Smoothing and Regression

WILEY SERIES IN PROBABILITY AND STATISTICS
APPLIED PROBABILITY AND STATISTICS SECTION

Established by WALTER A. SHEWHART and SAMUEL S. WILKS

A complete list of the titles in this series appears at the end of this volume.

Smoothing and Regression

Approaches, Computation, and Application

Edited by

MICHAEL G. SCHIMEK
Karl-Franzens-University of Graz, Austria
and University of Vienna, Austria

A Wiley-Interscience Publication
JOHN WILEY & SONS
New York • Chichester • Weinheim • Brisbane • Singapore • Toronto

This book is printed on acid-free paper. ♾

Published simultaneously in Canada.

Library of Congress Cataloging-in-Publication Data:

Smoothing and regression : approaches, computation, and application /
edited by Michael G. Schimek.
p. cm.
"A Wiley-Interscience Publication."
Includes bibliographical references.
ISBN 0-471-17946-9 (alk. paper)
1. Smoothing (Statistics) 2. Nonparametric statistics.
3. Regression analysis. I. Schimek, Michael G.
QA278.S64 2000
519.5′36—dc21 99-22017

Printed in the United States of America.

10 9 8 7 6 5 4 3 2 1

To my mother and the memory of my father

Contributors

Michel Delecroix, ENSAI, Rennes, France

Randall L. Eubank, Texas A&M University, College Station, Texas, USA

Ludwig Fahrmeir, Ludwig-Maximilians-Universität München, Munich, Germany

Jianqing Fan, University of North Carolina, Chapel Hill, North Carolina, USA and The Chinese University of Hong Kong, Hong Kong, PRC

Irene Gijbels, Université Catholique de Louvain, Louvain-la-Neuve, Belgium

Janet Grassmann, SONEM, Würzburg, Germany

Chong Gu, Purdue University, West Lafayette, Indiana, USA

Wolfgang Härdle, Humboldt-Universität zu Berlin, Berlin, Germany

Eva Herrmann, Technische Universität Darmstadt, Darmstadt, Germany

Sigbert Klinke, Humboldt-Universität zu Berlin, Berlin, Germany

Leonhard Knorr-Held, Ludwig-Maximilians-Universität München, Munich, Germany

Robert Kohn, Australian Graduate School of Management, Sydney, Australia

Thomas T. Kötter, SAP AG, Walldorf, Germany

Angelika van der Linde, Edinburgh University, Edinburgh, UK

Enno Mammen, Ruprecht-Karls-Universität Heidelberg, Heidelberg, Germany

Marlene Müller, Humboldt-Universität zu Berlin, Berlin, Germany

Guy P. Nason, University of Bristol, Bristol, UK

Douglas W. Nychka, National Center for Atmospheric Research, Boulder, Colorado, USA

Pascal Sarda, Université Paul Sabatier, Toulouse, France

Michael G. Schimek, Karl-Franzens-Universität Graz, Graz and Universität Wien, Vienna, Austria

David W. Scott, Rice University, Houston, Texas, USA

Bernard W. Silverman, University of Bristol, Bristol, UK

Jeffrey S. Simonoff, New York University, New York, USA

Michael Smith, University of Sydney, Sydney, Australia

Christine Thomas-Agnan, Université de Toulouse I, Toulouse, France

Berwin A. Turlach, The University of Western Australia, Perth, Australia

Gerhard Tutz, Ludwig-Maximilians-Universität München, Munich, Germany

Philippe Vieu, Université Paul Sabatier, Toulouse, France

Paul Yau, Australian Graduate School of Management, Sydney, Australia

Contents

Foreword

It is an honor to be asked to write a foreword for this volume, and it is my pleasure to congratulate the editor and the authors for a job well done.

This book is unique in that it brings together in one place a variety of points of view regarding nonparametric regression, smoothing and statistical modeling. These methods, as the editor notes, are becoming ubiquitous throughout the scientific, economic, environmental and social enterprises as ever increasing computational power and data collection ability become available.

The historical overviews here are important in tying together the different threads of the theory, which are widely scattered in the literature over time and in various journals. Many of the papers bring the reader up to date on the state of the art in the approaches discussed. Much useful computational and practical information, for example, discussion of useful software, is provided.

As I am sure the editor and authors will agree, this is a fun area to work in. There are many interesting and elegant theoretical questions, many interesting numerical and computational questions, and, finally, many interesting areas of application, where the statistician has the opportunity and satisfaction of working on "real world" scientific problems and contributing to the study of important scientific questions in a variety of areas.

The reader will probably agree with me that there is no one uniquely best method for all situations, but that substantive estimation and modeling problems may require the examination of several different approaches. Nevertheless there is abundant theory to give us guidance under various assumptions, although the last word has not been said on theoretical developments. This is particularly true in the multivariate case where data with very complex structure, as in environmental and demographic medical studies, is collected.

This book provides an important addition to the reading list of a modern nonparametric regression course. The extensive reference lists provide an excellent historical perspective, and a prime starting point for new researchers in the field. It will be a handy reference source for people wishing to obtain an overview of the various techniques, and pointers to practical issues in implementing the various methods. In addition, it is a fine summary of the state of the art as of the creation of the book.

Grace Wahba

Madison, December 1999

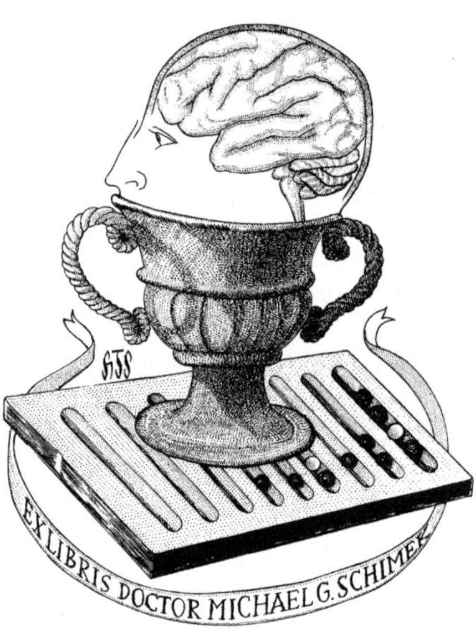
HJS
EX LIBRIS DOCTOR MICHAEL G. SCHIMEK

Preface

The idea for this volume can be traced back to the COMPSTAT Satellite Meeting on Smoothing, which took place in August 1994 in Semmering, Austria. I had the pleasure of organizing. During this workshop, it became clear that smoothing techniques are still a driving force in the development of nonparametric regression techniques. On the other hand they have become so diversified that it is really hard to keep up with all of the developments in modern regression analysis. In going from univariate to multivariate problems, the difference between competing methods is even more pronounced. During the past few years, important new approaches have emerged, such as the marginal integration method for the estimation of additive models. Markov chain Monte Carlo techniques have taken up a leading role in the Bayesian modeling of complicated data structures (dynamic, temporal, or spatial). Wavelets have become an important issue. Semiparametric regression is attracting more and more interest, and sliced inverse regression is offering new perspectives. Furthermore, all of these fascinating developments have been going on during the planning and writing of this book. We incorporated as much information on them as possible into this volume. References were last updated in late 1999.

What was the motivation to start an enterprise like this? There is certainly no lack of monographs in the field of smoothing techniques and nonparametric regression. Most of them are highly technical, some of them emphasize computational aspects or software products, and others embark on specific applications, but what they all have in common is that they specialize in one approach, say spline smoothing or local polynomial fitting. This volume attempts to bridge the gaps between these contributions in describing and occasionally comparing competing univariate and multivariate smoothing techniques. In addition, some of the subjects discussed in this book have not yet been covered in textbooks or monographs, such as smoothing under shape restrictions or of dependent data, or resampling methods for nonparametric regression, or vector generalized additive models.

What all of these topics have in common is that they are computationally demanding, some even requiring dynamic high-resolution graphics. Hence, in addition to defining and explaining them, it is only natural to discuss their numerical and software-related aspects. However, without the possibility of an appropriate implementation, such theoretical developments appear to be of limited value. Thus, in addition to explanations of their formal statistical background, all methods are discussed in the light of both computational efficiency and applicability for the analysis of real data. Smoothing techniques in regression are increasingly applied in the biosciences, in the environmental sciences, and in the fields of engineering and economics. They are also about to become relevant for many other fields, such as medical research and marketing. Last but not least, great efforts have been made in presenting graphics, which are of the utmost importance for any exploratory data analysis and the visualization of high-dimensional problems. Hence, also color plates were produced for this volume.

As a consequence, each regression approach included in the book is discussed from the following points of view: background theory, problems of computation and implementation, and examples of practical application.

The level of presentation is approximately the same throughout the volume, although some methods differ from others in terms of necessary technicalities, which are kept to a minimum to the greatest possible extent. Hence, the book addresses a wide audience: graduate students of statistics, applied mathematics, engineering, computer science and econometrics; scholars in the aforementioned fields; and applied researchers in these and other areas, such as biosciences, environmental sciences, medicine, psychometrics, and marketing. The wide range of examples in the volume is intended to further diversify the scope of application of nonparametric and semiparametric regression models and related techniques.

The emphasis of the volume is on multivariate regression problems. However, there are also introductory chapters that guide the reader through the most important univariate approaches, especially those of smoothing splines and of kernels. Other aspects of great importance are smoothing parameter choice, bandwidth selection, variance estimation, smoothing under functional constraints, and smoothing of autocorrelated data. A further topic of current interest is wavelet methods for regression. Some of these topics are essential for the understanding of multivariate regression problems.

The motivation behind the chapters on multivariate regression problems is to provide a state-of-the-art introduction to the large number of different approaches that are already established or are promising for the future of practical data analysis. Among these approaches are smoothing methods for discrete data (e.g., from contingency tables), local polynomial fitting, additive and generalized additive models (including alternating conditional expectations and additivity and variance stabilization), multivariate spline regression (e.g., thin-plate splines and interaction splines), multivariate and semiparametric kernel regression, spatial-process estimates as smoothers, resampling methods for nonparametric regression, multi-dimensional smoothing and visualization, projection pursuit regression, sliced inverse regression, nonparametric Bayesian dynamic and semiparametric models, and, finally, nonparametric Bayesian surface estimation. Competing algorithms are discussed throughout the book and are followed by software hints and examples using simulated as well as real data.

Due to the vast number of different techniques to be covered, the preparation of this volume demanded that experts in these very areas work together. As the book's editor, I designed it that way from the very beginning. To achieve this ambitious goal, I brought together researchers from many different countries and schools of thought, most of who would otherwise not be likely to write a book together. All agreed with me on the point that this project is solely feasible when there is a maximum of exchange among them. So authors commented on each others contributions and the final product is a unique piece of work with the same notation throughout. A crucial requirement was also that all authors obey a certain style of presentation and restrict themselves to a prespecified level of mathematical argumentation, avoiding the popular theorem–proof structure of statistics monographs. Proofs are either cited or given in an intuitive manner. The extensive list of references following each chapter should allow readers who desire further details, to trace down all of these technicalities. The specific format and the comprehensiveness of this volume are exactly what makes it special in the field of smoothing and regression.

I am most grateful to the enthusiastic support of all of the people who have helped to shape the book as it now stands. In addition to the authors and Stephen H. Quigley from John Wiley & Sons, Inc., there are many others I have to thank for stimulating discussions and for their most valuable reviews of chapters. In particular, I would like to name Dennis R. Cook, Dennis D. Cox, Naihua Duan, Sylvia Frühwirth–Schnatter, Peter Hall, Jeffrey D. Hart, M. C. Jones, James S. Marron, Jean D. Opsomer, Jürgen Pilz, James O. Ramsay, Burghart Seifert, Neil Shepherd, Stefan Sperlich, Joan G. Staniswalis, Alexander Tsybakov, Matthew P. Wand, and Thomas W. Yee.

Further, I wish to mention two of my former academic teachers, Prof. Leopold Schmetterer at the University of Vienna, Austria, who first introduced me to nonparametric concepts in statistics, and Prof. Bernard W. Silverman at the University of Bath, UK, who confronted me with smoothing techniques and computational statistics. Since my encounters with these individuals, I have been engaged in related problems, with a special emphasis on computing and applications in the biosciences and medicine.

Last but not least, I want to express my sincerest thanks to my mother, Ingrid, and my father, the late Prof. Herbert Toni Schimek, for their continuing support during my academic education. My father, who was a well-known Austrian designer and artist, not only shaped my eye for graphical detail but also stimulated my interest in science and technology. Both aspects turned out to be relevant to my later academic interests, as partly revealed in this book. Another, more direct contribution of my father is the exibris shown on page xvi, printed from a copperplate he engraved for me in 1979.

Michael G. Schimek

Graz, November 1999

Smoothing and Regression

CHAPTER 1

Spline Regression

Randall L. Eubank

Texas A&M University, College Station, Texas, USA

1.1 INTRODUCTION

Penalized least-squares regression and spline smoothing provide flexible data-fitting methodologies that have gained in popularity in recent years. While the basic spline-smoothing concept traces its origin back to Whittaker (1923), much of the modern development of smoothing splines and their variants is due to Grace Wahba. An introduction to her work, along with references to her more important papers, can be found in the Wahba (1990) monograph. Here, we give an overview of the work of Wahba and others as applied to the problem of univariate, nonparametric regression analysis.

Let us begin with an examination of the motivation behind spline smoothing. For this purpose, suppose that responses $y_1, \ldots, y_n$ have been observed at (nonstochastic) design points $t_1 < \cdots < t_n$ following the regression model

$$y_i = f(t_i) + \epsilon_i, \quad i = 1, \ldots, n, \tag{1.1}$$

where $f(\cdot)$ is an unknown regression function and $\epsilon_1, \ldots, \epsilon_n$ are zero-mean, uncorrelated random errors. The problem to be considered is the estimation of f from the observed data.

Classically, one of the most popular ways to estimate f has been through simple linear regression. Using this technique, one estimates $f(t)$ by $\hat{f}(t) = \hat{a} + \hat{b}t$, where $\hat{a}$ and $\hat{b}$ are the least-squares intercept and slope estimators, respectively, obtained by minimizing the *residual sum of squares*

$$\text{RSS}(g) = \sum_{i=1}^{n} (y_i - g(t_i))^2$$

Smoothing and Regression: Approaches, Computation, and Application
Edited by Michael G. Schimek
ISBN 0-471-17946-9

over all functions of the form $g(t) = a + bt$. This approach to estimation can be effective if f in Equation (1.1) is approximately linear, but it can also fail miserably. An example of such a failure is provided by the data in Figure 1.1. For this data, the simple linear-regression fit shown in the figure serves only to suggest the presence of a direct relationship as it smoothes out many other features of the data that might be attributed to the underlying regression curve.

To understand why linear regression fails in certain cases, it is useful to reconsider the Taylor-expansion argument that is often advanced as a motivation for the method. The basic premise of this argument is that if f has two derivatives, then for a point t that is close to some fixed point t_0, we have

$$f(t) = f(t_0) + f'(t_0)(t - t_0) + o(|t - t_0|^2)$$

as a result of Taylor's Theorem. Thus, for t close to t_0, f follows an approximate linear model with intercept $f(t_0) - f'(t_0)t_0$ and slope $f'(t_0)$. A global (i.e., over the entire design) linear approximation for f will therefore be satisfactory only when the slope of the regression function, $f'(t_0)$, is essentially constant and the error term $o(|t - t_0|^2)$ is uniformly small. The condition for a constant slope is clearly violated in Figure 1.1, which explains why simple linear regression has difficulties in fitting this data set. It now seems clear that to successfully fit data in general, we must alter our original estimation prescription and consider minimization of RSS(g) over functions g with slopes that vary over the design.

Also shown in Figure 1.1 is an estimator of f obtained by interpolating the data. For this estimator, we have allowed the ultimate slope flexibility, in that every two responses are connected by lines with their own individual slopes. For this estimator, RSS(g) = 0. However, the visual quality of the fit leaves much to be desired. Unlike the linear-regression fit, which uses too little of the information

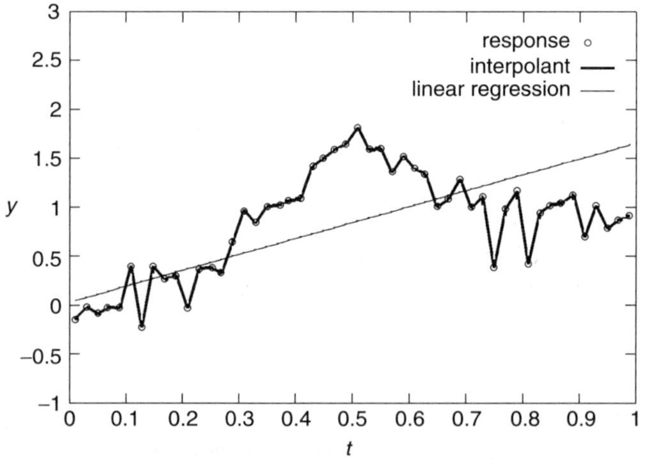

FIGURE 1.1 Linear regression and interpolation for a simulated data set.

in the data, this fit actually includes too much information. In this sense, it fails to provide a useful summary of the data and, more importantly, does not provide a satisfactory description of the basic trends or features in the data that could be due to the regression function in Equation 1.1 rather than to the random-noise component of the model.

The two "extreme" fits to the data in Figure 1.1 suggest that we need some compromise between fits with constant slopes and fits with completely flexible slopes. One way to accomplish this is to penalize functions whose slopes vary too rapidly. The rate of change in the slope of a function g is given by g'', and hence an overall measure of the change in slope of a potential fitted function is provided by

$$J(g) = \int_{t_1}^{t_n} g''(t)^2 \, dt.$$

Thus, a modified version of our original estimation criterion that incorporates a penalty for rapid changes in slope is

$$\mathrm{RSS}(g) + \lambda J(g), \qquad \lambda \geq 0, \tag{1.2}$$

which can now be minimized over, for example, all functions with two continuous derivatives. The value of the so-called smoothing parameter λ in (1.2) can be viewed as a measure of the importance that we place on the flexibility or lack of flexibility of the slope of a fit. If we take $\lambda = \infty$, then Equation (1.2) produces the constant slope linear regression fit while $\lambda = 0$ corresponds to an interpolatory fit with completely flexible slopes.

If $n \geq 2$, then it can be shown that there is a unique, computable minimizer f_λ for criterion (1.2), called a cubic smoothing spline. The specific form of f_λ and the reason for this particular name for the estimator are discussed more fully in the next section. For now, it suffices to say that the estimator is linear in the sense that for each estimation point t there are computable constants $\{h_i(t)\}_{i=1}^n$ such that

$$f_\lambda(t) = \sum_{i=1}^{n} h_i(t) y_i.$$

While use of the cubic smoothing-spline estimator may solve the problem of allowing for fits with variable slope, it also creates a new problem, namely, the determination of an appropriate value of the smoothing parameter for a given set of data. We would not expect the same value of λ to work for every data set. Indeed, if we agree that the "best" choice of λ is a value that minimizes the squared-error loss

$$L(\lambda) = n^{-1} \sum_{i=1}^{n} (f(t_i) - f_\lambda(t_i))^2,$$

then it is clear that a good choice of the smoothing parameter depends both on the unknown true regression curve as well as the inherent variability of the estimator. Since $L(\lambda)$ is unknown, we must estimate it in order to produce a data-driven choice for λ. One such estimator can be obtained from the *generalized cross-validation* (GCV) criterion of Craven and Wahba (1979), defined by

$$\mathrm{GCV}(\lambda) = n^{-1}\mathrm{RSS}(f_\lambda) \Big/ \left(1 - n^{-1}\sum_{i=1}^{n} h_i(t_i)\right)^2,$$

which provides an estimator of $L(\lambda) + \sigma^2$. It has been shown by Li (1986) that, under certain conditions, the minimizer $\hat{\lambda}$ of the GCV criterion satisfies $L(\hat{\lambda})/\inf_\lambda L(\lambda) \overset{P}{\to} 1$ as $n \to \infty$. This indicates that, at least in terms of the estimator's loss performance, one does as well asymptotically with the GCV choice of λ as could be done with the unknown, best possible value of the smoothing parameter.

The data in Figure 1.1 were actually simulated from Equation (1.1) using a uniform design of $n = 50$ points over the interval [0, 1], normal random errors with standard deviation $\sigma = 0.2$, and $f(t) = t + \exp\{-32(t - 0.5)^2\}$. As a result, we can compute the true loss in this special case. Values of $L(\lambda)$ as well as those for $\mathrm{GCV}(\lambda)$ are shown in Figure 1.2, from which we see that both criteria are minimized at about the same value of λ. Thus, GCV provides a good choice for the smoothing parameter for this data set. The fit corresponding to the GCV estimator of λ (dashed curve) is shown in Figure 1.3 along with the true regression function (solid curve). From this plot, the cross-validated smoothing spline can be seen to provide an excellent estimator of f that gives a substantial improvement over both the simple linear-regression estimator and the interpolation estimator in Figure 1.1.

In the next section, we discuss the general form and computation of a smoothing-spline estimator. Section 1.3 then focuses on a simple, special case

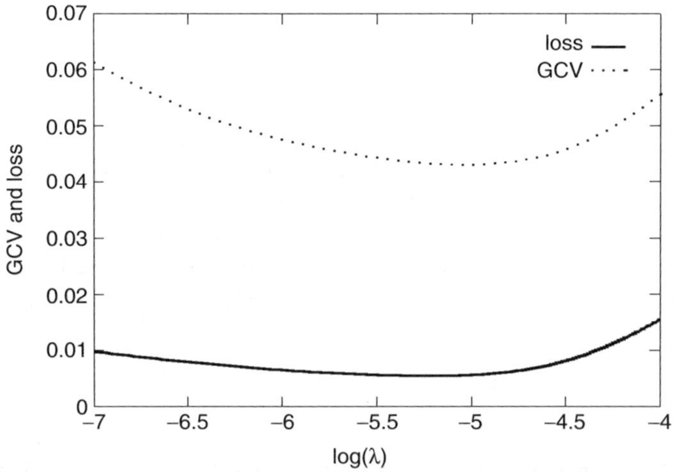

FIGURE 1.2 GCV and loss for a simulated data set.

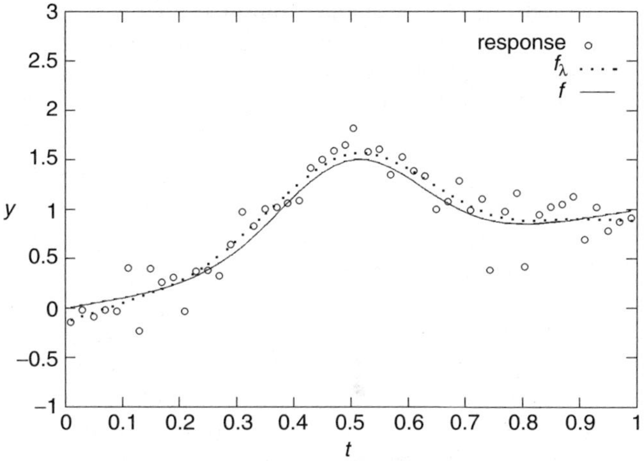

FIGURE 1.3 Smoothing spline fit to a simulated data set.

wherein it is easy to demonstrate the connection between smoothing splines and other nonparametric smoothers, such as Fourier series and kernel estimators. The large-sample frequentist properties of smoothing splines are discussed in Section 1.4 and Bayesian properties of the estimators and their frequentist implications are the subject of Section 1.5. In Section 1.6, we conclude with a discussion of how the penalized least-squares estimation paradigm that produced smoothing splines can be generalized to provide estimators for other problems.

1.2 GENERAL FORM OF THE ESTIMATOR

Our basic smoothing criterion, described in Equation (1.2), can be extended in several possible ways to produce other useful estimators of f. More generally, we could consider the estimation of f by a $2m$th-order smoothing spline f_λ obtained by minimizing

$$\sum_{i=1}^{n} w_i(y_i - g(t_i))^2 + \lambda \int_{t_1}^{t_n} g^{(m)}(t)^2 \, dt, \quad \lambda \geq 0, \tag{1.3}$$

where $w_1, \ldots, w_n$ are positive weights and m is a positive integer. Equation (1.3) is minimized over all functions g having $m - 1$ absolutely continuous derivatives and square integrable mth derivatives. The weights $w_1, w_2, \ldots, w_n$ in (1.3) can be used to place additional emphasis on the residuals from fitting the data in some important region of the design space, to adjust for heteroskedastic errors, or to produce an estimator when there are replicate samples (see, e.g. Eubank, 1988, 207ff).

If $n \geq m$, Equation (1.3) has a unique minimizer that is found to be a natural polynomial spline of order $2m$ with knots at the design points. More specifically, f_λ is a piecewise polynomial of order $2m$ (degree $2m - 1$) on $[t_1, t_n]$ with $2m - 2$ continuous derivatives at each of the join points, or knots, $t_1, \ldots, t_n$, where the different polynomial segments are tied together. The smoothing spline becomes a polynomial of order m outside of $[t_1, t_n]$ and therefore satisfies the so-called natural boundary conditions $g^{(m+j)}(t_1) = g^{(m+j)}(t_n) = 0$, $j = 0, \ldots, m - 1$, from which the term "natural spline" is derived. Thus, for example, when $m = 2$ in (1.3), f_λ is a piecewise cubic polynomial with two continuous derivatives that is linear outside of $[t_1, t_n]$.

The specific form of the estimator as a polynomial spline is, in many ways, immaterial. Its primary consequence is computational, in that, given a basis $x_1, \ldots, x_n$ for the $2m$th-order natural splines with knots at $t_1, \ldots, t_n$, we may substitute $g = \sum_{j=1}^n b_j x_j$ into Equation (1.3) and minimize with respect to $b_1, \ldots, b_n$ to get an explicit, computable form for the estimator. The result is that $f_\lambda = \sum_{j=1}^n b_{\lambda j} x_j$, where $b_\lambda = (b_{\lambda 1}, \ldots, b_{\lambda n})^T$ is obtained as the solution of

$$(X^T W X + \lambda\Omega)b = X^T W y, \tag{1.4}$$

where $X = \{x_j(t_i)\}_{i,j=1}^n$, $W = \text{diag}(w_1, \ldots, w_n)$, $\Omega = \{\int_{t_1}^{t_n} x_i^{(m)}(t) x_j^{(m)}(t)\, dt\}_{i,j=1}^n$, and $y = (y_1, \ldots, y_n)^T$. A clever choice for the spline basis can pay computational dividends, in that it can produce a band structure for the system described by Equation (1.4) that allows it to be solved in only $O(n)$ calculations. Examples of such bases are those described by Anselone and Laurent (1968) and Lyche and Schumaker (1973). Explicit FORTRAN code for solving a band-limited version of Equation (1.4) can be found in de Boor (1978, Chapter 14) for the most popular case of $m = 2$, or a cubic smoothing spline.

To select a good value of λ from the data, one can again use the GCV criterion, this time with the f_λ obtained from Equation (1.4). The residual sum of squares in our earlier form of the criterion is now replaced with the weighted residual sum of squares from Equation (1.3), and the values of $h_i(t_i)$ are taken to be the diagonal elements of the matrix $X(X^T W X + \lambda\Omega)^{-1} X^T W$. Efficient, $O(n)$ algorithms for computing $\{h_i(t_i)\}_{i=1}^n$ and, hence, the GCV criterion can be found in Hutchinson and de Hoog (1985, 1987).

To conclude this section, we consider the data in Figure 1.4, from Davidian et al. (1988). These data give responses to a drug as a function of its concentration, with four replicate responses being obtained at each concentration level. It is expected for data of this form, and is also obvious from the plot, that the response variances increase with drug concentration. Thus, if s_i^2 represents the sample variance for the four responses at the ith concentration, we should consider smoothing this data using Equation (1.3) with $w_i = s_i^{-2}$, $i = 1, \ldots, n$. A choice of $m = 2$ then gives a (weighted) cubic smoothing spline fit to the data. The resulting fit with the smoothing parameter selected using the GCV criterion is shown is Figure 1.4, along with the data. The smoothing-spline fit is found to be almost identical to a parametric fit for these data obtained by Davidian

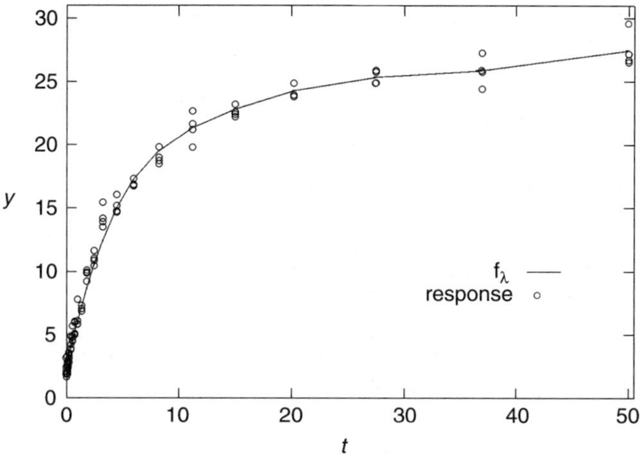

FIGURE 1.4 Smoothing spline fit to assay data.

et al. using a log-logistic model for the mean function and a log-linear (in the mean function) model for the variances. The close agreement between the parametric and nonparametric estimators in this case gives support for the postulated parametric model for the mean function.

Note that we can also "test" Davidian's et al. (1988) modeling assumption that the error variances are log linear in the response means by fitting $(\overline{y}_i, \log s_i^2)$, $i = 1, \ldots, n$, where $\overline{y}_i$ is the average of the four responses at the ith concentration. If one does so using the cubic smoothing spline from Equation (1.2), one finds that GCV chooses the value $\lambda = \infty$, which produces a simple linear fit. This finding also gives support to the modeling assumptions that were originally employed to fit the data.

1.3 THE LINEAR SMOOTHING SPLINE

A difficulty in understanding how smoothing splines process data stems from the absence of an easily interpretable closed form for the estimator. There is, however, one special case where (1.4) can be solved explicitly to provide an estimator that admits simple interpretations. This case and its consequences are the subject of this section.

Assume that $t_i = (2i - 1)/2n$, $i = 1, \ldots, n$, and take $m = 1$. Then, the minimizer of Equation (1.3) is known as a linear smoothing spline and (1.4) can be solved explicitly (Eubank, 1994) to see that

$$f_\lambda(t_i) = \overline{y} + 2\sum_{j=1}^{n-1} \frac{\hat{b}_j}{1 + \lambda\gamma_j} \cos(j\pi t_i), \qquad (1.5)$$

where $\overline{y}$ is the response average,

$$\hat{b}_j = n^{-1} \sum_{r=1}^{n} y_r \cos(j\pi t_r) \tag{1.6}$$

is the jth sample cosine Fourier coefficient, and

$$\gamma_j = \left[2n \sin\left(\frac{\pi j}{2n}\right)\right]^2, \quad j = 1, \ldots, n-1. \tag{1.7}$$

The oscillation properties of the functions $\cos(j\pi t)$ for increasing frequency j, along with the orthogonality properties

$$n^{-1} \sum_{i=1}^{n} \cos(j\pi t_i) = 0$$

and

$$\frac{2}{n} \sum_{i=1}^{n} \cos(j\pi t_i) \cos(k\pi t_i) = \delta_{jk},$$

show that the sample Fourier coefficients (1.7) provide a decomposition of the data into orthogonal frequency components. The $\hat{b}_j$s measure higher frequency aspects of the data as their indices increase. In general, one expects $\hat{b}_j$ for larger values of j to consist primarily of contributions from the random-error term in (1.1), while the information about a smooth regression function is expected to be concentrated in the lower frequency coefficients. From this viewpoint, the smoothing spline tends to use the data in a very natural way to produce an estimator of f. Specifically, since $\gamma_j \doteq (j\pi)^2$ for $j = o(n)$, f_λ uses all of the sample Fourier coefficients, but downweights or damps the ones with higher frequencies. The value chosen for λ then governs the degree of damping or smoothing.

The interpretation of the linear smoothing spline as a type of damped Fourier-series estimator can be extended to general smoothing splines through an appropriate choice of the natural-spline basis functions. It has been shown by Demmler and Reinsch (1975) that there is a natural-spline basis with oscillation properties similar to those for the cosine functions under which a smoothing spline admits a representation similar to that for the linear case. In fact, (1.6)–(1.7) are actually just a special case of the general Demmler–Reinsch representation.

Returning again to the linear smoothing spline, we see that by rearranging terms in (1.7) we can write $f_\lambda(t_i) = \sum_{r=1}^{n} h(t_i, t_r) y_r$, where

$$h(t, s) = n^{-1} \left[1 + 2 \sum_{j=1}^{n-1} \frac{\cos(j\pi t)\cos(j\pi s)}{1 + \lambda\gamma_j}\right].$$

If we approximate γ_j by $(j\pi)^2$ in this expression and replace the sum by an integral, we can argue heuristically that

$$nh(t, s) \doteq 2\int_0^\infty \frac{\cos(x\pi t)\cos(x\pi s)}{1+\lambda(\pi x)^2}\,dx = \frac{1}{2\sqrt{\lambda}}e^{-|t-s|/\sqrt{\lambda}}[1 + e^{-\min(t,s)/\sqrt{\lambda}}]$$

$$\approx \frac{1}{2\sqrt{\lambda}}e^{-|t-t|/\sqrt{\lambda}} \tag{1.8}$$

for fixed t as $\lambda \to 0$. Thus, the linear smoothing spline can also be viewed as being approximately a kernel-type estimator, with the kernel given by the Laplacian density $K(u) = \frac{1}{2}e^{-|u|}$, $-\infty < u < \infty$, and with a bandwidth $\sqrt{\lambda}$.

It is actually not difficult to make the argument leading to Equation (1.8) precise. One first uses the MacLaurin series for $\sin x$ to obtain approximation bounds, such as $0.4j\pi \le 2n\sin(j\pi/(2n)) \le j\pi$ and $|2n\sin(j\pi/(2n)) - j\pi| \le (j\pi)^3/(24n^2)$ for $1 \le j \le n-1$, and then combines these bounds with identity 1.445.2 on page 40 of Gradshteyn and Ryzhik (1980).

The connection between spline smoothing and kernel smoothing is not unique to the linear case. If we think of the t_i as being n-tiles of some continuous density p, then the general smoothing spline obtained from (1.3) admits a similar kernel approximation, with bandwidth $[\lambda/p(\cdot)]^{1/2m}$ and kernel K, obtained as the solution of a particular $2m$th-order differential equation (see Messer, 1991; Nychka, 1995; and Silverman, 1984). The kernel is of order $2m$ in the sense that $\int_{-\infty}^{\infty} K^2(u)\,du < \infty$; $\int_{-\infty}^{\infty} u^j K(u)\,du = 0$, $j = 0, \ldots, 2m-1$; and $\int_{-\infty}^{\infty} u^{2m}K(u)\,du \neq 0$. However, this approximation is valid only in the interior of $[t_1, t_n]$, and the approximate kernel degrades to order m near t_1 and t_n.

1.4 LARGE-SAMPLE EFFICIENCY

In this section, we discuss the large-sample properties of spline smoothers. Perhaps the best or strongest asymptotic results for smoothing splines can be found in Cox (1983), Rice and Rosenblatt (1983), and Speckman (1985). However, much of this work again traces its origins to seminal work by Grace Wahba, such as Wahba (1975).

The basic premise behind a large portion of the asymptotic analysis conducted in nonparametric regression involves assessment of an estimator's large-sample efficiency in terms of some type of performance measure. A standard choice for a performance criterion is the average squared-error risk, which has the form

$$R(\lambda) = n^{-1}\sum_{i=1}^{n} E(f_\lambda(t_i) - f(t_i))^2$$

for our smoothing-spline estimator f_λ. It is known (Nussbaum, 1985; Speckman, 1985; and Stone 1982) that, under various restrictions, the best rate of decay for the risk associated with any nonparametric estimator is of order $n^{-2m/(2m+1)}$ for estimation of regression functions with m derivatives. Thus, the goal in this type of analysis is to show that smoothing splines are asymptotically efficient in the sense that $\inf_{\lambda>0} R(\lambda) = O(n^{-2m/(2m+1)})$.

Analysis of the risk generally proceeds by first writing $R(\lambda) = V(\lambda) + B^2(\lambda)$, where

$$V(\lambda) = n^{-1} \sum_{i=1}^{n} \mathrm{Var}[f_\lambda(t_i)]$$

is the variance part of the risk and

$$B^2(\lambda) = n^{-1} \sum_{i=1}^{n} (Ef_\lambda(t_i) - f(t_i))^2$$

is the average squared bias. Arguments in Speckman (1985) can be used to show that for large n and small λ the variance term $V(\lambda)$ is well approximated by $C/(n\lambda^{1/2m})$, where C is a known constant that depends only on m and the design, and that $B^2(\lambda) \le \lambda \int_{t_1}^{t_n} f^{(m)}(t)^2\,dt$ (see also Eubank, 1988, Section 6.3.2). These equations give a large-sample upper bound for $R(\lambda)$ of the form

$$\frac{C}{n\lambda^{1/2m}} + \lambda \int_{t_1}^{t_n} f^{(m)}(t)^2\,dt,$$

from which we see that any choice of λ decaying at the rate $n^{-2m/(2m+1)}$ will ensure an optimal rate of decay for the smoothing-spline risk.

More exact expressions for the risk require an in-depth analysis of the properties of $B^2(\lambda)$, rather than just an upper bound on its size. This analysis tends to be somewhat complicated by its dependence on the boundary behavior of the regression function. To be somewhat more precise, it turns out that $B^2(\lambda)$ can decay to zero at a rate substantially faster than order λ if f satisfies some subset (or all) of the natural boundary conditions

$$f^{(m+j)}(t_1) = f^{(m+j)}(t_n) = 0, \quad j = 0, \ldots, m-1, \tag{1.9}$$

that are necessarily satisfied by the natural-spline basis functions used to construct f_λ. Arguments by Rice and Rosenblatt (1983) and Utreras (1987) have the consequence that if f has $2m$ derivatives and satisfies

$$f^{(m-1+j)}(t_1) = f^{(m-1+j)}(t_n) = 0, \quad j = 1, \ldots, k,$$

for some $0 \le k \le m-1$ (with $k = 0$ meaning that none of the conditions in (1.9) are satisfied), then $B^2(\lambda)$ can decay as fast as $\lambda^{(1+(2k+1)/2m)}$, and $B^2(\lambda) = O(\lambda^2)$ if f satisfies (1.9). This means, for example, that if we were fortunate enough that nature gave us a regression function with $2m$ (rather than m) derivatives and this function satisfied (1.9), then it would be possible for the smoothing-spline estimator to obtain a much faster rate of $O(n^{-4m/(4m+1)})$ for the risk.

Returning to the kernel approximation for f_λ discussed in Section 1.3, we can observe that the approximate form $C/(n\lambda^{1/2m})$ for $V(\lambda)$ is exactly what we would expect from a kernel estimator of order $2m$ with bandwidth $\lambda^{1/2m}$. However, even if f has $2m$ derivatives, the bias term $B^2(\lambda)$ decays at a rate of only $O(\lambda)$, rather than at the $O(\lambda^2)$ rate we would expect from a $2m$th-order kernel estimator, unless the natural boundary conditions are satisfied. The reason for this is that while smoothing splines behave like $2m$th-order kernel estimators in the interior of $[t_1, t_n]$, their parallel of the kernel estimator boundary correction is only of order m. Thus, for regression functions with more than the expected m derivatives that do not satisfy (1.9), the bias for f_λ is larger near t_1 and t_n than in the interior of the estimation interval. This entails that the bias part of the risk and, hence, the optimal level of smoothing tend to be dominated by the boundary behavior of f in such cases. From a data-analytic standpoint, this tendency has the consequence that, at least to the extent that the GCV estimator of λ estimates the minimizer of $R(\lambda)$, information from the boundary region may have an inordinately large influence on data-driven choices for the smoothing parameter. To compensate for this effect, the estimator may be boundary corrected along the lines of proposals in Eubank and Speckman (1991) and Oehlert (1992).

One reason for the problems that have been encountered in analyzing and understanding the large-sample properties of smoothing splines is the difficulty in obtaining simple large-sample approximations for the pointwise bias and variance of the estimators. Recently, Nychka (1995) has shown that it may be possible to develop such expressions through the use of Green's functions associated with spline-smoothing problems. In particular, his work can be used to show that, for the linear smoothing spline ($m = 1$),

$$f(t) - Ef_\lambda(t) = \frac{\lambda}{p(t)} f^{(2)}(t) + o(\lambda),$$

provided that f has a Hölder-continuous second derivative, and

$$\text{Var}(f_\lambda(t)) = \frac{\sigma^2}{4n(\lambda p(t))^{1/2}}(1 + o(1)).$$

These expressions are valid for t in any fixed interval that does not contain the boundary. They clearly show that the linear smoothing spline behaves like a second-order kernel smoother for estimation in the interior of $[t_1, t_n]$.

1.5 BAYESIAN MOTIVATION

The form of the smoothing-spline estimator that follows from (1.4) often leads parametric regression analysts to remark about possible connections between spline smoothing and ridge regression. One reason for the formal similarity of these two estimators derives from their common motivation as posterior means for certain Bayesian regression models. We explore the connection between Bayesian regresssion and spline smoothing in this section.

There are several elegant Bayesian regression models for which a smoothing spline is found to be an optimal estimator of the regression function (see, e.g. Kohn and Ansley, 1988; Wahba, 1978; Shiller, 1984; and van der Linde, 1993). Here, we deal only with the simple Bayesian model of Silverman (1985) and suppose that instead of using model (1.1), we collected our data from

$$y_i = \sum_{j=1}^{n} b_j x_j(t_i) + \epsilon_i, \quad i = 1, \ldots, n,$$

where $x_1, \ldots, x_n$ are the natural-spline basis functions and $\epsilon_1, \ldots, \epsilon_n$ are independent, normal errors with zero mean and variance σ^2. If we now take $b = (b_1, \ldots, b_n)^T$ to be the observed value of a random n-vector β with (prior) "density" proportional to $\exp\{-\lambda b^T \Omega b/(2\sigma^2)\}$, then we may readily check that $E[\beta|y] = b_\lambda$, where b_λ is the solution of (1.3). Thus, b_λ is the posterior mean of β for this Bayesian regression model. Accordingly, we can also see that the smoothing-spline estimator is the posterior mean of the regression function in this case. Note that the prior here is improper in that the matrix Ω has m zero eigenvalues corresponding to the polynomials $1, t, \ldots, t^{m-1}$ that are contained as a proper subset of the natural-spline basis. We may interpret this as meaning that smoothing splines can be viewed as a type of Bayesian polynomial regression estimator where the polynomial coefficients are modeled as having a diffuse normal prior and possible departures from a polynomial model are accounted for by fitting with flexible basis functions.

The connection between Bayesian regression and spline smoothing has considerable utility, since it provides a justification for the use of diagnostic and interval estimation techniques that parallel those used in frequentist, normal-theory linear regression. For example, it is easy to show that, under our Bayesian model, the ith residual $y_i - f_\lambda(t_i)$ has mean zero and variance $\sigma^2(1 - h_i(t_i))$, just as in ordinary least-squares regression. This suggests that one might conduct residual analysis of a smoothing-spline fit by following a linear-regression-type paradigm involving the examination of "standardized" residuals $(y_i - f_\lambda(t_i))/\sqrt{\sigma^2(1 - h_i(t_i))}$, $i = 1, \ldots, n$, and other related measures. It can actually be shown (see Craven and Wahba, 1979) that, for a smoothing-spline estimator,

$$y_i - f_\lambda^{(i)}(t_i) = (y_i - f_\lambda(t_i))/(1 - h_i(t_i)),$$

where $f_\lambda^{(i)}(t_i)$ is the smoothing-spline fit at t_i that is computed from the data set, with the observation for (t_i, y_i) deleted. This result can then be combined with the Bayesian residual variance identity to produce various types of deleted-observation influence diagnostics. Examples of these diagnostics are discussed in Eubank (1988, Section 5.5.5).

Perhaps the most important use of the Bayesian properties of smoothing splines has been for the development of Bayesian confidence, or, more correctly, prediction intervals by Wahba (1983). We develop these intervals by first observing that, under our Bayesian model, $\text{Var}[f(t_i)|y] = \sigma^2 h_i(t_i)$. This equation has the consequence that a $100(1-\alpha)\%$ prediction interval for the value of the mean function at t_i is given by

$$f_\lambda(t_i) \pm Z_{\alpha/2}\sigma\sqrt{h_i(t_i)}, \tag{1.10}$$

where $Z_{\alpha/2}$ is the $100(1-\alpha/2)$th percentage point of the standard normal distribution. Thus, (1.10) provides one way of producing interval estimators to accompany the smoothing-spline point estimator.

The derivation of the intervals (1.10) from a Bayesian perspective has prompted a number of investigations into the empirical and theoretical frequentist properties of these intervals under model (1.1). One of the most complete studies of this nature is that of Nychka (1988), who has shown that, under certain restrictions and with an optimally chosen level of smoothing, we have

$$n^{-1}\sum_{i=1}^{n} \text{prob}(|f_\lambda(t_i) - f(t_i)| \le Z_{\alpha/2}\sigma\sqrt{h_i(t_i)}) \doteq 1 - \alpha. \tag{1.11}$$

This states that under model (1.1) the average coverage for the Bayesian intervals across the design should be approximately the same as their nominal, pointwise coverage. While this condition is encouraging, it tells us very little about the pointwise coverage properties of any particular interval, since these properties can be unacceptably lower or higher than the nominal level and still satisfy (1.11).

Some intuitive notion of how the Bayesian intervals can be expected to work under model (1.1) can be obtained by referring back to the discussions in Sections 1.3 and 1.4, which suggested that smoothing splines are very close to being kernel estimators of order $2m$ with boundary corrections that are only of order m, unless the regression curve satisfies the natural boundary conditions in (1.9). In view of this situation, one might expect $(f_\lambda(t) - Ef_\lambda(t))/\sqrt{\text{Var}(f_\lambda(t))}$ to behave like a standard normal random variable for large n under mild restrictions on λ, since this behavior is exhibited by kernel-type estimators. If we now consider $(f_\lambda(t) - f(t))/\sqrt{\text{Var}(f_\lambda(t))}$ to be a potential pivotal quantity, we see that

$$\frac{f_\lambda(t) - f(t)}{\sqrt{\text{Var}(f_\lambda(t))}} = \frac{f_\lambda(t) - Ef_\lambda(t)}{\sqrt{\text{Var}(f_\lambda(t))}} + \frac{Ef_\lambda(t) - f(t)}{\sqrt{\text{Var}(f_\lambda(t))}},$$

and, consequently, the performance of standard frequentist confidence intervals depends on the bias of $f_\lambda(t)$ being small relative to its standard error. For typical boundary-corrected kernel estimators with optimal smoothing, the contribution of the bias term $(Ef_\lambda(t) - f(t))/\sqrt{\text{Var}(f_\lambda(t))}$ tends to be nonnegligible, thereby rendering $(f_\lambda(t) - f(t))/\sqrt{\text{Var}(f_\lambda(t))}$ unsuitable for a pivotal quantity without some type of bias correction. However, for smoothing-spline estimators of regression functions with more than m derivatives that do not satisfy the natural boundary conditions, the boundary "bias" properties of the estimator tend to cause automated smoothing methods to produce undersmoothed estimators with λ selected to be smaller than optimal. While this tendency has certain negative consequences for point estimation, it also entails that the bias of the estimator tends to be small relative to its standard error for estimation points away from the boundary. In such cases, frequentist intervals of the form

$$f_\lambda(t_i) \pm Z_{\alpha/2}\sqrt{\mathrm{Var}(f_\lambda(t_i))} \tag{1.12}$$

tend to be effective in the sense of having asymptotic coverage $1-\alpha$. Since one can show that $\mathrm{Var}(f_\lambda(t_i)) \leq \sigma^2 h_i(t_i)$, it follows that the Bayesian intervals (1.10) work at least as well as the intervals in (1.12) for such cases.

To illustrate the point of the preceding discussion, 1000 samples of 50 points each were generated from the model in (1.1), using uniformly spaced design points, normal errors with standard deviation $\sigma = 0.1$, and regression functions

$$f(t) = (27/4)^4[t^4(1-t)^8 + t^8(1-t)^4]/2$$

and

$$f^*(t) = 0.1g(t) + f(t),$$

where $g(t) = I(t \leq 0.1)(1-10t)^3 + I(t \geq 0.9)(10t-9)^3$, in which $I(A)$ is the indicator function for the event A. Note that functions f and f^* do and do not, respectively, satisfy the natural boundary conditions for $m = 2$ in Equation (1.9). Cubic smoothing splines were fit to each data set, and the Bayesian confidence intervals in Equation (1.10) were computed using a nominal level of 0.95. The observed average (over 1000 replications) coverage proportions at each of the 50 design points are shown in Figure 1.5. The solid curve at the top of the figure is a plot of the observed coverage proportions for f, and the solid curve below is a

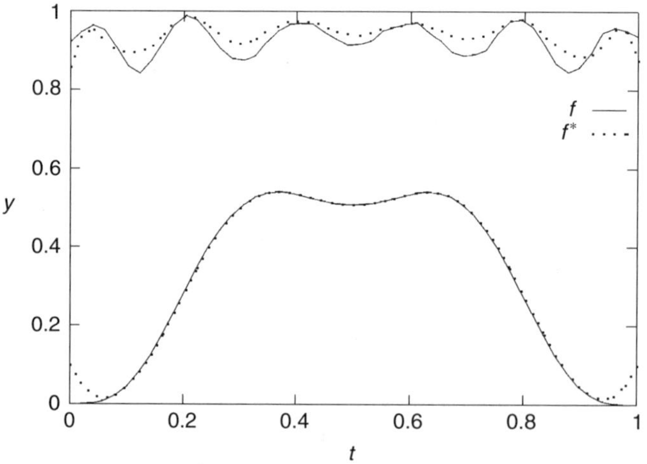

FIGURE 1.5 Empirical coverages of Bayesian intervals for f and f^*.

plot of the regression function. The same respective information for f^* is given by the dashed curves in the figure. From this figure, we see that the fact that f satisfies the natural boundary conditions has led to significant undercoverage for the Bayesian intervals. In contrast, the analogous results for f^* are much better away from the boundary region.

The relationship between spline smoothing and Bayesian estimation leads to a number of other useful results. For example, it can be used to develop efficient computational algorithms for smoothing splines via the Kalman filter. Discussions of this and other related issues can be found in Ansley and Kohn (1985) and Kohn and Ansley (1989).

1.6 EXTENSIONS AND IMPLEMENTATIONS

One of the more compelling features of the spline-smoothing technique is how easily it can be modified to produce estimators for other data-analysis problems. If we view the basic form of criterion (1.3) as

$$(\text{measure of fit}) + \lambda(\text{measure of roughness}),$$

then estimators for other settings can be readily produced by changing the definition of the (measure of fit) term and/or the (measure of roughness) term to more aptly suit the problem at hand. We illustrate this aspect of spline smoothing in this section through a few examples.

Suppose first that, instead of our original model (1.1), the mean function in our regression problem is the sum of a smooth function $\theta(\cdot)$ and a parametric linear component of the form $z^T\alpha$, where z is a known p-vector and α is an unknown vector of parameters. Then, given the obervations $(t_1, z_1^T, y_1), \ldots, (t_n, z_n^T, y_n)$, we might consider estimating both α and θ simultaneously by the minimizers with respect to a and g of

$$\sum_{i=1}^{n} w_i(y_i - z_i^T a - g(t_i))^2 + \lambda \int_{t_1}^{t_n} g^{(m)}(t)^2\, dt.$$

The resulting estimators exist and can be computed in a similar manner to the ordinary smoothing spline. This approach to estimation in semiparametric linear models is due to Wahba (1984), and the estimator is known as a partial smoothing spline.

Now suppose that, instead of the regression model in Equation (1.1), the responses come from a more general estimation setting wherein the ith response is assumed to have an exponential family density with corresponding log likelihood $l_i(y_i, \theta(t_i))$ that depends on some unknown function θ. We might then consider estimation of θ by the minimizer of a penalized log-likelihood criterion, such as

$$-\sum_{i=1}^{n} l_i(y_i, g(t_i)) + \lambda \int_{t_1}^{t_n} g^{(m)}(t)^2\, dt.$$

This estimator also exists and can be computed using the techniques discussed by O'Sullivan, Yandell, and Raynor (1986).

By modifying the roughness penalty, we can also produce parallels of smoothing splines that involve other interesting basis functions, such as exponential, rather than polynomial, splines. A rather general development of such smoothing splines is given by Kimeldorf and Wahba (1971). We can also alter the penalty to produce estimators for problems with multiple predictor variables. Specifically, suppose that we have two predictor variables s and t. We can obtain a bivariate smoothing spline by minimizing the residual sum of squares plus a penalty involving a roughness measure, such as

$$\int_{-\infty}^{\infty}\int_{-\infty}^{\infty}\left[\left(\frac{\partial^2 g(s,t)}{\partial s^2}\right)^2 + 2\left(\frac{\partial^2 g(s,t)}{\partial s \partial t}\right)^2 + \left(\frac{\partial^2 g(s,t)}{\partial t^2}\right)^2\right] ds\, dt.$$

The result is known as a Laplacian smoothing spline and it, as well as its higher dimensional generalizations, are discussed and illustrated by Wahba and Wendelberger (1980).

Finally, we should mention implementations of the standard smoothing-spline concept and its extensions. For ordinary splines, the most popular case of cubic smoothing splines, based on (1.3) in combination with *cross-validation* (CV) and GCV, is realized in the function `smooth.spline` in S–Plus (MathSoft, 1999). Alternatively, an equivalent degree of freedom can be specified. (See Chapter 2 for details.) S–Plus also offers a restricted form of B–splines (i.e., natural splines) in its function `ns`. Another implementation is found in XploRe4 (Härdle et al., 1999), wherein there is a library `spline` based on the Reinsch concept for cubic smoothing splines. The smoothing parameter can be chosen by a Monte Carlo technique that minimizes the CV score.

For the estimation of semiparametric linear models with partial smoothing splines, there is a macro implemented by Schimek (2000), which can be executed from S–Plus. In this macro, the smoothing parameter λ is chosen via an unbiased risk criterion. Other extensions are not readily available, but algorithmic hints in connection with S–Plus can be found in Green and Silverman (1994).

Acknowledgement

The author's research was supported in part by the National Science Foundation.

REFERENCES

Anselone, P. M., and Laurent, P. J. (1968). A general method for the construction of interpolating or smoothing spline functions, *Numer. Math.*, **12**, 66–82.

Ansley, C. F., and Kohn, R. (1985). Estimation, filtering and smoothing in state space models with incompletely specified initial conditions, *Ann. Statist.*, **13**, 1286–1316.

de Boor, C. (1978). *A Practical Guide to Splines,* Spinger–Verlag, New York.

Cox, D. D. (1983). Asymptotics for M-type smoothing splines, *Ann. Statist.*, **11**, 530–551.

Craven, P., and Wahba, G. (1979). Smoothing noisy day with spline functions: estimating the correct degree of smoothing by the method of generalized cross-validation, *Numer. Math.*, **31**, 377–403.

Davidian, M., Carroll, R. J., and Smith, W. (1988). Variance functions and the minimum detectable concentration in assays, *Biometrika*, **75**, 549–556.

Demmler, A., and Reinsch, C. (1975). Oscillation matrices with spline smoothing, *Numer. Math.*, **24**, 375–382.

Eubank, R. L. (1988). *Spline Smoothing and Nonparametric Regression*, Dekker, New York.

Eubank, R. L. (1994). A simple smoothing spline, *Amer. Statist.*, **48**, 103–106.

Gradshteyn, I. S., and Ryzhik, I. M. (1980). *Table of Integrals, Series and Products*, Academic Press, New York.

Green, P., and Silverman, B. W. (1994). *Nonparametric Regression and Generalized Linear Models: A Roughness Penalty Approach*, Chapman and Hall, London, UK.

Härdle, W., Klinke, S., and Müller, M. (1999). *XploRe—Academic Edition. The Interactive Statistical Computing Environment* (CD-ROM with Handbook). Springer–Verlag, Berlin, Germany.

Hutchinson, M. F., and de Hoog, F. R. (1985). Smoothing noisy data with spline functions, *Numer. Math.*, **47**, 99–106.

Hutchinson, M. F., and de Hoog, F. R. (1987). An efficient method for calculating smoothing splines using orthogonal transformations, *Numer. Math.*, **50**, 311–319.

Kimeldorf, G., and Wahba, G. (1971). Some results on Tchebycheffian spline functions, *J. Math. Anal. and Appl.*, **33**, 82–95.

Kohn, R., and Ansley, C. F. (1988). Equivalence between Bayesian smoothness priors and optimal smoothing for function estimation, in Spall, J. C. (ed.), *Bayesian Analysis of Time Series and Dynamic Models*, pp. 393–430. Dekker, New York.

Kohn, R., and Ansley, C. F. (1989). A fast algorithm for signal extraction, influence and cross-validation in state space models, *Biometrika*, **76**, 65–79.

Li, K. C. (1986). Asymptotic optimality of C_L and generalized cross-validation for ridge regression with application to spline smoothing, *Ann. Statist.*, **1**, 1101–1112.

Lyche, T., and Schumaker, L. L. (1973). Computation of smoothing and interpolating natural splines via local bases, *SIAM J. Numer. Anal.*, **10**, 1027–1038.

MathSoft (1999). *S-PLUS 2000: Guide to Statistics Vol. I*, MathSoft, Inc., Seattle, WA.

Messer, K. (1991). A comparison of a spline estimate to its equivalent kernel estimate, *Ann. Statist.*, **19**, 817–829.

Nussbaum, M. (1985). Spline smoothing in regression models and asymptotic efficiency in L_2, *Ann. Statist.*, **13**, 984–997.

Nychka, D. (1988). Bayesian confidence intervals for smoothing splines, *J. Amer. Statist. Assoc.*, **83**, 1134–1143.

Nychka, D. (1995). Splines as local smoothers, *Ann. Statist.*, **23**, 1175–1197.

Oehlert, G. W. (1992). Relaxed boundary smoothing splines, *Ann. Statist.*, **20**, 146–160.

O'Sullivan, F., Yandell, B. S., and Raynor, W. J. (1986). Automatic smoothing of regression functions in generalized linear models, *J. Amer. Statist. Assoc.*, **81**, 96–103.

Rice, J., and Rosenblatt, M. (1983). Smoothing splines: Regression, derivatives and deconvolution, *Ann. Statist.*, **11**, 141–156.

Schimek, M. G. (2000). Estimation and inference in partially linear models with smoothing splines, to appear in *Journal of Statistical Planning and Inference.*

Shiller, R. J. (1984). Smoothness priors and nonlinear regression, *J. Amer. Statist. Assoc.*, **79**, 609–615.

Silverman, B. W. (1984). Spline smoothing: The equivalent variable kernel method, *Ann. Statist.*, **12**, 896–916.

Silverman, B. W. (1985). Some aspects of the spline smoothing approach to nonparametric regression curve fitting (with discussion), *J. Roy. Statist. Soc. B*, **47**, 1–52.

Speckman, P. L. (1985). Spline smoothing and optimal rates of convergence in nonparametric regression models, *Ann. Statist.*, **13**, 970–983.

Stone, C. (1982). Optimal global rates of convergence for nonparametric regression, *Ann. Statist.*, **10**, 1040–1053.

Utreras, F. (1987). Boundary effects on convergence rates for Tikhonov regularization, *J. Approx. Theor.*, **54**, 235–249.

van der Linde, A. (1993). A note on smoothing splines as Bayesian estimators, *Statistics and Decisions*, **11**, 61–67.

Wahba, G. (1975). Smoothing noisy data by spline functions, *Numer. Math.*, **24**, 383–393.

Wahba, G. (1978). Improper priors, spline smoothing and the problem of guarding against model errors in regression, *J. Roy. Statist. Soc. B*, **40**, 364–372.

Wahba, G. (1983). Bayesian "confidence intervals" for the cross-validated smoothing spline, *J. Roy. Statist. Soc. B*, **45**, 133–150.

Wahba, G. (1984). Partial spline models for the semiparametric estimation of functions of several variables, in *Statistical Analysis of Time Series: Proceedings of the Japan, U.S. Joint Seminar*, pp. 319–329. Institute of Statistical Mathematics, Tokyo, Japan.

Wahba, G. (1990). *Spline Models for Observational Data*, SIAM, Philadelphia, PA.

Wahba, G., and Wendelberger, J. G. (1980). Some new mathematical methods for variational objective analysis using splines and cross validation, *Monthly Weather Review*, **108**, 1122–1143.

Whittaker, E. T. (1923). On a new method of graduation, *Proc. Edinburgh Math. Soc.*, **41**, 63–75.

CHAPTER 2

Variance Estimation and Smoothing-Parameter Selection for Spline Regression

Angelika van der Linde
Edinburgh University, Edinburgh, UK

2.1 INTRODUCTION AND SOME DEFINITIONS

Suppose noisy observations $y_d = (y(t_1), \ldots, y(t_n))$ of a smooth function f on a given domain S, and f is to be reconstructed from y_d. More formally,

$$y_d = f_d + \varepsilon_d, \qquad d = \{t_1, \ldots, t_n\} \subset S \subset \mathrm{IR}^p; \tag{2.1}$$

$$f_d = (f(t_1), \ldots, f(t_n)), \qquad f \text{ smooth}, \tag{2.2}$$

$$\varepsilon_d = (\varepsilon(t_1), \ldots, \varepsilon(t_n)), \qquad E(\varepsilon_d) = 0, \qquad \text{and} \qquad \operatorname{Cov} \varepsilon_d = \sigma^2 \Sigma, \tag{2.3}$$

where $\Sigma > 0$ is known, but the scale factor σ^2 of the error variance (error distribution) is unknown. As an example with simulated data, we consider a design with 50 equidistant points on $S = [0, 1]$. Observations y_d are generated by adding function values f_d from

$$f(t) = 0.5 + t + 3t^2 + \sin(5t)$$

and a realization of the random error $\varepsilon_d \sim N(0, 0.09I)$, where $\Sigma = I$. The simulated data are plotted in Figure 2.1.

Smoothing and Regression: Approaches, Computation, and Application
Edited by Michael G. Schimek
ISBN 0-471-17946-9

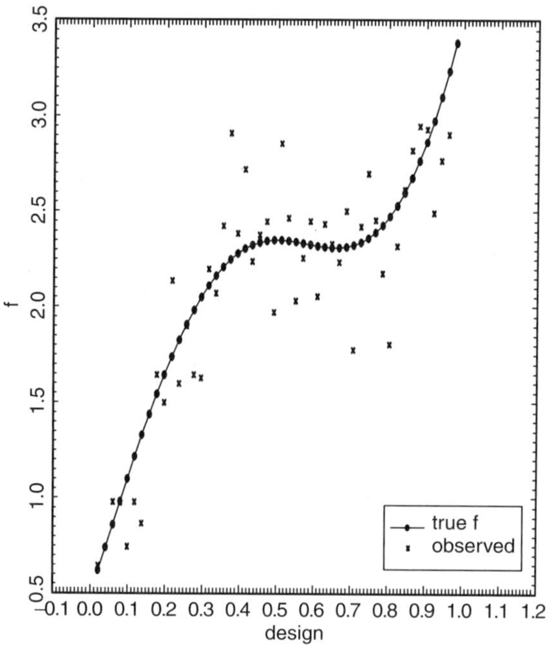

FIGURE 2.1 True function f with simulated observations y_d, where $\varepsilon_d \sim N(0, 0.09\mathrm{I})$. $n = 50$, and values of d are equidistant.

One way to estimate f is by using the smoothing spline $h^{\lambda}_{S(y_d)}$ to solve the optimization problem

$$(y_d - h_d)^T \Sigma^{-1}(y_d - h_d) + \lambda\, J(h) \to \min_{h \in \mathcal{H}}, \tag{2.4}$$

where $\mathcal{H}$ denotes a function space and $J(h)$ a roughness penalty term. (Compare this penalty term with that in Chapter 1.)

The smoothing parameter λ in Equation (2.4) is usually not given, but has to be estimated from the data. The main concern of this chapter is the estimation of the smoothing parameter λ and the scale factor σ^2 of the error variance, respectively the variance function. From the introductory presentation of the smoothing problem in Chapter 1, we know that the estimation of λ and σ is crucial for the fitting of a smoothing spline. To develop a further understanding, we work within an abstract setting in this chapter.

The function space $\mathcal{H}$ is a so-called *reproducing kernel Hilbert space* (RKHS). That is, $\mathcal{H}$ is a complete inner product space with all point evaluation functionals

$$\ell_t : \mathcal{H} \ni h \mapsto h(t) \in \mathbb{R}, \qquad t \in S,$$

being bounded. If ℓ_t is bounded then observing $h(t)$ is informative. Denoting the inner product in $\mathcal{H}$ by $\langle \cdot, \cdot \rangle$ any ℓ_t can then be represented by an element

$\tilde{K}(\cdot, t) \in \mathcal{H}$ such that

$$\ell_t(h) = \langle \tilde{K}(\cdot, t), h \rangle.$$

Since the representers $\tilde{K}(\cdot, t) \in \mathcal{H}$ are functions on S, there is a function

$$\tilde{K} = S \times S \ni (s, t) \mapsto \tilde{K}(s, t) \in \mathbb{R},$$

which is called the reproducing kernel of $\mathcal{H}$. Also, when $\mathcal{H}$ is addressed as an RKHS it is explicitly denoted as $\mathcal{H}(\tilde{K})$. $\mathcal{H}$ is chosen such that the penalty $J(h)$ is a seminorm on $\mathcal{H}$ with finite dimensional nullspace $\ker J = \{h | J(h) = 0\}$. The function space $\mathcal{H}$ can then be decomposed into a direct sum of the subspace of ultimately smooth functions $\ker J$ and an orthogonal complement $(\ker J)^{\perp}$, where the roughness penalty defines a norm. Accordingly, any function $h \in \mathcal{H}$ may be decomposed into $h = h_1 + h_2$. $\ker J$ as a finite dimensional space possesses a reproducing kernel K_1, and it is assumed that $(\ker J)^{\perp}$ does possess a reproducing kernel K, too. Then, by construction, we find that

$$\tilde{K} = K_1 + K$$
$$\mathcal{H}(\tilde{K}) = \ker J \oplus (\ker J)^{\perp} = \mathcal{H}(K_1) \oplus \mathcal{H}(K).$$

As an illustration, we find that the popular univariate cubic spline estimator of f results from the following choices:

$$S = [0, 1], \mathcal{H} = \{h | h, h' \text{ absolutely continuous}, J(h) = \int_0^1 (h''(t))^2 \, dt < \infty\}.$$

Here, the roughness penalty is defined in terms of the second derivative, and the ultimately smooth functions are just the linear functions

$$\ker J = \{h \in \mathcal{H} | h'' = 0 \text{ a.e.}\} = \{h \in \mathcal{H} | h(s) = a + bs, a, b \in \mathbb{R}\}.$$

Any linear function is determined by two values, for example, by $h(0)$ and $h'(0)$, and an inner product for linear functions may therefore be defined using these two values; thus,

$$\langle h, g \rangle_{K_1} = h(0)g(0) + h'(0)g'(0).$$

Any finite dimensional space possesses a reproducing kernel. Here,

$$K_1(t, s) = 1 + st,$$

such that $K_1(0, s) = 1$, $K_1'(\cdot, s) = s = K_1'(0, s)$, and thus

$$\begin{aligned} \langle h, K_1(\cdot, s) \rangle_{K_1} &= h(0)K_1(0, s) + h'(0)K_1'(0, s) \\ &= h(0) \qquad\qquad + h'(0)s = h(s). \end{aligned}$$

In particular, $h(t) = a + bt$, where $h(0) = c_1$ and $h'(0) = c_2$ is given by $a = c_1$ and $b = c_2$, respectively.

The complementary space to ker J does not contain any element of ker J apart from zero. It is obtained as

$$(\ker J)^{\perp} = \{h \in \mathcal{H} | h(0) = h'(0) = 0\},$$

where

$$\langle h, g \rangle_K = \int_0^1 h''(t) g''(t)\, dt.$$

A reproducing kernel K now has to satisfy

$$h(s) = \langle h, K(\cdot, s) \rangle_K = \int_0^1 h''(t) K''(t, s)\, dt.$$

Similarly,

$$h(s) = \int_0^1 h''(t) G(t, s)\, dt,$$

where G is Green's function, with respect to the differential operator D^2 ($D^2 h = h''$) and the initial values $h(0) = h'(0) = 0$, such that

$$G(t, s) = (t - s)_+.$$

Thus, the reproducing kernel is obtained by integrating Green's function:

$$K(s, t) = \int_0^1 (t - x)_+ (s - x)_+\, dx.$$

These ideas are easily generalized to higher order derivatives. Further examples for such a setting are given in Wahba (1990).

2.2 INTERPRETATION OF THE SMOOTHING PARAMETER

The smoothing spline obviously can be described as a penalized least-squares estimate of f, with λ determining the trade-off between data fit and smoothness. Other statistical characterizations of the pointwise estimates $h^{\lambda}_{S(y_d)}(t)$, $t \in S$, reveal that the basic modeling assumption about f is one of bounded roughness, that is

$$J(f) \le \tau^2, \tag{2.5}$$

and

$$\lambda = \sigma^2 / \tau^2. \tag{2.6}$$

The assumption of bounded roughness may be expressed by three distributional assumptions, which differ with respect to the amount of "randomization" in $f = f_1 + f_2$. "f nonrandom" corresponds to a minimax approach, "f_1 nonrandom, but f_2 random" yields the signal extraction approach, and "f random" underlies the Bayesian approach. If f_2 is considered to be random, then the bound of roughness τ^2 is introduced as a scale factor for its covariance. With

$$\begin{aligned} E(y(t)) &= f(t) \qquad t \in S \\ \mathrm{Cov}(y(t), y(s)) &= \mathrm{Cov}(\varepsilon(t), \varepsilon(s)) \qquad s, t \in S, \end{aligned} \tag{2.7}$$

$h^{\lambda}_{S(y_d)}(t)$ is known to be a minimax estimate:

$$\sup_{\substack{h \in \mathcal{H} \\ J(h) \leq \tau^2}} E(h(t) - h^{\lambda}_{S(y_d)}(t))^2 = \min_{c_t \in \mathbb{R}^n} \sup_{\substack{h \in \mathcal{H} \\ J(h) \leq \tau^2}} E(h(t) - c_t' y_d)^2. \tag{2.8}$$

(Compare Equation (2.8) to the work of Eubank, 1988, p. 192f and Li, 1982.) Alternatively, a Bayesian approach may be specified wherein

$$\begin{aligned} y(t) =& f_1(t) + f_2(t) + \varepsilon(t), \\ & f_1 \in \ker J,\ f_2 \in (\ker J)^{\perp},\ f = f_1 + f_2; \end{aligned} \tag{2.9}$$

the conditional distribution of y given f, $p^{y|f}$, is normal with moments as described by Equation (2.7) and the a prior distribution on $\mathcal{H}$ is diffuse on ker J and expresses preference of smooth functions in $(\ker J)^{\perp} \subset \mathcal{H}$. In particular,

$$\begin{aligned} p^{y_d|h_d} &= N(h_d, \sigma^2 \Sigma) \\ p^{h_d} &= N(0, \eta K_{1d} + \tau^2 K_d) \qquad \eta > 0, \end{aligned} \tag{2.10}$$

where K_d is obtained from the reproducing kernel of $(\ker J)^{\perp} \subset \mathcal{H}$ and thus is related to the seminorm $J(\cdot)$. Here, τ^2 describes the center of the prior distribution, with the degree of smoothness being specified a priori. Under these assumptions, we have

$$h^{\lambda}_{S(y_d)}(t) = \lim_{\eta \to \infty} E(h(t)|y_d). \tag{2.11}$$

For more details, see Eubank (1988, Chapter 5.5), Silverman (1985), van der Linde (1993, 1995b), and Wahba (1978, 1983).

Finally, the smoothing spline may be obtained as a realization of a smoothing process within a signal extraction approach. It is assumed that

$$\begin{aligned} y(t) &= x(t) + \varepsilon(t) \\ &= f_1(t) + z(t) + \varepsilon(t) \end{aligned} \tag{2.12}$$

and

$$E(x(t)) = f_1(t), \qquad E(z(t)) = 0,$$
$$\mathrm{Cov}(x(t), x(s)) = \mathrm{Cov}(z(t), z(s)) = \tau^2 K(s, t), \tag{2.13}$$

and ε and z are independent. K again denotes the reproducing kernel of $(\ker J)^{\perp} \subset \mathcal{H}$. Then, $h^{\lambda}_{S(y_d)}(t)$ is the realization of the best linear smoother $x^*(t)$ obtained from

$$E(x(t) - x^*(t))^2 = \min_{c_t \in \mathbb{R}^n} \{E(x(t) - \hat{x}(t))^2 | \hat{x}(t) = c_t' y_d, E(\hat{x}(t)) = f_1(t)\}. \tag{2.14}$$

For further reading, see Eubank (1984a, 1984b), Kimeldorf and Wahba (1970a, 1970b), Kohn and Ansley (1987, 1988), and Wecker and Ansley (1983).

The statistical characterizations of $h^{\lambda}_{S(y_d)}(t)$, $t \in S$, clearly show that τ^2 is a model scale parameter, σ^2 is a noise scale parameter, and λ^{-1} describes the ratio of signal to noise. The choice of λ (in units of σ^2) therefore may be thought of as a model choice, and the criteria of goodness of fit to be minimized in λ may depend on the distributional assumption used. The smoothing spline itself depends only on λ, and thus a decomposition of λ into τ^2 and σ^2 is not required. Therefore, most often, λ is estimated first and σ^2 is determined subsequently. However, alternative approaches, starting with an estimation of σ^2, have also been proposed and are described later in this chapter.

2.3 QUANTIFYING THE COMPLEXITY OF A SMOOTHING SPLINE

In the limiting case of $\lambda \to \infty$ ($\tau^2 \to 0$), the optimization problem (2.4) reduces to regression on ker J; in the case $\lambda \to 0$ ($\tau^2 \to \infty$), the resulting spline is wiggly, interpolating the observations. For the simulated data of our running example, one obtains for $\ln \lambda = -3.25$ a smoothing spline that is already a quite simple function, as can be seen in Figure 2.2.

For $\ln \lambda = -9.5$, the smoothing spline traces the noise around the curve, as depicted in Figure 2.3.

For an optimal value of $\ln \lambda = -6.06$, obtained from Equation (2.44), which is a criterion of robustified cross–validation (see later in this chapter), we observe in Figure 2.4 that the smoothing spline approximates the true function quite well.

For moderate λ, the smoothing spline is an interpolation spline in estimated function values; that is, it is the smoothest function in terms of $J(\cdot)$ that interpolates estimated function values in the design d, where

$$(h^{\lambda}_{S(y_d)})_d = \lim_{\eta \to \infty} E(h_d | y_d) \tag{2.15}$$

is obtained from (2.11). The estimated function values depend linearly on the observations, as can be seen from

$$(h^{\lambda}_{S(y_d)})_d = A(\lambda) y_d, \tag{2.16}$$

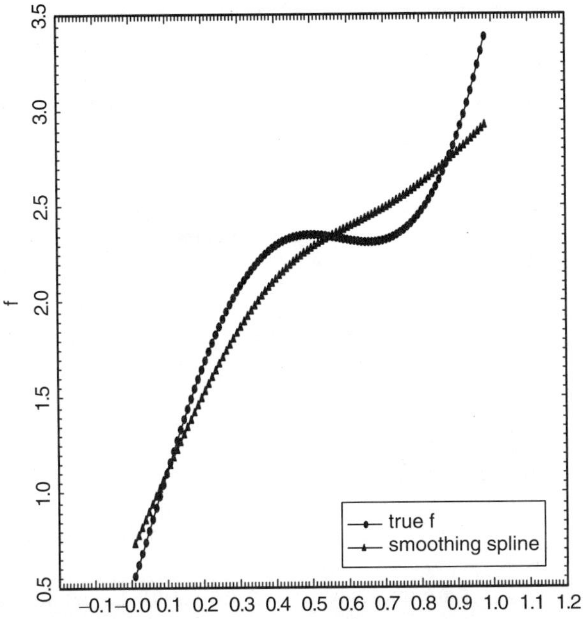

FIGURE 2.2 True function f and smoothing spline $h^{\lambda}_{S(y_d)}$ for $\ln \lambda = -3.25$.

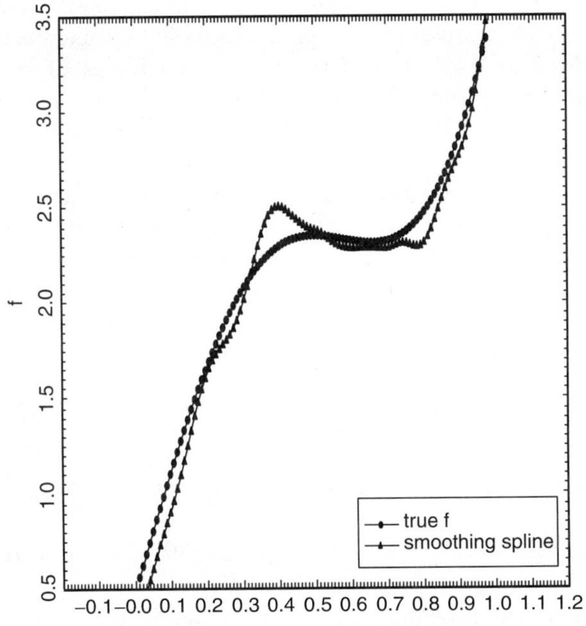

FIGURE 2.3 True function f and smoothing spline $h^{\lambda}_{S(y_d)}$ for $\ln \lambda = -9.5$.

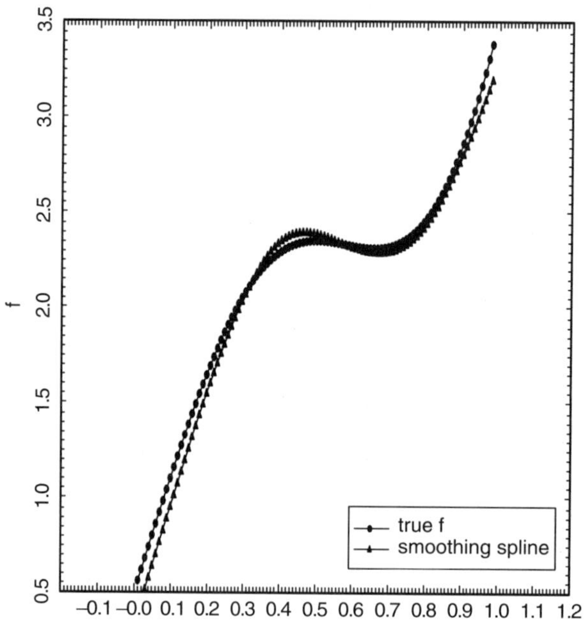

FIGURE 2.4 True function f and smoothing spline $h^{\lambda}_{S(y_d)}$ for $\ln \lambda = -6.06$.

which corresponds to Bayesian estimation in a linear model. The so-called hat matrix $A(\lambda)$ does not represent a projection matrix in $\mathbb{R}^n$, like the matrix resulting from Gauss–Markov estimation. Nevertheless, analogies of this method with usual procedures in linear models have been discussed. (See Eubank, 1984b; Green and Silverman, 1994, Chapter 3.3.4; and Hastie and Tibshirani, 1990, Chapter 3.5.)

Of particular interest has been the question of whether there is any quantity comparable to the number of parameters or to the number of degrees of freedom in parametric regression, since non-Bayesian estimation of σ^2 based on residuals in linear models usually requires an idea about the degrees of freedom. If, for example, $g(t) = (a_1(t), \ldots, a_p(t))^T \beta$, $a_i : S \to \mathbb{R}$, $\beta \in \mathbb{R}^p$; $y_d = A_d\beta + \varepsilon_d$; rank $(A_d) = p$; and $\operatorname{Cov} \varepsilon_d = \sigma^2\Sigma$, then the Gauss–Markov estimator of g_d is

$$\hat{g}_d^{GM} = A_d(A_d^T\Sigma^{-1}A_d)^{-1}A_d^T\Sigma^{-1}y_d,$$

where the number of parameters is given by $p = \text{rank } A_d(A_d^T\Sigma^{-1}A_d)^{-1}A_d^T$ and

$$\hat{\sigma}^2 = \|y_d - \hat{g}_d^{GM}\|^2_{\Sigma^{-1}}/(n-p) = \|\hat{\varepsilon}_d^{GM}\|^2_{\Sigma^{-1}}/(n-p).$$

Here, $\|\cdot\|$ denotes the Euclidian norm. If, in contrast, g_d is known, then

$$\hat{\sigma}^2 = \|y_d - g_d\|^2_{\Sigma^{-1}}/n = \|\varepsilon_d\|^2_{\Sigma^{-1}}/n.$$

The number of unknown parameters then describes the loss in degrees of freedom due to estimation of g_d, if estimated residuals $\hat{\varepsilon}_d^{GM}$, instead of ε_d, are used for estimating σ^2.

Generalizing to a nonparametric model, one estimates σ^2 by

$$\hat{\sigma}^2 = (y_d - \hat{f}_d)\Sigma^{-1}(y_d - \hat{f}_d)/(n - \text{edf}), \tag{2.17}$$

where edf denotes the *equivalent degrees of freedom*. Furthermore, some arguments have been suggested in order to analyze the analogous question for a nonparametric function f estimated by a smoothing spline.

2.3.1 Equivalent Degrees of Freedom

Equivalent degrees of freedom describe the amount of information in the data needed to estimate the residuals. Thus any definition of edf refers to a distributional assumption and a method of estimation.

It is useful to investigate as a basic regression model

$$y_d = h_d + \varepsilon_d, \tag{2.18}$$

with the distributional assumption $h_d \sim N(0, M)$, $M > 0$, and $\varepsilon_d \sim N(0, D)$, where $D > 0$, h_d and ε_d are independent,

$$\hat{h}_d = E(h_d | y_d), \tag{2.19}$$

$$\hat{\varepsilon}_d = y_d - \hat{h}_d = E(y_d - h_d | y_d) = E(\varepsilon_d | y_d). \tag{2.20}$$

Just as the parametric estimator for $g_d = A_d\beta$ results from a limiting case with $\text{Cov}\,\beta = \eta I$, $0 < \eta \to \infty$, the smoothing spline results from a partial limit $M = \eta K_{1d} + \tau^2 K_d$, with $0 < \eta \to \infty$. Let us define

$$\text{edf} := n - \text{tr}\,\text{Cov}\,\hat{\varepsilon}_d(\text{Cov}\,\varepsilon_d)^{-1}. \tag{2.21}$$

This amounts to summing up the roots of the determinantal equation

$$|\,\text{Cov}\,\hat{\varepsilon}_d - \rho D| = 0. \tag{2.22}$$

Using (2.20), we obtain

$$|D(M + D)^{-1}D - \rho D| = 0, \tag{2.23}$$

which is a well known expression in canonical correlation analysis (see, e.g., Rao, 1973, Chapter 8f). The roots ρ_i, $i = 1, \ldots, n$, are the squared canonical correlation coefficients quantifying the mutual linear predictability of ε_d and y_d.

Furthermore, we have

$$\begin{aligned}|\operatorname{Cov}\hat{\varepsilon}_d - \rho D| &= |\operatorname{Cov} E(\varepsilon_d | y_d) - \rho D| \\ &= |\operatorname{Cov} E(y_d - h_d | y_d) - \rho D| \\ &= |\operatorname{Cov}(y_d - E(h_d | y_d)) - \rho D| \\ &= |-\operatorname{Cov} h_d + \operatorname{Cov} E(h_d | y_d) + (1-\rho)D| \\ &= |-\operatorname{Cov}(h_d | y_d) + \kappa D| \end{aligned} \tag{2.24}$$

for $\kappa = 1 - \rho$. From (2.21), we obtain

$$\text{edf} = n - \sum_{i=1}^{n} \rho_i, \tag{2.25}$$

and from (2.24), we get

$$\text{edf} = \sum_{i=1}^{n} \kappa_i = \operatorname{tr} \operatorname{Cov}(h_d | y_d) D^{-1}, \tag{2.26}$$

which is the variation of h_d expressed as a variation of its estimate adjusted for noise.

In the special case of a parametric model, we have for the limit

$$\begin{aligned}\lim_{\eta\to\infty} |\operatorname{Cov}(h_d | y_d) - \kappa D| &= |\operatorname{Cov} \hat{g}_d^{GM} - \kappa\sigma^2\Sigma| \\ &= |\sigma^2 A_d (A_d^T \Sigma^{-1} A_d)^{-1} A_d^T - \kappa\sigma^2\Sigma| = 0, \end{aligned} \tag{2.27}$$

and

$$\text{edf} = \sum_{i=1}^{n} \kappa_i = \operatorname{tr} A_d (A_d^T \Sigma^{-1} A_d)^{-1} A_d^T \Sigma^{-1} = p. \tag{2.28}$$

Under the distributional assumptions for the signal extraction approach, we obtain for the limit

$$\lim_{\eta\to\infty} |\operatorname{Cov}(h_d | y_d) - \kappa D| = |\sigma^2 A(\lambda)\Sigma - \kappa\sigma^2\Sigma| = 0,$$

$$\text{edf} = \sum_{i=1}^{n} \kappa_i = \operatorname{tr} A(\lambda). \tag{2.29}$$

Even if the structure of f is not specified by a covariance, but is given only implicitly by the estimator $\hat{f}_d = A(\lambda) y_d$, the definition of edf in Equation (2.21) may nevertheless be used. Using the distributional assumptions in Equation (2.7) we obtain

$$\operatorname{Cov}\hat{\varepsilon}_d = \operatorname{Cov}((I - A(\lambda)) y_d) = \sigma^2 (I - A(\lambda)) \Sigma (I - A(\lambda))^T, \tag{2.30}$$

and

$$\begin{aligned}\text{edf} &= n - \operatorname{tr}(I - A(\lambda))\Sigma(I - A(\lambda))^T\Sigma^{-1} \\ &= \operatorname{tr}(2A(\lambda) - A(\lambda)\Sigma A(\lambda)^T\Sigma^{-1}).\end{aligned} \tag{2.31}$$

However, (2.24) and (2.26) do not hold; for example,

$$\operatorname{tr}\operatorname{Cov}\hat{h}_d(\sigma^2\Sigma)^{-1} = \operatorname{tr}A(\lambda)\Sigma A(\lambda)^T\Sigma^{-1}. \tag{2.32}$$

Thus, depending on the distributional assumptions, different values for edf result, even for the same estimator, numerically speaking. Using $\operatorname{tr}A(\lambda)$ in place of edf has been proposed by Wahba (1990, p. 63). The concept in (2.31) was used by Buckley et al. (1988) for estimating σ^2. The derivations in the literature, however, are based on heuristics and do not take into account distributional assumptions.

2.4 ESTIMATION OF σ^2

The estimation of the scale factor σ^2 of the error variance (error distribution) is essential for the diagnostics of the spline fit and is important for the determination of the necessary degree of smoothing, as we have seen so far. In this section, we discuss the various approaches to estimating σ^2 found in the literature.

2.4.1 The Minimax Approach

Buckley et al. (1988) investigated a minimax estimator according to the assumption of bounded roughness and the distributional assumptions (2.7). Actually, they argued for the setting of univariate cubic splines. They consider the class of estimators that are quadratic in the data to be nonnegative and unbiased if $f \in \ker J$; that is,

$$\hat{\sigma}^2 = \frac{y_d^T \Delta y_d}{\operatorname{tr}\Delta}, \tag{2.33}$$

where Δ is a symmetric, nonnegative definite matrix.

Their minimax estimator $\hat{\sigma}^2_{\text{BES}}$, is very close to a normalized residual sum of squares based on Speckman's (1985) estimator $\hat{f}_d^{\text{SP}} = A^{\text{SP}} y_d$ for $\Sigma = I$,

$$\hat{\sigma}^2_{\text{BES}} \approx \frac{(y_d - \hat{f}_d^{\text{SP}})^T(y_d - \hat{f}_d^{\text{SP}})}{\operatorname{tr}(I - A^{\text{SP}})^T(I - A^{\text{SP}})}. \tag{2.34}$$

This result may be taken as suggestive for the use of

$$\hat{\sigma}^2_{\text{MM}} = \frac{y_d^T(I - A(\lambda))^T\Sigma^{-1}(I - A(\lambda))y_d}{\operatorname{tr}(I - A(\lambda))^T\Sigma^{-1}(I - A(\lambda))\Sigma}, \tag{2.35}$$

which is an unbiased estimator for σ^2 if $f \in \ker J$ under the assumptions of Equation (2.7).

Estimators of the type in Equation (2.33) and, in particular, $\hat{\sigma}^2_{\text{MM}}$ may be considered to be a residual sum of squares divided by $(n - \text{edf})$. Here, the expression for edf from (2.31) is used, consistent with the distributional assumptions of the minimax approach. The simulated data in our running example yield $\hat{\sigma}^2_{\text{MM}} = 0.140$ as an estimate for the true $\sigma^2 = 0.09$ using $\ln \lambda = -6.06$. (See also Carter and Eagleson, 1992).

Calculating $\hat{\sigma}^2_{\text{MM}}$ requires computation of the smoothing parameter and of $A(\lambda)y_d$. An alternative estimator that is applicable in nonparametric regressions in general and is computationally simpler and yet quite efficient has been proposed by Hall et al. (1990). They investigated the difference-based estimators (for the univariate case wherein $S \subset \mathbb{R}$ and $\Sigma = I$)

$$\hat{\sigma}^2_D = (n-m)^{-1} \sum_{k=1}^{n-m} \left(\sum_{j=0}^{m} d_j y(t_{j+k}) \right), \tag{2.36}$$

where

$$\sum_{j=0}^{m} d_j = 0 \text{ and } \sum_{j=0}^{m} d_j^2 = 1,$$

and studied their asymptotic properties in terms of $\lim_{n\to\infty} E(\hat{\sigma}^2_{D,n} - \sigma^2)$. The estimator $\hat{\sigma}^2_D$ is again a special case of (2.33), as well as estimators based on a residual sum of squares with values of f locally approximated by values of a polynomial. Using a straight line as a local approximation, Gasser et al. (1986) proposed

$$\hat{\sigma}^2_G = (n-2)^{-1} \sum_{j=2}^{n-1} c_j^2 \hat{\varepsilon}_j^2, \tag{2.37}$$

where

$$\hat{\varepsilon}_j^2 = \left[y_{j-1} + \frac{y_{j+1} - y_{j-1}}{t_{j+1} - t_{j-1}} (t_j - t_{j-1}) - y_j \right]^2,$$

and

$$c_j^2 = \left[\left(\frac{t_{j+1} - t_j}{t_{j+1} - t_{j-1}} \right)^2 + \left(\frac{t_j - t_{j-1}}{t_{j+1} - t_{j-1}} \right)^2 + 1 \right]^{-1}.$$

Buckley and Eagleson (1989) further analyzed $\hat{\sigma}^2_G$ and showed that for an equally spaced design d,

$$E(\hat{\sigma}^2_G) = E\left(\frac{y_d^T \Delta_G y_d}{\operatorname{tr} \Delta_G} \right)$$

$$\approx \frac{f_d^T \Delta_G f_d}{\operatorname{tr} \Delta_G} + E\left(\frac{\varepsilon_d^T \Delta_G \varepsilon_d}{\operatorname{tr} \Delta_G}\right)$$

$$\approx \frac{n^{-4}}{6} J(f) + \sigma^2. \tag{2.38}$$

They suggested that $\hat{\sigma}_G^2$ be plotted as a function of n, to study its asymptotic behavior for $n \to \infty$ and that a value $\hat{\sigma}_{G,n_0}^2$ be chosen where the asymptote is reached (i.e., at infinity). Here, $J(f)$ again denotes the penalty yielding cubic splines. Buckley's and Eagleson's procedure works not only for Δ_G, but also for a more general Δ, such as that in Equation (2.33).

In order to generalize the asymptotic results obtained by Buckley et al. (1988) for estimators (2.34) and (2.35) to estimators based on kernels, Hall and Marron (1990) investigated, for the univariate case, for $\Sigma = I$, and for a set of estimators $\hat{f}$ for f of the form

$$\hat{f}(t_j) = \sum_{i=1}^{n} w_{ij} y(t_i),$$

where

$$\sum_{i=1}^{n} w_{ij} = 1$$

(which includes $h_{S(y_d)}^{\lambda}$) estimators for σ^2, namely,

$$\hat{\sigma}_H^2 = c^{-1} \sum_{j=1}^{n} \left(y(t_j) - \sum_{i=1}^{n} w_{ij} y(t_i) \right)^2, \tag{2.39}$$

where

$$c = n - 2\sum_{i=1}^{n} w_{ij} + \sum_{i,j=1}^{n} w_{ij}^2.$$

These estimators are unbiased if $f = 0$.

A good overview of some of the above mentioned estimators of σ^2, along with a simulation study of their performance, is given by Thomson et al. (1991). Finite sample properties have been studied by Seifert et al. (1993).

2.4.2 The Bayesian Approach

Within the empirical Bayesian approach of (2.9) and (2.10), estimation of the smoothing parameter λ as well as of the scale factor σ^2 (given λ) is based on the maximum likelihood concept with respect to the marginal distribution of the observations. Actually, because of the improper prior on ker J, transformed observations u are used:

$$u = R^T y_d. \tag{2.40}$$

The transformation matrix $R^T \in \mathrm{I\!R}^{(n-M)\times n}$, where $M = \dim \ker J$, is chosen such that $R^T h_{1d} = 0$ for $h_1 \in \ker J$. The resulting estimator is

$$\hat{\sigma}_B^2 = \frac{1}{n-M}\lambda(u^T(R^T K_d R + \lambda R^T \Sigma R)u). \tag{2.41}$$

For details, see Eubank (1988, Chapter 5.5.2).

Because the conditional distribution of y_d depends on two unknown parameters f_d and σ^2 and also on the hyperparameter τ^2, the fully Bayesian approach would require to specify priors for f_d, σ^2, and τ^2 in a hierarchical model. (See, e.g., Broemeling, 1985.) The posteriors then may be obtained from Markov chain Monte Carlo sampling. For related ideas, see Chapter 18 and Frühwirth–Schnatter (1994). For the special case of univariate cubic splines and $\Sigma = I$, Shiller (1984) showed that specifying a noninformative prior for σ^2 again yields (2.41) when τ^2 is fixed. Starting from

$$u = R^T y_d \sim N(0, R^T(\tau^2 K_d + \sigma^2\Sigma)R) \tag{2.42}$$

and applying some procedure for the estimation of variance components (such as REML or MINQUE; see, e.g., Christensen, 1987, Chapter XII), we may obtain an estimator for σ^2 without predetermining a value of λ.

2.4.3 The Signal Extraction Approach

In the signal extraction approach, the distribution of u is the same as the marginal distribution of the transformed observations in the Bayesian approach, and thus the same likelihood-based procedures may be applied. Furthermore, the discussion in Section 2.3 and, in particular, Equation (2.29), which gives edf $= \operatorname{tr} A(\lambda)$ under the signal extraction approach, show that Wahba's estimator

$$\hat{\sigma}_W^2 = \frac{y_d^T(I - A(\lambda))^T\Sigma^{-1}(I - A(\lambda))y_d}{\operatorname{tr}(I - A(\lambda))} \tag{2.43}$$

is strongly supported under the distributional assumptions (2.12) and (2.13). In our running example, the true value of $\sigma^2 = 0.09$ is estimated by $\hat{\sigma}_W^2 = 0.139$.

2.5 DETERMINATION OF λ

Our goal is to estimate $\lambda = \sigma^2/\tau^2$. Although most of the procedures that have been proposed to optimize λ are related to a distributional assumption — for example, with respect to asymptotic optimality properties — we would like an estimate to be robust to its derivation and to work well in the reconstruction of a function of bounded roughness no matter what the distributional assumptions are.

A suggestion that does not refer to a distributional assumption is the so-called *robustified cross-validation* (RCV) proposed by Robinson and Moyeed (1989).

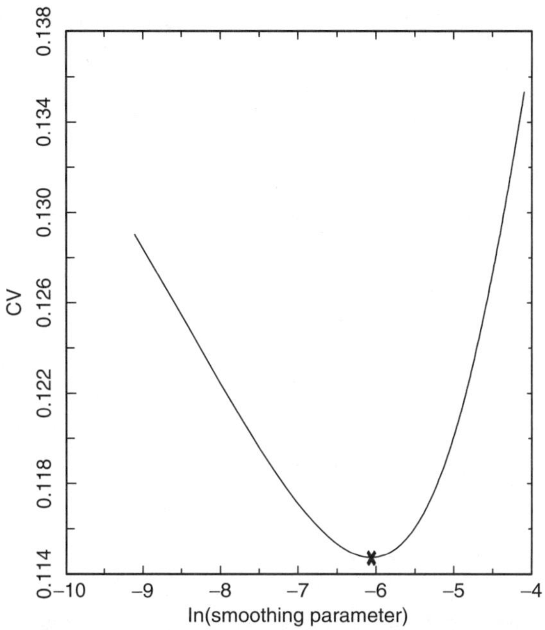

FIGURE 2.5 The criterion for robustified cross-validation in Equation (2.44) as a function of ln(λ).

The idea is simply that the estimated curve should not change very much if one observation is left out during estimation. After some algebraic manipulations, this procedure results in minimizing with respect to λ the criterion

$$\text{RCV}\ (\lambda) = n^{-1} \frac{1 + n^{-1}\,\text{tr}\,A(\lambda)^2}{(1 - n^{-1}\,\text{tr}\,A(\lambda))^2} \|(I - A(\lambda))y_d\|^2 \tag{2.44}$$

in case $\Sigma = I$. For a generalization to the case $\Sigma \neq I$, see van der Linde (1994).

The smoothing spline shown in Figure 2.4 results from minimizing RCV (λ). The criterion RCV as a function of $\ln \lambda$ is displayed in Figure 2.5.

Another natural criterion to optimize λ is the minimization of the estimation error $f_d - (h^{\lambda}_{S(y_d)})_d$, which may be measured by $\sigma^2\,\text{tr}\,A(\lambda)\Sigma$ under all of our distributional assumptions (i.e., signal extraction, Bayesian, and minimax), since as proven by van der Linde (1998)

$$\begin{aligned}
&\text{tr}\,\text{Cov}(x^*_d - x_d) \\
&= \text{tr} \lim_{\eta \to \infty} \text{Cov}(h_d | y_d) \\
&= \sum_{i=1}^{n} \min_{c_{t_i} \in \mathbb{R}^n} \sup_{J(h) \leq \tau^2} E(h(t_i) - c'_{t_i} y_d)^2 \\
&= \sigma^2\,\text{tr}\,A(\lambda)\Sigma.
\end{aligned} \tag{2.45}$$

However, $\sigma^2 \operatorname{tr} A(\lambda)\Sigma$ is the error occuring with respect to the class of regression functions of bounded roughness, and it is small when the class, that is, τ^2, is small. Therefore, in determining λ, one has to consider the estimation error for the one true function f, which is, of course, unknown.

For the assumption that f is nonrandom, most of the work has been done based on the distributional assumption specified in (2.7). The estimation (mean squared) error under this assumption is decomposed into bias and variance:

$$\begin{aligned} n^{-1}E(\|f_d - (h^{\lambda}_{S(y_d)})_d\|^2) &= n^{-1}E(\|f_d - A(\lambda)y_d\|^2) \\ &= n^{-1}\|(I - A(\lambda))f_d\|^2 + \sigma^2 n^{-1} \operatorname{tr} A(\lambda)\Sigma A(\lambda)^T. \quad (2.46) \end{aligned}$$

The idea, then, is that a good choice of λ yields a favorable compromise between bias and variance, since the bias increases and the variance decreases with the amount of smoothing. An unbiased estimate of the mean squared error is given by *goodness of fit* (GF)

$$\text{GF}\ (\lambda) = n^{-1}\|(I - A(\lambda))y_d\|^2 - \sigma^2 n^{-1} \operatorname{tr}(\Sigma - 2A(\lambda)\Sigma). \quad (2.47)$$

The process of plugging in an estimate for σ^2, such as $\hat{\sigma}^2_{\text{MM}}$ from (2.35), and minimizing (2.47) did not perform well in simulations by the author of this chapter. However, differentiating GF first for $\Sigma = I$ results in

$$\frac{d}{d\lambda} n^{-1}\|(I - A(\lambda))y_d\|^2 = -\sigma^2 2n^{-1} \frac{d}{d\lambda} \operatorname{tr} A(\lambda).$$

The result from plugging in $\hat{\sigma}^2_W = \|(I - A(\lambda))y_d\|^2 / \operatorname{tr}(I - A(\lambda))$ is then equivalent to

$$0 = \frac{d}{d\lambda}\left(\frac{\|(I - A(\lambda))y\|^2}{(n^{-1}\operatorname{tr}(I - A(\lambda)))^2}\right) = \frac{d}{d\lambda}\ \text{GCV}\ (\lambda), \quad (2.48)$$

which also results from an idea of cross-validation by Gu et al. (1992) that is explained later in this chapter.

Härdle (see discussion in Silverman, 1985, p. 37) has argued that in the case for which $\Sigma = I$, minimizing a criterion of goodness of fit of the form

$$\text{GF}_\psi(\lambda) = n^{-1}\|(I - A(\lambda))y_d\|^2 \psi(n^{-1} \operatorname{tr} A(\lambda)), \quad (2.49)$$

where

$$\psi(t) = 1 + 2t + \psi''(t_0)t^2,$$

is asymptotically equivalent to minimizing $n^{-1}\|f_d - A(\lambda)y_d\|^2$. Examples for this argument are given by Härdle and Marron (1985) and by Robinson and Moyeed (1989). Note that equation (2.49) includes *robustified cross-validation* (RCV) of (2.44) as well as *generalized cross-validation* (GCV) of (2.48).

2.5.1 Cross-Validation

A very popular and successful concept that was created by Craven and Wahba (1979) for the selection of λ is *cross-validation* (CV)—that is, predicting an observed value from the remaining observations and choosing the λ that yields the best prediction. If $\hat{y}^{\lambda}_{-i}(t_i)$ denotes the prediction of $y(t_i)$ when $y(t_i)$ is removed from the data, then the simplest CV criterion is

$$\text{CV }(\lambda) = n^{-1}\sum_{i=1}^{n}(y(t_i) - \hat{y}^{\lambda}_{-i}(t_i))^2. \tag{2.50}$$

For independent observations—that is, the case for which $\Sigma = I$—we have

$$\hat{y}^{\lambda}_{-i}(t_i) = h^{\lambda}_{S(y_d\setminus\{t_i\})}(t_i).$$

A computational simplification can be achieved by applying the deletion theorem (see Wahba, 1990, p. 50), giving

$$y(t_i) - h^{\lambda}_{S(y_d\setminus\{t_i\})}(t_i) = (1 - a_{ii}(\lambda))^{-1}(y(t_i) - h^{\lambda}_{S(y_d)}(t_i)), \tag{2.51}$$

where the values of $a_{ii}(\lambda)$ denote the elements in the diagonal of $A(\lambda)$. Substituting (2.51) in (2.50) and generalizing $a_{ii}(\lambda)$ to $n^{-1}\,\mathrm{tr}\,A(\lambda)$ then yields the criterion of generalized cross-validation

$$\text{GCV }(\lambda) = \frac{\|(I - A(\lambda))y_d\|^2}{(n^{-1}\,\mathrm{tr}(I - A(\lambda)))^2}, \tag{2.52}$$

already introduced in (2.48). A generalization to the case $\Sigma \neq I$ is discussed by Diggle and Hutchinson (1989) and by van der Linde (1995a). GCV for data with replications is analyzed in Gu et al. (1992).

The asymptotic performance of GCV for $\Sigma = I$ has been extensively studied (Kay, 1992; Li, 1985, 1986; Utreras, 1988; Wahba, 1979a, 1979b, 1990 (Chapter 4)). Some theorems relate GCV to the minimization of the mean squared error (2.46).

The *weak* GCV theorem states that there is a sequence $(\tilde{\lambda}_n)_{n\in\mathbb{N}}$ such that $\tilde{\lambda}_n$ minimizes $E(\mathrm{GCV}_n(\lambda))$ and

$$\lim_{n\to\infty} E(\|f_{d_n} - (h^{\tilde{\lambda}_n}_{S(y_{d_n})})_{d_n}\|^2)/\min_{\lambda} E(\|f_{d_n} - (h^{\lambda}_{S(y_{d_n})})_{d_n}\|^2) = 1. \tag{2.53}$$

The *strong* GCV theorem expresses that there is a sequence $(\hat{\lambda}_n)_{n\in\mathbb{N}}$ such that $\hat{\lambda}_n$ minimizes $\mathrm{GCV}_n(\lambda)$ and

$$\lim_{n\to\infty} E(\|f_{d_n} - (h^{\hat{\lambda}_n}_{S(y_{d_n})})_{d_n}\|^2/\min_{\lambda} E(\|f_{d_n} - (h^{\lambda}_{S(y_{d_n})})_{d_n}\|^2 = 1. \tag{2.54}$$

The weak GCV theorem also holds for the criterion GF_ψ given by (2.49), but the strong theorem does not. (For details, proofs, and convergence rates, see Wahba, 1990, Chapter 4.)

For the data in our example, the minimum of GCV (λ) yields $\ln\lambda = -6.34$ (see Figure 2.5), thus producing a smoothing spline very similar to that displayed in Figure 2.4. For small sample sizes, some evidence has been reported that the application of GCV (λ) may lead to undersmoothing. Heuristically, such an effect is likely to occur because CV (λ) as a sum of unstandardized residuals supports a minimization of σ^2. It may be helpful in this context to recall Geisser and Eddy's (1979) distinction of *low-structure selection* and *high-structure selection* of a model parameter. Low-structure selection is based on the deviations of observed values $y(t_i)$ to predicted values $\hat{y}^\lambda_{-i}(t_i)$, as is CV (λ). High-structure selection denotes a maximum-likelihood approach, according to which, for independent observations, the product of densities in $y(t_i)$ with unknown parameters is estimated based on $y_{d\setminus\{t_i\}}$ is maximized with respect to the model parameter. For dependent observations, it is suggested that the product of conditional densities of $y(t_i)$ given $y_{d\setminus\{t_i\}}$ be used. This process requires (2.7) to hold true and involves standardized residuals.

High structure cross-validation is closely related to penalized log likelihood. A result of Stone (1977) states that in the case of independent observations, the sum over i of the values of the log-likelihood function in $y(t_i)$ with parameters estimated by maximum likelihood based on $d \setminus \{t_i\}$ is asymptotically equivalent to *Akaike's information criterion* (AIC), which is the log-likelihood function in y_d with maximum-likelihood estimates of the parameters, penalized by the dimension of the vector of parameters. This result is not directly applicable to splines, since $h^\lambda_{S(y_{d\setminus\{t_i\}})}$ is a restricted, instead of a simple, maximum-likelihood estimate and σ^2 as an additional noise parameter has to be taken into account.

However, there is a parallel of AIC:

$$\mathrm{GF}_{\mathrm{AIC}}(\lambda) := n^{-1}\|(I - A(\lambda))y_d\|^2 + \exp(2n^{-1}\,\mathrm{tr}\,A(\lambda)). \tag{2.55}$$

Minimization of $\mathrm{GF}_{\mathrm{AIC}}$ looks like maximization of the log-likelihood function in y_d (for $\Sigma = I$), penalized by a measure of the complexity of the function that substitutes the number of unknown parameters.

2.5.2 Alternatives Based on Preestimated Values of σ^2

Finally, some procedures have been proposed based on some prior estimation of σ^2. Starting with the heuristics

$$E(\|\Sigma^{-1/2}(I - A(\lambda))y_d\|^2) = \sigma^2(n - \mathrm{edf}), \tag{2.56}$$

edf are specified; σ^2 is substituted for by an estimate that does not depend on λ, such as $\hat{\sigma}^2_D$ from (2.36) or $\hat{\sigma}^2_G$ from (2.37); and Equation (2.56) is solved for λ. (Compare with Hall and Titterington, 1987). Of course, a crucial point in this method is the determination of the correct edf. The relationship of this approach to other criteria in use when λ is estimated is discussed in Kay (1992).

A graphical procedure to determine λ from a plot of $\hat{\sigma}^2(\lambda)$ was proposed by Eagleson (1989). He argued that, in the case for which $\Sigma = I$, $\hat{\sigma}^2(\lambda)$ reaches an asymptote when the model is sufficiently complex or λ is as small as possible. The smoothing parameter to be singled out is the one at the "beginning" of the asymptote; that is, the model is chosen to be as simple as possible, but as complex as necessary. According to my experience, the procedure works only when the signal-to-noise ratio is large or when the function to be reconstructed is not hidden in noise.

2.5.3 The Bayesian Approach

Within the empirical Bayesian approach, the selection of the smoothing parameter is based mainly on maximizing the marginal likelihood of the observations. The likelihood function derived in Section 2.4.2 for transformed data $u = R^T y_d$ is the same as that for transformed data in the signal extraction approach.

Akaike (1980) proposed that the marginal log likelihood penalized by the number of hyperparameters be maximized. For splines, the number of hyperparameters is constantly equal to two, and thus his proposal results in a restricted maximum-likelihood estimation of two variance components, σ^2 and τ^2, or of λ and σ^2. A fully Bayesian approach allowing for informative priors for the variance components is discussed by Kitanides (1986). (See also the references discussed in Section 2.4.2.)

2.5.4 The Signal Extraction Approach

Within the signal extraction approach, maximum-likelihood methods to estimate variance components have been studied in the more general context of spatial statistics. In particular, the methods are known to be numerically crucial (Ripley, 1988, Chapter 2). Also in spatial statistics, low-structure CV criteria of goodness of fit have been discussed with conceptual difficulties found with respect to an appropriate standardization (Ripley, 1981, p. 57; Laslett et al., 1987).

The performance of the restricted maximum-likelihood estimate $\hat{\lambda}_{\text{ML}}$ of λ has been compared to $\hat{\lambda}_{\text{GCV}}$ that results from GCV. Wahba (1985) showed that under the distributional assumption of Equation (2.7) and for $\Sigma = I$, $\hat{\lambda}_{\text{ML}}$ performs asymptotically as well as $\hat{\lambda}_{\text{GCV}}$ if f is polynomial or indeed, a path of a stochastic process as specified in the signal extraction approach (2.13). Maximum-likelihood estimation, however, is inferior to GCV for $f \in C_p$, $p \in (1,2]$. Stein (1990) refined this investigation and showed for the special case where the process is Brownian motion and $\Sigma = I$ that, under the (correct) distributional assumptions of the signal extraction approach, $\hat{\lambda}_{\text{GCV}}$ has twice the asymptotic variance of $\hat{\lambda}_{\text{ML}}$.

A simulation study comparing the finite-sample behavior of CV and maximum-likelihood estimates of λ by Kohn et al. (1991) also indicates that $\hat{\lambda}_{\text{ML}}$ is superior.

2.6 ESTIMATION OF τ^2

There are two unknown parameters to be estimated: the prior scale factor τ^2 and the scale factor of the error distribution σ^2, or equivalently, the smoothing parameter $\lambda = \sigma^2/\tau^2$ and σ^2. Apart from the estimation of variance components, all of the aforementioned approaches deal with the estimation of λ and σ^2. Given estimates of $\hat{\lambda}$ and $\hat{\sigma}^2$, one trivially obtains, for the hyperparameter,

$$\hat{\tau}^2 = \hat{\sigma}^2/\hat{\lambda} \tag{2.57}$$

as an estimate for the bound of roughness or the size of the class of regression functions. Since spline smoothing may be looked at as smooth interpolation of estimated function values, $\hat{\tau}^2$ may be used to assess the interpolation error occuring for functions of bounded roughness.

However, experience from simulated examples (van der Linde, 1995a) indicates that, within an environment of $\hat{\lambda}$, $\hat{\sigma}^2(\hat{\lambda})$ often varies only slightly, and a change in $\hat{\lambda}$ is essentially reflected in $\hat{\tau}^2(\hat{\lambda})$ and may vary dramatically. Thus, $\hat{\tau}^2(\hat{\lambda})$ may not be reliable.

2.7 FINAL REMARKS

The discussion concerning the best estimator for the smoothing parameter is still going on, particularly with respect to generalizations. Most often, it is assumed that there are no replicates in the data. A paper by Gu et al. (1992) is the only study that deals with repeated observations for GCV. Further results of the comparison of the performance of GCV with maximum likelihood (for random f_2) would be of interest for this case.

Some attention has also been paid to correlated errors (Diggle and Hutchinson, 1989; Kohn et al., 1992; Schimek and Schmaranz, 1994; van der Linde, 1994). It might turn out to be difficult to tell particularly autoregressive or moving-average errors from the spline component, since discrete splines are closely related to such error structures (see Peele, 1979). For discussion of the problem of smoothing in the presence of dependent errors, see Chapter 6 in this volume.

For implementational issues of smoothing-parameter selection in spline regression we refer to Chapter 1 and the references therein.

REFERENCES

Akaike, H. (1980). Likelihood and the Bayes procedure, in Bernardo, J. M., et al. (eds.), *Bayesian Statistics*, pp. 143–146, University Press, Valencia, Spain.

Broemeling, L. D. (1985). *Bayesian Analysis of Linear Models*, Dekker, New York.

Buckley, M. J., Eagleson, G. K., and Silverman, B. W. (1988). The estimation of residual variance in nonparametric regression, *Biometrika*, **75**, 189–199.

Buckley, M. J., and Eagleson, G. K. (1989). A graphical method for estimating the residual variance in nonparametric regression, *Biometrika*, **76**, 203–210.

Carter, C. K., and Eagleson, G. K. (1992). A comparison of variance estimators in nonparametric regression, *J. Roy. Statist. Soc. B*, **54**, 773–780.

Christensen, R. (1987). *Plane Answers to Complex Questions*, Springer–Verlag, New York.

Craven, P., and Wahba, G. (1979). Smoothing noisy data with spline functions, *Numer. Math.*, **31**, 377–403.

Diggle, P., and Hutchinson, M. F. (1989). On spline smoothing with autocorrelated errors, *Austral. J. Statist.*, **31**, 166–182.

Eagleson, G. K. (1989). Curve estimation — whatever happened to the variance?, *Bulletin of the International Statistical Institute*, **53**, 535–551.

Eubank, R. L. (1984a). Approximate regression models and splines, *Commun. Statist. — Theor. Meth.*, **13**, 433–484.

Eubank, R. L. (1984b). The hat matrix for smoothing splines, *Statist. Prob. Letters*, **2**, 9–14.

Eubank, R. L. (1988). *Spline Smoothing and Nonparametric Regression*, Dekker, New York.

Frühwirth–Schnatter, S. (1994). Data augmentation and dynamic linear models, *J. Time Series Analysis*, **15**, 183–202.

Gasser, T., Sroka, L., and Jennen-Steinmetz, C. (1986). Residual variance and residual pattern in nonlinear regression, *Biometrika*, **73**, 625–633.

Geisser, S., and Eddy, W. F. (1979). A predictive approach to model selection, *J. Amer. Stat. Assoc.*, **74**, 153–160.

Green, P., and Silverman, B. W. (1994). *Nonparametric Regression and Generalized Linear Models*, Chapman and Hall, London, UK.

Gu, C., Heckmann, N., and Wahba, G. (1992). A note on generalized cross-validation with replicates, *Statist. Prob. Letters*, **14**, 283–287.

Hall, P., and Titterington, D. M. (1987). Common structure of techniques for choosing smoothing parameters in regression problems, *J. Roy. Statist. Soc. B*, **49**, 184–198.

Hall, P., Kay, J. W., and Titterington, D. M. (1990). Asymptotically optimal difference-based estimation of variance in nonparametric regression, *Biometrika*, **77**, 521–528.

Hall, P., and Marron, J. S. (1990). On variance estimation in nonparametric regression, *Biometrika*, **77**, 415–419.

Härdle, W., and Marron, J. S. (1985). Asymptotic nonequivalence of some bandwidth selectors in nonparametric regression, *Biometrika*, **72**, 481–484.

Hastie, T. J., and Tibshirani, R. J. (1990). *Generalized Additive Models*, Chapman and Hall, London, UK.

Kay, J. W. (1992). Asymptotic comparison factors for smoothing parameter choices in regression problems, *Statist. Prob. Letters*, **15**, 329–335.

Kimeldorf, G., and Wahba, G. (1970a). Spline functions and stochastic processes, *Sankhyā Ser. A*, **132**, 173–180.

Kimeldorf, G., and Wahba, G. (1970b). A correspondence between bayesian estimation of stochastic processes and smoothing by splines, *Ann. Math. Statist.*, **41**, 495–502.

Kitanides, P. K. (1986). Parameter uncertainty in estimation of spatial functions: Bayesian analysis, *Water Resources Research*, **22**, 499–507.

Kohn, R., and Ansley, C. F. (1987). A new algorithm for spline smoothing based on smoothing a stochastic process, *SIAM J. Sci. Stat. Comp.*, **8**, 33–34.

Kohn, R., and Ansley, C. F. (1988). Equivalence between Bayesian smoothness priors and optimal smoothing for function estimation, in Spall, J. C. (ed.), *Bayesian Analysis of Time Series and Dynamic Models*, Dekker, New York.

Kohn, R., Ansley, C. F., and Tharm, D. (1991). The performance of cross-validation and maximum likelihood estimators of spline smoothing parameters, *J. Amer. Statist. Assoc.*, **86**, 1042–1049.

Kohn, R., Ansley, C. F., and Wong, C. (1992). Nonparametric spline regression with autoregressive moving average errors, *Biometrika*, **79**, 335–346.

Laslett, G. M., McBratney, A. B., Pahl, P. J., and Hutchinson, M. F. (1987). Comparison of several spatial prediction methods for soil pH, *J. Soil Science*, **38**, 325–341.

Li, K.-C., (1982). Minimaxity of the method of regularization on stochastic processes, *Ann. Statist.*, **10**, 937–942.

Li, K.-C. (1985). From Stein's unbiased risk estimates to the method of generalized cross-validation, *Ann. Statist.*, **13**, 1352–1377.

Li, K.-C. (1986). Asymptotic optimality of C_L and generalized cross-validation in ridge regression with application to spline smoothing, *Ann. Statist.*, **14**, 1101–1112.

van der Linde, A. (1993). A note on smoothing splines as Bayesian estimates, *Statistics and Decisions*, **11**, 61–67.

van der Linde, A. (1994). On cross-validation for smoothing splines in the case of dependent observations, *Austral. J. Statist.*, **36**, 67–73.

van der Linde, A. (1995a). Computing the error for smoothing splines, *Computational Statistics*, **10**, 143–154.

van der Linde, A. (1995b). Splines from a Bayesian point of view, *Test*, **4**, 63–81.

van der Linde, A. (1998). Smoothing errors, *Statistics*, **31**, 91–114.

Peele, L. (1979). Characterizations of autoregressive prediction and moving average predictions as discrete-time spline interpolations, *Commun. Statist.—Theor. Meth.*, **8**, 35–46.

Rao, C. R. (1973). *Linear Statistical Inference and its Applications*, Wiley, New York.

Ripley, B. D. (1981). *Spatial Statistics*, Wiley, New York.

Ripley, B. D. (1988). *Statistical Inference for Spatial Processes*, Cambridge University Press, Cambridge, UK.

Robinson, T., and Moyeed, R. (1989). Making robust the cross-validatory choice of smoothing parameter in spline smoothing regression, *Commun. Statist. Theor. Meth.*, **18**, 523–539.

Schimek, M. G., and Schmaranz, K. G. (1994). Dependent error regression smoothing: A new method and PC program, *Comp. Statist. Data Anal.*, **17**, 457–464.

Seifert, B., Gasser, T., and Wolf, A. (1993). Nonparametric estimation of residual variance revisited. *Biometrika*, **80**, 373–383.

Shiller, R. J. (1984). Spline smoothing and optimal rates of convergence in nonparametric regression models, *Ann. Statist.*, **1**, 970–983.

Silverman, B. W. (1985). Some aspects of the spline smoothing approach to nonparametric regression curve fitting, *J. Roy. Statist. Soc. B*, **47**, 1–52 (with discussion).

Speckman, P. (1985). Spline smoothing and optimal rates of convergence in nonparametric regression models, *Ann. Statist.*, **13**, 970–983.

Stein, M. (1990). A comparison of generalized cross-validation and modified maximum likelihood for estimating the parameters of a stochastic process, *Ann. Statist.*, **18**, 1139–1157.

Stone, M. (1977). An asymptotic equivalence of choice of model by cross-validation and Akaike's criterion, *J. Roy. Statist. Soc. B*, **39**, 44–47.

Thomson, A. M., Kay, J. W., and Titterington, D. M. (1991). Noise estimation in signal restoration using regularization, *Biometrika*, **78**, 475–488.

Utreras, F. (1988). Convergence rates for multivariate smoothing spline functions, *J. Approx. Theor.*, **52**, 1–27.

Wahba, G. (1978). Improper priors, spline smoothing and the problem of guarding against errors in regression, *J. Roy. Statist. Soc. B*, **40**, 364–370.

Wahba, G. (1979a). How to smooth curves and surfaces with splines and cross-validation, in *Procedures of the 24th Conference on the Design of Experiments*, pp. 167–192, U.S. Army Research Office, HO 79-2, Research Triangle Park, NC.

Wahba, G. (1979b). Convergence rates of thin plate smoothing splines when the data are noisy, in Gasser, T., and Rosenblatt, M. (eds.), *Smoothing Techniques for Curve Estimation*, pp. 233–245, Springer–Verlag, Berlin, Germany.

Wahba, G. (1983). 'Bayesian confidence intervals' for the cross-validated smoothing spline, *J. Roy. Statist. Soc. B*, **45**, 133–150.

Wahba, G. (1985). A comparison of GCV and GML for choosing the smoothing parameter in the generalized spline smoothing problem, *Ann. Statist.*, **13**, 1378–1402.

Wahba, G. (1990). *Spline Models for Observational Data*, SIAM, Philadelphia, PA.

Wecker, W. E., and Ansley, C. F. (1983). The signal extraction approach to nonlinear regression and spline smoothing, *J. Amer. Stat. Assoc.*, **78**, 351–365.

CHAPTER 3

Kernel Regression

Pascal Sarda and Philippe Vieu
Université Paul Sabatier, Toulouse, France

3.1 INTRODUCTION

Given two real random variables X and Y, we denote the regression function of Y on X by

$$m(\cdot) = E[Y|X = \cdot]. \tag{3.1}$$

Our aim is to estimate the function m from a sample (X_i, Y_i), $i = 1, \ldots, n$ of independent realizations of (X, Y). Figure 3.1 shows a simulated example for which the sample is generated according to the relation

$$Y_i = -\exp(-20X_i^2) + X_i + X_i^2 I(X_i > 0) + \epsilon_i, \quad i = 1, \ldots, 100, \tag{3.2}$$

where the X_i's are drawn from the uniform distribution $U([-1, +1])$ and the ϵ_i are drawn from the normal distribution $N(0, 0.3)$.

We make repeated use of this data set througout this chapter to illustrate the behavior of kernel estimates on some finite sample. We use this example because it is typical of nonparametric problems, since it is impossible to define some reasonable *a priori* parametrization by looking only at the data cloud in Figure 3.1(a).

An outline of this chapter is as follows: First, we discuss the Nadaraya–Watson kernel regression estimate, which is defined from practical intuition in the next section. Then, we present some asymptotic results, paying particular attention to L_2 loss functions: Section 3.3 studies pointwise bias, Section 3.4 studies pointwise variance, and Sections 3.5 and 3.6 present asymptotic evaluations for the mean squared error (MSE) (pointwise and integrated, respectively). Our main

Smoothing and Regression: Approaches, Computation, and Application
Edited by Michael G. Schimek
ISBN 0-471-17946-9

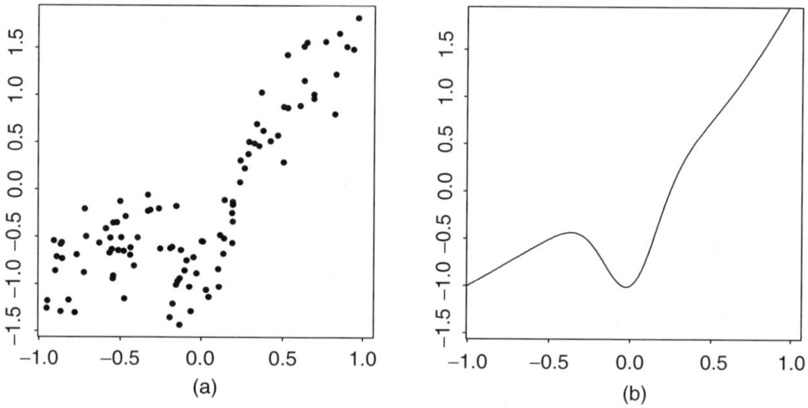

FIGURE 3.1 (a) The data cloud; (b) the true regression line.

results are Theorems 3.5 and 3.6, from which it can be observed that the regression estimate under consideration has some optimality properties. Similar results are given more briefly in Section 3.7, in which the quadratic loss function is changed to some L_∞ criterion. The final section, Section 3.8, is devoted to discussions of other functional estimation problems to which kernel methods can be applied in a similar manner.

3.2 THE NADARAYA–WATSON KERNEL REGRESSION ESTIMATE

Among the most popular nonparametric regression smoothers are the Nadaraya–Watson kernel estimates. This family of estimates was introduced by Nadaraya (1964) and Watson (1964) and traces back to the so-called regressogram from Tukey (1961). The regressogram uses the same basic ideas as the histogram does for density estimation. It consists in partitioning the set of values of X and then averaging the values of Y inside of each subinterval B_j, $j = 1, \ldots, J$. The resulting regression estimate is therefore a step function defined for any $x \in B_j$ by

$$\hat{m}_T(x) = \frac{\sum_{i=1}^{n} Y_i I(X_i \in B_j)}{\sum_{i=1}^{n} I(X_i \in B_j)}.$$

This is a typical nonparametric estimate in the sense that no parametric form needs to be assumed *a priori* for the unknown regression function m. However, this estimate has the drawback of needing both the number and the position of the intervals B_j to be chosen. From a theoretical point of view, this drawback can be seen through the bias of estimates near the bounds of each interval B_j. (See Härdle (1990) for a similar discussion on the histogram.) This drawback is also clear from an intuitive point of view, as can be seen in Figure 3.2, in which a

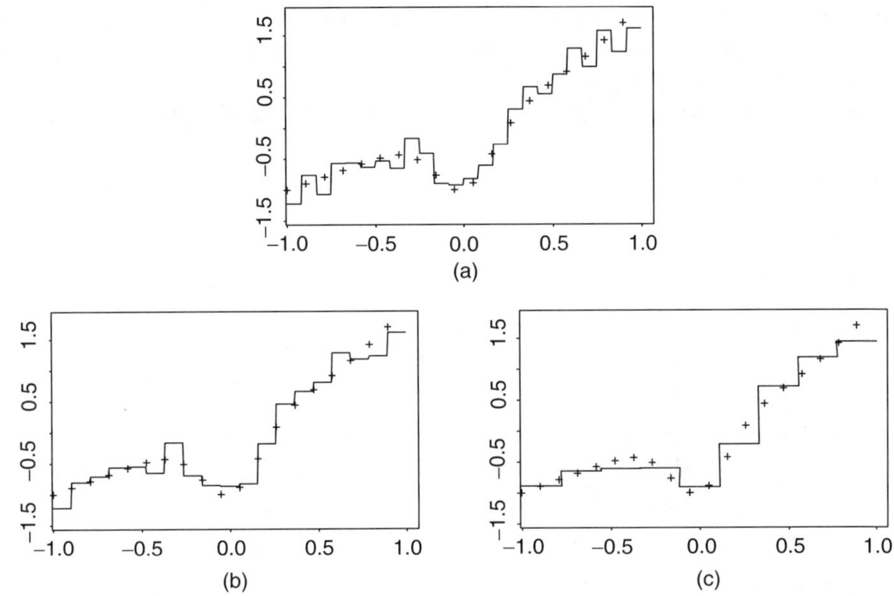

FIGURE 3.2 The regressogram (solid line) for 3 different bandwidths and true regression line (crosses): (a) $h = 0.08$; (b) $h = 0.1$; (c) $h = 0.22$.

regressogram is drawn for the data set presented in Figure 3.1. This regressogram was computed for intervals B_j with the same width h for three different values of h.

We note that small values of h (see Figure 3.2(a)) produce an undersmoothed estimate, while conversely, values of h that are too large (see Figure 3.2(c)) lead to an oversmoothed estimate. Apparently, Figure 3.2(b) presents a regressogram with a reasonable amount of smoothness that balances the trade-off between bias and variance. (See Section 3.5 for a precise quantification of these effects for the kernel estimate.) However, even the "good" value of h used in Figure 3.2 provides a quite rough estimate, because of the stepwise nature of the regressogram.

The first natural extension of the regressogram is a moving-window estimate. The idea is that a local average of Y values is still used, but the estimation at point x is based on a centered neighborhood of x. More precisely, it is defined, for any x, by

$$\frac{\sum_{i=1}^{n} Y_i I(X_i \in [x-h, x+h])}{\sum_{i=1}^{n} I(X_i \in [x-h, x+h])}.$$

This estimate has an advantage over the regressogram, since it depends on only one parameter, namely, the window width h. Clearly, this parameter has a great influence on the behavior of the estimate. This idea will be made more precise later on in this book. However, we may expect that once a reasonable value of h is chosen, the disadvantageous features previously observed for the regressogram no longer exist. (See Figure 3.2(b).)

Finally, this estimate can be generalized by using a weighted average by means of some kernel function K. This estimate is the Nadaraya–Watson kernel estimate, which is precisely defined by

$$\hat{m}_{\mathrm{NW}}(x) = \frac{\sum_{i=1}^{n} Y_i K((x - X_i)/h)}{\sum_{i=1}^{n} K((x - X_i)/h)}.$$

An interesting feature of this method is that it produces a continuous regression estimate, at least when a continuous kernel function K is used. The kernel K is a bounded and integrable real-valued function such that

$$\lim_{x \to \infty} |x| K(x) = 0.$$

Usually, the function K is taken to be a compactly supported symmetric probability density function, but using kernel functions that take negative values may be interesting in some sense to be made more precise later. To make ideas clearer, Figure 3.3 shows some of the most commonly used kernel functions, which are as follows:

$$\begin{aligned}
K_1(x) &= I(x \in [-0.5, +0.5]);\\
K_2(x) &= \frac{3}{4}(1 - x^2) I(x \in [-1, +0.1]);\\
K_3(x) &= \frac{15}{32}(3 - 10x^2 + 7x^4) I(x \in [-1, +0.1]);\\
K_4(x) &= \frac{1}{\sqrt{2\pi}} \exp(-x^2/2).
\end{aligned}$$

As before, h is some positive parameter that controls the smoothness of the estimate, as illustrated by Figure 3.4, where the kernel estimate is computed for the data set depicted in Figure 3.1(a) with the Epanechnikov kernel K_2 and for three different bandwidths.

As in Figure 3.2, we see the trade-off between bias and variance effects in Figure 3.4. The value $h = 0.05$ is too small, and the corresponding estimate has substantial variance, because too few data points are contained in the window. (See Section 3.4 for a theoretical description of these variance effects.) Conversely, the bandwidth $h = 0.5$ is too large, and it results in a large bias. (See Section 3.3 for discussion of the quantification of bias effects.) Finally, the bandwidth $h = 0.2$ appears to suitably balance these two effects. This value was

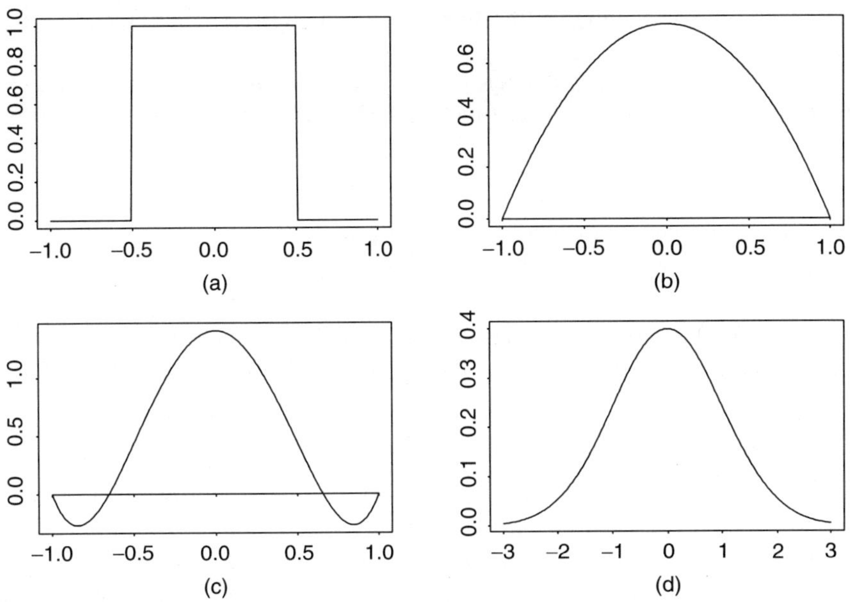

FIGURE 3.3 (a) Uniform kernel K_1; (b) Epanechnikov kernel K_2; (c) Kernel of order 4 K_3; (d) Normal kernel K_4.

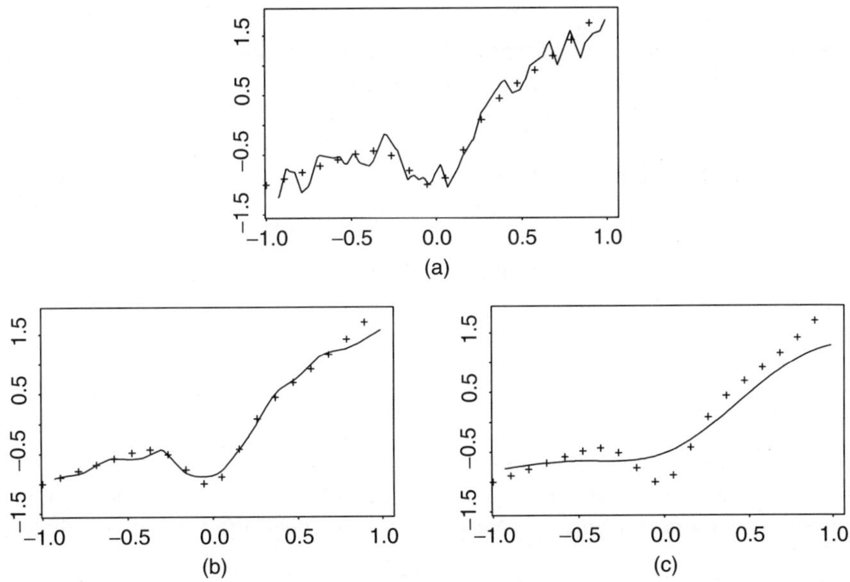

FIGURE 3.4 The kernel estimate (solid line) for 3 different bandwidths and the true regression line (crosses): (a) $h = 0.05$; (b) $h = 0.2$; (c) $h = 0.5$.

chosen because it minimizes the averaged squared error. (See Section 3.6 and Figure 3.6.)

In the remainder of this chapter, we concentrate on the kernel smoother. Obviously, since we take the kernel function to be the uniform density function on $[-\frac{1}{2}, +\frac{1}{2}]$, the results we present are still valid for the moving-window estimator.

3.3 POINTWISE BIAS PROPERTIES OF THE NADARAYA–WATSON ESTIMATE

In this section, we study the bias of the Nadaraya–Watson estimate at a fixed point x and see how it depends on the regularity of the regression function m.

3.3.1 Nonexistence of an Unbiased Estimate

The first property of nonparametric models is that, unlike for parametric models, there do not exist unbiased estimates. This property is summarized in the following theorem (see Collomb (1976) and references therein):

Theorem 3.1. *Let $\mathcal{D}$ be the set of distributions that are absolutely continuous with respect to the Lebesgue measure on $\mathbb{R}^2$ with continuous density $f_{X,Y}$ and such that*

$$\forall x \in \mathbb{R}, \int f_{X,Y}(x, y)\,dy \neq 0,$$

then there is no unbiased estimate of the regression function $m(x)$ (in the sense that there is no estimate m_n of m such that, for almost all x in $\mathbb{R}$ and all $f_{X,Y}$ in $\mathcal{D}$, $Em_n(x) = m(x)$).

Therefore, Theorem 3.1 shows that the generality of the nonparametric model results in a more difficult estimation problem, which is due to the fact that we are estimating an infinite number of parameters. From now on, we look for asymptotically unbiasedness properties. Later, in Theorem 3.2, we show that the kernel estimate is asymptotically unbiased at some fixed point x, and the rate of convergence is specified in Theorem 3.3.

3.3.2 Asymptotic Unbiasedness Property

Due to the previously described local feature of the kernel estimate, the continuity of functions to be estimated plays a prominent role in the statement of asymptotic properties of these estimates. More precisely, we need the assumption that

(H.1) the functions m and f are continuous at x and $f(x) \neq 0$. □

We introduce the next two conditions in order to make the presentation simpler. However, (H.2) can be weakened to the existence of absolute moments with order

$\beta > 2$, by using some truncation arguments (Mack and Silverman, 1982), whereas condition (H.3) can be suppressed (Collomb, 1976, p. 22). These two conditions are as follows:

(H.2) There exists a real M such that $|Y| < M < \infty$; □

(H.3) The kernel function K is nonnegative and compactly supported. □

The following condition on the bandwidth $h = h_n$ must also be introduced:

(H.4) $\lim_{n\to\infty} h = 0$ and $\lim_{n\to\infty} nh = \infty$. □

Theorem 3.2. *Suppose that conditions (H.1)–(H.4) are satisfied. Then the kernel estimate $\hat{m}_{\mathrm{NW}}$ is asymptotically unbiased at point x:*

$$\lim_{n\to\infty} E\hat{m}_{\mathrm{NW}}(x) - m(x) = 0.$$

Without loss of generality, we can assume that the kernel K integrates to unity. The proof of Theorem 3.2 is as follows: It is first shown that we have

$$\lim_{n\to\infty} E\hat{f}_{\mathrm{PR}}(x) - f(x) = 0, \tag{3.3}$$

where $\hat{f}_{\mathrm{PR}}$ is the usual Parzen–Rosenblatt kernel density estimate (Parzen, 1962; Rosenblatt, 1956), defined as

$$\hat{f}_{\mathrm{PR}}(x) = (nh)^{-1} \sum_{i=1}^{n} K((x - X_i)/h).$$

Then, a similar result is shown for the numerator of $\hat{m}_{\mathrm{NW}}$, namely,

$$\lim_{n\to\infty} E\hat{g}(x) - g(x) = 0, \tag{3.4}$$

where $g = mf$ and $\hat{g} = \hat{m}_{\mathrm{NW}}\hat{f}_{\mathrm{PR}}$. In a third step, we get the following decomposition:

$$E\hat{m}_{\mathrm{NW}}(x) = \frac{E\hat{f}_{\mathrm{PR}}(x)}{E\hat{g}(x)} + o((hn)^{-1}). \tag{3.5}$$

The decomposition in Equation (3.5) is based on the power-series development

$$\frac{1}{z} = 1 - (z-1) + \cdots + (-1)^p (z-1)^p + (-1)^{p+1}\frac{(z-1)^{p+1}}{z}$$

as applied to $z = \hat{f}_{\mathrm{PR}}(x)/E\hat{f}_{\mathrm{PR}}(x)$. Details on these computations are rather long and tedious and can be found in Collomb (1976).

Equations (3.3) and (3.4) are proven in the same way, and therefore we give only the proof for Equation (3.4). We have by equidistribution of the pairs (X_i, Y_i)

$$\begin{aligned} E\hat{g}(x) &= h^{-1} E Y_1 K((x - X_1)/h) \\ &= h^{-1} E[E[Y_1|X_1] K((x - X_1)/h)] \\ &= h^{-1} \int m(u) K((x-u)/h) f(u)\, du. \end{aligned}$$

Integrating by substitution, with $z = (x - u)/h$, we get

$$E\hat{g}(x) = \int K(z) g(x - zh)\, dz.$$

Because the kernel K integrates up to unity, we have

$$E\hat{g}(x) - g(x) = \int K(z)(g(x - zh) - g(x))\, dz. \tag{3.6}$$

The continuity of g and (H.4) ensure that, uniformly on z in the support of K, we have

$$\lim_{n\to\infty} g(x - zh) - g(x) = 0,$$

and this expression concludes the proof of (3.4).

3.3.3 Rate of Convergence of the Bias

The rate of convergence of the bias depends on smoothness assumptions on the regression function. We concentrate here on the simplest case, wherein the smoothness is characterized by the existence of continuous derivatives. Precisely, we assume that

(H.5) the functions m and f are twice continuously differentiable around x and $f(x) \neq 0$. □

The next condition on the kernel function K is necessary in order to take into account this new smoothness feature:

(H.6) The kernel K is symmetric. □

Theorem 3.3. *Under conditions (H.2)–(H.6), we have*

$$E\hat{m}_{\mathrm{NW}}(x) - m(x) = C_1 h^2 + o(h^2) + O((nh)^{-1}),$$

where

$$C_1 = \left(\int z^2 K(z)\, dz\right)\left(\frac{1}{2} m''(x) + \frac{m'(x) f'(x)}{f(x)}\right).$$

The original proof of Theorem 3.3 was made by Collomb (1976, 1977). It combines a similar result for density estimation, attributed to Parzen (1962) (see (3.8)), and another result for the function g (see Equation (3.7)) with the approximation (3.5). We just give the main steps of this proof here. Using a Taylor expansion in (3.6), we get

$$E\hat{g}(x) - g(x) = \frac{g''(x)}{2}\left(\int z^2 K(z)\,dz\right) h^2 + o(h^2). \tag{3.7}$$

The same kind of asymptotic evaluation also applies for the bias of the density estimate $\hat{f}_{\mathrm{PR}}$, namely,

$$E\hat{f}_{\mathrm{PR}}(x) - f(x) = \frac{f''(x)}{2}\left(\int z^2 K(z)\,dz\right) h^2 + o(h^2). \tag{3.8}$$

The result follows directly from these last two formulae together with (3.5).

The order for the bias depends only on the smoothness assumption made on the functions to be estimated. For instance, if, instead of using the assumptions in (H.5), we assume only that m and f are Lipschitz continuous with order α, and if we change suitably condition (H.6) on the kernel into

$$\left|\int z^\alpha K(z)\,dz\right| < \infty,$$

then we can get a bias of order h^α. The proof of this can be found, for instance, in Collomb (1976), but it is indeed very easy to see that it can be obtained directly by using a Lipschitz condition instead of the Taylor expansion in Equation (3.6).

In addition, it should be noted that the bias increases with h. These asymptotes agree with the intuitive aspects previously discussed with regard to Figure 3.3. As a result, one might consider choosing h to be as small as possible in order to get a smaller bias. However, we see in the next section that it is not so simple to get a smaller bias, because decreasing h also affects the variance of the estimate.

3.4 POINTWISE VARIANCE PROPERTIES OF THE NADARAYA–WATSON ESTIMATE

We now focus on the variance of the regression estimate at a fixed point x. To get an asymptotic evaluation of these variance effects, we need the following additional condition on the distribution of (X, Y):

(H.7) The conditional variance $\sigma^2(\cdot)$ is continuous at the point x. □

We now have the following theorem:

Theorem 3.4. *Under conditions (H.1)–(H.4) and (H.7), we have*

$$\mathrm{Var}(\hat{m}_{\mathrm{NW}}(x)) = C_2(nh)^{-1} + o((nh)^{-1}),$$

where

$$C_2 = \frac{\sigma^2(x)}{f(x)} \int K^2(z)\,dz.$$

The idea of the proof is, as for the bias term, to treat separately the numerator and the denominator of $\hat{m}_{\mathrm{NW}}$ and to go through the ratio via the expansion of $1/z$ at order 3. This proof is given precisely in Collomb (1976): we just present the main steps here. First, it must be shown that

$$\mathrm{Var}(\hat{f}_{\mathrm{PR}}(x)) = (nh)^{-1} f(x) \int K^2(z)\,dz + o((nh)^{-1}), \tag{3.9}$$

$$\mathrm{Var}(\hat{g}(x)) = (nh)^{-1} f(x)\phi(x) \int K^2(z)\,dz + o((nh)^{-1}), \tag{3.10}$$

and

$$\mathrm{Cov}(\hat{g}(x), \hat{f}_{\mathrm{PR}}(x)) = (nh)^{-1} g(x) \int K^2(z)\,dz + o((nh)^{-1}), \tag{3.11}$$

where $\phi(x) = \sigma^2(x) + m^2(x)$. We will only prove (3.10), since (3.9) and (3.11) follow along the same lines. (See Parzen (1962) for a proof of (3.9) and Collomb (1976) for a proof of Equation (3.11)). Because (X_i, Y_i) are i.i.d., we have

$$\mathrm{Var}(\hat{g}(x)) = (nh)^{-1} E\left(\frac{1}{h} Y^2 K^2((x - X)/h)\right) - n^{-1}\left(E\frac{1}{h} Y K((x - X)/h)\right)^2,$$

which gives us, together with Theorem 3.2 and (H.4),

$$\begin{aligned}\mathrm{Var}(\hat{g}(x)) &= (nh)^{-1} E\left(\frac{1}{h} Y^2 K^2((x - X)/h)\right) + o((nh)^{-1})\\ &= (nh)^{-1} \int \frac{1}{h}\phi(u) K^2((x - u)/h) f(u)\,du + o((nh)^{-1}).\end{aligned}$$

Using integration by substitution, with $z = (x - u)/h$, we get

$$\mathrm{Var}(\hat{g}(x)) = (nh)^{-1} \int \phi(x - zh) K^2(z) f(x - zh)\,dz + o((nh)^{-1}). \tag{3.12}$$

Now, the continuity of ϕ and f, together with (H.4), gives us, uniformly over z in the support of K,

$$\lim_{n\to\infty} (f\phi)(x - zh) - (f\phi)(x) = 0. \tag{3.13}$$

Equation (3.10) follows directly from (3.12) and (3.13). Finally, Theorem 3.4 follows from (3.9), (3.10), and (3.11), and the following decomposition, which can be proven using the development of $1/z$ at order 3 (see Collomb, 1976):

$$\begin{aligned}\mathrm{Var}(\hat{m}_{\mathrm{NW}}(x)) &= \frac{\mathrm{Var}(\hat{g}(x))}{(E\hat{f}_{\mathrm{PR}}(x))^2} - 4\frac{E\hat{g}(x)E((\hat{g}(x) - E\hat{g}(x))(\hat{f}_{\mathrm{PR}}(x) - E\hat{f}_{\mathrm{PR}}(x)))}{(E\hat{f}_{\mathrm{PR}}(x))^3} \\ &\quad + 3\,\mathrm{Var}(\hat{f}_{\mathrm{PR}}(x))\frac{(E\hat{g}(x))^2}{(E\hat{f}_{\mathrm{PR}}(x))^4} + o((nh)^{-1}).\end{aligned}$$

Note that the asymptotic variance does not depend on regularity conditions for the regression m as long as it is continuous at point x. Indeed, the variance is of the same order as the inverse of the number of X_i's falling into the interval $[x - h, x + h]$. (See, for example, Equation (2.8) in Stone (1981)). This identity comes from the local feature of the kernel procedure, since, for estimation at point x, we use only those X_i's. So, in practice, one would wish to take h to be as large as possible in order to have sufficient data in the window to reduce the variance. Of course, this procedure is in conflict with what we have observed for the bias; see the closing paragraph of Section 3.3.3. These asymptotic results agree with the intuitive arguments developed with respect to Figure 3.3.

3.5 TRADE-OFF BETWEEN BIAS AND VARIANCE: THE MEAN SQUARED ERROR

Because we have biased estimates (see Theorem 3.1), any suitable criterion to measure the accuracy of any nonparametric estimate should take into account both bias and variance effects. A natural way to take these effects into account is to look at the following *mean squared error* (MSE):

$$\mathrm{MSE}(\hat{m}_{\mathrm{NW}}(x)) = E(\hat{m}_{\mathrm{NW}}(x) - m(x))^2. \tag{3.14}$$

It is obvious that $\mathrm{MSE}(\hat{m}_{\mathrm{NW}}(x))$ is just the sum of the square bias and the variance. Therefore, we have the result stated in Theorem 3.5.

Theorem 3.5. *Under conditions (H.2)–(H.7), we have*

$$\mathrm{MSE}(\hat{m}_{\mathrm{NW}}(x)) = C_1^2 h^4 + C_2((nh)^{-1}) + o(h^4 + (nh)^{-1}), \tag{3.15}$$

where C_1 and C_2 are as defined in Theorems 3.3 and 3.4.

When we change the smoothness conditions on the regression function m and density f, we obtain different asymptotic evaluations for the MSE. In particular, if we replace condition (H.5) with a Lipschitz condition of order α and if we use the ideas developed at the end of Section 3.3.3, then we get a rate for MSE of order $O(h^{2\alpha}) + O((nh)^{-1})$. Also, if we replace condition (H.5) with condition (H.1) and use Theorem 3.2 instead of Theorem 3.3, we get pointwise mean squared convergence; however, no rate for the MSE can be specified under condition (H.1).

Let us further remark that the result in Theorem 3.5 remains valid when the MSE is changed into the pointwise *averaged squared errors* (ASE_x), defined as

$$\text{ASE}_x(\hat{m}_{\text{NW}}(x)) = \sum_{i=1}^{n} (\hat{m}_{\text{NW}}(X_i) - m(X_i))^2 w_x(X_i), \tag{3.16}$$

where w_x is some weight function with support centered around x and of length tending to zero as n goes to infinity, and when the o in Equation (3.15) is changed into o_P. Indeed, both the MSE and the ASE_x have been shown to be asymptotically equivalent (Marron and Härdle, 1986, and Vieu, 1991).

These asymptotes confirm what we have observed for the finite sample. (See Figure 3.3.) We can see that the performance of the estimate depends crucially on the value of the smoothing parameter h. A reasonable bandwidth is therefore necessary in order to balance bias and variance effects. Corollary 3.1 gives some insight into the asymptotic behavior of the optimal bandwidth at point x, in the sense of a minimal MSE.

Corollary 3.1. *Under conditions (H.2)–(H.7), if we use as bandwidth*

$$h_0(x) = C_3 n^{-1/5},$$

then we have

$$\text{MSE}(\hat{m}_{\text{NW}}(x)) = C_4 n^{-4/5} + o(n^{-4/5}), \tag{3.17}$$

where the constants C_3 and C_4 are defined by

$$C_3 = \left(\frac{C_2}{4C_1^2}\right)^{1/5}$$

and

$$C_4 = 5\left(\frac{C_2}{4}\right)^{4/5} C_1^{2/5}.$$

Minimizing the sum of the two main terms over h in Equation (3.15) leads to the bandwidth $h_0(x)$. Equation (3.17) is then obtained by plugging $h_0(x)$ in the asymptotic expression of the MSE.

To make things clearer, in the Figure 3.5 we show, for the simulated example presented in Section 3.1, the graph of ASE_x as a function of h for three special

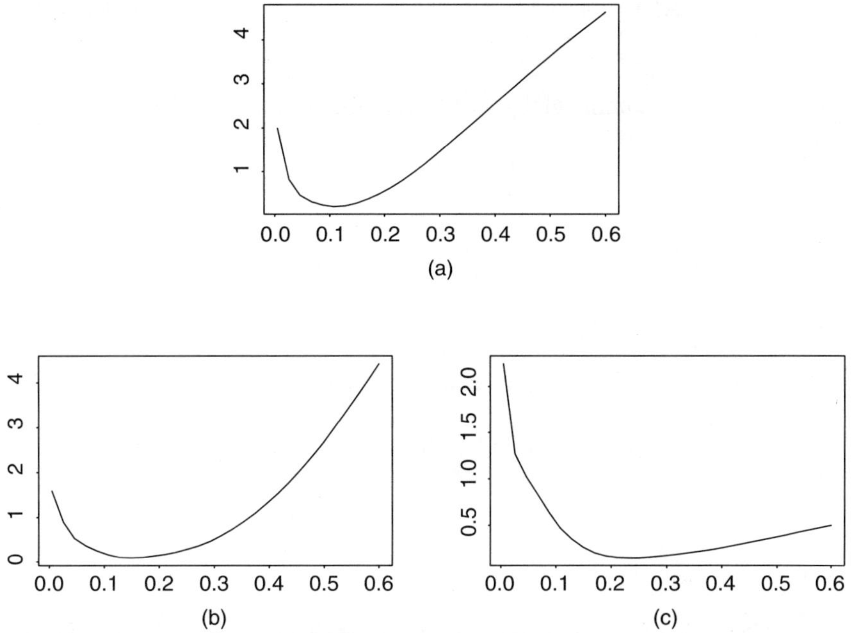

FIGURE 3.5 Pointwise averaged squared error as a function of h at different points x. (a) $x = 0$; (b) $x = 0.5$; (c) $x = -0.5$.

values of x. We have used the Epanechnikov kernel (see Figure 3.3(b)) as a kernel function and $w_x(t) = I(t \in [x - 0.3, x + 0.3])$ as a local weight function.

The convex shapes observed in these figures for a finite sample agree with the asymptotic expansion obtained in Theorem 3.5, but using ASE_x instead of MSE. Values of h that are too high or too small result in high values of ASE_x, because of increasing bias and increasing variance, respectively. Even if, for all x, we have the same general ASE_x feature, the corresponding optimal bandwidths $h_0(x)$ are quite different. For instance, in our example, these bandwidths are $h_0(0) \cong 0.11$, $h_0(0.5) \cong 0.15$, and $h_0(-0.5) \cong 0.23$. It is quite clear both from Theorem 3.3 and from intuition that the optimal bandwidth is smaller when the curvature of the underlying regression is important, and this relationship is exactly what we observe in the plots in Figure 3.5.

The rate of convergence given by (3.17) has been shown to be optimal over the class of twice continuously differentiable regression functions. (See Stone (1980).) This provides further theoretical support for the use of the Nadaraya–Watson kernel procedure.

In practice, the constant C_3 is not computable, since it depends on unknown quantities obtained through the other constants C_1 and C_2. However, Equation (3.17) provides a useful tool for building data-driven procedures for selecting the bandwidth, such as plug-in or cross-validation procedures. These data-driven bandwidth selectors are studied in detail in Chapter 4 of this volume.

3.6 GLOBAL RESULTS: MEAN INTEGRATED SQUARED ERROR PROPERTIES

In this section, we examine global versions of the results described in the previous sections. A natural global measure of accuracy for nonparametric estimates is the *mean integrated squared error* (MISE), defined as

$$\mathrm{MISE}(\hat{m}_{\mathrm{NW}}) = E\int (\hat{m}_{\mathrm{NW}}(x) - m(x))^2 w(x)\,dx, \tag{3.18}$$

where w is some positive weight function satisfying the following condition:

(H.8) The function w is bounded with compact support S. □

Moreover, we need global versions of the previous conditions for the distribution of (X, Y), namely,

(H.9) The functions m and f are twice continuously differentiable on an ϵ neighborhood of S and $\inf_{x\in S} f(x) \geq \xi > 0$; □

(H.10) The conditional variance $\sigma^2(\cdot)$ is continuous on an ϵ neighborhood of S. □

In the following theorem, we give the asymptotic behavior of the MISE.

Theorem 3.6. *Under conditions (H.2)–(H.4), (H.6), and (H.8)–(H.10), we have*

$$\mathrm{MISE}(\hat{m}_{\mathrm{NW}}) = C_5 h^4 + C_6((nh)^{-1}) + o(h^4 + (nh)^{-1}), \tag{3.19}$$

where

$$C_5 = \frac{1}{4}\left(\int z^2 K(z)\,dz\right)^2 \int \left(m''(x) + \frac{2m'(x)f'(x)}{f(x)}\right)^2 w(x)\,dx,$$

and

$$C_6 = \left(\int \frac{\sigma^2(x)}{f(x)} w(x)\,dx\right)\left(\int K^2(z)\,dz\right).$$

Equation (3.19) is a direct consequence of Theorem 3.5 by integration over S. The only thing to be checked is that the o in Equation (3.15) are uniform over S. This last point is quite obvious because all of the os involved in the proofs of Theorems 3.3 and 3.4 come from continuity arguments in functions m and f, which are indeed uniformly continuous over the compact S. For instance, it is easy to see that Equations (3.4) and (3.12) hold uniformly over x in S.

Similar results as in Theorem 3.6 hold when the smoothness conditions for the distribution of (X, Y) are changed. For instance, assuming only continuity of m and f, we get the L_2 convergence without a rate, whereas weakening condition (H.9) to some Lipschitz condition of order α leads to the rate $O(h^{2\alpha} + (nh)^{-1})$.

Asymptotic equivalences between MISE,

$$\text{ISE}(\hat{m}_{\text{NW}}) = \int (\hat{m}_{\text{NW}}(x) - m(x))^2 w(x)\, dx, \tag{3.20}$$

and

$$\text{ASE}(\hat{m}_{\text{NW}}) = \sum_{i=1}^{n} (\hat{m}_{\text{NW}}(X_i) - m(X_i))^2 w(X_i), \tag{3.21}$$

where ISE stands for *integrated squared error*, have been shown by Marron and Härdle (1986). Therefore, Theorem 3.6 remains true for these other two quadratic measures of accuracy.

The above asymptotic equivalences are interesting with respect to simulations, since it is easier to compute ASE than MISE or ISE. In Figure 3.6, we have drawn a plot of ASE as a function of h for the simulated data in Figure 3.1(a), using Epanechnikov's kernel function K_2.

The convex shape in the figure is as anticipated from Theorem 3.6. The right-hand side of the plot indicates a high ASE, because of substantial bias, while the left-hand side shows increasing ASE, due to variance effects. We have already observed the same trends in Figure 3.4. The optimal value of h for this example is $h_0 \cong 0.2$, and it leads to the kernel estimate previously presented in Figure 3.4(b).

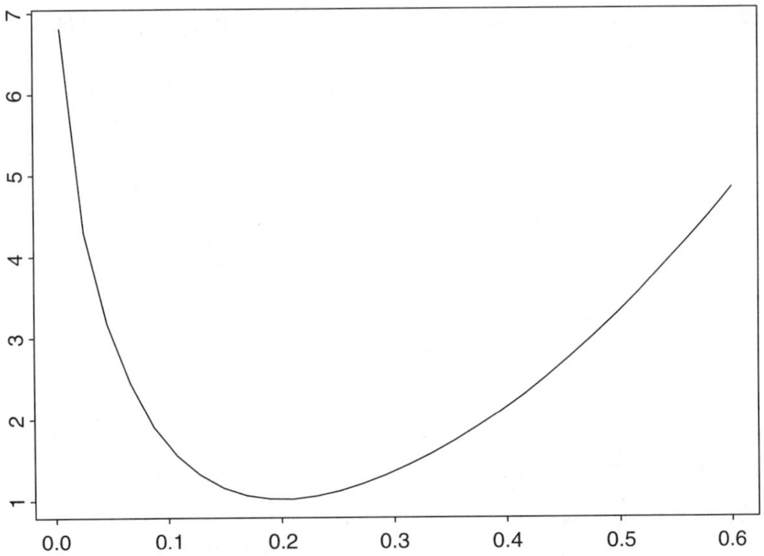

FIGURE 3.6 Averaged squared error as a function of h.

Corollary 3.2. *Under conditions (H.2)–(H.4), (H.6), and (H.8)–(H.10), and if we use as bandwidth*

$$h_0 = C_7 n^{-1/5},$$

we have

$$\text{MISE}(\hat{m}_{\text{NW}}(x)) = C_8 n^{-4/5} + o(n^{-4/5}), \tag{3.22}$$

where the constants C_7 and C_8 are defined by

$$C_7 = \left(\frac{C_6}{4C_5}\right)^{1/5}$$

and

$$C_8 = 5\left(\frac{C_6}{4}\right)^{4/5} C_5^{1/5}.$$

The proof of Corollary 3.2 follows the same lines as the proof of Corollary 3.1.

Just as with the pointwise error, we get the rate $n^{-4/5}$, which has been shown to be the global optimal rate of convergence by Stone (1982). The previous remarks also apply to the bandwidth h_0.

3.7 L_∞ CONVERGENCE PROPERTIES OF THE NADARAYA–WATSON ESTIMATE

3.7.1 Basic Tools

To state the L_∞ convergence results, we need two important tools, which are reported in Lemmas 3.1 and 3.2. The first tool is the Bernstein inequality. For clarity, we do not present this inequality in its more general form, because it is not necessary for its application to our problem. This inequality is given for weaker conditions by Hoeffding (1963).

Lemma 3.1. *Let $Z, Z_1, \ldots, Z_n$ be i.i.d. real random variables with zero mean. Assume that there exist some positive real numbers d and τ^2 such that*

$$|Z| \leq d$$

and

$$EZ^2 \leq \tau^2.$$

Then we have for any $\epsilon \in [0, \tau^2/d]$,

$$P\left[n^{-1}\left|\sum_{i=1}^{n} Z_i\right| > \epsilon\right] \leq 2\exp\left(\frac{-n\epsilon^2}{4\tau^2}\right).$$

Note that the variables and the bounds involved in Lemma 6.1 may depend on n. That means that this lemma remains valid when Z_i, τ, d, and ϵ are changed into $Z_{i,n}$, τ_n, d_n, and ϵ_n, respectively.

To state the almost-sure uniform convergence of the kernel estimate, we use the notion of almost-complete convergence, whose definition is given next. In Lemma 6.2, it is shown that this latter kind of convergence implies both convergence in probability and almost-sure convergence.

Let $(Z_n)_{n\in N}$ and Z be real, random variables, and let $(U_n)_{n\in N}$ be a sequence of real numbers. We define the sequence $(Z_n)_{n\in N}$ to converge almost completely to Z, and we note that $\lim_{n\to\infty} Z_n = Z$ is almost complete if, for any $\epsilon > 0$, we have

$$\sum_{n=1}^{\infty} P(|Z_n - Z| > \epsilon) < \infty.$$

Further, we say that $Z_n = O(U_n)$ is almost complete if there exists $\epsilon > 0$ such that we have

$$\sum_{n=1}^{\infty} P(|Z_n| > \epsilon U_n) < \infty.$$

Lemma 3.2. *(i) Assume that $(Z_n)_{n\in N}$ is a sequence of random variables such that $\lim_{n\to\infty} Z_n = Z$ is almost complete, then, we have*

$$\lim_{n\to\infty} Z_n = Z,$$

in probability, and

$$\lim_{n\to\infty} Z_n = Z,$$

a.s. (ii) Assume that $(Z_n)_{n\in N}$ is such that $Z_n = O(U_n)$, is almost complete; then, we have

$$Z_n = O(U_n),$$

in probability, and

$$Z_n = O(U_n),$$

a.s.

The proof of Lemma 6.2 consists in applying Borel-Cantelli's lemma. (See Bosq and Lecoutre (1987).)

3.7.2 Pointwise Almost Sure Consistency

As we did previously for quadratic accuracy measures we first state pointwise consistency results. Let x be a fixed point in $\mathbb{R}$. Next, Theorem 3.7 gives almost sure convergence under the continuity assumption for the functions to be estimated, whereas Theorem 3.8 specifies the rate of almost sure convergence under additional smoothness assumptions. To get almost sure pointwise consistency, we need the following conditions on the bandwidth and on the kernel function:

(H.11) K is compactly supported; □

(H.12) $\lim_{n\to\infty} h = 0$ and $\lim_{n\to\infty} \frac{nh}{\log n} = \infty$. □

Theorem 3.7. *Under conditions (H.1)–(H.2), (H.11), and (H.12), we have*

$$\lim_{n\to\infty} |\hat{m}_{\mathrm{NW}}(x) - m(x)| = 0, \tag{3.23}$$

a.s.

The proof of Theorem 3.7 treats the numerator and denominator of $\hat{m}_h(x)$ separately and uses the following decomposition to deal with the ratio:

$$\hat{m}_{\mathrm{NW}}(x) - m(x) = \frac{\hat{g}(x) - g(x)}{\hat{f}_{\mathrm{PR}}(x)} + (f(x) - \hat{f}_{\mathrm{PR}}(x))\frac{m(x)}{\hat{f}_{\mathrm{PR}}(x)}. \tag{3.24}$$

All we need to conclude our proof are the equations

$$\lim_{n\to\infty} |\hat{g}(x) - g(x)| = 0, \, a.c., \tag{3.25}$$

$$\lim_{n\to\infty} |\hat{f}_{\mathrm{PR}}(x) - f(x)| = 0, \, a.c., \tag{3.26}$$

$$\exists \delta > 0, \sum_{n=1}^{\infty} P[\hat{f}_{\mathrm{PR}}(x) \le \delta] < \infty, \tag{3.27}$$

since from Equations (3.24)–(3.27) we get the almost complete convergence of $\hat{m}_{\mathrm{NW}}(x)$, and using Lemma 3.2 we get (3.23). Next, we give only the proof of (3.25), since (3.26) follows along the same lines, whereas (3.27) is a direct consequence of (3.26) and (H.1). A complete proof of Theorem 3.7 can be found in Collomb (1984).

Let us now prove (3.25). Defining Z_i by

$$Z_i = \frac{1}{h}(Y_i K((x - X_i)/h) - EY_i K((x - X_i)/h)),$$

we have the following inequalities, for some real constants c_1 and c_2:

$$|Z_i| \le \frac{c_1}{h} \tag{3.28}$$

and

$$EZ_i^2 \le \frac{c_2}{h}. \tag{3.29}$$

Equation (3.28) follows directly from (H.2) and the boundedness assumption on K, whereas (3.29) is obtained following the same steps as those used to get (3.10). Now, we are able to apply Lemma 3.1 to the variables $z_1, \ldots, z_n$, and we have, for any ϵ,

$$P[|\hat{g}(x) - E\hat{g}(x)| > \epsilon] < 2\exp\left(\frac{-nh\epsilon^2}{4c_2}\right), \tag{3.30}$$

which gives us using (H.12),

$$\lim_{n\to\infty} |\hat{g}(x) - E\hat{g}(x)| = 0, a.c.$$

Equation (3.25) is a direct consequence of this last result and Equation (3.4).

3.7.3 Rate of Pointwise Almost-Sure Convergence

In this section, we give the rate of pointwise almost-sure convergence under smoothness assumptions on the distribution of (X, Y)—that is, the existence of higher order derivatives for the functions m, g, and f. For m and f, the following condition holds:

(H.13) The functions m and f are k-times continuously differentiable at point x, and $f(x) \neq 0$. □

We also introduce higher order kernels (see Gasser and Müller, 1979) defined by

(H.14) $\int z^j K(z)\,dz = 0$, for $j = 1, \ldots, k-1$, and $0 \neq \left|\int z^k K(z)\,dz\right| < \infty$. □

For $k > 2$, note that these kernels take on negative values. For instance, Figure 3.4(c) shows a fourth-order kernel. We are able to give the rates for both pointwise and uniform almost-sure convergence in such a case, whereas only the case of positive kernels (i.e., $k = 2$) is possible for quadratic errors. Indeed, this difference comes from the techniques used to deal with the ratio: While positivity of the kernel is necessary in order to treat Equation (3.5), we do not need such an assumption for the decomposition (3.24), which is proven via (3.27).

Theorem 3.8. *Under conditions (H.2) and (H.11)–(H.14), we have*

$$\hat{m}_{\text{NW}}(x) - m(x) = O(h^k + (\log n/nh)^{1/2}), a.s. \tag{3.31}$$

As Equations (3.24) and (3.27) have already been mentioned, the only points we have to state in order to prove Theorem 3.8 are

$$\hat{g}(x) - g(x) = O(h^k + (\log n/nh)^{1/2}), a.c., \tag{3.32}$$

and

$$\hat{f}_{\text{PR}}(x) - f(x) = O(h^k + (\log n/nh)^{1/2}), a.c. \tag{3.33}$$

As before, we prove only (3.32). For this proof, it is enough to show that

$$E\hat{g}(x) - g(x) = O(h^k) \tag{3.34}$$

and

$$\hat{g}(x) - E\hat{g}(x) = O((\log n/nh)^{1/2}), a.c. \tag{3.35}$$

To prove (3.34), it suffices to use a kth Taylor expansion on the term on the right-hand side of (3.6) and to use the same continuity argument, but on the kth derivative of g rather than on g itself. To prove Equation (3.35), it suffices to apply (1.30) with

$$\epsilon = \epsilon_0(\log n/nh)^{1/2}$$

which is possible because of the remark following Lemma 3.1.

If we change the smoothness assumptions on m and f, then the rate of convergence changes, too. For instance, if we add the restriction that the kth derivatives of m and f are Lipschitz continuous with order α, then the rate becomes $O(h^{k+\alpha} + (\log n/nh)^{1/2})$.

Corollary 3.3. *Under conditions (H.2) and (H.11)–(H.14), and if we use a bandwidth*

$$h_1(x) = C_9\left(\frac{n}{\log n}\right)^{-1/(2k+1)},$$

for some $C_9 > 0$, *then we have*

$$\hat{m}_{\mathrm{NW}}(x) - m(x) = O\left(\left(\frac{n}{\log n}\right)^{-k/(2k+1)}\right), a.s. \tag{3.36}$$

We omit the proof of this corollary, since it follows along the same lines as the proof of Corollary 3.1.

Note that these results are of the same kind as those obtained in Section 3.4 for quadratic loss, but here we do not have at hand a precise evaluation of the constants involved in the rates (3.31) and (3.36).

3.7.4 Uniform Almost-Sure Convergence

Let S be a compact subset of $\mathbb{R}$. We will now provide uniform versions of Theorem 3.8. The smoothness of the functions to be estimated is now determined by the following condition:

(H.15) The functions m and f are k-times continuously differentiable on an ϵ neighborhood of S, and $inf_{x\in S} f(x) \geq \xi > 0$. □

We also need the following condition on the kernel function:

(H.16) The kernel K is Lipschitz continuous of order $\beta > 0$. □

Theorem 3.9. *Under conditions (H.2), (H.11), (H.12), and (H.14)–(H.16), we have*

$$\sup_{x\in S} |\hat{m}_{\mathrm{NW}}(x) - m(x)| = O(h^k + (\log n/nh)^{1/2}), \ a.s. \tag{3.37}$$

The proof of this theorem can be found in Collomb (1984). It follows along the same lines as the proof of Theorem 3.8, but with some additional compactness arguments in order to deal with the supremum.

Corollary 3.4. *Under conditions (H.2), (H.11), (H.12), and (H.14)–(H.16), and if we use as bandwidth*

$$h_1 = C_{10} \left(\frac{n}{\log n} \right)^{-1/(2k+1)},$$

for some constant $C_{10} > 0$*, then we have*

$$\sup_{x\in S} |\hat{m}_{\mathrm{NW}}(x) - m(x)| = O\left(\left(\frac{n}{\log n} \right)^{-k/(2k+1)} \right), \ a.s. \tag{3.38}$$

The rate of convergence obtained in (3.38) has been shown to be the best one for the class of k-times continuously differentiable functions by Stone (1982). Note that the constant involved in this rate cannot be specified, which is not the case for Equations (3.38) and (3.22), for example.

Let us further remark that no assumption on the conditional variance $\sigma^2(\cdot)$ is needed to prove L_∞ properties. (In contrast, for instance, condition (H.7) was used for Theorem 3.6.) Indeed, to get an upper bound for the variance (see Equation (3.29)), we need only the existence of $\sigma^2(\cdot)$, which is ensured by condition (H.2), while continuity was necessary to specify the constant in (3.10).

3.8 COMPLEMENTARY BIBLIOGRAPHY

3.8.1 Literature on Nadaraya–Watson Estimates

The literature on Nadaraya–Watson estimates is quite extensive. In this chapter, we have presented only some basic consistency properties, and we refer for instance, to Collomb (1981, 1985) and to Härdle (1990) for further surveys of this topic. We have paid particular attention to L_2-type results, because exact asymptotic decompositions are feasible that allow for interesting developments. Similar results exist in L_∞ norm or in L_1 norm (see Krzyzak, 1985, and Wand, 1990), but they are less precise.

The most widely studied problem attached to kernel estimation is the bandwidth choice. Indeed, the importance of the bandwidth can be seen both practically from Figure 3.4 and theoretically from decompositions like those in (3.15) and (3.19). The data-driven bandwidths selectors are discussed later on in this book, and we just wish to mention here the key paper by Härdle and Marron (1985), as well as refer to Härdle (1990), Marron (1988), and Vieu (1993) for surveys on this subject. It has been pointed out in Figure 3.5 that choosing local bandwidths $h(x)$ has some advantages. Recent literature on this topic includes Fan et al. (1996), Schucany (1995), and Vieu (1991).

Because of the limited importance of kernel choice, the literature on it is far less extensive than that for bandwidth determination. However, several studies have tackled the problem of kernel choice. Most of these studies are devoted to seeking optimal kernels (in terms of the minimization of quadratic errors) in classes of fixed-order functions. (See condition (H.14).) We refer solely to Gasser and Müller (1979), Granovsky, Müller, and Pfeifer (1995), Jones and Wand (1992), Wand and Schucany (1990), and references therein. The choice of the kernel order has been treated by Berlinet (1993), Hall and Marron (1987), and Vieu (1999).

Further, Gasser and Müller (1979) have pointed out a boundary problem in kernel estimation: The bias at the edge of the support of the marginal density is larger than at interior points. This effect can be clearly discerned in Figure 3.4 near $x = 1$. Note that even with the theoretical optimal bandwidth (see Figure 3.4(b)) this problem is not overcome. A solution proposed by Gasser and Müller suggests applying nonsymmetric kernels at the edges of the estimation interval. Let us finally mention that asymmetric kernels are also used in a similar way to estimate the location and size of some breakpoints (Wu and Chu, 1993a, 1993b).

3.8.2 Literature on Using Kernels for Related Regression Problems

The first natural extension of the kernel theory consists in estimating a multivariate regression function m. When $\mathbf{X}$ is a vector of $\mathbb{R}^p$, $p \geq 1$, the regression function m can be estimated by $\hat{m}_{\text{NW}}$, in which K is a p-variate real-valued kernel function. Similar asymptotes are available. For instance, for a twice continuously differentiable function m, we get an L_2-rate of convergence of order $n^{-4/(4+p)}$ that generalizes Corollary 3.2. This rate is optimal (see Stone, 1982), but it gets worse when p increases. This well-known "curse of dimensionality" phenomenon can be dealt with by using, for instance, semiparametric regression models, as discussed by Stone (1985). See also the survey of Pelegrina et al. (1996). For a detailed discussion of kernel-based multivariate regression, see Chapter 12. Note that the recent work by Ferraty and Vieu (2000), proposes an extension of this kernel technique to the case where X is valued in an infinite dimensionality space.

Another generalization of the results discussed here is to consider nonindependent pairs (X_i, Y_i), in order to allow for application to time-series prediction. Starting with Collomb (1984), several extensions of Theorems 3.6 and 3.8 have been stated under several mixing conditions. We refer to Bosq (1996), Györfi et al. (1989), and Yoshihara (1994), and Chapter 6 for results of this nonparametric kernel prediction approach and to Doukhan (1994) for a general presentation on mixing models.

In a natural way, kernel regression techniques may be used in longitudinal analysis when we have, not just only one sample (X_i, Y_i), but rather, several samples $(X_{i,j}, Y_{i,j})$, $j = 1, \ldots, J$, observed on J individuals. Literature on this topic includes studies by Boularan et al. (1994, 1995a), Gasser et al. (1984), Hart and Wehrly (1993), and Nunez–Anton et al. (1999).

Asymptotic results are also available for the estimation of derivatives of m when one applies the derivative of $\hat{m}_{\text{NW}}$ as an estimate. (See Sarda and Vieu, 1988). Estimation of derivatives is of particular interest for at least two reasons. First of all, it allows us to estimate the constants involved in Corollaries 3.1 and 3.2 and then to estimate the optimal theoretical bandwidth. Second, it permits us to estimate particular points (see Boularan et al., 1995b) that can be used along longitudinal studies for classifying individuals.

3.8.3 Using Kernels for Other Functional Estimation Problems

The kernel techniques are not only useful for regression estimation, but they also can be used in any functional estimation field. For example, we have already mentioned, at several places in this chapter, the problem of density estimation. The Parzen–Rosenblatt kernel density estimate $\hat{f}_{\text{PR}}$ has been defined in Section 3.3.2, and some results have been shown in this chapter as preliminary steps in our proofs. (See Equations (3.3), (3.8), and (3.9).) More results can be found, for instance, in Devroye (1987), Izenman (1991), Scott (1992), Silverman (1986), and Wand and Jones (1995).

Kernel methods can be used to smooth the empirical cumulative distribution function. Hence, a kernel estimate of the distribution function $F(x)$ is defined as the cumulative function of the kernel density estimate $\hat{f}_{\text{PR}}$. In addition to its smoothness it turns out that this function also has nice efficiency properties. Papers on this topic include Azzalini (1981), Lejeune and Sarda (1992), Reiss (1981), and Shirahata and Chu (1992).

Combining a kernel estimate of a density function f together with some estimate of the corresponding cumulative distribution function F leads naturally to an estimate of the hazard function $\lambda(x) = f(x)/(1 - F(x))$. A survey on such nonparametric hazard estimators can be found in Hassani et al. (1986). Recent developments include Patil (1993a, 1993b), Youndjé et al. (1996), and references therein.

To conclude, let us mention that kernel techniques can also be used for estimating a conditional density, which provides an alternative tool for prediction via conditional mode estimation. See, for example, the contributions of Collomb et al. (1987), Youndjé et al. (1993, 1994), and Quintela del Rio and Vieu (1998). The conditional median can also be estimated via a kernel procedure (Gannoun, 1990).

3.8.4 Other Kernel-Based Regression Estimates

Other versions of Nadaraya–Watson estimates have been introduced in the literature. (For example, see Jones et al., 1994). Here, we just give definitions for those estimates that are of primary importance. We have dealt in previous sections of this chapter with random design (i.e., (X_i) is a sample of i.i.d. random variables). An alternative regression problem occurs when the design points (x_i) are known and satisfy $x_i = F^{-1}(i/n+1)$ for some known cumulative distribution function F. This case is refered to as the fixed-design regression problem.

The fixed-design regression model motivated the definition of the Priestley–Chao estimate (Priestley and Chao, 1972), given by

$$\hat{m}_{\text{PC}}(x) = \sum_{i=2}^{n}(x_i - x_{i-1})h^{-1}K\left(\frac{x-x_i}{h}\right)Y_i,$$

and the Gasser–Müller estimator (Gasser and Müller, 1979), given, for x restricted to some interval $[a, b]$, by

$$\hat{m}_{\text{GM}}(x) = \sum_{i=1}^{n}\int_{s_{i-1}}^{s_i} h^{-1}K\left(\frac{u-x_i}{h}\right)duY_i,$$

where $s_i = (x_i + x_{i+1})/2$, $i = 1, \ldots, n-1$; $s_0 = a$; and $s_n = b$.

Finally, another very important kernel-type regression estimate, which works both for fixed and for random design, is the local linear estimator $\hat{m}_{\text{LL}}$, given by

$$\hat{m}_{\text{LL}} = \sum_{i=1}^{n} h^{-1}K\left(\frac{x-X_i}{h}\right)Y_i\frac{s_2(x) - s_1(x)(x-X_i)}{s_2(x)s_0(x) - s_1^2(x)},$$

where

$$s_j(x) = \sum_{i=1}^{n}(x-X_i)^j h^{-1}K\left(\frac{x-X_i}{h}\right).$$

This estimator is a special case of the following local polynomial estimator (Stone, 1977, and Cleveland, 1979), which is defined for any x restricted to some interval $[a, b]$:

$$\hat{m}_{\text{LP}}(x) = \hat{P}(x).$$

Here, $\hat{P}$ is the polynomial of order p that minimizes

$$\sum_{i=1}^{n}(Y_i - P(X_i))^2 h^{-1}K\left(\frac{x-X_i}{h}\right).$$

A complete study of these estimates can be found in Fan and Gijbels (1996). These estimates are discussed in detail in Chapter 9 of this book.

3.8.5 Some Software Hints

A number of the techniques mentioned in this chapter are offered in the statistical environments S-Plus (MathSoft, 1999) and XploRe (Härdle et al., 1999), for Windows 98/NT, Linux, and UNIX platforms.

S-Plus contains a kernel-type scatterplot smoother in `ksmooth`. Available kernels are "box" (default and very crude), "triangle," "Parzen," and "normal." Other S-Plus code is provided by Härdle (1991) in the library `haerdle`, available from `statlib`, and in the library `sm` from Bowman and Azzalini (1997). XploRe also offers a Nadaraya–Watson kernel estimator. The default kernel is "quartic." Alternative kernels are "Gaussian" and "Epanechnikov." The relevant library is `KERNEL`. There are several options for bandwidth selection, such as cross-validation, plug-in, and rule-of-thumb approaches. (See Chapter 4 for an explanation of these approaches.) Macros can be found not only for regression, but also for derivative estimation.

Further references concerning relevant software for nonparametric regression estimation are provided by Simonoff (1996, p. 210ff).

REFERENCES

Azzalini, A. (1981). A note on the estimation of a distribution function and quantiles by kernel method, *Biometrika*, **68**, 326–328.

Berlinet, A. (1993). Hierarchies of higher order kernels, *Probability Theory and Related Fields*, **94**, 489–504.

Bosq, D. (1996). *Nonparametric Statistics for Stochastic Processes: Estimation and Prediction*, Springer–Verlag, Berlin, Germany.

Bosq, D., and Lecoutre, J. P. (1987). *Théorie de l'Estimation Fonctionnelle*, Economica, Paris, France.

Boularan, J., Ferré, L., and Vieu, P. (1994). Growth curves: A two-stage nonparametric approach, *J. Statist. Plann. and Infer.*, **38**, 327–350.

Boularan, J., Ferré, L., and Vieu, P. (1995a). A nonparametric model for unbalanced longitudinal data with application to geophysical data, *Computational Statistics*, **10**, 285–298.

Boularan, J., Ferré, L., and Vieu, P. (1995b). Location of particular points in nonparametric regression analysis, *Austral. J. Statist.*, **37**, 161–169.

Bowman, A. and Azzalini, A. (1997). *Applied Smoothing Techniques for Data Analysis.* Oxford University Press, Oxford, UK.

Cleveland, W. S. (1979). Robust locally weighted regression and smoothing scatterplots, *J. Amer. Statist. Assoc.*, **74**, 829–836.

Collomb, G. (1976). *Estimation Nonparamétrique de la Régression par la Méthode du Noyau.* Thesis at University Paul Sabatier, Toulouse, France.

Collomb, G. (1977). quelques propriétés de la méthode du noyau pour l'estimation nonparamétrique de la régression en un point fixé, *Comptes Rendus de l'Acad. des Sciences, Paris, Série A*, **285**, 289–292.

Collomb, G. (1981). Estimation nonparamétrique de la régression: revue bibliographique, *Inter. Statist. Review*, **49**, 75–93.

Collomb, G. (1984). Propriétés de convergence presque complète du prédicteur à noyau, *Z. Wahrscheinlichkeitstheorie verw. Gebiete*, **66**, 441–460.

Collomb, G. (1985). Nonparametric regression an up-to-date bibliography. *Statistics*, **2**, 309–324.

Collomb, G. Härdle, W., and Hassani, S. (1987). A note on prediction via estimation of the conditional mode function. *J. Statist. Plann. Infer.*, **15**, 227–236.

Devroye, L. (1987). *A Course on Density Estimation*, Birkhäuser, Boston, MA.

Doukhan, P. (1994). *Mixing: Properties and Examples*, Springer–Verlag, Berlin, Germany.

Fan, J., and Gijbels, I. (1996). *Local Polynomial Modeling and Its Applications*, Chapman and Hall, London, UK.

Fan, J., Gijbels, I., Hu, T. C., and Huang, L. S. (1996). A study of variable bandwidth selection for local polynomial regression, *Statistica Sinica*, **6**, 113–117.

Ferraty, F. and Vieu, Ph. (2000). Dimension fractale et estimation de la régression dans des espaces vectoriels semi-normés, to appear in *Comptes Rendus de l'Acad. des Sciences, Paris, Série I.*

Gannoun, A. (1990). Estimation non paramétrique de la médiane conditionnelle, *Pub. Instit. Statist. Univ. Paris*, **XXXV**, 11–22.

Gasser, T., and Müller, H. G. (1979). Kernel estimation of regression functions, in Gasser, T., and Rosenblatt, M. (eds.), *Smoothing Techniques for Curve Estimation*, pp. 23–68, Springer–Verlag, Heidelberg, Germany.

Gasser, T., Müller, H. G., Köhler, W., Molinari, L., and Prader, A. (1984). Nonparametric regression analysis of growth curves, *Annals of Statist.*, **12**, 210–219.

Granovsky, B. L., Müller, H. G., and Pfeifer, C. (1995). Some remarks on optimal kernel functions, *Statist. Decision*, **13**, 101–116.

Györfi, L., Härdle, W., Sarda, P., and Vieu, P. (1989). *Nonparametric Curve Estimation from Time Series*, Springer–Verlag, Berlin, Germany.

Hall, P., and Marron, J. S. (1987). Choice of kernel order in density estimation, *Ann. Statist.*, **16**, 161–173.

Härdle, W. (1990). *Applied Nonparametric Regression*, Cambridge University Press, Cambridge, UK.

Härdle, W. (1991). *Smoothing Techniques with Implementations in S*, Springer–Verlag, New York.

Härdle, W., Klinke, S., and Müller, M. (1999). *XploRe—Academic Edition. The Interactive Statistical Computing Environment* (CD-ROM with Handbook). Springer–Verlag, Berlin, Germany.

Härdle, W., and Marron, J. S. (1985). Optimal bandwidth selection in nonparametric regression function estimation, *Ann. Statist.*, **13**, 1465–1481.

Hart, J. D., and Wehrly, T. (1993). Consistency of cross-validation when data are curves, *Stoc. Proc. and Appl.*, **45**, 351–361.

Hassani, S., Sarda, P., and Vieu, P. (1986). Approche non paramétrique en théorie de la fiabilité: Revue bibliographique, *Revue de Statistique Appliquée*, **XXXV**, 653–666.

Hoeffding, W. (1963). Probability inequalities for sums of bounded random variables, *J. Amer. Statist. Assoc.*, **58**, 15–30.

Izenman, A. J. (1991). Recent developments in nonparametric density estimation, *J. Amer. Statist. Assoc.*, **413**, 205–223.

Jones, C., and Wand, M. P. (1992). Asymptotic effectiveness of some higher order kernels, *J. Statist. Plann. Infer.*, **31**, 15–21.

Jones, C., Davis, S. J., and Park, B. U. (1994) Versions of kernel type regression estimators, *J. Amer. Statist. Assoc.*, **89**, 825–839.

Krzyzak, A. (1985). The rates of convergence of kernel regression estimates and classification rules, *IEEE, Trans. Inform. Theory*, **32**, 668–679.

Lejeune, M., and Sarda, P. (1992). Smooth estimators of distribution and density functions, *Computat. Statist. Data Anal.*, **14**, 457–471.

Mack, Y. P., and Silverman, B. W. (1982). Weak and stong uniform consistency of kernel regression estimates, *Z. Wahrscheinlichkeitstheorie verw. Gebiete*, **61**, 405–415.

Marron, J. S. (1988). Automatic smoothing parameter selection: A survey, *Empirical Economics*, **13**, 187–208.

MathSoft, Inc. (1999). *S-PLUS 2000: Guide to Statistics*, Vol. I MathSoft, Inc., Seattle, WA.

Marron, J. S., and Härdle, W. (1986). Random approximations of some measure of accuracy in nonparametric estimation, *J. Multivariate Analysis*, **20**, 91–113.

Nadaraya, E. A. (1964). On estimating regression, *Theory Prob. Appl.*, **10**, 186–196.

Nunez-Anton, V., Rodriguez-Poo, J. and Vieu, P. (1999). Longitudinal data with nonstationary errors: a nonparametric three-stage approach. *Test*, **8**, 201–231.

Parzen, E. (1962). On estimation of probability density functions and mode, *Ann. Math. Statist.*, **33**, 1065–1076.

Patil, P. N. (1993a). On the least squares cross-validation bandwidth for hazard rate estimation, *Ann. Statist.*, **21**, 1792–1810.

Patil, P. N. (1993b). Bandwidth choice for nonparametric hazard rate estimation, *J. Statist. Plann. Infer.*, **35**, 15–30.

Pelegrina, L., Sarda, P., and Vieu, P. (1996). On multidimensional nonparametric regression, in Pratt, A. (ed.), *Proceedings in Computational Statistics*, pp. 149–160, Physica–Verlag, Heidelberg, Germany.

Priestley, M. B., and Chao, M. T. (1972). Nonparametric curve fitting, *J. Roy. Statist. Soc. Ser. B*, **34**, 385–392.

Quintela del Rio, A., and Vieu, P. (1997). A nonparametric conditional mode estimate, *Nonparametric Statistics*, **8**, 253–266.

Reiss, M. (1981). Nonparametric estimation of smooth distribution functions, *Scand. J. Statist.*, **8**, 116–119.

Rosenblatt, M. (1956). Remarks on some nonparametric estimates of a density function, *Ann. Math. Statist.*, **27**, 832–837.

Sarda, P., and Vieu, P. (1988). Vitesse de convergence d'estimateur à noyau de la règression et de ses dérivées. *Comptes Rendus de l'Acad. des Sciences, Paris, Série I*, **306**, 83–86.

Schucany, W. R. (1995). Adaptive bandwidth choice for kernel regression, *J. Amer. Statist. Assoc.*, **90**. 535–540.

Scott, D. W. (1992). *Multivariate Density Estimation: Theory, Practice and Visualisation.* Wiley, New-York.

Shirahata, S., and Chu, I. S. (1992). Integrated square errors of kernel type estimators of distribution functions, *Ann. Inst. Statist. Math.*, **44**, 579–591.

Silverman, B. W. (1986). *Density Estimation for Statistics and Data Analysis*. Chapman and Hall, London, UK.

Simonoff, J. (1996). *Smoothing Methods in Statistics*, Springer–Verlag, New York.

Stone, C. J. (1977). Consistent nonparametric regression (with discussion), *Ann. Statist.*, **5**, 595–645.

Stone, C. J. (1981). Optimal rates of convergence for nonparametric regression, *Ann. Statist.*, **9**, 1348–1360.

Stone, C. J. (1982). Optimal global rates of convergence for nonparametric regression, *Ann. Statist.*, **10**, 1040–1053.

Stone, C. J. (1985). Additive regression and other nonparametric models, *Ann. Statist.*, **13**, 689–705.

Tukey, J. W. (1961). Curves as parameters, and touch estimation, *Proceedings of the 4th Symposium on Mathematics, Statistics and Probability*, pp. 681–694, Berkeley, CA.

Vieu, P. (1991). Nonparametric regression: Local optimal bandwidth choice, *J. Roy. Statist. Soc., Ser. B*, **53**, 453–464.

Vieu, P. (1993). Bandwidth selection for kernel regression: A survey, in Härdle, W., and Simar, L., *Computer Intensive Methods in Statistics*, pp. 134–149, Physica–Verlag, Heidelberg, Germany.

Vieu, P. (1999). Multiple kernel procedure: an asymptotic support, *Scand. J. Statist.*, **26**, 61–72.

Wand, M. P. (1990). On exact L_1 rates of convergence in nonparametric kernel regression, *Scand. J. of Statist.*, **17**, 251–256.

Wand, M. P., and Jones, M. C. (1995). *Kernel Smoothing*, Chapman and Hall, London, UK.

Wand, M. P., and Schucany, W. (1990). Gaussian based kernels, *Canadian J. Statist.*, **18**, 197–204.

Watson, G. S. (1964). Smooth regression analysis, *Sankhyā Ser. A*, **26**, 359–372.

Wu, J. S., and Chu, C. K. (1993a). Kernel type estimators of jump points and values of a regression function, *Ann. Statist.*, **21**, 1545–1566.

Wu, J. S., and Chu, C. K. (1993b). Nonparametric function estimation and bandwidth selection for discontinuous regression functions, *Statistica Sinica*, **3**, 557–576.

Yoshihara, K. (1994). *Weakly Dependent Stochastic Sequences and their Applications: Volume IV: Curve Estimation Based on Weakly Dependent Data*, Sanseido, Tokyo, Japan.

Youndjé, E., Sarda, P., and Vieu, P. (1993). Estimateur à noyau d'une densité conditionnelle: choix de fenêtre pour des données dépendantes, *Comptes Rendus Acad. des Sciences, Paris, Serie I*, **316**, 935–938.

Youndjé, E., Sarda, P., and Vieu, P. (1994). Choix de paramètre de lissage en estimation de densité conditionnelle, *Publi. de l'Inst. Statist. Univ. Paris*, **XXXVIII**, 57–80.

Youndjé, E., Sarda, P., and Vieu, P. (1996). Optimal smooth hazard estimates, *Test*, **5**, 379–394.

CHAPTER 4

Variance Estimation and Bandwidth Selection for Kernel Regression

Eva Herrmann

Technische Universität Darmstadt, Darmstadt, Germany

4.1 INTRODUCTION

Methods for nonparametric regression and variance estimation are an important subject of statistical research with practical relevance. In this chapter, we give an overview of a variety of methods for nonparametric variance estimation and of automatic bandwidth selection rules for kernel regression estimation. Therefore, we concentrate on the main ideas of, differences as well as and similarities between, such methods.

First, consider the nonparametric regression model

$$Y_i = m(x_i) + \varepsilon_i, i = 1, \ldots, n, \tag{4.1}$$

where $Y_1, \ldots, Y_n$ are observable data variables with respect to the design variables $x_1, \ldots, x_n$. The error variables or residuals, $\varepsilon_1, \ldots, \varepsilon_n$; their variance function σ^2, where $\mathrm{Var}(\varepsilon_i) = \sigma^2(x_i)$, $i = 1, \ldots, n$; and the regression function m are unknown.

Typically, nonparametric estimators of the regression function m perform a locally weighted average of the data; see also Chapters 1, 3 and 9 of this volume. The region of the local averaging and hence the amount of smoothness of the estimated regression function are controlled by a smoothing parameter that is called *bandwidth* for kernel-type regression estimators.

When choosing the bandwidth, one has to make a standard compromise between bias and variance effects. Thus, most of the objective and data-dependent rules for bandwidth selection need an explicit estimator of the residual variance.

Smoothing and Regression: Approaches, Computation, and Application
Edited by Michael G. Schimek
ISBN 0-471-17946-9

Besides, inference about the regression also requires knowledge about the residual variance.

In the remaining part of this introduction, we pose some standard assumptions for our nonparametric regression model and the kernel regression estimators used here. In the next section, we describe and compare several methods for nonparametric estimation of a constant residual variance and of a variance function σ^2. In Section 4.3, we compare data-dependent bandwidth selectors that have been proposed for kernel regression estimation.

The design variables $x_1, \ldots, x_n$ either should stem from a fixed-design setting and satisfy $x_i = F^{-1}(i/n+1)$, $i = 1, \ldots, n$, where F is a cumulative distribution function of a design density f, or should be ordered variables from a sample of random variables with common design density $f = f_X$. For random design, we assume that (x_i, Y_i) stem from a bivariate sample of independent and identically distributed random variables with the same distribution as a random variable (X, Y) that satisfies $m(x) = E(Y|X = x)$ and $\sigma^2(x) = \text{Var}(Y|X = x)$. Additionally, we assume that the design density f has support $[a, b]$, where $a, b \in \mathbb{R}$, $a < b$. Variance and regression estimation are restricted to this compact interval.

The error variables are assumed to be independent random variables such that $E(\varepsilon_i) = 0$ and $\text{Var}(\varepsilon_i) = \sigma^2(x_i)$, $i = 1, \ldots, n$. Unless otherwise stated, we assume that the design density f, the variance function σ^2, and the regression function m have at least two continuous derivatives on the interval $[a, b]$. Additionally, the design density is assumed to be strictly positive on $[a, b]$, and the second derivative should satisfy a γ-Lipschitz condition for some $\gamma > 0$ in case of random design. For some results, we pose further conditions on these functions. In particular, we consider the important case of constant variance in more detail, since it is assumed for several methods of variance estimation and data-adaptive bandwidth selection.

Several kernel-type regression estimators have been introduced in the previous chapter. Here, we assume a kernel K of order $k > 1$, which is a Lipschitz continuous function on $\mathbb{R}$ with support $[-1, 1]$, whose jth moments

$$\mu_j = \mu_j(K) = \int_{-1}^{1} x^j K(x)\, dx, \; j = 0, \ldots, k,$$

satisfy $\mu_0 = 1$; $\mu_j = 0$, $j = 1, \ldots, k-1$; and $\mu_k \neq 0$. We use the abbreviation

$$K_h(x) = \frac{1}{h} K\left(\frac{x}{h}\right),$$

where $h > 0$ is the bandwidth, throughout this chapter. Some examples and illustrations of such kernels have already been given in Chapter 3. Our assumption of compact support excludes some popular kernel functions, for example, the Gaussian density. Most of the results presented in the next sections are also valid for such rapidly vanishing kernel functions, but they have been proved explicitly only for the more restrictive assumption of compact support.

The most important kernel regression estimators are the Nadaraya–Watson estimator $\hat{m}_{\text{NW}}$ (Nadaraya, 1964, and Watson, 1964; see also the Chapter 3 for several consistency properties), the Priestley–Chao estimator $\hat{m}_{\text{PC}}$ (Priestley and Chao, 1972), and the Gasser–Müller estimator $\hat{m}_{\text{GM}}$ (Gasser and Müller, 1979). Another kernel regression estimator, based on a previous linear interpolation of the data points, is proposed in Clark (1977).

These estimators have to be modified in the boundary region $[a, a+h) \cup (b-h, b]$ by using boundary kernels; see, for example, Müller (1988, Section 4.3), for these modifications. Optimal choices of boundary kernels are discussed, for example, in Fan and Gijbels (1996, Section 3.4.2).

Other kernel-type regression estimators are the local polynomial estimators $\hat{m}_{\text{LP}}$ of order p (Cleveland, 1979, and Stone, 1977, see also Chapter 9 in this volume), which are defined for each $x \in [a, b]$ by solving a locally weighted least-squares problem. It is appropriate to use $K_h(\cdot - x)$ with a nonnegative second-order kernel K as a local weight function. The local polynomial estimator is well defined only if at least $p+1$ different design points lie in $(x-h, x+h)$, this condition is always assumed.

The different kernel-type regression estimators are compared in detail in several papers; see, for example, the recent monographs of Fan and Gijbels (1996, Section 2.3), Simonoff (1996, Section 5.2), and Wand and Jones (1994, Section 5.6) for overviews and further references. Only some particularly important asymptotic equivalences are sketched below.

For a fixed design and $t \in (a, b)$, the Priestley–Chao estimator $\hat{m}_{\text{PC}}$, the Gasser–Müller estimator $\hat{m}_{\text{GM}}$, and the local linear estimator $\hat{m}_{\text{LL}}$ are asymptotically equivalent if the same second-order kernel is used. The local polynomial estimator $\hat{m}_{\text{LP}}$ for $p \geq 1$ with weighting kernel K is asymptotically equivalent with the Priestley–Chao estimator $\hat{m}_{\text{PC}}$ and the Gasser–Müller estimator $\hat{m}_{\text{GM}}$ if these estimators use a $(p+1)$th-order kernel K^*, where

$$K^*(x) = K(x) \cdot (c_0 + \cdots + c_{p-1}x^{p-1}).$$

(Compare Müller, 1987.) Thereby, the coefficients $c_0, \ldots, c_{p-1}$ are uniquely determined by the moment conditions. The kernel K^* is referred to as the equivalent kernel for local polynomial estimation. Such an asymptotic equivalence does not hold for random design, where the local polynomial estimators have a better conditional asymptotic variance. Nevertheless, typically, the variance itself of the local polynomial estimator is infinite. Recent modifications of the Priestley–Chao and Gasser–Müller estimators for random design have been proposed by Hall and Turlach (1997) and Herrmann (1996) that make them competitive and asymptotically at least as good as local polynomial estimators. The modifications follow ideas already sketched by Chu and Marron (1991) and Jones et al. (1994). The modified Gasser–Müller estimator proposed by Herrmann (1996) is defined as

$$\hat{m}_{\text{MGM}}(x, h) = \sum_{i=1}^{n} \int_{s_{i-1}}^{s_i} K_h(x-u)\, du Y_i,$$

using kernel quantile estimators $s_i = \hat{F}^{-1}(i + 0.5/n + 1)$, $i = 0, \ldots, n$, and is used in the following examples.

In the next sections, we assume a kernel-type regression estimator $\hat{m}$ that is based on a kernel K of even order k. For the fixed-design case, a modified or ordinary Priestley–Chao estimator; a modified or ordinary Gasser–Müller estimator, with boundary modifications; or a local polynomial estimator $\hat{m}_{\mathrm{LP}}$ with $p = k - 1$ and a suitable weighting kernel leading to the equivalent kernel K may be used; these estimators are all asymptotically equivalent. For equidistant design, the Nadaraya–Watson estimator $\hat{m}_{\mathrm{NW}}$ can be used, too.

For random design, we assume that a modified Priestley–Chao estimator, a modified Gasser–Müller estimator, or a local polynomial estimator with $p = k - 1$ and a suitable weighting kernel leading to the equivalent kernel K is used. For uniform random design, the Nadaraya–Watson estimator may also be used. Again, all of these estimators are asymptotically equivalent.

For these cases, the following results on mean and variance will hold at least conditionally on $x_1, \ldots, x_n$. For the modified Priestley–Chao and the modified Gasser–Müller estimators, they even hold unconditionally.

For each $\delta > 0$, we have

$$\sup_{x \in [a+\delta, b-\delta]} \left| E(\hat{m}(x, h)|x_1, \ldots, x_n) - m(x) - \frac{h^k}{k!} m^{(k)}(x) \mu_k(K) \right| = o_p(h^k)$$

and

$$\sup_{x \in [a,b]} \left| E(\hat{m}(x, h)|x_1, \ldots, x_n) - m(x) \right| = O_p(h^k)$$

for all k-times continuously differentiable regression functions m.

Additionally, we obtain

$$\sup_{x \in [a+\delta, b-\delta]} \left| \operatorname{Var}(\hat{m}(x, h)|x_1, \ldots, x_n) - \frac{\sigma^2(x)}{n f(x)} \int_{-1}^{1} \{K(y)\}^2 \, dy \right| = o_p(n^{-1} h^{-1})$$

and

$$\sup_{x \in [a,b]} \operatorname{Var}(\hat{m}(x, h)|x_1, \ldots, x_n) = O_p(n^{-1} h^{-1}).$$

4.2 NONPARAMETRIC VARIANCE ESTIMATORS

Knowledge of an approximate value for the residual variance is not only of interest in itself and important for the assessment of nonparametric regression estimators, but is also necessary for several data-driven bandwidth selectors, as can be seen in the next section.

In the main part of this section, we consider nonparametric estimators of a constant residual variance σ^2 for the underlying regression model (4.1). All of the estimators described here stem from some (typically normalized) estimated residual sum of squares. Since the residuals are estimated mostly by linear functions, the variance estimators are given by symmetric positive semidefinite quadratic forms of the data variables $Y_1, \ldots, Y_n$. If we use matrix and vector notation, the variance estimators can be written as

$$\hat{\sigma}^2 = \mathbf{Y}^T \mathbf{Q} \mathbf{Y},$$

where $\mathbf{Q}$ is a suitable symmetric and positive semidefinite matrix and $\mathbf{Y}^T = (Y_1, \ldots, Y_n)$. Let $\mathbf{m}^T = (m(x_1), \ldots, m(x_n))$, $\mathbf{1}^T = (1, \ldots, 1)$, and let $\mathbf{D}$ denote the diagonal matrix with the same diagonal elements as $\mathbf{Q}$. If we assume identically distributed residuals with variance σ^2, $|m_3| = |E(\varepsilon_i^3)| < \infty$, and $m_4 = E(\varepsilon_i^4) < \infty$, then we obtain

$$E(\hat{\sigma}^2 | x_1, \ldots, x_n) = \mathbf{m}^T \mathbf{Q} \mathbf{m} + \sigma^2 \operatorname{tr}(\mathbf{Q})$$

and

$$\operatorname{Var}(\hat{\sigma}^2 | x_1, \ldots, x_n) = 2\sigma^4 \operatorname{tr}(\mathbf{Q}^2) + 4\sigma^2 \mathbf{m}^T \mathbf{Q}^2 \mathbf{m} + 4m_3 \mathbf{1}^T \mathbf{D} \mathbf{Q} \mathbf{m} + (m_4 - 3\sigma^4) \operatorname{tr}(\mathbf{D}^2).$$

Such quadratic variance estimators are considered, for example, by Buckley et al. (1988) from a theoretical point of view. Buckley et al. assumed additionally fixed design and normally distributed residuals. Therefore, the last two terms in the above formula for the variance vanish. Moreover, they considered the class M_C of all regression functions m with absolutely continuous second derivative satisfying

$$\int_a^b \{m''(x)\}^2 \, dx < C$$

and the class S of all quadratic and nonnegative variance estimators that are unbiased as linear regression functions. They derived an almost explicit form of the minimax mean squared-error optimal variance estimator with respect to these classes. Thereby, in general, the minimax estimator depends on C, n, σ^2, and the design. See also Fujioka (1994) and Ullah and Zinde-Walsh (1992) for related work. Asymptotic considerations of Buckley et al., however, show, for example, that for equidistant design, the minimax risk can be approximated by

$$\inf_{\hat{\sigma}^2 \in S} \sup_{m \in M_C} E(\hat{\sigma}^2 - \sigma^2)^2 = 2n^{-1}\sigma^4(1 + O(n^{-7/9})).$$

This approximation can be seen as a benchmark for variance estimators of constant variance, which are discussed in the following subsections. We describe the two most important types of such variance estimators. Also, extensions of both types of variance estimators for the estimation of a smooth variance function are sketched in Section 4.2.4. Some of the variance estimators described below are also discussed in Chapter 2 of this volume. Further reviews on estimation of a constant variance function can be found in Dette et al. (1998b) and Green and Silverman (1994, Section 3.4).

4.2.1 Variance Estimators Based on Differencing

Removing trend effects by differencing is a classical approach, for example, in time-series analysis. Suppose that a constant variance is to be estimated in our regression model, under the assumptions given in Section 4.1. The basic idea is to estimate the ith residual by some kind of differencing scheme, given by

$$\sum_{k=-\ell}^{r-\ell} d_{k,i} Y_{i+k},$$

where $0 \leq \ell \leq r$ and $r \geq 1$. Note that ℓ only indicates which of the residuals is estimated by the weighting scheme; it does not specify a further parameter of the estimation method. Natural assumptions on the differencing weights are

$$\sum_{k=-\ell}^{r-\ell} d_{k,i} = 0 \text{ and } \sum_{k=-\ell}^{r-\ell} d_{k,i}^2 = 1. \tag{4.2}$$

The assumptions lead to variance estimators, defined as

$$\hat{\sigma}^2 = \sum_{i=\ell+1}^{n+\ell-r} \frac{1}{n-r} \left(\sum_{k=-\ell}^{r-\ell} d_{k,i} Y_{i+k} \right)^2,$$

that are unbiased for constant regression functions and asymptotically unbiased for Lipschitz continuous regression functions. Let us introduce an $(n-r) \times n$ matrix $\mathbf{D}_r$, defined as

$$\mathbf{D}_r = \begin{pmatrix} d_{-\ell,\ell+1} & \cdots & d_{r-\ell,\ell+1} & & \mathbf{0} \\ & \ddots & & \ddots & \\ \mathbf{0} & & d_{-\ell,n+\ell-r} & \cdots & d_{r-\ell,n+\ell-r} \end{pmatrix}$$

Then, the above variance estimators can also be written in quadratic form as $\hat{\sigma}^2 = \mathbf{Y}^T \mathbf{Q} \mathbf{Y}$, where $\mathbf{Q} = (1/n-r)\mathbf{D}_r^T \mathbf{D}_r$.

Rice (1984) proposed that the uniquely defined estimator

$$\hat{\sigma}_1^2 = \frac{1}{2(n-1)} \sum_{i=1}^{n-1} (Y_i - Y_{i-1})^2$$

be used, which satisfies Equation (4.2) for $r = 1$. The variance estimator of Gasser et al. (1986) relies on a differencing scheme with $r = 2$, which is unbiased even for linear regression functions. Thus, the estimated residuals can be

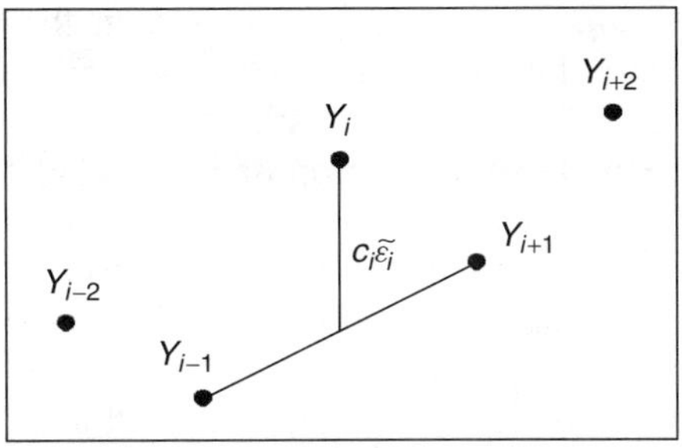

FIGURE 4.1 Difference scheme of the estimated residuals for the variance estimator of Gasser et al. (1986).

interpreted up to normalization as differences between the data and the line that joins the two neighboring points, compare Figure 4.1. Note that this method is equivalent, up to normalization, to estimating residuals from fitting a least-squares line to successive triples of points, an approach that has already been sketched by Rice (1984). This approach yields the estimated residuals given by

$$\tilde{\varepsilon}_i = \sum_{k=-1}^{1} d_{k,i} Y_{i+k} = \frac{1}{c_i}\left(Y_i - \frac{x_{i+1} - x_i}{x_{i+1} - x_{i-1}} Y_{i-1} - \frac{x_i - x_{i-1}}{x_{i+1} - x_{i-1}} Y_{i+1}\right),$$

$$i = 2, \ldots, n-1,$$

where the normalizing factor c_i is defined as

$$c_i^2 = 1 + \left(\frac{x_{i+1} - x_i}{x_{i+1} - x_{i-1}}\right)^2 + \left(\frac{x_i - x_{i-1}}{x_{i+1} - x_{i-1}}\right)^2. \qquad (4.3)$$

The estimated residuals can also be rewritten as divided differences:

$$\tilde{\varepsilon}_i c_i = -\frac{(x_{i+1} - x_i)(x_i - x_{i-1})}{x_{i+1} - x_{i-1}} \left(\frac{Y_{i+1} - Y_i}{x_{i+1} - x_i} - \frac{Y_i - Y_{i-1}}{x_i - x_{i-1}}\right). \qquad (4.4)$$

The underlying idea can be extended in a straightforward manner to fit higher order polynomials and use more neighboring points. Thereby, a differencing scheme for $r \geq 1$ can be defined uniquely by posing the additional conditions

$$\sum_{k=-\ell}^{r-\ell} d_{k,i}(x_{i+k} - x_i)^j = 0, \qquad j = 0, \ldots, r-1,$$

and

$$\sum_{k=-\ell}^{r-\ell} d_{k,i}^2 = 1.$$

The resulting differencing scheme again depends on the design points. These variance estimators are unbiased for all polynomial regression functions of an order of at most $r-1$. We denote these estimators by $\hat{\sigma}_r^2$.

Buckley and Eagleson (1989) have proposed a graphical method to assess the amount of bias of related variance estimators. The graphical method can be used to decide which $r \in \{1, 2, 3\}$ should be used. Therefore, Buckley and Eagleson modified $\hat{\sigma}_r^2$ for nonequidistant design by a different averaging formula of the squared estimated residuals in order to obtain a simpler form of the asymptotic bias. Used for nonequidistant design, this modification typically increases the variance of the single estimators. It also weights down those squared estimated residuals whose design points are close together. Hence, bias also typically increases. Nevertheless, for equidistant or nearly equidistant design, it is appropriate to choose the degree r of $\hat{\sigma}_r^2$. Chan (1991) derived an optimal design for $\hat{\sigma}_1^2$ that equals equidistant design for $x_2, \ldots, x_{n-1}$, wherein x_1 and x_n should be chosen slightly closer to x_2 and x_{n-1}, respectively. A simplified version of Gasser et al. (1986) weighting scheme of the variance estimator $\hat{\sigma}_2^2$ has been proposed by Altman and Paulson (1993).

An interesting minimax result of such estimators for general $r > 0$ was proved by Seifert et al. (1993). They used a discrete version of the restriction assumption given in Buckley et al. (1988) on the regression function, namely,

$$\mathbf{m}^T (\Delta^{(p)})^T \mathbf{W} \boldsymbol{\Delta}^{(p)} \mathbf{m} \leq \mathbf{C}\sigma^2, \tag{4.5}$$

where $\mathbf{W}$ is some positive definite weighting matrix and $\boldsymbol{\Delta}^{(p)}$ is a matrix that yields the pth-order divided differences. Seifert et al. (1993) showed that the variance estimator $\hat{\sigma}_p^2$ is the unique minimax mean squared-error optimal variance estimator with respect to all quadratic variance estimators based on differencing schemes satisfying (4.2) for $r \leq p$, and all regression functions for which inequality (4.5) is true for arbitrary but fixed $\mathbf{C}$, $\mathbf{W}$ and sample size $n > p$.

Hence, $\hat{\sigma}_2^2$ seems to be a good choice that well suits the concept of second-order kernel estimation or local linear estimation. Nevertheless, $\hat{\sigma}_2^2$ does not reach the minimax risk proposed by Buckley et al. (1988). For equidistant design and normally distributed error variables, the bias is asymptotically negligible, and the variance of $\hat{\sigma}_2^2$ satisfies

$$\text{Var}(\hat{\sigma}_2^2) = \frac{35}{9n}\sigma^4(1 + O(n^{-1})),$$

in contrast to the asymptotic minimax value of $2n^{-1}\sigma^4$ obtained by Buckley et al. (1988).

An improvement can be given only by increasing r. Hall et al. (1990) analyzed differencing schemes with equal differencing weights $d_{k,i} = d_k$ for all i from an asymptotic point of view. This analysis can be seen as an approach to improving Rice (1984) estimator $\hat{\sigma}_1^2$ by increasing r, without the aim of estimating exactly unbiased variances for nonconstant polynomial regression functions. Hall et al.

proposed to use such variance estimators based on differencing schemes with fixed r that optimize the asymptotic mean squared error. This approach can be seen as equivalent to minimizing

$$\sum_{j=-\ell}^{r} \left(\sum_{k=\max\{-\ell,-\ell-j\}}^{\min\{r-\ell,r-\ell-j\}} d_k d_{j+k} \right)^2 .$$

Hall et al. (1990) gave a table of the resulting weighting schemes for $1 \leq r \leq 10$. For example, $r = 2$, the resulting variance estimator is given by

$$\hat{\sigma}^2_{\text{HKT},2} = \frac{1}{n-2} \sum_{i=2}^{n-1} \left\{ \frac{1}{4}(\sqrt{5}+1)Y_{i-1} - \frac{1}{2}Y_i - \frac{1}{4}(\sqrt{5}-1)Y_{i+1} \right\}^2 .$$

An analogous proposal for weighting schemes with $r = 3$ can be found in Seifert et al. (1993). This proposal optimizes the asymptotic mean integrated squared error with respect to all variance estimators based on differencing schemes with $r = 3$ that are unbiased for linear regression functions. Seifert et al. also compared some of these estimators in a simulation example. The results of this simulation demonstrated that all of these modifications, especially the modified estimator of Hall et al. (1990), may suffer from high bias for small sample sizes; thus, they are better for very large sample sizes. Another simulation study that compared these estimators, with similar results, is reported in Dette et al. (1998a); see also Dette et al. (1998b).

Hall et al. (1990) demonstrated that their resulting differencing schemes tend to become spiky for increasing values of r; that is, most of the weight is concentrated in a single observation. Thus, for increasing values of r, variance estimators based on the asymptotically optimal differencing schemes are highly related to variance estimators for which the residuals are estimated as difference between the observation Y_i and a suitable estimated regression curve. Such variance estimates are discussed in the next subsection.

4.2.2 Variance Estimators Based on Curve Fitting

A different approach is to estimate residuals after some previous regression estimation step. A variance estimator is obtained as a weighted average of those squared residuals. Early references for such approaches are Breiman and Meisel (1976), who used piecewise linear fits with an adaptive number of pieces for previous curve fitting, and Cleveland (1979), who used a robust variant of local polynomial estimation for regression estimation.

Note that the motivation of the variance estimators based on differencing is slightly different. Typically, they do not aim to approximate the residuals, especially the squared residuals, very well. Instead, they first estimate a functional of the variance and then normalize this estimator.

In this subsection, we assume again that a constant variance should be estimated and that a kernel-type regression estimator $\hat{m}(\cdot, h)$ is used in the first step. Therefore, the bandwidth should satisfy $h = o(1)$ and $n^{-1}h^{-1} = o(1)$. The variance estimator then becomes

$$\hat{\sigma}_h^2 = \frac{1}{cn} \sum_{i=1}^{n} (Y_i - \hat{m}(x_i, h))^2,$$

where the normalizing factor c may be defined such that the variance estimator is unbiased for constant regression functions. If $s_{ij}(h)$ are the entries of the smoothing matrix $\mathbf{S}(h)$, which is often called the *hat matrix*, we obtain

$$\hat{m}(x_i, h) = \sum_{j=1}^{n} s_{ij}(h) Y_j$$

and

$$\hat{\sigma}_h^2 = \frac{1}{cn} \mathbf{Y}^T (\mathbf{I} - \mathbf{S}(h))^T (\mathbf{I} - \mathbf{S}(h)) \mathbf{Y}.$$

Then, the normalizing factor is defined as follows:

$$\begin{aligned} c &= \frac{1}{n} \operatorname{tr}\{(\mathbf{I} - \mathbf{S}(h))^T (\mathbf{I} - \mathbf{S}(h))\} \\ &= 1 - \frac{2}{n} \sum_{i=1}^{n} s_{ii}(h) + \frac{1}{n} \sum_{i=1}^{n} \sum_{j=1}^{n} s_{ij}(h) s_{ji}(h). \end{aligned} \tag{4.6}$$

From the definition of the kernel regression estimators, it follows that $c = 1 + O_p(n^{-1}h^{-1})$; see also Rice (1984), for example. Hence, the normalizing factor can be neglected asymptotically.

The obvious problem that occurs when using such estimators is the choice of the bandwidth. A representation of the asymptotically optimal bandwidth with respect to the mean squared error was derived for the Nadaraya–Watson regression estimator, $k = 2$, and equidistant design by Hall and Marron (1990) and for random design for the local polynomial estimator of odd order $p = k - 1$ by Ruppert et al. (1995). Let K be a kernel of order k or the equivalent kernel for the local polynomial estimator. Then, if m is at least k-times continuously differentiable, the asymptotically optimal bandwidth has the form

$$h_\sigma = \left(\frac{\sigma^4 C_1(K)(b-a)}{n^2 \left(\int \{m^{(k)}(x)\}^2 f(x)\, dx \right)^2} \right)^{1/(4k+1)}, \tag{4.7}$$

as long as $\int \{m^{(k)}(x)\}^2 f(x)\, dx \neq 0$. The kernel constant $C_1(K)$ is given by

$$C_1(K) = \frac{(k!)^4 \int \{(K * K)(y) - 2K(y)\}^2\, dy}{2k\{\mu_k(K)\}^4},$$

where $*$ denotes the convolution and, again, $\mu_k(K) = \int y^k K(y)\, dy$.

If this bandwidth is compared with the asymptotically optimal bandwidth for regression function estimation (see Equations (4.15) and (4.16) in Section 4.3), it is obvious that the bandwidth used here in the context of variance estimation must be chosen to be somewhat smaller than for regression estimation itself, at least asymptotically. Obviously, the formula of the asymptotically optimal bandwidth h_σ gives no practical advice on how to choose the bandwidth right now, since it depends on unknown quantities. Nevertheless, an estimator of this bandwidth can be obtained by applying plug-in ideas, that is, by inserting suitable estimates of these quantities into the formula for h_σ. Alternatively, a data-adaptive bandwidth designed for regression function estimation itself can be used. This suboptimal—and in our context, at least asymptotically, oversmoothing—choice has been proposed by several authors and leads to variance estimators that reach the asymptotically optimal bound of Buckley et al. (1988) for equidistant design and normally distributed error variables.

Another possibility is to transform an arbitrary bandwidth rule for estimating the regression function itself in a way that is motivated by the asymptotically optimal bandwidths. The relationship between the asymptotically optimal bandwidths given by Equations (4.7) and (4.16), with the weight function $\tilde{w}(x) = 1$ for all $x \in [a, b]$, results in

$$h_\sigma = \left(\tilde{h}_a^{4k+2} \frac{2k \int \{(K * K)(y) - 2K(y)\}^2 \, dy}{(b-a) \left[\int \{K(y)\}^2 \, dy \right]^2} \right)^{1/(4k+1)} . \tag{4.8}$$

Inserting a bandwidth rule for estimating the regression function, which is consistent for $\tilde{h}_a$ on the right-hand side, yields a bandwidth rule that is consistent for h_σ and hence an asymptotically optimal bandwidth in the previous step of variance estimation.

Several kinds of such variance estimators can be found in the literature; see also Chapter 2 in this volume for a discussion of several proposals of such estimators in the context of spline smoothing.

Some of these estimators can be used quite analogously for kernel regression estimation, and these are sketched next. Wahba (1983) uses a smoothing spline regression estimator wherein the smoothing parameter is selected by a generalized cross-validation criterion (see also Section 4.3.1) and a slightly different normalizing factor

$$\tilde{c} = \frac{1}{n} \operatorname{tr}(\mathbf{I} - \mathbf{S}(h)). \tag{4.9}$$

In Wahba (1983), this normalizing constant is interpreted as an equivalent degree of freedom, in analogy to parametric regression estimation. Wahba also described some simulation results wherein the variance estimator sometimes suffered from a high bias, especially if the regression function estimator was severely undersmoothed. Again in the context of smoothing-spline regression estimation, Carter and Eagleson (1992) demonstrated that the variance estimator with the normalizing constant in Equation (4.6) is less sensitive with respect to undersmoothed regression estimators in comparison with the variance estimator proposed by Wahba (1983).

Ruppert et al. (1995) proposed a more sophisticated variance estimator, based on a local polynomial regression estimator; see also Wand and Jones (1994, Section 5.8). Ruppert et al. used a variance estimator with a normalizing constant as in Equation (4.6) a smoothing matrix $\mathbf{S}(\hat{h})$ that stems from local polynomial estimation of order $p = k - 1$. A plug-in bandwidth $\hat{h}$ is used, which estimates the asymptotically optimal bandwidth h_σ, given in Equation (4.7), by inserting previous estimates of σ^2 and $\int \{m^{(k)}(x)\}^2 f(x)\,dx$. Both values are obtained from a previous regression estimator $\hat{m}^Q$, which stems from a piecewise quartic fit for $k \le 4$. The number of pieces, N, is selected by Mallows's criterion (Mallows, 1973). Then, the estimators of σ^2 and $I_k = \int \{m^{(k)}(x)\}^2 f(x)\,dx$ are given by

$$\sigma_Q^2(N) = \frac{1}{n - 5N} \sum_{i=1}^{n} \{Y_i - \hat{m}^Q(x_i)\}^2$$

and

$$\hat{I}_k^Q(N) = \frac{1}{n} \sum_{i=1}^{n} \{(\hat{m}^Q)^{(k)}(x_i)\}^2.$$

4.2.3 Comparison of the Different Approaches

Variance estimators based on differences and those based on prior curve fitting are constructed from a different point of view. If the underlying regression curve is relatively smooth and if the sample size is reasonably large, then the variance estimators discussed in the previous subsection are preferable. Nevertheless, for small sample sizes, these estimators can suffer from a large bias that is negative for nearly interpolating regression estimates and positive if a considerable amount of smoothing is involved and $\int \{m^{(k)}(x)\}^2 f(x)\,dx$ is relatively large. Hence, if one wants to guard against a wide class of regression functions, variance estimators based on differencing, especially the estimator $\hat{\sigma}_2^2$ of Gasser et al. (1986), seem to be more appropriate. Then, the increase in variance of a factor of about 2 is negligible in comparison to the decrease in bias. Nevertheless, even these variance estimators may have a large positive bias for a very small variance and extremely variable regression functions. Therefore, even robust versions of variance estimators based on differencing may be considered for such cases; see, for example, Thompson et al. (1991). The simulation example described next should illustrate the previous remarks.

We fix $n = 201$ observations on an equidistant design, with normally distributed error variables and two different regression functions, which are shown in Figure 4.2. We then show sample medians and the region between 10% and 90% sample quantiles from 500 replications of several variance estimators for several values of the variance ranging between 0.0001 and 1. First, we compare Gasser et al. (1986) variance estimator $\hat{\sigma}_2^2$ with the variance estimator

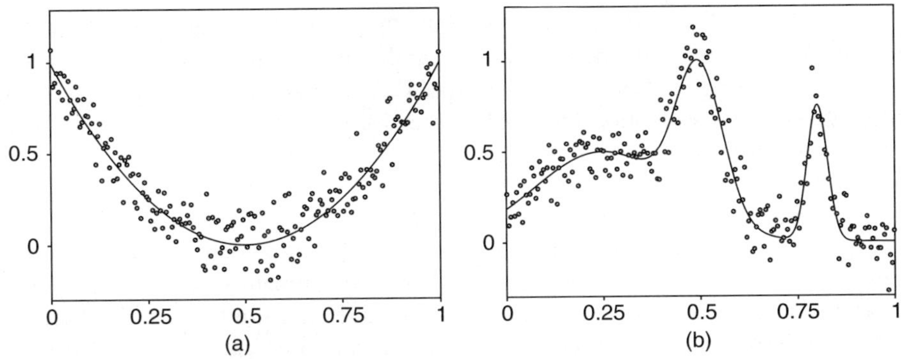

FIGURE 4.2 Two regression functions m_1 in (a) and m_2 in (b) used in the simulation example with sample data sets for $\sigma^2 = 0.01$. (a) m_1 a slowly varying quadratic function; (b) m_2 a superposition of three exponential peaks.

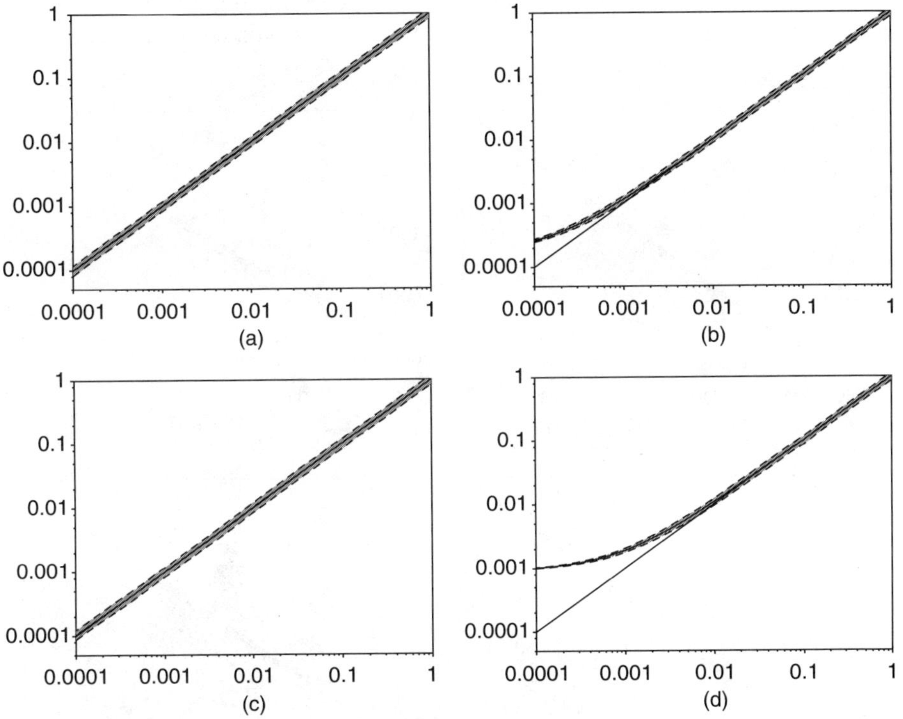

FIGURE 4.3 Sample quantiles for the variance estimators $\hat{\sigma}_2^2$ on the left and $\hat{\sigma}_{\mathrm{HKT},2}^2$ on the right. (a) and (b) Results for the regression function m_1; (c) and (d) results for the regression function m_2.

of Hall et al. (1990) which is denoted by $\hat{\sigma}^2_{\text{HKT},2}$ in Figure 4.3, using $r = 2$. Results for a variance estimator $\hat{\sigma}^2_W$ comparable to the one proposed by Wahba (1983), but using a Gasser–Müller-type kernel regression estimator with a generalized cross-validation bandwidth (see Section 4.3.1) and the normalizing constant from Equation (4.9) and for Ruppert et al.'s (1995) variance estimator $\hat{\sigma}^2_{\text{RSW}}$ are given in Figure 4.4. Thereby, we use the second-order Epanechnikov kernel, given by $K(x) = \frac{3}{4}(1 - x^2)_+$..

An interesting simulation study has been reported by Thompson et al. (1991). They compared eight different methods. Variance estimators based on previous ridge regression estimators with smoothing parameters from a generalized cross-validation criterion with different normalizing constants and the variance estimator $\hat{\sigma}^2_{\hat{r}}$ from Section 4.2.1 with an iteratively adapted degree $\hat{r}$ of differencing behave equally well overall as long as the underlying regression function is relatively smooth. For rapidly variable regression functions, bias-trimming versions of $\hat{\sigma}^2_{\hat{r}}$ or modifications using edge-detecting algorithms can lead to drastic improvements.

Hence, if one feels that bias effects from the underlying regression function are only of minor importance, then variance estimators based on regression estimators

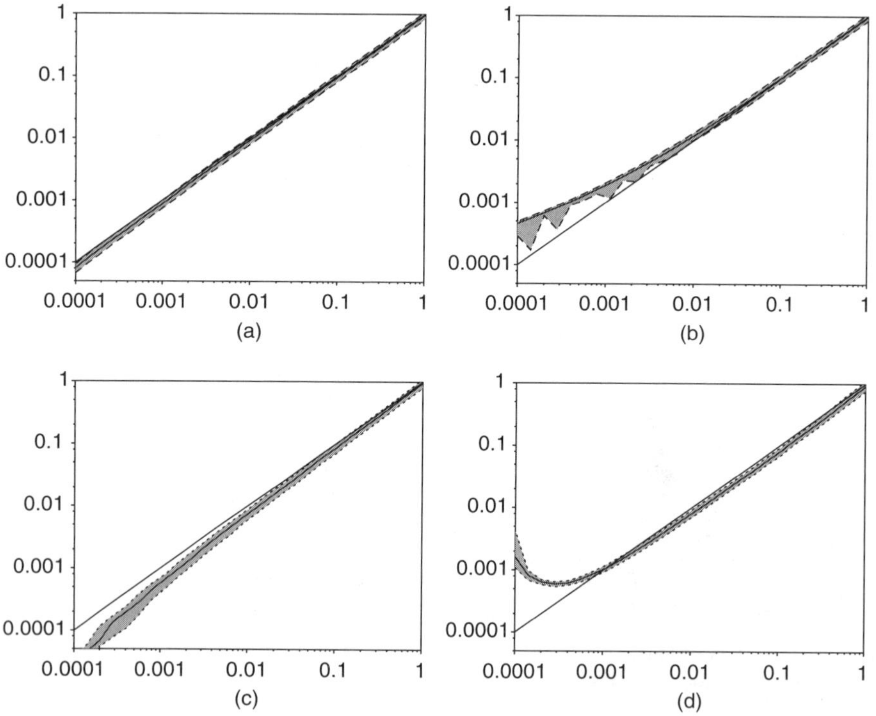

FIGURE 4.4 Sample quantiles for the variance estimators $\hat{\sigma}^2_W$, shown in the plot on the left, and $\hat{\sigma}^2_{\text{RSW}}$, shown in the plot on the right. (a) and (b) Results for the regression function m_1; (c) and (d) results for the regression function m_2.

are most suitable. If variable regression functions may occur, then a difference-based variance estimator seems to be more appropriate, and even further robustifications, like the bias-trimmed estimator of Thompson et al. (1991), should be considered. Additionally, they are easy to apply since no further smoothing parameters have to be selected.

4.2.4 Variance Function Estimation

Each of the variance estimators discussed in the previous subsections can simply be generalized to variance function estimation by performing some local averages instead of averaging the overall normalized squared estimated residuals. In this subsection, we sketch some straightforward modifications of the constant variance estimators discussed previously with kernel-type weights. We also discuss the asymptotically optimal choice of weights and give further references on this topic.

First, we look at variance estimators based on differencing schemes that satisfy (4.2) and

$$\sum_{k=-\ell}^{r-\ell} d_{k,i}(x_{i+k} - x_i)^j = 0,\ j = 0, \ldots, s-1, \tag{4.10}$$

where some $1 \leq s \leq r$ and $r \geq 1$. Estimated squared residuals give

$$\left(\sum_{k=-\ell}^{r-\ell} d_{k,i} Y_{i+k} \right)^2 = \left(\sum_{k=-\ell}^{r-\ell} d_{k,i} \varepsilon_{i+k} \right)^2 + O_p(n^{-s})$$

as long as the regression function m is s-times continuously differentiable. If the variance function is Lipschitz continuous, we obtain the new regression problem

$$Y_i' = \sigma^2(x_i) + \tilde{\varepsilon}_i,\ i = \ell + 1, \ldots, n - r + \ell, \tag{4.11}$$

where

$$Y_i' = \left(\sum_{k=-\ell}^{r-\ell} d_{k,i} Y_{i+k} \right)^2$$

and $E(\tilde{\varepsilon}_i) = 0$, neglecting $O_p(n^{-1})$ remainder terms. Hence, general methods for nonparametric regression estimation can be used for estimating the variance function. If general bandwidth rules, which are discussed in the next section, are applied here, then the r-dependence of the residuals has to be taken into account. Variance function estimators based on differences have been proposed by Müller and Stadtmüller (1987b) and have also been used by Herrmann (1997). They are also discussed in Müller and Stadtmüller (1993).

Müller and Stadtmüller (1993) assumed equidistant design and compared several variance estimators. They showed that very weak assumptions on the smoothness of m are sufficient to obtain rates of convergence for the variance function estimation that depend only on the smoothness of σ^2. They also derived the asymptotically optimal bandwidth for kernel estimators of σ^2 using a kernel K of order k for variance estimators based on differencing schemes with $d_{k,i} = d_k$ with respect to the mean squared error. Assuming equidistant design, a Lipschitz continuous regression function m, a k-times continuously differentiable variance function σ^2, and a Lipschitz continuous moment function m_4, where $E(\varepsilon_i^4) = m_4(x_i)$, the asymptotically optimal local bandwidth is given by

$$h_a(x) = \left(\frac{(k!)^2 (m_4(x) - 3\sigma^4(x) + 2D\sigma^4(x)) \int \{K(y)\}^2\, dy}{2kn\{(\sigma^2)^{(k)}(x)\}^2 \{\mu_k(K)\}^2} \right)^{1/(2k+1)},$$

for $(\sigma^2)^{(k)}(x) \neq 0$, where

$$D = \sum_{j=-\ell}^{r} \left(\sum_{k=\max\{-\ell,-\ell-j\}}^{\min\{r-\ell,r-\ell-j\}} d_k d_{j+k} \right)^2;$$

see Müller and Stadtmüller (1993, Lemma 3.1 and Equation (5.19) therein). Again, the weighting scheme of Hall et al. (1990) optimizes D and hence the asymptotic variance, whereas, roughly speaking, differencing schemes satisfying (4.10), with $r = s$, minimize bias effects that stem from the underlying regression function m.

Asymptotic properties of variance function estimators based on prior curve fitting can be analyzed even more simply. In analogy to Equation (4.11), they can be written as estimators of the regression problem

$$Y_i' = \sigma^2(x_i) + \tilde{\varepsilon}_i,\ i = 1, \ldots, n, \tag{4.12}$$

where $Y_i' = (Y_i - \hat{m}(x_i, h_1))^2$, and uncorrelated residuals, where $E(\tilde{\varepsilon}_i) = 0$ and $E(\tilde{\varepsilon}_i^2) = m_4(x_i)$, neglecting remainder terms of the order $O_p(n^{-1/2} h_1^{-1/2} + h_1^{k_1})$ as long as the regression function estimator uses a kernel K_1 of order k_1 and if m is at least k_1-times continuously differentiable. If σ^2 is estimated by a kernel-type regression estimator using a kernel K of order k_2; if σ^2 is at least k_2-times continuously differentiable; and if there is a Lipschitz continuous function m_4, where $E(\varepsilon_i^4) = m_4(x_i)$, then an asymptotically optimal local bandwidth for this second regression problem is given by

$$h_{2,a}(x) = \left(\frac{(k_2!)^2 (m_4(x) - \sigma^4(x)) \int \{K(y)\}^2\, dy}{2k_2 n\{(\sigma^2)^{(k_2)}(x)\}^2 f(x) \{\mu_{k_2}(K)\}^2} \right)^{1/(2k_2+1)}$$

as long as $(\sigma^2)^{(k)}(x) \neq 0$. This follows from the asymptotic expression for conditional bias and variance in Ruppert et al. (1997) for local polynomial estimation and random design. For Gasser–Müller-type regression estimators and equidistant design it can already be found in Müller and Stadtmüller (1993). An asymptotic representation of the optimal local bandwidth h_1 in the previous regression estimation step is

$$h_{1,a}(x) = \left(\frac{\sigma^4(x) C_1(K_1)(b-a)}{n^2 \{m^{(k_1)}(x)\}^2 f(x)} \right)^{1/(4k+1)};$$

also given in Ruppert et al. (1997) for local polynomial estimation, again assuming a k_1-times continuously differentiable regression function m, a kernel K_1 of order k_1, and $m^{(k_1)}(x) \neq 0$. This is exactly a local form of (4.7).

Hall and Carroll (1989) showed that an α-Lipschitz continuity condition on m, where $\alpha \geq k_2/(2k_2+2)$, suffices to ensure that errors which stem from estimating m become asymptotically negligible. Note, however, that Müller and Stadtmüller (1993) proved that variance function estimators that are based on differencing require even fewer smoothness assumptions on m.

Now, we briefly discuss data-adaptive bandwidth selection of both required bandwidths. Neumann (1994) proved the consistency of variance function estimators for which global leave-one-out cross-validation bandwidths are used for both h_1 and h_2; see also Section 4.3.1. Neumann assumed a Gasser–Müller-type kernel regression estimator and a fixed design.

Ruppert et al. (1997) proposed the use of local plug-in bandwidth selection rules for regression estimation — for example, the rules of Fan and Gijbels (1995) or Ruppert (1997) for local polynomial regression estimation — in both bandwidth selection problems; see also Section 4.3.4.

The approach of Fan and Gijbels (1995) (see also Fan and Gijbels, 1996, Section 4.3), is slightly different. They proposed the use of the same local bandwidth for both smoothing procedures. This bandwidth is defined as a minimizer of a locally weighted penalized residual sum-of-squares criterion; see also Section 4.3.4. The variance estimator is then given as the locally weighted residual sum of squares itself, which is stated in a normalized form.

4.2.5 Some Further Aspects

The variance estimation procedures that are based on curve fitting can easily be generalized to multivariate regression problems. This approach has already been included in the variance estimation approach of Breiman and Meisel (1976), who used piecewise linear fits with an adaptive number of pieces for multivariate regression estimation. Extensions to higher dimensions are also discussed in Ruppert et al. (1997).

Variance estimators based on differences for bivariate regression estimation have been proposed by Hall et al. (1991) and Herrmann et al. (1995). Variance estimation and estimation of more general functions of variance and covariances become much more difficult for regression models that allow dependent error variables. Typical nonparametric estimators of such functions can again be divided in those based on differences and those requiring preliminary curve fitting. An overview of methods used in the context of bandwidth selection can be found in Quintela del Rio (1996). Chapter 6 of this volume provides a full discussion of such regression models for dependent data. Note that variance estimators can fail to be consistent in the case of dependent errors. This is especially true for fixed-design estimators and estimators based on differences.

4.3 BANDWIDTH CHOICE FOR KERNEL REGRESSION ESTIMATORS

The performance of kernel regression estimators depends heavily on the choice of the bandwidth, as already noted in Section 4.1. In this section, we give an overview of some important and popular bandwidth selection methods, concentrating on such methods that are proposed for regression. For a comparison of and an overview on bandwidth selectors for the related case of kernel density estimation, we refer to the recent monographs of Fan and Gijbels (1996, Sections 4.10 and 4.12), Scott (1992, Section 6.5), Simonoff (1996, Subsection 3.1.3), and Wand and Jones (1994, Chapter 3).

A natural error criterion for regression function estimators is the *integrated squared error* (ISE). For a kernel regression estimator with bandwidth h, the ISE is defined as

$$\mathrm{ISE}(h) = \int_a^b \{\hat{m}(x, h) - m(x)\}^2\, dx.$$

A straightforward generalization is given by a *weighted version of ISE*, denoted by WISE, which is defined as

$$\mathrm{WISE}(h) = \int_a^b \{\hat{m}(x, h) - m(x)\}^2 w(x)\, dx,$$

where w is a nonnegative and smooth weight function. We now assume that this weight function is twice continuously differentiable. There are two important motivations to use such a weight function. Firstly, w may be chosen in such a way that it is constant in a suitable inner part and equals 0 in an appropriate boundary region of $[a, b]$. Such a choice weights down the errors in the boundary region, which typically dominate the error in the inner part. Hence, also from a practical point of view, it is often more appropriate to optimize, for example, a global bandwidth with respect to the performance of the corresponding regression estimator in the inner part. Such a weight function also drastically simplifies theoretical considerations; even so, however, it may not be necessary.

Secondly, the weight function may be chosen as the design density f itself, so that WISE is a continuous analog of and asymptotically equivalent to an *averaged squared error* (ASE) defined as

$$\mathrm{ASE}(h) = \frac{1}{n}\sum_{i=1}^{n}\{\hat{m}(x_i, h) - m(x_i)\}^2.$$

Both proposals can also be combined. In that case, a *weighted ASE* criterion (WASE) defined as

$$\text{WASE}(h) = \frac{1}{n}\sum_{i=1}^{n} \tilde{w}(x_i)\{\hat{m}(x_i, h) - m(x_i)\}^2 \tag{4.13}$$

where $\tilde{w}$ is a nonnegative weight function, is asymptotically equivalent to the WISE where the weight function w is given by

$$w(x) = \tilde{w}(x) f(x).$$

In the next two subsections, we describe several methods for global bandwidth selection.

4.3.1 Bandwidth Selectors Based on a Residual Sum-of-Squares Criterion

Here, we consider several classical bandwidth selectors that are based essentially on the residual sum of squares. Most of the procedures described here can be used for a general linear regression estimator that depends on a smoothing parameter. Hence, we also sketch some methods that have been proposed for spline regression estimation. Some of these methods have already been discussed in Chapters 1 and 2 of this book.

Here, it is appropriate again to use vector and matrix notation. Let $\mathbf{Y} = (Y_1, \ldots, Y_n)^T$ and $\hat{\mathbf{m}}(h) = (\hat{m}(x_1, h), \ldots, \hat{m}(x_n, h))^T$ and denote by $\tilde{\mathbf{W}}$ the diagonal matrix with diagonal elements $\tilde{w}(x_1), \ldots, \tilde{w}(x_n)$. Let $\mathbf{S}(h)$ again be the smoothing matrix with entries $s_{ij}(h)$ and $\hat{\mathbf{m}}(h) = \mathbf{S}(h)\mathbf{Y}$. The weighted residual sum of squares is defined by

$$\frac{1}{n}\sum_{i=1}^{n} \tilde{w}(x_i)\{Y_i - \hat{m}(x_i, h)\}^2 = \frac{1}{n}(\mathbf{Y} - \hat{\mathbf{m}}(h))^T \tilde{\mathbf{W}}(\mathbf{Y} - \hat{\mathbf{m}}(h)).$$

Hence, bandwidth selectors based on a residual sum-of-squares criterion aim to optimize the WASE criterion given by Equation (4.13). Obviously, this residual sum-of-squares criterion has to be modified, since it is to be optimal for interpolating regression estimates.

In *cross-validation* (CV) criteria (Clark, 1975; Stone, 1974), $\hat{m}(x_i, h)$ is substituted by a regression estimator $\tilde{m}(x_i, h)$ that does not use the ith observation Y_i. A simple approach is to define

$$\tilde{m}(x_i, h) = (\hat{m}(x_i, h) - s_{ii}(h)Y_i)/(1 - s_{ii}(h)),$$

where $s_{ii}(h) \neq 1$, but see, for example, Neumann (1994) for a different proposal for the Gasser–Müller-type kernel regression estimator. The CV bandwidth h_{CV} is then defined as a bandwidth that minimizes

$$\text{CV}(h) = \frac{1}{n}\sum_{i=1}^{n} \tilde{w}(x_i)(Y_i - \tilde{m}(x_i, h))^2 = \frac{1}{n}\sum_{i=1}^{n} \tilde{w}(x_i)\left(\frac{Y_i - \hat{m}(x_i, h)}{1 - s_{ii}(h)}\right)^2.$$

The CV bandwidth h_{CV} often yields good approximations of h_{WASE}, the optimal bandwidth with respect to the WASE itself. Nevertheless, practical experience and also theoretical considerations, which are discussed later, show that h_{CV} suffers from an unsatisfactory high variability and sometimes tends to choose very small bandwidths.

These drawbacks have motivated the *generalized cross-validation* (GCV) approach of Craven and Wahba (1979) and Wahba (1977) in the context of smoothing-spline estimation. In this method, the adjusting constants $1 - s_{ii}(h)$ are replaced by their weighted mean $1/n\,\mathrm{tr}(\tilde{\mathbf{W}} - \tilde{\mathbf{W}}\mathbf{S}(h))$. This expression gives the GCV criterion

$$\mathrm{GCV}(h) = n \sum_{i=1}^{n} \tilde{w}(x_i) \left(\frac{y_i - \hat{\mathbf{m}}(x_i, h)}{\mathrm{tr}(\tilde{\mathbf{W}} - \tilde{\mathbf{W}}\mathbf{S}(h))} \right)^2 = n \frac{(\mathbf{Y} - \hat{\mathbf{m}}(h))^T \tilde{\mathbf{W}} (\mathbf{Y} - \hat{\mathbf{m}}(h))}{\{\mathrm{tr}(\tilde{\mathbf{W}} - \tilde{\mathbf{W}}\mathbf{S}(h))\}^2}.$$

A different approach has been proposed by Rice (1984); see also Mallows (1973) for similar ideas. For constant variance, it is straightforward to prove that

$$\begin{aligned} &E((\mathbf{Y} - \hat{\mathbf{m}}(h))^T \tilde{\mathbf{W}} (\mathbf{Y} - \hat{\mathbf{m}}(h)) | x_1, \ldots, x_n) \\ &= E\left(\sum_{i=1}^{n} \tilde{w}(x_i)(m(x_i) - \hat{m}(x_i, h))^2 \middle| x_1, \ldots, x_n \right) + \sigma^2(\mathrm{tr}(\tilde{\mathbf{W}}) - 2\,\mathrm{tr}(\tilde{\mathbf{W}}\mathbf{S}(h))). \end{aligned}$$

Hence, the criterion

$$\mathrm{R}(h) = \frac{1}{n}(\mathbf{Y} - \hat{\mathbf{m}}(h))^T \tilde{\mathbf{W}} (\mathbf{Y} - \hat{\mathbf{m}}(h)) + \frac{\hat{\sigma}^2}{n}(2\,\mathrm{tr}(\tilde{\mathbf{W}}\mathbf{S}(h)) - \mathrm{tr}(\tilde{\mathbf{W}})),$$

which is often called the *unbiased risk criterion*, is a natural modification of the residual sum of squares. The bandwidth selector of Rice (1984) is simply the bandwidth h_R that minimizes this criterion.

Similar ideas are behind the bandwidth estimators that are defined as minimizers of the criteria described next. They are stated only for the most important case $\tilde{\mathbf{W}} = \mathbf{I}$, and we do not consider modifications for degenerate situations here. The criteria are as follows:

- *Akaike's finite prediction error* (AFPE) (Akaike, 1970)

$$\mathrm{AFPE}(h) = \frac{1}{n}(\mathbf{Y} - \hat{\mathbf{m}})^T (\mathbf{Y} - \hat{\mathbf{m}}) \frac{n + \mathrm{tr}(\mathbf{S}(h))}{n - \mathrm{tr}(\mathbf{S}(h))};$$

- *Akaike's information criterion* (AIC) (Akaike, 1974)

$$\mathrm{AIC}(h) = \log((\mathbf{Y} - \hat{\mathbf{m}})^T (\mathbf{Y} - \hat{\mathbf{m}})) + \frac{2}{n}\,\mathrm{tr}(\mathbf{S}(h));$$

- *Shibata's criterion* (SC) (Shibata, 1981)

$$\mathrm{SC}(h) = \frac{1}{n}(\mathbf{Y} - \hat{\mathbf{m}})^T (\mathbf{Y} - \hat{\mathbf{m}}) \left(1 + \frac{2}{n}\,\mathrm{tr}(\mathbf{S}(h))\right);$$

- another criterion proposed by Rice (1984), defined as

$$\mathrm{T}(h) = \frac{(\mathbf{Y} - \hat{\mathbf{m}})^T(\mathbf{Y} - \hat{\mathbf{m}})}{n - 2\,\mathrm{tr}(\mathbf{S}(h))},$$

- and a modified AIC criterion by Hurvich et al. (1998), defined as

$$\mathrm{AIC}_c(h) = \log((\mathbf{Y} - \hat{\mathbf{m}})^T(\mathbf{Y} - \hat{\mathbf{m}})) + \frac{2\,\mathrm{tr}(\mathbf{S}(h)) + 2}{n - \mathrm{tr}(\mathbf{S}(h)) - 2}.$$

All of these criteria can be shown to be asymptotically equivalent for bandwidths that are already roughly of the same and optimal order. For such cases, $1/n\,\mathrm{tr}(\mathbf{S}(h)) \to 0$, and the assumption can be proven by simple Taylor expansion of the criteria with respect to $1/n\,\mathrm{tr}(\mathbf{S}(h))$; see, for example, Rice (1984). Only the CV rule needs some further considerations, which can be found in Härdle et al. (1988) for the equidistant design case and the Nadaraya–Watson estimator with second-order kernel. Also, the one-sided CV rule of Hart and Yi (1998) fits in this concept. Nevertheless, it is important in practice that these criteria differ, especially for very small bandwidths, and hence protect, more or less, against choosing bandwidths that lead to severe undersmoothing; see also Hurvich et al. (1998) for more details and several simulation examples. Chui (1990) gives further motivation by a deeper analysis of the asymptotic distribution of the cross-validation bandwidth rule that is skewed to the left. Hart and Yi (1998) prove analogous results for their one-sided CV bandwidth with less skew.

Härdle et al. (1988), however, have proven for equidistant design, four-times continuously differentiable regression functions, and the Nadaraya–Watson kernel estimator that all of these bandwidth selectors have a relatively high variance of the order $n^{-1/5}$ for second-order kernels and, even worse, that they are negatively correlated with the optimal bandwidth h_{WASE}.

Figure 4.5 compares the ASE and some of the criteria functions for the simulation example with regression functions m_1 and m_2 from Figure 4.2. The minimal values are marked by vertical lines in Figure 4.5. A similar concept is the integrated residual sum-of-squares criterion of Fan and Gijbels (1995); see also Fan and Gijbels (1996, Section 4.5). This criterion is an integrated version of a locally weighted penalized residual sum-of-squares criterion and should also give a rough estimate of the bandwidth. Since it depends on a local approach, we discuss it in Section 4.3.4.

4.3.2 Bandwidth Selectors Based on Higher Order Fitting

A different approach for estimating the WASE or the WISE uses previous estimates of the regression function or its derivatives and of the residual variance and inserts them in asymptotic representations of these error criteria. Versions of such bandwidth selectors have been proposed for kernel regression and for local

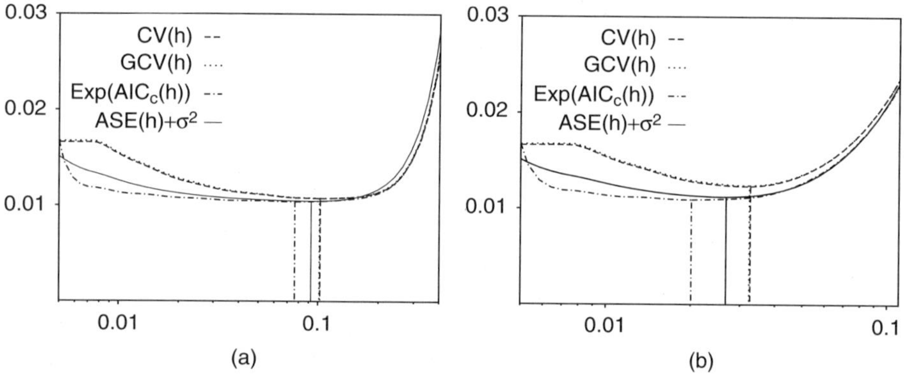

FIGURE 4.5 Comparison of true and estimated ASEs for several of the criteria functions. The situation for m_1 in the simulation examples of Figure 4.2 is shown in (a), and likewise for the situation for m_2 in (b).

polynomial estimation. (See also Chapter 9 in this volume.) Our description of estimators avoids the restriction of a specific kernel-type regression estimation.

A natural approximation of the WISE is its conditional expectation, which is the conditional mean integrated squared error

$$E(\text{WISE}(h)|x_1, \ldots, x_n) = \int \left[\{m(x) - E(\hat{m}(x, h)|x_1, \ldots, x_n)\}^2 + \text{Var}(\hat{m}(x, h)|x_1, \ldots, x_n)\right] w(x)\, dx.$$

If the kernel estimator is given as a weighted sum

$$\hat{m}(x, h) = \sum_{i=1}^{n} s_i(x, h) Y_i,$$

then we obtain

$$E(\text{WISE}(h)|x_1, \ldots, x_n) = \int \left[\left\{ m(x) - \sum_{i=1}^{n} s_i(x, h) m(x_i) \right\}^2 + \sum_{i=1}^{n} s_i^2(x, h) \sigma^2(x_i) \right] w(x)\, dx. \quad (4.14)$$

Inserting estimators for the unknown functions m and σ^2 leads to an estimated conditional mean integrated squared error that can be viewed as an estimation of WISE itself. We refer to such bandwidth selectors as plug-in bandwidth selectors from here on. These bandwidth selectors have the great advantage of allowing us to optimize the estimation of σ^2 and especially the estimation of m for the assessment of the kernel estimator using a kernel K and a bandwidth h. This feature stands in contrast to the criteria that rely on the residual sum of squares, for which only estimates of m with the same kernel and bandwidth can be used.

If the choice of the kernel is already optimized with respect to the smoothness assumptions on m, then there is little to gain by using these bandwidth estimators. Nevertheless, since the real smoothness of m is unknown, there is typically enough space for improvement at this intermediate stage of bandwidth selection. Hence, we may use higher order kernels and a higher order of local polynomial estimation for a previous estimate of m, respectively, without incorporating the problems of higher order approximation at the final kernel regression estimation.

Härdle et al. (1992) have given rates of consistency for this approach; they assumed equidistant design and constant variance. An estimate of (4.14) is obtained by replacing m with a higher order kernel estimator for m and by replacing the unknown variance with a $\sqrt{n}$ consistent variance estimator. Only a slight further modification is performed by subtracting a known diagonal bias term from the estimated squared bias. If the order of the second kernel estimator is large enough, and if the regression function possesses enough continuous derivatives, then estimators $\hat{h}$ of the $E(\text{WISE}|x_1, \dots, x_n)$ optimal bandwidth h_0 may be derived in such a way that $n^{1/2}(\hat{h} - h_0)/h_0$ is asymptotically normally distributed.

A further simplification stems from approximating $E(\text{WISE}(h)|x_1, \dots, x_n)$ in Equation (4.14) by the leading asymptotic terms, given by

$$\frac{h^{2k}}{(k!)^2}\{\mu_k(K)\}^2 \int w(x)\{m^{(k)}(x)\}^2\,dx + \frac{1}{nh}\int \{K(y)\}^2\,dy \int w(x)\{\sigma^2(x)/f(x)\}\,dx,$$

if m is k-times continuously differentiable and $\int w(x)\{m^{(k)}(x)\}^2\,dx \neq 0$. The new approximation error is typically negligible in comparison with the difficulty of estimating m. This approximation can be minimized explicitly, and the resulting asymptotically optimal bandwidth is then given by

$$h_a = \left(\frac{(k!)^2 \int w(x)\{\sigma^2(x)/f(x)\}\,dx \int \{K(y)\}^2\,dy}{2kn\{\mu_k(K)\}^2 \int w(x)\{m^{(k)}(x)\}^2\,dx} \right)^{1/(2k+1)}. \tag{4.15}$$

Note that the asymptotically optimal bandwidth of the weighted average squared error has the form

$$\tilde{h}_a = \left(\frac{(k!)^2 \int \tilde{w}(x)\sigma^2(x)\,dx \int \{K(y)\}^2\,dy}{2kn\{\mu_k(K)\}^2 \int \tilde{w}(x)\{m^{(k)}(x)\}^2 f(x)\,dx} \right)^{1/(2k+1)}, \tag{4.16}$$

where m is k-times continuously differentiable and $\int \tilde{w}(x)\{m^{(k)}(x)\}^2 f(x)\,dx \neq 0$. Neglecting first the minor problem of variance estimation, we have to estimate the kth derivative of m. In this case, we may again use kernel-type estimators. The Priestley–Chao and Gasser–Müller estimators can simply be adapted to derivative estimation by using derivative kernels and derivative boundary kernels; see, for example, Fan and Gijbels (1996, Section 3.2) and Gasser and Müller

(1984). At least in the interior, such estimates are exactly the kth derivatives of kernel estimators of an order $k' > k$. The local polynomial estimator can also be modified for derivative estimation if we define

$$\hat{m}_{\mathrm{LP}}^{(k)}(x, h) = \hat{P}^{(k)}(x),$$

where $\hat{P}$ is the locally adapted polynomial, which has to be of order $p \geq k$. Hence, we use a higher order approximation here. Again, we have to select a bandwidth for this higher order approximation. If we denote by $\hat{m}_k(x, g)$ a suitable kernel-type estimator of $m^{(k)}(x)$ of order $k + 2$ using a bandwidth g, then we can define a bandwidth for the original problem as

$$h(g) = \left(\frac{(k!)^2 \int w(x)\{\hat{\sigma}^2(x)/f(x)\}\, dt \int \{K(y)\}^2\, dy}{n 2k\{\mu_k(K)\}^2 \int w(x)\{\hat{m}_k(x, g)\}^2\, dx} \right)^{1/(2k+1)}.$$

The new bandwidth selection problem of selecting g is less difficult than the original problem, as demonstrated in Figure 4.6. This figure shows the ASE for the simulation examples of Figure 4.2 and 4.5 and compares the average squared error as a function of h using the corresponding kernel estimator $\hat{m}(\cdot, h)$ of order $k = 2$ and as a function of g using a kernel estimator $\hat{m}(\cdot, h(g))$. Here, Gasser–Müller-type kernel regression estimators with an Epanechnikov kernel and the variance estimator of Gasser et al. (1986) are used. Obviously, the intermediate estimation step simplifies the bandwidth selection problem, but, of course, a bandwidth must have been previously chosen.

A simple iterative procedure has been proposed by Gasser et al. (1991) for Gasser–Müller-type kernel regression estimation with $k = 2$, equidistant design, and constant variance. They proposed that Gasser et al. (1986) variance estimator

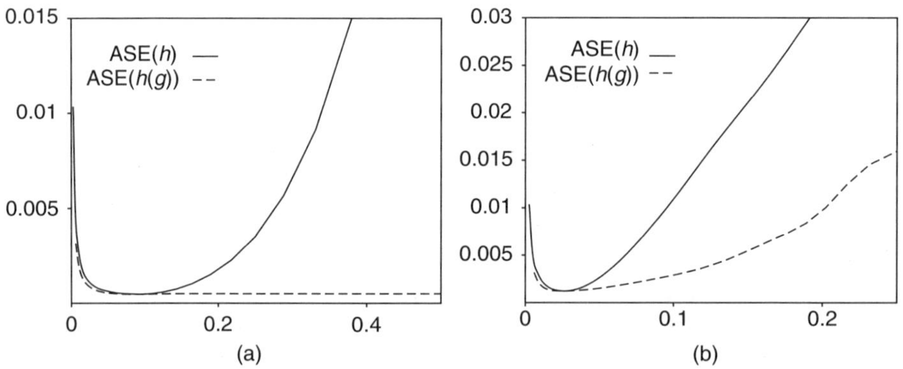

FIGURE 4.6 The ASE as a function of h using a kernel estimator $\hat{m}(\cdot, h)$ in (a) and of g using the corresponding kernel estimator $\hat{m}(\cdot, h(g))$ in (b).

$\hat{\sigma}_2^2$ be used (see Section 4.2.1) and that $\hat{h}_0 = n^{-1}(b-a)$, $\hat{g}_i = \hat{h}_{i-1}n^{1/10}$, and

$$\hat{h}_i = \left(\frac{\hat{\sigma}_2^2 \int w(x)\,dx \int \{K(y)\}^2\,dy}{n\{\mu_2(K)\}^2 \int w(x)\{\hat{m}_2(x,\hat{g}_i)\}^2\,dx} \right)^{1/5}$$

for $i \geq 1$. This procedure is stopped after 11 iteration steps, resulting in a bandwidth rule $\hat{h} = \hat{h}_{11}$. Gasser et al. (1991) proved the consistency of their method using only rather weak smoothness assumptions of m''. Additionally, they proved that there is no real breakdown for regression functions without continuous second derivatives, even though the iterative plug-in bandwidth becomes inconsistent with respect to the optimal bandwidth.

A general and slightly modified version of this iterative plug-in bandwidth is described in Herrmann (1997). This version also adapts heteroscedasticity and nonequally spaced design. In addition, it can be used for random design, because of the asymptotic equivalence of appropriate kernel estimators for random design. First, the bandwidth is again defined for the Gasser–Müller-type kernel regression estimator and its modification for random design. (See Herrmann, 1996.) Then, $\int w(x)\{\sigma^2(x)/f(x)\}\,dx$ is estimated by going back to the more precise expression for the variance of the Gasser–Müller-type regression estimator,

$$n\sum_{i=1}^{n} w(x_i)\sigma^2(x_i)(s_i - s_{i-1})^2 \approx \frac{nh}{\int \{K(y)\}^2\,dy}\left(\sum_{i=1}^{n} w(x_i)\sigma^2(x_i)s_i(x,h)^2\right), \tag{4.17}$$

where $s_0, \ldots, s_n$ stem from the definition of the kernel estimator. For the estimation, $\sigma^2(x_i)$ is replaced by normalized squared residuals $\tilde{\varepsilon}_i^2$ of the variance estimation approach of Gasser et al. (1986) (see Equations (4.4) and (4.3)) with a suitable modification for $i = 1$ and $i = n$. The bandwidth choice of the kth derivative is again performed iteratively through $\hat{h}_0 = (k-1)(b-a)n^{-1}$ and

$$\hat{h}_i = \left(\frac{(k!)^2 \sum_j w(x_j)\tilde{\varepsilon}_j^2(s_j - s_{j-1})^2 \int \{K(y)\}^2\,dy}{2k\{\mu_2(K)\}^2 \int w(x)\{\hat{m}_2(x,\hat{g}_i)\}^2\,dx} \right)^{1/(2k+1)}$$

using $\hat{g}_i = c\hat{h}_{i-1}n^{2/\{(2k+1)(2k+3)\}}$ and $(k+1)(2k+1)$ iteration steps. An analogous iteration formula that depends only on the linear weighting scheme of the regression estimator is given by

$$\hat{h}_i = \left(\frac{(k!)^2 \hat{h}_{i-1} \sum_j \sum_\ell w(x_j)\tilde{\varepsilon}_j^2 s_j(x_\ell,h)^2}{(2k)n\{\mu_2(K)\}^2 \int w(x)\{\hat{m}_2(x,\hat{g}_i)\}^2\,dx} \right)^{1/(2k+1)},$$

using the same bandwidths $\hat{g}_i = c\hat{h}_{i-1}n^{2/\{(2k+1)(2k+3)\}}$ and, again, $(k+1)(2k+1)$ iteration steps. The resulting estimator is $\hat{h} = \hat{h}_{(k+1)(2k+1)}$. The constant c is chosen as $c = 1.1(1 + 0.05(k-2))$, a choice that is motivated from practical experience. This iterative procedure performs quite well in several simulations, and some empirical results are given in Herrmann (1997) and Hurvich et al. (1998).

Note that the asymptotic motivation of these bandwidth selectors is not as restrictive as it seems to be at first glance. For example, the asymptotically optimal bandwidth h_a may not approximate the optimal bandwidth for regression functions with overlaying structure. If there is a rough overall variability on the one hand and a fine variability structure on the other hand, then the asymptotically optimal bandwidth always adapts mainly the fine variability structure; even so, it might be favorable only to adapt the overall structure for finite sample size. Hence, the asymptotically optimal bandwidth typically leads to an undersmoothed regression estimator. On the other hand, estimating the optimal bandwidth via plug-in ideas can also result in estimators of the kth derivative that smooth away the fine variability structure and instead adapt the overall variability. This explains why using plug-in bandwidths often yields better results than using h_a itself for small sample size. Gijbels and Mammen (1998) proved that simple—that is, noniterative—plug-in bandwidths, in fact, tend to oversmooth fine structure, even in intermediate situations, in which this effect is not desirable. Starting with a very small bandwidth in the iterative procedure can guard against this defect; see, for example, the simulation examples and illustrations in Herrmann (1997). We now sketch another proposal for global plug-in bandwidth selection.

The procedure of Ruppert et al. (1995) has been proposed for random design, local linear regression estimation, and constant variance, and Ruppert et al. used the ASE as an error criterion. Their variance estimator has already been discussed in Section 4.2.2. The asymptotically optimal bandwidth for estimating $\int \{m''(x)\}^2 f(x)\,dx$ with

$$\frac{1}{n}\sum_{i=1}^{n}\{\hat{m}_2(x_i, g)\}^2$$

using a local polynomial derivative estimator $\hat{m}_2$ of order $p = 3$ is given by

$$g_a = \left(-\frac{12\sigma^2(b-a)\displaystyle\int \{K_2(y)\}^2\,dy}{n\displaystyle\int m''(x)m^{(4)}(x)f(x)\,dx\mu_4(K_2)} \right)^{1/7}$$

if $\int m''(x)m^{(4)}(x)f(x)\,dx < 0$ and by

$$g_a = \left(\frac{30\sigma^2(b-a)\displaystyle\int \{K_2(y)\}^2\,dy}{n\displaystyle\int m''(x)m^{(4)}(x)f(x)\,dx\mu_4(K_2)} \right)^{1/7}$$

if $\int m''(x)m^{(4)}(x)f(x)\,dx > 0$. Here, K_2 denotes the equivalent derivative kernel, which is assumed to satisfy $\mu_4(K_2) = \int y^4 K_2(y)\,dy > 0$. Ruppert et al. also gave a table of the necessary kernel constants for the most popular kernels. Now, the same blocked quartic fit $\hat{m}^Q$ from the first estimation step of the variance estimator is used again to estimate g_a by taking $\hat{\sigma}_Q^2$ as previous estimate of σ^2 and

$$\frac{1}{n}\sum_{i=1}^{n}(\hat{m}^Q)''(x_i)(\hat{m}^Q)^{(4)}(x_i)$$

as a previous estimate of $\int m''(x)m^{(4)}(x)f(x)\,dx$. The resulting bandwidth $\hat{g}$ is used for estimating $\int\{m''(x)\}^2 f(x)\,dx$. Hence, the plug-in bandwidth for the original problem is given by

$$\hat{h} = \left(\frac{\hat{\sigma}^2_{\mathrm{RSW}}(b-a)\displaystyle\int\{K(y)\}^2\,dy}{\displaystyle\sum_{i=1}^{n}\{\hat{m}_2(x_i,\hat{g})\}^2\{\mu_2(K)\}^2}\right)^{1/5}.$$

Simulation results are reported in Ruppert et al. (1995) and Hurvich et al. (1998). Since all asymptotically optimal bandwidths are stated in Ruppert et al. (1995), the bandwidth selector can be extended straightforwardly to local polynomial estimation using polynomials of a general odd order p.

Certainly, all of these plug-in approaches are highly related. Note that an estimation of the optimal bandwidth g may also be incorporated in the iterative plug-in estimator by first estimating an optimal constant $c_a = g_a/h_a n^{-2/\{(2k+1)(2k+3)\}}$; see, for example, Engel et al. (1994) for such an approach in the case of an iterative plug-in bandwidth selector for kernel density estimation. It is also straightforward to modify the iterative plug-in procedure for estimating with respect to a WASE criterion and to adapt the plug-in estimator of Ruppert et al. (1995) for estimating with respect to a WISE criterion.

Another plug-in-type estimator has been proposed by Fan and Gijbels (1995). It will be discussed in its local form in Section 4.3.4.

4.3.3 Comparison of the Different Methods

In sum, one can deduce from both simulation results and asymptotic theory that plug-in bandwidths have a significantly smaller variance than bandwidth selectors, which use a penalized residual sum-of-squares criterion. Therefore, using plug-in bandwidths very often results in better estimators. Additionally, the iterative plug-in estimator and the plug-in bandwidth selector of Ruppert et al. (1995) do not need a numerical minimization step and can be computed quite fast. They can even be modified in order to select bandwidths of regression derivative estimation. Even though they use the formula of the asymptotically optimal bandwidth, they are, according to asymptotic theory, better in estimating the sample optimal bandwidths h_{WISE} and h_{WASE} than are the bandwidth rules discussed in Section 4.3.1. This advantage follows from the negative correlation of the asymptotic joint distribution of such estimators and the finite-sample optimal ones proved in Härdle et al. (1988). Plug-in bandwidths may also have a negative correlation in the asymptotic distribution, as proved by Herrmann (1994) for the iterative plug-in rule of Gasser et al. (1991). Because of the small variance, this negative correlation is less important for plug-in bandwidths than for the cross-validation-like bandwidth selectors, which are considered in Härdle et al. (1988). Therefore, asymptotic theory indicates that plug-in bandwidths are asymptotically equivalent with the asymptotically optimal bandwidths in estimating h_{WISE} or h_{WASE}, whereas bandwidth selectors based on some residual sum-of-squares criteria exhibit worse behavior, at least for equidistant design, constant variance, second-order kernels, and four-times continuously differentiable regression functions. In spite of the asymptotic theory, Hurvich et al. (1998) have summarized several simulation situations wherein their modified AIC rule behaves almost as well as plug-in bandwidths.

Hall and Johnstone (1992) derived bandwidth selectors whose joint asymptotic distribution with the optimal bandwidth not only shows positive correlations, but is even asymptotically efficient. They also gave an example of such an estimator. However, for cases involving finite samples, this estimator does not lead to satisfactory simulation results. Also, note that differences between the performance of bandwidth selectors are less important for kernel order $k \geq 4$.

Typically, plug-in bandwidths are preferable in many situations. Nevertheless, they have more problems in estimating the optimal bandwidth in situations wherein very good regression estimators exist, namely, for very large sample sizes, very smooth regression functions, and very small residual variances. In particular, they have difficulties in estimating relatively large optimal bandwidths — for example, when the regression function is nearly linear. They also depend on further parameters, for example, the constant c for the iterative plug-in rule and the maximal number of blocks and the polynomial order for the previous blocked fit of Ruppert et al. (1995) rule. The proposals for plug-in bandwidths given in this chapter do a good job in many situations, but may lead to a significant bias under certain circumstances.

4.3.4 Local Bandwidth Choice

The global rules discussed may be generalized for local bandwidth choice. First of all, the bandwidth selectors based on a residual sum of squares can be adapted locally by using a local weighting scheme for the sum of squares instead of averaging overall; see, for example, Vieu (1991) for such a proposal in the case of the cross-validation criterion. The local residual sum-of-squares criterion of Fan and Gijbels (1995) (see also Chapter 9 of this volume), is similar. It is proposed mainly for a previous step of bandwidth selection for local polynomial estimation. In this case, the normalized local residual sum of squares

$$\hat{\sigma}_{\mathrm{FG}}(x, h) = \frac{1}{c} \sum_{i=1}^{n} (Y_i - \hat{P}(x_i))^2 \frac{1}{h} K\left(\frac{x - x_i}{h}\right)$$

is viewed as a variance estimator, where $\hat{P}$ again denotes the locally fitted polynomial. It uses the same weighting kernel K and the same bandwidth h for both the local weighting and the local polynomial estimation of order p. Denoting by $s_1(x, h), \ldots, s_n(x, h)$ the linear weights of the regression estimator $\hat{m}(x, h) = \sum s_i(x, h) Y_i$, we define the local residual sum-of-squares criterion as

$$\mathrm{RSC}(x, h) = \hat{\sigma}^2_{\mathrm{FG}}(x, h) \left(1 + (p + 1) \sum_{i=1}^{n} s_i^2(x, h)\right).$$

A global criterion IRSC($[c, d]$) for $x \in [c, d] \subset [a, b]$ is simply obtained by integration of RSC($\cdot, h$) on $[c, d]$. The resulting bandwidths are consistent with the optimal bandwidths only for $p = 1$ and the Gaussian weighting kernel. Otherwise, one must use the adjusting constants that have been proposed using ideas that are detailed in Müller and Stadtmüller (1987a) and Müller (1988). Similarly, the resulting bandwidth can be adjusted for estimating derivatives of an order less than p.

Very much related is a proposal by Cleveland and Loader (1996) that uses a local version of Rice's (or Mallows's) criterion for local polynomial estimation. Again, the locally weighted residual sum of squares is based on a single polynomial fit. Cleveland and Loader also apply the same weighting kernel K and bandwidth h for local weighting of the residual sum of squares and for the local polynomial estimation. They proposed that Gasser et al. (1986) estimator $\hat{\sigma}_2^2$ be used as a nonparametric variance estimator for an equidistant design example.

Another possibility for modifying a global residual sum-of-squares criterion has been proposed by Fan et al. (1996), mainly for cases involving kernel density estimation. Fan et al. concentrated on bandwidth functions that are interpolating cubic splines for a finite set of knots. Then, a leave-one-out cross-validation criterion is minimized numerically over the finite number of bandwidths that are associated with the knots.

Several proposals for local bandwidth choice modify global bandwidth selectors by using plug-in ideas. For equidistant design and constant variance, Müller and Stadtmüller (1987a) started with the bandwidth $\hat{h}_R$ that stems from minimizing Rice's criterion and multiplied it by a factor for local adaptation. This factor was obtained from the analogous relation for the asymptotically optimal bandwidths by inserting estimators for the derivatives of the regression function using higher order kernels. Also for equidistant design and constant variance, Staniswalis (1989) estimated the mean squared error by using a prior curve estimate. In this case, a regression estimator with bandwidth $\hat{h}_R$ was also used.

Similarly, Fan and Gijbels (1995) proposed a modification of the bandwidths obtained from their residual sum-of-squares criterion. For global bandwidth choice, they first fitted a polynomial of order $p+2$ locally and adjusted the IRSC($[a, b]$) optimal bandwidth for estimating the $(p+1)$th derivative. The fit of the resulting polynomial of order $p+2$ and the local variance estimator $\hat{\sigma}^2_{\mathrm{FG}}(x, h_{\mathrm{IRSC}})$ were inserted into Equation (4.14). Then, the proposed global bandwidth selector yielded the bandwidth that minimizes the estimated mean integrated squared error.

For local bandwidth selection, Fan and Gijbels (1995) split the smoothing interval $[a, b]$ in $k = [n/(10 \log n)]$ subintervals I_k of the same length. For each interval, the local polynomial estimator of order $p+2$ was used for finding the IRSC(I_k) optimal bandwidth, which was again adjusted for estimating the $(p+1)$th derivative. Then, the bandwidth step function was smoothed using the length of the subintervals as a smoothing parameter. A local polynomial estimator of order $p+2$ was obtained for estimating the integrated squared error on each subinterval. The minimizing bandwidths again formed a bandwidth step function that was slightly resmoothed and used for the final local polynomial estimation of order p.

The asymptotic distribution of a simple local plug-in approach for the Nadaraya–Watson kernel regression estimator and random design has already been derived in Tsybakov (1987). Nevertheless, no practical proposals for bandwidth selection in the previous derivative estimation step were given in that paper.

The plug-in approach of Schucany (1995) assumes fixed design and constant variance. It estimates the mean squared error for a kernel regression estimator of order k. If $s_1(x, h), \ldots, s_n(x, h)$ again denote the kernel weights, then the exact mean squared error in this case is given by

$$E(\hat{m}(x, h) - m(x))^2 = \left\{ m(x) - \sum_{i=1}^{n} s_i(x, h) m(x_i) \right\}^2 + \sigma^2 \sum_{i=1}^{n} s_i^2(x, h).$$

The variance part of this equation is estimated by using the $\sqrt{n}$ consistent variance estimator of Hall et al. (1990) and by approximating $\sum_{i=1}^{n} s_i^2(x, h)$ by $1/nh \int \{K(y)\}^2 \, dy$. The bias part can be approximated by

$$m(x) - \sum_{i=1}^{n} s_i(x, h) m(x_i) \approx B_x h^k + C_x h^{k+1}, \tag{4.18}$$

which is used as a motivation for the following steps. For fixed bandwidth, the bias part is estimated by inserting a previous curve estimator of order $k+2$ using the same bandwidth. This step is repeated for several different bandwidths $h_1, \ldots, h_\ell$ that are chosen in such a way that they are equidistant and of the same order. For $k = 2$ and $n \geq 100$, Schucany (1995) proposed that $h_i = i/8n^{-1/7}$, $i = 1, \ldots, 7$, be used. Next, the coefficients B_x and C_x were estimated by least-squares estimation, viewing the right-hand side of Equation (4.18) as a parametric model for a function of the bandwidth. The resulting local bandwidth estimator is then given by

$$\hat{h}(x) = \left(\frac{\hat{\sigma}^2 \int \{K(y)\}^2 \, dy}{2kn\hat{B}_x} \right)^{1/(2k+1)},$$

which approximates the bandwidth that minimizes the estimated mean squared error. Also, a modification of the estimated bias has been proposed in regions near inflection points. In this case, the bandwidth $\hat{h}$ is chosen as a numerical solution of

$$2(B_x\hat{h}^k + C_x\hat{h}^{k+1})(kB_x\hat{h}^{k-1} + (k+1)C_x\hat{h}^k) = \frac{\hat{\sigma}^2 \int \{K(y)\}^2 \, dy}{n\hat{h}^2}.$$

Ruppert (1997) proposed a local empirical bias bandwidth rule for local polynomial estimation of order $p = k - 1$ that is based on ideas that are quite similar to those of Schucany (1995). Since it uses several user-specified parameters and several estimating steps, we provide only a rough sketch of it here. For a fixed grid of bandwidths, the mean squared error and especially the bias are again estimated by fitting a polynomial as a function of h. Then, an approximate and local parametric model for the expectation of the regression estimator is given by $\hat{m}(x, h) = a_0 + a_k h^k + \cdots + a_{k+\tau-1} h^{k+\tau-1}$ for some $\tau \geq 1$. The regression estimator of a fixed number of bandwidths near h, which are taken from the same fixed grid of bandwidths, is used to fit this model for estimating the bias by $\hat{a}_k h^k + \cdots + \hat{a}_{k+\tau-1} h^{k+\tau-1}$. The variance is estimated by $\hat{\sigma}^2(x) \sum_{i=1}^n s_i^2(x, h)$, where a local variance estimator from a previous curve estimator is used as an estimator of the regression problem given by Equation (4.12). Thereby, the same bandwidth selection rule is proposed, but with a scale model for the new variance. A previous scale estimator stems from a blocked fit. Afterwards, the estimated mean squared errors of the original regression estimator are interpolated to a finer grid of bandwidths and smoothed as a function of x. For each x, the least local minimum of the smoothed estimated error criterion is taken as the bandwidth rule, which is again resmoothed.

The local modifications of Brockmann et al. (1993) and Herrmann (1997) of the global iterative plug-in bandwidth are quite simple and fast in comparison with the proposals discussed previously. In particular, they do not use any numerical minimization procedure. Here, we describe only the version by Herrmann (1997), which is stated for our general regression setting. Additionally, it has been proposed for the Gasser–Müller kernel estimator for fixed design and can also be used for the modified and even the unmodified Gasser–Müller kernel estimator for random design. We also sketch the application for a local polynomial estimator. This estimator adds a single local iteration step to the global iterative plug-in bandwidth estimator $\hat{h}$. Suppose that we use a kernel estimator of order k. Then a local variance estimator is obtained as a local variant of Equation (4.17) by using

$$n \sum_{i=1}^{n} \tilde{\varepsilon}_i^2 (s_i - s_{i-1}) \int_{s_{i-1}}^{s_i} K_{\hat{h}}(x-u)\,du,$$

where $\tilde{\varepsilon}_i^2$ are the normalized squared residuals from the variance estimator $\hat{\sigma}_2$ of Gasser et al. (1986). Note that

$$n(s_i - s_{i-1}) \int_{s_{i-1}}^{s_i} K_h(x-u)\,du \int \{K(y)\}^2\,dy$$

approximates the local kernel weight $s_i(x,h)^2 h$ for the convolution kernel-type estimator and can also be replaced by it. Defining bandwidths

$$g_1 = 0.86(1 + 0.05(k-2))\hat{h} n^{4/\{(2k+1)(2k+5)\}}$$

and

$$g_2 = 1.4(1 + 0.05(k-2))\hat{h} n^{2/\{(2k+1)(2k+3)\}},$$

we obtain a local estimator $\tilde{m}_k^2(x, g_1, g_2)$ of the squared kth derivative as a smoothed version of a squared kernel derivative estimator $\hat{m}_k(x, g_1)$ using a derivative kernel of order $k+2$. The resmoothing is done by

$$\tilde{m}_k^2(x, g_1, g_2) = \int \tilde{K}_{g_2}(x-u)\{\hat{m}_k(u, g_1)\}^2\,du$$

with a positive and symmetric weighting kernel $\tilde{K}$. The resmoothing procedure has the nice advantage of adapting automatically to single zeros of the kth derivative; see Herrmann (1997). The resulting local plug-in bandwidth is now obtained by

$$\hat{h}(x) = \left(\frac{(k!)^2 \sum_i \tilde{\varepsilon}_i^2 (s_i - s_{i-1}) \int_{s_{i-1}}^{s_i} K_{\hat{h}}(x-u)\,du \int \{K(y)\}^2\,dy}{2k\tilde{m}_k^2(x, g_1, g_2)\{\mu_k(K)\}^2} \right)^{1/(2k+1)}.$$

An analogous formula that uses the weighting scheme only is given by

$$\hat{h}(x) = \left(\frac{(k!)^2 \hat{h} \sum_i \tilde{\varepsilon}_i^2 s_i(x, \hat{h})^2}{2kn\tilde{m}_k^2(x, g_1, g_2)\{\mu_k(K)\}^2} \right)^{1/(2k+1)}.$$

A comparative simulation study of several local bandwidth rules has yet to be undertaken. Some simulations indicate that there is a small advantage of local bandwidth rules in comparison with analogous global rules for moderate to large sample sizes. Asymptotic considerations on optimal rates of the bandwidths selectors of Wang and Gasser (1996) show again, however, that local bandwidths are more difficult to estimate than global ones.

4.3.5 Further Aspects and Some Software Hints

Several bandwidth selectors can be or have already been generalized to multivariate regression problems. (See the discussion in Chapter 12 of this book.) Also, generalizations for the estimation of kernel derivatives are quite straightforward, at least for plug-in estimators.

Nearly all bandwidth selectors proposed and discussed behave worse for dependent error variables. This is especially true for fixed design. As a result, several modifications have been proposed; see, for example, Quintela del Rio (1996) for an overview and comparison of several methods. Further references can also be found in Simonoff (1996, p. 203). Chapter 6 in this volume provides a detailed discussion of regression estimation under dependence.

Finally, we give some information concerning the implementation of bandwidth selection techniques. General versions of a FORTAN 77 subroutine for the fast iterative global and local plug-in rules discussed previously can be obtained from the Internet at `http://www.unizh.ch/biostat`. Quite comfortable MATLAB and S-Plus interface routines are provided there as well. A Gasser–Müller-type kernel regression estimator that uses the modified version for random design can be used with a user-provided bandwidth or, alternatively, with the global or local iterative plug-in bandwidth for estimating a univariate regression function and its first and second derivatives. In this manner, fast updating algorithms have been adopted. (See, for example, Gasser and Kneip, 1989, and Seifert et al., 1994).

S-Plus interface routines for local polynomial estimation and bandwidth choice, as described in Ruppert et al. (1995), and related routines for nonparametric kernel density estimation, as discussed in Wand and Jones (1994), together with descriptions of these routines, can be found on the S archive of the statlib server `http://lib.stat.cmu.edu/S/` of the American Statistical Association by using the keyword `kernel`. This Web page also provides several other smoothing routines that are related to nonparametric regression estimation.

REFERENCES

Akaike, H. (1970). Statistical prediction information, *Ann. Inst. Statist. Math.*, **22**, 203–217.

Akaike, H. (1974). A new look at statistical model identification, *IEEE Trans. Automat. Control.*, **19**, 716–723.

Altman, N. S., and Paulson, C. P. (1993). Some remarks about the Gasser–Sroka–Jennen-Steinmetz variance estimator, *Comm. Statist., Theory Methods*, **22**, 1045–1051.

Breiman, L., and Meisel, W. S. (1976). General estimates of the intrinsic variability of data in nonlinear regression models, *J. Amer. Statist. Assoc.*, **71**, 301–307.

Brockmann, M., Gasser, T., and Herrmann, E. (1993). Locally adaptive bandwidth choice for kernel regression estimators, *J. Amer. Statist. Assoc.*, **88**, 1302–1309.

Buckley, M. J., and Eagleson, G. K. (1989). A graphical method for estimating the residual variance in nonparametric regression, *Biometrika*, **76**, 203–210.

Buckley, M. J., Eagleson, G. K., and Silverman, B. W. (1988). The estimation of residual variance in nonparametric regression, *Biometrika*, **75**, 189–199.

Carter, C. K., and Eagleson, G. K. (1992). A comparison of variance estimators in nonparametric regression, *J. Roy. Statist. Soc. Ser. B*, **54**, 773–780.

Chan, L.-Y. (1991). Optimal design for estimation of variance in nonparametric regression using first order differences, *Biometrika*, **78**, 926–929.

Chu, C. K., and Marron, J. S. (1991). Choosing a kernel regression estimator, *Statist. Sci.*, **6**, 404–436.

Chui, S.-T. (1990). Why do bandwidth selectors tend to choose smaller bandwidths and a remedy, *Biometrika*, **77**, 436–441.

Clark, R. M. (1975). A calibration curve for radiocarbon dates, *Antiquity*, **49**, 251–266.

Clark, R. M. (1977). Non-parametric estimation of a smooth regression function, *J. Roy. Statist. Soc. Ser. B*, **39**, 107–113.

Cleveland, W. S. (1979). Robust locally weighted regression and smoothing scatterplots, *J. Amer. Statist. Assoc.*, **74**, 829–836.

Cleveland, W. S., and Loader, C. (1996). Smoothing by local regression: Principles and methods, in Härdle, W., and Schimek, M. G. (eds.), *Statistical Theory and Computational Aspects of Smoothing*, pp. 10–49, Physica–Verlag, Heidelberg, Germany.

Craven, P., and Wahba, G. (1979). Smoothing noisy data with spline functions, *Numer. Math.*, **31**, 377–403.

Dette, H., Munk, A., and Wagner, T. (1998a). Estimating the variance in nonparametric regression by quadratic forms—What is a reasonable choice? *J. Roy. Statist. Soc. Ser. B*, **60**, 751–764.

Dette, H., Munk, A., and Wagner, T. (1998b). A review of variance estimators with extensions to multivariate nonparametric regression models, in Ghosh, S. (ed.), *Multivariate Design and Sampling*, Dekker, New York.

Engel, J., Herrmann, E., and Gasser, T. (1994). An iterative bandwidth selector for kernel estimation of densities and their derivatives, *J. Nonparametr. Statist.*, **4**, 21–34.

Fan, J., and Gijbels, I. (1995). Data-driven bandwidth selection in local polynomial fitting: Variable bandwidth and spatial adaptation, *J. Roy. Statist. Soc. Ser. B*, **57**, 371–394.

Fan, J., and Gijbels, I. (1996). *Local Polynomial Modeling and Its Application–Theory and Methodologies*, Chapman and Hall, New York.

Fan, J., Hall, P., Martin, M., and Patil, P. (1996). On local smoothing of nonparametric curve estimators, *J. Amer. Statist. Assoc.*, **91**, 258–266.

Fujioka, T. (1994). The max-mse's of minimax estimators of variance in nonparametric regression, *Hiroshima Math. J.*, **24**, 573–581.

Gasser, T., and Kneip, A. (1989). Discussion of Buja, A., Hastie, T., and Tibshirani, R., Linear smoothers and additive models, *Ann. Statist.*, **17**, 532–535.

Gasser, T., Kneip, A., and Köhler, W. (1991). A flexible and fast method for automatic smoothing, *J. Amer. Statist. Assoc.*, **86**, 643–652.

Gasser, T., and Müller, H.-G. (1979). Kernel estimation of regression functions, in Gasser, T., and Rosenblatt, M. (eds.), *Smoothing Techniques for Curve Estimation*, pp. 23–68, Springer-Verlag, Berlin, Germany.

Gasser, T., and Müller, H.-G. (1984). Estimating regression functions and their derivatives by the kernel method, *Scand. J. Statist.*, **11**, 171–185.

Gasser, T., Sroka, L., and Jennen-Steinmetz, C. (1986). Residual variance and residual pattern in nonlinear regression, *Biometrika*, **73**, 625–633.

Gijbels, I., and Mammen, E. (1998). On local adaptivity of kernel estimates with plug-in local bandwidth selectors, *Scand. J. Statist.*, **25**, 503–520.

Green, P. L., and Silverman, B. W. (1994). *Nonparametric Regression and Generalized Linear Models*, Chapman and Hall, London, UK.

Hall, P., and Carroll, R. H. (1989). Variance function estimation in regression: The effect of estimating the mean, *J. Roy. Statist. Soc. Ser. B*, **51**, 3–14.

Hall, P., and Johnstone, I. (1992). Empirical functionals and efficient smoothing parameter selection, *J. Roy. Statist. Soc. Ser. B*, **54**, 475–530.

Hall, P., Kay, J. W., and Titterington, D. M. (1990). Asymptotically optimal difference based estimation of variance in nonparametric regression, *Biometrika*, **77**, 521–528.

Hall, P., Kay, J. W., and Titterington, D. M. (1991). On estimation of noise variance in two-dimensional signal processing, *Adv. Appl. Prob.*, **23**, 476–495.

Hall, P., and Marron, J. S. (1990). On variance estimation in nonparametric regression, *Biometrika*, **77**, 415–419.

Hall, P., and Turlach, B. (1997). Enhancing convolution and interpolation methods for nonparametric regression, *Biometrika*, **84**, 779–790.

Härdle, W., Hall, P., and Marron, J. S. (1988). How far are automatically chosen regression smoothing parameters from their optimum? (with discussion), *J. Amer. Statist. Assoc.*, **83**, 86–101.

Härdle, W., Hall, P., and Marron, J. S. (1992). Regression smoothing parameters that are not far from their optimum, *J. Amer. Statist. Assoc.*, **87**, 227–233.

Hart, J. D., and Yi, S. (1998). One-sided cross-validation, *J. Amer. Statist. Assoc.*, **93**, 620–631.

Herrmann, E. (1994). Asymptotic distribution of bandwidth selectors in kernel regression estimation, *Statist. Papers*, **35**, 17–26.

Herrmann, E. (1996). *On the Convolution Type Kernel Regression Estimator*, Technical Report, Technische Universität Darmstadt, Darmstadt, Germany.

Herrmann, E. (1997). Local bandwidth choice in kernel regression estimation, *J. Comput. Graph. Statist.*, **6**, 35–54.

Herrmann, E., Wand, M. P., Engel, J., and Gasser, T. (1995). A bandwidth selector for bivariate kernel regression, *J. Roy. Statist. Soc. Ser. B*, **57**, 171–180.

Hurvich, C. M., Simonoff, J. S., and Tsai, C.-L. (1998). Smoothing parameter selection in nonparametric regression using an improved AIC criterion, *J. Roy. Statist. Soc. Ser. B*, **60**, 271–293.

Jones, M. C., Davies, S. J., and Park, B. U. (1994). Versions of kernel-type regression estimators, *J. Amer. Statist. Assoc.*, **89**, 825–832.

Mallows, C. (1973). Some comments on c_p, *Technometrics*, **15**, 661–675.

Müller, H.-G. (1987). Weighted local regression and kernel methods for nonparametric curve fitting, *J. Amer. Statist. Assoc.*, **82**, 231–238.

Müller, H.-G. (1988). *Nonparametric Analysis of Longitudinal Data*, Springer–Verlag, Berlin, Germany.

Müller, H.-G., and Stadtmüller, U. (1987a). Variable bandwidth kernel estimators of regression curves, *Ann. Statist.*, **15**, 182–201.

Müller, H.-G., and Stadtmüller, U. (1987b). Estimation of heteroscedasticity in regression analysis, *Ann. Statist.*, **15**, 610–635.

Müller, H.-G., and Stadtmüller, U. (1993). On variance function estimation with quadratic forms, *J. Statist. Plann. Infer.*, **35**, 213–231.

Nadaraya, E. A. (1964). On estimating regression, *Theory Probab. Appl.*, **10**, 133–141.

Neumann, M. H. (1994). Fully data-driven nonparametric variance estimators, *Statistics*, **25**, 189–212.

Priestley, M. B., and Chao, M. T. (1972). Nonparametric curve fitting, *J. Roy. Statist. Soc. Ser. B*, **34**, 385–392.

Quintela del Rio, A. (1996). Comparison of bandwidth selectors in nonparametric regression under dependence, *Comput. Statist. Data Anal.*, **21**, 563–580.

Rice, J. (1984). Bandwidth choice for nonparametric regression, *Ann. Statist.*, **12**, 1215–1230.

Ruppert, D. (1997). Empirical-bias bandwidths for local polynomial nonparametric regression and density estimation, *J. Amer. Statist. Assoc.*, **92**, 1049–1063.

Ruppert, D., Sheather, S. J., and Wand, M. P. (1995). An effective bandwidth selector for local least squares regression, *J. Amer. Statist. Assoc.*, **90**, 1257–1270.

Ruppert, D., Wand, M. P., Holst, U., and Hössjer, O. (1997). Local polynomial variance function estimation, *Technometrics*, **39**, 262–273.

Schucany, W. R. (1995). Adaptive bandwidth choice for kernel regression, *J. Amer. Statist. Assoc.*, **90**, 535–540.

Scott, D. W. (1992). *Multivariate Density Estimation: Theory, Practice, and Visualization*, Wiley, New York.

Seifert, B., Brockmann, M., Engel, J., and Gasser, T. (1994). Fast algorithms for nonparametric curve estimation, *J. Comput. Graph. Statist.*, **3**, 192–213.

Seifert, B., Gasser, T., and Wolf, A. (1993). Nonparametric estimation of residual variance revisited, *Biometrika*, **80**, 373–383.

Shibata, R. (1981). An optimal selection of regression variables, *Biometrika*, **68**, 45–54.

Simonoff, J. (1996). *Smoothing Methods in Statistics*, Springer–Verlag, New York.

Staniswalis, J. G. (1989). Local bandwidth selection for kernel estimates, *J. Amer. Statist. Assoc.*, **84**, 284–287.

Stone, C. J. (1977). Consistent nonparametric regression (with discussion), *Ann. Statist.*, **5**, 595–645.

Stone, M. (1974). Cross-validatory choice and assessment of statistical predictions (with discussion), *J. Roy. Statist. Soc. Ser. B*, **36**, 111–147.

Thompson, A. M., Kay, J. W., and Titterington, D. M. (1991). Noise estimation in signal restoration using regularization, *Biometrika*, **78**, 475–488.

Tsybakov, A. B. (1987). On the choice of the bandwidth in kernel nonparametric regression, *Theory. Probab. Appl.*, **32**, 142–148.

Ullah, A., and Zinde-Walsh, V. (1992). On the estimation of residual variance in nonparametric regression, *J. Nonparametric Statistics*, **1**, 263–265.

Vieu, P. (1991). Nonparametric regression: Optimal local bandwidth choice, *J. Roy. Statist. Soc. Ser. B*, **53**, 453–464.

Wahba, G. (1977). A survey of some smoothing problems and the methods of generalized cross-validation for solving them, in Krisnaiah, P. R. (ed.), *Applied Statistics*, pp. 507–523, North Holland, Amsterdam, Netherlands.

Wahba, G. (1983). Bayesian "confidence intervals" for the cross-validated smoothing spline, *J. Roy. Statist. Soc. Ser. B*, **45**, 133–150.

Wand, M. P., and Jones, M. C. (1994). *Kernel Smoothing*, Chapman and Hall, London, UK.

Wang, K., and Gasser, T. (1996). Optimal rate for estimating local bandwidth in kernel estimation of regression functions, *Scand. J. Statist.*, **23**, 303–312.

Watson, G. S. (1964). Smooth regression analysis, *Sankhyā Ser. A*, **26**, 359–372.

CHAPTER 5

Spline and Kernel Regression under Shape Restrictions

Michel Delecroix
ENSAI, Rennes, France

Christine Thomas-Agnan
Université de Toulouse I, Toulouse, France

5.1 INTRODUCTION

Shape restrictions on a functional parameter, such as positivity, monotonicity, and convexity, arise in a variety of statistical models. In econometric models, for instance, a cost function must be homogeneous of degree one, nondecreasing, and concave in the price of inputs in order to be consistent with the economic theory of production. Hildreth (1954) considered the problem of estimating production functions and Engel curves, and Matzkin (1991) studied the estimation of monotone and concave utility functions. For applications to life testing and reliability, Marshall and Proschan (1965) have studied the estimation of distributions with monotone failure rates. Other applications to meteorology and computerized tomography are described in Wahba (1982), and to biometrics in Kelly and Rice (1990). Little can be said about the parametric approach, which usually yields an optimization (maximum likelihood) that is subject to constraints on the parameters, as, for example, in Hawkins (1994), and suffers from the classical drawbacks of parametric models. As Dent (1973, p. 131) noted, "Prior specification is a weakening of the position most researchers would like to assume." Let us also quote Ramsay (1988, p. 425): "Attempting to impose monotonicity on polynomials quickly becomes unpleasant."

In this chapter, we present different approaches that can be found in the literature on nonparametric models for the problem of incorporating the information

Smoothing and Regression: Approaches, Computation, and Application
Edited by Michael G. Schimek
ISBN 0-471-17946-9

on shape in an estimation procedure. Some of these approaches are based on kernel smoothers, whereas others are based on splines. As there have been very few propositions based on other smoothing techniques, like local polynomials (see Chapter 9 of this book), we therefore concentrate on splines and kernels. Some of the approaches that we discuss are specific to a particular kind of shape constraint, while others are more flexible. For each method, we describe the principle and the main properties and give a precise account of the computational aspects. It is, of course, impossible to give all of the details for each of the methods we present, but we hope that a schematic description will convince the reader of the feasibility of the approaches and enable him or her to choose the most appropriate one for any particular application. With a clear idea of the difficulties involved with each approach, the reader will then have to refer to the original text describing the method for its actual implementation.

First, we rapidly sketch the precursors to such methods, like the approaches discussed by Hildreth (1954) and Brunk (1955, 1958). Mukerjee (1988) found a way to force the Parzen–Rosenblatt kernel smoother to satisfy a monotonicity constraint. But there have been more attempts to deal with the problem by using splines. For example, Dierckx (1980), for convexity, and Ramsay (1988), for monotonicity, used constrained least-squares splines. Villalobos and Wahba (1987) used smoothing-spline estimates and incorporated a discretized version of the shape restriction by including it as constraints in the optimization problem that yields the smoothing spline. Mammen (1991b) proposed a least-squares procedure in a class of functions satisfying some light smoothness condition and a shape restriction, and the solution turned out to be a least-squares spline with knot points that depend on the observations. Lastly, Delecroix et al. (1996) introduced a two-step procedure consisting of a smoothing step followed by an "isotonizing" step. In the first step, a smoothing-spline estimate or a convolution-type kernel estimate can be used. The second step is a projection in a Hilbert space onto the cone of restrictions.

A related but different problem, that of estimating departures from monotonicity of a regression function, has been studied by Cuesta-Albertos et al. (1994, 1995).

Although we concentrate in this chapter on the regression model, a lot of work has also been done in the context of density estimation under monotonicity restriction (Birgé, 1987a,b, 1997; Groeneboom, 1985) or unimodality restriction (Bickel and Fan, 1996; Prakasa Rao, 1969; Wang, 1995). Furthermore, Grenander (1956) introduced a monotone density estimator that inspired later work by Birgé (1989) and Groeneboom et al. (1996).

5.2 DESCRIPTION OF THE MAIN METHODS

For description purposes, we adopt as a common setting the regression model described next and come back later to the specificity of each method with respect to adaptability to the model. Let $(X_1, Y_1), \ldots, (X_n, Y_n)$ be i.i.d. bivariate random variables for which we are given one observation, denoted by (x_i, y_i).

The functional parameter m to be estimated is $m(x) = E(Y_i|X_i = x)$, which is the regression function of Y_i on X_i. In the fixed-design case, the explanatory variable is denoted by a lowercase letter, and we have $Y_i = m(x_i) + \epsilon_i$, where ϵ_i are realizations of n random variables with zero mean and covariance structure defined by $E(\epsilon_i) = 0$ if $i \neq j$ and σ^2 otherwise.

Besides models for the estimation of regression functions, one finds in the literature models in which other types of functional parameters have to be estimated with constraints. For example, let us mention the median regression in Cryer et al. (1972), the percentile regression model in Casady and Cryer (1976), the dichotomous choice model in Matzkin (1991), and the models of others like Groeneboom and Jongbloed (1993) and Ferraty (1997).

The shape restriction can usually be formulated by the fact that m belongs to a cone C in a function space. The most frequent restrictions, such as positivity, monotonicity, and convexity fall into this framework. The same is true for piecewise monotonicity, piecewise convexity, and unimodality. Note, however, that for unimodality with unknown mode, the resulting cone is not convex.

5.2.1 The Precursors

Motivated by economic problems, Hildreth (1954) introduced an estimation procedure for the concavity restriction. This procedure consists of a least-squares estimate that is subject to discretized concavity restrictions. Let C_d be the discrete cone of restrictions in

$$\mathbb{R}^n\colon \left\{(z_1, \ldots, z_n)\colon \frac{z_{i+1} - z_i}{x_{i+1} - x_i} \geq \frac{z_{i+2} - z_{i+1}}{x_{i+2} - x_{i+1}}, i = 1, \ldots, n-2\right\}.$$

Hildreth's estimate Y_i^* at X_i is the solution of

$$\min_{(z_1,\ldots,z_n)\in C_d} \sum_{i=1}^{n} (y_i - z_i)^2. \tag{5.1}$$

He proposed an iterative procedure to compute this estimate. This estimate, extended between observation points to be a piecewise linear function, falls into the class of least-squares spline estimates. (See Section 5.2.3.1.) A consistency result has been given by Hanson and Pledger (1976). Several authors have proposed various algorithms for its computation, including Dent (1973), Fraser and Massam (1989), and Holm and Frisen (1985). A multivariate version of the algorithm has been studied by Holloway (1979) and Ruud (1995), the latter in the case of additional monotonicity constraints.

But Hildreth's (1954) estimation procedure has not been as popular as the one introduced a little later for the monotonicity restriction by Brunk (1958, 1970). Brunk's procedure also consists in a least-squares procedure corresponding to maximum likelihood when the data are normal. Let C_d be the discrete cone of restrictions in $\mathbb{R}^n\colon \{(z_1, \ldots, z_n)\colon z_1 \leq z_2 \leq \cdots \leq z_n\}$. Then let Y_i^* be solution of the minimization problem

$$\min_{(z_1,\dots,z_n)\in C_d} \sum_{i=1}^{n} (y_i - z_i)^2. \tag{5.2}$$

This problem has a unique solution that can be expressed by the two min–max formulas. Let $X_{(1)}, \dots, X_{(n)}$ be the order statistics corresponding to $X_1, \dots, X_n$ and $Y_{[i]}$ be the observation at the point $X_{(i)}$. Then

$$Y_i^* = \max_{s\le i} \min_{t\ge i} \sum_{j=s}^{t} Y_{[j]}/(t-s+1) \tag{5.3}$$

$$Y_i^* = \min_{s\ge i} \max_{t\le i} \sum_{j=s}^{t} Y_{[j]}/(t-s+1). \tag{5.4}$$

Another expression for Y_i^* is obtained through the cumulative sum diagram of $Y_{[i]}$, that is, the set of points $\{(k, \sum_{j=1}^{k} Y_{[j]}), 0 \le k \le n\}$. Then Y_i^* is the left slope at X_i of the greatest convex minorant of the cumulative sum diagram on $[0, n]$.

None of these expressions, however, is optimal for actual computation of Y_i^*, and a popular algorithm for this computation is the *pool adjacent violators* (PAV) algorithm. The proof that the PAV algorithm yields the solution to Equation (5.2) can be found, for example, in Barlow et al. (1972) and in Robertson et al. (1988). In addition, a similar algorithm for isotonic median regression can be found in Menendez and Salvador (1987) and Shamberg (1991).

In Brunk's approach however, no attempt is made to extend smoothly the estimation to values of x between observation points. The simplest extension, which consists of constant steps between the values of X_i, has been studied extensively, but this estimate has been criticized for its lack of smoothness, the presence of too many flat spots, and its rate of convergence, which is slower than nonparametric rates.

Cryer et al. (1972), adapted Brunk's idea to the estimation of the median regression function (i.e., the median of the conditional distribution of Y_i given X_i) under monotonicity restriction and Casady and Cryer (1976), extended this idea to the estimation of monotone percentiles. Their idea for estimating the median regression function is to replace the arithmetic average in the max–min formulas by a median. It is interesting to note that it also corresponds to replacing the sum of squares in (5.2) by the sum of absolute deviations.

These early nonparametric estimators are not a solution to the monotone and concave, respectively, smoothing problems because of the lack of smooth extensions.

5.2.2 The Isotonized Kernel

Mukerjee (1988), presented a way to force the Parzen–Rosenblatt kernel smoother to satisfy a monotonicity constraint. (For details on kernel smoothing, the reader is referred to Chapter 3 in this volume.) The idea is to use a two-step procedure, wherein a smoothing step is followed by an isotonizing step, or the reverse. (See also Delecroix et al., 1993, and Friedman and Tibshirani, 1984.) If isotonization is performed last and only at observation points, then the problem of extension between observation points remains. On the other hand, if smoothing is performed last, then the difficulty is to preserve the order restriction acquired at the first step. Mukerjee proved that this difficulty is overcome by choosing a log-concave density as a kernel. More precisely, after Y_i^* are found by Brunk's procedure, the isotonized kernel is defined by

$$\hat{m}(x) = \frac{\sum_{i=1}^{n} K((x - X_{(i)})/h_n)Y_i^*}{\sum_{i=1}^{n} K((x - X_{(i)})/h_n)} \tag{5.5}$$

for x such that the denominator is positive, and if necessary, it is extended first by continuity, then by linear interpolation in the possible gaps inside the interval $(X_{(1)}, X_{(n)})$, and finally as a continuous function with constant values in each of the outer intervals. In Equation (5.5), h_n is the bandwidth and K is a log-concave density (i.e., $K(\cdot) = \exp(C(\cdot))$, where C is a concave function). This definition includes in particular the case of the uniform, the normal, and the Laplacian kernels.

Mammen (1991a) compared this estimator with the one obtained by interchanging the smoothing step and the isotonizing step and proved that the two estimators are asymtotically first-order equivalent. Higher order stochastic expansion terms depend on the smoothness of the kernel.

Mukerjee and Stern (1994) proposed an alternative method for isotonizing a kernel smoother. This method is computationally efficient and can be adapted to the case of multivariate regression functions.

5.2.3 The Family of Constrained Splines

Retaining the same advantages as polynomials (e.g., linearity in the coefficients, ease of computation), spline functions provide the statistician with a flexible set of functions with which to build regression models and also with a better local sensitivity to coefficient values. Let us recall that an mth-order polynomial spline on (a, b) with simple knots $\xi_1, \dots, \xi_k$ that are strictly in (a, b) is a function f that satisfies the following requirements:

(a) f coincides with a polynomial of degree less than or equal to $m - 1$ between knots and on (a, ξ_1) and (ξ_k, b).

(b) f and its derivatives up to order $m - 2$ are continuous on (a, b).

The space $\mathcal{S}$ of splines of order m on (a, b) with knots $\xi_1, \dots, \xi_k$ has dimension $p = m + k$. A convenient basis for computational purposes is provided by the normalized B-splines (see Schumaker, 1981), because their support is located in a range of $m + 1$ adjacent knots and because of the existence of recurrence schemes for their evaluation. (For more details on splines, see Chapter 1 of this book.)

5.2.3.1 Least-Squares Splines

A least-squares spline estimate is a linear combination f of a given set of spline functions that solves a minimization problem of the form

$$\min_{s \in \mathcal{S}} \sum_{i=1}^{n} (y_i - f(x_i))^2. \tag{5.6}$$

Weights can be introduced in the least squares if necessary. The smoothing parameter p is the dimension of the spline space $\mathcal{S}$.

Positivity of the first or second derivative at a given point t of a function f in such a spline space can thus be written as positivity of a linear combination of the coordinates $a_1, \ldots, a_p$ of f with respect to a basis. A discretized version of the monotonicity (and concavity) restrictions are then obtained by choosing a finite number of discretization points t_j and results in a finite number of linear constraints on these coefficients. It is quite natural to incorporate these constraints in the estimation procedure by including them as linear constraints directly in the minimization problem that defines the least-squares spline. This approach is found in Dierckx (1980), Holm and Frisen (1985), Mammen (1991b), and Ramsay (1988).

Dierckx's (1980) focus is on local convexity and local concavity restrictions; that is, the mean function is known to be convex or concave on some predetermined subintervals. He chose to work with cubic splines ($m = 4$) for the following reason: Because the second derivative of a cubic spline is a linear function between adjacent knots, satisfaction of a discrete constraint on the sign of the second derivative at the knots (same sign) implies automatic satisfaction of the global constraint (between knots). Dierckx presented an algorithm that makes use of the Theil–Van de Panne procedure for the resulting quadratic programming problem. The estimate is $\hat{m}(x) = \sum_{j=-3}^{k} c_j^* N_j(x)$, where the values of c_j^* solve the quadratic problem

$$\min_{\sum_{j=-3}^{k} d_{j,l} c_j e_j \leq 0} \sum_{i=1}^{k} \left(y_i - \sum_{j=-3}^{k} c_j N_j(x_i) \right)^2. \tag{5.7}$$

Here, N_j, $j = -3 \cdots k$, denotes the basis of normalized B-splines and e_j determines the kind of constraints according to the following values:

(i) $e_j = 1$ if the function is locally concave at the knot ξ_i;
(ii) $e_j = -1$ if the function is locally convex at the knot ξ_i;
(iii) $e_j = 0$ if the function is unrestricted at the knot ξ_i.

The numbers $d_{j,l}$ are determined by the values of the second derivatives of the basis splines at the knots and are given by the simple formulas

$$d_{j,l} = 0 \quad \text{if} \quad j \le l-4 \quad \text{or} \quad j \ge l \tag{5.8}$$

$$d_{l-3,l} = \frac{6}{(t_{l+1} - t_{l-2})(t_{l+1} - t_{l-1})} \tag{5.9}$$

$$d_{l-1,l} = \frac{6}{(t_{l+2} - t_{l-1})(t_{l+1} - t_{l-1})} \tag{5.10}$$

$$d_{l-2,l} = -(d_{l-3,l} + d_{l-1,l}). \tag{5.11}$$

An automatic allocation of knots, which are a priori different from the observation points, is proposed based on a given upper limit for the sum of squared residuals.

For the monotonicity restriction, Ramsay (1988) constructed a basis of so-called I-splines that has the following properties: Given a sequence of knots,

(i) each I-spline of order m is an indefinite integral of a corresponding B-spline of order m;
(ii) each I-spline is monotone;
(iii) any linear combination of I-splines with positive coefficients is monotone.

It is unclear, though, whether positivity of the coefficients characterizes monotonicity in the span of I-splines. Note that an I-spline of order m is a piecewise polynomial of degree m. An explicit formula for calculating I-splines is given. The I-spline procedure for a regression problem then consists of a least-squares minimization under positivity constraints on the coefficients. It is interesting that Brunk's estimator as extended to be piecewise linear coincides with the I-spline procedure of order one with a knot at each data point. However, the number of knots that is recommended in the I-spline procedure is usually much smaller than the number of observation points.

Holm and Frisen (1985) proposed a stepwise method to compute the piecewise linear (second order) least-squares spline with knots at each data point for the estimation of nondecreasing convex functions. The respective cases of decreasing convex, decreasing concave, and nondecreasing concave functions can be treated similarly, as well as can the cases of functions that have such characteristics on given subintervals. With the aim of estimating, for example, sigmoid functions, one can then consider the cases of functions that shift from one of these characteristics to another at unknown breakpoints. Holm and Frisen proposed to calculate the sum of squared residuals for all solutions corresponding to all possible shifting points (chosen among data points) and then to select the solution that yields the smallest sum of squared residuals.

Mammen (1991b) considered least-squares estimation of a regression in the class of functions whose $(m-1)$st derivative consists of at most k monotone pieces. This regression encompasses the respective cases of piecewise monotone ($m = 1$) and piecewise convex or concave ($m = 2$) functions. As the least-squares estimators in this class are not uniquely defined, Mammen further restricted the class to its intersection with the set of polynomial splines of order m. With a given bound on the $(m-1)$st derivative, the least-squares estimator in this class is then unique. The minimizing spline has data-dependent knots, and Mammen proposed an active set algorithm for their estimation. When $m = 1$ and k is known, the estimator coincides with Brunk's estimator, extended to be piecewise constant between observation points.

5.2.3.2 *Smoothing Splines*

In contrast to least-squares splines, smoothing D^m-splines are defined as the unique solution of the minimization problem

$$\min_{f \in H^m(a,b)} \sum_{i=1}^{n} (y_i - f(x_i))^2 + \lambda \int_a^b (f^{(m)})^2 \, dt, \tag{5.12}$$

where λ is a positive smoothing parameter that balances the trade-off between fidelity to the data on one side and smoothness on the other side. Interpolating splines satisfy the interpolation conditions $f(x_i) = y_i$, $i = 1, \ldots, n$, while minimizing the integral term. (The interpolating splines appear as the limit of smoothing splines when the smoothing parameter tends to zero.) Smoothing splines (and interpolating splines as well) can be shown to be polynomial splines of order $2m$ with knots at the data points x_i, $i = 1, \ldots, n$. (See Chapter 1 in this volume.)

It is clear that incorporating the shape restrictions in the minimization problem appears to be a natural way, as in the case of least-squares splines, to isotonize the estimates. Wright and Wegman (1980) gave an existence and characterization theorem for an isotonized version of splines between smoothing splines and interpolating splines. They proposed to minimize the integral term while satisfying some approximate interpolation conditions of the form $\alpha_i \leq f(x_i) \leq \beta_i$, $i = 1, \ldots, n$. The shape restriction was not discretized in their result, so their estimate reached a global satisfaction of monotonicity or convexity, respectively, on (a, b). They proved that the solution is a polynomial spline of order $2m$ with knots at the data points and possibly at a countable number of knots elsewhere. However, the drawback to this method is that no computing algorithm was mentioned, except for the case of the discretized restriction.

Utreras (1985, 1991) gave an existence and uniqueness theorem for smoothing splines constrained to belong to a closed convex subset satisfying some minor regularity assumption. A characterization was derived for the general case, but is not really computationally workable. A more precise characterization was provided for the case of monotone smoothing splines (see also Michelli and Utreras, 1988).

Villalobos and Wahba (1987) derived a computable representation of the minimizer of Equation (5.12) subject to a finite number of inequality constraints, in a multivariate setup. The resulting quadratic programming problem is then supposed to be solved numerically by available nonlinear programming algorithms. Laurent (1980) provides a specific algorithm to compute this type of constrained splines.

Elfving and Andersson (1988) derived a characterization for smoothing splines constrained to be convex or convex and monotone over a whole interval rather than at a discrete set of points. They provided a computing algorithm for this type of problem. Turlach (1997) developed another such computing algorithm.

5.2.3.3 Hybrid Splines

In the least-squares spline approach, the tuning of the amount of smoothing is performed by the choice of the number of knots and their location, whereas in the smoothing-spline approach, it is done by the choice of the parameter λ, as the knots are automatically located at each observation point. In the former case, the drawback is "computational cost which increases enormously with the number of knots," together with "the chance of missing the optimal location, because of an inappropriate initial position" in the case of automatic knot allocation, according to Dierckx (1993, p. 75). A hybrid formulation consists in minimizing the penalized least squares in Equation (5.12) in an approximating subspace of splines with given knots that are a priori different from (and fewer than) the observation points. (See Wahba, 1990, Chapter 7.) This formulation thus allows a compromise between the choice of λ and the number of knots and their location; this compromise can be made either subjectively from examination of graphical displays or by automatic data-driven techniques, such as cross-validation. This approach has been used, for example, by Kelly and Rice (1990) for the monotone estimation of dose response curves. They enforced monotonicity by restricting the coefficients in the B-spline basis to be nondecreasing at the optimization step.

Before we close this section on splines, we must refer the reader to the approach proposed by Ramsay (1998).

5.2.4 The Projection Method

Friedman and Tibshirani (1984) investigated a two-step procedure. The first step is a local smoothing by the k-nearest neighbors, and the second step is an application of the PAV algorithm to the outcome of the first step. However, Friedman and Tibshirani said nothing about the extension between observation points, which is necessary, since the second step is a projection of the discrete data onto a discrete cone.

A similar scheme (i.e., smoothing first and then projecting) was investigated later by Delecroix et al. (1993). However, in this case, the projection took place in an infinite dimensional space in order to eliminate the problem of extension between observation points, and the shape restriction was not limited to monotonicity, but could be any restriction described by a cone. The necessity to discretize the cone remained, however, for computability reasons. The projection method described by Delecroix et al. is based on the following simple principle: Choose a Hilbert space endowed with a given norm, and find an initial estimate that lies in this function space. If the cone of restrictions is closed and convex, which is usually the case, then it is at least theoretically possible to project any element of the Hilbert space onto this cone. Therefore, to get an estimate in the cone, it is enough to project the initial estimate onto the cone. If the true mean function belongs to the cone, one is thus guaranteed, by the contracting properties of projections, to be closer to the true mean than was the original estimate. If, moreover, the initial estimate happens to be consistent (in a mean-square sense, or almost surely), then so is its projected version (in the same sense). The same is true if the projection happens onto a larger cone, which is the case for the usual discretizations.

The difficulties in applying this simple principle are, on one hand, to find concrete spaces and estimates in which these conditions are satisfied and, on the other hand, actually to compute the projections. The latter difficulty is solved by first discretizing the cone as in the spline method and then solving a quadratic optimization problem with linear constraints. As for the Hilbert space and the initial estimates, possible choices are given. One can take, for example, a Sobolev space $\mathcal{H}^m(0,1)$ endowed with the norm

$$\|u\| = \int_0^1 u(t)^2\,dt + \frac{1}{\tau^{2m}} \int_0^1 u^{(m)}(t)^2\,dt, \tag{5.13}$$

where τ is a parameter regulating the balance between our interest in the estimation precision of the function itself and of its mth derivative. One can use a convolution-type kernel estimate or a smoothing-spline estimate in the first step to get Sobolev norm consistency, but the use of any other smoother is possible only if the estimate lies in the Sobolev space, which is not very restrictive. It is not possible to give all the details of this procedure here (see Delecroix et al., 1993), but let us summarize the steps of the procedure, for the monotonicity constraint, for example:

(1) Calculate an initial estimate $\hat{m}(\cdot)$ and choose its smoothing parameter;
(2) choose the number p of discretization points t_k for the cone and their location;
(3) choose the parameter τ;
(4) calculate the first derivative of the initial estimate at the discretization points $\hat{m}'(t_k)$;
(5) compute the solution α_i^* of the quadratic program

$$\min_{\alpha_1 \geq 0, \ldots, \alpha_p \geq 0} \sum_{i,j=1}^{p} \alpha_i \alpha_j d(t_i, t_j) - 2 \sum_{i=1}^{p} \alpha_i \hat{m}'(t_i), \tag{5.14}$$

where the function $d(t, s)$ is known;
(6) compute the final estimate by

$$\tilde{m}(x) = \hat{m}(x) - \sum_{i=1}^{p} \alpha_i^* h(t_i, x), \tag{5.15}$$

where the function $h(s, t)$ is known.

Step (5) may not be the most efficient way to compute this projection, however, an adaptation of the Boyle and Dykstra (1986) procedure for calculating projections onto the intersection of convex sets is worthy of investigation.

5.3 A COMPARATIVE VIEW

The choice of an isotonizing estimator must be based not only on the characteristics of the particular application, but also on the motivation for isotonizing and on the availability of software. There are basically two main motivations for choosing an isotonizing estimator. The first one is to improve the efficiency of the estimate by using this additional information. The second one is to constrain the estimate with respect to the shape restriction that the underlying function is known to satisfy, for visual reasons or for further treatment. For example, with respect to the first motivation, let us quote Wright and Wegman (1980, p. 1033): "It is in the context of a relatively sparse data set that the added knowledge of isotonicity will allow for the relatively largest improvement in efficiency." For the second motivation, monotonicity is often essential in variable transformation problems, because of the need to use the inverse transformation to describe effects in terms of the original variable.

5.3.1 Spline versus Kernel

The fact that most of the isotone estimates are based on splines rather than on kernels probably stems from the fact that splines are defined as a solution of a minimization problem. Therefore, enforcing the restrictions at the minimization step appears to be a natural solution. Among the methods that we have covered in this chapter, only the Mukerjee (1988) estimate is kernel based, although the projection method can take a kernel estimate as the initial estimate. Let us now mention a few other advantages of B-spline decomposition. Kelly and Rice (1990) stressed that if the coefficients in the B-spline bases are nonnegative, then so is the function, and if the sequence of coefficients is nondecreasing, then so is the function. Dierckx (1980, Section 2.3.1) made use of the fact that with a cubic B-spline decomposition, enforcing the constraint at the knots is equivalent to enforcing the full constraint for convexity.

5.3.2 Model Adaptability

All of the estimates that we have described in this chapter have been presented in the regression framework, but some of them can easily be adapted to other types of models as well. For instance, those estimates that are defined by minimization of a criterion or estimates for the projection method can be adapted to other types of models. Let us give some specific examples. Kruskal (1965) analyzed factorial experiments with monotone transformations. Like Brunk's (1958) estimates, these estimates are derived by minimizing, under monotonicity restrictions, a criterion that is suitable for the transformation model. Friedman and Tibshirani (1984) used their monotone smoothing procedure to find an optimal transformation for the response in linear regression. Ramsay (1988) illustrated the virtues of monotone spline response variable transformation in nonlinear regression, dose response models, principal components, and canonical correlation. Ferraty (1997) used the projection method to estimate monotone transformations. On the other hand, the isotonized kernel estimate does not seem easy to adjust to a model other than regression.

5.3.3 Types of Constraints

The most general shape restriction can be expressed by the fact that the functional parameter belongs to a closed cone. Let us mention that some cones of interest are not convex, for example, the set of unimodal functions with unknown mode and, similarly, the set of piecewise monotone or piecewise convex–concave functions with unknown breakpoints. (See Demetriou, 1995, and Reboul, 1998.) This factor must be checked before using projections that are properly defined only in the convex case. Among the estimators that we have presented in this chapter, those that are geared towards only one type of constraint are, for the monotonicity restriction, Mukerjee's (1988) isotone kernel and Ramsay's (1988) monotone splines and, for the convexity restriction, Dierckx's (1980) least-squares splines. Constrained smoothing splines and the projection method are open to very general types of constraints, including mixtures of monotonicity, positivity, and convexity.

5.3.4 Discretized Constraint versus Global Constraint

A limited number of estimates achieve global satisfaction of the constraint over the whole interval: Mukerjee's and Ramsay's for monotonicity, Dierckx's for convexity, and Elfving and Andersonn's for convexity and monotone convexity. The other estimates satisfy a discretized constraint. However, in practice, the enforcement of this lighter constraint often results in global satisfaction. Wright and Wegman (1980, p. 1034) wrote, "In general, unless there is a very pathological situation, a finite number of constraints will do to make the spline isotone." For the projection method, it is possible to allocate iteratively the discretization points by inspection in the regions of nonsatisfaction of the constraints until a reasonable aspect is reached.

5.3.5 Smoothness of the Estimate

Some of the methods described in this chapter are shape-constrained smoothers, whereas others are shape-constrained nonparametric estimators. Although Dent (1973, p. 131) wrote that "between observation points there is nothing (statistically) preferred in having a nonlinear curve rather than a linear segment," in many situations, smoothness is just as desirable as isotonicity. Another motivation for smoothing is described in the next section.

5.3.6 Asymptotic Results

Before we consider asymptotic efficiency, let us mention a finite-sample result that Kelly (1989) found for a pure Gaussian regression model: The absolute error of each component of the isotonic regression estimator is stochastically smaller than that of the usual maximum-likelihood estimator of the regression.

Asymptotic behavior has not been studied completely for all of the isotone estimates we have discussed throughout this chapter. Here, we cite the available results, without providing details on the specific assumptions.

Brunk's estimator has an asymptotic distribution with a normalizing factor of $n^{1/3}$ at points of positive slope. Smoothing an isotonized estimator results in improved convergence rates that are similar to nonparametric convergence rates. The isotonizing step consists in local averaging on random neighborhoods that are smaller than the those in the smoothing step. A modification of Brunk's estimator with $n^{2/5}$ convergence rate is discussed by Wright (1982). Mammen (1991b) discussed a similar result that was found with a normalizing factor of $n^{2/5}$ in the convex–concave case.

With respect to almost-sure and uniform-norm cases, we refer the reader to the consistency results of Hanson and Pledger (1976) for Hildreth's estimate, Wright and Wegman (1980) for their isotone spline, and Mukerjee (1988) for the isotonized kernel. Mukerjee also provided an asymptotic normality result. The monotone median estimate (see Cryer et al., 1972) has been shown to be consistent in terms of probability and uniform norm. Mammen (1991b) has provided a consistency result with respect to the L^2-norm and probability for his least-squares splines. Consistency with respect to the mean square and the Sobolev norm for the projection method has already been mentioned in this chapter. Utreras (1985) proved a consistency theorem for constrained smoothing splines for convergence with respect to the mean square and the Sobolev norm. Mammen and Thomas-Agnan (1998) showed that constrained smoothing splines achieve optimal rates in shape-restricted Sobolev classes. Durot (1996) proved convergence in probability and asymptotic normality of the L_1-error for Brunk's estimator.

Kiefer (1982) attempted to find a general pattern for asymptotic behavior in estimation problems with order restrictions, in spite of the great diversity of models, performance criteria, type of probabilistic convergence, and classes of functions concerned in all the results. His conclusion after examining several examples was, as shown by Utreras (1985), that in all encountered cases, the rate of convergence for smooth functions cannot be improved upon by the additional information of order restriction; however, the constants in the upper bounds may be decreased, which can have a bearing on the actual improvement for finite samples.

5.3.7 Choice of Smoothing Parameter

In most cases, the classical methods for choosing the smoothing parameter for cases under constraints can be used in the same way as for cases without constraints, but in some cases, an adaptation is needed. For example, Wahba (1982) proposed an extension of the generalized cross-validation method to the constrained case for smoothing splines. Kelly and Rice (1990, p. 1072) observed that "the monotonicity constraint is a powerful smoothing device as well." In fact, in our simulations, it appeared that the MISE as a function of the smoothing parameter usually had a wider zone of stability around the optimal parameter in the constrained case, making its optimal choice easier.

5.3.8 Further Questions

Beyond the problem of estimation, the question of testing for monotonicity or convexity seems to be of great interest for some scientists. A small number of papers on this problem are currently available, including those of Bowman et al. (1996), Diack and Thomas-Agnan (1998), Schlee (1980), and Yatchew (1992). Hence, there is need for further research. Testing for $m = m_0$ in the context of isotonic regression is addressed in Tocquet (1998).

The problem of multivariate versions of the constrained estimates would also find many domains of application, but it has not been widely investigated. It seem to us that, as in other domains of multivariate statistical modeling, the solution goes through dimension reduction techniques that allow the application of the one-dimensional methods.

5.4 EXAMPLES

Because of the large number of different estimates, a systematic comparison of simulations is an ambitious enterprise. Instead, we selected four estimates in the monotonicity constraint case and four in the convexity constraint case and computed them on simulated regression data sets. For each case, two of the four estimates were ordinary estimates and two were isotone estimates. From our study, we saw that a more systematic comparison is difficult also because, on the examples we computed, we observed that there is no method that seems to be uniformly best, but for each method, there exists a data set for which that particular method is the best.

For the monotonicity case, we thus compared

- the *Parzen–Rosenblatt kernel* smoother (PRK)
- the *Mukerjee isotonized kernel* (MUK)
- the *ordinary smoothing spline* (OSS)
- the *projected smoothing spline* (PSS)

For the convexity case, we compared

- the *ordinary least-squares spline* (OLS)
- the *Dierckx constrained least-squares spline* (DLS)
- the *ordinary smoothing spline* (OSS)
- the *projected smoothing spline* (PSS)

For each isotonized estimate, we chose the optimal smoothing parameters by computing the actual mean integrated square error, with the exception of the number of discretization points for projections, which was set to half the sample size. For comparison purposes, in the upcoming subsection, we present the graphs of selected examples together with the values of the corresponding MISE, and we also show the ordering of the four estimates with respect to this criteria. The ordering is given in increasing order of the MISE. The sample size is denoted by n, the standard deviation of the noise by σ, and the true function by u. The computations are performed on a mainframe using MATLAB software, which has a convenient toolbox for computing splines.

5.4.1 Monotonicity

For the PRK and the MUK estimates, the bandwidth is denoted by h. For the OSS estimates, the parameter is $\rho = 1/(\lambda + 1) \in (0, 1)$. For the projection, the number of discretization points is p, and the norm parameter is τ and is choosen among the following values: {0.5, 1, 3, 5, 7}.

Example 5.1 is presented in Figure 5.1 and corresponds to the following values:

- $u(x) = \exp(-(x + 4))$,
- $n = 10$,

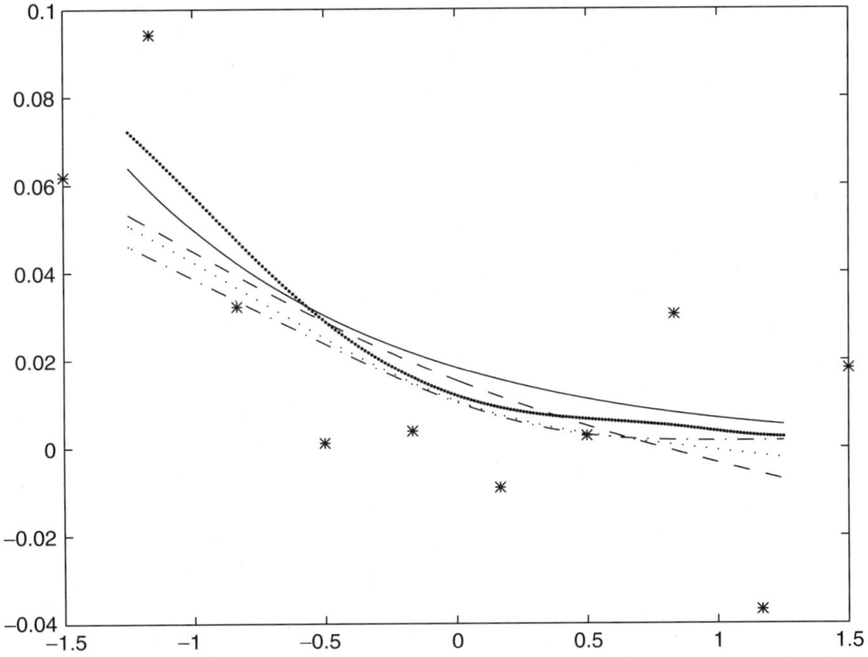

FIGURE 5.1 Monotonicity: Observations and function estimates of Example 5.1.

- $\sigma = 0.04$, and
- $p = 10$.

For the splines, the optimal parameters are $\rho = 0.3$ for the OSS estimate, and $\rho = 0.4$ and $\tau = 3$ for the PSS estimate. The corresponding MISEs are 0.0078 for the OSS estimate and 0.0052 for the PSS estimate. For the kernels, the optimal bandwidth is $h = 1.75$ for the MUK estimate and $h = 2$ for the PRK estimate. The associated MISEs are 0.4060 for the MUK estimate and 0.133 for the PRK estimate. In terms of increasing MISE, the order is PSS $\leq$ OSS $\leq$ MUK $\leq$ PRK, with an MISE ratio of 0.66 for OSS vs. PSS, of 0.71 for OSS vs. MUK and 0.78 for MUK vs. PRK.

Example 5.2 is displayed in Figure 5.2 and corresponds to the following values:

- $u(x) = -0.5x$,
- $n = 30$,
- $\sigma = 1$, and
- $p = 30$.

For the splines, the optimal parameters are $\rho = 0.25$ for the OSS estimate and $\rho = 0.92$ and $\tau = 0.5$ for the PSS estimate. The associated MISEs are 5.7711 for

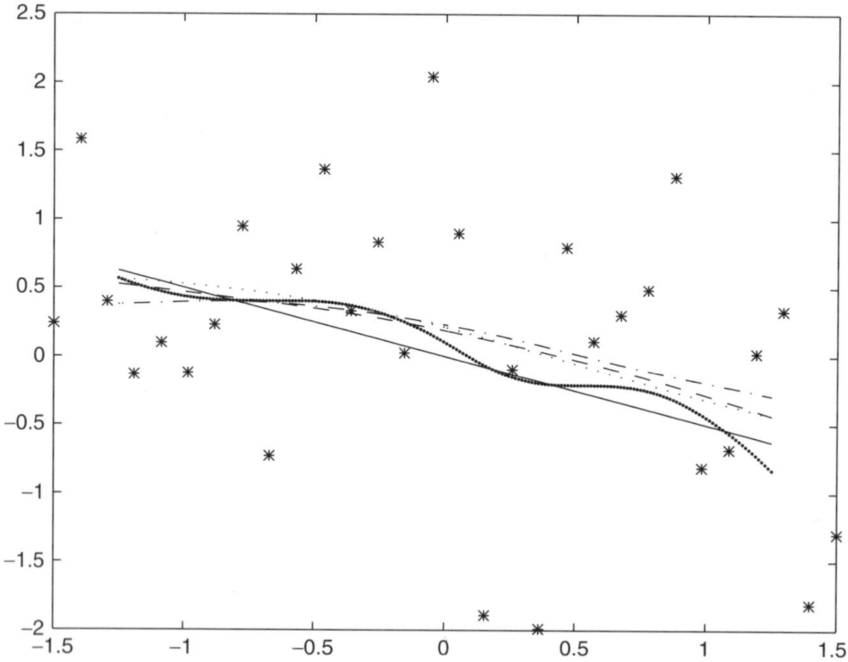

FIGURE 5.2 Monotonicity: Observations and function estimates of Example 5.2.

the OSS estimate and 2.2274 for the PSS estimate. For the kernels, the optimal bandwidth is $h = 1.69$ for the MUK estimate and $h = 1.43$ for the PRK estimate. The associated MISEs are 5.8062 for the MUK estimate and 10.146 for the PRK estimate. In terms of increasing MISE, the order is PSS $\leq$ OSS $\leq$ MUK $\leq$ PRK, with an MISE ratio of 0.386 for OSS vs. PSS, 0.994 for OSS vs. MUK, and 0.572 for MUK vs. PRK.

Example 5.3, presented in Figure 5.3, is a variant of Example 5.2, with $\sigma = 0.7$. For the splines, the optimal parameters are $\rho = 0.69$ for the OSS estimate and $\rho = 0.80$ and $\tau = 0.5$ for the PSS estimate. The corresponding MISEs are 1.6859 for the OSS estimate and 1.6835 for the PSS estimate. For the kernels, the optimal bandwidth is $h = 1.16$ for the MUK estimate and $h = 1.43$ for the PRK estimate. The associated MISEs are 1.0420 for the MUK estimate and 2.4284 for the PRK estimate. In terms of increasing MISE, the order is MUK $\leq$ PSS $\leq$ OSS $\leq$ PRK, with an MISE ratio of 0.6189 for MUK vs. PSS, 0.9986 for PSS vs. OSS, and 0.6942 for OSS vs. PRK.

5.4.2 Concavity

For the OSS and the PSS estimates we keep the same notations for concavity as for monotonicity. For the least-squares splines, the parameter is the number k of knots that are evenly distributed on the interval (x_1, x_n).

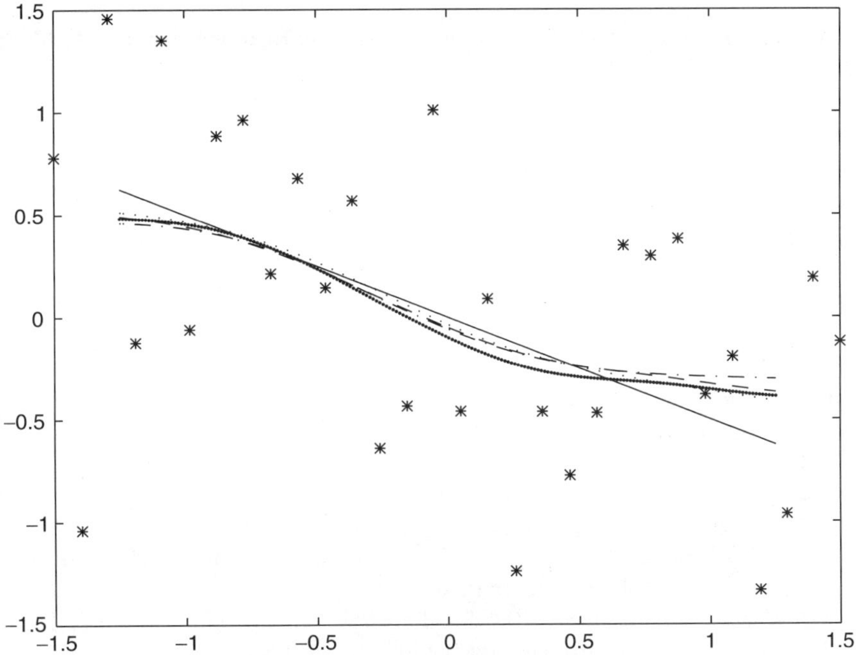

FIGURE 5.3 Monotonicity: Observations and function estimates of Example 5.3.

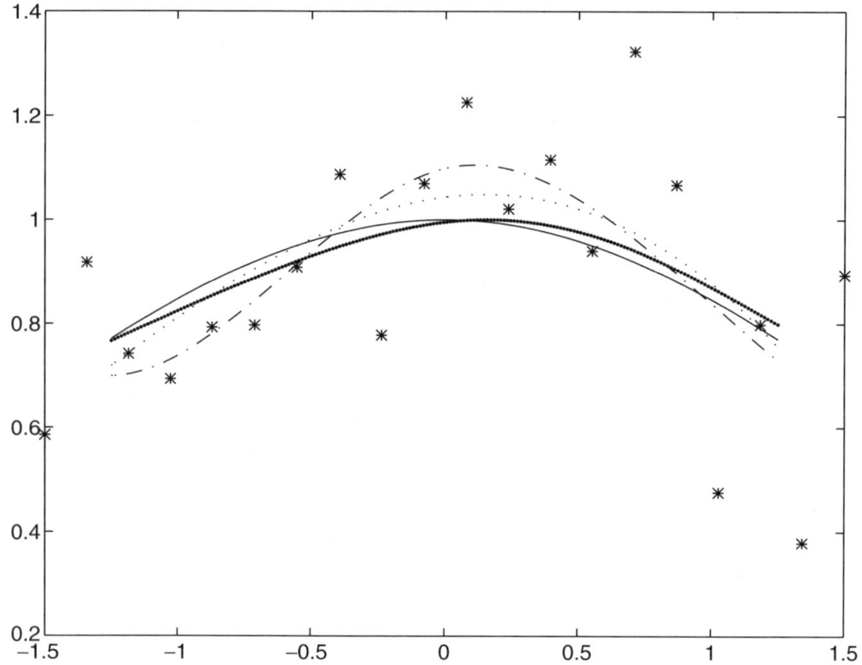

FIGURE 5.4 Concavity: Observations and function estimates of Example 5.4.

Example 5.4 is presented in Figure 5.4 and corresponds to the following values:

- $u(x) = \exp(-x^2/6)$,
- $n = 20$,
- $\sigma = 0.2$, and
- $p = 10$.

For the smoothing splines, the optimal parameters are $\rho = 0.60$ for the OSS estimate and $\rho = 0.60$ and $\tau = 3$ for the PSS estimate. The associated MISEs are 0.0909 for the OSS estimate and 0.0908 for the PSS estimate. For the least-squares splines, the optimal number of knots is $k = 1$ for the DLS estimate and $k = 1$ for the SPR estimate. The associated MISEs are 0.3550 for the DLS estimate and 0.1.0859 for the OLS estimate. In terms of increasing MISE, the order is PSS $\leq$ OSS $\leq$ DLS $\leq$ OLS, with an MISE ratio of 0.999 for PSS vs. OSS, 0.256 for OSS vs. DLS, and 0.327 for DLS vs. OLS.

Example 5.5, presented in Figure 5.5, is a variant of Example 5.4 with $n = 10$. For the smoothing splines, the optimal parameters are $\rho = 0.83$ for the OSS estimate and $\rho = 0.53$ and $\tau = 0.5$ for the PSS estimate. The associated MISEs are 0.9423 for the OSS estimate and 0.2705 for the PSS estimate. For the least-squares splines, the optimal number of knots is $k = 2$ for the DLS estimate and

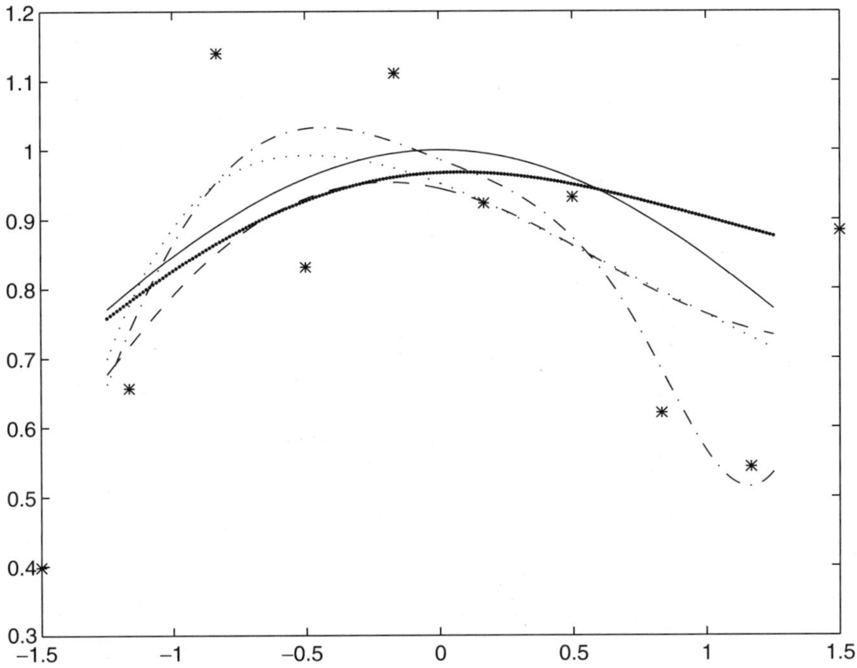

FIGURE 5.5 Concavity: Observations and function estimates of Example 5.5.

$k = 3$ for the SPR estimate. The corresponding MISEs are 0.877 for the DLS estimate and 3.3641 for the OLS estimate. In terms of increasing MISE, the order is PSS $\leq$ DLS $\leq$ OSS $\leq$ OLS, with an MISE ratio of 0.308 for PSS vs. DLS, 0.93 for DLS vs. OSS, and 0.28 for OSS vs. OLS.

Even though it is impossible to draw any definitive conclusions about the relative performance of the methods from the analysis of examples involving single data sets only, we can say the following: We have based the comparison on the MISE criterion, but we think that it is also necessary to study MISE for at least the first derivative in the monotonicity case and the second derivative in the convexity case to account for the role of derivatives in these constraints. The gain is more satisfactory visually and in regard to the constraint than in terms of the MISE, however the reduction in the MISE can be substantial. For obvious reasons, the largest improvements are obtained for functions that are close to the boundary of the cone, that is, in the monotonicity case, in regions where the function is flat and, in the convexity case, in regions where the function is close to linear. Finally, we should say that the gains are more significant in the case of convexity as compared with the case of monotonicity.

5.4.3 Testing Concavity on Real Data

Human capital theory predicts the concavity of the logarithm of wage as a function of experience. A real data set due to Simioni (1995) produces a curve of

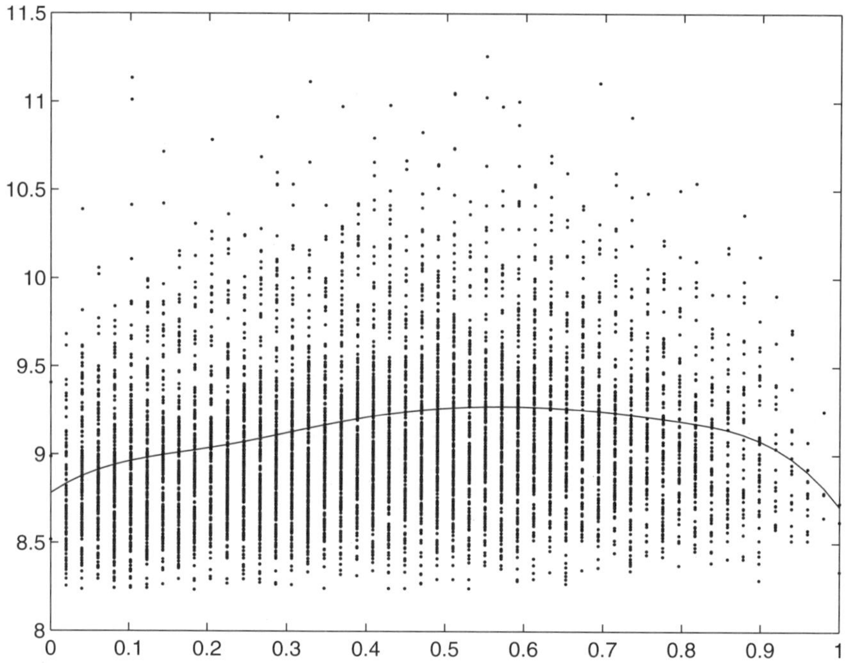

FIGURE 5.6 Logarithm of wage as a function of experience: Observations and spline estimate.

low curvature. Diack and Thomas-Agnan (1998) could show that their test was able to confirm concavity as predicted by economic theory. The data set and the unconstrained hybrid cubic spline estimate, on which their test is based, are displayed in Figure 5.6.

5.5 SOFTWARE HINTS

We are not aware of nonparametric regression software that would allow one to set up constraints. However some of the methods can be easily implemented in such packages. In order to let the reader judge for themselves we give sample MATLAB code here for the computation of the projected smoothing spline and Dierckx spline in the case of concavity. To run the following `M` files the working directory should contain the two function files `d2x.m` and `ps2.m` at the end of the MATLAB code.

Projected Smoothing Spline Code

```
% The sample size is no
% The data set is (xo,yo)
% The true regression function is u
global tau h sigma no np xo yo
sigma=0.5; tau=7; h=0.01;
no=100; np=no/2; p=0.9;
```

```
% DISCRETIZATION POINTS
pas=(xo(no)-xo(1)-2*h)/(np-1);
t=xo(1)+h+(0:(np-1))*pas;
% SCALAR PRODUCTS MATRIX
for i=1:np
    for j=1:np
             xmat(i,j)=ps2(t(i)-t(j));
    end
end
% OPTIMIZATION=ALPHA
p1=fnder(csaps(xo,yo,p),2);
c=-fnval(p1,t);
b=zeros(1,np);
A=-eye(np);
alpha=qp(xmat,c,A,b);
% CONSTRAINED AND UNCONSTRAINED ESTIMATES
xd=linspace(-1.25,1.25,200);% points for evaluation of
the true function and the two estimates
y=u(xd);
uchap=csaps(xo,yo,p,xd);
for i=1:length(xd)
    utild(i)=uchap(i)-d2x(xd(i)*ones(1,np)-t)*alpha;
end
graf=plot(xd,y,xd,uchap,'y-',xd,utild,'y.')
print -dps graf
```

Dierckx Spline Code

```
sigma=0.5;
graf0=plot(xo,yo,'y*',xo,u(xo),'y-');
title('Raw Data and True function');
print -dps graf0
% INTERIOR KNOTS
k=4;
zt=zeros(1,k+2);
t=zeros(1,k+8);
zt=linspace(-1.5,1.5,k+2);
t(1)=-1.5;t(2)=-1.5;t(3)=-1.5;
t(1,4:k+5)=zt;
t(k+6)=1.5;t(k+7)=1.5;t(k+8)=1.5;
% MATRIX B
B=spcol(t,4,xo);
% MATRIX D
D=zeros(k+4,k+2);
for i=1:(k+4)
    for j=1:(k+2)
         if (j-i==0)
              pro=(t(j+4)-t(j+1))*(t(j+4)-t(j+2));
              D(k+5-i,k+3-j)=6/pro;
           elseif (j-i==-2)
```

```
              pro=(t(j+5)-t(j+2))*(t(j+4)-t(j+2));
              D(k+5-i,k+3-j)=6/pro;
           elseif (j-i==-1)
              D(k+5-i,k+3-j)=-6*((1/((t(j+4)-t(j+1))*(t(j+4)-
t(j+2))))+...(1/((t(j+5)-t(j+2))*(t(j+4)-t(j+2)))));
            end
    end
end
% OPTIMIZATION
c=qp(B'*B,-yo*B,D',zeros(1,k+2));
c1=qp(B'*B,-yo*B,[],[]);
% ESTIMATOR
sp=spmak(t,c');
sp1=spmak(t,c1');
xd=linspace(-1.25,1.25,200);% points for evaluation of
the true function and the two estimates
udier=fnval(sp,xd);
u1=fnval(sp1,xd);
yd=u(xd);
plot(xd,yd,xd,udier,'y.',xd,u1,'y-.')
```

The Two Function Files

```
function y=d2x(x)
% second derivative at x
global tau;
y=(tau^3/6)*(exp(-tau*abs(x))-2*exp(-tau*abs(x)/2)...
.*cos(tau*abs(x)*sqrt(3)/2));
function y=ps2(x)
% scalar product of second derivative
global tau;
y=(tau^5/6)*(exp(-tau*abs(x)) + 2*(exp(-tau*abs(x)/2)...
.*cos((tau*abs(x)*sqrt(3)/2)+(pi/3))));
```

REFERENCES

Barlow, R., Bartholomew, D., Bremner, J., and Brunk, H. (1972). *Statistical Inference under Order Restrictions*, Wiley, New York.

Bickel, P. J., and Fan, J. (1996). Some problems on the estimation of unimodal densities, *Statistica Sinica*, **6**, 23–45.

Birgé, L. (1986). Estimating a density under order restrictions: Nonasymptotic minimax risk, *Ann. Statist.*, **15**, 995–1012.

Birgé, L. (1987). On the risk of histograms for estimating decreasing densities, *Ann. Statist.*, **15**, 1013–1022.

Birgé, L. (1989). The Grenander estimator: A nonasymptotic approach, *Ann. Statist.*, **17**, 1532–1549.

Birgé, L. (1997). Estimation of unimodal densities without smoothness assumptions, *Ann. Statist.*, **25**, 970–980.

Bowman, A. W., Jones, M. C., and Gijbels, I. (1996). Testing monotonicity of regression, preprint.

Boyle, J. P., and Dykstra, R. L. (1986). A method for finding projections onto the intersection of convex sets in Hilbert spaces, in Dykstra, R. L., Robertson, T. and Wright, F. T. (eds.), *Advances in Order Restricted Statistical Inference*, pp. 28–47, Springer–Verlag, Berlin, Germany.

Brunk, H. D. (1955). Maximum likelihood estimates of monotone parameters, *Ann. Math. Statist.*, **26**, 607–616.

Brunk, H. D. (1958). On the estimation of parameters restricted by inequalities, *Ann. Math. Statist.*, **29**, 437–454.

Brunk, H. D. (1970). Estimation of isotonic regression, in Puri, M. L. (ed), *Nonparametric Techniques in Statistical Inference*, pp. 177–195, Cambridge University Press, Cambridge, UK.

Casady, R. J., and Cryer, J. D. (1976). Monotone percentile regression, *Ann. Statist.*, **4**, 532–541.

Cryer, J. D., Robertson, T., Wright, F. T., and Casady, R. J. (1972). Monotone median regression, *Ann. Math. Statist.*, **43**, 1459–1469.

Cuesta-Albertos, J. A., Dominguez-Menchero, J. S., and Matran-Bea, C. (1994). Some stochastics on monotone functions, *J. Computat. Appl. Math.*, **55**, 165–182.

Cuesta-Albertos, J. A., Dominguez-Menchero, J. S., and Matran-Bea, C. (1995). Consistency of L_p best monotone approximations, *J. Stat. Plann. Infer.* **47**, 295–318.

Delecroix, M., Simioni, M., and Thomas-Agnan, C. (1993). A shape preserving smoother: Simulation study, *Computational Statistic*, **10**, 155–175.

Delecroix, M., Simioni, M., and Thomas-Agnan, C. (1996). Functional estimation under shape restrictions, *J. Nonparametric Statistics*, **6**, 69–89.

Demetriou, I. C. (1995). Discrete piecewise monotonic approximation by a stricly convex distance function, *Math. Computat.*, **64**, 17–180.

Dent, W. (1973). A note on least-squares fitting of functions constrained to be either nonnegative, nondecreasing or convex, *Management Science*, **20**, 130–132.

Diack, C., and Thomas-Agnan, C. (1998). A nonparametric test of the nonconvexity of regression, *J. Nonparametric Statist.*, **9**, 335–362.

Dierckx, P. (1980). An algorithm for cubic spline fitting with convexity constraints, *Computing*, **24**, 349–371.

Dierckx, P. (1993). *Curve and Surface Fitting with Splines*, Oxford Science Publications, Oxford, UK.

Durot, C. (1996). *Sharp Asymptotics for Isotonic Regression*, Technical Report Université de Paris-Sud, Paris, France.

Elfving, T., and Andersson, L. E. (1988). An algorithm for computing constrained smoothing spline functions, *Numerische Mathematik*, **52**, 583–595.

Ferraty, F. (1997). Estimations de transformations optimales en ACP curvilineaire, *Rev. Statistique Appliquee*, **XLV(1)**, 5–39.

Fraser, D. A. S., and Massam, H. (1989). A mixed primal-dual bases algorithm for regression under inequality constraints: Application to concave regression, *Scand. J. Statist.*, **16**, 65–74.

Friedman, J., and Tibshirani, R. (1984). The monotone smoothing of scatterplots, *Technometrics*, **26**, 243–250.

Grenander, U. (1956). On the theory of mortality measurement, *Scand-Aktuarietidskr*, **39**, 125–153.

Groeneboom, P. (1985). Estimating a monotone density, in Le Cam, L. M., and Olshen, R. A. (eds.), *Conference in Honor of Jerzy Neyman and Jack Kiefer*, 539–555.

Groeneboom, P., and Jongbloed, G. (1993). *Isotonic Estimation and Rates of Convergence in Wicksell's Problem*, Technical Report 93–39, Faculty of Technical Mathematics and Informatics, Delft University of Technology, Delft, Netherlands.

Groeneboom, P., Hooghiemstra, G., and Lopuhaa, H. P. (1996). *Asymptotic Normality of the L1-error of the Grenander Estimator*, Technical Report, Faculty of Technical Mathematics and Informatics, Delft University of Technology, Delft, Netherlands.

Hanson, D. L., and Pledger, G., (1976). Consistency in concave regression, *Ann. Statist.*, **4**, 1038–1050.

Hawkins, D. (1994). Fitting monotonic polynomials to data, *Computational Statistics*, **9**, 233–247.

Hildreth, C. (1954). Point estimates of ordinates of concave functions, *J. Amer. Statist. Assoc.*, **49**, 598–619.

Holloway, C. A. (1979). On the estimation of convex functions, *Oper. Research*, **27**, 401–407.

Holm, S., and Frisen, M. (1985). *Nonparametric Regression with Simple Curve Characteristics*, Technical Report 4, Department of Statistics, University of Goteborg, Goteborg, Sweden.

Kelly, R. (1989). Stochastic reduction of loss in estimating normal means by isotonic regression, *Ann. Statist.*, **17**, 937–940.

Kelly, C., and Rice, J. (1990). Monotone smoothing with application to dose-response curves and the assessment of synergism, *Biometrics*, **46**, 1071–1085.

Kiefer, J. (1982). Optimum rates for nonparametric density and regression estimates under order restrictions, in Kallianpur, G., Krishnaiah, P. R., Ghosh, J. K. (eds.), *Statistics and Probability: Essays in Honor of C. R. Rao*, pp. 419–428, North Holland, Amsterdam, Netherlands.

Kruskal, J. B. (1965). Analysis of factorial experiments by estimating monotone transformations of the data, *J. Royal Statist. Soc.*, **27**, 251–263.

Laurent, P. J. (1980). *An Algorithm for the Computation of Spline Functions with Inequality Constraints*, Technical Report I.M.A.G. 335, Grenoble, France.

Mammen, E. (1991a). Estimating a smooth monotone regression function, *Ann. Statist.*, **19**, 724–740.

Mammen, E. (1991b). Nonparametric regression under qualitative smoothness assumptions, *Ann. Statist.*, **19**, 741–759.

Mammen, E., and Thomas-Agnan, C. (1999). Smoothing splines and shape restrictions, *Scand. J. Statist.*, **26**, 239–252.

Marshall, A. W., and Proschan, F. (1965). Maximum likelihood estimation for distributions with monotone failure rate, *Ann. Math. Statist.*, **36**, 69–77.

Matzkin, R. L. (1991). Semiparametric estimation of monotone and concave utility functions for polychotomous choice models, *Econometrica*, **59**, 1315–1327.

Menéndez, J. A., and Salvader, B. (1987). An algorithm for isotonic median regression, *Computat. Statist. Data Anal.*, 399–406.

Michelli, C. A., and Utreras, F. (1988). Smoothing and interpolation in a convex subset of a Hilbert space, *SIAM J. Sci. Stat. Comput.*, **9**, 728–747.

Mukerjee, H. (1988). Monotone nonparametric regression, *Ann. Statist.*, **16**, 741–750.

Mukerjee, H., and Stern, S. (1994). Feasible nonparametric estimation of multiargument monotone functions, *J. Amer. Statist. Assoc.*, **89**, 77–80.

Prakasa Rao, B. L. S. (1969). Estimation of a unimodal density, *Sankhyā Ser.*, **31**, 23–36.

Ramsay, J. O. (1988). Monotone regression splines in action, *Statist. Sci.*, **3**, 425–461.

Ramsay, J. O. (1998). Estimating smooth monotone functions, *J. Roy. Statist. Soc. B*, **60**, 365–375.

Reboul, L. (1998). Estimation sous contraintes de forme et applications à la fiabilité, Thesis, Université de Paris-Sud, Paris, France.

Robertson, T., Dykstra, R. L., and Wright, F. T. (1988). *Order Restricted Statistical Inference*, Wiley, New York.

Ruud, P. (1995). *Restricted Least-Squares Subject to Monotonicity and Concavity Constraints, Computationally Intensive Methods in Econometrics*, 7th World Congress of the Econometric Society, Tokyo, Japan.

Schlee, W. (1980). Nonparametric test of the monotony and convexity of regression, *Nonparametric Statistical Inference*, **2**, 823–836.

Schumaker, L. (1981). *Spline Functions: Basic Theory*, Wiley, New York.

Stromberg, U. (1991). An algorithm for isotonic regression with arbitrary convex distance function, *Computat. Statist. Data Anal.*, **11**, 205–219.

Tocquet, A. S. (1998). Construction et étude de tests en régression, Thesis, Université de Paris-Sud, Paris, France.

Turlach, B. (1997). Constrained smoothing splines revisited, Technical Report SRR 008-97, Australian National University, Canberra, Australia.

Utreras, F. (1985). Smoothing noisy data under monotonicity constraints: Existence characterization and convergence rates, *Numerische Mathematik*, **47**, 611–625.

Utreras, F. (1991). The variational approach to shape preservation, in Laurent, P. J., Le Mehaute, A., and Schumaker, L. (eds.), *Curves and Surfaces*, pp. 461–476, Academic Press, Boston, MA.

Villalobos, M., and Wahba, G. (1987). Inequality constrained multivariate smoothing splines with application to the estimation of posterior probabilities, *J. Amer. Statist. Assoc.*, **82**, 239–248.

Wahba, G. (1982). Constrained regularization for ill posed linear operator equations, with applications in meteorology and medecine, Gupta, S. S., and Berger, J. O. (eds.), *Statistical Decision Theory and Related Topics III*, **2**, 383–418.

Wahba, G. (1990). Spline models for observational data, CBMS-NSF Conference Series in Applied Mathematics 59, SIAM, Philadelphia, PA.

Wang, Y. (1995). The L1-theory of estimation of monotone and unimodal densities, *J. Nonparametric Statistics*, **4**, 249–261.

Wright, F. T. (1982). Monotone regression estimates for grouped observations, *Ann. Statist.*, **10**, 278–286.

Wright, I. W., and Wegman, E. J. (1980). Isotonic, convex and related splines, *Ann. Statist.*, **8(5)**, 1023–1035.

Yatchew, A. J. (1992). Nonparametric regression tests based on least-squares, *Econometric Theory*, **8**, 435–451.

CHAPTER 6

Spline and Kernel Regression for Dependent Data

Robert Kohn
The Australian Graduate School of Management, Sydney, Australia

Michael G. Schimek
Karl-Franzens-Universität Graz, Graz, Austria

Michael Smith
University of Sydney, Sydney, Australia

6.1 INTRODUCTION

In previous chapters, univariate spline and kernel smoothing were introduced and selection techniques, such as cross-validation, for the smoothing parameter, or bandwidth, were discussed. An important assumption made throughout was that the errors are independent and identically distributed (i.i.d.). This chapter deals with observations collected in temporal order and shows that the regression curve estimates can be adversely affected if the errors are autocorrelated, but are treated as if they were independent. In particular, if the explanatory variable is time, in which case the regression function represents a trend, then ignoring the autocorrelation in the errors can lead to poor function estimates because the regression function is confounded with the error structure. In general, the effective signal-to-noise ratio can be increased by taking account of the time dependence of the errors, leading to improved function estimates.

Consider the model

$$y_t = f(x_t) + e_t, \tag{6.1}$$

Smoothing and Regression: Approaches, Computation, and Application
Edited by Michael G. Schimek
ISBN 0-471-17946-9

where $y_1, \ldots, y_n$ is a sequence of observations collected at design points x_t (forming the independent variable) and e_t is the error sequence. If the x_t are equally spaced and in temporal order, then $f(x_t)$ is the trend. We assume that e_t is a stationary zero mean autocorrelated error sequence that can be modeled as an autoregression of maximal order p. That is, the AR(p) process

$$e_t = \phi_1 e_{t-1} + \cdots + \phi_p e_{t-p} + \epsilon_t, \tag{6.2}$$

where the errors $\epsilon_1, \ldots, \epsilon_n$ are independently distributed $N(0, \sigma^2)$. Once we choose to model the errors e_t as a linear process, little loss in generality occurs if we assume that the errors follow an autoregressive process, because most linear processes can be approximated by an autoregressive model of a sufficiently high order. (For a general introduction to time-series models, see Chatfield (1996).)

In order to distinguish between a trend f that, by assumption, is changing slowly and the autoregressive process e_t, we suppose that e_t is stationary, which means that all the roots of the autoregressive polynomial $\phi(B) = 1 - \phi_1 B - \cdots - \phi_p B^p$ lie outside the unit circle. One way to enforce stationarity is to parameterize the autoregressive coefficients $\phi_1, \ldots, \phi_p$ in terms of the partial autocorrelations $\psi_1, \ldots, \psi_p$, as in Monahan (1984). The partial autocorrelations provide a one-to-one transformation of the autoregressive coefficients within the region of stationarity.

To illustrate how autoregressive errors can confound a trend, consider an example with $n = 100$ observations generated from the model of Equation (6.1); in that case, the regression function $f(x_t) = 1280x_t^4(1 - x_t)^4$ for $t = 1, 2, \ldots, 100$, and $x_t = t/100$. The errors e_t are generated using a first-order autoregressive process (i.e., AR(1)) with $\phi_1 = 0.3713$. The regression function is estimated in two ways—first, as if the errors were independent, and second, supposing that the errors follow a AR(1) process, but with ϕ_1 and σ^2 assumed unknown. The estimation is carried out using the Bayesian regression approach discussed in detail later in the chapter, but qualitatively similar results are obtained using other methods of estimation. Figure 6.1(a) shows that the estimate obtained assuming that the errors are independent is rougher than the true function, whereas the estimate that takes account of the autocorrelation in the errors (see Figure 6.1(b)) is more satisfactory.

Schimek (1992) carried out simulation experiments for different trend functions and sample sizes, using autoregressive and moving-average error models that have appeared frequently in the literature. He reported improved trend recovery for most AR(1) cases when the correct error model was used. He also observed undersmoothing for positive autocorrelation and oversmoothing for negative autocorrelation when the errors were assumed to be independent, but this did not occur in all cases. The quality of the trend estimate also depends on the sample size and on the type of underlying trend model. Most researchers who report on effects due to unspecified AR(1) errors used periodic curves (e.g., Diggle and Hutchinson, 1989), which may explain some of the contradictory findings regarding the necessary degree of smoothing when other trends are studied.

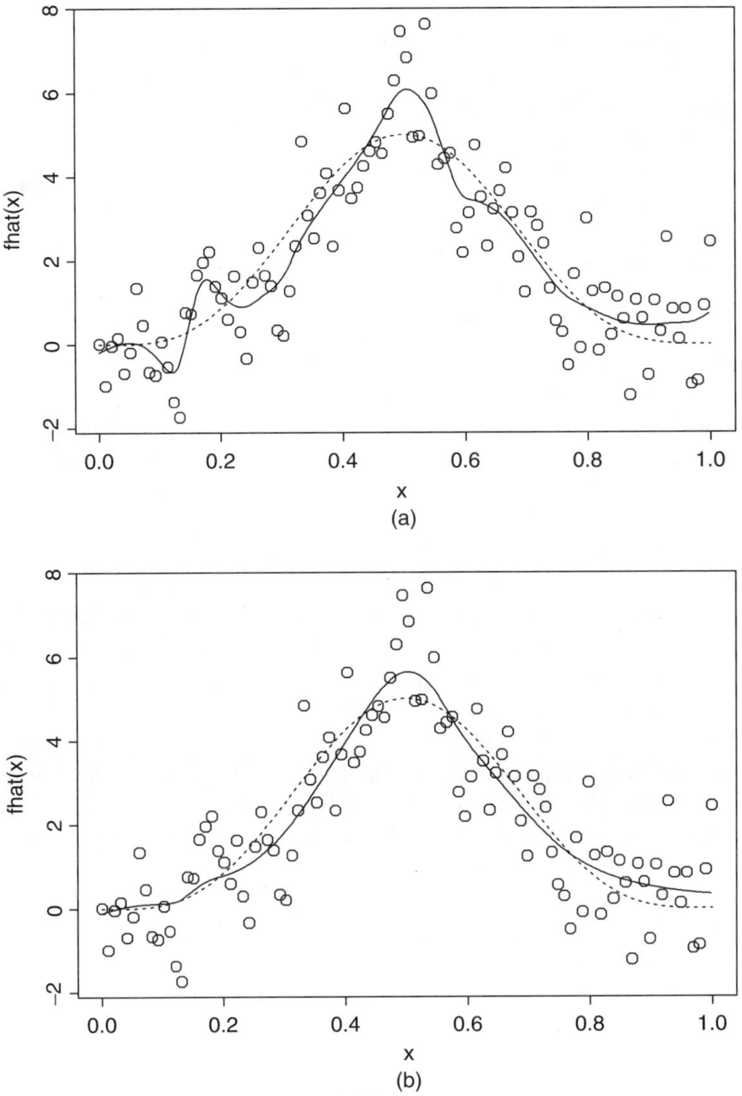

FIGURE 6.1 (a) Independent errors; (b) AR(1) errors ($\phi = 0.3713$).

Similar effects occur with AR(2) error processes. The effect of autocorrelation on the trend estimate for a first-order moving average MA(1) error models is qualitatively similar to that for AR(1) error models, that is, undersmoothing for positive autocorrelation and oversmoothing for negative autocorrelation — but is less pronounced than for AR models. The sample size n also affects the quality of the curve estimate. Independently of the error specification, the necessary degree of smoothing increases substantially as the number of observations increases.

Asymptotic results (e.g., Altman, 1990) cannot be expected to work well for time series with fewer than 200 observations, as noted in Schimek (1992). In conclusion, it seems that global smoothing parameter (bandwidth) adaptations are not necessarily sufficient to compensate for violations of independence, and there is a nonnegligible influence due to certain features of the trend function. Consequences of this finding — especially for global bandwidth selection — are discussed throughout this chapter.

The rest of the chapter is organized as follows: Early work that tried to correct for autocorrelation of the data under the assumption of a known covariance (autocorrelation) matrix is reviewed. There, the emphasis is on spline smoothing and, to a lesser extent, kernel smoothing. Then two directions of research are discussed; one concentrates on *cross-validation* (CV) and the other on so-called plug-in methods, with the primary goal of improving kernel estimation under time dependence. At least two different approaches can be distinguished for the method of CV. The first is the popular leave-one-out approach introduced for independent observations (see Chapters 2 and 4) constituting what is now called *time-series cross-validation* (TSCV). The second approach applies a state space (Kalman filter) representation of CV. Research undertaken so far goes in the direction of "automatic" bandwidth adaptation, applying either CV or plug-in techniques. In Section 6.4, we present a recent Bayesian alternative to traditional kernel and smoothing spline approaches to nonparametric regression with dependent data. This approach decomposes the function as a linear combination of basis functions and applies a Bayesian hierarchical model to explicitly parameterize the possibility that each of the regression coefficients is exactly equal to zero. The error distribution is modeled as a stationary autoregressive process, but one in which the possibility that each of the partial autocorrelations is zero is also explicitly parameterized. The entire model is estimated simultaneously using Markov chain Monte Carlo techniques, and the effect is to provide simultaneous estimates of the function f, along with a parsimonious autoregressive process in which the order is also estimated. Lastly, the use of smoothing when the observations are dependent is demonstrated with two examples of real data estimated by the Bayesian approach.

6.2 APPROACHES FOR A KNOWN AUTOCORRELATION FUNCTION

6.2.1 Adaptations of Smoothing with a Cubic Spline

In discussing Silverman's (1985) paper on cubic smoothing splines, Diggle (1985) considered the following regression problem for autocorrelated data:

$$y_t - f(x_t) = \phi[y_{t-1} - f(x_{t-1})] + e_t.$$

The time points x_t are equally spaced, and the parameter ϕ represents a stationary $AR(1)$ error process. The function f can be estimated by a weighted least-squares procedure, conditional on y_1 and fixed ϕ ($|\phi| < 1$ for stationarity). Let $\lambda > 0$ be the smoothing parameter for a cubic smoothing spline. Further we have entries $d_{tt'}$ of the inverted $n \times n$ covariance matrix $\mathbf{D}^{-1}$ (i.e., $t = t' = (1, 2, \cdots, n)$). Then the penalized sum-of-squares equation is

$$\mathrm{SS}(f) = \sum_{t=2}^{n} [(y_t - f(x_t)) - \phi(y_{t-1} - f(x_{t-1}))]^2 + \lambda \int (f''(x))^2 \, dx,$$

with $d_{tt'} = 0$ for $t \neq t'$. (Because of undefined element d_{11} not time reversible.)

Schimek (1988) introduced a spline-related smoothing technique for arbitrary autoregressive and moving-average errors. He replaced the ordinary cubic spline by a covariance-weighted version, which behaves in a manner similar to that of standard smoothing splines (with λ measured on the same scale as in the case of i.i.d. errors).

Under the assumptions made earlier, the approach is characterized by a discrete generalized sum-of-squares equation with a quadratic roughness penalty

$$\mathrm{SS}(f) = \sum_{t=1}^{n} \sum_{t'=1}^{n} (y_t - f(x_t)) d_{tt'} (y_{t'} - f(x_{t'})) + \lambda \sum_{t=2}^{n-1} (\Delta^2 f(x_t))^2 \qquad (6.3)$$

and with

$$\Delta^2 = f(x_{t+1}) - 2f(x_t) + f(x_{t-1}).$$

Let $d_{tt'}$ again be the elements of $\mathbf{D}^{-1}$, where $\mathbf{D} \neq \mathbf{I}$ is the covariance matrix of size $n \times n$ representing a (stationary) autoregressive or (invertible) moving-average error process with independently distributed $N(0, \sigma_\epsilon^2)$ random shocks ϵ. The λ parameter is defined as usual. In matrix notation (compare with Diggle and Hutchinson, 1989), Equation (6.3) is

$$\mathrm{SS}^d(f) = (y - f)^T \mathbf{D}^{-1} (y - f) + \lambda f^T \mathbf{Q} f, \qquad (6.4)$$

where y is an n-dimensional column vector of observations y_t, f is an n-dimensional column vector of curve estimates $f(x_t)$, and

$$\mathbf{Q} = \begin{pmatrix} 1 & -2 & 1 & & & & & & 0 \\ -2 & 5 & -4 & & 1 & & & & \\ 1 & -4 & 6 & & -4 & & 1 & & \\ & \ddots & \ddots & \ddots & \ddots & \ddots & \ddots & \ddots & \\ & & 1 & & -4 & & 6 & -4 & 1 \\ & & & & 1 & & -4 & 5 & -2 \\ 0 & & & & & & 1 & -2 & 1 \end{pmatrix}.$$

is an $n \times n$ banded and symmetric operator matrix.

Minimizing Equation (6.4) with respect to f yields

$$\begin{aligned}\hat{f} &= (\mathbf{D}^{-1} + \lambda\mathbf{Q})^{-1}\mathbf{D}^{-1}y \\ &= \mathbf{A}_\lambda y,\end{aligned}$$

where $\mathbf{A}_\lambda$ is the so-called hat matrix. The function $\hat{f}$ is a generalized least-squares estimator that is related to ridge regression. Under the assumption that $\lambda > 0$ for all n, the asymptotic unbiasedness of the estimator $\hat{f}$ can be specified by $\hat{f}_n(t) \to f_n(t)$ for $n \to \infty$ and (independently) $\lambda \to 0$. The mathematical expectation of $\hat{f}_n - f_n$, neglecting the dependence on x, is

$$E(\hat{f}_n - f_n) = (\mathbf{A}_\lambda - \mathbf{I}_n)f_n.$$

This requires that $E(\hat{f}_n - f_n) = 0$, which is fulfilled if $\mathbf{A}_\lambda \to \mathbf{I}_n$ for $n \to \infty$ and $\lambda \to 0$. We thus consider $\mathbf{A}_\lambda - \mathbf{I}_n$ and apply the identity $\mathbf{I} = \mathbf{B}_\lambda^{-1}\mathbf{B}_\lambda$, where $\mathbf{B}_\lambda = (\mathbf{D}_n^{-1} + \lambda\mathbf{Q}_n)$ is a nonsingular matrix. We calculate

$$\begin{aligned}\mathbf{A}_\lambda - \mathbf{I}_n &= \mathbf{B}_\lambda^{-1}\mathbf{D}_n^{-1} - \mathbf{B}_\lambda^{-1}\mathbf{B}_\lambda \\ &= -\lambda(\mathbf{D}_n^{-1} + \lambda\mathbf{Q}_n)^{-1}\mathbf{Q}_n \\ &= -\lambda(\mathbf{I} + \lambda\mathbf{D}_n\mathbf{Q}_n)^{-1}\mathbf{D}_n\mathbf{Q}_n. \end{aligned} \tag{6.5}$$

For the linear stochastic error processes under consideration, we assume weak stationarity, so $S = \sum_{v=0}^{\infty} |\gamma_v|$ is finite with $E(e_t, e_{t+v}) = \gamma_v$. Then, for any n and all t, we have $\sum_{t'=1}^{n} |d_{tt'}| < 2S$, making $\|\mathbf{D}_n\|$ uniformly bounded by $2S$. $\mathbf{Q}_n$ is uniformly bounded over n because of its symmetric band structure. For n tending to infinity, Equation (6.5) converges to a null matrix, since $\|\mathbf{D}_n\mathbf{Q}_n\|$ is uniformly bounded over n and, by assumption, λ tends to zero.

There are $O(n)$ algorithms for AR(m), as well as MA(m) errors, as long as $n \gg m$. For applications to evoked potential series and intervention analysis, see Schimek et al. (1991) and Schimek (1994), respectively. For the implementation of the algorithm and more applications, see Schimek and Schmaranz (1994), who also provide the computer program DERS, written in C for PCs under MS-Windows.† DERS is an exploratory tool that allows analysts to examine even long time series for departures from the usual independence assumption when standard smoothing—not necessarily spline smoothing—should be applied. No such routines take account of serial correlation in Xplore and S-Plus, the two most popular software packages for nonparametric regression.

Diggle and Hutchinson (1989) suggested a procedure for cubic spline smoothing in the presence of errors of which the autocorrelation function is given by

$$v_{tt'} = \exp(-\rho|x_t - x_{t'}|)$$

† DERS can be obtained from M. G. Schimek.

for some known parameter $\rho > 0$. Allowing for distinct, but not necessarily equally spaced, design points, we see that the penalized sum-of-squares equation (in matrix notation) takes the form

$$\mathrm{SS}(f) = (y - f)^T \mathbf{V}^{-1}(y - f) + \lambda \int (f''(x))^2 \, dx, \qquad (6.6)$$

where $\mathbf{V}$ is a correlation matrix replacing the inverse of Silverman's (1985) diagonal weighting matrix. For a known autocorrelation, $\mathbf{V}$ is banded. Diggle and Hutchinson (1989) show that, for the tridiagonal case (i.e., AR(1)), Equation (6.6) can be minimized in $O(n)$ operations. For this purpose, we require a symmetric tridiagonal matrix $\mathbf{W} = \mathbf{V}^{-1}$ consisting of nonzero elements $w_{11} = 1/(1 - a_1^2)$, $w_{tt} = 1 + a_{t-1}^2/(1 - a_{t-1}^2) + a_t^2/(1 - a_t^2)$ for $t = 2, 3, \ldots, n-1$, $w_{nn} = 1/(1 - a_{n-1}^2)$, and $w_{t+1,1} = w_{t,t+1} = -a/(1 - a_t^2)$ for $t = 1, 2, \ldots, n-1$, where $a_t = \exp(-\rho(x_{t+1} - x_t))$. Then B-spline-based estimation can be carried out (de Boor, 1978). This approach is more limited than that of Schimek (1988), because error processes other than AR(1) cannot be handled. Diggle and Hutchinson (1989) suggested using the *generalized cross-validation* (GCV) criterion

$$\mathrm{GCV}(\lambda) = (y - \hat{f})^T \mathbf{V}^{-1}(y - \hat{f})/[\mathrm{tr}(\mathbf{I} - \mathbf{A})]^2$$

to determine the smoothing parameter λ in the presence of a known autocorrelation structure. Empirical findings reported for known AR(1) error processes compared with independent (white-noise) errors are similar to those obtained by Schimek's (1988) discrete spline approach.

Diggle and Hutchinson (1989) also proposed a penalized maximum-likelihood estimator for unknown ρ. The likelihood function to be maximized is

$$L(\lambda) = -n \ln[\mathbf{R}(\lambda)] + 2n \ln(1 - d/n),$$

under multivariate normality of y, where $\mathbf{R}(\lambda) = (y - f)^T \mathbf{V}^{-1}(y - f)$, $d = \mathrm{tr}(\mathbf{A}_\lambda)$, and $\mathbf{A}_\lambda$ is again the hat matrix. We can also allow for more general penalties on d and include ρ in the estimation by maximizing

$$L(\lambda, \rho) = -n \ln[\mathbf{R}(\lambda, \rho)] - \ln |\mathbf{V}(\rho)| - \zeta(n, d(\lambda, \rho)),$$

where, for each n, ζ is an increasing function of d. Under the technical assumption that ρ is small and in the vicinity of the true value, and noting that $|\mathbf{V}(\rho)| = (1 - a^2)^{n-1}$ under the assumptions that $f(\rho) = a/(1 - a^2)$ and $\zeta(n, d(\hat{\lambda}, \rho))$ is approximately constant, we have

$$L(\hat{\lambda}, \rho) \simeq -n \ln(a) = n\rho.$$

These assumptions — as a matter of fact, not very realistic — are necessary to obtain a solution in the case of an unknown first-order autocorrelation structure. In simulations, Diggle and Hutchinson (1989) applied different penalties, such as $\zeta(n, d) = d \ln(n)$ due to Schwarz (1978). The results are not promising. One reason is given by van der Linde (1994), who showed that the penalties cannot be motivated by the heuristics of the usual "low-structure" CV techniques introduced for independent observations (i.e., techniques based on the deletion theorem) by Geisser and Eddy (1979). Recently, an approach similar to that of Diggle and Hutchinson (1989) was taken by Wang (1998), who also assumed specific (temporal as well as spatial) parametric forms of the correlation matrix. Apparently, he obtained reasonable results for certain polynomial splines.

6.2.2 Adaptations of Kernel Smoothing

Altman (1990) studied ways to adapt kernel smoothing for stationary AR errors based on the widely used kernel estimator of Priestley and Chao (1972).

For regularly spaced design points t/n (for $t = 1, 2, \ldots, n$), a stationary error sequence with covariance function

$$E(e_{n,t}, e_{n,t'}) = \sigma^2 \rho_n(|t - t'|)$$

is assumed, where the variance σ^2 is independent of the sample size n and the correlation function $\rho_n(k)$. Let $\mathbf{D}_n$ be the covariance matrix of the errors. This definition allows the autocorrelation among the errors to vary with both the interval between the time points and the sample size. The interesting case here is $\rho_n(t) = \rho(t)$, which means a constant error process, regardless of how close the time points become. Two further absolute summability conditions for the sequence of correlation functions are required by Altman (1990): $\sum_t^{k/2} |\rho_n(t)|$ converges as n and $k \to \infty$, and $\sum_t^{k/2} t\rho_n(t) = o(k)$ as n and $k \to \infty$. Under the first condition, the sum of correlations is

$$S_\rho = \lim_{n,k \to \infty} S_{\rho_n}(k),$$

where

$$S_{\rho_n}(k) = \sum_{t=1}^{k/2} \rho_n(t).$$

Under the assumption that the errors have finite second moments, the asymptotic properties of the kernel estimator can be assessed by the *mean squared error* (MSE)

$$\text{MSE}(x, h, n) = E(\hat{f}_{h,n}(x) - f(x))^2,$$

where h is the bandwidth of the kernel. The MSE of the Priestley and Chao kernel and the asymptotic form of its bias can be derived when the design points become dense on the interval $h/2 < x < 1 - h/2$ for $h \to 0$ and $nh \to \infty$. Under the foregoing conditions, correlation does not affect the bias, but does affect the variance to a considerable extent. In fact, the behavior of the variance term as a function of h depends on both the correlation and the kernel.

For a mean function f with r derivatives and a kernel K of order r under the usual assumptions of summability and equally spaced design points, the following result for the MSE is due to Altman (1990):

$$\text{MSE}(x, h, n) = (h^r f^{(r)}(x) s_K / r!)^2 + \sigma^2 (N_k / nh)(1 + 2S_\rho) + o(1/nh) + o(h^{2r}).$$

Here, $s_K = \int x^r K(x)\,dx$ is the rth moment of K (the first $r - 1$ moments are zero), and N_K is the squared norm of K. This result can be used to calculate the asymptotically optimal bandwidth h_n and to demonstrate the consistency of the estimator. Comparing the correlated error process with variance σ^2 to the i.i.d. error sequence and assuming that $S_r > 0$ (< 0) reveals that h_n is greater (smaller) under correlation than under i.i.d. The MSE achieved at the optimal bandwidth is again greater (smaller), which corresponds to undersmoothing (oversmoothing) for the positive (negative) AR(1) parameters mentioned earlier in the chapter. With respect to the choice of the bandwidth, an appropriate value will result in a small value of the MSE. Altman's idea is to correct selection criteria such as CV for correlation.

In general, there are two approaches to adjusting selection criteria: a direct method to make CV more nearly unbiased for the *expected squared predictor error* (ESPE) and an indirect method transforming the residuals to make them less correlated. For equidistant time points and a known correlation function ρ_n with a corresponding correlation matrix $\mathbf{R}_n$, CV can be adjusted by

$$r^2_{CV,\rho}(t, h, n) = \frac{r^2(t, h, n)}{\left[1 - \sum_{t'=-(nh/2)}^{nh/2} w_{h,n}(t/n, t + t')\rho_n(t')\right]^2},$$

where $r^2(t, h, n)$ denotes the squared residuals and $w_{h,n}$ are the Priestley–Chao kernel weights. In the case of GCV, the adjustment is

$$r^2_{\text{GCV},\rho}(t, h, n) = \frac{r^2(t, h, n)}{(1 - 1/n \text{ tr } \mathbf{W}_{h,n}\mathbf{R}_n)^2},$$

where $\mathbf{W}$ is the matrix of the kernel weights.

The indirect method that was used by Schimek (1988) and Diggle and Hutchinson (1989) is based on an appropriate transformation of the residuals.

Altman (1990) also carried out simulations using a uniform and a spline equivalent kernel. She considered the trend function $f(x) = \cos(3.15\pi x)$ for $x \in [0, 1]$, with the errors generated from a stationary AR(1) process with $\rho(1) \in [-0.9, 0.9]$. Altman considered two sample sizes: $n = 128$ and $n = 256$. The results can be summarized as follows:

(1) Autocorrelation has a substantial effect on bandwidth selection methods.
(2) Kernel regression performs reasonably well for data with correlated errors if the selection criterion is suitably corrected.

(3) Overestimation of the correlations turns out to be less serious than underestimation.
(4) For the assumed variance of $\sigma^2 = 1$, the choice of kernel does not matter much.
(5) Under the wrong assumption of i.i.d. errors, CV and GCV fail even for the larger sample size.
(6) Undersmoothing and oversmoothing were observed in the same way as reported earlier for cubic smoothing splines.

6.3 APPROACHES FOR AN UNKNOWN AUTOCORRELATION FUNCTION

Until recently, it was unclear whether the selection of h could be done independently of the user. Hart (1991, p. 184), who devoted a great deal of research to that question, writes, "It should be clear that dependence in the data calls for a good deal of interaction between the analyst and the data." In this section, we move from adaptations of the estimator or the cross-validation criterion for a known serial correlation to more sophisticated, "automated" techniques for an unknown covariance function. We can differentiate between plug-in methods and modifications of the CV score. Here we consider leave-one-out and maximum-likelihood approaches based on the state-space model for the latter case. Although "automated," these approaches are still limited from theoretical as well as practical points of view for reasons that are explained later.

6.3.1 The Plug-In Technique

Herrmann et al. (1992) developed a plug-in technique for the kernel estimator introduced by Gasser and Müller (1979). Let W be the kernel function and h the bandwidth. Herrmann et al.'s proposal adapts the plug-in estimator for independent observations put forward in Gasser et al. (1991), because the latter method breaks down for serially correlated errors. The goal was to obtain a method that is robust for a broad variety of correlation structures.

The assumptions are as usual: equally spaced time points and stationarity. Further, $E(e_t, e_{t+\nu}) = \gamma_\nu$ for the errors e_t. The mean-integrated-squared-error criterion for a kernel estimator $\hat{f}$ with bandwidth h is

$$\text{MISE}(h) = E\left[\int_0^1 v(t)\{f(t) - \hat{f}(t, h)\}^2\, dt\right],$$

where v is a twice continuously differentiable function with support $[\delta, 1-\delta]$ and $v(t) > 0$ for all $t \in (\delta, 1-\delta)$ and some $\delta > 0$. These assumptions on v are made to avoid boundary problems during the bandwidth selection process. The optimal bandwidth h_M minimizes the MISE. For the purpose of estimation, let

h_A be the bandwidth minimizing the asymptotic MISE. Then, for a second-order kernel, we have

$$h_A = \left(C_1 S \Big/ \left[n C_2^2 \int_0^1 v(t)\{f''(t)\}^2\, dt \right]^{1/5} \right) \tag{6.7}$$

with

$$C_1 = \int_0^1 v(t)\, dt \int_{-1}^1 \{W(x)\}^2\, dx,\ C_2 = \int_{-1}^1 x^2 W(x)\, dx \text{ and } S = \sigma^2 + 2\sum_{\nu=1}^{\infty} \gamma_\nu .$$

This equation, due to Hart (1991), demonstrates the impact of error correlation on model estimation. Under the wrong assumption of i.i.d. errors, h becomes too small when $S > \sigma^2$. This phenomenon is often experienced in practice when high-frequency noise is dominated by low-frequency noise. In the case of no correlation, we have $S = \sigma^2$.

Plug-in estimators for h_M apply Equation (6.7) on the basis of known constants C_1 and C_2 (which are themselves based on estimators for σ^2 and $\int v(t)\{f''(t)\}^2\, dt$). Such estimators are known in the literature. An estimator $\hat{S}_M$ for S is also required. Herrmann et al. (1992) provide one that is unbiased for m-order dependent errors (residuals) and a linear regression function f. Moreover, their error is consistent when $m \to \infty$ and $m^2/n \to 0$ as $n \to \infty$ for residuals that are not necessarily m-dependent, but that satisfy some mixing condition (i.e., AR or MA processes) for smooth regression functions. Now it is necessary to choose m, which is a kind of tuning or secondary parameter. The authors propose two heuristic methods, for which they provide some asymptotic backing, that are suitable in different situations. One turns out to be appropriate for correlation structures dominated by positive autocorrelation (i.e., $S > \sigma^2$), and the other is for more complex situations. Similar heuristics are used for the choice of smoothing parameter in the exploratory context of DERS (Schimek and Schmaranz, 1994). Simulations reported in Herrmann et al. (1992) make clear that the plug-in technique, in combination with the preceding two methods, works reasonably well for low-order, but not all, AR/MA error processes. But the approach is certainly not an "automated" one, as the statistician has to intervene in many instances.

6.3.2 Leave-One-Out Cross-Validation

There have been a number of attempts to adapt the idea of leave-one-out (generalized) CV to time-dependent data. The usual approach is based on risk estimation in which serial correlation is taken into account. (See Chiu (1989), Hart (1991), and Wu and Chu (1994), who use kernel estimation.) Consider the *mean average squared error* (MASE) as the risk criterion,

$$\text{MASE}(h) = n^{-1} \sum_{t=1}^{n} E(\hat{f}_{t,h} - f_t)^2,$$

where $f_t = E(x_t)$. Given an estimate $\hat{M}(h)$ of MASE(h), the bandwidth h can be selected to minimize $\hat{M}(h)$. The problem is how to estimate those terms that depend on the underlying serial correlation. Without a crude estimation of the trend—and estimating the trend is our primary goal—we cannot evaluate the correlation structure. This leads to a kind of circularity, because our need to estimate risk (MASE) is a result of our ignorance of the trend. To overcome this methodological dilemma, a technical parameter has to be specified. Hence, the foregoing approaches are neither automatic nor objective in the light of a critical perspective.

Time-series cross-validation (TSCV), as proposed by Hart (1994), seems to overcome the dilemma. Essentially, this technique is an application of the prequential method of Dawid (1984). In standard leave-one-out CV, the prediction of x_j is based on the data $x_1, \ldots, x_{j-1}, x_{j+1}, \ldots, x_n$; in TSCV, the prediction of x_j depends only on the previous data $x_1, \ldots, x_{j-1}$. The idea of the latter is to produce a good prediction of the future in a time series.

Let us assume again kernel smoothers of the Gasser–Müller form, yielding estimates $\hat{f}_{t,h}$ and boundary kernels used to modify the estimates when $(t - 0.5)/n \in [0, h)$ or $(t - 0.5)/n \in (1 - h, 1]$ (see Gasser and Müller, 1979, and Müller, 1991 for details). Hart (1994) modified CV to TSCV for equidistant time points and the usual stationary AR(p) error assumption. For that purpose, he introduced a parametric probability model M_θ for the process $\{\epsilon_j = x_j - f_j\}$. A predictor of x_j can be of the form

$$\hat{x}_j(h, \theta) = \tilde{f}_h^{(j)} + g_\theta(\hat{\epsilon}_1, \ldots, \hat{\epsilon}_{j-1}),$$

where $\tilde{f}_h^{(j)}$ is a kernel estimate of f_j that uses solely the data $x_1, \ldots, x_{j-1}$, $\hat{\epsilon}_t = x_t - \tilde{f}_{t,h}^{(j)}$ for $t = 1, \ldots, j-1$ and $g_\theta(\hat{\epsilon_1}, \ldots, \hat{\epsilon}_{j-1})$ is the optimal mean squared error predictor of ϵ_j given $\epsilon_1, \ldots, \epsilon_{j-1}$ under the probability model M_θ. In TSCV, one can choose h and θ by minimizing the prediction error criterion

$$P(h, \theta) = \sum_{t=2}^{n} (\hat{x}_t(h, \theta) - x_t)^2.$$

Under the AR error assumption, it is possible to find analytically, for each h, the value $\hat{\theta}(h)$ that minimizes $P(h, \theta)$ with respect to θ. We obtain a TSCV curve $P(h) = P(h, \theta(h))$ that can be plotted and examined as in the case of independent data. Hart (1994, 1996) proposed to evaluate the TSCV bandwidth by a boundary-type kernel smoother because only the data on the left of a time point j are relevant. For the final trend estimation $\tilde{f}_h^{(j)}$ based on a kernel K, a bandwidth

$$C_{K,L} = \left\{ \frac{J_K (s_L^2)^2}{J_L (s_K^2)^2} \right\}^{1/5}, \tag{6.8}$$

is suggested, where L is a boundary kernel used in the computation of $\tilde{f}_h^{(j)}$,

$$s_K^2 = \int x^2 K(x)\, dx, \quad s_L^2 = \int x^2 L(x)\, dx,$$

$$J_K = \int K^2(x)\,dx,$$

and

$$J_L = \int L^2(x)\,dx.$$

Equation (6.8) is motivated by the fact that $C_{K,L}$ is free of unknown parameters.

Hart (1994) also studied the relationship between the prediction error criterion $P(h, \theta)$ and the mean average squared error MASE(h) for the special case of AR(1) errors. His main result was that, when the fitted error model matches the true error model, minimizing $P(h, \theta)$ is asymptotically equivalent to minimizing MASE(h). Further, TSCV can be made robust to the misspecification of the error model by using a kernel L such that $L(0) = 0$ (see Müller, 1991). The resulting TSCV curve has an expected value that is approximately proportional to MASE(h) plus a constant, even when the correlation model is inadequate. This could also be the reason for the relatively good results obtained via the plug-in technique of Herrmann et al. (1992), which is based on a MISE criterion (related to MASE). There is no doubt that TSCV outperforms leave-one-out (generalized) CV, because the latter is characterized by a total lack of robustness (see Hart, 1991). Quintela del Rio (1996) examined several bandwidth selectors for the Gasser–Müller kernel estimator in an extensive Monte Carlo study. Among others, he compared the above mentioned plug-in technique with several modifications of CV (e.g., due to Altman, 1990). His main finding was that all plug-in techniques do very well over a wide range of error structures and trend functions, whereas even modified leave-one-out CV (not TSCV, which was not in the study) comes last. Quintela del Rio (1996) also demonstrated the crucial role of technical (secondary) parameters in bandwidth selection.

6.3.3 The State-Space Approach

Kohn et al. (1992) provided a quasi-Bayesian method to estimate the trend observed with autocorrelated noise, using a state-space framework to efficiently implement the computation. They modeled the trend by means of a smoothness prior and the autocorrelated errors as an autoregressive moving-average model. The trend was estimated by its posterior mean, which is a cubic smoothing spline. The smoothing parameter and the autoregressive moving-average parameters were estimated by maximizing the likelihood of the stochastic model for the observations. The authors also provided an alternative GCV technique for estimating the unknown parameters. Therefore, the approach is almost automatic, except for the assumption that the orders of the autoregression and the moving average are known. The technique requires the x_t to be in temporal order (as in all the previous approaches), which effectively means that it is useful only for modeling the trend.

Let us consider Equation (6.1) and take the prior for the trend function to be

$$f(x) = \beta_0 + \beta_1 x + 1/\alpha^{1/2} \int_0^x (x - s)W(ds), \qquad (6.9)$$

where $W(s)$ is a Wiener process with $W(0) = 0$ and $\text{Var}(W(1)) = 1$. Kohn et al. (1992) assumed a diffuse prior for $\beta = (\beta_0, \beta_1)^T$; that is, $\beta \approx N(0, kI_2)$ with $k \to \infty$. This is the smoothness prior used by Wahba (1978). To motivate that prior for the trend, we note that f is smooth and has a smooth first derivative, but its second derivative is diffuse because it has infinite variance. We also note that the level and the slope of f at $x = 0$ are diffuse because $f(0) = \beta_0$ and $f^{(0)}(0) = \beta_1$. That is, we assume no prior information about the level and the slope of the trend at $x = 0$ or about the second derivative of the trend. By Kohn and Ansley (1983), the posterior mean of f for this prior (i.e., $\lim_{k\to\infty} E(f(x)|y;k)$) is a cubic smoothing spline. This suggests using the posterior mean of f as the estimate of the trend and makes the prior for f attractive because it is not only smooth, but also capable of accommodating a variety of functional forms. Kohn et al. modeled the errors as an autoregressive moving-average series, but for conciseness, we confine the discussion to an AR error term of the form of Equation (6.2) with the order of the autoregression assumed known. In practice, p represents an upper bound on the expected order of the AR process. To obtain an estimate of the trend, it is also necessary to estimate the variance α and the autoregressive parameters $\phi = (\phi_1, \ldots, \phi_p)^T$.

Kohn et al. used maximum likelihood to estimate α and ϕ; they obtained the posterior mean of f on the basis of these parameter estimates. The maximum-likelihood method of estimating α and ϕ is equivalent to estimating them by their posterior modes using flat priors. This gives a model-based approach for solving the nonparametric problem and is therefore different from most previous approaches.

To make the computation of the parameter estimates and the posterior mode efficient, Kohn et al. expressed both the prior (Equation (6.9)) for f and the AR process (Equation (6.2)) as a state-space model and used efficient $O(n)$ state-space filtering and smoothing routines to evaluate the likelihood and the GCV function to obtain the posterior mean of f.

In the case of independent errors, Wahba (1978) showed that using a Bayesian approach with the function estimated by the posterior mean is equivalent to estimating the regression function by a penalized least-squares method, as in Equations (6.3) and (6.4). This result extends to the autocorrelated-error case discussed here and provides a further motivation for a Bayesian approach.

6.4 A BAYESIAN APPROACH TO SMOOTHING DEPENDENT DATA

Smith and Kohn (1996) introduced an alternative approach to data-driven regression smoothing for independent data based on a Bayesian hierarchical model estimated using a Markov-chain Monte Carlo sampling scheme. Smith et al. (1998) extended this approach to estimate the autoregressive error regression model of Equations (6.1) and (6.2). They examined both the case where the independent variable is restricted to be in temporal order and the more general case where the independent variable is unrestricted. They also employed a parsimonious autoregressive structure that explicitly accounts for the possibility that each of the p partial autocorrelations $\psi_1, \ldots, \psi_p$ is exactly equal to zero. Both papers represent the function as a linear combination of regression spline terms.

This section summarizes the approach suggested by Smith et al. (1998) and demonstrates how bases other than regression splines can be employed successfully in such a framework. It also discusses how the "one at a time" sampling steps employed by these authors can be replaced with the more efficient "focused sampling" steps introduced for binary-indicator-based hierarchical Bayesian models by Wong et al. (1997).

6.4.1 Functional Representation

The unknown regression function f in Equation (6.1) is represented as a linear combination of basis terms b_i, for $i = 1, \ldots, r$, augmented with an intercept and linear term, so that

$$f(x_t) = \beta_r + \beta_{r-1}x_t + \sum_{i=1}^{r-2} \beta_i b_i(x_t, k_i). \tag{6.10}$$

Here, $\beta = (\beta_1, \ldots, \beta_r)^T$ are unknown regression coefficients, while $\min(x_t) < k_i < \max(x_t)$ is the location of the basis term b_i, sometimes called a *knot*. These coefficients can be placed at all the unique design points, although Smith and Kohn (1996) demonstrated that a smaller grid of knots is all that is required for the accurate estimation of many functions. Popular choices for smooth basis terms b_i include regression splines (Friedman and Silverman, 1989), natural splines (Wahba, 1990), and a wide variety of radial basis functions (Powell, 1987). (See Wong et al. (1997) for an extensive comparison of a variety of smooth bases and Smith and Kohn (1998) for discontinuous and other nonsmooth bases.)

6.4.2 A Bayesian Hierarchical Dependent-Error Linear Model

Decomposing the function outlined in Equation (6.10) enables the regression of Equation (6.1) to be rewritten as the linear model $y = \mathbf{X}\beta + e$. Here, $y = (y_1, \ldots, y_n)^T$, $\beta = (\beta_1, \ldots, \beta_r)^T$, $e = (e_1, \ldots, e_n)^T$, and $\mathbf{X}$ is an $n \times r$ design matrix. We model the uncertainty associated with the significance (or otherwise) of each of the basis terms using a vector of binary indicator variables $\gamma = (\gamma_1, \ldots, \gamma_r)^T$, where $\gamma_i = 1$ iff $\beta_i \neq 0$ and $\gamma_i = 0$ iff $\beta_i = 1$. We can therefore rewrite the linear model conditional on γ as

$$y = \mathbf{X}_\gamma \beta_\gamma + e. \tag{6.11}$$

Here, β_γ is the vector of regression coefficients of the "active" basis terms (that is, the β_i such that $\gamma_i = 0$), while X_γ is a matrix of the corresponding columns of X.

To facilitate selecting the AR order, indicator variables $\kappa = (\kappa_1, \ldots, \kappa_s)^T$ are introduced to denote the nonzero partial autocorrelations $\psi = (\psi_1, \ldots, \psi_p)^T$. In the same spirit as in selecting subsets for regression parameters, a conditional prior $\psi|\kappa, \sigma^2$ can be defined, where $\psi_1, \ldots, \psi_p$ are a priori conditionally independent, with

$$\psi_i|\kappa_i = 0, \sigma^2 \text{ identically equal to zero}$$

$$\psi_i|\kappa_i = 1, \sigma^2 \sim \text{ uniform } (-1, 1).$$

The maximal order p of the autoregression and the prior probabilities $p(\kappa_i = 1), i = 1, \ldots, p$, are prescribed by the user. In all the empirical work presented here, a descending prior on the order is used, where $p(\kappa_1 = 1) = 0.5$, $p(\kappa_2 = 1) = 0.4$, $p(\kappa_3 = 1) = 0.3$, $p(\kappa_4 = 1) = 0.2$, and $p(\kappa_i = 1) = 0.1$ for $i = 5, \ldots, p$. This prior favors lower order autoregressions, although, as will be demonstrated by the example in Section 6.5.1, the empirical results appear to be insensitive to the prior.

The formulation in terms of ψ allows the error variance to be explicitly calculated as $\text{Var}(e) = \sigma^2 \mathbf{\Omega}_\psi$, where $\mathbf{\Omega}_\psi$ is a matrix depending only upon ψ and the data y. An important aspect of such a hierarchical linear model is the choice of prior for β. It is sufficient to consider the prior conditional on the other model parameters (Mitchell and Beauchamp, 1988). In particular, conditional on γ, inactive elements of β are exactly zero, while we take a proper prior on β_γ that is a fraction of the likelihood (O'Hagan, 1995), so that

$$p(\beta_\gamma|\gamma, \sigma^2, \psi, \kappa) \propto p(y|\gamma, \sigma^2, \psi, \beta_\gamma, \kappa)^{1/n},$$

which results in the prior

$$\beta_\gamma|\gamma, \sigma^2, \psi, \kappa \sim N((\mathbf{X}_\gamma^T \mathbf{\Omega}_\psi^{-1} \mathbf{X}_\gamma)^{-1} \mathbf{X}_\gamma^T \mathbf{\Omega}_\psi^{-1} y, n\sigma^2 (\mathbf{X}_\gamma^T \mathbf{\Omega}_\psi^{-1} \mathbf{X}_\gamma)^{-1}).$$

In effect, this places a prior on β_γ with variance equal to the weighted least-squares estimate, scaled up by a factor of n.

Lastly, we set the prior $p(\sigma^2) \propto 1/\sigma^2$ and make all subsets equally likely a priori by taking $p(\gamma) = 2^{-r}$. In addition, the parameters κ, γ, and σ^2 are assumed independent a priori.

6.4.3 Estimation Using a Markov Chain Monte Carlo Method

Smith et al. (1998) discuss at length a Markov chain Monte Carlo sampling scheme for estimating the preceding Bayesian model. We employ this sampling scheme here, but with the additional feature that, in generating the binary indicators γ_i and κ_i, we use focused sampling, as opposed to more traditional one-at-a-time sampling. The sampling scheme is outlined as follows:

Step (0): Choose the initial state $(\kappa^{[0]}, \psi^{[0]}, \beta^{[0]}, \gamma^{[0]})^T$.

Sequentially iterate the following steps:
Step (1): Generate an iterate from $\beta|\kappa, \psi, \gamma, y$.
Step (2): Select $\mathcal{S} \subset \{1, 2, \ldots, s\}$ in a random manner.
Step (3): Repeat the following K_1 times:
Generate iterates from $\kappa_i, \psi_i|\kappa_{j\neq i}, \psi_{j\neq i}, \gamma, \beta, y$ for $i \in \mathcal{S}$.
Step (4): Select $\mathcal{C} \subset \{1, 2, \ldots, m\}$ in a random manner.
Step (5): Repeat the following K_2 times:
Generate iterates from $\gamma_i|\gamma_{j\neq i}, \kappa, \psi, y$ for $i \in \mathcal{C}$.

This scheme is a focused sampling version of the sampler presented in Smith et al. (1998), who derive the conditional distributions at steps (1), (3), and (5) and indicate how to generate indicators from them. Steps (2) and (4) select a random subset of binary indicator variables $\kappa_{i\in\mathcal{S}}$ and $\gamma_{i\in\mathcal{C}}$ from which indicators should be generated in each iteration. The sets $\mathcal{S}$ and $\mathcal{C}$ are chosen to consist of the indices of the binary indicators that were active in the previous iteration, along with a randomly selected subset of the indices of the remaining previously inactive indicators. In this sense, the method is focused because it generates indicators that are most likely to be active and only those that are unlikely to be active with a low probability. (See Wong et al. (1997) and Smith and Kohn (1998) for a more in-depth discussion of focused sampling.)

After an initial warm-up period, the focused sampler produces iterates $\{\beta^{[j]}, \kappa^{[j]}, \psi^{[j]}, \gamma^{[j]}\}_{j=1}^{J}$ upon which inference can be based. We estimate $E(\beta|y)$, $E(\psi|y)$, and $E(\kappa_i|y) = p(\kappa_i = 1|y)$ using the mixture and histogram estimates:

$$\hat{\beta} = \frac{1}{J}\sum_{j=1}^{J} E(\beta|\gamma^{[j]}, \psi^{[j]}, \kappa^{[j]}, y), \qquad \hat{\psi} = \frac{1}{J}\sum_{j=1}^{J} \psi^{[j]};$$

$$\hat{\kappa}_i = \frac{1}{J}\sum_{j=1}^{J} p(\kappa_i = 1|\psi_{k\neq i}^{[j]}, \kappa_{k\neq i}^{[j]}, \gamma, \beta, y), \qquad i = 1, \ldots, s.$$

The estimate $\hat{\phi}$ of the AR parameters is obtained from $\hat{\psi}$, as there is a one-to-one relationship between the two, while the functional estimate $\hat{f} = \mathbf{X}\hat{\beta}$. We use a warm-up of 2000 iterates and a sampling period of $J = 2000$ iterations.

6.5 APPLICATIONS OF SMOOTHING DEPENDENT DATA

Smith et al. (1998) undertook extensive simulations comparing the Bayesian approach outlined in Section 6.4, based on a sampler that generates the γ_i one-at-a-time, with a cubic regression spline basis for the function f. They found that, for a variety of test functions and signal-to-noise ratios, when $x_t = t/n$, the Bayesian approach constantly outperformed the smoothing spline approach of Kohn et al. (1992) with both maximum-likelihood and GCV-based smoothing parameters, as well as Hart's (1994) TSCV-based procedure. Smith et al. (1998)

also examined the more general case when x_t was not ordered in time. They empirically demonstrated that there can be large gains in efficiency in accounting for an AR error structure, compared with incorrectly treating the errors as independent. They also demonstrated that modeling the errors as a longer-lag-length autoregression with parsimonious partial autocorrelation representation results in little loss in efficiency over employing the correct AR error model.

In the rest of this section, we demonstrate the potential use of smoothing dependent data with two real-world examples. The first is a model for the real average cost of Victorian hotel rooms, while the second concerns the estimation of the daily and weekly behavior of electrical load data.

6.5.1 Victorian Hotel Room Data

This example estimates the trend in the real average cost of a room for motels, hotels, and guest houses in Victoria, Australia, during the period from January 1980 to June 1995. The following monthly data[‡] were obtained:

`TAKINGS`: Total monthly room takings.

`NIGHTS`: Total number of nights room was occupied.

`CPI`: Monthly consumer price index in Melbourne, Victoria, normalized to equal 1 in January 1980.

We used these data to calculate the real average cost of a room, $y =$ `TAKINGS/(NIGHTS*CPI)`, measured in January 1980 Australian dollars. The regression model outlined in Equations (6.1) and (6.2) was fit to the data, with independent variable $x_t = t$ and a parsimonious AR error structure, as outlined in Section 6.4. To account for cyclical and seasonal behavior, the maximal lag length was $p = 24$, which corresponds to two years of monthly data. The basis used was radial, with terms of the form $b_i(x_t, k_i) = z^2 \log(z)$, where $z = \|x_t - k_i\|$ is the Euclidean distance between x_t and k_i; such a basis is commonly called a *thin-plate spline*. Knots are placed at all the observations, so that $k_i = x_i$, for $i = 1, \ldots, n$, which results in $r = n + 2$.

Figure 6.2 plots y against time, along with the estimated trend $\hat{f}$. The first point to notice is the substantial drop in the real average cost of a room between 1983 and 1989, which was a period of substantial deregulation and reform in the hospitality industry and, more generally, throughout the Australian economy. Table 6.1 contains the estimates of the autoregressive coefficients $\hat{\phi}$, partial autocorrelations $\hat{\psi}$, and lag indicators $\hat{\kappa}$. These capture the highly seasonal nature of the hospitality industry, with positive autocorrelations at 12- and 24-month lags, both of which have a posterior probability of being nonzero of 1 (up to three decimal

[‡] The data were originally collected by the Australian Bureau of Statistics and were provided to us by R. Hyndman. The data can currently be found at the Time Series Library at the following URL: `www.maths.monash.edu.au/~hyndman/`.

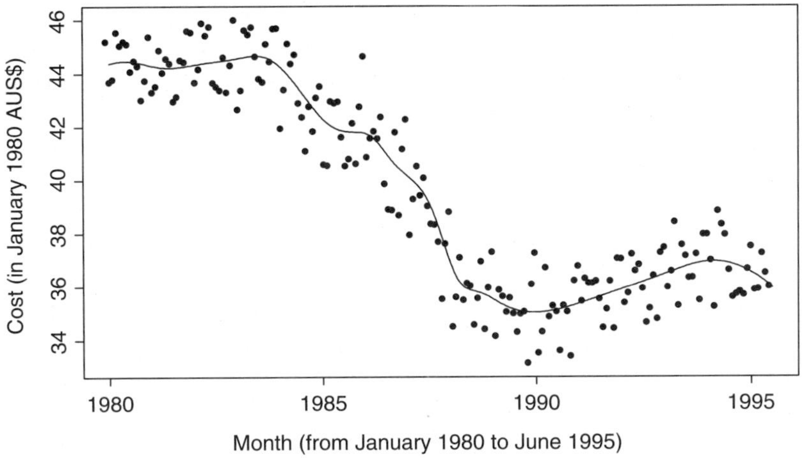

FIGURE 6.2 Observations and estimated trend for the cost of a Victorian hotel room.

TABLE 6.1 Parameter Estimates for Victorian Hotel Room Data

Lag i	$\hat{\phi}_i$	$\hat{\psi}_i$	$\hat{\kappa}_i$	Lag i	$\hat{\phi}_i$	$\hat{\psi}_i$	$\hat{\kappa}_i$
1	0.044	**0.145**	**0.520**	13	−0.007	−0.001	0.040
2	0.023	**−0.004**	**0.101**	14	0.016	0.000	0.036
3	0.001	0.003	0.080	15	0.003	0.001	0.037
4	0.004	0.001	0.052	16	−0.001	−0.001	0.039
5	−0.002	0.000	0.041	17	0.002	0.001	0.036
6	−0.002	0.013	0.071	18	0.003	0.003	0.042
7	−0.002	0.000	0.034	19	0.003	0.002	0.045
8	0.000	0.000	0.039	20	−0.003	−0.001	0.042
9	−0.004	0.001	0.044	21	0.001	0.002	0.040
10	−0.030	0.008	0.085	22	−0.013	0.000	0.033
11	0.012	**0.127**	**0.501**	23	−0.023	0.002	0.037
12	0.344	**0.813**	**1.000**	**24**	0.576	**0.576**	**1.000**

places). The figures in bold in Table 6.1 are for those partial autocorrelations that have a nonnegligible probability of being nonzero. Notice that the representation is parsimonious, with most partial autocorrelations having a high posterior probability of being exactly equal to zero.

6.5.2 Electrical Load Data

Another common use of regression smoothing when the data are dependent is to remove diurnal or other periodic effects from data where shorter term autocorrelation also exists. Examples arise naturally in environmental data and operational data from public utilities. To illustrate, we consider estimating the daily and weekly periodic components of the New South Wales electricity system load from a three-week window of data arising during the period from May 31, 1993, to June 20, 1993. Such load data have been studied by many authors, including Harvey and Koopman (1993) and Hsu and Yang (1995), and we refer the reader to these works for comprehensive studies.

Half-hourly observations (so that $n = 1008$) were obtained on the following variables:

L: Total system load (megawatts) in New South Wales.

TOD $\in [0, 1)$: The time of the day of the observations ($0 =$ midnight).

TOW $\in [0, 1)$: The time of the week of the observations ($0 =$ midnight on Monday morning).

Figure 6.3(a) plots the three weeks of load data, L, against the time of the week, TOW; strong daily and weekly periodic effects are apparent. The regression model used is

$$\mathsf{L}_t = \beta_1 + \beta_2 t + f_1(\mathsf{TOD}_t) + f_2(\mathsf{TOW}_t) + e_t,$$

where the errors are modeled with an autoregression of maximal order $p = 12$. The signal is modeled as an additive combination of two periodic quadratic reproducing kernel bases (Luo and Wahba, 1997) with knots placed at all possible observations, so that

$$f_1(\mathsf{TOD}) = \sum_{i=1}^{48} \beta_i^1 R(\mathsf{TOD}, i/48),$$

$$f_2(\mathsf{TOW}) = \sum_{i=1}^{336} \beta_i^2 R(\mathsf{TOD}, i/336),$$

where $R(x, k) = ((|x - k| - \frac{1}{2})^2 - \frac{1}{12})$. In the notation of Equation (6.10), the ith basis function is $b_i(x) = R(x, k_i)$, where k_i is the knot. Notice that, because the basis terms are periodic on [0,1), so are the functions f_1 and f_2. Estimating such additive models is a trivial extension of the univariate regression model, where the design matrix X in Section 6.4 is simply constructed from the basis terms of both f_1 and f_2.

Figures 6.3(b) and 6.4 plot the estimates $\hat{f}_2$ and $\hat{f}_1$, respectively, while Figure 6.3(c) plots $\hat{f}_1 + \hat{f}_2$ against TOW. These two functions explain a large amount of the variation in the total system load. Table 6.2 outlines the estimates of the AR coefficients $\hat{\phi}$, partial autocorrelations $\hat{\psi}$, and lag indicators $\hat{\kappa}$. These estimates indicate that a strong AR process occurs in the data, even when daily and weekly periodicity is controlled for directly by the periodic functions f_1 and f_2. The figures in bold in Table 6.2 are for partial autocorrelations that have a nonnegligible probability of being nonzero. All the partial autocorrelations of order six and above appear highly likely to be exactly zero, so that an AR(5)

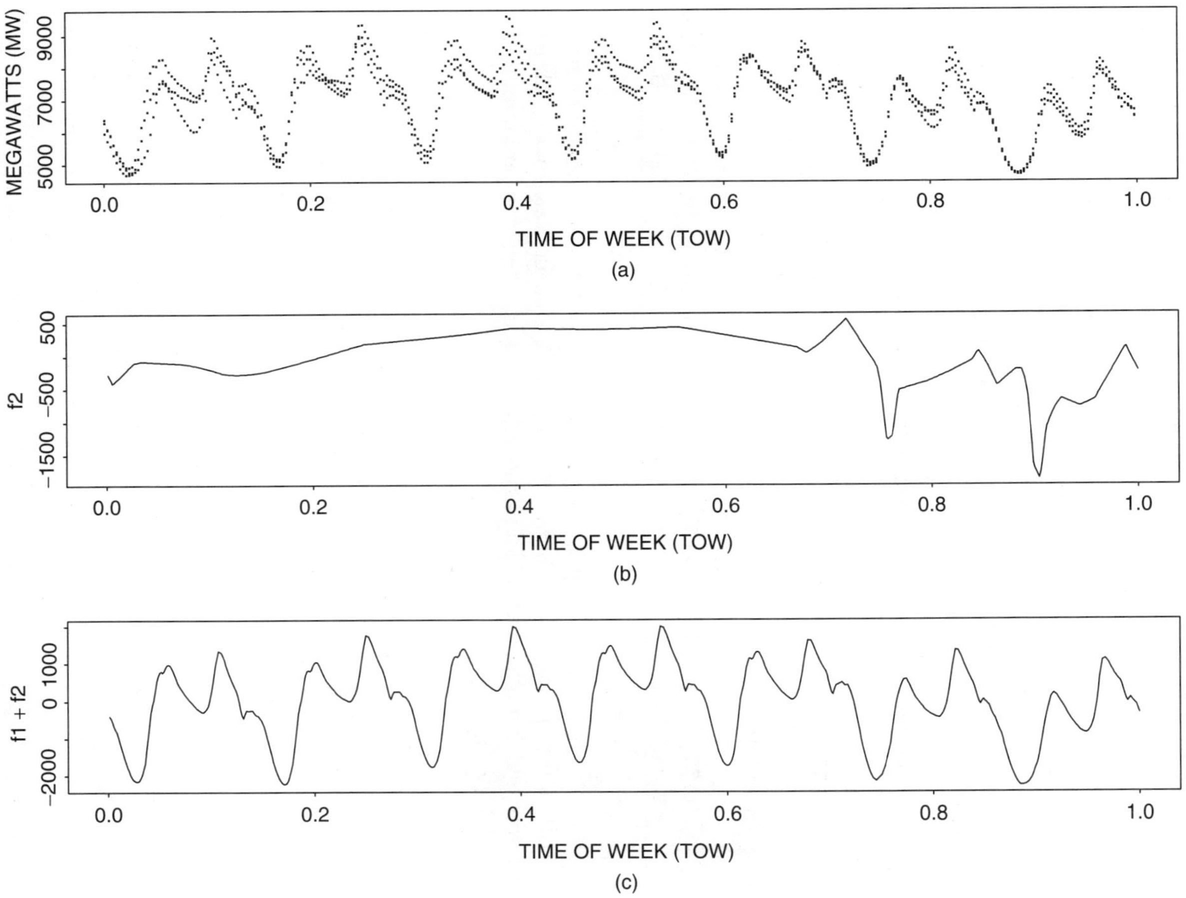

FIGURE 6.3 (a) New South Wales system load data; (b) function estimate of weekly periodic component; (c) function estimate of periodic weekly and periodic daily component.

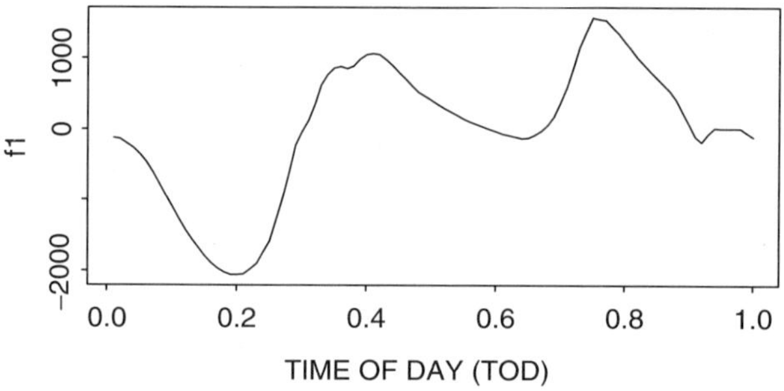

FIGURE 6.4 Function estimate of periodic daily component.

TABLE 6.2 Parameter Estimates for Electrical Load Data

Lag i	$\hat{\phi}_i$	$\hat{\psi}_i$	$\hat{\kappa}_i$	Lag i	$\hat{\phi}_i$	$\hat{\psi}_i$	$\hat{\kappa}_i$
1	1.536	**0.988**	**1.000**	7	0.005	0.000	0.006
2	−0.553	**−0.555**	**1.000**	8	−0.002	0.000	0.004
3	0.087	−0.002	0.041	9	0.011	0.000	0.002
4	−0.244	0.000	0.015	10	−0.031	0.000	0.001
5	0.158	**0.158**	**0.963**	11	0.020	0.020	0.173
6	−0.003	0.000	0.004	12	0.000	0.000	0.001

model for the errors appears appropriate. Such short-term autocorrelation (two and a half hours) is induced by a variety of weather-related variables, such as temperature and sunlight, and social phenomena, such as television programming and special events.

Acknowledgements

M. G. Schimek expresses his thanks to the Australian Graduate School of Management for financial support of, and hospitality during, his research visit in September 1997, while working on this chapter.

REFERENCES

Altman, N. S. (1990). Kernel smoothing of data with correlated errors, *J. Amer. Statist. Assoc.*, **85**, 749–759.

Chatfield, C. (1996). *The Analysis of Time Series: An Introduction*, 5th ed. Chapman and Hall, London, UK.

Chiu, S.-T. (1989). Bandwidth selection for kernel estimate with correlated noise, *Statist. Probab. Lett.*, **8**, 347–354.

Dawid, A. P. (1984). Statistical theory: The prequential approach (with discussion), *J. Roy. Statist. Soc. A*, **147**, 278–292.

De Boor, C. (1978). *A Practical Guide to Splines*, Springer-Verlag, New York.

Diggle, P. J. (1985). Discussion of "Some aspects of the spline smoothing approach to non-parametric regression curve fitting" by B. W. Silverman, *J. Roy. Statist. Soc., B*, **47**, 28–29.

Diggle, P. J., and Hutchinson, M. F. (1989). On spline smoothing with autocorrelated errors, *Austral. J. Statist.*, **31**, 166–182.

Friedman J. H., and Silverman, B. W. (1989). Flexible parsimonious smoothing and additive modeling, *Technometrics*, **31**, 3–39.

Gasser, T., Kneip, A., and Köhler (1991). A flexible and fast method for automatic smoothing, *J. Amer. Statist. Assoc.*, **86**, 643–652.

Gasser, T., and Müller, H. G. (1979). Kernel estimation of regression functions. In Gasser, T., and Rosenblatt, M. (eds.), *Smoothing Techniques for Curve Estimation*, Springer-Verlag, New York, 23–68.

Geisser, S., and Eddy, W. F. (1979). A predictive approach to model selection, *J. Amer. Statist. Assoc.*, **74**, 153–160.

Hart, J. D. (1991). Kernel regression with time series errors, *J. Roy. Statist. Soc., B*, **53**, 173–187.

Hart, J. D. (1994). Automated kernel smoothing of dependent data using time series cross-validation, *J. Roy. Statist. Soc., B*, **56**, 529–542.

Hart, J. D. (1996). Some automated methods of smoothing time-dependent data, *Nonparametric Statistics*, **6**, 115–142.

Harvey, A., and Koopman, S. (1993). Forecasting hourly electricity demand using time-varying splines, *J. Amer. Statist. Assoc.*, **88**, 1228–1253.

Herrmann, E., Gasser, T., and Kneip, A. (1992). Choice of bandwidth for kernel regression when residuals are correlated, *Biometrika*, **79**, 783–795.

Hsu, Y., and Yang, C. (1995). Electrical load forecasting, in Murray, A. (ed.), *Applications of Neural Networks*, pp. 157–189, Kluwer Academic Publishers, Boston, MA.

Kohn, R., and Ansley, C. F. (1983). On the smoothness properties of the best linear unbiased estimate of a stochastic process observed with noise, *Ann. Statist.*, **11**, 1011–1017.

Kohn, R., Ansley, C. F., and Wong, C. M. (1992). Nonparametric spline regression with autoregressive moving average errors, *Biometrika*, **79**, 335–346.

Luo, Z., and Wahba, G. (1997). Hybrid adaptive splines, *J. Amer. Statist. Assoc.*, **92**, 107–116.

Mitchell, T., and Beauchamp, J. (1988). Bayesian variable selection in linear regression, *J. Amer. Statist. Assoc.*, **83**, 1023–1036.

Monahan, J. (1984). Full Bayesian analysis of ARMA time series models, *J. Econometrics*, **21**, 307–331.

O'Hagan, A. (1995). Fractional Bayes factors for model comparison (with discussion), *J. Roy. Statist. Soc., B*, **57**, 99–138.

Müller, H. G. (1991). Smooth optimum kernel estimators near endpoints, *Biometrika*, **78**, 521–530.

Powell, M. J. D. (1987). Radial basis functions for multivariate interpolation: A review, in Mason, J., and Cox, M. (eds.), *Algorithms for Approximation*, pp. 143–167, Oxford University Press, Oxford, UK.

Priestley, M. B., and Chao, M. T. (1972). Non-parametric function fitting, *J. R. Statist. Soc., B.*, **34**, 385–392.

Quintela del Rio, A. Q. (1996). Comparison of bandwidth selectors in nonparametric regression under dependence, *Comput. Statist. Data Anal.*, **21**, 563–580.

Schimek, M. G. (1988). A roughness penalty regression approach for statistical graphics, in Edwards, D. and Raun, N.E. (eds.) COMPSTAT 1988. Proceedings in Computational Statistics. Physica-Verlag, Heidelberg, Germany, pp. 37–43.

Schimek, M. G. (1992). Serial correlation in spline smoothing: A simulation study. *Computational Statistics*, **7**, 309–327.

Schimek, M. G. (1994). Single case studies: The time series vs. the smoothing regression approach, in Fischer, G., and Laming, D. (eds.), *Contributions to Mathematical Psychology, Psychometrics and Methodology*, pp. 345–357, Springer–Verlag, New York.

Schimek, M. G., Millner, M. M., and Kubik, W. (1991). Non-parametric estimation of evoked potential signals: Methods and problems, in Dvorak, I., and Holden, A. V. (eds.), *Mathematical Approaches to Brain Functioning Diagnostics*, pp. 195–208, Manchester University Press, Manchester, UK.

Schimek, M. G., and Schmaranz, K. G. (1994). Dependent error regression smoothing: A new method and PC program, *Computat. Statist. Data Anal.*, **17**, 457–464.

Schwarz, G. (1978). Estimating the dimension of a model, *Ann. Statist.* **6**, 461–464.

Silverman, B. W. (1985). Some aspects of the spline smoothing approach to nonparametric regression curve fitting (with discussion), *J. Roy. Statist. Soc., B*, **47**, 1–52.

Smith, M., and Kohn, R. (1996). Nonparametric regression using Bayesian variable selection, *J. Econometrics*, **75**, 317–344.

Smith, M., and Kohn, R. (1998). Nonparametric estimation of irregular functions with independent or autocorrelated errors, in Dey, D., Muller, P., and Sinha, D. (eds.), *Practical Nonparametric and Semiparametric Bayesian Statistics*, pp. 157–179, Springer-Verlag, New York.

Smith, M., Wong, C., and Kohn, R. (1998). Additive nonparametric regression with autocorrelated errors, *J. Roy. Statist. Soc., B*, 311–331.

van der Linde, A. (1994). On cross-validation for smoothing splines in the case of dependent observations, *Austral. J. Statist.*, **36**, 67–73.

Wahba, G. (1978). Improper priors, spline smoothing and the problem of guarding against errors in regression, *J. Roy. Statist. Soc., B*, **40**, 364–372.

Wahba, G. (1990). *Spline Models for Observational Data*, SIAM, Philadelphia, PA.

Wang, Y. (1998). Smoothing spline models with correlated random errors, *J. Amer. Statist. Assoc.*, **93**, 341–348.

Wong, F., Hansen, M., Kohn, R., and Smith, M. (1997). *Focused Sampling and Its Application to Nonparametric Regression*, preprint.

Wu, J. S., and Chu, C. K. (1994). Nonparametric estimation of a regression function with dependent observations, *Stoch. Proc. Appl.*, **50**, 149–160.

CHAPTER 7

Wavelets for Regression and Other Statistical Problems

Guy P. Nason and Bernard W. Silverman
Department of Mathematics, University of Bristol, Bristol, UK

7.1 INTRODUCTION

This chapter provides an introductory review, largely based on the work of Nason and Silverman (1994), of the use of the *discrete wavelet transform* (DWT) for statistical purposes and a guide to a publicly available package of routines (see Nason, 1993) for the statistical language S. The package is called `wavethresh` and is available from the StatLib archive. This chapter also includes discussion of two subsequent developments of wavelet regression methodology using material drawn from Nason (1996) and Johnstone and Silverman (1997).

A gentle introduction to wavelet methods has been provided by Strang (1993). For a more detailed discussion, the reader is referred, for example, to Chui (1992) and Daubechies (1992). The statistical aspects of the `wavethresh` package are mainly due to Donoho and Johnstone (1994). In this chapter, we concentrate on the discrete wavelet transform. This transform is based on filtering ideas that have been discussed extensively in the engineering literature. Vaidyanathan (1990) and Vetterli and Herley (1992) have provided detailed surveys and numerous references on the DWT. Also, some other specific references are mentioned in Section 7.6.

We do not claim that wavelets are useful in all statistical curve and surface estimation problems. The general aim of this review is to widen interest in and access to wavelet methods so that they can be tried and tested in practice and that a mature view of their usefulness and potential may thereby be obtained.

Smoothing and Regression: Approaches, Computation, and Application
Edited by Michael G. Schimek
ISBN 0-471-17946-9

Standard linear regression techniques formulate a model of the response in terms of some explanatory variables. If a polynomial regression is appropriate, then orthogonal polynomials may be useful. Although the resultant variables may be more complicated, the regression coefficients of the polynomial variables are independent. This independence is desirable, especially when the modeling is of a changing process when modeling terms have to be added or dropped. Alternatively, linear regression may be extended by using nonparametric regression methods, such as spline smoothing. Most of the aforementioned methods are linear in the response, but for wavelet regression with thresholding, this is not the case. Wavelet regression is potentially highly locally adaptive; it retains the independence of regression coefficients, like orthogonal polynomials and can, under the right conditions, provide an insightful interpretation of the data.

This chapter is set out as follows: A brief discussion of the wavelet transform is given in Section 7.2. The three sections thereafter give a tutorial guide to the `wavethresh` software. In Section 7.3, the one-dimensional transform and its inverse are discussed, and an explanation is given of the way in which the wavelet decomposition is held as an object in S. Section 7.4 covers various thresholding procedures for smoothing and for data compression. Section 7.6 gives precise details on the implementation of the software. Section 7.7 gives details on how to retrieve and install the `wavethresh` software.

7.2 WAVELET EXPANSIONS

7.2.1 Fourier and Wavelet Series

We first set the scene by remembering Fourier series. Given a function f defined on $[-\pi, \pi]$, we can represent f in terms of the Fourier basis $\{\exp(inx)\}_{n=-\infty}^{\infty}$ as

$$f(x) = \sum_{-\infty}^{\infty} c_n \exp(inx), \tag{7.1}$$

where the *Fourier coefficients* are computed by

$$c_m = (2\pi)^{-1} \int_{-\pi}^{\pi} f(x) \exp(-imx)\, dx. \tag{7.2}$$

Since $\exp(inx) = \cos(nx) + i \sin(nx)$, the Fourier series in Equation (7.1) can be regarded as an expansion of f in terms of sine and cosine functions. We regard the series expansion as a transform, taking a function f into a set of coefficients c_n.

The wavelet transform is in some, but not all, ways similar to the Fourier transform. Given a function f, we wish to expand that function in terms of some orthonormal basis functions ψ_ν. For example, the Fourier series expansion uses the orthonormal system $(2\pi)^{-1/2} \exp(i\nu x)$ (on $[-\pi, \pi]$). Wavelet expansions are orthogonal series expansions wherein the basis functions are constructed in an intriguing way. Here, we write our general wavelet basis element as ψ_{jk}; notice that there are two subscripts, not just one. This is because the wavelet basis functions are all dilations and translates of a single function called the *mother wavelet* ψ. The *wavelet* ψ_{jk} is obtained from the mother wavelet by shrinking by a factor of 2^j and translating by $2^{-j}k$, to obtain

$$\psi_{jk}(x) = 2^{j/2}\psi(2^j x - k). \tag{7.3}$$

Here, the j subscript represents the dilation number, or resolution level, and the k subscript represents the translation number. The scale factor $2^{j/2}$ normalizes ψ_{jk} so that $||\psi_{jk}|| = ||\psi||$.

For certain choices of ψ, the set of functions ψ_{jk} forms an orthonormal basis for all functions in $L^2(\mathrm{I\!R})$, and therefore we shall the wavelets $\{\psi_{jk}\}$ to approximate functions.

The continuous wavelet series representation is

$$f(x) = \sum_{jk} f_{jk}\psi_{jk}(x), \tag{7.4}$$

where the *wavelet coefficients* are found in the usual way, by

$$\begin{aligned} f_{jk} &= \int_{-\infty}^{\infty} f(x)\psi_{jk}(x)\,dx \\ &= \langle f, \psi_{jk}\rangle, \end{aligned} \tag{7.5}$$

where $\langle\cdot,\cdot\rangle$ denotes the inner product. Clearly, given a function, we wish to compute its wavelet coefficients. The software described later in this chapter computes wavelet coefficients, but uses a discretized version of Equation (7.5).

7.2.2 What Can Wavelets Offer?

The question of what wavelets can offer is discussed in detail by Strang (1993), so we say relatively little about this topic here. In the Fourier transform of a function f on $(-\infty, \infty)$,

$$\widehat{f}(\omega) = \int_{-\infty}^{\infty} f(t)\exp(-i\omega t)\,dt,$$

wherein we usually identify t with time and ω with frequency. To obtain information about a particular frequency, we have to integrate over the whole domain of f from $t = -\infty$ to $t = \infty$, even though we might want to know the frequency behavior of f only over a particular time period. The wavelet transform is one way by which such local frequency information can be obtained. Wavelets provide *time–frequency localization* in that the coefficient f_{jk} gives information about the function f near the time point $2^{-j}k$ and near a frequency proportional to 2^j. Daubechies (1992) has provided more details on and an example of this method.

A simple example of a wavelet basis is the Haar basis, generated from the mother wavelet

$$\psi(x) = \begin{cases} -1 & 0 \le x \le \frac{1}{2}, \\ 1 & \frac{1}{2} \le x \le 1, \\ 0 & \text{otherwise.} \end{cases} \tag{7.6}$$

It is easy to convince oneself that wavelets derived from the Haar mother wavelet form an orthonormal system by observing that

1. wavelets with different translate numbers, but on the same scale, do not have intersecting supports;
2. wavelets on different scales either have nonintersecting supports or, if they do not, one wavelet takes the value $-k$ and then k over a set wherein the other wavelet is constant (for some k); and
3. $\langle \psi_{jk}, \psi_{jk} \rangle = 1$ for all j, k.

The major impetus behind the theory of wavelets was the realization that there are many other families of wavelets, arising from mother functions more regular than the Haar function in Equation (7.6). Some of the useful properties of wavelets are exhibited by the Haar wavelet basis, but in many contexts, more regular wavelets are useful.

Release 2.2 of the `wavethresh` software, the latest release at the time of this writing, includes two families of wavelets. Both of these families are attributable to Daubechies (1988), and we refer to them as the "extremal phase" or "least asymmetric" wavelets. The wavelets in each of the families have compact support. As in Daubechies (1988), we index each mother wavelet in a family by N. For each family, the regularity of the mother wavelet (and hence the regularity of all the derived wavelets) is proportional to N. If you look forward to Figure 7.4 you can see a picture of Daubechies's "extremal phase" wavelet for $N = 2$. Figure 7.4 shows an irregular wavelet, but as N increases the wavelet becomes smoother. In the `wavethresh` software, the number N is identified with the `filter.number` argument.

Wavelets usually possess another interesting property, that of "vanishing moments." We say that a wavelet has L vanishing moments if, for each $l = 0, \ldots, L-1$, a wavelet ψ satisfies

$$\int x^l \psi(x)\, dx = 0.$$

This property has important implications for applications. Daubechies (1992) provides a good example which shows that this property ensures that fine-scale wavelet coefficients are large only where a function or its derivatives have singularities. This property promotes the use of wavelets for compression techniques, because the wavelets provide *sparse* representations of functions.

Another often-quoted benefit of wavelet methods is that wavelet bases are capable of representing various classes of functions more efficiently than, say, Fourier bases. (See Donoho and Johnstone (1994), for example.) For example, take functions that are only piecewise continuous; that is they contain some discontinuities. You will need many Fourier basis functions to represent the discontinuities accurately, and the effect of the basis functions will be global. Wavelets, on the other hand, are able to represent the discontinuities more efficiently and, at the same time, are local and do not affect the representation elsewhere.

The discrete wavelet transform is fast, in principle, faster even than the fast Fourier transform. However, both the standard FFT and the discrete wavelet transform operate only on data sets that contain 2^M observations (for some M). There are some methods of overcoming this difficulty, but these methods are not implemented in the current release of `wavethresh`.

7.3 THE DISCRETE WAVELET TRANSFORM IN S

This section assumes that the discrete wavelet transform software is already installed. If you want instructions for retrieving, unpacking, and setting up the software, see Appendix A at the end of this chapter. The transform software is an implementation of the method described by Mallat (1989b), but uses the compactly supported orthonormal wavelets as described by Daubechies (1988, 1992). We discuss our implementation in greater detail in Section 7.6.

7.3.1 Accessing the Software

We assume that the `wavethresh` software is installed somewhere on your system (UNIX or Windows). For the purposes of our discussions we need to invent a directory name so that we can refer to the location of the software. On your system, the directory where the software is stored will most likely be different than the one we use here. The directory that we have invented is `/stats/WAVELETS/DISTRIB2.2`. As such, accessing the `wavethresh` software is made possible after the following function has been executed in S:

```
> attach("/stats/WAVELETS/DISTRIB2.2/.Data")
```

Expert users may wish to use the `search()` or `objects()` functions to verify that the `wavethresh` directory has been attached. Nonexpert users can simply use the `wvrelease()` function to verify this:

```
> wvrelease()
S wavelet software, release 2.2, installed
Copyright Guy Nason 1993
```

If you do not get this message (or something like it), then the wavethresh software is not present. If you want to use the software often, then you may want to attach the software every time you start up S. This is accomplished by means of the .First() function. The following lines form a plausible .First() function:

```
> .First <- function(){
 attach("/stats/WAVELETS/DISTRIB2.2/.Data")
 wvrelease()
 }
```

7.3.2 Getting Help with the Software

You can get help on any of the wavelet functions or objects by using the S help facility. For example, to obtain help on the wd wavelet decomposition function, simply type

```
> help(wd)
```

You may also be able to use the S-Plus help.start() function, which starts an interactive help viewer. See your local documentation to see if this function is available. Typing "wavelet" as a help search topic lists all of the help items available for the wavelet software.

7.3.3 Preparing an Example

We assume at this point that you have started S and have attached the wavethresh software. You will probably want to try out the wavelet techniques on your own data. However, for the purposes of this tutorial, we create some simulated data to use as an example. We create a sampled version of the contrived function

$$y(x) = \begin{cases} 4x^2(3-4x) & \text{for } x \in \left[0, \frac{1}{2}\right] \\ \frac{4}{3}x(4x^2-10x+7)-\frac{3}{2} & \text{for } x \in \left[\frac{1}{2}, \frac{3}{4}\right] \\ \frac{16}{3}x(x-1)^2 & \text{for } x \in \left[\frac{3}{4}, 1\right] \end{cases}$$

and sample it 512 times in the interval [0, 1]. Figure 7.1 shows y plotted against x.

The wavethresh software contains a function called example.1() that computes x and y automatically for you:

```
> x <- example.1()
> y <- x$y
> x <- x$x
```

This sampled function can be viewed using the S plot() function. (First start up your favorite graphics device, such as X11(), openlook(), or motif()).

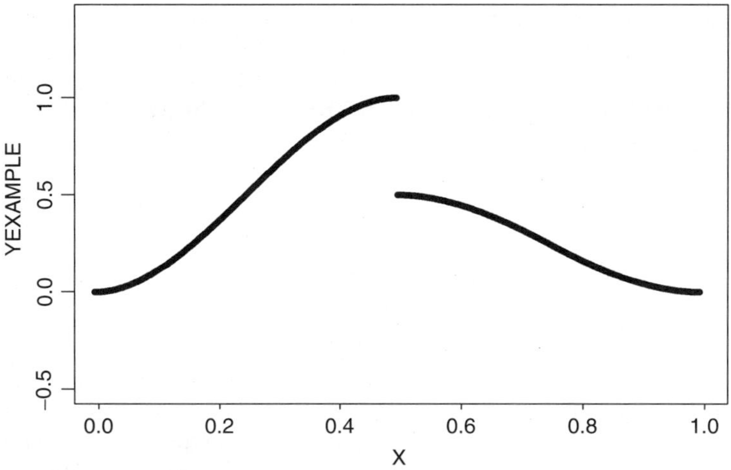

FIGURE 7.1 A plot of the function y versus x.

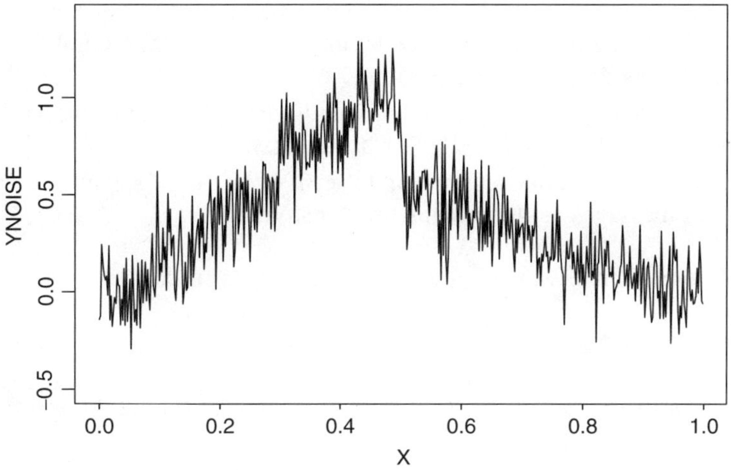

FIGURE 7.2 The sampled function with Gaussian white noise.

We now add some Gaussian white noise to y to create the vector `ynoise` by

```
> ynoise <- y + rnorm(512, s=0.15)
> plot(x, ynoise, type="l")
```

This function is plotted in Figure 7.2.

7.3.4 Applying the Discrete Wavelet Transform

The function for a discrete wavelet transform is `wd`, which stands for *wavelet decomposition*. This function can take several arguments; a complete explanation is given in the `wd` help page. The first argument, `data`, is the vector to which you wish to apply the transform. The length of this vector, N, must be a power of 2 (i.e., let $N = 2^M$), which is why we previously specified 512 as the length of y. The remaining arguments specify the type of wavelet that is used, the regularity or smoothness of the wavelet, and the method of handling the transform at the boundaries. For Release 2.2 of `wavethresh`, the arguments are defined as follows:

`data` is the vector that you wish to transform.

`filter.number` determines the regularity of the wavelet. The wavelets get smoother as `filter.number` increases. (See Daubechies, 1992.)

`family` determines which family of wavelets is used. At present, there are two families. They are both compactly supported orthonormal wavelets and are defined in Daubechies (1992). The family names are `DaubExPhase` and `DaubLeAsymm`, and, contrary to their names, the second of these families is less asymmetric than the first! Ten members of the `DaubExPhase` family are available (indexed by `filter.numbers` from 1 to 10), as are seven members of the `DaubLeAsymm` family (indexed by `filter.number` from 4 to 10). Note that the first wavelet in the `DaubExPhase` family is the Haar wavelet, as described in Equation (7.6).

`bc` represents the boundary handling conditions. This argument can be set to `periodic` (the default value) or `symmetric`.

`verbose` holds a value of true or false; if it is true, then the `wd` functions prints messages as it performs the transform.

We apply the discrete wavelet transform to the `ynoise` data by

```
> ywd <- wd(ynoise)
```

The `wd()` function applies `filter.number=2` for the `DaubExPhase` family by default.

7.3.5 Wavelet Decomposition Objects

The S object `ywd` is an example of a `wd.object`, and it has class `wd`. The `wd` class objects are lists, and you can find out what they contain by using the `names()` function or by looking at the `wd.object` help page.

There are various methods for this class of object. Probably the simplest is the `summary() method`. For example,

```
> summary(ywd)
Levels:  9
Length of original:  512
Filter was:  Daub cmpct on ext. phase N=2
Boundary handling:  periodic
```

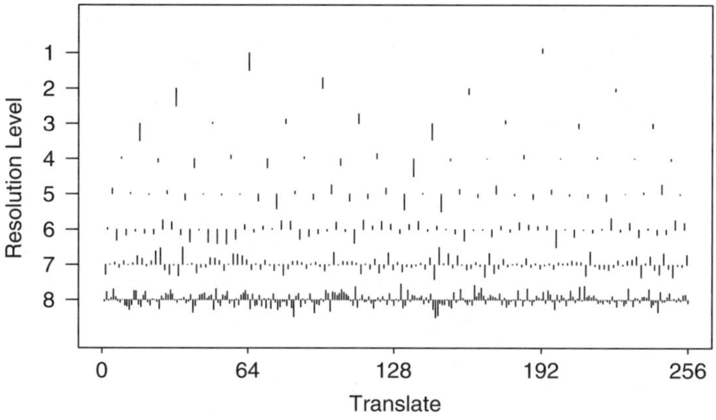

FIGURE 7.3 Wavelet coefficients for the `ynoise` data.

tells us that the length of the original data was 512 samples, the transform was performed using $N = 2$ Daubechies compactly supported wavelets from the `DaubExPhase` family, the decomposition consists of 9 levels, and the boundary handling was periodic. The number of levels refers to the number of levels of smoothed data and wavelet coefficients (and therefore, $2^{\texttt{levels}}$ = the length of the data).

Another method is `plot()`, which plots the wavelet coefficients in the manner proposed by of Donoho and Johnstone (1994). This method is applied by

```
> plot(ywd)
```

and results in the plot shown in Figure 7.3. Roughly speaking, the horizontal axis of Figure 7.3 corresponds to spatial position along the x-axis of Figure 7.1. It is possible to change the x-axis of Figure 7.3 so that it depicts the actual values of x rather than the translates. The vertical axis of Figure 7.3 represents a frequencylike quantity, the coefficients at the bottom are "high-frequency" information, and the "frequency" decreases as you move up the axis. So, the coefficients on the left-hand side of Figure 7.3 correspond to the left-hand side of the function plotted in Figure 7.2, and the right-hand sides in both figures correspond. The reason that the x-axis in Figure 7.3 is labelled from 0 to 256 is that the number of wavelet coefficients at the highest resolution level (at the bottom of Figure 7.3) is exactly half the number of original data points, and the number of coefficients decreases by half at each level. At the highest level, the coefficient with translation number k is plotted at position $k + \frac{1}{2}$.

7.3.6 Exact Reconstruction

From `ywd`, we can exactly reconstruct the original sampled function. In this section, we do this and compare our result to the original function. The function that performs the *inverse discrete wavelet transform* (IDWT) is `wr()`, which stands for *wavelet reconstruction*. So, to reconstruct, we make the following command:

```
> ywr <- wr(ywd)
```

Note that it is not necessary to specify the type of wavelet, because `wr()` works it out. The `wr()` function can return a `wd` class object, but by default, it produces a vector; this vector is the reconstructed function at the highest possible resolution level. We can then plot the reconstructed values by

```
> plot(x, ywr, type="l")
```

and we should get exactly the same plot as in Figure 7.2. To check that the reconstruction is exactly the same, up to numerical error, we can subtract the original function from the reconstructed function and look at the error:

```
> max(abs(ywr - ynoise))
[1] 1.109357e-11
```

As you can see the error is small.

7.3.7 What do the Daubechies Wavelets Look Like?

The orthonormal basis functions involved with the Fourier transform are familiar, since they are just sine and cosine functions. For the Daubechies' wavelets however, things are not so simple, and there is no known closed-form formula for the wavelets. The wavelets are the result of a process which says that we want functions that are orthonormal, have good time–frequency localization properties, and relate to each other by dilation and translation. The nearest one can get to a closed-form expression is an infinite-product expression for the Fourier transform of the wavelets. All of this is more fully explained in Daubechies (1992).

In any event, we do not really need to have a closed-form formula, since we have an algorithm that performs the inverse discrete wavelet transform. To get a picture of a wavelet, all we need to do to arrange for a wavelet series to have only one nonzero wavelet coefficient. Then we apply the inverse wavelet transform to this series and plot the reconstruction. The `wavethresh` software has this option built into the `draw` function. If we take our previous `ywd wd.object` and issue the command

```
> draw(ywd, enhance=F)
```

then we should see a picture of Daubechies's $N = 2$ "extremal phase" wavelet for `filter.select=2`, as in Figure 7.4. The Daubechies's wavelets in `wavethresh`

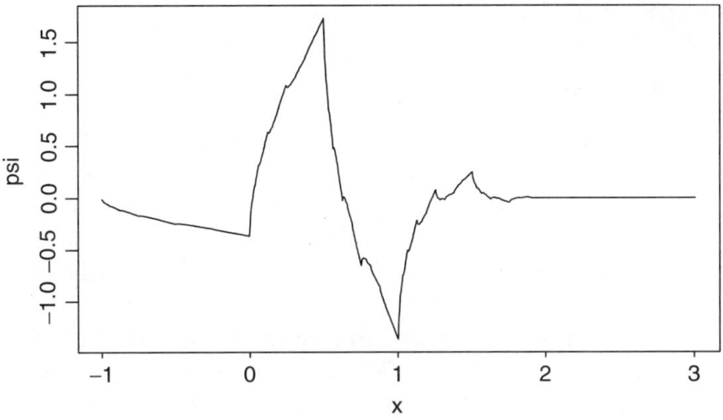

FIGURE 7.4 Daubechies's "extremal phase" wavelet $N = 2$.

are all compactly supported. The support is related to the `filter.number`, or N in Daubechies's notation. The support for the "extremal phase" wavelets is $[N - 1, N]$. So, the support of the wavelet in Figure 7.4 is $[-1, 2]$, but you should notice that the size of the wavelet is not large over the whole of that interval. For example, the significant parts of the Daubechies's $N = 2$ wavelet are contained within the interval [0, 1.2], which we would term the *effective support* of the wavelet. For the smoother wavelets, the effective support is much less than the actual support. The `enhance` option of `draw()` draws the wavelet on the effective support. In the `wavethresh` package, we define the effective support in the following way: Suppose that $\psi(x)$ is the wavelet under consideration. Define z_0 to be the maximum absolute value of $\psi(x)$ multiplied by the `draw()` argument `efactor`. Then define the set

$$A_{\texttt{efactor}} = \{x : |\psi(x)| > z_0\}.$$

We define the effective support of ψ to be the smallest interval (or square region for two dimensions) containing $A_{\texttt{efactor}}$. The argument `efactor` is under user control, but the default value of 0.05 seems to work well.

For an example, try the following commands, which show the difference between enhanced and nonenhanced pictures:

```
> draw.default(filter.number=10, enhance=F)
> draw.default(filter.number=10)
```

You may also have guessed that the `draw` function is generic and that methods exist for one- and two-dimensional wavelets. We will look at some two-dimensional pictures of the wavelet later in this chapter.

7.4 WAVELET SHRINKAGE

7.4.1 Thresholding Methods

We now move on to the statistical techniques of wavelet regression and smoothing. Following the work of Donoho and Johnstone (1994), we implement a thresholding function. The thresholding function is generic and is called `threshold()`. The idea behind thresholding is the removal of small wavelet coefficients, considered to be noise. This leaves the large coefficients in the `wd` object, which can then be used to estimate the signal after reconstruction.

There are many ways to threshold. To threshold using our software, you have to choose a thresholding "policy." These policies specify how the threshold is computed, as discussed later. Once the threshold has been computed, it is applied to the coefficients either as a hard or soft threshold, as specified by the argument `type`. Finally, each of the policies can be applied separately to each level, or a policy can be applied to a group of levels simultaneously; this factor is controlled by the `by.level` and `levels` arguments, respectively.

The policies are as follows:

`universal`: The threshold is computed as

$$\lambda = s\sqrt{2\log M},$$

where M is the number of data points, or equivalently, the number of wavelet coefficients, and s is an estimate of the variation of the coefficients (on the standard deviation scale). The `dev` argument allows the user to replace the default measure of variation (`var`) by his or her own choice—for example, by the mean absolute deviation (`mad`). This type of thresholding was proposed by Donoho and Johnstone (1994).

`manual`: The threshold is supplied by the user.

`probability`: The user supplies a probability `value` p. The threshold is then the pth quantile of the coefficients.

The types of threshold are as follows:

`hard`: The coefficients compared to the threshold(s). If a coefficient is smaller in absolute magnitude than the threshold, it is removed; otherwise, it is left alone ("keep" or "kill").

`soft`: The coefficients are modified by the formula

$$d_{jk}^{\text{new}} = \text{sgn}(d_{jk})(|d_{jk}| - \lambda)_+,$$

where λ is the threshold.

7.4.2 Example of Thresholding

Here, we apply thresholding to the coefficients illustrated in Figure 7.3. We use the universal policy:

```
> threshy <- threshold(ywd)
> plot(threshy)
```

The threshy object is of class wd, and so plot uses the plot.wd function to plot the wavelet coefficients, as in Figure 7.5. If you compare Figure 7.5 to Figure 7.3, you will see that many of the smaller coefficients have disappeared. We can now reconstruct the function corresponding to the thresholded coefficients by

```
> yrecon <- wr(threshy)
> plot(x, yrecon, type="l", xlab = "x", ylab =
                "Reconstructed function")
```

The plot should look like Figure 7.6 which is a slightly "smoother" plot than Figure 7.2 and should look a bit more like Figure 7.1. However, the plot in Figure 7.6 still looks jagged. Figure 7.7 shows what happens if a smoother wavelet is used. We now repeat the decomposition, shrinkage, and reconstruction using the Daubechies wavelet selected with filter.select=4. The command that we issue to S is

```
> plot(x, wr(threshold(wd(ynoise, filter.number=4))),
xlab="x", ylab="Reconstructed function", type="l")
```

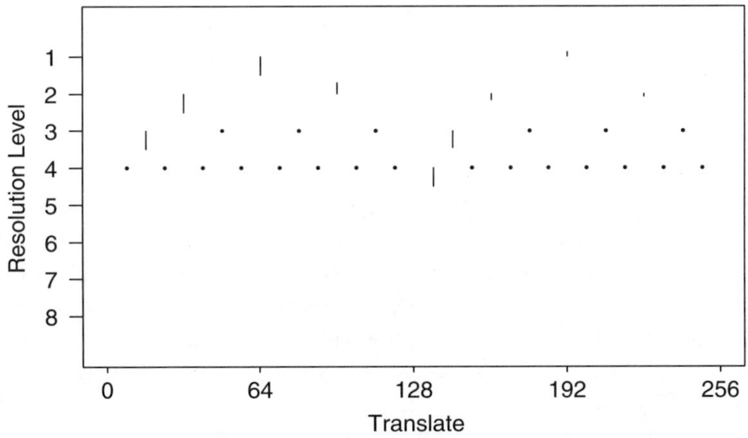

FIGURE 7.5 Wavelet coefficients after universal thresholding.

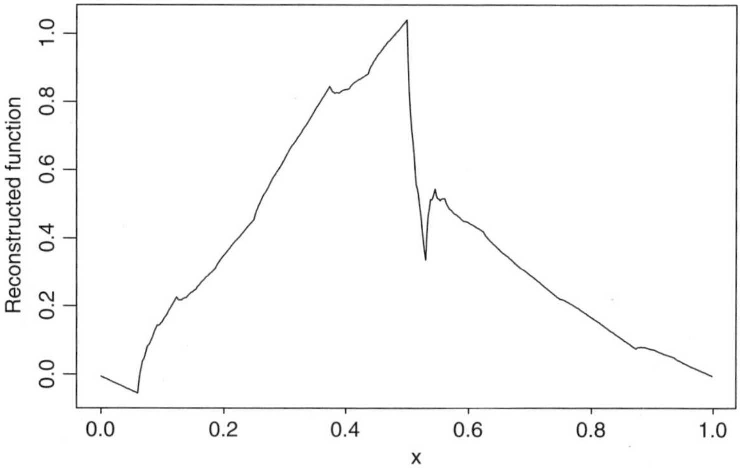

FIGURE 7.6 Reconstructed function after universal thresholding.

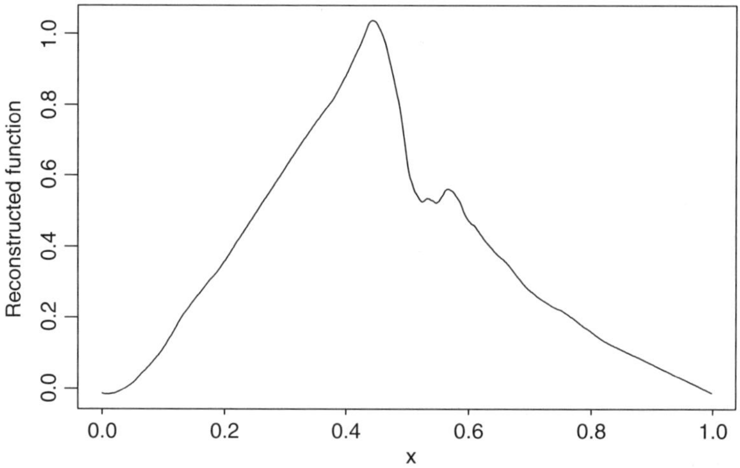

FIGURE 7.7 Reconstruction with a smoother wavelet and universal thresholding.

As you can see in Figure 7.7, the reconstruction looks better. You may want to try the procedure again, but for different wavelets, and see what happens. The default action of `threshold()` is to select the universal threshold of Donoho and Johnstone (1994), but you can also select your own threshold value. To do this, you need to tell the `threshold()` function that you want to set the threshold—do this by setting `policy="manual"` and supplying the value of the threshold by `value=`t, where t is your threshold. It is useful to know what threshold has been used, and setting the `verbose=T` argument causes the thresholding information to be printed.

The next subsection briefly refers to some recently developed methods for choosing thresholds for denoising. These are not available in `wavethresh` version 2.2, but will be present in later versions of the software. Some of the methods are available in the S + WAVELETS toolkit by Bruce and Gao (1996) and also in the WaveLab software for MATLAB developed by the Stanford group (available on the Internet from `stat.stanford.edu`).

7.4.3 SureShrink Methods

SureShrink is a technique that chooses a threshold by minimizing the Stein unbiased estimate of risk for thresholded estimates and was developed by Donoho and Johnstone (1995). Given n noisy wavelet coefficients $x = x_i$, the SURE criterion is given by

$$\text{SURE}(t; x) = n - 2 \cdot \#\{i : |x_i| \leq t\} + \sum_{i=1}^{n} (|x_i| \wedge t)^2,$$

where $x \wedge y$ is equal to the minimum of x and y. For SURE wavelet shrinkage estimation, the threshold that minimizes the SURE criterion is used. It can be found using an $O(n \log n)$ algorithm. The SureShrink method is not only intuitively and computationally appealing, but it also satisfies an important and remarkable adaptivity result found by Donoho and Johnstone (1995) which states that the SureShrink estimator asymptotically achieves the optimal rate of convergence for estimation of function simultaneously over large and diverse classes of functions. The reason that the result is remarkable is that the SureShrink estimator knows nothing beforehand about the underlying function, but attains the optimal rate of convergence nevertheless. Indeed, over certain function classes, SureShrink is still optimal, whereas linear methods cannot obtain that rate. The SureShrink procedure in Donoho and Johnstone (1995) was designed to operate level by level —that is, to assign a threshold to every level of coefficients within a wavelet transform. This technique turns out to be especially useful when dealing with correlated noise (see Section 7.5.)

7.4.4 Cross-Validation

Cross-validation is another well-known statistical technique that has been used to find good thresholds. A twofold cross-validation technique was used by Nason (1996) to choose good thresholds. Many "classical" cross-validation techniques use a *leave-one-out* strategy. That is, a point is left out of a data set, the remaining points are used to form an estimate using a particular threshold, and then the missing point's value is predicted and compared to the left-out point's actual value. Each point is left out in turn, and the prediction errors for each point are averaged. The average prediction error is then numerically minimized over all possible thresholds in order to find the "optimal" threshold. As in SureShrink, cross-validation attempts to minimize the L_2 risk of an estimator $||\hat{f} - f||_2^2$.

Unfortunately, the fast DWT requires a dyadic number of data points to work, and so the leave-one-out technique cannot be implemented.

Nason's twofold cross-validation technique works well for independently distributed data by using a *leave-half-out* method. This procedure systematically leaves out the odd-indexed data points, forms a wavelet shrinkage estimate using the even-indexed points and a given threshold, and computes the prediction error in using the even-indexed estimator to predict the odd points. The roles of the even and odd points are reversed, and an estimate of the L_2 risk is formed from the odd-predict-even and the even-predict-odd prediction errors. This estimate is then numerically minimized over all thresholds to produce the optimal threshold.

One advantage of such a simple cross-validation technique is that it is easy to adapt to different definitions of risk. For example, the prediction error can be measured using L_∞ or L_1 norms, and in some applied problems (e.g., image denoising), these norms can give better results. However, the twofold cross-validation does not fare well when applied to correlated data, even if a level-dependent threshold is used: the SureShrink method is far superior. Also superior to the twofold cross-validation is the method based on generalized cross-validation described by Weyrich and Warhola (1994).

Recently, much attention has been paid to Bayesian methods of coefficient estimation. Some of these methods bestow a prior distribution on each wavelet coefficient and then use the coefficient data to produce an *a posterior* estimate of the wavelet coefficients. For further information, see Abramovich et al. (1998) or Chipman et al. (1997). Another interesting and related approach is taken by Crouse et al. (1998), who use hidden Markov models to estimate coefficients.

7.4.5 An Example from Anaesthesiology

The example we discuss in this subsection arises from data collected by inductance plethysmography and is discussed in more detail in Nason (1996). A plethysmograph is an apparatus for measuring variations in the size of parts of the body. In this experiment, the inductance plethysmograph consists of a coil of wire encapsulated in a belt. A radio-frequency carrier signal is passed through the wire, and size variations change the inductance of the coil which can be detected as a change in voltage. When properly calibrated the output voltage of the inductance plethysmograph is proportional to the change in volume of the part of the body under examination.

It is of both clinical and scientific interest to discover how anaesthetics and analgesics may alter normal breathing patterns postoperatively. Sensors exist that measure blood oxygen saturation, but by the time that they indicate critically low levels of oxygen in the blood, the patient is often apnoeic (i.e., ceases breathing) and in considerable danger. It is possible for a nurse to continually observe a patient, but this is expensive, prone to error, and requires training. In this example, the plethysmograph was arranged around the chest and abdomen of a set of patients and used to measure the flow of air during breathing. The data recordings were made by the Department of Anaesthesia at the Bristol Royal Infirmary after the patients had undergone surgery under general anaesthetic. Figure 7.8(a) shows a section of plethysmograph recording lasting approximately 80 seconds (4096 data points). The two main sets of regular oscillations correspond to normal breathing. The disturbed behaviour in the center of the plot, where the normal breathing pattern disappears, corresponds to the patient vomiting.

Figures 7.8(b), (c), and (d) show the reconstructions via universal, SureShrink, and cross-validation thresholds. At present, signals such as the one in Figure 7.8(a) are classified using neural networks. Wavelet shrinkage can act as a dimension reducer, compressing the signal into few coefficients, so that

- the subsequent neural networks may be smaller and simpler and
- the classification rates are better as the noise is removed.

For the purposes of dimension reduction, it is possible that any of the shrinkage procedures discussed here are adequate. Although the cross-validation method best retains the sharpness of the peaks, it is still noisy in places, whereas the universal procedure effectively suppresses the noise, but has slightly more rounded peaks.

7.4.6 New Algorithms and Other Statistics

Threshold selection is a key component of wavelet shrinkage, but other aspects can be just as important (for example, the concept of primary resolution, which ultimately determines the actual scale that wavelet shrinkage is performed on — see Hall and Nason (1997) and Hall and Patil (1995) for further information). One key component is the type of wavelet transform itself. Little is known about how the choice of mother wavelet influences estimation for a general curve estimation problem, although in some cases it is clear that certain degrees of smoothness are useful — for example, in survival function estimation, where the ratio of estimators is required (see Antoniadis et al., 1999).

Additionally, other algorithms exist that may prove more useful for curve estimation and statistics in a more general fashion. These algorithms include the stationary wavelet transform, which is also known as the nondecimated wavelet transform. It is well known that the DWT is not translation invariant (or equivariant). This means that the above estimators that rely on the DWT are sensitive to choice of origin, and in practice, the success of estimation depends on whether or not the interesting features of a function are located near to coefficients in the DWT that are fixed to a certain dyadic grid. This problem could be overcome by shifting the data; performing wavelet shrinkage, as discussed previously; and unshifting the estimate. If this procedure is performed for all possible shifts, then, up to the resolution of the data, all possible interesting features in the underlying function can be estimated well by at least one of the shifted wavelet transforms. Coifman and Donoho (1995) demonstrated that a computationally efficient algorithm (TI) exists to perform all possible n shifts in only $O(n \log n)$ time, not the $O(n^2)$ time that might be expected from a naive implementation.

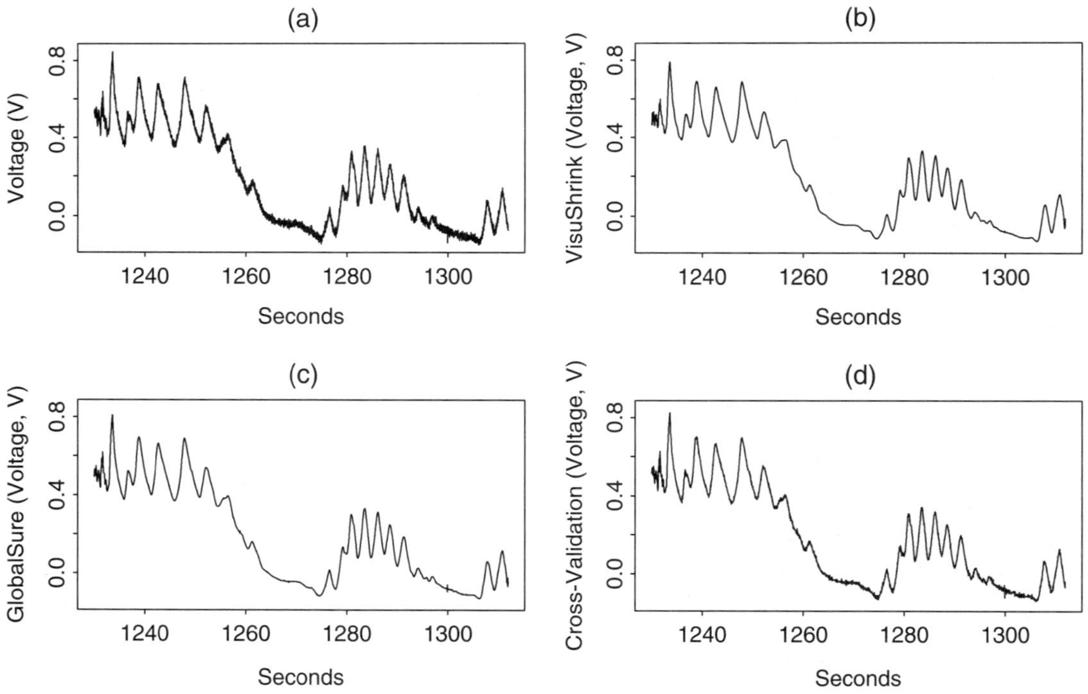

FIGURE 7.8 (a) Section of inductance plethysmograph recording. Reconstructions using Daubechies' least asymmetric wavelet $N = 6$ with thresholds t: (b) Universal: $t = 0.048$; (c) SureShrink: $t = 0.025$; (d) Cross-validation: $t = 0.010$.

Such an algorithm produces a table of coefficients that represents the data with respect to many possible shifted wavelet bases. The table is an overdetermined system, and there is no unique inverse. However, two methods can be used to reconstruct an estimate from thresholded data: basis selection and basis averaging. (For a detailed description see Nason and Silverman, 1995). Coifman and Donoho's translation-invariant shrinkage produces excellent results with basis averaging with, say, SureShrink for the shrinkage part. Nason and Silverman (1995) have discussed the potential of the algorithm both for curve estimation and for obtaining time-scale representations of stochastic time series. (See also von Sachs et al., 1996).

Also of statistical interest is the wavelet packet transform (see, for example, Wickerhauser, 1994), which has many potential uses, such as in dimension reduction and classification Coifman and Saito, 1994; Learned and Willsky, 1995; Saito 1994; and Saito and Coifman, 1996, the stationary wavelet packet transform (Cohen et al., 1997, and Nason et al., 1997), and the multiple wavelet transform (where there is more than one mother and father wavelet; see Downie and Silverman, 1996).

7.5 ESTIMATORS FOR DATA WITH CORRELATED NOISE

7.5.1 Level-Dependent Thresholding

Johnstone and Silverman (1997) have considered the extension of the wavelet thresholding methodology to data with correlated noise. One of the appeals of wavelets in this context is that the wavelet transform often decorrelates the noise process. For example, Figure 7.9 shows the wavelet transform of a process with long-range correlation. Because of the correlation structure, the variance of the wavelet coefficients is no longer the same at each level, but decreases as the resolution level increases. For this particular noise structure, it is also the case that the correlations between coefficients, both within and across levels, are small (see Figure 7.10). So, assuming multivariate normality, the wavelet transform maps a highly dependent series to an array of approximately independent coefficients whose variances depend only on their level.

How should one deal with a wavelet array of coefficients wherein the noise is of this kind? A natural approach, investigated in detail by Johnstone and Silverman (1997), is to use level-dependent thresholding, using for each coefficient a threshold that is proportional to its standard deviation. Since, for a stationary noise process, the variances are constant within levels, these standard deviations can be estimated from data considering the wavelet coefficients level by level rather than all together.

7.5.2 An Example from Neurophysiology

Johnstone and Silverman (1997) have presented an example from neurophysiology, using data supplied by Eisenberg and Levis. The context of the data is the main issue in molecular physiology of the detection and measurement of the

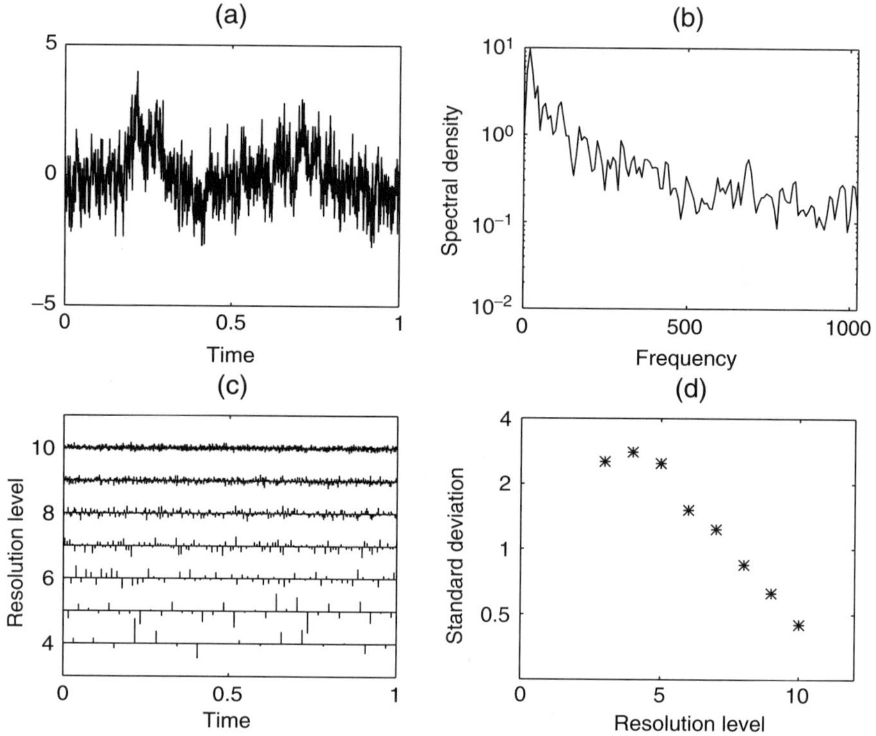

FIGURE 7.9 (a) A generated process exhibiting long-range correlation; (b) estimated spectral density of the data shown in (a); (c) discrete wavelet transform of the series using Daubechies' nearly symmetric wavelet of order 8, with coefficients scaled relative to the largest coefficient at all levels; (d) sample standard deviations (plotted on a logarithmic scale) of the various levels of the discrete wavelet transform shown in (c). Reproduced from Johnstone and Silverman (1997).

picoamp currents that flow through the single-membrane channels that control movement in and out of cells. The data generated by Eisenberg and Levis, part of which is shown in Figure 7.11, is intended to represent most of the relevant challenges in processing such single-channel data. The data consists of a step function randomly switching between values of 0 ("off") and 1 ("on"), measured in the presence of additive, correlated noise.

This generated example differs in kind from the simulated data often considered in statistical studies, in that its underlying model is carefully selected by practitioners directly involved in routine collection and analysis of real data. The obvious advantage of using a generated data set, rather than an actual data set obtained in practice, is that in the case of a real data set, the "truth" is not known, and so it is impossible to quantify the performance of the various methods.

Various approaches were applied to the estimation of the original (0 or 1) signal from these data. The first approach was to classify the original data pointwise, regardless of the time structure. Three wavelet-based approaches were used, all

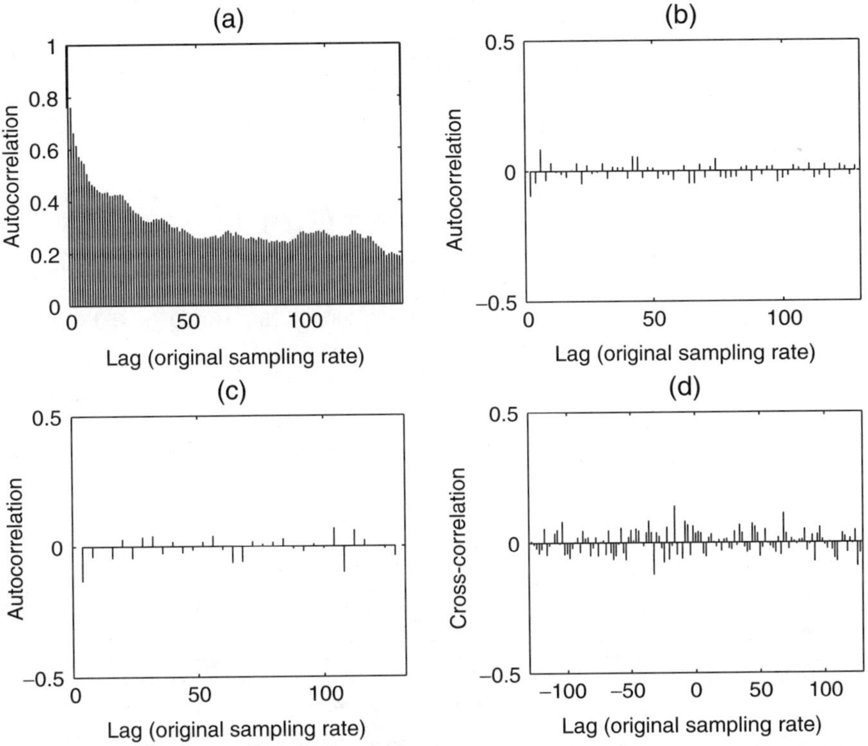

FIGURE 7.10 (a) Sample autocorrelation function of the series illustrated in Figure 7.9(a); (b) sample autocorrelation function of wavelet coefficients at level 10 of the discrete wavelet transform of the data shown in (a); (c) sample autocorrelation function of wavelet coefficients at level 9; (d) sample cross-correlations of coefficients between levels 9 and 10. Reproduced from Johnstone and Silverman (1997).

based on the level-dependent thresholding method for correlated data. Median absolute deviation estimates were used to give an estimate $\hat{\sigma}_j$ of the standard deviation of the wavelet coefficients at each level j. The methods used are given in Table 7.1. The *translation-equivariant*, or *stationary*, wavelet transform uses a redundant basis in which there is the same number of coefficients at each level, and it can be calculated in $O(n \log n)$. It contains the ordinary discrete wavelet transform coefficients for every choice of time origin. (See Coifman and Donoho (1995) or Nason and Silverman (1995) for further discussion.) The Stein unbiased risk (SURE) method is based on the work of Stein (1981) and is described in detail by Johnstone and Silverman (1997).

Table 7.1 compares the error rates of the various methods we have discussed thus far on 10 successive segments of length 4096 from the original data record. Also shown is the error rate of the current special-purpose method developed specifically by Eisenberg and Levis for this problem. While translation-invariant level dependent wavelet thresholding is not yet quite as good as the special-purpose

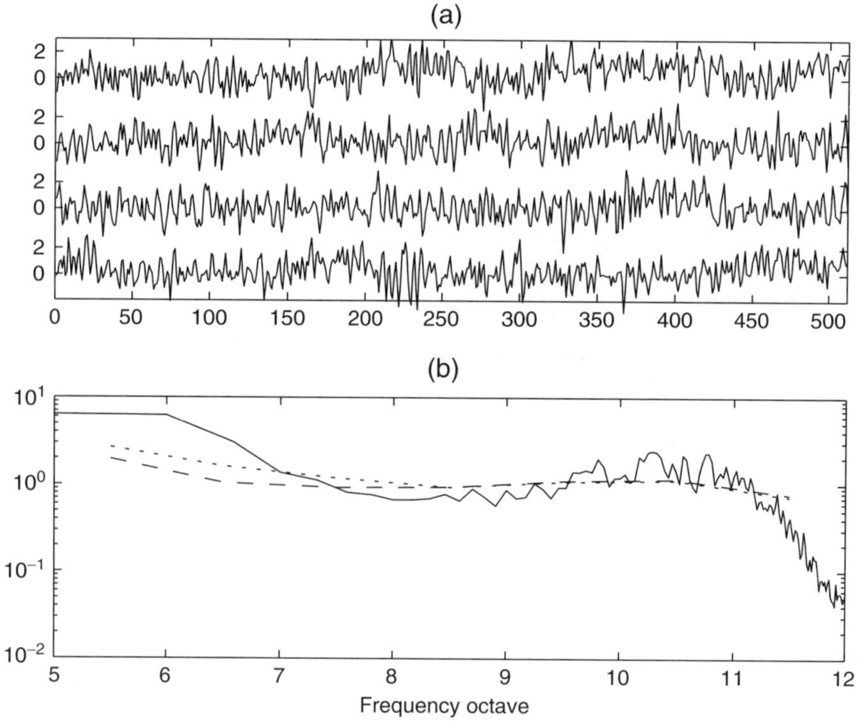

FIGURE 7.11 (a) A subset of 2048 data points (2025–4072) from a record of 100,000 points provided by Eisenberg and Levis, with successive segments of length 512 shown from bottom to top; (b) ——— represents an estimate of the power spectral density of the observed data y (a sequence of length 4096, with data points 1001–5096), with both amplitude and frequency shown on logarithmic scales, frequency coded according to frequency octaves; the right-hand endpoint is 50 kHz. The median absolute deviation – – – and standard deviation ······ of the Haar wavelet coefficients of the observed data are plotted versus resolution level (plus 0.5) on the same vertical scale. Reproduced from Johnstone and Silverman (1997).

TABLE 7.1 Comparison of Various Methods for the Neurophysiology Data[a]

Method	Errors	Percent
(i) Pointwise classification of raw data	1180.3	28.8%
(ii) Wavelet thresholding, thresholds $\hat{\sigma}_j\sqrt{2\log n}$ at levels $j \geq 6$	643.1	15.7%
(iii) Same as (ii), but using translation-invariant wavelet transform	350.7	8.6%
(iv) Same as (iii), but using SURE thresholds at levels $j = 6,7$, and 8	95.9	2.3%
(v) Special-purpose algorithm provided by Eisenberg and Levis	82.2	2.0%

[a]The performances shown are the averages over 10 successive segments, each of length 4096. Method (i) estimates the signal to be zero if the observed value is less than 0.5 and unity otherwise.

filtering method, it is encouraging that relatively straightforward adaptation of the general wavelet thresholding prescription achieves reasonable results in this special context. Also, the special-purpose method has no promise of working on any other kind of data, while the methodology that we have proposed is part of a very generally applicable toolkit.

7.6 IMPLEMENTATION OF THE WAVELET TRANSFORM

This section is a reference for the `wavethresh` implementation of the discrete wavelet transform (DWT) and the inverse discrete wavelet transform (IDWT) software. The algorithm follows exactly that described by Mallat (1989b), and this section should be read with it. The DWT algorithm as described by Mallat (1989b) is a special case of a two-channel subband coder using the conjugate quadrature filters of Smith and Barnwell (1986). Vaidyanathan (1990) has provided a comprehensive survey and comparison of many filtering methods, including subband coders; other significant contributions include those of Vetterli (1984), Mintzer (1982, 1985) on filter design, and Smith and Eddins (1990) on subband coding for images.

To proceed, we assume that we are considering expansions in terms of a family of wavelets

$$\psi_{jk}(x) = 2^{j/2}\psi(2^j x - k)$$

for some mother wavelet ψ. In what follows, the coefficients of such expansions are written as vectors d^j, where j represents the resolution level; the individual elements of these vectors give the coefficients at various translations k.

7.6.1 Computing the DWT

As in Mallat (1989b), we begin with a set of $N = 2^M$ data

$$c_0^M, \ldots, c_{N-1}^M. \tag{7.7}$$

In all that follows, in contrast to the double-subscript notation c_{jk} used previously, the superscript denotes the resolution level and the subscript represents the coefficient within that level. The reason for promoting the j to a superscript is that the k subscript may get a little more complicated! The algorithm consists of $M = \log_2 N$ stages, and the superscript M signifies that this is the original data at the highest resolution level. At each stage, we produce two vectors, a "smooth" vector c and a vector d comprising the "detail" at that level, both of which are described by the upcoming formulae.

We do not consider ways of extending data sets of other sizes to length 2^M, but natural possible approaches are periodic extension and symmetric reflection (both are considered in detail by Smith and Eddins, 1990), as well as padding out with zeroes, boundary value replication, and antisymmetric reflection. Note that the choice of boundary conditions for the DWT itself is a separate matter and is discussed next and in Section 7.3.4.

At each stage, new c's and d's are produced from the previously smoothed data c by using finite impulse response filters h and g, as in Equations (7.9) and (7.16). The filters h and g are intimately related by the relation

$$g(n) = (-1)^n h(1-n), \tag{7.8}$$

and they are known as quadrature mirror filters (see Vaidyanathan, 1990, for further information on filter terminology and design). It is because of this relation that `filter.select()` needs to know only one half of the pair (h in our case) for a particular wavelet family. The filter h is a smoothing filter, and the filter g is a highpass filter. If periodic boundary conditions are used for the DWT, the lengths of the new c and d are exactly half the length of the previous c, so the total length of c and d at each stage remains the same throughout the algorithm. If symmetric boundary reflection is used in the DWT, then some extra numbers are required at either end of the c and d filters. This explains why the lengths of c's and d's obtained by `accessC()` and `accessD()` functions are not necessarily exact powers of two when the `boundary=T` option is supplied. An increase in the filter length produces a corresponding increase in the number of these "extras."

In order to allow different boundary methods to be used, we define the sequences $f[j]$ and $l[j]$ to represent the first and last indices for a particular sequence at a particular level. So, for example, if we say that a sequence s at level 3 has $f[3] = -4$ and $l[3] = 3$, then the sequence elements are

$$s^3_{-4}, s^3_{-3}, \ldots, s^3_2, s^3_3.$$

For the original data in Equation (7.7), we have $f[M] = 0$ and $l[M] = N - 1$. With periodic boundary handling, it is simple to predict the values of $f[j]\, and l[j]$ for $j = 0, \ldots, M-1$, since the sequence of c and d halves exactly at each step. However, for symmetric end-reflection boundary handling, we need to develop some inequalities that specify exactly where the start and finish for each sequence are; we do so in the next subsection. Sometimes, we subscript f and l with C or D to make it clear which sequence we are referring to. For both periodic and symmetric boundary handling, the values for f_C, l_C, f_D, and l_C are usually precomputed by the `first.last()` function.

7.6.2 Level j to Level $j-1$

We now describe the decomposition of a sequence at level j into a sequence at level $j-1$. A decomposition means extracting a "smoother" signal at a lower resolution and extracting the signal detail at that resolution. The first decomposition step starts at level M and produces level $M-1$.

At level j, we have data

$$c^j_{f[j]}, \ldots, c^j_{l[j]},$$

where $f[j] \leq 0 \leq l[j]$. The smoothing operation to the next level, $j-1$, is achieved with the filter

$$h(n), \quad 0 \leq n \leq N_h - 1,$$

that has N_h nonzero coefficients. The smoothing formula is a convolution followed by dyadic decimation, as in Mallat (1989b):

$$c_k^{j-1} = \sum_{n \in \mathbb{Z}} h(n-2k) c_n^j \tag{7.9}$$

$$= \sum_{m \in \mathbb{Z}} h(m) c_{m+2k}^j$$

$$= \sum_{m=0}^{m=N_h-1} h(m) c_{m+2k}^j. \tag{7.10}$$

To identify which k we can compute, and to obtain exact reconstruction, the summation indices must satisfy the following inequalities (for symmetric end reflection):

$$0 \leq m \leq N_h - 1; \tag{7.11}$$

$$f[j] \leq m + 2k \leq l[j]. \tag{7.12}$$

From Equation (7.12), we can write

$$\tfrac{1}{2}(f[j] - m) \leq k \leq \tfrac{1}{2}(l[j] - m). \tag{7.13}$$

Using Equations (7.11) and (7.13), we can obtain the range of the coefficient indices for the next level:

$$\tfrac{1}{2}(f[j] - N_h + 1) \leq k \leq \tfrac{1}{2} l[j].$$

Thus, the coefficient indices for the next level are

$$f[j-1] = \lceil \tfrac{1}{2}(f[j] - N_h + 1) \rceil \tag{7.14}$$

and

$$l[j-1] = \lfloor \tfrac{1}{2} l[j] \rfloor, \tag{7.15}$$

where $\lceil x \rceil$ is the smallest integer greater than or equal to x and $\lfloor x \rfloor$ is the largest integer less than or equal to x. Equations (7.14) and (7.15) are built into the `first.last()` function.

The detail d is computed as

$$d_k^{j-1} = \sum_{n \in \mathbb{Z}} g(n-2k) c_n^j. \tag{7.16}$$

Equation (7.8) follows from Daubechies (1992, Equation 5.1.34). Hence, combining Equations (7.16) and (7.8), we have

$$d_k^{j-1} = \sum_{n\in\mathbb{Z}} (-1)^n h(2k+1-n) c_n^j$$
$$= \sum_{m=0}^{m=N_h-1} (-1)^{1-m} h(m) c_{2k+1-m}^j. \quad (7.17)$$

The inequalities are similar. From Equations (7.17) and (7.11), we have

$$f_D[j-1] = \lceil \tfrac{1}{2}(f_C[j]-1)\rceil \quad (7.18)$$

and

$$l_D[j-1] = \lfloor \tfrac{1}{2}(l_C[j]+N_h-2)\rfloor, \quad (7.19)$$

where f_C and l_C are the first and last indices, respectively, for the data c at level j (see Equations (7.14) and (7.15)), and f_D and l_D are the corresponding indices for the detail d and are also built into `first.last`. Equations (7.10) and (7.17) are the actual formulae built into the `wd` function. The decomposition process is illustrated in Figure 7.12.

7.6.3 The Result of the DWT

At the end, of the algorithm we obtain

$$DWT\{c^M\} = \{c_0^0, d_0^0, d_0^1, d_1^1, \ldots, d_0^{M-1}, \ldots, d_{N/2-1}^{M-1}\} \quad (7.20)$$

Most of the coefficients are the detail d, except for c_0^0, which was the last smoothed data to be produced and is a weighted total of all the data (application

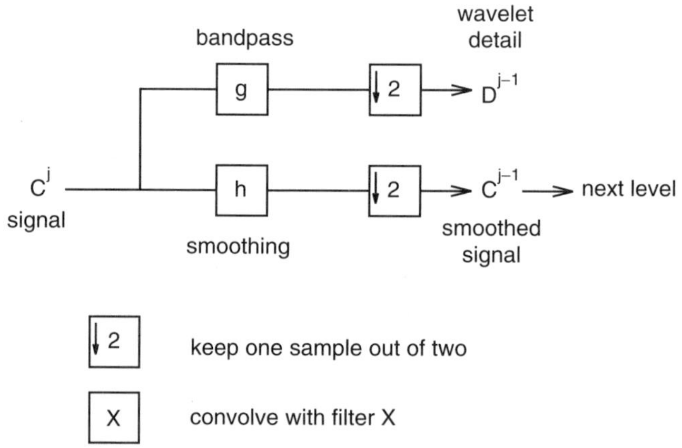

FIGURE 7.12 Decomposition of discrete signal c^j into smoothed signal c^{j-1} and wavelet coefficients d^{j-1} by use of conjugate filters h and g (adapted from Mallat, 1989b).

of many smoothing filters). If the decomposition was produced using the Haar basis, then this c_0^0 is exactly the sample mean multiplied by the square root of the number of original data points. The collection of coefficients in Equation (7.20) plus all of the levels of smoothed data make up the *wavelet decomposition object*. The transform that we actually compute is really more closely related to the expansion

$$f(x) = \sum_k c_{0k}\phi_{0k}(x) + \sum_{j\geq 0}\sum_k d_{jk}\psi_{jk}(x), \tag{7.21}$$

rather than to the expansion in Equation (7.4). The functions ϕ_{0k} are simply the integer translates of a scaling function ϕ. The set $\{\phi_{0k}, \psi_{jk}\}_{(j,k)\in\mathbb{Z}^2}$ also forms a basis for $L^2(\mathbb{R})$ for appropriate choice of ϕ and ψ.

7.6.4 Computing the IDWT

To reconstruct, we begin with the lowest resolution coefficients and work up towards full resolution (or whatever level we desire). So, given the wavelet decomposition in Equation (7.20), we begin the reconstruction using c_0^0 and d_0^0 to produce c_0^1 and c_1^1. We then use these c and the d at the same resolution level to produce the c at the next resolution level. The general idea of a step in this reconstruction is shown in Figure 7.13.

7.6.5 Level j to Level $j+1$

The formula for obtaining the c from the previous level is

$$c_n^{j+1} = \sum_{k\in\mathbb{Z}} h(n-2k)c_k^j + \sum_{k\in\mathbb{Z}} g(n-2k)d_k^j,$$

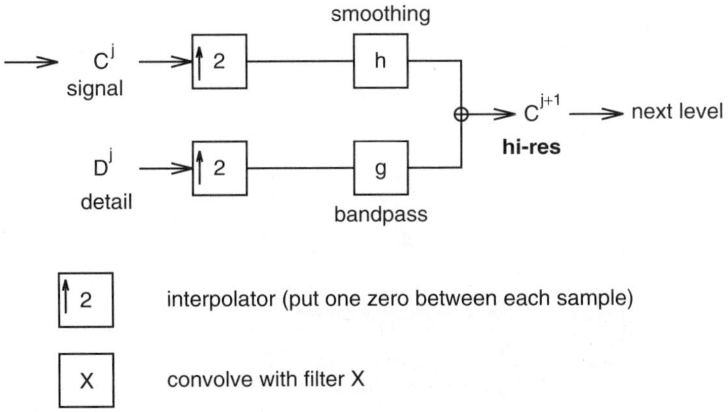

FIGURE 7.13 Reconstruction step: producing a higher resolution signal c^{j+1} from smoothed signals c^j and d^j from filters h and g (adapted from Mallat, 1989b).

where h and g are as before. Since there are only a finite number of h's, we can place limits on the indices of summation; furthermore, we may rewrite g in terms of h, because of Equation (7.8) and obtain

$$c_n^{j+1} = \sum_{2k\geq n+1-N_h}^{2k\leq n} h(n-2k)c_k^j + \sum_{2k\geq n-1}^{2k\leq N_h+n-2} h(1+2k-n)d_k^j.$$

We know which n of the c to compute, since these were worked out in the decomposition part.

7.6.6 The DWT of Images

The theory for the two-dimensional transform is a straightforward extension of that for the one-dimensional transform. Mallat (1989b) also gives details of how to compute the DWT and its inverse on an image. (See also Nason and Silverman, 1994). We give the details of the algorithm in Figure 7.14 for the two-dimensional decomposition and in Figure 7.15 for the reconstruction. The transform described in Mallat (1989b) is based on a separable multiresolution approximation (see Mallat, 1989a, or Daubechies 1992) wherein the two-dimensional scale function written as

$$\Phi(x, y) = \phi(x)\phi(y),$$

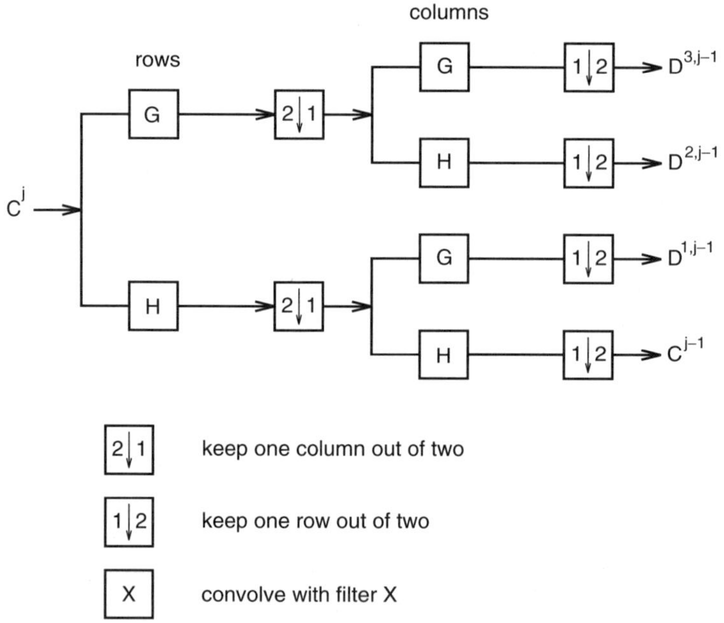

FIGURE 7.14 Decomposition of an image. The algorithm is based on one-dimensional convolutions along rows and columns of the image (adapted from Mallat, 1989b).

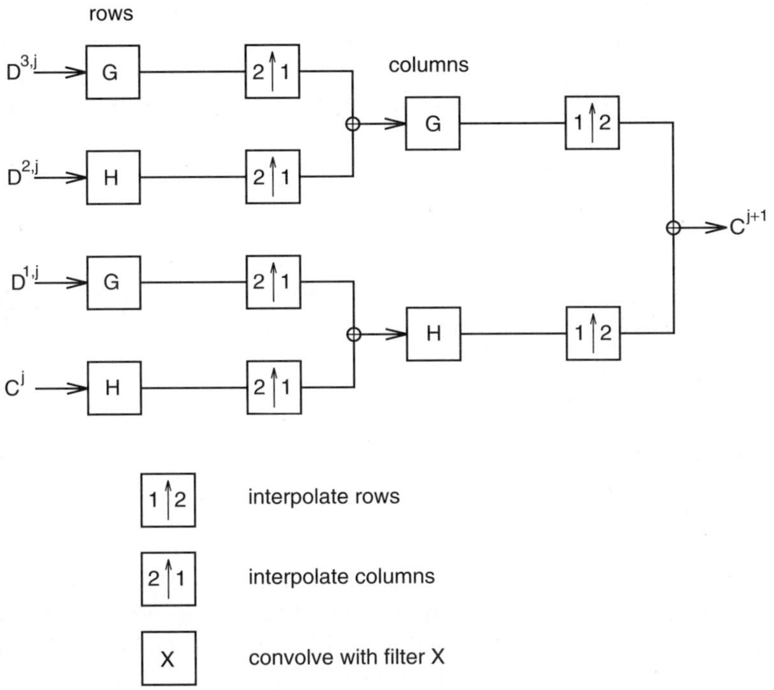

FIGURE 7.15 Reconstruction of an image (adapted from Mallat, 1989b).

where $\phi(x)$ is a one-dimensional scale function from a univariate multiresolution approximation. This time, there are three mother wavelets, not one (as in the univariate case):

$$\Psi^1(x, y) = \phi(x)\psi(y); \quad \Psi^2(x, y) = \psi(x)\phi(y);$$
$$\Psi^3(x, y) = \psi(x)\psi(y).$$

If we begin with an n^2 image at level j, then we obtain four sets of information. First, we obtain a$(n)/2^2$ image representing the smoothed data at level $j-1$; the other three sets are wavelet coefficients corresponding to the $\{\Psi^1, \Psi^2, \Psi^3\}$ wavelet functions, and there are $(n/2)^2$ coefficients in these "images" as well. Due to the separable nature of the decomposition, each of the details in the wavelet images is oriented horizontally, vertically, and diagonally. This can also be seen from Figure 7.14, which is a diagram depicting the decomposition of an image.

7.7 HOW TO OBTAIN AND INSTALL THE WAVELET SOFTWARE

The `wavethresh` package is available, along with other statistical software, from the StatLib archive. StatLib is a statistical software archive wherein the software is freely obtainable via email and anonymous FTP. The package is also available from the Statistics Group's Web server at the University of Bristol:

```
http://www.stats.bris.ac.uk/pub/reports/Wavelets/Wavelets.html
```

To obtain the `wavethresh` software via anonymous FTP, type

```
% ftp lib.stat.cmu.edu
```

Then respond to the prompt for a login id with `statlib`, and use your full e-mail address as the password. Then type

```
ftp> cd S
```

to enter the S software directory and type

```
ftp> get wavethresh
```

and you should then see messages informing you that the software is being retrieved. The StatLib archive is also reproduced at other locations; for example, users in the United Kingdom can find a StatLib mirror at `hensa.unix.ac.uk`. The software is distributed as a shar archive file that we call `wavethresh`. You can unpack the software anywhere you like, but it is a good idea to create a particular subdirectory for it. A good name for a subdirectory is `WAVELET`; so, `cd` to a directory of your choosing and type

```
% mkdir WAVELET
```

Then copy the distribution file (`wavethresh`) to this directory and then `cd` to it. To unpack the distribution, type

```
%  sh wavethresh
```

You can now delete the distribution file `wavethresh`, if you like. The archive is unpacked into a subdirectory named `DISTRIB2.2`. If you `cd` into this directory and take a look at the files there, you should see the following files (although your `ls` function may list them differently):

```
total 88
   2 Copyright                      4 StoIRS.c
   7 ImageDecomposeStep.c           2 conbar.c
   2 Makefile                       4 convolve.c
   2 README                         3 wavedecomp.c
   3 Sconvolve.c                    1 wavelet.h
  53 Source-0.S                     3 waverecons.c
   2 StoIDS.c
```

Most of these files are the C code to perform the transforms and have to be compiled. The `Source-0.S` file contains all of the S functions in ASCII format. In addition, there is a subdirectory `.Data`, revealed by typing `ls -a` (under UNIX).

Setting Up
Before you can use any of the software, you must perform some initialization tasks. The `Makefile` is set up for using S-Plus; you may wish to edit and change this file to S. You might also need to modify other aspects of the `Makefile`, but it should be correct for most UNIX systems. Assuming that you are still in the `DISTRIB2.2` directory, make the software by typing

```
% make all
```

This command sets up the S functions and constructs the C object code. Type

```
% make cleanobj
```

if you want to clean up.

If you use raw S, then `dyn.load` may not work for you, in which case you need to replace the call to `dyn.load` in the `maybe.load` function by `dyn.load2`. At present, the `Makefile` contains the relevant modifications for a DEC machine running Ultrix V4.2A, although if may give you a guide for running the software on other machines.

Acknowledgements

We would like to thank Iain Johnstone for assistance and encouragement with the software and this chapter. We would also like to thank Howard Grubb, Emma McCoy, Peter Clifford, and Martin Maechler for suggesting improvements and making contributions to the `wavethresh` package, and we would like to thank Andrew Black of the Bristol Royal Infirmary for supplying the anaesthesiology data.

Both authors were supported in part by U.K. EPSRC grant KR70236. The material for this chapter was completed while B. W. Silverman was a Fellow at the Center for Advanced Study in the Behavioral Sciences at Stanford University. He is grateful for the support of National Science Foundation grant SBR-9601236.

REFERENCES

Abramovich, F., Sapatinas, F., and Silverman, B. W. (1998). Wavelet thresholding via a Bayesian approach, *J. Roy. Statist. Soc. B*, **60**, 725–749.

Antoniadis, A., Grégoire, G., and Nason, G. P. (1999). Density and hazard rate estimation for right censored data using wavelet methods, *J. Roy. Statist. Soc. B*, **61**, 63–84.

Bruce, A., and Gao, H.-Y. (1996). *Applied Wavelet Analysis with S-Plus*, Springer–Verlag, New York.

Chipman, H., Kolacyzk, E. D., and McCulloch, R. E. (1997). Adaptive Bayesian wavelet shrinkage, *J. Amer. Statist. Ass.*, **92**, 1413–1421.

Chui, C. K. (1992). *An Introduction to Wavelets*, Academic Press, London.

Cohen, I., Raz, S., and Malah, D. (1997). Orthonormal and shift-invariant wavelet packet decomposition, *Sig. Proc.*, **57**, 251–270.

Coifman, R. R., and Donoho, D. L. (1995). Translation-invariant de-noising, in Antoniadis, A., and Oppenheim, G. (eds.), *Wavelets and Statistics: Proceedings of the XVth Rencontres Franco-Belges des Statisticiens*, pp. 125–150, Springer-Verlag, Berlin.

Coifman, R. R., and Saito, N. (1994). Constructions of local orthonormal bases for classification and regression, *Comptes Rendus Acad. Sci. Paris*, **319**, 191–196.

Crouse, M. S., Nowak, R. D., and Baraniuk, R. G. (1998). Wavelet-based signal processing using hidden Markov models, *IEEE Trans. Sig Proc.*, **46**, 886–902.

Daubechies, I. (1988). Orthonormal bases of compactly supported wavelets, *Comms. Pure Appl. Math.*, **41**, 909–996.

Daubechies, I. (1992). *Ten Lectures on Wavelets*, Society for Industrial and Applied Mathematics, Philadelphia, PA.

Donoho, D. L., and Johnstone, I. M. (1994). Ideal spatial adaptation by wavelet shrinkage, *Biometrika*, **81**, 425–455.

Donoho, D. L., and Johnstone, I. M. (1995). Adapting to unknown smoothness via wavelet shrinkage, *J. Amer. Statist. Assoc.*, **90**, 1200–1224.

Downie, T. R., and Silverman, B. W. (1996). *The Discrete Multiple Wavelet Transform and Thresholding Methods*. Technical Report, Department of Mathematics, University of Bristol, Bristol, UK.

Hall, P., and Nason, G. P. (1997). On choosing a non-integer resolution level when using wavelet methods, *Statistics and Probability Letters*, **34**, 5–11.

Hall, P., and Patil, P. (1995). Formulae for mean integrated squared error of nonlinear wavelet-based density estimators, *Ann. Statist.*, **23**, 905–928.

Johnstone, I. M., and Silverman, B. W. (1997). Wavelet threshold estimators for data with correlated noise, *J. Roy. Statist. Soc. B*, **59**, 319–351.

Learned, R. E., and Willsky, A. S. (1995). A wavelet packet approach to transient signal classification, *App. Comput. Harm. Anal.*, **2**, 265–278.

Mallat, S. G. (1989a) Multiresolution approximations and wavelet orthonormal bases of $L^2(R)$. *Trans. Amer. Math. Soc.*, **315**, 69–87.

Mallat, S. G. (1989b). A theory for multiresolution signal decomposition: The wavelet representation. *IEEE Trans. Pattn. Anal. Mach. Intell.*, **11**, 674–693.

Mintzer, F. (1982). On half-band, third-band and *N*th band FIR filters and their design, *IEEE Transactions on Acoustics, Speech and Signal Processing*, **30**, 734–738.

Mintzer, F. (1985). Filters for distortion-free two-band multirate filter banks, *IEEE Transactions on Acoustics, Speech and Signal Processing*, **33**, 626–630.

Nason, G. P. (1993). *The* `WaveThresh` *Package: Wavelet Transform and Thresholding Software for S.* Available from the StatLib archive.

Nason, G. P. (1996). Wavelet shrinkage using cross-validation, *J. Roy. Statist. Soc. B*, **58**, 463–479.

Nason, G. P., Sapatinas, T., and Sawczenko, A. (1997). Regression modeling of time-series using the stationary wavelet packet transform, preprint.

Nason, G. P., and Silverman, B. W. (1994). The discrete wavelet transform in S, *J. Comput. Graph. Statist.*, **3**, 163–191.

Nason, G. P., and Silverman, B. W. (1995). The stationary wavelet transform and some statistical applications, in Antoniadis, A., and Oppenheim, G. (eds.), *Wavelets and Statistics*, pp. 281–300, Springer-Verlag, Berlin, Germany.

Saito, N. (1994). *Local Feature Extraction and Its Applications Using a Library of Bases*, Ph.D. thesis, Yale University, New Haven, CT.

Saito. N., and Coifman, R. R. (1996). On local feature-extraction for signal classification, *Zeitschrift für Angewandte Mathematik und Mechanik*, **76**, 453–456.

Smith, M. J. T., and Barnwell, T. P. (1986). Exact reconstruction techniques for tree-structured subband coders, *IEEE Transactions on Acoustics, Speech and Signal Processing*, **34**, 434–441.

Smith, M. J. T., and Eddins, S. L. (1990). Analysis/synthesis techniques for subband image coding. *IEEE Transactions on Acoustics, Speech and Signal Processing*, **38**, 1446–1456.

Stein, C. (1981). Estimation of the mean of a multivariate normal distribution, *Ann. Statist.*, **9**, 1135–1151.

Strang, G. (1993) Wavelet transforms versus Fourier transforms, *Bulletin (New Series) of the American Mathematical Society*, **28**, 288–305.

Vaidyanathan, P. P. (1990). Multirate digital filters, filter banks, polyphase networks, and applications: A tutorial, *Proceedings of the IEEE*, **78**, 56–93.

Vetterli, M. (1984). Multi-dimensional sub-band coding: Some theory and algorithms, *Signal Processing*, **6**, 97–112.

Vetterli, M., and Herley, C. (1992). Wavelets and filter banks: Theory and design, *IEEE Transactions on Signal Processing*, **40**, 2207–2232.

Von Sachs, R., Nason, G. P., and Kroisandt, G. (1998). Spectral representation and estimation for locally-stationary wavelet processes, in Dubuc, S. (ed.), *CRM Proceedings and Lecture Notes*. **18**, 381–397.

Weyrich, N., and Warhola, G. T. (1994). *De-Noising Using Wavelets and Cross-Validation.* Technical report AFIT/EN/TR/94-01, Department of Mathematics and Statistics, Air Force Institute of Technology, AFIT/ENC Wright-Patterson Air Force Base, OH.

Wickerhauser, M. V. (1994). *Adapted Wavelet Analysis from Theory to Software*, A. K. Peters, Wellesley, MA.

CHAPTER 8

Smoothing Methods for Discrete Data

Jeffrey S. Simonoff
New York University, New York, USA

Gerhard Tutz
Ludwig–Maximilians Universität München, Munich, Germany

8.1 INTRODUCTION

The statistical analysis of discrete data has a long history, the modern era having begun with Pearson's (1900) paper on χ^2 goodness-of-fit tests for contingency tables. For purposes of our discussions, it is useful to classify such data into two broad categories:

1. *Categorical data*, includes both data with nominal categories (that is, categories that do not have a natural ordering) and data with ordered categories.
2. *Discrete regression data*, where the target variable is categorical and the predictor variable(s) are either categorical or continuous.

In contrast to the situation of continuous data, in which the benefits of smoothing (in the form of a scatter plot smoother, for example) are clear, the applicability of smoothing methods to discrete data is less clear. The notion of functions having similar values for "close" observations is the basis of smoothing, and this concept needs to be generalized to discrete data in order for the analyst to apply smoothing methods.

In this chapter, we motivate a discussion of, and describe, the application of smoothing methods to discrete data. Section 8.2 treats categorical data, with

Smoothing and Regression: Approaches, Computation, and Application
Edited by Michael G. Schimek
ISBN 0-471-17946-9

"closeness" naturally defined for data classified into ordered categories; in that situation, observations that fall, for example, into adjacent categories are closer to each other than those that fall into categories that are farther apart. Section 8.3 addresses discrete regression data, for which observations are defined as "close" on the basis of both their categories in the target variable and their values for the predictor variable(s).

The unifying concept for all of the methods discussed here is their close conceptual connection to nonparametric regression methods for continuous data. As will be seen, estimation problems can be viewed as regression problems, and theoretical results are very similar to those derived for nonparametric regression estimators. Smoothing methods for categorical data also provide a useful structure for the development of effective density estimation techniques.

8.2 SMOOTHING CONTINGENCY TABLES

8.2.1 Univariate Categorical Data

We begin with a discussion of univariate categorical data. The data takes the form of counts $\{n_j,\ j = 1, \ldots, k\}$, with $\sum_j n_j = n$. The data are typically represented using a multinomial distribution with sample size n and underlying probability vector π, so that the log likelihood equals $\sum_{j=1}^{k} n_j \log(n\pi_j)$ (omitting constants).

The most commonly used estimator of π is the frequency estimator, $p_j = n_j/n$. This estimator, which is uniform-minimum-variance unbiased, is the nonparametric maximum likelihood estimator and is very accurate when the sample is large enough. As $\inf_j n\pi_j \to \infty$, $p = \{p_1, \ldots, p_k\}$ is consistent and (as is the case with most maximum-likelihood estimators) is best asymptotically normal, with $E(p_j/\pi_j) = 1$ and

$$\mathrm{Var}(p_j/\pi_j) = (1 - \pi_j)/(n\pi_j). \tag{8.1}$$

Thus, if all the expected cell counts $n\pi_j$ are large, little meaningful improvement on the frequency estimator is possible.

Equation (8.1) shows how the frequency estimator can fail. If the expected cell counts are small, p is highly variable as an estimator of π. (Example 1, given later in this section, illustrates this property.) Modifications of p aim at reducing this variability. Many authors have suggested adding a flattening constant α to each cell, yielding

$$\hat{\pi}_j = \frac{n_j + \alpha}{n + k\alpha}.$$

This is a *shrinkage estimator* in the sense of Stein (1962), in that it shrinks the frequency estimates towards the uniform vector. Fienberg and Holland (1970) justified a particular data-based choice of α through an empirical Bayes argument using a Dirichlet prior. The estimator also can be written as a weighted sum of the frequency estimates (that is, as a *kernel probability estimator*), since

$$\hat{\pi}_j = \sum_{i=1}^{k} K(j|i) p_i, \tag{8.2}$$

where

$$K(j|i) = \begin{cases} \dfrac{n+\alpha}{n+k\alpha}, & \text{if } i = j \\ \dfrac{\alpha}{n+k\alpha}, & \text{if } i \neq j \end{cases}.$$

An equivalent form of the kernel function is

$$K_h(j|i) = \begin{cases} h, & \text{if } i = j \\ (1-h)/(k-1) & \text{if } i \neq j \end{cases}, \tag{8.3}$$

where $h = (n+\alpha)/(n+k\alpha)$ (Aitchison and Aitken, 1976).

This kernel estimator is appropriate for nominal data; that is, data classed into categories for which there is no natural ordering. Many categorical variables have categories that are naturally ordered, and it is sensible to take advantage of that ordering. Such variables can arise as a discretization of a continuous variable or as a set of inherently discrete, but ordered, categories (e.g., the Likert scale, with its categories "strongly disagree," "disagree", "neutral", "agree", and "strongly agree"). Discrete random variables, such as those with binomial or Poisson distributions, also can be viewed in this way, with the integer values of the random variable representing the categories.

In each of these situations, a reasonable hypothesis is that the probability that an observation will fall into a given cell is similar to the probability that it will fall into a nearby cell. That is, the underlying probability vector π is smooth, with π_i not too different from π_j for i close to j. This suggests smoothing over neighboring cells to obtain a more accurate probability estimate in a given cell. The kernel formulation (8.2) allows this if smoothness is incorporated into the kernel function K. If the weights $K(j|i)$ decrease smoothly as $|i-j|$ increases, the kernel estimate $\hat{\pi}_j$ "borrows" information from nearby cells to improve the estimation of π_j. Aitchison and Aitken (1976), Habbema et al. (1978), Titterington (1980), Wang and Van Ryzin (1981), and Aitken (1983), proposed different possible weighting schemes. More generally, weights can be derived by integrating any continuous, symmetric unimodal density function with little effect on the accuracy of the estimation (Burman, 1987; Hall and Titterington, 1987). See Simonoff (1995) for a more detailed discussion of these estimators.

Viewed in this manner, the kernel probability estimator is apparently the discrete analog of the kernel density estimator for continuous data, viz.,

$$\hat{f}(x) = \frac{1}{nh} \sum_{j=1}^{n} K\left(\frac{x - x_j}{h}\right)$$

(Rosenblatt, 1956; Parzen, 1962), where h is the bandwidth and the kernel function K satisfies $\int K(u)\,du = 1$, $\int uK(u)\,du = 0$, and $\int u^2K(u)\,du > 0$.

An even better correspondence is with kernel regression estimation. Since $E(p_j) = \pi_j$, estimating π_j coincides with estimating the expectation of the frequency estimator, for which the literature treating continuous responses can be applied (Eubank, 1988; Härdle, 1990; Wand and Jones, 1995; Simonoff, 1996a, Chapters 5 and 6). The kernel probability estimator (8.2) is a kernel regression estimator with predictor values j/k and response values p_j, $j = 1, \ldots, k$. (Note that in this context the response values are not independent.)

Kernel estimation can be viewed as fitting a constant locally to π. A more flexible approach is to allow higher order local polynomials to be used in the estimation of π. One way this can be done is by utilizing weighted least-squares polynomial fitting, weighting by the kernel function K (Stone, 1977; Cleveland, 1979; Fan, 1992; Fan and Gijbels, 1996). In that case, in fitting a local tth-order polynomial, the cell probability estimate $\hat{\pi}_i$ is the constant term of the minimizer $\hat{\beta}$ of

$$\sum_{j=1}^{k}\left[p_j - \left\{\beta_0 + \cdots + \beta_t\left(\frac{i}{k} - \frac{j}{k}\right)^t\right\}\right]^2 K\left(\frac{i/k - j/k}{h}\right). \tag{8.4}$$

As noted earlier, the frequency estimator is not problematic when the expected cell counts are large. Sparse asymptotics (Fienberg and Holland, 1973; Simonoff, 1983), which require the number of cells to become infinite as the sample size becomes infinite, model the occurrence of large tables with relatively few observations in each cell, for which smoothing can be beneficial. However, under sparse asymptotics in which the sample size n grows at the same rate as the number of cells k, the frequency estimator is generally not consistent, since the expected cell counts do not get large as n and k get large. Squared-error analysis is based on the *mean sum of squared errors*, $\text{MSSE}(\hat{\pi}) = E[\sum_i(\hat{\pi}_i - \pi_i)^2]$, with $k \times \text{MSSE}$ approaching zero corresponding to consistency. The frequency estimator satisfies $\text{MSSE}(p) = O(N^{-1})$, so under these sparse asymptotics, $k \times \text{MSSE} \not\to 0$ (Hall and Titterington, 1987).

In contrast, the local polynomial estimator $\hat{\pi}$ is sparse asymptotic consistent, with properties very similar to those for nonparametric regression estimation. Assume, for mathematical convenience, that the vector π is generated from an underlying density f on $[0, 1]$ possessing $t + 1$ uniformly continuous derivatives through the relation

$$\pi_i = \int_{(i-1)/k}^{i/k} f(u)\,du. \tag{8.5}$$

(In this way, the smoothness of π can be represented by using derivatives of f.) Let $h \to 0$ with $nh \to \infty$. Then, for t odd, and assuming regularity conditions on K, the MSSE for the tth-degree local polynomial estimator satisfies

$$\text{MSSE}(\hat{\pi}) = \frac{1}{k}\left[\frac{h^{t+1}\mu_{t+1}(K_{(t)})}{(t+1)!}\right]^2 \int f^{(t+1)}(u)^2\,du + \frac{R(K_{(t)})}{nhk} + o[h^{2t+2}k^{-1} + (nhk)^{-1}], \tag{8.6}$$

where $K_{(t)}$ is a $(t+1)$th-order kernel function (this means that $\int K_{(t)}(u)\,du = 1$, $\mu_l(K_{(t)}) \equiv \int u^l K_{(t)}(u)\,du = 0$ for $l = 1, \ldots, t$, and $\mu_{t+1}(K_{(t)}) \neq 0$) and $R(g) \equiv \int g^2(u)\,du$ for any function g (Aerts et al., 1997a).

The first two terms in (8.6) correspond to squared bias and variance terms, respectively. The squared bias is a direct function of the bandwidth h, as would be expected, since $h = 0$ corresponds to the frequency estimator, which is unbiased. The variance is an inverse function of h, since using a larger bandwidth results in smoothing over more cells (and hence, more observations). From a practical point of view, it is unlikely that much improvement in the estimates would be seen for $t > 1$ or 2.

The bandwidth that minimizes the leading terms of (8.6) is

$$h_0 = \left[\frac{(t+1)(t!)^2 R(K_{(t)})}{2n\mu_{t+1}(K_{(t)})^2 \int f^{(t+1)}(u)^2\,du}\right]^{1/(2t+3)}, \tag{8.7}$$

which shows that the optimal MSSE converges to zero at the rate of $O(n^{-(2t+2)/(2t+3)}k^{-1})$, which is optimal. (See Hall and Titterington, 1987.) If a linear function is fit locally, the rate is thus $O(n^{-4/5}k^{-1})$. This rate holds without putting any additional conditions on the cell probabilities for i close to either 1 or k (the boundaries) and without modifying the estimates for these cells, a property not shared by kernel estimators (Burman, 1987; Hall and Titterington, 1987). Dong and Simonoff (1994) adapted work on boundary-corrected kernel regression estimation to derive boundary-corrected kernel probability estimators that yield similar rates of MSSE convergence for boundary cells. Generally speaking, taking t even in local polynomial estimation does not affect the accuracy of interior cells asymptotically, but leads to degradation of the MSSE rate for boundary cells unless boundary correction is performed, giving a theoretical advantage to local polynomial estimators with odd t.

The bandwidth h, which controls the level of smoothing, determines the properties of the local polynomial estimator. Too large a value of h gives an oversmoothed estimate, in which interesting structure can be smoothed over, while too small a value of h gives an undersmoothed estimate, with too much variability in the cell probability estimates ($h = 0$ corresponds to the frequency estimates). The plug-in approach to bandwidth selection substitutes an estimate for $\int f^{(t+1)}(u)^2\,du$ into (8.7), giving a choice for h. A simpler version of this approach substitutes the (known) value of $\int f^{(t+1)}(u)^2\,du$ for a "reference" density (say, Gaussian) into (8.7), although this choice is not consistent for h_0 if the wrong density is chosen. Another possibility is to use *cross-validation* (CV), wherein h is chosen to minimize an unbiased estimate of MSSE, viz.,

$$\sum_{j=1}^{k} \hat{\pi}_j^2 - \frac{2}{n} \sum_{j=1}^{k} n_j \hat{\pi}_j^{(j)},$$

where $\hat{\pi}_j^{(j)}$ is the value of $\hat{\pi}_j$ if one observation is deleted from the jth cell. Hall and Titterington (1987) demonstrated sparse asymptotic consistency for the kernel estimate based on CV, and showed that the CV bandwidth is asymptotically optimal. The close connection with nonparametric regression also suggests trying methods designed for local polynomial regression bandwidth selection, as in Ruppert et al. (1995) or Hurvich et al. (1998), for example.

Example 8.1: Trouble Spots of Automobiles
The April issue of *Consumer Reports* magazine is devoted to a description of new automobiles. The magazine also reports reliability figures for previous versions of automobiles being offered for sale that year, based on the experiences of readers who respond to an annual questionnaire. Reliability figures are given for 14 potential problem areas of the auto, such as the engine, the cooling system, the electrical system, and so on. Table 8.1 gives the number of potential problem areas (out of 14) for which more than 5% of the owners reported serious problems

TABLE 8.1 Number of Trouble Spots for 1992 Models of Medium-Sized Autos and Coupes

Make and Model	Number of Trouble Spots	Make and Model	Number of Trouble Spots
BMW 3-series	5	Mercury Cougar	7
Buick Century	5	Mercury Sable	8
Buick Regal	4	Mitsubishi Diamante	3
Buick Skylark	8	Mitsubishi Galant	5
Cadillac Eldorado	9	Nissan Maxima	3
Chevrolet Beretta	8	Oldsmobile Achieva	7
Chevrolet Corsica	8	Oldsmobile Ciera	5
Chevrolet Lumina	6	Oldsmobile Cutlass Supreme	6
Ford Taurus	8	Pontiac Grand Am	8
Ford Thunderbird	7	Pontiac Grand Prix	5
Honda Accord	2	Saab 900	5
Infiniti G20	2	Saab 9000	5
Lexus ES300	2	Subaru Legacy	3
Lexus SC300	2	Toyota Camry	4
Mazda 626	5		

(over the 12-month period from April 1994 through March 1995) for 29 1992 medium-sized autos and coupes equipped with an automatic transmission, as reported in the April 1996 issue. We will term this number the number of trouble spots for a given model of automobile. Considering the April 1996 issue as a realization of an annual (reasonably stationary) stochastic process, we use these data to estimate the distribution of probabilities of the number of trouble spots for a "typical" medium-sized auto (which could be applied to four-year-old autos in 1997, for example).

Figure 8.1 summarizes these data. The $\times$ values are the frequency estimates for the probability that there are j trouble spots, $j = 0, \ldots, 14$. The dotted lines connecting the frequency estimates emphasize that this is a distribution with ordered categories. The local polynomial estimator amounts to fitting a smooth curve through the $\times$ values. The $\circ$ values (connected by solid lines) are local quadratic estimates, wherein the kernel function is the tricube function

$$K(u) = 0.35(1 - u^3)^3, \quad -1 \leq u \leq 1,$$

and the bandwidth is chosen such that the kernel covers 60% of the cells (chosen by visual inspection). The smoothed probability estimates highlight the dominant pattern of most autos having between two and nine trouble spots, peaking at around four to eight trouble spots, while smoothing over the high variability of the frequency estimates.

The local quadratic estimate in Figure 8.1 is still not satisfactory, despite its pleasing smooth representation of the probabilities. The estimated probabilities do not reflect the sharp drop in the probability of having 10 or more trouble spots; indeed, there are no autos in the sample with that many trouble spots. (Presumably, such unreliable auto models are no longer being produced.) Even worse, 4 of the 15 cells (corresponding to the tails of the distribution) have negative estimated probabilities, which is, of course, undesirable.

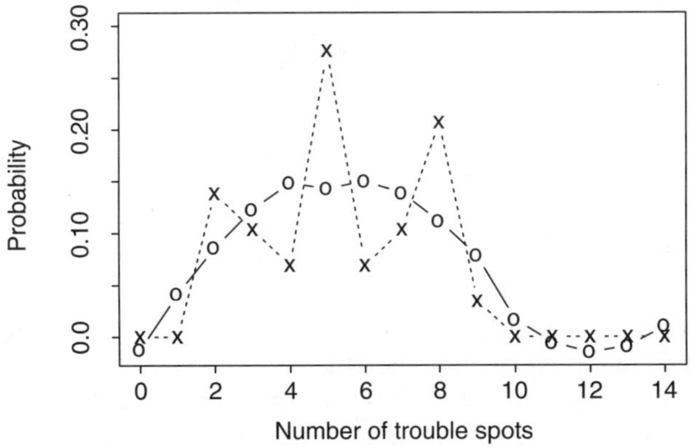

FIGURE 8.1 Number of trouble spots for 1992 models of medium sized autos and coupes ($\times$: frequency estimates, $\circ$: local quadratic estimates).

The reason for these difficulties is that the minimization of (8.4) is based on a local least-squares criterion, which is appropriate for regression data, but not tables of counts. In this context, it is more appropriate to use the multinomial likelihood function than to use a function based on least squares (see Simonoff, 1996b; Staniswalis, 1989; Tibshirani and Hastie, 1987). It is helpful to formulate the problem in terms of a Poisson model, since the multinomial distribution for $\{n_j\}$ can be derived by treating the n_js as independent Poisson variates and then conditioning on $\sum n_j = n$. The log-likelihood is then

$$\sum_{j=1}^{k}[n_j \log(n\pi_j) - n\pi_j]$$

(ignoring constants). The canonical link for the Poisson distribution is the log link (McCullagh and Nelder, 1989), so the local log-likelihood (Tibshirani and Hastie, 1987; Firth et al., 1991; Fan et al., 1995) has the form

$$L' = \sum_{j=1}^{k} \left\{ n_j \left[\beta_0 + \cdots + \beta_t \left(\frac{i}{k} - \frac{j}{k} \right)^t \right] - \exp\left[\beta_0 + \cdots + \beta_t \left(\frac{i}{k} - \frac{j}{k} \right)^t \right] \right\} K\left(\frac{i/K - j/K}{h} \right) \tag{8.8}$$

for cell i. The *local polynomial likelihood estimate* is then $\exp(\hat{\beta}_0)$, where $\hat{\beta}$ is the maximizer of (8.8). This guarantees that the estimate will be nonnegative and also has the effect of fitting locally exponential polynomial forms, which are more sensible than polynomial forms, to the density.

Example 8.2: Trouble Spots of Automobiles (Continued)
Figure 8.2 gives the local quadratic likelihood estimate for the automobile data from Table 8.1, using the same bandwidth as in Figure 8.1 (+ connected by solid lines). While the general pattern is similar to that of Figure 8.1, the unappealing aspects of the least-squares-based estimate are no longer present. The probability estimates are all nonnegative, and the virtually zero probability of an auto having 10 or more trouble spots is now evident.

Consumer Reports also reports the average reliability of all motor vehicles of a given model and year. For example, the average 1992 automobile had six trouble spots. If all models followed this same reliability profile, then the numbers in Table 8.1 could be modeled as a random sample from a binomial distribution based on 14 trials (the problem areas) and the probability of any one problem area being a trouble spot equal to 6/14. Figure 8.2 gives the corresponding probabilities for this binomial random variable (dots connected by dotted lines). It is apparent that the observed probability vector is more variable than that of the binomial; that is, there is extrabinomial variation. (A Pearson χ^2-goodness-of-fit statistic testing the fit of a binomial distribution equals 29.55 on 14 degrees of freedom, strongly rejecting a binomial fit ($p < 0.01$)).

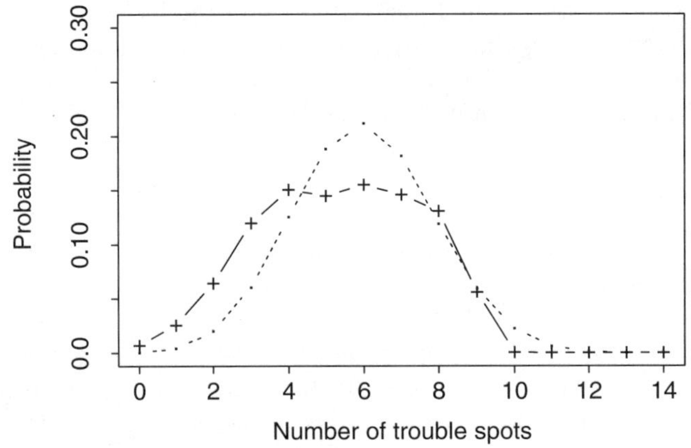

FIGURE 8.2 Number of trouble spots for 1992 models of medium-sized autos and coupes (+: local quadratic likelihood estimates, ·: estimates based on binomial trials).

Extrabinomial variation (also termed overdispersion) often arises when data that come from separate populations are combined, and that is the case here, as 16 of the autos are produced by US manufacturers and 13 by non-US manufacturers. The two subgroups are apparently different, as the average number of trouble spots for US-manufactured autos is 6.5, while that for non-US-manufactured autos is 3.67. Figure 8.3 gives the probability function for this mixture of binomials,

$$\frac{16}{29} \times \text{Binomial}\left(14, \frac{6.5}{14}\right) + \frac{13}{29} \times \text{Binomial}\left(14, \frac{3.67}{14}\right),$$

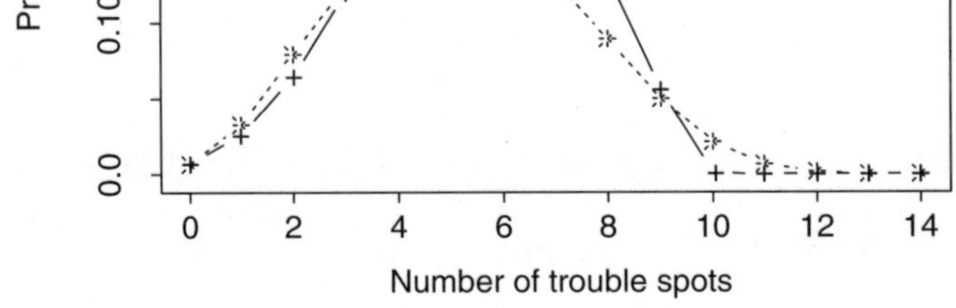

FIGURE 8.3 Number of trouble spots for 1992 models of medium-sized autos and coupes (+: local quadratic likelihood estimates, $*$: estimates based on mixture of binomial trials).

shown as $*$ connected by dotted lines, along with the local quadratic likelihood estimates. The figure supports the hypothesis that the mixture is the source of the extrabinomial variation, as does a Pearson χ^2-goodness-of-fit test, which equals 13.22 on 12 degrees of freedom ($p = 0.35$).

8.2.2 Contingency Tables

The problems associated with sparse multinomials are magnified when one moves to higher dimensions (contingency tables). Since the number of cells in a table increases multiplicatively with the dimension of the table, higher dimensional tables are more likely to be sparse. Fortunately, smoothing methods extend directly to higher dimensions as well. Unfortunately, however, so does the so-called *curse of dimensionality*, which refers to the need for progressively larger sample sizes in higher dimensions to achieve comparable accuracy.

Local polynomial estimators generalize to d-dimensional tables in the same way that they do for nonparametric regression with multiple predictors. For example, a local linear estimator $\hat{\pi}_{ij}$ for the probability of falling in the (i, j)th cell of an $R \times C$ two-dimensional table $\{n_{ij}\}$ is $\hat{\beta}_0$, where $\hat{\beta}$ is the minimizer of

$$\sum_{k=1}^{R}\sum_{l=1}^{C}\left[p_{kl} - \beta_0 - \beta_1\left(\frac{i}{R} - \frac{k}{R}\right) - \beta_2\left(\frac{j}{C} - \frac{l}{C}\right)\right]^2 K_{h_R,h_C}(i, j, k, l, R, C) \tag{8.9}$$

and $K_{h_R,h_C}(i, j, k, l, R, C)$ is a two-dimensional kernel function (typically, a symmetric bivariate density function) with h_R and h_C the smoothing parameters for rows and columns, respectively.

Aerts et al. (1997b) studied the properties of local polynomial estimation for sparse multidimensional contingency tables. For general d, define K_d: $\mathbb{R}^d \to \mathbb{R}$ to be the kernel function, often a d-variate probability density function, and let H be a nonsingular $d \times d$ matrix that contains the bandwidths to be used for smoothing. (H is often a diagonal matrix, as in (8.9), but need not be.) A popular technique for generating K_d is from a univariate kernel K using a product kernel:

$$K_d(u) = \prod_{j=1}^{d} K(u_j). \tag{8.10}$$

(More general formulations for K_d also are possible.) Assume that all second partial derivatives of f are piecewise continuous and square integrable, and assume regularity conditions on K_d. Define $h > 0$, and let the $d \times d$ matrix $\mathbf{A}$ satisfy $\mathbf{H} = h\mathbf{A}$, where $\mathbf{A}$ has a unit determinant. (This does not force a constant bandwidth for all dimensions, as $\mathbf{A}$ allows full generality of $\mathbf{H}$.) Then, if $h \to 0$ and $nh^dk \to \infty$ as $n \to \infty$ (where k is the total number of cells in the table), the *asymptotic* MSSE for the local linear estimator is

$$\text{AMSSE} = \frac{R(K_d)}{nh^dk} + \frac{h^4}{4k}\int \{\text{tr}[\mathbf{A}\mathbf{A}^T\nabla^2 f(u)]\}^2\, du_1 \cdots du_d, \tag{8.11}$$

where $\nabla^2 f(u)$ is the $d \times d$ Hessian matrix

$$\nabla^2 f(u)_{ij} = \frac{\partial^2 f(u)}{\partial u_i \partial u_j}.$$

Aerts et al. (1997b) also described generalization of these results to higher order local polynomial estimators.

The higher the dimension of the table, the more slowly the variance converges to zero, because of the relative lack of local information in higher dimensions. The variance is $O[(nh^d k)^{-1}]$, reflecting the curse of dimensionality. So, for example, for a two-dimensional table wherein R and C both become infinite at a rate proportional to $\sqrt{n}$ (so that the total number of cells in the table increases at the same rate as the sample size), the optimal MSSE rate is $O(n^{-2/3}k^{-1})$ (Burman, 1987), which is attained when $h = O(n^{-1/6})$ for both rows and columns, compared with the optimal $O(n^{-4/5}k^{-1})$ rate in one dimension.

Just as was true in one dimension, a more appropriate version of this estimator is based on local likelihood, rather than local least squares. Equation (8.8) generalizes in the obvious way. For example, the local linear likelihood estimator for a two-dimensional table is $\exp(\hat{\beta}_0)$, where $\hat{\beta}_0$ is the constant term of the maximizer of

$$\sum_{k=1}^{R}\sum_{l=1}^{C}\left\{n_{kl}\left[\beta_0 + \beta_1\left(\frac{i}{R} - \frac{k}{R}\right) + \beta_2\left(\frac{j}{C} - \frac{l}{C}\right)\right] - \exp\left[\beta_0 + \beta_1\left(\frac{i}{R} - \frac{k}{R}\right)\right.\right.$$
$$\left.\left. + \beta_2\left(\frac{j}{C} - \frac{l}{C}\right)\right]\right\} K_{h_R,h_C}(i, j, k, l, R, C). \qquad (8.12)$$

Example 8.3: Bond Rating

Researchers into the characteristics of urban environments have long been aware of the importance of large organizations in determining a city's health (both fiscally and from the point of view of quality of life); see, for example, Perrow (1991). Historically, researchers focused attention on the for-profit sector, but the public and not-for-profit sectors are also important, particularly because downsizing has reduced the size of many for-profit organizations in recent years. Table 8.2 gives a cross-classification of the fiscal health of 57 of the 75 largest U.S. cities, as measured by *Standard and Poor's* 1994 rating for their general-obligation bonds (the other 18 cities do not have such bonds) and by the number of not-for-profit organizations among a city's top 10 employers, according to Chamber of Commerce lists published in 1995 (Abzug et al., 1999). Bond rating, which is ordered from AAA (the top rating) to BB (the lowest rating for the cities studied), has been shown to be a good proxy for a city's general fiscal health (Przybylski et al., 1996), and the number of organizations among the top 10 employers is a measure of the importance of the not-for-profit sector in the life

TABLE 8.2 Standard and Poor's Ratings of General Obligation Bonds by the Number of Not-for-Profits Among the Top Ten Employers for 57 US Cities

Bond	Number of Not-for-Profits Among Top 10 Employers					
Rating	0	1	2	3	4	5
AAA	0	4	2	0	0	0
AA+	0	2	2	2	0	0
AA	4	9	5	2	1	0
AA−	1	1	4	0	0	0
A+	0	0	0	1	1	0
A	0	1	1	4	1	0
A−	1	0	2	0	0	0
BBB+	0	0	0	0	0	0
BBB	1	0	1	1	0	0
BBB−	0	0	0	0	0	2
BB+	0	0	0	0	0	0
BB	0	0	0	1	0	0

of the city. Thus, the joint probability distribution of the two variables describes the connection between the prominence of the not-for-profit sector and fiscal health. The 1995 data can be considered a realization of the underlying process, so the distribution shown could apply to the current employment situation in large U.S. cities.

The table is quite sparse: 12×6 with 57 observations. Table 8.3 is the local quadratic likelihood estimate of the expected cell counts produced by using a product tricube kernel covering the closest 35% of the cells and treating the cell positions as being on an equispaced grid. (The expected cell counts are given so that a direct comparison with Table 8.2 is possible; the estimated probabilities are the expected cell counts divided by 57.) The bandwidth was chosen by visual inspection. It is apparent that expected counts are generally concentrated in the upper left of the table, corresponding to higher ratings and fewer not-for-profits among the top 10 employers. The table also shows a clear relationship between the bond rating and the number of not-for-profits. A higher bond rating is associated with a lower number of not-for-profits among the top 10 employers. In particular, the "ideal" number of not-for-profits is apparently unity, with zero or two not-for-profits being roughly similar thereto. On the other hand, as the number of not-for-profits among the top ten employers gets larger, the probability of a lower bond rating also increases.

There are several possible reasons for this pattern. First, not-for-profits do not contribute to a city's tax base, so very large not-for-profit organizations constitute a potential drain on government resources, compared with comparably sized for-profit organizations. Further, most large not-for-profit organizations are hospitals and health organizations, and given the strong political pressures on health-care

TABLE 8.3 Local Quadratic Likelihood Estimated Cell Counts for US City Bond Rating Table

Bond	Number of Not-for-Profits Among Top 10 Employers					
Rating	0	1	2	3	4	5
AAA	0.51	2.08	1.82	0.46	0.05	0.00
AA+	1.27	4.85	3.64	1.05	0.17	0.00
AA	1.72	5.60	3.97	1.45	0.33	0.00
AA−	1.25	2.61	2.91	1.70	0.55	0.00
A+	0.60	0.74	1.79	1.74	0.79	0.01
A	0.35	0.33	1.13	1.48	0.71	0.01
A−	0.36	0.18	0.84	0.97	0.36	0.02
BBB+	0.44	0.07	0.64	0.47	0.14	0.14
BBB	0.37	0.03	0.41	0.22	0.06	0.46
BBB−	0.19	0.05	0.20	0.20	0.09	0.51
BB+	0.05	0.03	0.14	0.22	0.14	0.42
BB	0.01	0.01	0.16	0.28	0.18	0.24

providers in the US in recent years, a large dependence on such employers in a city could be viewed as risky by the bond market. Finally, a larger number of not-for-profits among the largest employers also reflects a smaller number of large employers from the public and for-profit sectors, potentially indicating a less healthy municipal government (that is, one unable to provide many services beyond basic police and fire protection and road repair) and a less healthy business sector. On the other hand, the existence of one or two (but not more) large universities or hospitals in a city can be a sign of the city's prominence and is associated with relatively higher bond rating.

These data also can be viewed as regression data, with the bond rating being the (ordinal categorical) response variable and the number of not-for-profits among the top 10 employers the (ordinal categorical) predictive variable. Smoothing methods for this type of data are the focus of Section 8.3.

8.2.3 Connections to Regression and Density Estimation

As noted earlier, and as Equations (8.4) and (8.9) show, categorical data and contingency table smoothing are operationally very similar to nonparametric regression, with the cell relative frequencies being the response values and the cell indices being the predictor values. This close connection is reflected in the theoretical properties of the estimates, as the MSSE results (8.6) and (8.11) equal the corresponding local polynomial-regression integrated mean-squared error values divided by k.

Categorical data and contingency table smoothing are also closely connected to density estimation, through (8.5), with an estimate of a cell probability p_i giving a density estimate for $f(x_i)$. If the bins are so narrow that only one observation can fall into any bin (except for exact duplications, to the resolution of the data), the resultant regression estimator corresponds exactly to a density estimator for continuous data. For example, the univariate local linear regression–based probability estimator that is the constant term of the minimizer of (8.4) with $t = 1$ becomes equivalent to the constant term of the minimizer of

$$\int \left[n^{-1} \sum_{i=1}^{n} \delta(u - x_i) - \beta_0 - \beta_1(x - u) \right]^2 K\left(\frac{x-u}{h}\right) du$$

as the bins narrow, substituting i/k for x and j/k for u and taking δ to be the Dirac delta function. This local polynomial estimator is a generalized jackknifing continuous-boundary kernel density estimator (Jones, 1993).

Similarly, as the bins narrow, (8.8) becomes approximately

$$\sum_{i=1}^{n} K\left(\frac{x-x_i}{h}\right) \log[f(x_i)] - n \int K\left(\frac{x-u}{h}\right) f(u)\, du$$

(ignoring constants, changing to a continuous scale by substituting x for i/k, taking the first sum to be over observations instead of cells, and approximating the second sum with an integral). This equation is identical to the local likelihood for density estimation proposed in Hjort and Jones (1996) and Loader (1996) and has the same advantages over the usual kernel density estimators as do those of the local likelihood probability estimators over kernel probability estimators, including automatic boundary bias correction, improved bias (particularly in the tails), and guaranteed nonnegativity.

8.2.4 Computational Issues

Kernel and local polynomial regression estimation are available in many statistical packages. Equations (8.2) and (8.4) demonstrate that categorical data smoothing can be accomplished on the basis of the least-squares criterion, with the response variable being the frequency estimates and the predicting variable the cell number divided by the number of cells. Thus, these estimates can be easily calculated for one-dimensional data by using, for example, any of the packages Data Desk, MINITAB, NCSS, S-Plus, SPSS, Stata, Systat, XLISP-STAT, and XploRe. JMP, SAS/INSIGHT, and S-Plus can be used for multidimensional contingency tables, since they provide code for loess, a multiple predictor local polynomial method. Such probability matrices are estimated using d predictor variables based on an equispaced design corresponding to the cell index in a given dimension divided by the number of cells in that dimension.

Local likelihood-based estimators can be calculated with the use of a package that allows smoothing-based generalized likelihood modeling. Such packages include GLIM and S-Plus.

8.3 SMOOTHING APPROACHES TO CATEGORICAL REGRESSION

8.3.1 Kernel Regression Estimation

Let us start with the simple case of a dichotomous response, where the data are given by (y_i, x_i), $i = 1, \ldots, n$, with $y_i \in \{0, 1\}$ and x_i denoting a univariate explanatory variable. In this case, the expectation of the response is given by $E(y_i|x_i) = P(Y = 1|x_i)$. Thus, estimating $\pi(x) = P(Y = 1|x)$ coincides with estimating the expectation, for which we can draw on the literature on continuous responses, as was done in Section 8.2.

Kernel Regression for a Dichotomous Response

The kernel-based regression smoother applied to dichotomous response data has the form

$$\hat{\pi}(x) = \hat{P}(Y = 1|x) = \sum_{i=1}^{n} y_i \omega_h(x, x_i), \tag{8.13}$$

where $\omega_h(x, x_i)$ is a weight function reflecting the distance between the evaluation value x of the predictor and observation x_i. For continuous x, weights may be of the Gasser–Müller, Priestley–Chao, or Nadaraya–Watson type (see Chapter 3). The last-named one is given by

$$\omega_h(x, x_i) = K_h(x|x_i) \Big/ \sum_{j=1}^{n} K_h(x|x_j),$$

where $K_h(x|x_i) = K((x - x_i)/h)$ is a continuous, symmetric unimodal kernel function. Müller and Schmitt (1988) have considered Gasser–Müller estimates of the form (8.13) in quantal bioassay and give asymptotic results, see also Staniswalis and Cooper (1988).

In the case of a categorical explanatory variable, discrete kernels may be used instead of continuous ones. In this situation, $K((x - x_i)/h)$ is replaced by a discrete kernel function — for example the Aitchison-and-Aitken kernel (8.3). This function treats all categories where $x \neq x_i$ as equivalent and is therefore less appropriate for a variable with ordered categories, for which a kernel that reflects the ordering is more sensible.

In real data sets, there is only a limited number $g \leq n$ of distinct values of the explanatory variables. For a categorical explanatory variable x, the number of design points or outcomes g is fixed from the beginning, so a grouped version of (8.13) is more efficient with respect to computation time. Let $x_{(1)} < \cdots < x_{(g)}$ denote the distinct values of the explanatory variable, and let $p_{(i)}$ denote the relative frequency of outcome $Y = 1$ at fixed point $x_{(i)}$ based on $m_{(i)}$ observations. Then (8.13) has the form

$$\hat{P}(Y = 1|x) = \sum_{i=1}^{g} p_i \tilde{\omega}_h(x, x_{(i)}), \tag{8.14}$$

with weights $\tilde{\omega}_h(x, x_{(i)}) = m_{(i)}\omega(x, x_{(i)})$. The estimate (8.14) is a weighted sum of relative frequencies wherein the weights contain the number of observations used in the computation of the relative frequencies. Since the number of local observations, $m_{(i)}$, may vary strongly (if, for example, different sampling schemes were used for different i), it is important to include that number in the weighting procedure. Note that this issue was not relevant in Section 8.2, where the multinomial character of the contingency table implied a fixed total sample size for the table.

A Nadaraya–Watson Type of Estimator for a Multivariate Response

Estimate (8.14) was derived straightforwardly from the continuous kernel regression approach. This is because of the univariate nature of dichotomous data. If the number of response categories exceeds two, we have to deal with the multivariate case, and the usual kernel regression must be modified. One approach is to consider each category of y as a dichotomous response. This assumption is appropriate, however, only for nominal y. If y is ordinal, the ordering of response categories should be taken into account. As is shown next, this may be done in the spirit of the Nadaraya–Watson kernel estimator.

Accordingly, let Y now be multicategorical; that is, $Y_i \in \{1, \ldots, k\}$ with $y_i^T = (y_{i1}, \ldots, y_{ik})$ denoting dummy variables with $y_{ir} = 1$ if $Y_i = r$ and $y_{ir} = 0$ otherwise. The response probability $\pi_r(x) = P(Y = r|x)$ may be rewritten (abusing notation in the obvious way) as $P(Y = y|x) = p(y, x)/p(x)$, where $p(y, x)$ and $p(x)$ denote the densities for (y, x) and x, respectively. The Nadaraya–Watson estimate is usually constructed by substituting estimates for $p(y, x)$ and $p(x)$. The common kernel density estimate for $p(x)$ has the form

$$\hat{p}(x) = \frac{1}{n}\sum_{i=1}^{n} K_{h_1}^{(x)}(x|x_i).$$

For the joint probability $p(y, x)$, one can use the product kernel estimate

$$\hat{p}(y, x) = \frac{1}{n}\sum_{i=1}^{n} K_{h_1}^{(x)}(x|x_i)K_{h_2}^{(y)}(y|y_i)$$

as in (8.10), where $K^{(x)}$ is a kernel for the explanatory variable and $K^{(y)}$ is a kernel for the dependent variable y. Depending on the character of the explanatory variable, $K^{(x)}$ may be a continuous or a discrete kernel, whereas a discrete kernel will be appropriate for $K^{(y)}$. Substituting $\hat{p}(x)$ and $\hat{p}(y, x)$ for $p(x)$ and $p(y, x)$, respectively, yields the discrete kernel regression estimator

$$\hat{\pi}_y(x) = \hat{P}(Y = y|x) = \sum_{i=1}^{n} K_{h_2}^{(y)}(y|Y_i)\omega_{h_1}(x, x_{(i)}) \tag{8.15}$$

with the usual Nadaraya–Watson weights

$$\omega_{h_1}(x, x_i) = \frac{K_{h_1}^{(x)}(x|x_i)}{\sum_{j=1}^{n} K_{h_1}^{(x)}(x|x_j)}.$$

The grouped version of (8.15) is given by

$$\hat{\pi}_r(x) = \sum_{i=1}^{g} p_{ir} K_{h_2}^{(y)}(y|Y_i)\tilde{\omega}_{h_1}(x, x_{(i)}),$$

where p_{ir} denotes the relative frequency of category r at measurement point $x_{(i)}$, $i = 1, \ldots, g$.

In (8.15) two smoothing concepts are used, namely, smoothing over the response categories $1, \ldots, k$ and smoothing across the explanatory variables. For the first type of smoothing, the amount of smoothing is determined by h_2; for the second type, h_1 is the determining parameter. Which of the two smoothing concepts is more important depends on the regression. In the dichotomous case ($k = 1$), not much is gained from smoothing across the response categories. Thus, the limiting case $h_2 \to 0$, which yields the simpler estimate (8.13), does equally well in most cases. If, however, the number of categories k is large compared with the range of x (i.e., the number of distinct values g), then smoothing across response categories becomes important. In the extreme case where no covariate is present ($g = 1$), smoothing across response categories is actually the *only* smoothing concept involved. In this case, (8.15) yields the usual kernel estimator for categorical data described in Section 8.2.

The smoothing parameters h_1 and h_2 in (8.15) control the smoothness of the final regression estimate, so their choice is crucial. A common method of choosing them is to use cross-validation, where an unbiased estimate of the mean integrated squared error of the estimator is derived by successively dropping each observation from the sample (see Chapter 4). The cross-validation bandwidths then minimize the cross-validation score,

$$[1 - \hat{\pi}(y|x)]^2 + \sum_{\tilde{y} \neq y} \hat{\pi}(\tilde{y}|x)^2. \tag{8.16}$$

Minimization of (8.16) leads to consistent estimates of $\pi(y|x)$ under regularity conditions for all discrete kernels that are continuous in h_1 (Tutz, 1991).

Example 8.4: Bond Rating (Continued)

Now let us consider the bond-rating data as a regression problem with the bond rating as response variable and the number of not-for-profits as regressor. Since the number of ordinal response categories is rather high in comparison to the values of the predictor variable, the double-smoothed discrete kernel regression estimator (8.15) is used. Cross-validation yields $h_1 = 0.7$ and $h_2 = 0.9$, based on the discrete Habbema kernel, where the kernel weights are proportional to $h^{(i-j)^2}$. Thus, cross-validation indicates that smoothing across response categories is useful. Figure 8.4 and Table 8.4 give the estimated response probabilities. From the Figure, it is seen that, with increasing number of not-for-profits, the probability is shifted to higher response categories (lower ranking). The same effect may be seen from Table 8.3, where the cell counts are estimated by treating the data as a contingency table. Although the basic results are the same in Tables 8.3 and 8.4, there are some differences in the estimates because there are basic differences between the approaches underlying those tables. Table 8.3 is based on the local quadratic likelihood method of estimating cell counts, whereas the direct kernel regression that yields Table 8.4 is based on a local constant least-squares fit. Moreover, for the direct kernel estimate, the smoothing parameter was chosen by least-squares cross-validation, a criterion that has rather high sample variability and in the present case yields rather weak smoothing, resulting in some bumpiness of the curve in Figure 8.4. The cell counts in Table 8.3 were estimated on the basis of bandwidths chosen by visual inspection, resulting in a smoother pattern; this could have also been done in Figure 8.4 and Table 8.4 by increasing h_1 and h_2, respectively. Despite these differences, the most important point is the similarity of the implications from either formulation of the problem.

8.3.2 Local Polynomial Modeling

As described in Section 8.2, local polynomial approaches to contingency table estimation have many advantages over kernel-based estimators (which may be considered a special case thereof). This effect occurs in the regression context

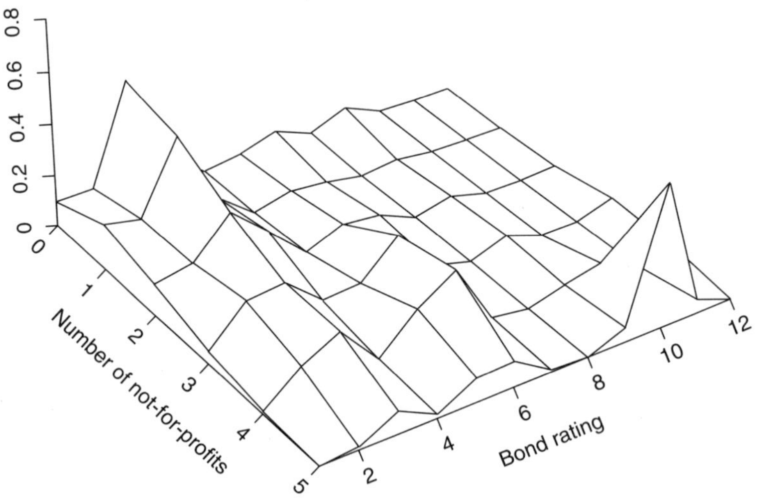

FIGURE 8.4 Bond rating as a function of the number of not-for-profits among the top 10 employers, based on direct regression kernel estimate.

TABLE 8.4 Estimated Probabilities for Bond Rating Depending on the Number of Non-for-Profits

Bond Rating	Number of Not-for-Profits Among Top 10 Employers					
	0	1	2	3	4	5
AAA	0.095	0.181	0.125	0.046	0.009	0.000
AA+	0.097	0.147	0.143	0.149	0.093	0.013
AA	0.472	0.419	0.297	0.207	0.202	0.089
AA−	0.137	0.126	0.169	0.084	0.039	0.017
A+	0.011	0.013	0.032	0.091	0.180	0.095
A	0.028	0.049	0.090	0.233	0.277	0.101
A−	0.071	0.035	0.074	0.052	0.028	0.009
BBB+	0.014	0.005	0.011	0.009	0.004	0.000
BBB	0.069	0.021	0.043	0.063	0.044	0.058
BBB−	0.007	0.002	0.004	0.007	0.075	0.558
BB+	0.000	0.000	0.001	0.005	0.011	0.056
BB	0.000	0.000	0.011	0.053	0.039	0.003

also. The first step is to replace the kernel smoothing over values of x with local polynomial smoothing.

For simplicity of notation, we consider in the following the grouped-data form of the estimator, where $x_1 < \cdots < x_g$ denote the distinct values of the explanatory variable. At measurement point x_i, the number of observations is m_i, yielding the vector of observations $n_i^T = (n_{i1}, \ldots, n_{ik})$, $\sum_{r=1}^{k} n_{ir} = m_i$, and the vector of relative frequencies $p_i^T = (p_{i1}, \ldots, p_{ik})$.

For the dichotomous case, with $k = 2$, local polynomial least squares fitting at target point x is given by

$$\sum_{j=1}^{g} [p_{j1} - \beta_0 - \cdots - \beta_t (x - x_j)^t]^2 K_h(x|x_i). \tag{8.17}$$

The constant term of the minimizer of (8.17) yields the estimate $\hat{\pi}_1(x) = \hat{P}(Y = 1|x)$. Obviously, the kernel-type estimate (8.14) is a special case in which the degree of the polynomial is given by $t = 0$. Moreover, if x refers to the margin of a contingency table, (8.17) is equivalent to (8.4). However, here x may also be a continuous explanatory variable.

As noted earlier, least-squares fitting is not the best choice for categorical response data, since heteroscedasticity (which is a basic trait of categorical data) is not taken into account and probability estimates can be negative. Local polynomial likelihood fitting implicitly takes care of these problems. For simplicity, the Poisson likelihood based on the log-link function may be used. With $n_{ir} \sim P(m_i \pi_r(x_i))$, $r = 1, 2$, the local polynomial likelihood for the estimation of $\pi_r(x_i)$

is given by

$$\sum_{j=1}^{g}\{n_{jr}[\beta_0+\cdots+\beta_t(x_i-x_j)^t]-\exp[\beta_0+\cdots+\beta_t(x_i-x_j)^t]\}K_h(x_i|x_j). \tag{8.18}$$

Maximization of (8.18) yields $m_i\hat{\pi}_r(x_i)=\exp(\hat{\beta}_0)$, and thus, $\hat{\pi}_r(x_i)=\exp(\hat{\beta}_0)/m_i$. Of course, in the dichotomous case, it is sufficient to maximize (8.18) only once for $\pi(x_i)=\pi_1(x_1)$. There are, however, two disadvantages of using (8.18). First, since $\hat{\pi}_r(x_i)$ is based on m_i, $\pi_r(x_i)$ may be estimated only at measurement points $x_1,\ldots,x_g$, where $m_i\neq 0$. Second, a very large h yields estimates with $\hat{\beta}_0$ constant for all i, but this implies probability estimates of the form $\hat{\pi}_r(x_i)=\text{constant}/m_i$, which is not a constant. Note that these issues do not come up when one uses the Poisson likelihood in the contingency table context, since $g=1$ and $m_1=n$ in that case.

An alternative is to use the original binomial distribution underlying the binary response variable. With $n_i=n_{i1}$ and $\pi_1(x_i)=\pi(x_i)$, one has $n_i\sim B(m_i,\pi(x_i))$, and the binomial log-likelihood function at x_i is given by

$$n_i\log\frac{\pi(x_i)}{1-\pi(x_i)}+m_i\log(1-\pi(x_i))=n_i\theta(x_i)-m_i\log(1+\exp(\theta(x_i))),$$

where $\theta(x_i)=\log\{\pi(x_i)/[1-\pi(x_i)]\}$ is the natural parameter. Thus, assuming the canonical link $\log\{\pi(x_i)/[1-\pi(x_i)]\}=z^T\beta$, one obtains, for the estimation of $\pi(x)$, the local polynomial likelihood

$$\sum_{j=1}^{g}\{n_j[\beta_0+\cdots+\beta_t(x-x_j)^t]$$
$$-m_j\log(1+\exp[\beta_0+\cdots+\beta_t(x-x_j)^t])\}K_h(x|x_j). \tag{8.19}$$

The estimator for $\pi(x)$ is given by $\log\hat{\pi}(x)/(1-\hat{\pi}(x))=\hat{\beta}_0$, yielding $\hat{\pi}(x)=\exp(\hat{\beta}_0)/(1+\exp(\hat{\beta}_0))$. Equation (8.19) may be used directly to estimate $\pi(x)$ for any value x.

A disadvantage of the local polynomial likelihood approach is that the estimation of $\hat{\beta}_0,\ldots,\hat{\beta}_t$ requires iterative procedures. For each target value x, some form of iterative proportional fitting or Fisher scoring has to be performed. (This is also true for local polynomial likelihood contingency table fitting, but only one iteration cycle is necessary, as that situation corresponds to only one predictor value x; for details of the iterative fit for the local polynomial likelihood approach, see, e.g., McCullagh and Nelder, 1989.)

It is noteworthy that for the special case $t=0$, estimation based on (8.19) also yields the simple Nadaraya–Watson estimator applied to the relative frequencies $P(Y=1|x)=\sum_j\tilde{\omega}_h(x,x_j)p_j$. Thus, for $t=0$, one obtains a simple linear estimator that may be computed directly, without iterations.

In the multicategorical case, where $k > 2$, the kernel estimator (8.15) uses additional smoothing over response categories. Local polynomial fitting in this case then refers to using a local polynomial for both response y and explanatory variable x. On the basis of a Poisson likelihood, one considers $n_{ir} \sim P(m_i\pi_{ir})$, and the estimation of $\pi_r(x_i)$ results from maximization of the local polynomial (linear-case) likelihood

$$\sum_{j=1}^{g}\sum_{s=1}^{k}\left\{n_{js}\left[\beta_0+\beta_y\left(\frac{r}{k}-\frac{s}{k}\right)+\beta_x(x_i-x_j)\right]\right.$$
$$\left.-\exp[\beta_0+\beta_y\left(\frac{r}{k}-\frac{s}{k}\right)+\beta_x(x_i-x_j)]\right\}K_{1,h_1}\left(\frac{r}{k}\Big|\frac{s}{k}\right)K_{2,h_2}(x_i|x_j), \quad (8.20)$$

yielding $m_i\hat{\pi}_{ir} = \exp(\hat{\beta}_0)$, and therefore, $\hat{\pi}_{ir} = \exp(\hat{\beta}_0)/m_i$. If $h_1 \to 0$, then (8.20) reduces to local polynomial fitting over x, and for $k = 2$, one obtains (8.18). As in the case of the double-smoothed kernel regression estimator (8.15), the importance of smoothing over y and x depends on the number of measurement points g, the number of response categories k, and the number of observations n. For categorical x, equation (8.20) leads to the smoothing of contingency tables as given in equation (8.12), the only difference being that the regression estimates are estimates of the conditional probability given x_i and are normalized by m_i rather than n. As noted earlier, the use of the Poisson likelihood is most appropriate when m_i is constant for all i.

A thorough investigation of the asymptotic behavior in the dichotomous case has been undertaken by Fan et al. (1995). (For an earlier investigation, see also Staniswalis and Cooper, 1988 and Staniswalis, 1989)). Fan et al. (1995) show that, for odd degree, the asymptotic bias is of the same order of magnitude near the boundary as in the interior of the function. For even degree, the boundary bias dominates the interior bias. The asymptotic variance is $O(n^{-1}h^{-1})$, whereas the order of the bias decreases with the degree of the polynomial (e.g., it is h^2 for linear, and h^4 for quadratic, fit). However, for a globally optimal bandwidth, local linear fits often seem to be more effective than quadratic fits. The bandwidths that minimize the asymptotic mean squared error are of order $n^{-1/(2t+3)}$ for t odd and $n^{-1/(2t+5)}$ for t even.

Example 8.5: Failure of Newly Founded Firms

For econometricians, the factors that determine the success of newly founded firms are of interest from a scientific viewpoint. People who work in credit scoring have a more practical viewpoint, because the failure of firms usually means problems with credit. We consider a sample of 1701 Bavarian firms that were founded between January 1985 and December 1986 (see Brüderl et al., 1992) and that were observed for at least three years following their foundation. The response is dichotomous, with $y = 0$ signifying that the firm was still in business after three years and $y = 1$ signifying failure of the firm within the first three years. We restrict the consideration to one predictor variable, namely, the age of the founder, who also was the owner of the firm. Age as a substitute for experience has been shown to have some effect upon the survival of the firm, and a preliminary investigation indicates that the effect is stable across subgroups, built from variables that incorporate the size of the firm or the invested capital.

Figure 8.5(a) shows the local relative frequencies of failure and the fit of two versions of the parametric logit model. In the simple version the predictor is linearly specified, and in the second version the model includes a quadratic term. Clearly, both models fit well only in the medium range of age. For low and high values of the predictor, the fit is very unsatisfying. In Figure 8.5, the local polynomial likelihood fit based on (8.19) and the canonical logit model are given. The figure shows local polynomial likelihood smoothers of various degrees, namely, the simple local likelihood smoother ($t = 0$) in (b), the linear smoother ($t = 1$) in (c), and the quadratic smoother ($t = 2$) in (d). On the basis of the Epanechnikov kernel, the smoothing parameter is chosen by maximizing the AIC criterion (see, Hastie and Tibshirani, 1990, pp. 158ff), because problems in the iterative estimation procedure occurred with the use of CV. This criterion yields $h = 0.5$, $h = 0.75$, and $h = 4.00$ for $t = 0, 1, 2$ respectively (if the age variable is normalized to have a standard deviation of unity). Plainly, local polynomial likelihood gives an improved fit at the left and the right boundary. While the simple local likelihood smoother and the linear smoother are rather

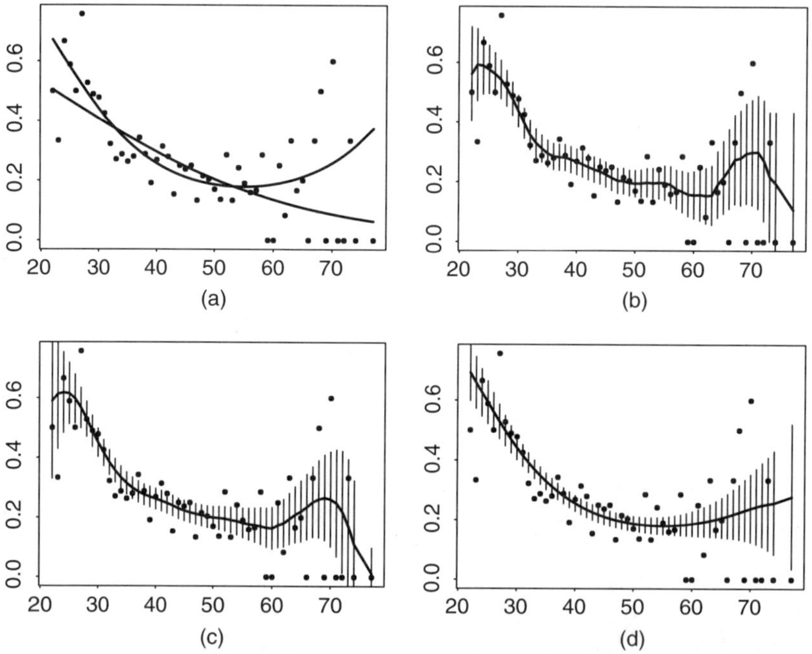

FIGURE 8.5 Failure of newly founded firms as a function of the age of the person in charge. (a) Linear and quadratic logit model; local polynomial likelihood, (b) $t = 0$, (c) $t = 1$, (d) $t = 2$.

similar, the quadratic estimator yields a very smooth curve that is quite similar to the simple parametric model with a quadratic term. This is, of course, due to the large smoothing parameter, which makes the fit almost a global quadratic one.

Moreover, in Figure 8.5, variability bands are given that are computed by 1.96 times the estimated standard deviation. The standard deviation is based on the first term in equation (8.30), which gives an approximation for the more general case of an ordered response variable. Although the intervals are rather wide for low and high ages (which is an effect of local sample sizes in the regions), between 25 and 60 years of age the probability of failure is obviously decreasing with increasing age. Thus, one should be cautious in founding a firm if one has little experience.

8.3.3 Penalizing the Smooth Component

The approaches considered so far are almost model free, in the sense that a parametric model is used only in the specification of the link function for the Poisson or binomial model. Since these models are employed only locally (with respect to y and x), almost no structure is assumed. An alternative approach that also uses a link function and, therefore, a generalized linear model is based on penalizing functions. Let us consider briefly the structure of (univariate) generalized linear models. We restrict the discussion to the case of a dichotomous response.

A generalized linear model for observations (y_i, x_i), $y_i \in \{0, 1\}$, has the form

$$\pi(x_i) = h(z_i^T \beta),$$

where z_i is a design vector composed from the x_i and β is a parameter vector. The likelihood contribution of the ith (grouped) observation p_i is given by

$$l_i(\beta) = n_i \log \frac{\pi(x_i)}{1 - \pi(x_i)} + m_i \log(1 - \pi(x_i)).$$

Simple derivation shows that the score function of the ith observation is given by

$$s_i(\beta) = \frac{\partial l_i(p_i; \beta)}{\partial \beta} = z_i d_i(\beta) \frac{p_i - \pi(x_i)}{\sigma_i^2(\beta)},$$

where $\sigma_i^2(\beta) = \pi(x_i)(1 - \pi(x_i))/m_i$ is the covariance of p_i evaluated with parameter β and $d_i(\beta) = \partial h(z_i^T \beta)/\partial \eta$ is the derivative of the response function. The score equation $s(\beta) = \sum_{i=1}^{g} s_i(\beta) = 0$ may be solved by iterative Fisher scoring, yielding the updated estimate

$$\beta^{(p+1)} = \beta^{(p)} + F^{-1}(\beta^{(p)}) s(\beta^{(p)}),$$

where $F(\beta) = \sum_{i=1}^{g} [d_i(\beta)^2 / \sigma_i^2(\beta)] z_i z_i^T$ denotes the Fisher matrix.

Fisher scoring may be rewritten as an iteratively weighted least-squares estimate (McCullagh and Nelder, 1989). Defining the pseudo-observations $\tilde{y}_i(\beta) = z_i^T\beta + (y_i - \pi_i(\beta))/d_i(\beta)$, one obtains

$$\hat{\beta}^{(p+1)} = (\mathbf{Z}^T\mathbf{W}^{(p)}\mathbf{Z})^{-1}\mathbf{Z}^T\mathbf{W}^{(p)}\tilde{y}^{(p)}, \tag{8.21}$$

where $\mathbf{Z}^T = [z_1, \ldots, z_g]$ is the transposed total design matrix, $\mathbf{W}^{(p)} = \text{diag}\,(d_i^2(\beta^{(p)})/\sigma_i^2(\beta^{(p)}))$ is a weight matrix, and $\tilde{y}^{(p)} = (\tilde{y}_1(\beta^{(p)}), \ldots, \tilde{y}_g(\beta^{(p)}))$ is the vector of pseudo observations.

Smoothing proceeds by replacing the parameterized predictor $\eta_i = z_i^T\beta$ with the general predictor $\eta_i = f(x_i)$. In the model

$$\pi(x_i) = h(f(x_i)),$$

the parameters are replaced by $f(x_i)$, where $f(\cdot)$ is a sufficiently smooth function. Estimation is then based on maximizing the penalized log-likelihood

$$\sum_{i=1}^{g} l_i(y_i; f(x_i))/m_i - \tfrac{1}{2}\lambda \int (f''(u))^2\, du, \tag{8.22}$$

where l_i is the likelihood contribution and the second term is a roughness penalty in terms of f'', which denotes the second derivative of f. (See Simonoff, 1983, 1987, for related estimators in the contingency table context.) The maximization of the penalized maximum likelihood criterion (8.22) yields a cubic spline function. The criterion may be transformed into

$$\sum_{i=1}^{g} l_i(y_i; f(x_i))/m_i - \tfrac{1}{2}\lambda f^T\mathbf{K}f, \tag{8.23}$$

where $f^T = (f(x_1), \ldots, f(x_g))$ is the vector of the predictor values and $\mathbf{K}$ is the penalty matrix that corresponds to spline smoothing. (See, e.g., Green and Silverman, 1994, p. 12, or Fahrmeir and Tutz, 1994, p. 154.)

The iterative Fisher scoring (8.21) now has the very simple form

$$f^{(p+1)} = \mathbf{S}^{(p)}\tilde{y}^{(p)}, \tag{8.24}$$

with the smoother matrix $\mathbf{S}^{(p)} = (\mathbf{W}^{(p)} + \lambda\mathbf{K})^{-1}\mathbf{W}^{(p)}$ and the weighting matrix $\mathbf{W}^{(p)}$ from (8.21). Therefore, one has an iterative estimator that is essentially a linear smoother of the real-valued pseudo observations. Note that this Fisher scoring procedure is not restricted to the spline criterion (8.22): In (8.24), any smoother matrix $\mathbf{S}$ may be used. The corresponding penalized maximum likelihood criterion is given by (8.23), with the roughness penalty matrix $\mathbf{K} = \mathbf{S}^- - \mathbf{I}$, where $\mathbf{S}^-$ is a generalized inverse of S. The procedure considered here connects linear smoothing of continuous responses with categorical-response data smoothing by applying the former type of smoothing to the adjusted variables or pseudo observations $\tilde{y}_i$. This procedure is also the kernel of generalized additive models (GAMs), as considered by Hastie and Tibshirani (1990).

8.3.4 Model-Based Smoothing for Ordinal Data

The (univariate) penalization criterion (8.22) cannot be used directly on an ordinal response variable. Although the penalizing approach may be extended to multi-category response data, in the following discussion we will consider a local likelihood approach that is easy to use by basing the analysis on existing ordinal models. The ordinal scale level of the response is exploited by locally specifying an ordinal regression model, where "locally" means now with respect to explanatory variable x only. Incorporating a model into the smoothing process has the big advantage that the smoothed estimates may also be used for diagnostic checks of the model.

Local Fitting of Ordinal Models

Let us start by sketching the structure of ordinal models. The most commonly used such model is the proportional-odds model, which has been considered extensively by McCullagh (1980). The model assumes that for observations (Y_i, z_i),

$$P(Y_i \leq r|x_i) = G(\gamma_{0r} + z_i^T \gamma), \quad r = 1, \ldots, k, \tag{8.25}$$

where $-\infty = \gamma_{00} < \gamma_{01} < \cdots < \gamma_{0k} = \infty$, in which the constants γ_{0r} may be interpreted as thresholds of a latent variable, and where G is the logistic cumulative distribution function (see McCullagh, 1980.) Thus, the response probabilities $\pi_R(x_i) = P(Y_i = r|x_i)$ are given by $\pi_r(x) = G(\hat{\gamma}_{or} + z_i^T \gamma) - G(\gamma_{o,r-1} + z_i^T \gamma)$. The influence variable z_i is deliberately chosen to be different from our explanatory variable x_i, since, in what follows z_i is constructed from x_i, but is not necessarily equal to it. By including Y_i, the form (8.25) hides the fact that the response is vector valued. The actual response is given by the vector of relative frequencies $p_i^T = (p_{i1}, \ldots, p_{iq})$, $q = k - 1$, where p_{ik} is omitted because of the restriction that $p_{ik} = 1 - p_{i1} - \cdots - p_{iq}$. By examining this q-dimensional response, the underlying structure of model (8.25) is that of a multivariate generalized linear model. As outlined in Fahrmeir and Tutz (1994), these models have the form

$$\pi_i = h(\mathbf{Z}_i \beta), \quad i = 1, \ldots, n, \tag{8.26}$$

where $\pi_i^T = (\pi_1(z_i), \ldots, \pi_q(z_i))$ denote the response probabilities ($q = k - 1$), $\mathbf{Z}_i$ is a design matrix, $h = (h_1, \ldots, h_q)$:$\mathrm{IR}^q \to \mathrm{IR}^q$ denotes the multivariate response function, and β is the parameter vector. For model (8.25), one obtains

$$\mathbf{Z}_i = \begin{bmatrix} 1 & & & z_i^T \\ & \ddots & & z_i^T \\ & & 1 & z_i^T \end{bmatrix},$$

with the rth component of the response funtion $h_r(\eta_1, \ldots, \eta_q) = G(\eta_r) - G(\eta_{r-1})$, where $\eta_0 = -\infty$ and the total-parameter vector $\beta^T = (\gamma_{01}, \ldots, \gamma_{0q}, \gamma^T)$. For alternative multivariate generalized linear models and the estimation theory that is sketched next, see, for example, Fahrmeir and Tutz (1994).

Parametric models like (8.26) are usually fitted by using a maximum-likelihood technique. The multinomial log-likelihood contribution of observation (p_i, z_i) is given by

$$l_i(\beta) = \sum_{r=1}^{g} n_{ir} \log \frac{\pi_r(z_i)}{1 - \sum_{s=1}^{q} \pi_s(z_i)} + m_i \log \left(1 - \sum_{s=1}^{q} \pi_s(z_i)\right),$$

where the $\pi_r(z_i)$, $r = 1, \ldots, q$, are determined by the model—for instance, model (8.25). The corresponding score function is given by

$$s_i(\beta) = \frac{\partial l_i}{\partial \beta} = \mathbf{Z}_i^T D_i(\beta) \Sigma_i^{-1}(\beta)(y_i - \mu_i(\beta)),$$

where $D_i(\beta) = \partial h(\eta_i)/\partial \eta$ is the derivative of $h(\eta)$ evaluated at $\eta_i = \mathbf{Z}_i \beta$ and $\Sigma_i(\beta) = \text{Cov}\,(p_i)$ denotes the covariance of the observation p_i, given parameter β. Iterative Fisher scoring with score function $s(\beta) = \sum_{i=1}^{n} s_i(\beta)$ now has the form

$$\hat{\beta}^{(p+1)} = \hat{\beta}^{(p)} + \mathbf{F}^{-1}(\beta^{(p)}) s(\beta^{(p)}), \tag{8.27}$$

where the Fisher matrix is given by

$$\mathbf{F}(\beta) = \sum_{i=1}^{n} F_i(\beta) = \sum_{i=1}^{n} \mathbf{Z}_i^T D_i(\beta) \Sigma_i^{-1}(\beta) D_i^T(\beta) \mathbf{Z}_i.$$

Comparison of this material with that of Section 8.3.3 shows that the given forms are generalizations of the corresponding likelihood contributions, score functions, and Fisher matrices in the binary case.

To obtain smoothed fits of p_i, given the explanatory variable x_i, one applies the local likelihood principle to the observations of the simple model

$$\pi_1(x_i) + \cdots + \pi_r(x_i) = G(\gamma_{0r}), \tag{8.28}$$

which implies that only the constants are incorporated (as $\beta^T = (\gamma_{01}, \ldots, \gamma_{0q})$ and $z_i = 0$). The corresponding local likelihood to be maximized for estimating $\pi_r(x)$, $r = 1, \ldots, q$, is given by

$$\sum_{i=1}^{g} l_i(\beta) \omega(x, x_i), \tag{8.29}$$

yielding the weighted-score equation

$$s_{(x)}(\beta) = \sum_i s_i(\beta) \omega(x, x_i) = 0,$$

which may be solved by an iterative fitting procedure

$$\hat{\beta}^{(p+1)} = \hat{\beta}^{(p)} + \mathbf{F}_{(x)}^{-1}(\beta^{(p)}) s_{(x)}(\beta^{(p)})$$

with the weighted Fisher matrix $\mathbf{F}_{(x)}(\beta) = \sum_{i=1}^{n} \omega(x, x_i) F_i(\beta)$. Maximization of (8.29) yields the local likelihood estimates $\hat{\pi}_r(x) = G(\hat{\gamma}_{0r}) - G(\hat{\gamma}_{0,r-1})$.

Kauermann and Tutz (2000) and Tutz and Kauermann (1997) show that, under regularity conditions and fixed r, the local likelihood estimate at fixed point x is consistent and asymptotically normally distributed. If x is continuous and $\beta(x)$ is a smooth function of x, one assumes that the smoothing parameter approaches infinity with $h = \rho n^{-\alpha}$, $\alpha > 0$. Then the variance of the local likelihood estimate $\hat{\beta} = \hat{\beta}(x)$ satisfies

$$\operatorname{Var} \hat{\beta} = \left\{ \sum_i \omega_\gamma(x, x_i) F_i(\beta) \right\}^{-1} m(K) + O(n^{-1-\alpha}) + O(n^{-2+2\alpha}), \qquad (8.30)$$

with the local Fisher matrix $F_i(\beta) = D_i^T(\beta) \Sigma_i^{-1} D_i^T(\beta)$ and $m(K) = K(0)^{-1} \int K(z)^2\, dz$ depending only on the kernel function. The mean squared error satisfies $\text{MSE} = O(n^{-1+\alpha}) + O(n^{-4\alpha})$, implying consistency when $0 < \alpha < 1$. The estimate is asymptotically normally distributed for $\frac{1}{5} < \alpha < 1$, with $V^{-1/2}(\hat{\beta} - \beta) \sim N(0, I)$ where $V = \operatorname{Var}(\hat{\beta})$.

If the explanatory variable is discrete, taking $\omega_h(x, x_i) = O(n^{-2\alpha})$ yields similar results, with consistency for $\alpha > 0$ and asymptotic normality for $\alpha > 1/4$. Although alternative forms of the asymptotic variance are available, the form (8.30) has the advantage of being easily computable. Estimated variances can be combined with asymptotic normality so that error bands $\hat{\beta}_i \pm \sqrt{\widehat{\operatorname{Var}} \hat{\beta}_i}$ can be constructed for each component of β. Kauermann and Tutz (2000) provide more details of this estimator, also describing a bias-corrected form. A simpler weighted least-squares approach has been considered by Tutz and Kauermann (1998).

The extension to local polynomial likelihood is straightforward. Locally fitting the model

$$\pi_1(x_i) + \cdots + \pi_r(x_i) = G(\gamma_{0r} + \cdots + (x - x_i)^t \gamma_t) \qquad (8.31)$$

(i.e., enriching the local model (8.28) by including higher order polynomial terms $z_i^T = (x - x_i, \ldots, (x - x_i)^t)$ yields the local polynomial likelihood estimates $\hat{\pi}_r(x) = G(\hat{\gamma}_{0r}) - G(\hat{\gamma}_{0,r-1})$. The combination of (8.31) and (8.29) gives the model-based estimate in which smoothing over explanatory variables is done by local polynomial likelihood techniques and smoothing over the response is based on an ordinal model.

Local Parametric Modeling as a Diagnostic Tool
The parametric proportional-odds model for $Y_i|x_i$ (8.25) can be written

$$\pi_1(x_i) + \cdots + \pi_r(x_i) = G(\gamma_{0r} + x_i\gamma), \quad r = 1, \ldots, q. \tag{8.32}$$

Smoothed estimates can be used to assess the fit of this model. In a first step, a smoothed estimate can be compared graphically with the model fit. A more refined technique that allows the identification of areas of the explanatory variable wherein the parametric model fails is based on model (8.32) itself. Let model (8.32) be fit locally by maximizing (8.29) with respect to $\beta^T = (\gamma_{01}, \ldots, \gamma_{0q}, \gamma)$. Note that in (8.32) the explanatory variable is not centered at x. Therefore, the use of the weight function $\omega(x, x_i)$ in (8.29) yields local estimates $\hat{\gamma}_{01}(x), \ldots, \hat{\gamma}_{0q}(x)$, $\hat{\gamma}(x)$, which may be interpreted as the local parameters of model (8.32) at x. Actually, the parameter functions $\hat{\gamma}_{01}(x), \ldots, \hat{\gamma}_{0q}(x), \hat{\gamma}(x)$ that one obtains may be plotted and thus compared with the "fixed model" estimates based on the (unsmoothed) fit of model (8.32). (For an example, see Tutz, 1995).

Example 8.6: Duration of Unemployment
An important variable used in predicting the duration of a person's unemployment is the age of the unemployed person. We will consider the trichotomous response of short-term unemployment ($\leq$6 months), medium-term unemployment (between 7 and 12 months) and long-term unemployment ($>$12 months). The data are from the Socioeconomic Panel in Germany (Hanefeld, 1987) and refer to 675 unemployed men with German citizenship.

We fit the proportional-odds model (8.25) locally with polynomials of degree $t = 0, 1$, based on the Epanechnikov kernel. Smoothing parameters are chosen according to CV, yielding $h = 0.75$ and $h = 1.0$ for $t = 0$ and 1, respectively. Figure 8.6 shows the resulting curves with local relative frequencies given in the plots. Clearly, with increasing age of the unemployed person, the probability is shifted mostly toward long-term unemployment. Medium-term unemployment is rather constant, but long-term unemployment increases steeply starting at about 45 years of age. The local linear fit, which has less bias at the boundary, shows a stronger decrease in short-term unemployment and a stronger increase in long-term unemployment for older people.

8.3.5 Multiple Predictors and Additive Models

Local polynomial likelihood fitting generalizes directly to the case of multiple predictors. For $k = 2$ and observations $(y_i, x_{i1}, \ldots, x_{ip})$, the local linear likelihood at target value $x_0^T = (x_{01}, \ldots, x_{0p})$ is given by

$$\sum_{j=1}^{g} \{n_{j1}[\beta_0 + \beta_1(x_{01} - x_{j1}) + \cdots + \beta_p(x_{0p} - x_{jp})] \\ - \exp[\beta_0 + \beta_1(x_{01} - x_{j1}) + \cdots + \beta_p(x_{0p} - x_{jp})]\}\omega(x_0, x_j), \tag{8.33}$$

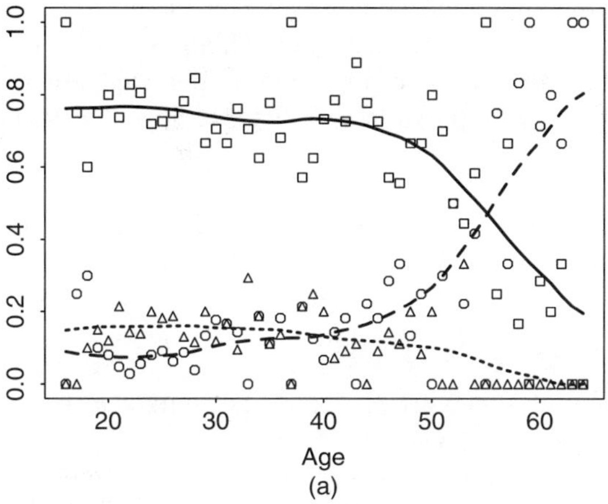

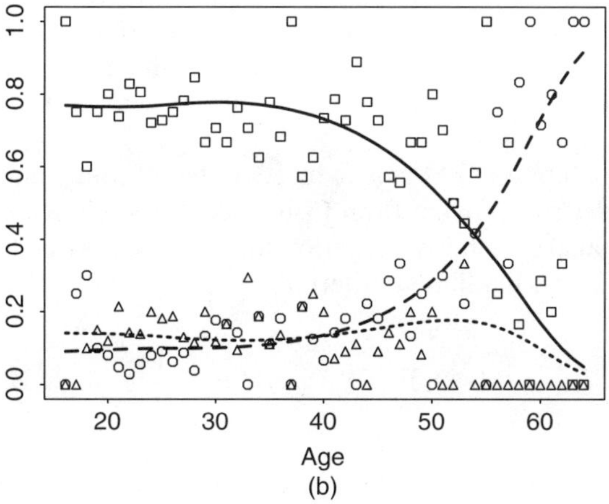

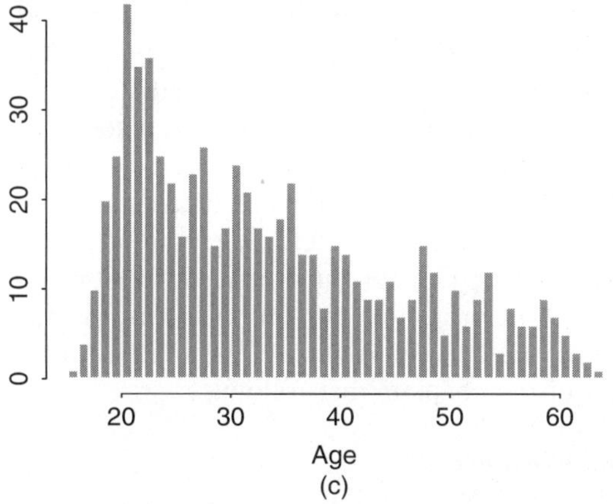

FIGURE 8.6 Locally fitted proportional-odds model for Unemployment Data with (a) local (constant) likelihood and (b) local linear fitting. Relative frequencies as a function of age are given as squares (short-term unemployment), triangles (medium-term unemployment), and circles (long-term-unemployment). (c) The local sample sizes from which the relative frequencies are determined.

with the weight function determined by kernel products $\omega(x_0, x_j) = \prod_{s=1}^{p} K_{h_s}(x_{0s}|x_{js})$. For a two-dimensional predictor, maximization of (8.33) yields a smoothed surface that may be presented graphically. For higher dimensional data, graphical display is possible only by "slicing" the surface by fixing the values of all but two of the predictors, which are then plotted. More seriously, in higher dimensions, the curse of dimensionality has to be considered.

One way of reducing these problems is to generalize categorical regression to GAMs, where one assumes that

$$\pi(x_i) = h(f_1(x_{i1}) + \cdots + f_p(x_{ip})), \tag{8.34}$$

with univariate smooth functions $f_1, \ldots, f_p$ and a fixed response function h—for example, the logistic distribution function.

The strength of this approach lies in the easy presentation of the fit. The smooth functions f_i may be presented separately, which, hopefully, provides insight into the multivariate-data problem. Of course, some caution is appropriate: If the additive model (8.34) is not correct, the resultant estimates are difficult to interpret and use. This approach is only sketched here; a detailed discussion is given elsewhere in the book.

The fitting problem for GAMs is reduced to that of fitting an additive model to the pseudo observations arising from Fisher scoring. (See Section 8.3.3.) Consequently, there is also an intuitive criterion that is maximized when fitting GAMs, namely, the penalized likelihood criterion

$$\sum_{i=1}^{g} l_i(y_i, f_1(x_{i1}), \ldots, f_p(x_{ip}))/m_i - \frac{1}{2}\sum_{i=1}^{p} \lambda_i f_i^T \mathbf{K}_i f_i,$$

where $\lambda_1, \ldots, \lambda_p$ are smoothing parameters for the components, $f_i^T = (f_i(x_{i1}), \ldots, f_i(x_{ip}))$ are the values of the ith component, and $\mathbf{K}_i$ is a smoothing matrix built from $x_{1i}, \ldots, x_{gi}$. An appropriate choice of $\mathbf{K}_i$ yields spline smoothing in the backfitting algorithm of additive models. Additive and generalized additive models have been outlined in a series of papers by Hastie and Tibshirani (1985, 1986, 1987) and are covered in Hastie and Tibshirani (1990). For an extension to the multicategorical case, see Yee and Wild (1996). Alternative approaches to handle the multiple-predictor case are semiparametric (Green and Yandell, 1985; O'Sullivan et al., 1986; Severini and Staniswalis, 1994; Carroll et al., 1997), projection pursuit (Diaconis, 1983; Roosen and Hastie, 1994) and single index models (Horowitz, 1993, 1998). All these approaches are considered in some of the chapters that follow.

Example 8.7: Chronic Bronchitis

In an occupational epidemiology multicenter study, the focus was on the dependence of the development of chronic bronchitis on the average concentration of dust at the workplace and the time since beginning of exposure to the disease. (For a detailed description of the data, see Küchenhoff and Ulm, 1997.) The

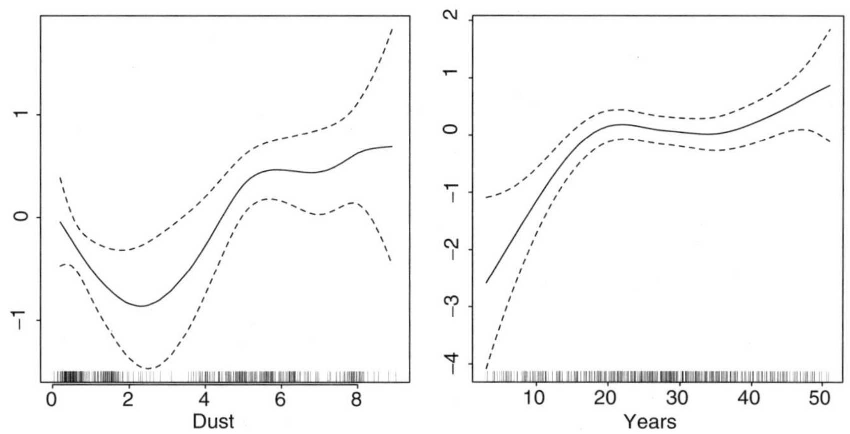

FIGURE 8.7 Effects of concentration of dust and exposure time on probability of chronic bronchitis for the generalized additive logit model.

analysis is based on 920 workers, all of whom are smokers. A generalized additive logit model based on spline smoothing was fitted, with $y = 1$ signifying the occurrence of bronchitis. Figure 8.7 shows the additive components, together with error bands provided by the software package S-Plus. The components signal a generally increasing probability of bronchitis with increasing duration of the exposure. The effect of dust concentration shows a dip at about 2.5 mg/m^3 of dust, but with rather wide error bands in this area due to a small number of observations between 2 and 4 mg/m^3. For higher concentrations, the probability of bronchitis is strictly increasing. The increase only for higher concentrations may suggest that there is a threshold value of dust concentration above which damage in the form of increased probability of bronchitis occurs. Küchenhoff and Ulm (1997) give various approaches to the modeling of threshold values.

In Figure 8.8 two fits are shown: the two-dimensional effects based on the additive model and, for comparison, the local linear two-dimensional fit of f(dust, years) based on nearest neighborhood, with the smoothing parameter chosen such that it uses about 50% of the data. The figure reveals that the gap of observations between 1.5 and 4 mg/m^3 is smoothed by the nearest-neighborhood fit, with the dip around 2.5 mg/m^3 being much less prominent. There is no strong indication of interaction between dust and duration of the exposure, a finding that is also supported by the fact that an interaction term in the simple linear logit model is nonsignificant.

8.3.6 Computational Issues

Simple kernel regression for dichotomous responses according to (8.13) reduces to smoothing of relative frequencies and may be done by the packages mentioned in Section 8.2.4. More advanced methods may be found in the packages S-Plus and XploRe. Local polynomial modeling for dichotomous responses can be done with loess weights using the function `gam` of S-Plus,

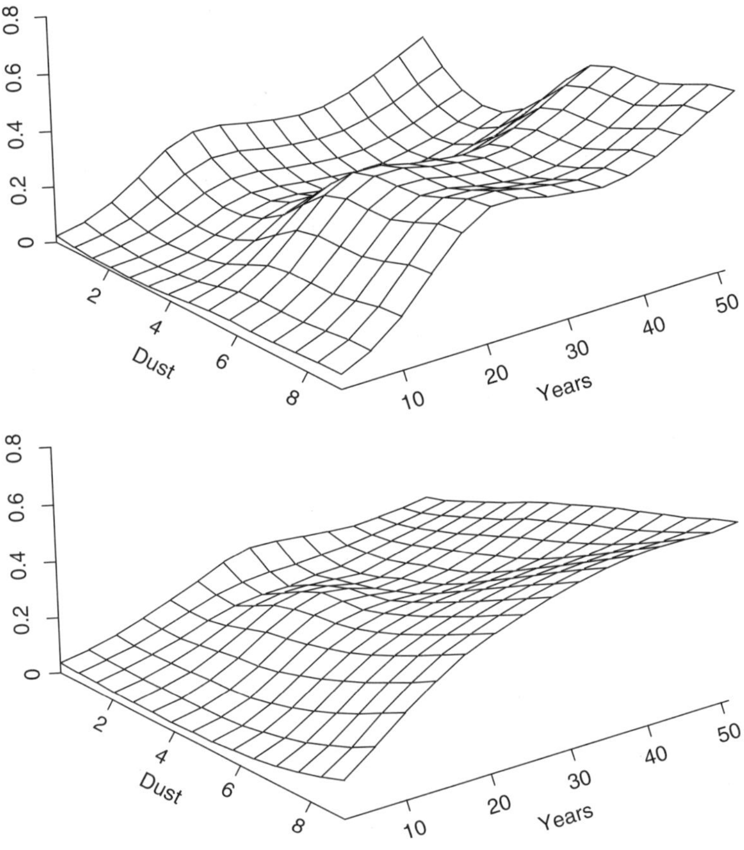

FIGURE 8.8 Generalized additive logit model (top) and local linear fit of f(dust, years) with nearest neighborhood to probability of chronic bronchitis.

which fits GAMs. For the model-based smoothing of ordinal data, an S-Plus macro is available from Silke Edlich, Technische Universität Berlin (`http://stat.cs.tu-berlin.de/˜monster`).

8.4 CONCLUSION

In this chapter, we have described smoothing methods for discrete data. The inherent heteroscedasticity of count data, combined with the restriction of probabilities to be in [0, 1], makes least-squares approaches to such data less natural than Poisson, binomial, and multinomial likelihood approaches. The general idea of generalized modeling (which can be extended to include GAMs) makes it possible to construct smoothed probability estimates that are flexible enough to highlight interesting and possibly unexpected structure that might be missed in a parametric modeling, while still being true to the inherent nominal or ordinal nature of the discrete data. The local likelihood formulation also brings out the similarities of the contingency table and discrete regression problems.

Acknowledgement

G. Tutz's work was supported in part by Deutsche Forschungsgemeinschaft (Projekt VARKOFF).

REFERENCES

Abzug, R., Ahlstrom, D., and Simonoff, J. S. (1999). The organizational landscape of cities: A comparative geography, preprint.

Aerts, M., Augustyns, I., and Janssen, P. (1997a). Smoothing sparse multinomial data using local polynomial fitting, *J. Nonparametric Statistics*, **8**, 127–147.

Aerts, M., Augustyns, I., and Janssen, P. (1997b). Local polynomial estimation of contingency table cell probabilities, *Statistics*, **30**, 127–148.

Aitchison, J., and Aitken, C. G. G. (1976). Multivariate binary discrimination by the kernel method, *Biometrika*, **63**, 413–420.

Aitken, C. G. G. (1983). Kernel methods for the estimation of discrete distributions, *J. Statist. Comput. Simul.*, **16**, 189–200.

Brüderl, J., Preisendörfer, P., and Ziegler, R. (1992). Survival chances of newly founded business organizations, *American Sociological Review*, **57**, 227–242.

Burman, P. (1987). Smoothing sparse contingency tables, *Sankhyā A*, **49**, 24–36.

Carroll, R. J., Fan, J., Gijbels, I., and Wand, M. P. (1997). Generalized partially linear single index models, *J. Amer. Statist. Assoc.*, **91**, 477–489.

Cleveland, W. S. (1979). Robust locally weighted regression and smoothing scatterplots, *J. Amer. Statist. Assoc.*, **74**, 829–836.

Diaconis, P. (1983). *Projection Pursuit for Discrete Data*, Technical Report, Department of Statistics, Stanford University, CA.

Dong, J., Simonoff, J. S. (1994). The construction and properties of boundary kernels for sparse multinomials, *J. Comput. Graph. Statist.* **3**, 57–66.

Eubank, R. L. (1988). *Spline Smoothing and Nonparametric Regression*, Dekker, New York.

Fahrmeir, L., and Tutz, G. (1994). *Multivariate Statistical Modeling Based on Generalized Linear Models*, Springer–Verlag, New York.

Fan, J. (1992). Design-adaptive nonparametric regression, *J. Amer. Statist. Assoc.*, **87**, 998–1004.

Fan, J., and Gijbels, I. (1996). *Local Polynomial Modeling and its Applications*, Chapman and Hall, London, UK.

Fan, J., Heckman, N. E., and Wand, M. P. (1995). Local polynomial kernel regression for generalized linear models and quasi-likelihood functions, *J. Amer. Statist. Assoc.* **90**, 141–150.

Fienberg, S. E., and Holland, P. W. (1970). Methods for eliminating zero counts in contingency tables, in Patilgerman, G. P. (ed.), *Random Counts on Models and Structures*, pp. 233–260, Pennsylvania State University Press, University Park, PA.

Fienberg, S. E., and Holland, P. W. (1973). Simultaneous estimation of multinomial cell probabilities, *J. Amer. Statist. Assoc.*, **68**, 683–691.

Firth, D., Glosup, J., and Hinkley, D. V. (1991). Model checking with nonparametric curves. *Biometrika*, **78**, 245–252.

Green, D. J., and Silverman, B. W. (1994). *Nonparametric Regression and Generalized Linear Models: A Roughness Penalty Approach.* Chapman and Hall, London, UK.

Green, P. J., and Yandell, B. S. (1985). Semi-parametric generalized linear models, in Gilchrist, R., Francis, B. J., and Whittaker, J. (eds.), *Generalized Linear Models*, pp. 44–55. Springer–Verlag, Berlin, Germany.

Habbema, J. D. F., Hermans, J., and Remme, J. (1978). Variable kernel density estimation in discriminant analysis, in Corstner, L. C. A., and Hermands, J. (eds.), *COMPSTAT Proceedings*, pp. 178–185. Physica–Verlag, Vienna, Austria.

Hall, P., and Titterington, D. M. (1987). On smoothing sparse multinomial data, *Austral. J. Statist.*, **29**, 19–37.

Hanefeld, U. (1987). *Das sozio-oekonomische Panel*, Campus, Frankfurt a. M., Germany.

Härdle, W. (1990). *Applied Nonparametric Regression*, Cambridge University Press, Cambridge, UK.

Hastie, T., and Tibshirani, R. (1985). Comments on "projection pursuit," *Ann. Statist.*, **13**, 502–508.

Hastie, T., and Tibshirani, R. (1986). Generalized additive models (with discussion), *Statist. Sci.*, **1**, 297–318.

Hastie, T., and Tibshirani, R. (1987), Generalized additive models: Some applications, *J. Amer. Statist. Assoc.* **82**, 371–386.

Hastie, T., and Tibshirani, R. (1990). *Generalized Additive Models*, Chapman and Hall, London, UK.

Hjort, N. L., and Jones, M. C. (1996). Locally parametric nonparametric density estimation, *Ann. Statist.*, **24**, 1619–1647.

Horowitz, J. L. (1993). Semiparametric and nonparametric estimation of quantal response models, in Maddala, G. S., Rao, C. R., and Vinod, H. D. (eds.), *Handbook of Statistics*, Volume 11. Elsevier Science Publisher, Amsterdam, The Netherlands.

Horowitz, J.L. (1998). *Semiparametric Methods in Econometrics*, Springer–Verlag, New York.

Hurvich, C. M., Simonoff, J. S., and Tsai, C.-L. (1998). Smoothing parameter selection in nonparametric regression using an improved Akaike information criterion, *J. Roy. Statist. Soc. B*, **60**, 271–293.

Jones, M. C. (1993). Simple boundary correction for kernel density estimation. *Statist. Comput.*, **3**, 135–146.

Kauermann, G., and Tutz, G. (2000). Local likelihood estimation in varying coefficient models including additive bias correction. To appear in *Journal of Nonparametric Statistics*.

Küchenhoff, H., and Ulm, K. (1997). Comparison of statistical method for assessing threshold limiting values in occupational epidemiology, *Comput. Statist.*, **12**, 249–264.

Loader, C. R. (1996). Local likelihood density estimation, *Ann. Statist.*, **24**, 1602–1618.

McCullagh, P. (1980). Regression model for ordinal data (with discussion), *J. Roy. Statist. Soc. B*, **42**, 109–127.

McCullagh, P., and Nelder, J. A. (1989). *Generalized Linear Models* (2d ed.), Chapman and Hall, London, UK.

Müller, H.-G., and Schmitt, T. (1988). Kernel and probit estimates in quantal bioassay, *J. Amer. Statist. Assoc.*, **83**, 750–759.

O'Sullivan, F., Yandell, B. S., and Raynor, W. J. (1986). Automatic smoothing of regression functions in generalized linear models, *J. Amer. Statist. Assoc.*, **81**, 96–103.

Parzen, E. (1962). On estimation of probability density function and mode, *Ann. Math. Statist.*, **33**, 1065–1076.

Pearson, K. (1900). On a criterion that a system of deviations from the probable in the case of a correlated system of variables is such that it can be reasonably supposed to have arisen in random sampling, *Philos. Mag., Ser. 5*, **50**, 157–175.

Perrow, C. (1991). A society of organizations, *Theory and Society*, **20**, 725–762.

Przybylski, M., Littlepage, L., and Rosentraub, M. S. (1996). Philanthropy, nonprofits, and the fiscal health of cities, *Nonprofit Voluntary Sector Quart.*, **25**, 14–39.

Roosen, C. B., and Hastie, T. (1994). Automatic smoothing spline projection pursuit, *J. Comput. Graph. Statist.* **3**, 235–248.

Rosenblatt, M. (1956). Remarks on some nonparametric estimates of a density function, *Ann. Math. Statist.*, **27**, 832–837.

Ruppert, D., Sheather, S. J., and Wand, M. P. (1995). An effective bandwidth selector for local least squares, *J. Amer. Statist. Assoc.*, **90**, 1257–1270.

Severini, T. A., and Staniswalis, J. G. (1994). Quasi-likelihood estimation in semiparametric models, *J. Amer. Statist. Assoc.*, **89**, 501–511.

Simonoff, J. S. (1983). A penalty function approach to smoothing large sparse contingency tables, *Ann. Statist.*, **11**, 208–218.

Simonoff, J. S. (1987). Probability estimation via smoothing in sparse contingency tables with ordered categories, *Statist. Probab. Lett.*, **5**, 55–63.

Simonoff, J. S. (1995). Smoothing categorical data, *J. Statist. Plann. Inf.*, **47**, 41–69.

Simonoff, J. S. (1996a). *Smoothing Methods in Statistics*, Springer–Verlag, New York.

Simonoff, J. S. (1996b). Smoothing-based modeling of contingency tables, in *1996 Proceedings of the ASA Statistical Education Section*, pp. 8–17.

Staniswalis, J. G. (1989). The kernel estimate of a regression function in likelihood-based models, *J. Amer. Statist. Assoc.*, **84**, 276–283.

Staniswalis, J. G., and Cooper, V. D. (1988). Kernel estimates of dose-response, *Biometrics*, **44**, 1103–1119.

Stein, C. (1962). Confidence sets for the mean of a multivariate normal distribution (with discussion), *J. Roy. Statist. Soc. B*, **24**, 265–296.

Stone, C. J. (1977). Consistent nonparametric regression (with discussion). *Ann. Statist.*, **5**, 595–645.

Tibshirani, R., and Hastie, T. (1987). Local likelihood estimation, *J. Amer. Statist. Assoc.*, **82**, 559–568.

Titterington, D. M. (1980). A comparative study of kernel-based density estimates for categorical data, *Technom.*, **22**, 259–268.

Tutz, G. (1991). Consistency of cross-validatory choice of smoothing parameters for direct kernel estimates, *Computational Statistics*, **4**, 295–314.

Tutz, G. (1995). Smoothing for categorical data: Discrete kernel regression and local likelihood approaches, in Bock, H. H., and Polasek, W. (eds.), *Data Analysis and Information Systems*, pp. 261–271, Springer–Verlag, Berlin, Germany.

Tutz, G. and Kauermann, G. (1997). Local estimators in multivariate generalized linear models with varying coefficients, *Computational Statistics*, **12**, 193–208.

Tutz, G. and Kauermann, G. (1998). Locally weighted least squares in categorical varying coefficient models, in Galata, R. and Küchenhoff, H. (eds.), *Econometrics in Theory and Practice. Festschrift for Hans Schneeweiß*, pp. 119–130, Physica–Verlag, Heidelberg, Germany.

Wand, M. P., and Jones, M. C. (1995). *Kernel Smoothing*, Chapman and Hall, London.

Wang, M. C., and van Ryzin, J. (1981). A class of smooth estimators for discrete distributions, *Biometrika*, **68**, 301–309.

Yee, T. W., and Wild, C. J. (1996). Vector generalized additive models, *J. Royal Statist. Soc. B*, **58**, 481–493.

CHAPTER 9

Local Polynomial Fitting

Jianqing Fan
University of North Carolina, Chapel Hill, North Carolina, USA and The Chinese University of Hong Kong, PRC

Irene Gijbels
Université Catholique de Louvain, Louvain-la-Neuve, Belgium

9.1 INTRODUCTION

Nonparametric regression is about data-analytic approaches to regression problems arising from many statistical applications. The aim of these model-free nonparametric approaches is to release the restrictive assumptions imposed on the form of a regression function in favor of one that describes the given data well. They are powerful in detecting possible nonlinearities and in exploring fine functional relationships. In addition they provide useful diagnostic tools for parametric regression models. Local polynomial fitting is a powerful nonparametric method and can easily be understood and implemented.*

9.1.1 Nonlinear Regression

There is a surge of interest in nonlinear regression in the literature. With the advancement of modern science and technology, there are many situations in which a linear relationship can not adequately describe the structural dependence between some variables. Consider, for example, the data presented in Figure 9.1, in which the relationship between the concentration of nitric oxides in engine

* All calculations in this chapter are based on prototype software written by the authors which is not publically available.

Smoothing and Regression: Approaches, Computation, and Application
Edited by Michael G. Schimek
ISBN 0-471-17946-9

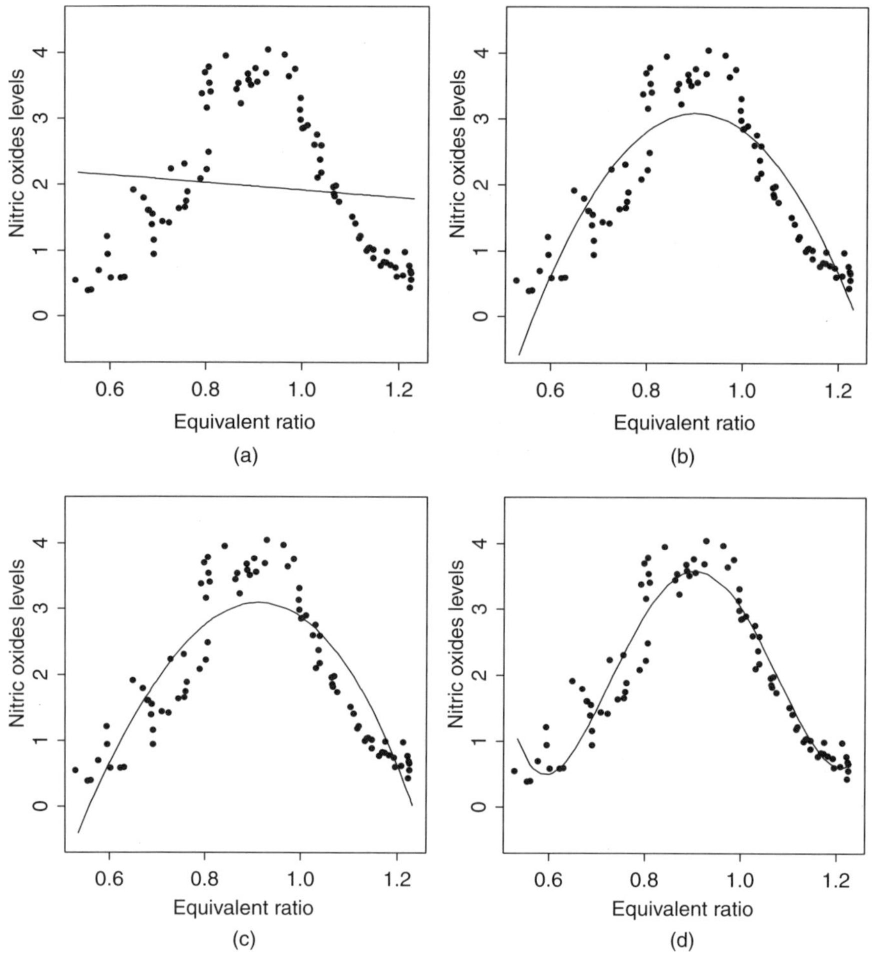

FIGURE 9.1 Polynomial fits to data on the burning of ethanol. Presented are the scatterplots of the concentration of nitric oxides versus the equivalence ratio, along with the fitted polynomial regression functions; (a) linear; (b) quadratic; (c) cubic; (d) quartic.

exhaust (dependent variable) and the equivalence ratio (independent variable), a measure of the richness of the mixture of air and ethanol, is depicted for the burning of ethanol in a single-cylinder automobile test engine. From inspection, one can see that the relationship between the concentration of nitric oxides and the equivalence ratio is highly nonlinear. What can we do with such a nonlinear data set? Traditionally, trial-and-error methods on polynomial regression are often employed. Figure 9.1 presents the resulting fits by using four different degrees of polynomials. One can easily see that all of the resulting fits have large biases. As they have homogeneous degrees of smoothness, polynomial functions are not effective in modeling many nonlinear functions. Examples of this inadequacy are depicted by the Ethanol Data in Figure 9.1.

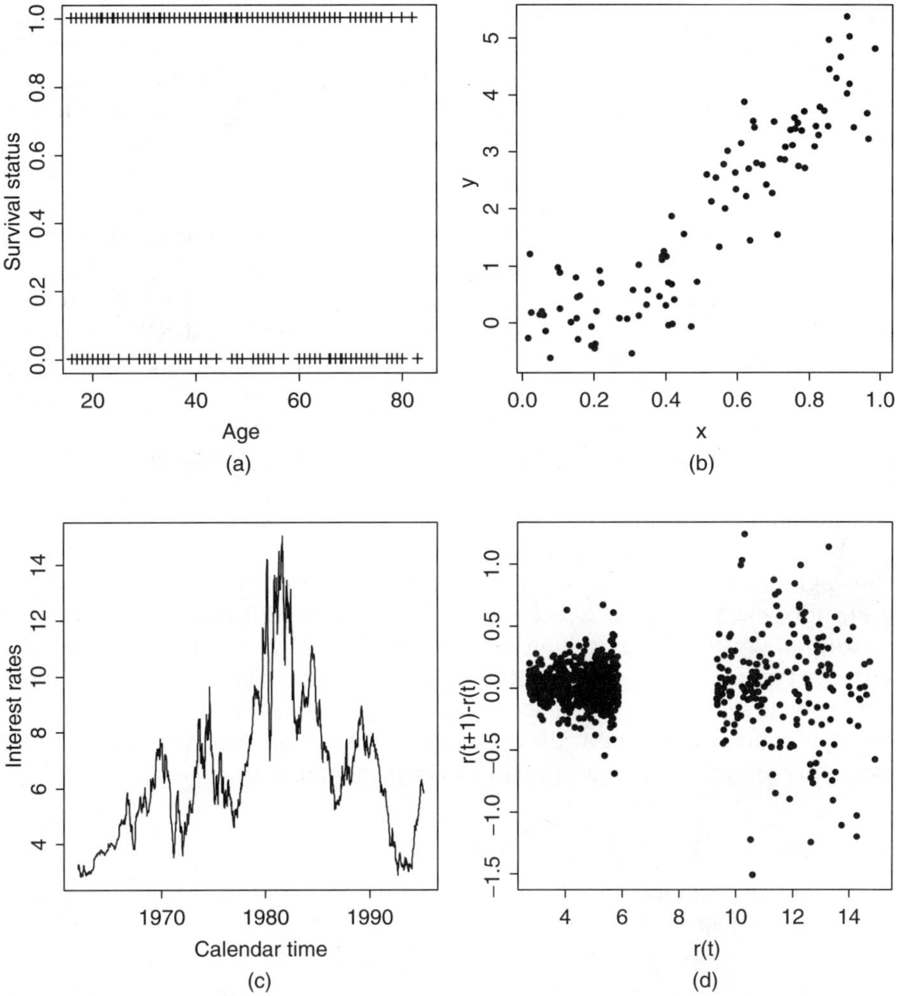

FIGURE 9.2 Limitations of scatterplot visualization. (a) Binary response data (burn injuries) whose mean regression function is hard to image without a nonparametric smoother; (b) a simulated data set for which the discontinuity of the mean regression function at $x = 0.5$ is hard to detect without a derivative estimator; (c) the yields $\{r_t\}$ of a 12-month US Treasury Bill; (d) the scatterplot for one-step ahead prediction of $\{r_t\}$ versus $\{r_{t+1} - r_t\}$, whose conditional variance function is hard to picture by inspection.

9.1.2 Why is Smoothing Needed?

The nonlinearity of the ethanol data is detected with the assistance of scatterplot visualization. However, this approach, while simple, has serious limitations. For one thing, it cannot process data beyond the domain of our visualization. Consider, for example, the data presented in Figure 9.2(a), which were collected from 435 adults between the ages 17 and 85 suffering from burn injuries. The Burn Data were collected at the University of Southern California General Hospital Burn Center and contain information on several variables. The binary response variable is taken to be unity for those who survived their burn injuries and zero otherwise, and the age of the victim is taken as the covariate. One has a difficult time tracing the mean response curve via scatterplot visualization. Figure 9.2(b) presents a simulated set of data of size 100 from the model $Y = 4X^2 + I(X > 0.5) + \varepsilon$, where $\varepsilon \sim N(0; 0.5^2)$ is independent of $X \sim$ uniform[0, 1]. From scatterplot visualization, one can hardly detect a jump at the point $x = 0.5$. For this purpose, one needs to visualize the derivative curve of the regression function, which is beyond the scope of this chapter. Figure 9.2(c) concerns the yields of the 12-month U.S. Treasury Bill from the second market rates (on Fridays), which are annualized using a 360-day year of bank interest and quoted on a discount basis. The data consist of 1,735 weekly observations, from January 5, 1962, to March 31, 1995. Let r_t denote the interest rate on week t. Figure 9.2(d) gives the scatterplot for the one-step ahead prediction, where the points $\{(r_t, r_{t+1} - r_t)\}$ are presented. It is difficult to picture the volatility function $\sqrt{\mathrm{Var}(r_{t+1}|r_t)} = \sqrt{\mathrm{Var}(r_{t+1} - r_t|r_t)}$ without the aid of a smoothing technique.

The scatterplot visualization method especially loses its appeal when dealing with high-dimensional data. It is difficult to visualize data beyond two dimensions using this method. Further, because of the intrinsic difficulty called *curse of dimensionality* in analyzing high-dimensional data, high-dimensional regression surfaces typically are approximated by lower dimensional functions and lower dimensional smoothers, which serve as building blocks and are iteratively used for high-dimensional data analyses. In this respect, the scatterplot visualization method is not useful, because it can hardly be automated.

9.1.3 Local Polynomial Regression

While the data presented in Figures 9.1 and 9.2 show different kinds of nonlinearity, they admit one common feature: The underlying curves to be estimated are smooth (except for a jump in Figure 9.2(b)). What this means is that the underlying function can be locally approximated by a linear function. Let h be the size of the local neighborhood. Then for a given point x_0, we can locally model the data as

$$Y_i = a(x_0) + b(x_0)X_i + \text{error}, X_i \in x_0 \pm h,$$

where $a(x_0)$ and $b(x_0)$ are two local parameters. This model naturally leads to the following local regression problem: Find $\widehat{a}(x_0)$ and $\widehat{b}(x_0)$ to minimize

$$\sum_{i=1}^{n} \{Y_i - a(x_0) - b(x_0)X_i\}^2 I\left(\frac{|X_i - x_0|}{h} \leq 1\right). \tag{9.1}$$

The resulting estimator of the regression function at the point x_0 is thus $\widehat{m}(x_0) = \widehat{a}(x_0) + \widehat{b}(x_0)x_0$. As x_0 runs through a grid of points in a region of interest, we obtain the estimated curve.

An improved version of the above idea is to include a smooth weight function, which weighs down the contribution from an observation X_i away from x_0. Let K be a nonnegative unimodal function with support in $[-1, 1]$, so that Equation (9.2), below uses only the data in the strip $x_0 \pm h$. The weighted local regression now becomes the minimization of

$$\sum_{i=1}^{n}\{Y_i - a(x_0) - b(x_0)X_i\}^2 K\left(\frac{X_i - x_0}{h}\right). \tag{9.2}$$

When K is the uniform density function on $[-1, 1]$, (9.2) reduces to (9.1). In general, the weight function K does not have to be compactly supported, as long as it decays fast enough to eliminate the impact of remote data points.

As an illustration of the above local linear regression approach, consider the Ethanol Data presented in Figure 9.1. The window size h is taken to be 0.051. To estimate the regression curve at the point $x_0 = 0.8$, we fit a line through the data in the neighborhood $x_0 \pm 0.051$ with a weight function $K(t) = (1 - t^2)_+$, where the subscript "+" denotes the positive part of the function, so that K has support $[-1, 1]$, and obtain the estimated value $\widehat{m}(0.8) = 2.86$ (see Figure 9.3). Suppose now that we wish to estimate the regression function at another point $x_0 = 0.88$; we then fit another line using the data in the slided window 0.88 ± 0.051. The whole curve is obtained by estimating the regression function in a grid of points. For example, the curve in Figure 9.3 was obtained by running 101 local linear regressions, associated with 101 grid points from 0.535 to 1.232.

To put this basic idea in a more general statistical framework, let us assume that the observed data $\{(X_i, Y_i), i = 1, \ldots, n\}$ are a random sample from a certain population (X, Y) with conditional mean function $m(x) = E(Y|X = x)$ and conditional variance function $\sigma^2(x) = \text{Var}(Y|X = x)$. For the ease of comprehension,

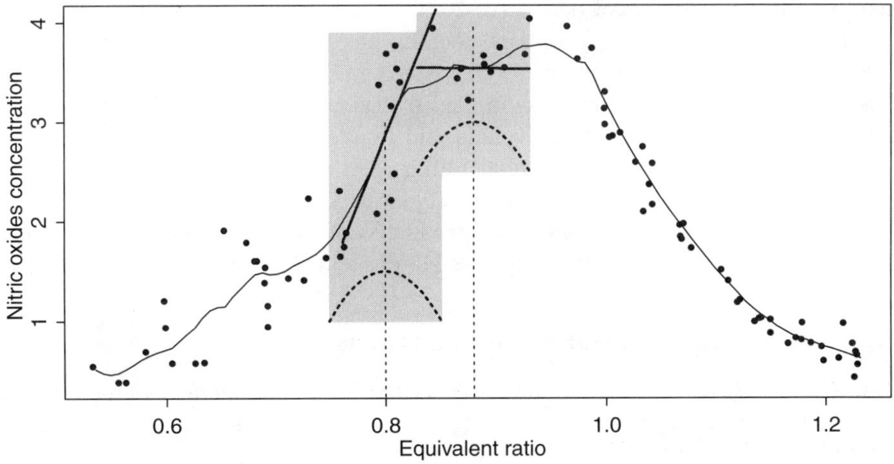

FIGURE 9.3 Illustration of the local linear fit. For each given point x_0, a line is fitted through the data contained in the strip $x_0 \pm 0.051$, using the weight function indicated at the bottom of the strip.

we write

$$Y = m(X) + \sigma(X)\varepsilon, \tag{9.3}$$

where $E(\varepsilon|X = x) = 0$ and $\text{Var}(\varepsilon|X = x) = 1$. For a given point x_0, if $m(x_0)$ has a pth-order derivative, then by Taylor's expansion,

$$\begin{aligned} m(x) &\approx m(x_0) + m'(x_0)(x - x_0) + \cdots + \frac{m^{(p)}(x_0)}{p!}(x - x_0)^p \\ &= \beta_0 + \beta_1(x - x_0) + \cdots + \beta_p(x - x_0)^p, \end{aligned} \tag{9.4}$$

for x in a neighborhood of x_0, where $\beta_j = m^{(j)}(x_0)/j!$. Let h be the size of the neighborhood. Since the local model in Equation (9.4) is valid for $X_i \in x_0 \pm h$, we consider the following *local polynomial regression* problem: Minimize with respect to $\beta_0, \ldots, \beta_p$

$$\sum_{i=1}^{n}\{Y_i - \beta_0 - \beta_1(X_i - x_0) - \cdots - \beta_p(X_i - x_0)^p\}^2 K\left(\frac{X_i - x_0}{h}\right). \tag{9.5}$$

Let $\widehat{\beta}_0, \ldots, \widehat{\beta}_p$ be the resulting fitted coefficients. Then it is clear that $\widehat{\beta}_\nu$ is an estimator for $\beta_\nu = m^{(\nu)}(x_0)/\nu!$, or equivalently,

$$\widehat{m}_\nu(x_0) = \nu!\widehat{\beta}_\nu, \qquad \nu = 0, \ldots, p, \tag{9.6}$$

is an estimator for $m^{(\nu)}(x_0)$. Note that when $p = 1$, the above local polynomial regression problem reduces to the *local linear regression* procedure, and β_0 and β_1 are just a reparametrization of the local parameters a and b in Equation (9.2). We prefer the parameterization in Equation (9.4), because these parameters represent a multiple of the local derivatives.

The idea of using polynomials locally has a long history. Among the early references are Henderson (1916), Macaulay (1931), Spencer (1904), and Woolhouse (1870). For a more complete overview of the long history of local regression, see, for example, Cleveland and Loader (1996). Early papers on the systematic study of local polynomial fitting are those by Cleveland (1979) and Stone (1977). Recent books that extensively embark on this subject include Wand, Fan and Gijbels (1996), Simonoff (1996), and Wand and Jones (1995). For some recent developments in local fitting, see Härdle and Schimek (1996).

9.1.4 Kernel Function and Bandwidth Parameter

The function weight K in Equation (9.5) is called a *kernel function*. In the context of nonparametric regression, it does not have to integrate up to unity; Any multiple of K results in the same weighting scheme. In the local regression setting, we assume throughout this chapter that K is a nonnegative function. Commonly used kernel functions include the following:

Gaussian kernel: $K(t) = (\sqrt{2\pi})^{-1} \exp(-t^2/2)$

Symmetric Beta family: $K(t) = \dfrac{1}{\text{Beta}(1/2, \gamma + 1)}(1 - t^2)_+^{\gamma}, \gamma = 0, 1, 2, \ldots$

Note that the symmetric Beta kernels have support $[-1, 1]$. When $\gamma = 0, 1, 2$, and 3, the resulting kernels are called, respectively, the *uniform*, the *Epanechnikov*, the *biweight*, and the *triweight* kernel function. Figure 9.4 depicts these kernel functions.

The parameter h in Equation (9.5) is called a *bandwidth*, or a *smoothing parameter*. It can be vaguely understood as the size of a local neighborhood. However, this interpretation can be difficult when the kernel K has unbounded support, such as is the case for a Gaussian kernel. The bandwidth parameter is confounded with the effective support of the kernel K. For example, using h_0 with a uniform kernel function is effectively about the same as using $2h_0$ with a triweight kernel function, because the effective support of the latter is about half as wide as that of the uniform kernel function. Because of this confounding effect, we take the kernel function K to be fixed when we discuss bandwidth selection. The normalization of a kernel can be defined via the concept of *canonical kernels*, as in Marron and Nolan (1988).

To demonstrate the effect of bandwidth selection, Figure 9.5 summarizes three estimated curves using local linear fits with the Epanechnikov kernel and bandwidths $h = 0$, 0.051, and ∞. When $h = 0$, there is only one data point in the neighborhood of a design point X_i, and the least-squares estimate, while degenerate, gives the fitted value to be the observed data. Hence, the resulting estimate is basically the linear interpolation. Such an estimate, which appears wiggly, is an example of an undersmoothed curve. As h increases, the size of the local

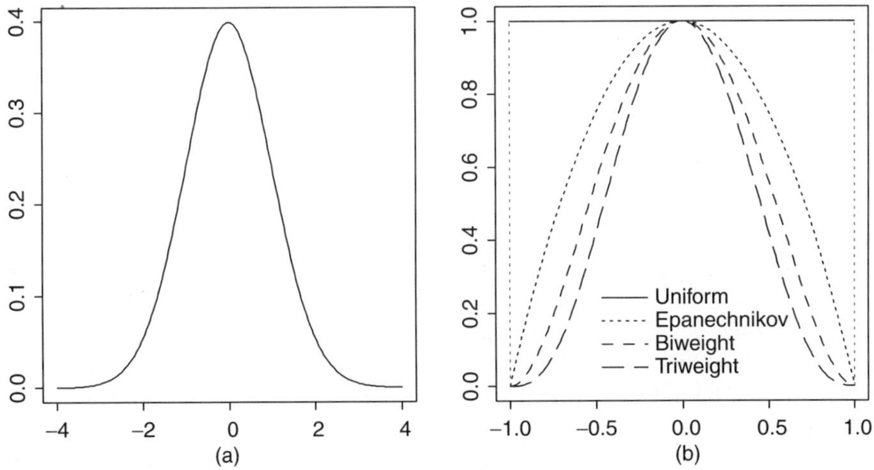

FIGURE 9.4 Commonly used kernels. (a) Gaussian kernel; (b) symmetric Beta family of kernels that are renormalized to have a maximum height of unity.

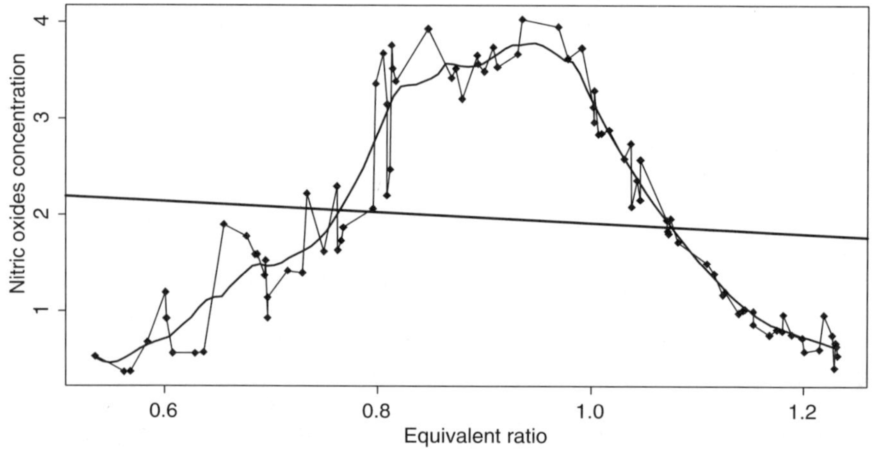

FIGURE 9.5 Local linear fit with bandwidth $h = 0$ (thin curve), 0.051 (solid curve), and ∞ (thick line).

neighborhood increases and so does the potential modeling bias. On the other hand, the number of local data points also increases, reducing the variance of the estimated coefficients. When $h = \infty$, the resulting local model becomes a global model, and hence the resulting estimate is a straight line, which has a large modeling bias for the ethanol data set. This kind of estimate, which ignores the features of the underlying curve, is called an oversmoothed estimate. As h ranges from 0 to ∞, we can obtain a family of estimated curves, ranging from the most complex curve (interpolation) to the simplest curve (line), and there are many interesting functions lying between these two extreme situations. Thus, the family of potential nonlinear functions produced by a local polynomial fit is rich.

A satisfactory estimated function should achieve a balance between the bias, which requires a small bandwidth, and the variance, which requires a large bandwidth. An example of this balance is the curve for $h = 0.051$ in Figure 9.5. There are two possible methods for choosing an appropriate amount of smoothing. A data-driven approach consists of choosing a bandwidth that minimizes the risk of an estimated curve. Such an automatic bandwidth selection method avoids subjectivity and is very useful for multivariate regression analysis. A subjective choice of the bandwidth can be thought of in a similar way, except that the trade-off now relies on the visualization. The drawback of this approach is that it cannot be automated and can be difficult to use when the data are out of the domain of visualization, as explained in Section 9.1.2.

Local modeling techniques provide parsimony to parametric modeling in the sense that, when the bandwidth is large, the local model becomes a global model. Therefore, the boundary between parametric and nonparametric modeling is not clear, particularly when one takes the bandwidth into consideration. The key difference is that using a parametric method is tantamount to using bandwidth $h = \infty$, while in nonparametric modeling, global modeling is the only possible candidate.

9.2 PROPERTIES OF LOCAL POLYNOMIAL FITTING

The basic idea of local polynomial fitting is very simple: In a neighborhood of x_0, a polynomial of degree p is fitted to the data, and the $p+1$ coefficients of the fit lead to estimates for $\beta_\nu = m^{(\nu)}(x_0)/\nu!$, $\nu = 0, \ldots, p$, where $m^{(\nu)}(x_0)$ denotes the νth derivative of the regression function $m(\cdot)$ evaluated at the point x_0. The neighborhood of the point x_0 is determined by two parameters: a symmetric kernel function K and a bandwidth parameter h.

In this section, we summarize some properties of the local polynomial regression estimator $\widehat{m}_\nu(x_0)$ of $m^{(\nu)}(x_0)$, defined by (9.6). When the design density $f_X(\cdot)$ has a finite support, one needs to study separately the behavior of the local polynomial regression estimator for points x_0 in the interior of the support of the design density and at the boundary of the support. It is well known (see, e.g., Gasser and Müller, 1979, and Härdle, 1990) that for kernel-type estimators, the behavior in the interior and at the boundary can be quite different. In the first two sections we focus on estimation in the interior, while the behavior at the boundary is discussed in the third section.

9.2.1 Bias and Variance

Since the model in Equation (9.3) involves the random design $X_1, \ldots, X_n$, one can consider conditional bias and variance expressions if given $X_1, \ldots, X_n$ or unconditional bias and variance if not. The question of whether to study the conditional mean squared error $E[\{\widehat{m}_\nu(x_0) - m^{(\nu)}(x_0)\}^2|\mathcal{X}]$, where $\mathcal{X} = (X_1, \ldots, X_n)$, or the unconditional mean squared error $E[\{\widehat{m}_\nu(x_0) - m^{(\nu)}(x_0)\}^2]$ is of a philosophical nature and relates to the decision to focus more on the behavior of the estimators for the design points at hand or more on the general behavior of the estimators for all possible design realizations from the design density f_X. We do not discuss this issue here; instead, we emphasize the conditional expressions for bias and variance, given $(X_1, \ldots, X_n)$, and complement this discussion by making some remarks on unconditional results.

Let x_0 be a point at the interior of the support of the design density f_X. Recall that the estimators $\widehat{\beta}_\nu$, $\nu = 0, \ldots, p$, for $\beta_\nu = m^{(\nu)}(x_0)/\nu!$, $\nu = 0, \ldots, p$, are obtained from minimizing (9.5) with respect to $\beta_0, \ldots, \beta_p$. For the derivations shown next, it is more convenient to use matrix notation. We denote the design matrix of the least-squares problem in (9.5) by

$$\mathbf{X} = \begin{pmatrix} 1 & (X_1 - x_0) & \cdots & (X_1 - x_0)^p \\ \vdots & \vdots & & \vdots \\ 1 & (X_n - x_0) & \cdots & (X_n - x_0)^p \end{pmatrix},$$

and let

$$\mathbf{y} = \begin{pmatrix} Y_1 \\ \vdots \\ Y_n \end{pmatrix}, \boldsymbol{\beta} = \begin{pmatrix} \beta_0 \\ \vdots \\ \beta_p \end{pmatrix}, \quad \text{and} \quad \widehat{\boldsymbol{\beta}} = \begin{pmatrix} \widehat{\beta}_0 \\ \vdots \\ \widehat{\beta}_p \end{pmatrix}.$$

Further, denote by

$$\mathbf{W} = \text{diag}\{K_h(X_i - x_0)\}$$

the diagonal matrix containing the weights, where $K_h(\cdot) = K(\cdot/h)/h$ is a rescaled version of the kernel K. The weighted least-squares problem in (9.5) can then be expressed as

$$\min_{\boldsymbol{\beta}} (\mathbf{y} - \mathbf{X}\boldsymbol{\beta})^T \mathbf{W} (\mathbf{y} - \mathbf{X}\boldsymbol{\beta}),$$

where $\mathbf{A}^T$ denotes the transpose of a vector or a matrix $\mathbf{A}$. The solution to this minimization problem follows from simple least-squares theory and is given by

$$\widehat{\boldsymbol{\beta}} = (\mathbf{X}^T\mathbf{W}\mathbf{X})^{-1}\mathbf{X}^T\mathbf{W}\mathbf{y}.$$

Here, we implicitly assume that the matrix $\mathbf{X}^T\mathbf{W}\mathbf{X}$ is invertible. When there are at least $(p+1)$ data points in the neighborhood around the point x_0, that is, the number of distinct points in the set

$$\{X_i : K_h(X_i - x_0) > 0\}$$

is at least $(p+1)$—then the matrix $\mathbf{X}^T\mathbf{W}\mathbf{X}$ is typically invertible.

The exact conditional bias and variance expressions for the estimated vector $\widehat{\boldsymbol{\beta}}$ of $\boldsymbol{\beta} = (\beta_0, \ldots, \beta_p)^T = (m(x_0), \ldots, m^{(p)}(x_0)/p!)^T$ are easily obtained from the above explicit expression for the estimator. Introducing the following notation:

$\mathbf{m} = (m(X_1), \ldots, m(X_n))^T$	$n \times 1$ vector
$\mathbf{r} = \mathbf{m} - \mathbf{X}\beta$	$n \times 1$ vector of residuals
$S_n = \mathbf{X}^T\mathbf{W}\mathbf{X}$	$(p+1) \times (p+1)$ matrix Y
$\Sigma = \text{diag}\{K_h^2(X_i - x_0)\sigma^2(X_i)\}$	$n \times n$ diagonal matrix
$S_n^* = \mathbf{X}^T\Sigma\mathbf{X}$	$(p+1) \times (p+1)$ matrix,

the expressions for the conditional bias and conditional variance are as follows:

$$E(\widehat{\boldsymbol{\beta}}|\mathcal{X}) = S_n^{-1}\mathbf{X}^T\mathbf{W}\mathbf{m} = \boldsymbol{\beta} + S_n^{-1}\mathbf{X}^T\mathbf{W}\mathbf{r}$$

$$\text{Var}(\widehat{\boldsymbol{\beta}}|\mathcal{X}) = S_n^{-1}S_n^*S_n^{-1}. \tag{9.7}$$

Seifert and Gasser (1996) investigated in detail the finite-sample variance of local polynomial regression estimators. They found that no upper bound can be given for the conditional variance and that the unconditional variance can be infinite when a kernel with compact support is used. The difficulties apparently arise in regions where the design is sparse and when the design realization has a small variance and/or a skew empirical distribution. Seifert and Gasser proposed two small-sample modifications of local polynomial regression estimators that they showed to have a good finite-sample behavior. When kernels with infinite support were used, such as a Gaussian kernel, the degree of such problematic finite-sample behavior was reduced.

Asymptotic expressions for the conditional bias and variance are useful, for example, for the bandwidth selection problem. We first derive an asymptotic expression for the conditional variance, which can be obtained by establishing asymptotic expressions for the matrices S_n and S_n^* that appear in the exact variance expression. Note that the matrix $S_n = (S_{n,j+l})_{0 \le j,l \le p}$, where

$$S_{n,j} = \sum_{i=1}^{n} K_h(X_i - x_0)(X_i - x_0)^j, \tag{9.8}$$

and that the matrix $S_n^* = (S_{n,j+l}^*)_{0 \le j,l \le p}$, where

$$S_{n,j}^* = \sum_{i=1}^{n} K_h^2(X_i - x_0)(X_i - x_0)^j \sigma^2(X_i). \tag{9.9}$$

Therefore, one of our tasks is to find asymptotic expressions for both types of element $S_{n,j}$ and $S_{n,j}^*$. Before doing so, we introduce some more notation:

$$\begin{aligned}
\nu_j &= \int u^j K(u)\,du; & \nu_j &= \int u^j K^2(u)\,du;\\
S &= (\mu_{j+l})_{0 \le j,l \le p}; & c_p &= (\mu_{p+1}, \ldots, \mu_{2p+1})^T\\
\widetilde{S} &= (\mu_{j+l+1})_{0 \le j,l \le p}; & \widetilde{c}_p &= (\mu_{p+2}, \ldots, \mu_{2p+2})^T;\\
S^* &= (\nu_{j+l})_{0 \le j,l \le p}; & H &= \operatorname{diag}(1, h, \ldots, h^p).
\end{aligned}$$

For the quantity defined in (9.8), we find that

$$\begin{aligned}
S_{n,j} &= ES_{n,j} + O_P\{\sqrt{\operatorname{Var}(S_{n,j})}\}\\
&= nh^j \int u^j K(u) f_X(x_0 + hu)\,du + O_P(\sqrt{nE\{K_h^2(X_1 - x_0)(X_1 - x_0)^{2j}\}})\\
&= nh^j\{f_X(x_0)\mu_j + o(1) + O_P((nh)^{-1/2})\}\\
&= nh^j f_X(x_0)\mu_j\{1 + o_P(1)\},
\end{aligned} \tag{9.10}$$

where we assume that $h \to 0$, $nh \to \infty$, $f_X(\cdot)$ is continuous in a neighborhood of x_0, and K has a compact support. In general, (9.4) holds for kernel functions with light tails. See Parzen (1962) for a reasonable set of conditions on K. For the matrix S_n, we then obtain

$$S_n = nf_X(x_0)HSH\{1 + o_P(1)\}. \tag{9.11}$$

Using similar arguments, we derive the asymptotic behavior of S_n^*. From (9.9) and (9.10), it is already clear that

$$S_{n,j}^* = nh^{j-1} f_X(x_0)\sigma^2(x_0)\nu_j\{1 + o_P(1)\},$$

provided that $\sigma^2(\cdot)$ is continuous in a neighborhood of x_0. This expression yields

$$S_n^* = nh^{-1} f_X(x_0)\sigma^2(x_0) H S^* H\{1 + o_P(1)\}. \tag{9.12}$$

Equations (9.11) and (9.12) lead to the following asymptotic expression for the conditional variance, provided that $f_X(x_0) > 0$:

$$\mathrm{Var}(\widehat{\boldsymbol{\beta}}|\mathcal{X}) = \frac{\sigma^2(x_0)}{f_X(x_0)nh} H^{-1}S^{-1}S^*S^{-1}H^{-1}\{1 + o_P(1)\}. \tag{9.13}$$

Assuming that $m^{(p+1)}(\cdot)$ is continuous in a neighborhood of x_0 and using the Taylor expansion for the vector of residuals $\mathbf{r}$, we can express the conditional bias as

$$\begin{aligned} S_n^{-1}\mathbf{X}^T\mathbf{W}\mathbf{r} &= S_n^{-1}\mathbf{X}^T\mathbf{W}[\beta_{p+1}(X_i - x_0)^{p+1} + o_P\{(X_i - x_0)^{p+1}\}]_{1\le i\le n} \\ &= S_n^{-1}\{\beta_{p+1}c_n + o_P(nh^{p+1})\}, \end{aligned} \tag{9.14}$$

where $\beta_{p+1} = m^{(p+1)}(x_0)/(p+1)!$ and $c_n = (S_{n,p+1}, \ldots, S_{n,2p+1})^T$. Using (9.10) and (9.11), we find that this leads to the asymptotic expression for the conditional bias:

$$\mathrm{Bias}(\widehat{\boldsymbol{\beta}}|\mathcal{X}) = H^{-1}S^{-1}c_p\beta_{p+1}h^{p+1}\{1 + o_P(1)\}. \tag{9.15}$$

The asymptotic conditional variance of $\widehat{\beta}_\nu = \widehat{m}_\nu(x_0)/\nu!$ is obtained by taking the $(\nu+1)$th diagonal element of the right-hand side of (9.13). The asymptotic conditional bias of $\widehat{m}_\nu(x_0)/\nu!$ is given by the $(\nu+1)$th element of the vector provided on the right-hand side of (9.15). Here, one needs to be careful however. When K is symmetric, the $(\nu+1)$th element of $S^{-1}c_p$ is zero when $p-\nu$ is even, and therefore, the leading bias term is zero. This result is valid, but not informative. In order to find the asymptotic conditional bias for $\widehat{\beta}_\nu$ when $p-\nu$ is even, it suffices to use Taylor expansions up to a higher order in the above derivation for the asymptotic bias. A higher order Taylor expansion will of course require more stringent conditions on the functions $f_X(\cdot)$ and $m(\cdot)$ and on the bandwidth h. We do not elaborate further on this point. More details can be found in Fan and Gijbels (1996, Theorem 3.1, p. 62).

The asymptotic expressions for the conditional bias and variance of $\widehat{m}_\nu(x_0)$ are summarized in the Theorem 9.1, which was derived by Fan (1992) and Ruppert and Wand (1994). For simplicity, we assume that the kernel function is symmetric, with a bounded support.

Theorem 9.1. *Assume that $f_X(x_0) > 0$; $f_X(\cdot)$, $m^{(p+1)}(\cdot)$, and $\sigma^2(\cdot)$ are continuous in a neighborhood of x_0; and K has a compact support. If $h \to 0$ such that $nh \to \infty$, then the asymptotic conditional variance of $\widehat{m}_\nu(x_0)$ is given by*

$$\mathrm{Var}\{\widehat{m}_\nu(x_0)|\mathcal{X}\} = e_{\nu+1}^T S^{-1} S^* S^{-1} e_{\nu+1} \frac{(\nu!)^2\sigma^2(x_0)}{f_X(x_0) n h^{1+2\nu}} + o_P(n^{-1}h^{-(1+2\nu)}), \quad (9.16)$$

where $e_{\nu+1} = (0, 0, \ldots, 0, 1, 0, \ldots, 0)^T$ denotes the unit vector of dimension $p+1$ with a value of unity at position $\nu+1$.

For $p - \nu$ odd the asymptotic conditional bias of $\widehat{m}_\nu(x_0)$ is

$$\mathrm{Bias}\{\widehat{m}_\nu(x_0)|\mathcal{X}\} = e_{\nu+1}^T S^{-1} c_p \frac{\nu!}{(p+1)!} m^{(p+1)}(x_0) h^{p+1-\nu} + o_P(h^{p+1-\nu}). \quad (9.17)$$

For $p - \nu$ even, and under the additional assumptions that $f'_X(\cdot)$ and $m^{(p+2)}(\cdot)$ are continuous in a neighborhood of x_0 and $nh^3 \to \infty$, the asymptotic conditional bias is given by

$$\mathrm{Bias}\{\widehat{m}_\nu(x_0)|\mathcal{X}\} = e_{\nu+1}^T S^{-1} \widetilde{c}_p \frac{\nu!}{(p+2)!} \left\{ m^{(p+2)}(x_0) + (p+2) m^{(p+1)}(x_0) \frac{f'_X(x_0)}{f_X(x_0)} \right\}$$
$$\times h^{p+2-\nu} + o_P(h^{p+2-\nu}). \quad (9.18)$$

Further, it can be shown under certain conditions that the estimator $\widehat{m}_\nu(x_0)$ is asymptotically normal:

$$\{\mathrm{AV}_{\nu,p}(x_0)\}^{-1/2}\{\widehat{m}_\nu(x_0) - m^{(\nu)}(x_0) - \mathrm{Ab}_{\nu,p}(x_0)\} \xrightarrow{D} \mathrm{N}(0; 1). \quad (9.19)$$

Here, $\mathrm{AV}_{\nu,p}(x_0)$ and $\mathrm{Ab}_{\nu,p}(x_0)$ are, respectively, the leading terms in the asymptotic conditional variance and bias given previously.

9.2.2 Equivalent Kernels

The estimator $\widehat{\beta}_\nu$ of $m^{(\nu)}(x_0)/\nu!$ can be rewritten as a type of kernel estimator:

$$\widehat{\beta}_\nu = e_{\nu+1}^T \widehat{\boldsymbol{\beta}} = e_{\nu+1}^T S_n^{-1} \mathbf{X}^T \mathbf{W} \mathbf{y} = \sum_{i=1}^n W_\nu^n \left(\frac{X_i - x_0}{h} \right) Y_i. \quad (9.20)$$

Here, $W_\nu^n(t) = e_{\nu+1}^T S_n^{-1}\{1, th, \ldots, (th)^p\}^T K(t)/h$. Note that the definition of the weight function $W_\nu^n(\cdot)$ depends on the design points and on x_0. This dependence reflects the fact that local polynomial regression estimators adapt automatically to the design and to the location x_0 (either in the interior or near the boundary of the support).

It is easily seen that the kernel weights W_ν^n satisfy some discrete moment conditions. Indeed, for $0 \le \nu$ and $q \le p$, we find that

$$\sum_{i=1}^{n}(X_i-x_0)^q W_\nu^n\left(\frac{X_i-x_0}{h}\right) = e_{\nu+1}^T S_n^{-1}\sum_{i=1}^{n}(X_i-x_0)^q \begin{pmatrix} 1 \\ X_i - x_0 \\ \vdots \\ (X_i-x_0)^p \end{pmatrix} K_h(X_i-x_0)$$

$$= e_{\nu+1}^T S_n^{-1} S_n e_{q+1} = \delta_{\nu,q}.$$

Consequently, the finite-sample bias is zero when using local polynomial fitting of degree p to estimate polynomials of order up to p.

The above representation of the local polynomial regression estimator as a conventional kernel estimator has an asymptotic counterpart. This can easily be found by substituting the asymptotic expression in (9.11) for the matrix S_n into the definition of $W_\nu^n(t)$, yielding an asymptotic expression for the kernel function in (9.20):

$$W_\nu^n(t) = \frac{1}{nh^{\nu+1} f_X(x_0)} e_{\nu+1}^T S^{-1}(1, t, \ldots, t^p)^T K(t)\{1 + o_P(1)\}.$$

This expression in turn leads to the following asymptotic expression for the estimator:

$$\widehat{\beta}_\nu = \frac{1}{nh^{\nu+1} f_X(x_0)} \sum_{i=1}^{n} K_\nu^*\left(\frac{X_i - x_0}{h}\right) Y_i\{1 + o_P(1)\}.$$

Here,

$$K_\nu^*(t) = e_{\nu+1}^T S^{-1}(1, t, \ldots, t^p)^T K(t); \qquad (9.21)$$

which is called the *equivalent kernel*. It is easily seen, using similar arguments as before, that this equivalent kernel satisfies the moment conditions

$$\int u^q K_\nu^*(u)\, du = \delta_{\nu,q}, \quad 0 \le \nu, q \le p,$$

which make up the asymptotic counterpart of the discrete moment conditions on $W_\nu^n(\cdot)$ discussed previously. A picture of the biweight kernel and some corresponding equivalent kernels is provided in Figure 9.6. See Section 3.2.2 of Fan and Gijbels (1996) for more details on equivalent kernels and their relationship with other kernels discussed in the literature.

The asymptotic conditional bias and variance of the local polynomial regression estimator provided in Theorem 9.1 can be rewritten in terms of the equivalent kernel K_ν^*. For the asymptotic conditional variance, we have

$$\text{Var}\{\widehat{m}_\nu(x_0)|\mathcal{X}\} = \int K_\nu^{*2}(t)\, dt \frac{(\nu!)^2\sigma^2(x_0)}{f_X(x_0) nh^{1+2\nu}} + o_P\left(n^{-1}h^{-(1+2\nu)}\right). \qquad (9.22)$$

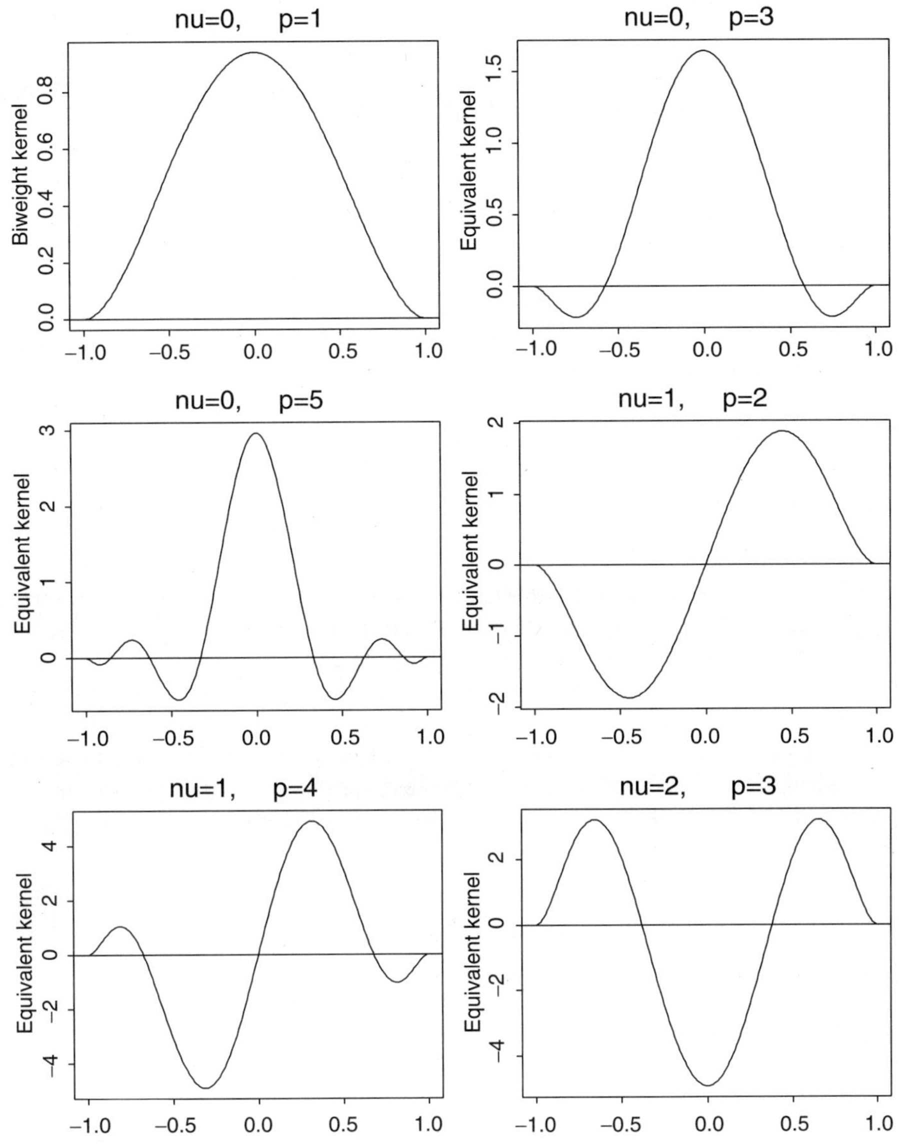

FIGURE 9.6 Biweight kernel and corresponding equivalent kernels for some values of p and ν.

The asymptotic conditional bias of $\widehat{m}_\nu(x_0)$ can be reexpressed as

$$\text{Bias}\{\widehat{m}_\nu(x_0)|\mathcal{X}\} = \left\{\int t^{p+1} K_\nu^*(t)\,dt\right\} \frac{\nu!}{(p+1)!} m^{(p+1)}(x_0) h^{p+1-\nu} + o_P(h^{p+1-\nu}) \tag{9.23}$$

for $p - \nu$ odd and as

$$\text{Bias}\{\widehat{m}_\nu(x_0)|\mathcal{X}\} = \left\{\int t^{p+2} K_\nu^*(t)\,dt\right\} \frac{\nu!}{(p+2)!} \left\{ m^{(p+2)}(x_0) + (p+2) m^{(p+1)}(x_0) \times \frac{f_X'(x_0)}{f_X(x_0)} \right\} h^{p+2-\nu} + o_P(h^{p+2-\nu}) \tag{9.24}$$

for $p - \nu$ even.

Note that the leading term in the asymptotic conditional bias has a more complex structure for $p - \nu$ even. Furthermore, the term depends on the design density $f_X(x_0)$ and its derivative function $f_X'(x_0)$. This rather complex structure of the leading term is unpleasant to work with when, for example, doing bandwidth selection based on L_2-type measures of performance. (See also Section 9.3.2.)

As an illustration, we look at four special cases: local constant ($p = 0$) and local linear ($p = 1$) fitting for estimating the regression function ($\nu = 0$) and local linear ($p = 1$) and local quadratic ($p = 2$) fitting for estimating the first derivative of the regression function ($\nu = 1$). Table 9.1 lists, in the second column, the expression for the equivalent kernel for each of these cases. In the last column, the equivalent kernels for boundary points are provided. These are discussed in the next subsection, where one can also find the expressions for the $\mu_{j,c}$'s (see (9.25)). Table 9.2 presents the leading terms in the asymptotic conditional bias and variance.

We close this section by remarking that, with respect to making a decision on theoretical terms, local polynomial fits possess various optimality properties. See Stone (1982), Fan (1992), and Fan and Gijbels (1996, Sections 3.4 and 3.5) for discussions.

9.2.3 Estimation at the Boundary

In this section, we briefly discuss the behavior of the local polynomial regression estimator $\widehat{m}_\nu(x_0)$ for x_0 at the boundary of the support of the design density $f_X(\cdot)$. Assume, for simplicity, that this support is [0, 1] and that we want to estimate the

TABLE 9.1 Some Examples of Equivalent Kernels

ν	p	$K_\nu^*(t)$	$K_{\nu,c}^*(t)$
0	0	$K(t)$	$\mu_{0,c}^{-1}K(t)$
0	1	$K(t)$	$(\mu_{2,c} - \mu_{1,c}t)K(t)/(\mu_{0,c}\mu_{2,c} - \mu_{1,c}^2)$
1	1	$\mu_2^{-1}tK(t)$	$(-\mu_{1,c} + \mu_{0,c}t)K(t)/(\mu_{0,c}\mu_{2,c} - \mu_{1,c}^2)$
1	2	$\mu_2^{-1}tK(t)$	a[a]

[a] $a = \dfrac{-\mu_{1,c}\mu_{4,c} + \mu_{2,c}\mu_{3,c} + t(\mu_{0,c}\mu_{4,c} - \mu_{2,c}^2) - t^2(\mu_{0,c}\mu_{3,c} - \mu_{1,c}\mu_{2,c})}{(\mu_{0,c}\mu_{2,c} - \mu_{1,c}^2)\mu_{4,c} - \mu_{2,c}^3 - \mu_{0,c}\mu_{3,c}^2 + 2\mu_{1,c}\mu_{2,c}\mu_{3,c}}.$

TABLE 9.2 Leading Terms in the Asymptotic Conditional Bias and Variance

ν	p	Leading Term in Bias	Leading Term in Variance
0	0	$\mu_2 \frac{1}{2}\left\{m''(x_0) + 2m'(x_0)\frac{f'_X(x_0)}{f_X(x_0)}\right\} h^2$	$\nu_0 \frac{\sigma^2(x_0)}{f_X(x_0)nh} \equiv V_0$
0	1	$\mu_2 \frac{1}{2} m''(x_0)h^2$	V_0
1	1	$\mu_2^{-1}\mu_4 \frac{1}{6}\left\{m'''(x_0) + 3m''(x_0)\frac{f'_X(x_0)}{f_X(x_0)}\right\} h^2$	$\mu_2^{-2}\nu_2 \frac{\sigma^2(x_0)}{f_X(x_0)nh^3} \equiv V_1$
1	2	$\mu_2^{-1}\mu_4 \frac{1}{6} m'''(x_0)h^2$	V_1

regression function or its derivatives at left boundary points $x_0 = ch$ or at right boundary points $x_0 = 1 - ch$, where $c > 0$. If the kernel function K has support $[-1, 1]$, then real boundary points are those for which $c < 1$, whereas for $c > 1$ we have interior points.

The asymptotic conditional bias and variance expressions for the estimator when $x_0 = 1 - ch$ (right boundary points) are derived in a way analogous to that for interior points. Essentially, the only difference is in the asymptotic moment calculations. This results for example in a different asymptotic behavior of the finite-sample moments $S_{n,j}$, defined in (9.8). For these moments, we now obtain

$$S_{n,j} = nh^j f_X(1-)\mu_{j,c}\{1 + o_P(1)\}, \tag{9.25}$$

where $\mu_{j,c} = \int_{-\infty}^{c} u^j K(u)\, du$, which is a truncated version of the moment μ_j. Compare (9.25) with (9.10). This difference in the asymptotic behavior of $S_{n,j}$ for boundary points carries through to the definition of the equivalent kernel, which is now given by

$$K^*_{\nu,c}(t) = e^T_{\nu+1} S_c^{-1}(1, t, \ldots, t^p)^T K(t), \tag{9.26}$$

where $S_c = (\mu_{j+\ell,c})_{0\le j,\ell\le p}$. Note that the only difference between the definition of the equivalent kernel for interior points and that for boundary points is the replacement of the matrix S by the matrix S_c.

The asymptotic conditional variance for right boundary points $x_0 = 1 - ch$ is given by

$$\mathrm{Var}\{\widehat{m}_\nu(x_0)|\mathcal{X}\} = \int_{-\infty}^{c} K^{*2}_{\nu,c}(t)\, dt \frac{(\nu!)^2\sigma^2(1-)}{f_X(1-)nh^{1+2\nu}} + o_P(n^{-1}h^{-(1+2\nu)}), \tag{9.27}$$

and the asymptotic conditional bias

$$\mathrm{Bias}\{\widehat{m}_\nu(x_0)|\mathcal{X}\} = \left\{\int_{-\infty}^{c} t^{p+1} K^*_{\nu,c}(t)\, dt\right\} \frac{\nu!}{(p+1)!} m^{(p+1)}(1-)h^{p+1-\nu} + o_P(h^{p+1-\nu}), \tag{9.28}$$

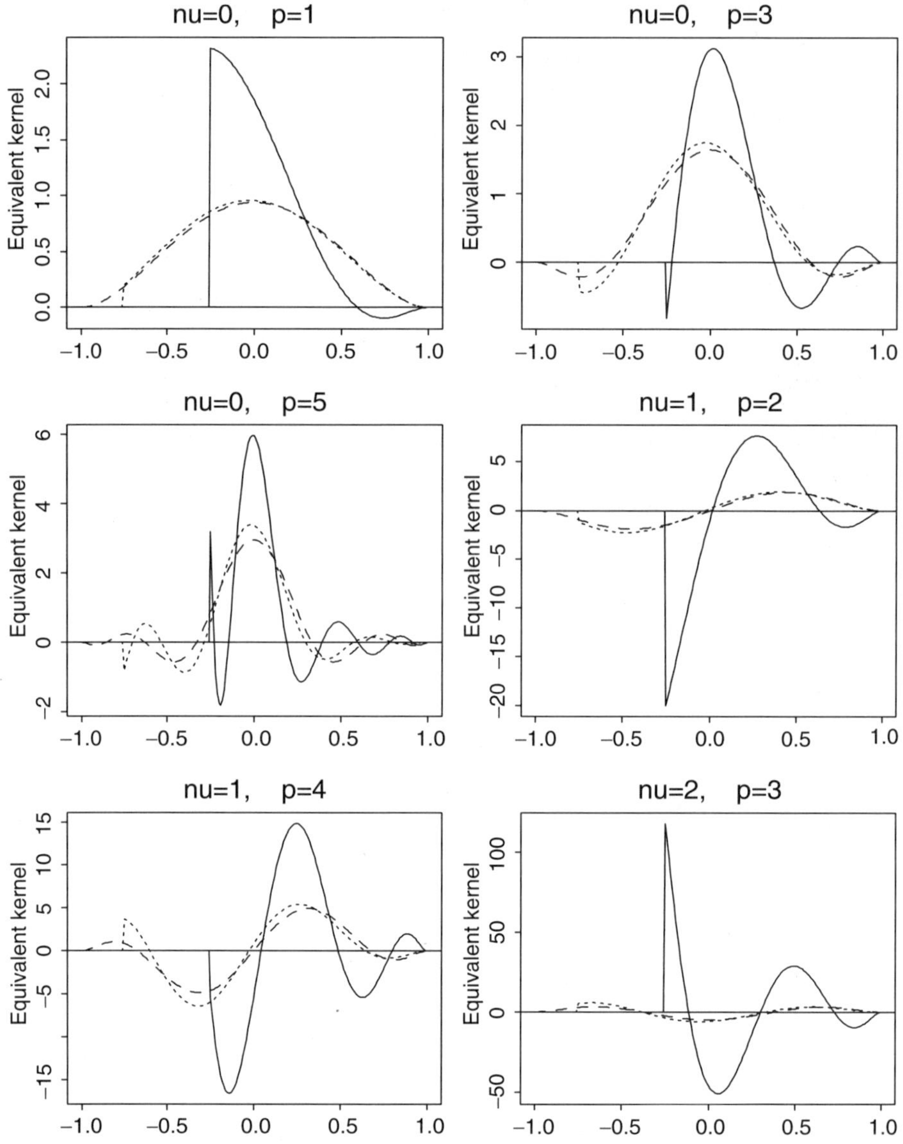

FIGURE 9.7 Biweight kernel and corresponding equivalent kernels for right boundary points $x_0 = 1 - ch$ with $c = 0.25$ and $c = 0.75$, and for interior points $c \geq 1$.

is informative for both $p - \nu$ odd and even, since now the leading term in (9.28) is not necessarily zero for $p - \nu$ even.

For left boundary points $x_0 = ch$, similar derivations can be made. Now the truncated moments $\nu_{j,c} = \int_{-c}^{+\infty} u^j K(u)\, du$ appear in Equation (9.25), and with these truncated moments, the equivalent kernel is defined as in (9.26). The expressions for the asymptotic conditional bias and variance are similar to those provided in (9.27) and (9.28), but with the appropriate definition of the equivalent kernel and with the right-hand limits $m^{(p+1)}(0+)$, $\sigma^2(0+)$, and $f_X(0+)$ instead of the left-hand limits at the point 1.

Table 9.1 provides expressions for equivalent kernels at the boundary for some values of p and ν. Figure 9.7 presents the biweight kernel and some of the corresponding equivalent kernels at two right boundary points $x_0 = 1 - ch$, where $c = 0.25$ and 0.75, and at interior points, where $c \geq 1$.

Note that for $p - \nu$ odd the asymptotic conditional bias and variance are of the same order at the boundary and in the interior. Compare the expressions in (9.22) and (9.23) with the expressions in (9.27) and (9.28), respectively. This comparison implies that, for $p - \nu$ odd, local polynomial regression estimators have the same L_2 rate of convergence at the boundary and in the interior. In this sense, this estimation procedure for $p - \nu$ odd adapts automatically to the boundary of the support of the design density. Note, however, that the constant factors, determined by the kernel function K, p, and ν, in the leading terms of the asymptotic conditional bias and variance are different for the boundary region than for the interior, and so there are certain boundary effects that are noticed. A discussion of these constant factors can be found in Fan and Gijbels (1992).

For $p - \nu$ even the rate of convergence of the local polynomial regression estimator is different at the boundary than in the interior; compare expressions (9.24) and (9.28). For a local constant fit, for example, the conditional bias term is of order h^2 for points in the interior and order h for points at the boundary.

9.3 CHOICE OF BANDWIDTH

Local polynomial fitting requires a choice for the bandwidth parameter h, the degree p of the polynomial, and the kernel function K. It is well known that the choice of the kernel function is of less importance in kernel smoothing. This also holds true for local polynomial fitting. For a thorough discussion on optimal choices for the kernel function, see Fan and Gijbels (1996, Chapter 3), Gasser et al. (1985), and Granovsky and Müller (1991). Here, we discuss only the choice of the bandwidth and the degree of the polynomial p. This section is devoted to bandwidth selection. See also Chapter 4 in this book.

9.3.1 Different Types of Bandwidths

As already discussed in Section 9.1, the choice of the bandwidth parameter h can have a big influence on the final smoothing result and hence is quite important. We can distinguish between different types of bandwidths. The bandwidth h can be chosen to be the same constant over all the range of x in $m(x)$. Such a bandwidth is called a *constant*, or *global*, bandwidth and is a reasonable choice when the unknown regression function shows approximately the same smoothness over all the estimation range.

When the regression function shows a different behavior in different regions—for example, when it is nearly linear in one region and changes rather rapidly in some other region—it may be advisable to use different bandwidths in these different regions. In the latter case, the bandwidth h changes with the location x and is denoted by $h(x)$. This bandwidth is referred to as *local variable* bandwidth. A bandwidth can also change with the data point X_i, resulting in $h(X_i)$, which is called a *global variable* bandwidth.

9.3.2 Optimal Bandwidths

When talking about *optimal* bandwidths, we first need to specify the criterion that is used for evaluating the performance of an estimator. Such an evaluation can be based on any appropriate distance function between the unknown regression function and the estimated function. Examples include the L_1, L_2, and L_∞ distances, as well as the Hellinger distance. For mathematical convenience, we often focus on the L_2-loss function. A measure of local performance of the estimator is then the conditional *mean squared error* (MSE), defined as

$$\begin{aligned}\text{MSE}_{\nu,p}(x_0;h) &= E[\{\widehat{m}_\nu(x_0) - m^{(\nu)}(x_0)\}^2|\mathcal{X}] \\ &= [\text{Bias}\{\widehat{m}_\nu(x_0)|\mathcal{X}\}]^2 + \text{Var}\{\widehat{m}_\nu(x_0)|\mathcal{X}\}, \end{aligned} \tag{9.29}$$

and a measure of global performance is the conditional *mean integrated squared error* (MISE), defined as

$$\begin{aligned}\text{MISE}_{\nu,p}(h) &= \int E[\{\widehat{m}_\nu(x) - m^{(\nu)}(x)\}^2|\mathcal{X}]w(x)\,dx \\ &= \int [\text{Bias}\{\widehat{m}_\nu(x)|\mathcal{X}\}]^2 w(x)\,dx + \int \text{Var}\{\widehat{m}_\nu(x)|\mathcal{X}\}w(x)\,dx, \end{aligned} \tag{9.30}$$

where $w(\cdot)$ is a positive weight function.

The ideal optimal global bandwidth is then defined as the bandwidth that minimizes the conditional MISE:

$$h^o_{\nu,p} = \text{argmin}_h \text{MISE}_{\nu,p}(h). \tag{9.31}$$

The ideal optimal local variable bandwidth is the bandwidth that minimizes the conditional MSE:

$$h^o_{\nu,p}(x_0) = \text{argmin}_h \text{MSE}_{\nu,p}(x_0;h). \tag{9.32}$$

Approximations of these ideal optimal bandwidths are obtained by minimizing the asymptotic expressions for the MSE and the MISE, denoted by AMSE and AMISE, respectively, which follow directly from the asymptotic expressions for bias and variance provided in Section 9.2.1. Since the case $p - \nu$ even leads to more complex expressions for the asymptotic conditional bias, it is convenient to focus on the case $p - \nu$ odd. In Section 9.4, we discuss in more detail the choice between odd or even $p - \nu$ and argue in favor of the case $p - \nu$ odd.

Minimization of the asymptotic conditional MSE (i.e., AMSE) for $p - \nu$ odd leads to the theoretical optimal local variable bandwidth

$$\begin{aligned} h_{\nu,p}^{\text{opt}}(x_0) &= \text{argmin}_h \text{AMSE}_{\nu,p}(x_0; h) \\ &= C_{\nu,p}(K) \left[\frac{\sigma^2(x_0)}{\{m^{(p+1)}(x_0)\}^2 f_X(x_0)} \right]^{1/(2p+3)} n^{-1/(2p+3)}, \end{aligned} \tag{9.33}$$

where the constant factor $C_{\nu,p}(K)$ depends only on K, p, and ν and is given by

$$C_{\nu,p}(K) = \left[\frac{\{(p+1)!\}^2 (2\nu+1) \displaystyle\int K_\nu^{*2}(t)\,dt}{2(p+1-\nu) \left\{ \displaystyle\int t^{p+1} K_\nu^*(t)\,dt \right\}^2} \right]^{1/(2p+3)}. \tag{9.34}$$

These constant factors are easy to calculate and are tabulated for certain choices of K, p, and ν in Fan and Gijbels (1996, Table 3.2 p. 67). In particular, when $p = 1$ and $\nu = 0$, the constant $C_{\nu,p}(K)$ is 0.776, 1.351, 1.719, and 2.036 for the Gaussian, uniform, Epanechnikov, and biweight kernel, respectively.

The theoretical optimal constant or global bandwidth is obtained by minimizing the asymptotic conditional MISE (i.e., AMISE) and is given by

$$\begin{aligned} h_{\nu,p}^{\text{opt}} &= \text{argmin}_h \text{AMISE}_{\nu,p}(h) \\ &= C_{\nu,p}(K) \left[\frac{\displaystyle\int \sigma^2(x) w(x) / f_X(x)\,dx}{\displaystyle\int \{m^{(p+1)}(x)\}^2 w(x)\,dx} \right]^{1/(2p+3)} n^{-1/(2p+3)}. \end{aligned} \tag{9.35}$$

Such theoretical optimal bandwidths cannot be directly used in practice, since they depend on the unknown quantities $m^{(p+1)}(\cdot)$, $\sigma^2(\cdot)$, and $f_X(\cdot)$ and hence cannot be calculated. In the next section, we briefly discuss some practical data-driven bandwidth choices, which consistently estimate $h_{\nu,p}^{\text{opt}}$ or $h_{\nu,p}^{\text{opt}}(x_0)$.

9.3.3 Data-Driven Bandwidths

9.3.3.1 Rule-of-Thumb Global Bandwidth Selector

A simple data-driven global bandwidth selector is obtained by using (9.35) and considering a reference parametric model for the unknown regression function $m(\cdot)$. One can, for example, take $m(\cdot)$ to be a polynomial. For the degree of the polynomial, one could, for instance, take $p+3$, which would imply that $m^{(p+1)}(\cdot)$ is a quadratic function, leaving a certain flexibility for fitting functions with a curvature. More precisely, consider the parametric model

$$m(x) = \alpha_0 + \alpha_1 x + \cdots + \alpha_{p+3} x^{p+3}, \qquad (9.36)$$

and take $w(\cdot)$ in (9.35) to be equal to $f_X(\cdot) w_0(\cdot)$ for a given positive weight function $w_0(\cdot)$. We fit the parametric model (9.36) to the data and obtain the fit

$$\check{m}(x) = \check{\alpha}_0 + \check{\alpha}_1 x + \cdots + \check{\alpha}_{p+3} x^{p+3},$$

as well as $\check{\sigma}^2$, the standardized residual sum of squares from this parametric fit. Substituting the estimated quantities

$$\check{m}^{(p+1)}(x) = (p+1)!\check{\alpha}_{p+1} + (p+2)!\check{\alpha}_{p+2} x + \frac{(p+3)!}{2} \check{\alpha}_{p+3} x^2$$

and $\check{\sigma}^2$ into (9.35) leads to the estimated expression

$$C_{\nu,p}(K) \left[\frac{\check{\sigma}^2 \int w_0(x)\, dx}{n \int \{\check{m}^{(p+1)}(x)\}^2 w_0(x) f_X(x)\, dx} \right]^{1/(2p+3)}.$$

Note that the denominator can be written as $nE[\{\check{m}^{(p+1)}(X)\}^2 w_0(X)]$, which can be estimated by

$$\sum_{i=1}^{n} \{\check{m}^{(p+1)}(X_i)\}^2 w_0(X_i).$$

Hence, a *rule-of-thumb* (ROT) estimator for $h_{\nu,p}^{\text{opt}}$ in (9.35) is given by

$$\check{h}_{\text{ROT}} = C_{\nu,p}(K) \left[\frac{\check{\sigma}^2 \int w_0(x)\, dx}{\sum_{i=1}^{n} \{\check{m}^{(p+1)}(X_i)\}^2 w_0(X_i)} \right]^{1/(2p+3)}. \qquad (9.37)$$

The above data-driven bandwidth selector is of the same nature as the rule-of-thumb bandwidth selector in density estimation using a normal density as a reference (see, e.g., Silverman, 1986). It is a rather crude bandwidth selector, but it gives a first idea of an appropriate magnitude for the bandwidth parameter. We illustrate the use of this bandwidth selector in Section 9.4.

9.3.3.2 Plug-In Global Bandwidth Selector

The theoretical global bandwidth in (9.35) contains two unknown quantities, namely, $\int \sigma^2(x)w(x)/f_X(x)\,dx$ and $\int \{m^{(p+1)}(x)\}^2 w(x)\,dx$, and hence the basic task is to estimate these unknown quantities. Ruppert et al. (1995) proposed nonparametric estimators for these quantities and substituted them into (9.35), leading to a direct plug-in-type of bandwidth selectors. They focus on local linear regression estimation ($p = 1$ and $\nu = 0$) under homoscedastic errors ($\sigma^2(x) = \sigma^2$ for all x). Here, we briefly discuss their procedure.

Taking $w(x) = f_X(\cdot)I\{x \in [a, b]\}$, where $[a, b] \subset \mathbb{R}$, the theoretical optimal global bandwidth is given by

$$h_{0,1}^{\text{opt}} = C_{0,1}(K)\left[\frac{\sigma^2(b-a)}{n\int_a^b \{m''(x)\}^2 f_X(x)\,dx}\right]^{1/5}, \tag{9.38}$$

where the unknown quantities to be estimated are

$$\theta_{22} \equiv \int_a^b \{m''(x)\}^2 f_X(x)\,dx \quad \text{and} \quad \sigma^2.$$

Note that $\theta_{22} = E[\{m''(X)\}^2 I\{X \in [a, b]\}]$, and hence an appropriate estimator is

$$\widehat{\theta}_{22} = \frac{1}{n}\sum_{i=1}^{n}\{\widehat{m}_2(X_i; g)\}^2 I\{X_i \in [a, b]\},$$

where $\widehat{m}_2(\cdot; g)$ is obtained from a local cubic fit using a pilot bandwidth g. Ruppert et al. (1995) derived the theoretical optimal choice for the pilot bandwidth g, which is obtained by minimizing the AMSE. They found that

$$g_{\text{AMSE}} \simeq C(K)\left[\frac{\sigma^2(b-a)}{n|\theta_{24}|}\right]^{1/7}, \tag{9.39}$$

where $\theta_{24} = \int m''(x)m^{(4)}(x)\,dx$ and the definition of the constant factor $C(K)$ depends on the sign of the quantity θ_{24}, but its definition involves only the kernel function K. Note that the theoretical optimal bandwidth g_{AMSE} is of order $n^{-1/7}$, compared with the order $n^{-1/5}$ for the theoretical optimal bandwidth $h_{0,1}^{\text{opt}}$, and involves the unknown quantity θ_{24}.

The unknown variance σ^2 is estimated by

$$\widehat{\sigma}^2 = (\text{df})^{-1} \sum_{i=1}^{n} \{Y_i - \widehat{m}(X_i; \lambda)\}^2, \tag{9.40}$$

where df is the *degrees of freedom* (see Ruppert et al. (1995) for the exact expression for df) and $\widehat{m}(\cdot; \lambda)$ is a local linear regression estimator using a pilot bandwidth λ. The theoretical optimal value for λ can be shown to be

$$\lambda_{\text{AMSE}} = C^*(K) \left[\frac{\sigma^4(b-a)}{n^2\theta_{22}^2} \right]^{1/9}, \tag{9.41}$$

where $C^*(K)$ is a constant factor that depends only on the kernel function K.

Since both g_{AMSE} and λ_{AMSE} depend on unknown quantities themselves, we need to find some strategies to deal with this problem. Ruppert et al. (1995) proposed three plug-in bandwidth selection strategies. For brevity, we only mention one of them:

1. Divide the sample $X_1, \ldots, X_n$ into N subsamples, and obtain least-squares quartic fits of $m(\cdot)$ and σ^2 using the subsamples. Use these least-squares quartic fits to get preliminary estimators for θ_{24} and σ^2.
2. Use the preliminary estimators from Step 1 to estimate g_{AMSE} and θ_{22}.
3. Estimate λ_{AMSE} by using the preliminary estimator for σ^2 from Step 1 and the estimator for θ_{22} from Step 2. Obtain the final estimator of σ^2 from (9.40).
4. Substitute the estimators for θ_{22} and σ^2 into (9.38), and obtain the estimator for $h_{0,1}^{\text{opt}}$.

For a detailed description of this and other plug-in global bandwidth selection procedures, see Ruppert et al. (1995).

9.3.3.3 Residual Squares and Preasymptotic Substitution Bandwidth Selectors

The rule-of-thumb and plug-in bandwidth selection procedures are *post-asymptotic substitution methods*, since they both focus on the theoretical optimal global bandwidth in (9.35), which minimizes the asymptotic conditional mean integrated squared error. This theoretical optimal bandwidth is a first-order approximation of the ideal optimal global bandwidth in (9.31), which minimizes the finite-sample conditional mean integrated squared error.

An attempt to stay closer to the finite-sample level and hence to avoid going to the asymptotic immediately consists of trying to estimate the mean squared error and minimize this estimate or an integrated version of it. This approach results in bandwidth selectors that rely less on asymptotes and hopefully perform better overall. The bandwidth selectors discussed in this subsection are based on this idea and are called *preasymptotic substitution methods*.

We first describe how to obtain an estimator $\widehat{\text{MSE}}_{\nu,p}(x_0;h)$ of $\text{MSE}_{\nu,p}(x_0;h)$. However, we only outline the estimation procedure. For more details refer to Fan and Gijbels (1995a, 1996). As is clear from (9.39), we need to provide estimates for the conditional bias and variance given in (9.7). Note that the only unknown quantity in the conditional bias is the vector of residuals, **r**, and that the conditional variance depends on the unknown variance function $\sigma^2(\cdot)$. So, the essential task is to estimate these two unknown quantities.

Recall that the vector of residuals in the conditional bias is

$$\mathbf{r} = \mathbf{m} - \mathbf{X}\boldsymbol{\beta} = \begin{pmatrix} m(X_1) - \sum_{j=1}^{p} \beta_j (X_1 - x_0)^j \\ \vdots \\ m(X_n) - \sum_{j=1}^{p} \beta_j (X_n - x_0)^j. \end{pmatrix}$$

Using a Taylor expansion of order a, for some given positive integer a, we can approximate the ith element of this vector by

$$\beta_{p+1}(X_i - x_0)^{p+1} + \cdots + \beta_{p+a}(X_i - x_0)^{p+a}. \tag{9.42}$$

The rate of convergence of the final bandwidth selector depends on the choice of the integer a. See Huang (1995) or Fan and Gijbels (1996) for a discussion of this issue. For simplicity of presentation, we take $a = 2$. Using the approximation in (9.42) the conditional bias can be approximated as follows

$$S_n^{-1}\mathbf{X}^T\mathbf{W}\mathbf{r} \approx S_n^{-1} \begin{pmatrix} \beta_{p+1}S_{n,p+1} + \beta_{p+2}S_{n,p+2} \\ \vdots \\ \beta_{p+1}S_{n,2p+1} + \beta_{p+2}S_{n,2p+2} \end{pmatrix},$$

where the elements $S_{n,j}$ are defined in (9.8). An estimator for the right-hand side of the above expression is

$$S_n^{-1} \begin{pmatrix} \widehat{\beta}_{p+1}S_{n,p+1} + \widehat{\beta}_{p+2}S_{n,p+2} \\ \vdots \\ \widehat{\beta}_{p+1}S_{n,2p+1} + \widehat{\beta}_{p+2}S_{n,2p+2} \end{pmatrix}, \tag{9.43}$$

where $\widehat{\beta}_{p+1}$ and $\widehat{\beta}_{p+2}$ are the estimated regression coefficients from a local polynomial fit of degree $p+2$. This fitting step requires a pilot bandwidth h^*, and we suggest an adaptive choice of this pilot bandwidth later on.

The conditional variance in (9.7) depends on the unknown function $\sigma^2(\cdot)$. Assuming local homoscedasticity, we can approximate the conditional variance by

$$S_n^{-1}(\mathbf{X}\mathbf{W}^2\mathbf{X})S_n^{-1}\sigma^2(x_0).$$

An estimator for the quantity $\sigma^2(x_0)$ is the normalized weighted residual sum of squares resulting from a local polynomial fit of order $p+2$ using a pilot bandwidth h^*. More explicitly, $\sigma^2(x_0)$ can be estimated by

$$\widehat{\sigma}^2(x_0) = \frac{\sum_{i=1}^{n}(Y_i - \widehat{Y}_i)^2 K_{h^*}(X_i - x_0)}{\mathrm{tr}\{\mathbf{W}^* - \mathbf{W}^*\mathbf{X}^* S_n^{*-1}\mathbf{X}^{*T}\mathbf{W}^*\}}, \tag{9.44}$$

where $\mathbf{X}^*$, $\mathbf{W}^*$, and S_n^* are defined similarly to $\mathbf{X}$, $\mathbf{W}$, and S_n, respectively, but now for a local polynomial fit of order $p+2$ using a pilot bandwidth h^*. Note that the estimator in (9.40) is similar in nature to the estimator in (9.44), but the former is a global type of estimator, which is intuitively explained by the assumption of homoscedasticity.

The estimator for the conditional variance is then given by

$$S_n^{-1}(\mathbf{X}\mathbf{W}^2\mathbf{X})S_n^{-1}\widehat{\sigma}^2(x_0). \tag{9.45}$$

Estimators for the conditional bias and the variance, respectively, of the estimator $\widehat{m}_\nu(x_0) = \nu!\widehat{\beta}_\nu$ are then given by

$$\begin{aligned}\widehat{b}_{\nu,p}(x_0) &= \nu! \times \text{ the } (\nu+1)\text{th element of the estimated vector in (9.43)}\\ \widehat{V}_{\nu,p}(x_0) &= (\nu!)^2 \times \text{ the } (\nu+1)\text{th diagonal element of the}\\ &\qquad \text{estimated matrix in (9.45).}\end{aligned} \tag{9.46}$$

These expressions lead to the following estimator for the conditional MSE of $\widehat{m}_\nu(x_0)$:

$$\widehat{\mathrm{MSE}}_{\nu,p}(x_0) = \widehat{b}^2_{\nu,p}(x_0) + \widehat{V}_{\nu,p}(x_0). \tag{9.47}$$

Note that, in marked contrast to the postasymptotic substitution methods, (9.46) provides estimates of conditional bias and variance, which are the raw materials for constructing confidence intervals. See Fan and Gijbels (1996, Section 4.4) and Fan et al. (1998).

We next describe a procedure for an adaptive choice of the pilot bandwidth h^*. Consider the *residual sum of squares* RSC (criterion)

$$\mathrm{RSC}(x_0; h) = \widehat{\sigma}(x_0)\{1 + (p+1)V\}, \tag{9.48}$$

where $\widehat{\sigma}^2(\cdot)$ is defined similarly to (9.44), but for a local polynomial fit of degree p, and where V is the first diagonal element of the matrix $S_n^{-1}(\mathbf{X}\mathbf{W}^2\mathbf{X})S_n^{-1}$. It can be shown, under certain conditions, that the conditional expectation of the statistic $\mathrm{RSC}(x_0; h)$ can be approximated up to the first order as

$$E\{\mathrm{RSC}(x_0; h)|\mathcal{X}\} \approx \sigma^2(x_0) + C_p\beta^2_{p+1}h^{2p+2} + (p+1)a_0\frac{\sigma^2(x_0)}{f_X(x_0)nh},$$

where $a_0 = \int K_0^{*2}(t)\,dt$ and $C_p = \mu_{2p+2} - c_p^T S^{-1} c_p$. From this expression, it follows, heuristically, that the minimizer of $E\{\mathrm{RSC}(x_0; h)|\mathcal{X}\}$ is approximately equal to the minimizer of the right-hand side of the above expression, which is

$$h_0(x_0) = \left\{ \frac{a_0 \sigma^2(x_0)}{2 C_p \beta_{p+1}^2 f_X(x_0) n} \right\}^{1/(2p+3)}. \tag{9.49}$$

Comparing $h_0(x_0)$ with the theoretical optimal variable bandwidth in (9.33), we find that the two quantities relate as follows:

$$h_{\nu,p}^{\mathrm{opt}}(x_0) = \mathrm{adj}_{\nu,p} h_0(x_0). \tag{9.50}$$

Here, the adjusting constant is given by

$$\mathrm{adj}_{\nu,p} = \left[\frac{(2\nu+1) C_p \int K_\nu^{*2}(t)\,dt}{(p+1-\nu) \left\{ \int t^{p+1} K_\nu^*(t)\,dt \right\}^2 \int K_0^{*2}(t)\,dt} \right]^{1/(2p+3)}. \tag{9.51}$$

See Fan and Gijbels (1995a) and Fan and Gijbels (1996, p. 120) for some tabulated values of these adjusting constants. In particular, when $p = 1$ and $\nu = 0$, the adjusting constant is 0.8941 for the Epanechnikov kernel and 1 for the Gaussian kernel.

From the above considerations, it is easily seen that an appropriate data-driven global bandwidth selector for estimating $m^{(\nu)}(\cdot)$ on an interval $[a, b]$ by local polynomial fitting of degree p is

$$\widehat{h}_{\nu,p}^{\mathrm{RSC}} = \mathrm{adj}_{\nu,p} \mathrm{argmin}_h \int_{[a,b]} \mathrm{RSC}(y; h)\,dy. \tag{9.52}$$

The performance of this data-driven global bandwidth estimator has been studied in Fan and Gijbels (1995a). It is this data-driven bandwidth selector that is used as a pilot bandwidth in the refined bandwidth selection procedure that we describe next.

Here, we discuss only the data-driven global bandwidth selection procedure. An analogous procedure, using similar ingredients, exists for local variable bandwidth selection. For details on the data-driven bandwidth selectors based on the residual sum-of-squares criterion refer to Fan and Gijbels (1995a).

The algorithm for calculating the *refined global bandwidth selector* is as follows:

1. *Pilot estimation.* A polynomial of degree $p+2$ is fitted, using the pilot bandwidth $h^* = \widehat{h}_{p+1,p+2}^{\mathrm{RSC}}$ defined in (9.52). This local polynomial fit provides the estimates $\widehat{\beta}_{p+1}$, $\widehat{\beta}_{p+2}$, and $\widehat{\sigma}^2(x_0)$.

2. *Bandwidth selection.* The estimator $\widehat{\text{MSE}}_{\nu,p}(\cdot; h)$ is calculated via (9.47), and then the refined global bandwidth selector is given by

$$\widehat{h}^{\text{R}}_{\nu,p} = \text{argmin}_h \int_{[a,b]} \widehat{\text{MSE}}_{\nu,p}(y; h)\, dy. \tag{9.53}$$

3. *Estimation of* $\mathbf{m}^{(\nu)}(\cdot)$. The bandwidth $\widehat{h}^{\text{R}}_{\nu,p}$ is used to fit a polynomial of degree p and to obtain $\widehat{m}_\nu(\cdot)$.

We also refer to the bandwidth in (9.53) as the global preasymptotic substitution bandwidth selector. As an example the local linear fit for the Ethanol Data presented in Figure 9.3 was obtained using the refined global bandwidth selector in (9.53).

9.3.3.4 Empirical-Bias Bandwidth Selector

Ruppert (1997) proposed a data-driven local variable bandwidth selector that is based on the idea of modeling the bias term. The basic principle is the same as that used in the previous paragraphs: Estimate the conditional mean squared error and then minimize this estimate to obtain the bandwidth selector.

The main difference between Ruppert's (1997) approach and the preasymptotic substitution method discussed previously above is in the estimation of the conditional bias of $\widehat{m}_\nu(x_0)$. In Ruppert's approach, the bias is estimated empirically by calculating $\widehat{m}_\nu(x_0; h) = \widehat{m}_\nu(x_0)$ on a grid of h values and then modeling in an appropriate way the behavior of $\widehat{m}_\nu(x_0; h)$ as a function of h. The appropriate model is dictated by the asymptotic expression for the conditional bias.

Ruppert (1997) proposed the following procedure for estimating the bias of $\widehat{m}_\nu(x_0; h_0)$: Let $J_b > 1$ be an integer, and let $h_0^1, h_0^2, \ldots, h_0^{J_b}$ be in a neighborhood of h_0. Calculate $\widehat{m}_\nu(x_0; h_0^l)$ for $l = 1, \ldots, J_b$. Then for some integer $a \geq 1$, fit the model

$$\text{d}_0(x_0) + \text{d}_{p+1-\nu}(x_0)h^{p+1-\nu} + \cdots + \text{d}_{p+a-\nu}(x_0)h^{p+a-\nu} \tag{9.54}$$

to the "data" $\{(h_0^l, \widehat{m}_\nu(x_0; h_0^l)\colon l = 1, \ldots, J_b\}$ using ordinary least squares. This results in the fit

$$\widehat{\text{d}}_0(x_0) + \widehat{\text{d}}_{p+1-\nu}(x_0)h^{p+1-\nu} + \cdots + \widehat{\text{d}}_{p+a-\nu}(x_0)h^{p+a-\nu}.$$

The form of the parametric model in (9.54) is inspired by the asymptotic expressions. See, for example, (9.23) and (9.24) and higher order asymptotic expressions in, Huang (1995). An estimator for the bias is then

$$\widehat{\text{d}}_{p+1-\nu}(x_0)h^{p+1-\nu} + \cdots + \widehat{\text{d}}_{p+a-\nu}(x_0)h^{p+a-\nu}. \tag{9.55}$$

9.3.3.5 Other Bandwidth Selectors

A *nearest-neighbor* type of bandwidth is a useful "reparametrization" of the constant bandwidth h such that there are about $2k$ data points in each neighborhood, where $k > 0$ is a given integer. Let $X_{(1)} \leq X_{(2)} \leq \cdots \leq X_{(n)}$ denote the ordered observations $X_1, \ldots, X_n$, and let $X_{(l)}$ be the design point that is closest to the point x_0. Consider the bandwidth

$$\widehat{h}(x_0) = \frac{X_{(l+k)} - X_{(l-k)}}{2}. \tag{9.56}$$

The discrete parameter k could be selected using, for example, cross-validation. (See, e.g., Fan and Gijbels, 1996, p. 151.)

There are a few variations on the nearest-neighborhood type of bandwidth. A k-nearest neighborhood of a point refers to the neighborhood that contains k data points that are nearest to the given point. The bandwidth used in the `loess` procedure, provided in S-Plus, is of the nearest neighbor type (Cleveland, 1979 and Cleveland et al., 1992).

9.4 CHOICE OF THE DEGREE

What is the effect of the choice of the degree of the polynomial on the estimation procedure? To get an idea about this, we can first look at the asymptotic expressions for the conditional bias and variance provided in (9.22)–(9.24). As already mentioned in Section 9.2.2, the asymptotic conditional bias has a simpler structure when $p - \nu$ is odd than when $p - \nu$ is even. This is an important factor for the bandwidth selection problem. Looking at the orders of the terms, we can conclude that the order of the asymptotic variance does not depend on p. The order of the asymptotic conditional bias, however, does increase with p, and hence this bias reduces when p is larger. But what about the constant factors in the asymptotic conditional variance? How do they change with p? An investigation of these constant factors revealed that they remain the same when passing from an even-order ($p - \nu$ even) fit to the consecutive odd-order fit ($p - \nu + 1$ odd) (Ruppert and Wand, 1994, and Fan and Gijbels, 1995b). In other words, when $p - \nu$ is even and K is symmetric, the local polynomial fit of degree p uses one parameter less than the local polynomial fit of degree $p + 1$, without saving on the asymptotic variance. But the extra parameter enables the local polynomial fit of order $p + 1$ to reduce the bias of the estimate significantly, particularly at the boundary. Further, the constant factors in the asymptotic conditional variance increase with the degree, since more parameters have to be fitted. (See Fan and Gijbels, (1996, Table 3.3 p. 79.) For example, for estimating the regression function ($\nu = 0$), the asymptotic variance of the local quadratic fit is 2.08 or 1.69 times as large as that of the local linear fit when the Epanechnikov kernel or the Gaussian kernel is used, respectively.

Based on the above purely asymptotic facts, one could conclude that a local polynomial fit with odd $p - \nu$ is preferable to a local polynomial fit with even $p - \nu$. Keeping in mind that a higher order local polynomial fit involves more parameters, it is recommended to use $p = \nu + 1$ or $p = \nu + 3$ in practice. Of course, one should also consider the influence of the degree p on the finite-sample performance of the estimators, but this factor is very case-dependent, and no precise information is available on the finite-sample behavior.

So far, we have assumed that, on the whole, the estimation-region local polynomial fits of the same order are carried through. However, when the unknown curve has different degrees of smoothness in different regions — for example, a flat part in a certain region and a rapidly changing part in another region — it might be more reasonable to adapt the degree of the local polynomial fit to the point x_0 under consideration. The bandwidth h is assumed to be fixed here.

An adaptive procedure for choosing the degree of the polynomial fit locally is easy in principle: In a fixed neighborhood of x_0, choose the order of fit such that the estimated mean squared error $\widehat{\text{MSE}}_{\nu,p}(x_0; h)$ is the smallest. Let us describe the procedure more precisely. Denote by

$$x_j = x_L + j\Delta, \qquad j = 0, \ldots, n_{\text{grid}},$$

the grid points at which the estimated curve is to be evaluated, where x_L is the first grid point. Suppose that we consider only local polynomial fits up to order R. For a given bandwidth h, the algorithm for an adaptive choice of the order of fit is as follows:

Step 1. For all orders $p(\nu < p \leq R)$ and for all grid points, obtain $\widehat{\text{MSE}}_{\nu,p}(x_j; h)$.

Step 2. For each order and for each grid point, calculate the smoothed estimated MSE by taking the weighted local average of the estimated MSE in the neighboring $2[h/\Delta] + 1$ grid points.

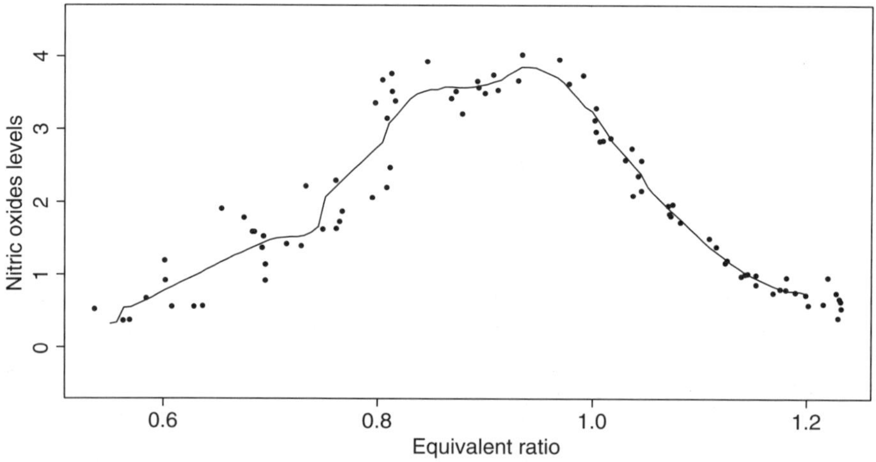

FIGURE 9.8 Adaptive-order local fit for the Ethanol Data using the rule-of-thumb bandwidth selector.

Step 3. For each grid point x_j, choose the order p_j that has the smallest smoothed estimated MSE, fit a polynomial of degree p_j locally, and obtain the estimator for $m^{(\nu)}(x_j)$.

Simulation studies in Fan and Gijbels (1995b) revealed that with such an adaptive choice of the degree of the polynomial, the problem of bandwidth selection is less crucial. Hence, a crude rule-of-thumb data-driven bandwidth selector, such as the one described in Section 9.3.3.1, could be sufficient.

As an illustration of the adaptive-order-estimation procedure, we depict in Figure 9.8 the adaptive order local fit for the Ethanol Data, using the bandwidth $1.5 \times \check{h}_{\mathrm{ROT}}$ defined in (9.37) with $p = 1$ and $\nu = 0$.

9.5 LOCAL MODELING

The idea of local polynomial regression extends readily to other settings where the loss function can be nonquadratic. Further, the family of local models does not have to be the family of polynomial functions. It can be any favored family of parametric models. We first give a general framework for the local modeling and then illustrate it by a few examples. More details can be found in Fan and Gijbels (1996, Chapter 5).

9.5.1 A General Framework

Suppose that we are interested in estimating a population function $\theta(\cdot)$ based on the observed data $\{(X_i, Y_i), i = 1, \ldots, n\}$ from a certain population (X, Y). The function $\theta(\cdot)$ can be, for example, the logit transform of the conditional probability function for the bivariate data given in Figure 9.2(a) in the context of generalized linear models, or the hazard regression function in the proportional hazards model, or a conditional quantile function. In parametric modeling, the function $\theta(x)$ is modeled by a parametric family $\theta(x) = g(x; \beta)$, and one tries to minimize (or maximize) an objective function

$$L(\beta) = \sum_{i=1}^{n} l\{X_i, Y_i, g(X_i; \beta)\}. \tag{9.57}$$

Usually, l is a discrepancy loss function or the log-likelihood function of an individual observation. The local modeling method aims simply to relax the global parametric model assumption to the local parametric model assumption. The benefit of the local modeling method is that it is applicable to any smooth function $\theta(\cdot)$, not just to the family $g(x; \beta)$. Hence, the modeling bias can be reduced significantly.

For a given point x_0, the *local modeling* method consists simply of replacing the objective function in (9.57) by its local version

$$L(\beta(x_0)) = \sum_{i=1}^{n} l\{X_i, Y_i, g(X_i; \beta(x_0))\}K\left(\frac{X_i - x_0}{h}\right). \tag{9.58}$$

Optimizing (9.58) yields an estimate $\widehat{\beta}(x_0)$ of the local parameter $\beta(x_0)$, and an estimate of the function $\theta(\cdot)$ at the point x_0 is simply $\widehat{\theta}(x_0) = g\{x_0; \widehat{\beta}(x_0)\}$.

Which choice of the objective function should one use? What are the relations between the population (X, Y) and the functions l and θ? Since the local estimate $\widehat{\beta}(x_0)$ optimizes (9.58), the estimate $g\{x_0; \widehat{\beta}(x_0)\}$ should converge to its population version, namely,

$$\theta(x_0) = \text{argmin}_a E\{l(X, Y, a)|X = x_0\}. \tag{9.59}$$

Here, argmin should be replaced by argmax if l is a log-likelihood function. In other words, the estimator $\widehat{\theta}(x_0)$ is a consistent estimator of the function $\theta(x_0)$ if $h \to 0$ in such a way that $nh \to \infty$.

One can choose a preferred parametric model $g(x; \beta)$ in the local model. An advantage of doing so is that the local model is parsimonious to the global parametric model in the sense that when $h = \infty$, the local model reduces to the global model. In other words, the global parametric model $g(x; \beta)$ is embedded in the family of local models indexed by h. This method can provide useful diagnostic tools for checking the goodness of fit of the parametric model $g(x; \beta)$.

In practice, the function $g(x; \beta)$ is often chosen to be a polynomial function:

$$g(x; \beta) = a_0 + a_1 x + \cdots + a_p x^p.$$

For each given x_0, we can reparametrize the function as

$$g(x; \beta) = \beta_0(x_0) + \beta_1(x_0)(x - x_0) + \cdots + \beta_p(x_0)(x - x_0)^p. \tag{9.60}$$

The theoretical justification of this reparametrization comes from Taylor's expansion. With such a reparametrization and by suppressing the dependence of the βs on x_0, (9.58) becomes

$$\sum_{i=1}^{n} l\{X_i, Y_i, \beta_0 + \beta_1(X_i - x_0) + \cdots + \beta_p(X_i - x_0)^p\}K\left(\frac{X_i - x_0}{h}\right). \tag{9.61}$$

Let $\widehat{\beta}_j$, $j = 0, \ldots, p$, optimize (9.61). Then, as in (9.6),

$$\widehat{\theta}_\nu(x_0) = \nu!\widehat{\beta}_\nu, \quad \nu = 0, \ldots, p, \tag{9.62}$$

estimates the function $\theta^{(\nu)}(x_0)$.

Existing software for fitting parametric models can easily be used to compute $\widehat{\beta}_\nu$, $\nu = 0, \ldots, p$. For each given x_0, the objective function in (9.58) or (9.61) is just a weighted version of (9.57) for the parametric model. Thus, the local parameters β can easily be computed by using the software for optimizing (9.57). This attribute is another advantage of the local modeling approach.

It is clear that local polynomial regression estimation is a specific case of the above local modeling framework with a quadratic loss function $l(x, y, \theta) = (y - \theta)^2$. An extension of the ideas for estimating bias and variance in Section 9.3 to the above local modeling framework can be found in Fan et al. (1998), wherein methods for selecting bandwidths and constructing confidence intervals are also proposed. A closely related framework is the local estimating equation method introduced by Carroll et al. (1998), who proposed the use of the empirical-bias method to select bandwidths.

9.5.2 Quantile Regression and Robust Regression

Conditional quantiles are very useful and informative summary statistics and have important applications for constructing predictive intervals and detecting heteroscedasticity. In particular, the conditional median regression function is more informative than the conditional mean regression function when the error distribution is asymmetric.

Let $G(y|x)$ be the conditional distribution function of Y, given $X = x$. Then the *conditional α-quantile function* is $\xi_\alpha(x) = G^{-1}(\alpha|x)$. This function minimizes (9.59) with the loss function $l(x, y, \theta) = l_\alpha(y - \theta)$, where

$$l_\alpha(t) = |t| + (2\alpha - 1)t. \tag{9.63}$$

Figure 9.9(a) presents a few examples of functions $l_\alpha(t)$. Thus, to estimate the conditional quantile function, we need to minimize

$$\sum_{i=1}^{n} l_\alpha\{Y_i - \beta_0 - \beta_1(X_i - x_0) - \cdots - \beta_p(X_i - x_0)^p\}K\left(\frac{X_i - x_0}{h}\right), \tag{9.64}$$

and the resulting estimator for $\xi_\alpha^{(\nu)}(x_0)$ is simply $\nu!\widehat{\beta}_\nu$.

As an illustration, consider the 12-month U.S. Treasury Bill Data presented in Figure 9.2(d). Figure 9.6(b) depicts the conditional median, conditional 10th percentile, and the conditional 90th percentile. The fan shape of the conditional quantiles indicates that the variability gets larger as the interest rate increases. The conditional 10th and 90th percentiles form a 80%-predictive interval. For example, given the current interest rate of 10%, with an 80% probability, the difference of the next week's rate and this week's rate falls in the interval $[-0.373\%, 0.363\%]$.

For robust estimation of the regression function, one can simply replace the loss function in (9.64) by an outlier-resistant loss function, such as

$$l(t) = \begin{cases} t^2/2 & \text{when } |t| \le c, \\ c|t| - c^2/2 & \text{when } |t| > c; \end{cases}$$

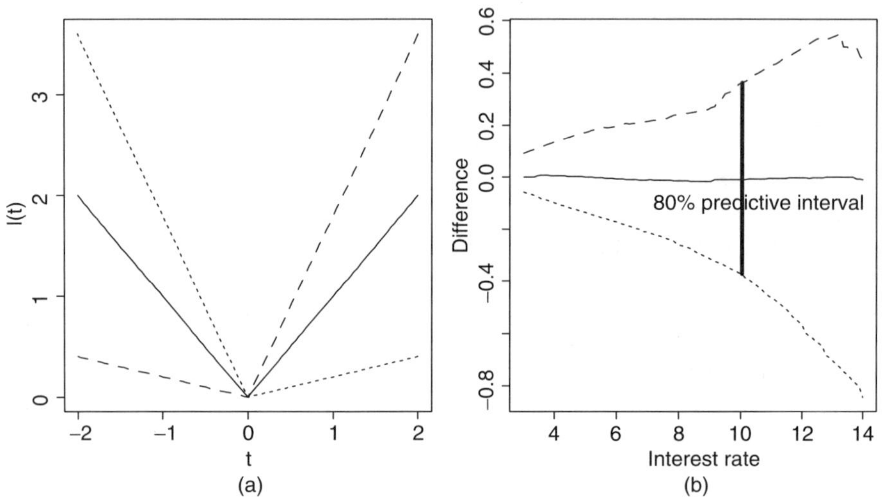

FIGURE 9.9 Quantile regression. (a) Asymmetric loss function for quantile regression with $\alpha = 0.1$ (short-dashed lines), 0.5 (solid lines), and 0.9 (long-dashed lines), (b) conditional quantiles for the data presented in Figure 9.2(d); short-dashed curve $\alpha = 0.1$; solid curve $\alpha = 0.5$; long-dashed curve —$\alpha = 0.9$. The vertical bar indicates the 80%-predictive interval at the point $x = 10$.

in other words, take the derivative of $l(t)$ to be Huber's ψ function: $\psi_c(t) = \max\{-c, \min(c, t)\}$. When the conditional distribution of Y, given $X = x$, is symmetric about the regression function $m(x)$, then the resulting estimates are consistent for all $c \geq 0$. Another useful robust procedure is LOWESS, introduced by Cleveland (1979), which reduces the influence of outliers by an iterative reweighted least-squares scheme with weights proportional to the residuals from the previous iteration.

There is a large amount of literature on nonparametric quantile regression and robust regression. Härdle and Gasser (1984) and Tsybakov (1986) considered, respectively, local constant and local polynomial fitting. Other contributions in this area include those of Chaudhuri (1991), Hall and Jones (1990), Koenker et al. (1992), and Truong (1989). For further references, see Fan and Gijbels (1996, Sections 5.5 and 5.7).

9.5.3 Generalized Linear Models

Generalized linear models (GLM) are an extension of the normal linear regression model. They contain a family of useful stochastic distributions that can be used to deal with both discrete and continuous responses. In particular, it is assumed that the conditional distribution of Y, given X, belongs to the canonical exponential family

$$f(y|x) = \exp([\theta(x)y - b\{\theta(x)\}]/a(\phi) + c(y, \phi)), \tag{9.65}$$

for some known functions $a(\cdot)$, $b(\cdot)$, and $c(\cdot, \cdot)$. The parameter $\theta(\cdot)$ is called the canonical parameter, and ϕ is called the dispersion parameter. These two parameters are related to the conditional mean and conditional variance as follows:

$$m(x) = E(Y|X = x) = b'\{\theta(x)\},$$
$$\mathrm{Var}(Y|X = x) = a(\phi)b''\{\theta(x)\}. \tag{9.66}$$

The function $(b')^{-1}$, which links the mean regression function to the canonical parameter function $\theta = (b')^{-1}(m)$, is called a canonical link. More details on GLMs can be found in McCullagh and Nelder (1989). In the parametric model, the canonical function $\theta(x)$ is typically modeled as a linear function or, more generally, as a polynomial function of x. The aim of the local modeling approach is to release this assumption and to search for a function that describes the data well.

As an example, consider the bivariate data in Figure 9.2(a). It is reasonable to assume that the conditional distribution of Y, given $X = x$, is a Bernoulli distribution with probability of success $p(x)$. Then its conditional density, with respect to the counting measure, can be written as

$$p(x)^y\{1 - p(x)\}^{1-y} = \exp[y\theta(x) - b\{\theta(x)\}], \tag{9.67}$$

where $\theta(x) = \log p(x)/1 - p(x)$ is the logit transformation of $p(x)$ and $b(\theta) = \log\{1 + \exp(\theta)\}$. The canonical link is the logit transformation $\log\{t/(1 - t)\}$.

For the model in (9.65), the function $l(x, y, \theta)$ is the log-likelihood function, which is proportional to $\theta y - b(\theta)$. Thus, the local likelihood in (9.61) reduces to

$$\sum_{i=1}^{n}[\{\beta_0 + \beta_1(X_i - x_0) + \cdots + \beta_p(X_i - x_0)^p\}Y_i - b\{\beta_0 + \beta_1(X_i - x_0)$$
$$+ \cdots + \beta_p(X_i - x_0)^p\}]K\left(\frac{X_i - x_0}{h}\right). \tag{9.68}$$

Let $\widehat{\beta}_\nu$, $\nu = 0, \ldots, p$, maximize (9.68). Then, an estimator of the function $\theta^{(\nu)}(x)$ is simply $\widehat{\theta}_\nu(x) = \nu!\widehat{\beta}_\nu$.

We now use the Burn Data in Figure 9.2(a) to illustrate the application of the local modeling approach to the generalized linear models setting. Of interest is the estimation of the probability of surviving given the age of victims. We employed the local linear fit with the Epanechnikov kernel function and chose a nearest-neighbor type of bandwidth with a span of 0.4 of the data range. The results are summarized in Figure 9.10. Note that the conditional probability function must be monotonic for the parametric linear model, whereas for the local linear model, the conditional probability function can be any curve. The former model can overstate the probability of survival for the younger group and for the senior group. The solid curves in Figure 9.10 suggest that the conditional probability function is unimodal, which is reasonable in the current context.

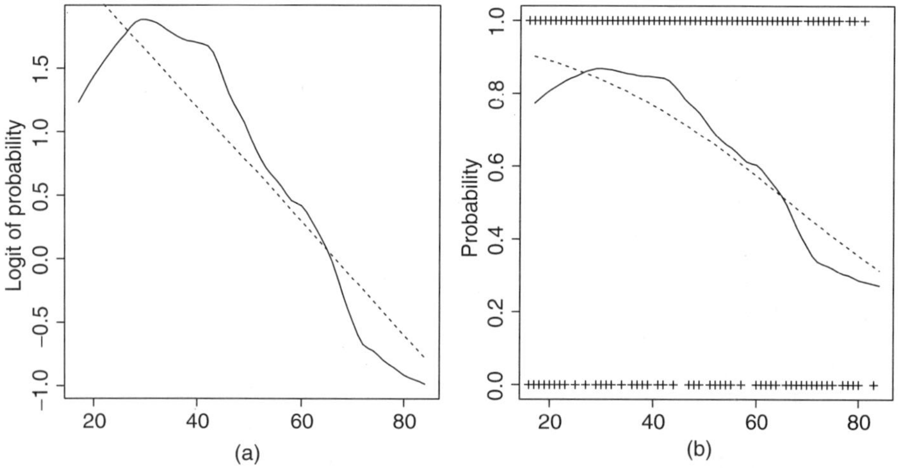

FIGURE 9.10 Illustration of the local likelihood approach for the Burn Data. (a) Estimated logit transform of the conditional probability. (b) Estimated conditional probability. Solid curve: local modeling with about 40% of the data points in each local neighborhood; dashed curve: global parametric logit linear model.

For literature on nonparametric generalized linear models, see Fan et al. (1995), Green and Yandell (1985), Gu (1990), Severini and Staniswalis (1994), Stone (1994), and Tibshirani and Hastie (1987), and Chapter 8 in this book.

9.5.4 Proportional Hazards Model

A useful model for studying the effect of a covariate X on a survival time T is the proportional hazards model. Let $\lambda(t|X)$ be the *conditional hazard rate* function, defined by

$$\lambda(t|X) = \lim_{\Delta \downarrow 0} \frac{P(t \le T < t + \Delta | T \ge t, X)}{\Delta},$$

and let $\lambda_0(t)$ be the *baseline hazard function*, representing the hazard risk at a certain given value of the covariate, say zero. The *proportional hazards model* assumes that the covariate x inflates the hazard risk by a factor of $\exp\{\theta(x)\}$, namely,

$$\lambda(t|x) = \lambda_0(t) \exp\{\theta(x)\}. \tag{9.69}$$

The function $\theta(x)$ is called the *hazard regression function* and characterizes the risk contribution of the covariate at value x. The model in (9.69) with time-dependent covariates has been introduced by Cox (1972).

A salient feature of survival analysis is that the survival time is often not fully observable, due, for example, to the termination of a study or early withdrawal of subjects. For the ith individual, the observed data are the minimum $Z_i = \min(T_i, C_i)$ of the survival time T_i and the *censoring time* C_i, as well as the censoring indicator $\delta_i = I(T_i \leq C_i)$, indicating whether the observed time is the survival time, along with the covariate X_i. In other words, the available data are

$$\{(X_i, Z_i, \delta_i) : i = 1, \cdots, n\},$$

which are assumed to be a random sample from a population (X, Z, δ). Based on these censored data, we are interested in estimating the hazard regression function $\theta(x)$ and the baseline hazard function $\lambda_0(t)$. For a thorough introduction to survival analysis, see recent books by Andersen et al. (1993) and Fleming and Harrington (1991).

In the parametric analysis, a linear form $\theta(x) = \beta x$ is imposed on the hazard regression function. The local modeling approach aims to remove this restriction and is applicable to any smooth hazard regression function. For simplicity of discussion, we restrict the covariate to be univariate. For multivariate settings, a dimensionality reduction technique, such as additive models, should be incorporated. (See Chapter 10 in this book.)

A commonly used technique for estimating the hazard regression function is the *partial likelihood* technique introduced by Cox (1975). Let $t_1^o < \cdots < t_N^o$ denote the ordered observed failure times. Let (j) provide the label for the item failing at t_j^o, so that the covariates associated with the N failures are $X_{(1)}, \ldots, X_{(N)}$. Denote by $R_j = \{i : Z_i \geq t_j^o\}$ the risk set at a time instantaneously before t_j^o. Then, the log-partial likelihood in our context is given by

$$\sum_{j=1}^{N} \left[\theta(X_{(j)}) - \log\left(\sum_{k \in R_j} \exp\{\theta(X_k)\} \right) \right]. \tag{9.70}$$

Substituting the parametric form of $\theta(\cdot)$ into (9.70) yields a maximum partial likelihood estimate of the hazard regression function.

We now apply the local modeling technique to estimate the hazard regression function $\theta(\cdot)$. For a given x_0, we approximate $\theta(x)$ by

$$\theta(x) \approx \beta_0 + \cdots + \beta_p (x - x_0)^p, \tag{9.71}$$

for x in a neighborhood of x_0. Let

$$\beta = (\beta_1, \ldots, \beta_p)^T \text{ and } \mathbf{X}_j = \{(X_j - x_0), \ldots, (X_j - x_0)^p\}^T.$$

Then the *local partial likelihood* is

$$\sum_{j=1}^{N} K_h(X_{(j)} - x_0) \left[\mathbf{X}_{(j)}^T \beta - \log\left\{ \sum_{k \in R_j} \exp(\mathbf{X}_k^T \beta) K_h(X_k - x_0) \right\} \right], \tag{9.72}$$

where $K_h(\cdot) = K(\cdot/h)/h$. The kernel function is imposed to ensure that the model in (9.71) is used only locally. See Fan et al. (1997) for a derivation of the local partial likelihood in (9.72). When the kernel function is uniform and the bandwidth is of the nearest-neighbor type, the local likelihood in (9.72) is the one introduced by Tibshirani and Hastie (1987). For a related approach based on the local likelihood, see Gentleman and Crowley (1991).

The function value $\theta(x_0)$ is not directly estimable, since (9.72) does not depend on the intercept β_0. However, the derivative functions are directly estimable. Let $\widehat{\beta}(x_0)$ be the maximum local log-partial likelihood estimate that maximizes (9.72). An estimate $\widehat{\theta}_\nu(x_0)$ of $\theta^{(\nu)}(x_0)$ is then given by $\nu!\widehat{\beta}_\nu(x_0)$.

We impose the condition $\theta(0) = 0$ for identifiability. With this extra constraint, the function $\theta(x)$ can be estimated by

$$\widehat{\theta}(x) = \int_0^x \widehat{\theta}'(t)\,dt, \tag{9.73}$$

where $\widehat{\theta}'(t) = \widehat{\theta}_1(t)$ is the derivative estimator. In practice, the function $\widehat{\theta}_1(x)$ is often evaluated at either grid points or the design points. Assume that $\widehat{\theta}_1(x_j) = \widehat{\beta}_1(x_j)$ are computed at points $\{x_0, \cdots, x_m\}$. Then, $\widehat{\theta}(x_i)$ can be approximated by the trapezoidal rule

$$\widehat{\theta}(x_i) = \sum_{j=1}^{i} (x_j - x_{j-1})(\widehat{\beta}_{1,j} + \widehat{\beta}_{1,j-1})/2,$$

where $\widehat{\beta}_{1,j} = \widehat{\beta}_1(x_j)$. Once $\widehat{\theta}(\cdot)$ is estimated, we can use Breslow's (1974) type of estimator for the *cumulative hazard function* $\Lambda_0(t) = \int_0^t \lambda_0(u)\,du$:

$$\widehat{\Lambda}_0(t) = \sum_{j=1}^{N} \left[\sum_{i \in \mathcal{R}_j} \exp\{\widehat{\theta}(X_i)\} \right]^{-1} I\{t_j \le t\}. \tag{9.74}$$

A kernel-smoothing technique can then be employed to obtain an estimate of $\lambda_0(t)$ via

$$\widehat{\lambda}_0(t) = \int W_g(t - x) d\widehat{\Lambda}_0(x), \tag{9.75}$$

where $W_g(\cdot) = W(\cdot/g)/g$, with W a kernel function and, g a bandwidth. With the estimated hazard regression function and the estimated baseline hazard function, we can construct the estimated hazard risk function of individuals with covariate x via (9.69)

$$\widehat{\lambda}(t|x) = \widehat{\lambda}_0(t)\exp\{\widehat{\theta}(x)\}.$$

Using the relationship between the *survivor function* and the hazard risk function, we find that individuals with the covariate value x have an estimated survivor function

$$\widehat{P}(T > t|X = x) = \exp\left\{ -\int_0^t \widehat{\lambda}(u|x)\,du \right\}.$$

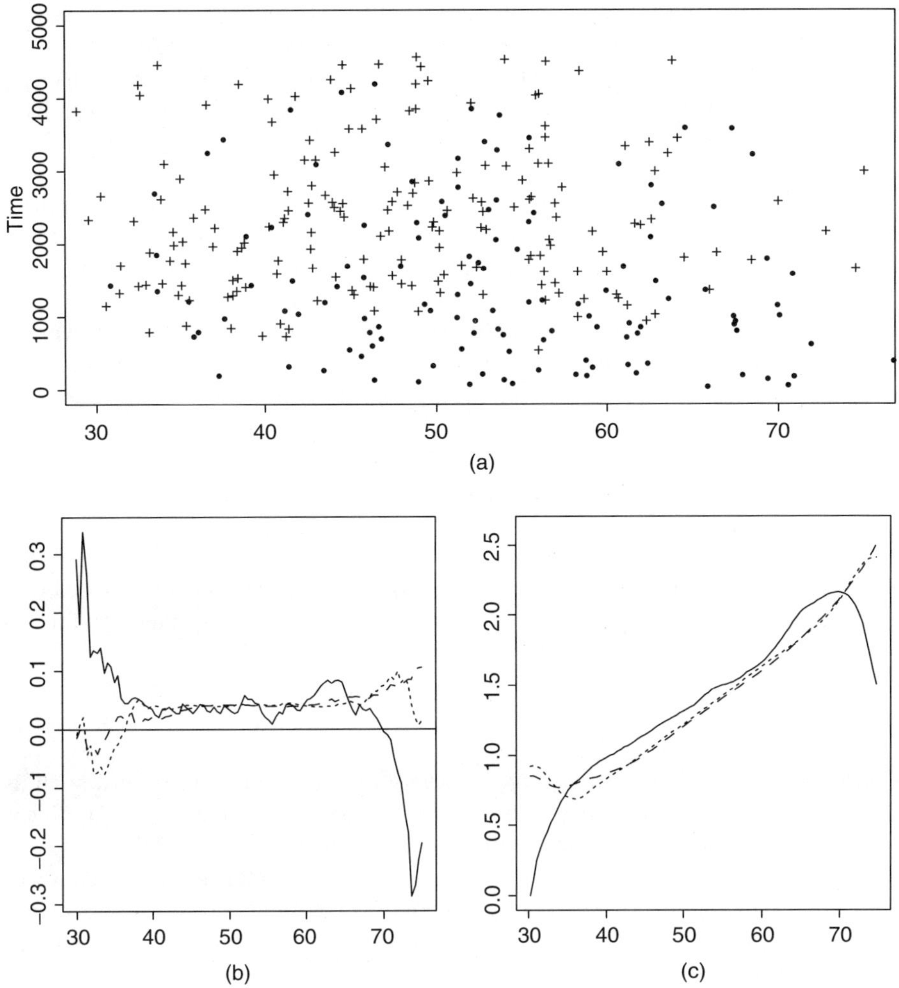

FIGURE 9.11 Local partial-likelihood estimation of the hazard regression function. (a) Observed survival time versus age, where "+" indicates censored observations; (b) estimated derivative function $\theta'(\cdot)$; (c) estimated hazard regression function $\theta(\cdot)$: solid curve, bandwidth = 10; short-dashed curve, bandwidth = 20; long-dashed curve, bandwidth = 30.

From this function, we can easily get an estimate of the individuals' mean survival time and the median survival time. For example,

$$\widehat{E}(T|X=x) = \int_0^\infty \widehat{P}(T > t|X=x)\,dt,$$

where the integral exists. This is the area of the survivor curve.

We conclude this section with an analysis of the Primary Biliary Cirrhosis (PBC) Data that can be found in Fleming and Harrington (1991). PBC is a rare but fatal chronic liver disease of unknown cause. The analysis is here based on the data collected at the Mayo Clinic between January 1974 and May 1984. Of 312 patients who participated in the randomized trial, 187 cases were censored. In our analysis, we take the time, in days, between registration and death, liver transplantation, or the time of the study (July 1986) as the response and the ages of the patients as a covariate. The observed data are presented in Figure 9.11(a). The local partial-likelihood method in (9.72) with $p = 2$ was employed for three different bandwidths $h = 10$, 20, and 30. The estimated hazard regression function and its derivative function are respectively given in Figure 9.11(c) and (b). Note that since the hazard regression function is identifiable only within a constant, the curves in Figure 9.11(c) are normalized to have the same average height, so that they can be better compared. Figure 9.11(c) reveals the fact that it is reasonable to model linearly the hazard regression function of covariate age.

9.6 SOME MORE APPLICATIONS

This section discusses briefly additional applications of local polynomial fitting. We first consider how to estimate a conditional variance and how to detect a discontinuity of a regression curve, and then we show how to estimate conditional densities.

9.6.1 Estimation of Variance

Conditional variance functions have a variety of statistical applications, such as measuring the volatility or risk in finance, monitoring the reliability in nonlinear prediction, identifying homoscedastic transforms in regression, and monitoring the signal-to-noise ratios in quality control of experimental design. Because of their important applications in finance, where data are often dependent, we formulate our problems in the stochastic setup. For simplicity of presentation, we focus only on local linear fitting.

Let $\{(X_i, Y_i)\}$ be a two-dimensional, strictly stationary process having the same joint distribution as (X, Y). Let $m(x) = E(Y|X = x)$ and $\sigma^2(x) = \text{Var}(Y|X = x)$ be, respectively, the regression function and the conditional variance function. Our approach is based on the residuals of the local linear fit. Let $\widehat{m}_{h_1,K}(\cdot)$ be the local linear fit of $m(\cdot)$ using a kernel K and a bandwidth h_1. Consider the squared residuals

$$\widehat{r}_i = \{Y_i - \widehat{m}_{h_1,K}(X_i)\}^2. \tag{9.76}$$

Note that the conditional variance function can be expressed as

$$\sigma^2(x) = E[\{Y - m(X)\}^2|X = x],$$

which is the regression function of the squared residuals. Therefore, a natural procedure is to run a local linear fit on the squared residuals. Let $\widehat{\sigma}^2_{h_2,W}(x)$ be the local linear fit based on the data $\{(X_i, \widehat{r}_i), i = 1, \ldots, n\}$, using a bandwidth h_2 and a kernel W. It has been shown by Fan and Yao (1998) and Ruppert et al. (1997) that the estimator $\widehat{\sigma}^2_{h_2,W}(x)$ performs as well as the ideal estimator, which is a local linear fit to the true squared residuals

$$\{(X_i, \{Y_i - m(X_i)\}^2), i = 1, \cdots, n\}$$

using the same bandwidth h_2 and the same kernel W. In particular, the bias of the estimator $\widehat{\sigma}^2_{h_2,W}(x)$ is of order $O(h_2^2 + h_1^4 + (nh_1)^{-1})$, though the bias of $\widehat{m}_{h_1,K}$ is of order $O(h_1^2)$, and the variance of $\widehat{\sigma}^2_{h_2,W}(x)$ is of order $O\{(nh_2)^{-1}\}$.

This discussion suggest that if the bandwidth h_1 is of order $n^{-1/5}$, then the residual-based *conditional variance estimator* performs asymptotically as well as the ideal one. In particular, the optimal bandwidth for estimating the mean regression function is permitted to be used for computing the residuals. This leads to the data-driven conditional variance estimator discussed next.

Let $\widehat{h}(X_1, \ldots, X_n; Y_1, \ldots, Y_n)$ be a data-driven bandwidth selection rule for local linear regression based on the data $(X_1, Y_1), \ldots, (X_n, Y_n)$. This rule can be in one case a cross-validation bandwidth rule and in another case one of the bandwidth selection methods discussed in Section 9.3.3. Our data-driven version of conditional variance estimation is as follows:

Step 1. Use bandwidth $h_1 = \widehat{h}(X_1, \ldots, X_n; Y_1, \ldots, Y_n)$ in local linear regression to obtain the estimate $\widehat{m}_{h_1,K}(X_i)$ for $i = 1, \ldots, n$.

Step 2. Compute squared residuals $\widehat{r}_i = \{Y_i - \widehat{m}_{h_1,K}(X_i)\}^2$, $i = 1, \ldots, n$.

Step 3. Apply bandwidth $h_2 = \widehat{h}(X_1, \ldots, X_n; \widehat{r}_1, \ldots, \widehat{r}_n)$ in local linear regression to obtain $\widehat{\sigma}^2_{h_2,W}(\cdot)$.

To illustrate the usefulness of this automatic method, consider the yields of a 12-month US Treasury Bill. We use the refined global bandwidth selector of Fan and Gijbels (1995a) and the Epanechnikov kernel. Figure 9.12(a) gives the estimated mean regression function for the data in Figure 9.12(d). The bandwidth $\widehat{h}_1 = 3.99$ was chosen. Figure 9.12(b) depicts the estimated conditional standard deviation (the volatility function) and the conditional variance function. Further the bandwidth $\widehat{h}_2 = 3.63$ was selected. Visual inspection suggests that the volatility function should be a power function. Indeed, the correlation coefficient between $\{\log(x_j)\}$ and $\{\log(\widehat{\sigma}(x_j))\}$ is 0.997, where x_j, $j = 1, \ldots, 201$, are grid points in the interval [3, 14]. Fitting a line through the data $\{(\log(x_j), \log\{\widehat{\sigma}(x_j)\}), j = 1, \ldots, 201\}$, we obtain the estimate

$$\widehat{\sigma}(x) = 0.0154x^{1.3347}.$$

This estimate is presented as a thick-dashed curve in Figure 9.12(b). This case is an example where the nonparametric analyses yield a good parametric model $\sigma(x) = \alpha x^{\beta}$. Based on the linear regression of the data $(\log(X_i), \log(\widehat{r}_i))$, one can also obtain directly an estimate of α and β.

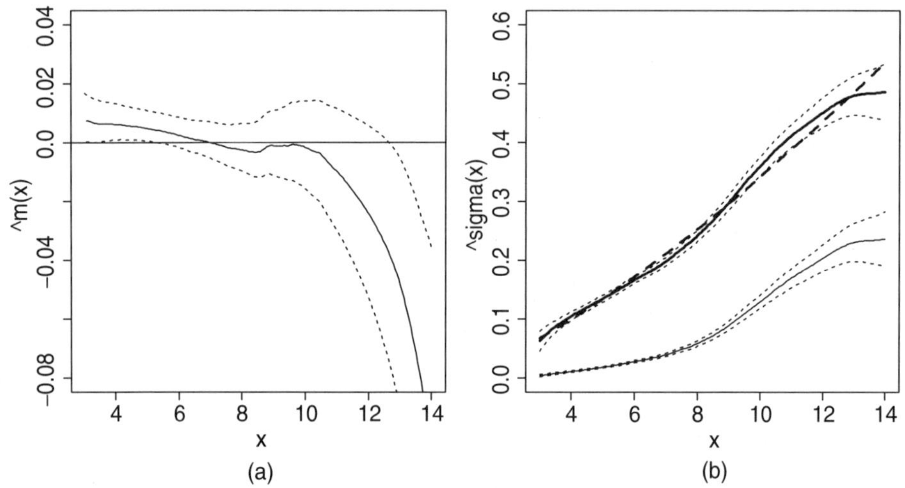

FIGURE 9.12 The regression function and the volatility for the 12-month US Treasury Bill Data. (a) The estimated mean regression function. (b) The estimated volatility function (thick curve), and the estimated conditional variance function (thin curve) for the data in Figure 9.2(d). The two dashed curves around a solid curve indicate one standard error above and below the estimated mean regression function.

9.6.2 Change-Point Detection

Jump points arise for example when the treatment effects change suddenly, without warning or planning. Examples of this kind occur in the context of medical monitoring and quality control. The statistical problem can be abstracted as follows.

Assume that the bivariate data $(X_1, Y_1), \ldots, (X_n, Y_n)$ are a random sample from a population (X, Y) with conditional mean function m, which is smooth except for a few number of jump discontinuities. For simplicity, we assume that there is only one discontinuity point, also called a *change point*.

One approach to this problem is to regard the change point as the location where the derivative function $|m'(\cdot)|$ is maximized. Thus, a natural approach is first to estimate the derivative curve and then to find the maximizer of the absolute value of the estimated derivative function. Let $D(x, h)$ be a derivative estimator resulting from a local polynomial fit of order p with bandwidth h and kernel K. For simplicity, assume that the support of K is $[-1, 1]$. This idea translates into the following estimating scheme: Plot the function $|D(\cdot, h)|$ for a range of values of h, and identify the jump as the point x, in the vicinity of which $|D(x, h)|$ is consistently large for a range of values of h. More precisely, let $\widetilde{x}(h)$ be the global maximum of the function $|D(\cdot, h)|$ and

$$\widetilde{x}_-(h) = \sup_{h_1 \in [h, \eta_n]} \{\widetilde{x}(h_1) - 2h_1\}, \quad \widetilde{x}_+(h) = \inf_{h_1 \in [h, \eta_n]} \{\widetilde{x}(h_1) + 2h_1\}, \tag{9.77}$$

for $h \leq \eta_n$, where $\eta_n > 0$ is a prescribed number that tends to zero more slowly than $n^{-1} \log n$. Let $\widetilde{h}$ denote the minimum of values h such that $\widetilde{x}_-(h) \leq \widetilde{x}_+(h)$. The proposed jump-point estimator is $\widehat{x}_0 = \widetilde{x}(\widetilde{h})$. This idea is due to Gijbels et al. (1995).

Gijbels et al. (1995) also proposed a further refinement of the above concept. For a given bandwidth h, pretend that the change point lies in the interval $\widetilde{x}(h) \pm 2h$ and that the regression function is a step function on this interval. Then find the unknown location of the jump such that it minimizes the residual sum of squares, using only the data in the strip $\widetilde{x}(h) \pm 2h$. The resulting estimator is a refinement of the estimator $\widetilde{x}(h)$. In particular, we can take $h = \widetilde{h}$ to yield a refinement of $\widehat{x}_0$.

Another intuitive method is a one-sided kernel approach developed by Müller (1992). The idea as extended to the local polynomial setting works as follows: Let K_- be a kernel function supported on $[-1, 0]$. Let $\widehat{m}_-(x, h)$ be a local polynomial fit using the bandwidth h and the kernel K_-. Note that the estimator $\widehat{m}_-(x, h)$ uses only the local data on the left-handside of the point x. Analogously, let K_+ be a kernel function supported on $[0, 1]$ and $\widehat{m}_+(x, h)$ be a local polynomial fit using the bandwidth h and the kernel K_+. Then, $\widehat{m}_+(x, h)$ uses only the data on the right-hand side of the point x. At the smooth locations, the estimates $\widehat{m}_-(x, h)$ and $\widehat{m}_+(x, h)$ are about the same, since both are consistent estimates of $m(x)$. At the discontinuity point, however, they estimate, respectively, the left limit and the right limit of the function m at the point x. Thus, a natural estimator is the location such that the difference function $|\widehat{m}_+(x, h) - \widehat{m}_-(x, h)|$ is maximized. The bandwidth for detecting the change point is typically much smaller than the optimal bandwidth for curve estimation. For a wavelet-based method, see Wang (1995).

To illustrate these two approaches, consider the simulated data in Figure 9.2(b). We apply a local linear fit with the Epanechnikov kernel. Figure 9.13(a) presents the absolute value of the estimated derivative curve for a wide range of bandwidths, ranging from 0.05 to 0.3. Though the bandwidths have been increased by a factor of 6, the location of the maximum remains about the same: All of the bandwidths are close to the true value. Using the bandwidth $h = 0.05$, we obtain the jump point $\widehat{x} = 0.49$. Figure 9.13(b) depicts the difference curves $|\widehat{m}_+(x, h) - \widehat{m}_-(x, h)|$ for two bandwidths $h = 0.1$ and $h = 0.3$. Since the kernel is one-sided, we do not have enough data at the boundary points to fit a line. Thus, we only compute the difference curves for $x \in [0.1, 0.9]$. The choice $h = 0.3$ gives a correct estimate $\widehat{x} = 0.51$ of the change point, while the choice $h = 0.1$ results in a wrong estimate of 0.23 for the location of the change point. This example shows two advantages of the derivative method: The estimator of the jump point is less sensitive to the choice of the bandwidth, and sparsity of the local data points occurs less often.

9.6.3 Estimation of Conditional Densities

In the univariate setting, the population mean, while easier to estimate, is far less informative than the probability density function. Likewise, in the regression

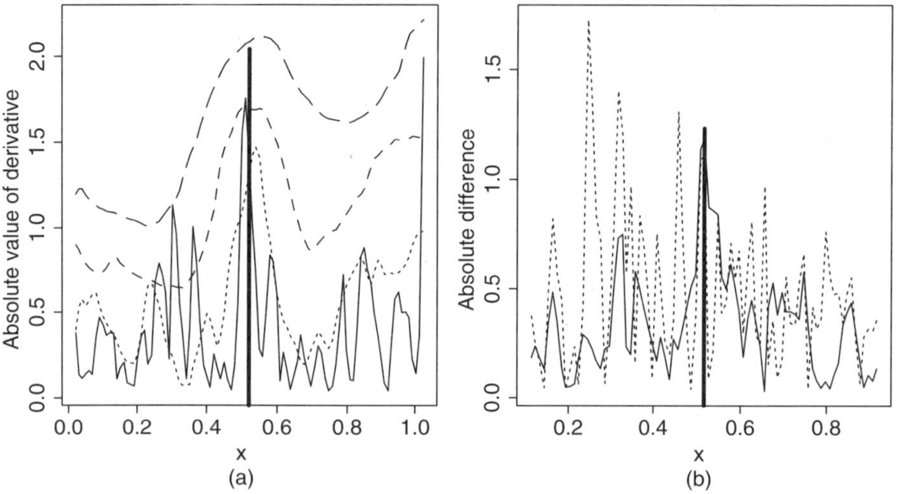

FIGURE 9.13 Change-point detection. (a) Maximum derivative method. Presented is the absolute value of the estimated derivative curve with bandwidths $h = 0.3$ (long-dashed curve), 0.2 (dashed curve), 0.1 (short-dashed cure), and 0.05 (solid curve). (b) One-sided kernel method. Presented are $|\widehat{m}_+(x, h) - \widehat{m}_-(x, h)|$ with bandwidths $h = 0.3$ (solid curve) and 0.1 (dashed curve).

setting, the regression function is far less informative than the conditional probability density function. The conditional density function provides information about not only its center (the regression function), but also the degree of spread from the center. If the shape of the conditional density estimator appears multimodal or highly skewed, it is not appropriate to summarize the density via the conditional mean regression function.

Let $(X_1, Y_1), \ldots, (X_n, Y_n)$ be a random sample from the population (X, Y) with the conditional density $g(y|x)$. Note that

$$E\{K_{h_2}(Y - y)|X = x\} \approx g(y|x) \quad \text{as } h_2 \to 0, \tag{9.78}$$

where $K_h = K(\cdot/h)/h$ is a rescaled kernel function. Thus, $g(y|x)$ can be regarded as the approximate regression function of the variable $K_{h_2}(Y - y)$ on X. Considerations of this nature lead to the following local polynomial regression problem:

$$\sum_{i=1}^{n} \left\{ K_{h_2}(Y_i - y) - \sum_{j=0}^{p} \beta_j (X_i - x)^j \right\}^2 W_{h_1}(X_i - x) \tag{9.79}$$

for a given bandwidth h_1 and a kernel function W. Let $\{\widehat{\beta}_j(x, y), j = 0, \ldots, p\}$ be the solution of the least-squares problem. Then, from Section 9.1.3, an estimator of $g^{(\nu)}(y|x) = (\partial^\nu g(y|x)\partial y^\nu)$ is $\nu!\widehat{\beta}_\nu(x, y)$. We write $\widehat{g}(y|x) = \widehat{\beta}_0(x, y)$ as the estimator of the conditional density.

The bandwidth selection problem appears somewhat challenging for this problem. For a given bandwidth h_2, the estimator $\widehat{g}(y|x)$ is derived from a local polynomial fit. Hence, any bandwidth selection method in Section 9.3.3 can be used to select h_1. For selection of h_2, a simple method is to apply a rule-of-thumb bandwidth selector, such as the normal reference rule (see Silverman, 1986, p. 45)

$$\widehat{h}_2 = \left[\frac{8\pi^{1/2} \int K^2(x)\,dx}{3\left\{\int x^2 K(x)\,dx\right\}^2} \right]^{1/5} S_y n^{-1/5}, \tag{9.80}$$

where S_y is the sample standard deviation of $Y_1, \ldots, Y_n$. When K is the Gaussian kernel, $\widehat{h}_2 = 1.06 S_y n^{-1/5}$; for the Epanechnikov kernel, $\widehat{h}_2 = 2.34 S_y n^{-1/5}$. The latter method is proposed in Fan et al. (1996), where one can also find many interesting numerical examples.

With the estimated conditional density function, one can derive many statistical estimators. For instance, the mean regression function can simply be estimated by

$$\widehat{m}(x) = \int y \widehat{g}(y|x)\,dy.$$

It can be shown that this estimator is the same as the local polynomial regression estimator when the kernel function K has mean zero. Similarly, one can derive estimates for the conditional variance and conditional quantile function.

Acknowledgments

This project was supported by a direct grant from the Faculty of Science of The Chinese University of Hong Kong. J. Fan was supported partially by NSF grant DMS-9504414 and NSA Grant 96-1-0015. This chapter was written while I .Gijbels was visiting the Department of Statistics of The Chinese University of Hong Kong. She gratefully acknowledges the Department of Statistics for its hospitality and support. I. Gijbels was partially supported by Projet d'Actions de Recherche Concertées (No. 93/98-164) from the Belgian Government and by research grant No. 15.001.95F of the National Science Foundation (FNRS), Belgium.

REFERENCES

Andersen, P. K., Borgan, Ø., Gill, R. D., and Keiding, N. (1993). *Statistical Models Based on Counting Processes*, Springer–Verlag, New York.

Breslow, N. E. (1974). Covariance analysis of censored survival data, *Biometrics*, **30**, 89–99.

Carroll, R. J., Ruppert, D., and Welsh, A. H. (1998). Local estimation equations. *J. Amer. Statist. Assoc.*, **93**, 214–227.

Chaudhuri, P. (1991). Nonparametric estimates of regression quantiles and their local Bahadur representation, *Ann. Statist.*, **19**, 760–777.

Cleveland, W. S. (1979). Robust locally weighted regression and smoothing scatterplots, *J. Amer. Statist. Assoc.*, **74**, 829–836.

Cleveland, W. S., Grosse, E., and Shyu, W. M. (1992). Local regression models, in Chambers, J. M., and Hastie, T. J. (eds.), *Statistical Models in S*, pp. 309–376, Wadsworth & Brooks, CA.

Cleveland, W. S., and Loader, C. (1996). Smoothing by local regression: principles and methods, in Härdle, W., and Schimek, M. G. (eds.), *Statistical Theory and Computational Aspects of Smoothing*, pp. 10–49, Physica-Verlag, Heidelberg, Germany.

Cox, D. R. (1972). Regression models and life-tables (with discussion), *J. Roy. Statist. Soc. B*, **34**, 187–220.

Cox, D. R. (1975). Partial likelihood, *Biometrika*, **62**, 269–276.

Fan, J. (1992). Design-adaptive nonparametric regression, *J. Amer. Statist. Assoc.*, **87**, 998–1004.

Fan, J., Farmen, M., and Gijbels, I. (1998). Local maximum likelihood estimation and inference, *J. Roy. Statist. Soc. B*, **60**, 591–608.

Fan, J., and Gijbels, I. (1992). Variable bandwidth and local linear regression smoothers, *Ann. Statist.*, **20**, 2008–2036.

Fan, J., and Gijbels, I. (1995a). Data-driven bandwidth selection in local polynomial fitting: Variable bandwidth and spatial adaptation, *J. Roy. Statist. Soc. B*, **57**, 371–394.

Fan, J., and Gijbels, I. (1995b). Adaptive order polynomial fitting: Bandwidth robustification and bias reduction, *J. Comp. Graph. Statist.*, **4**, 213–227.

Fan, J., and Gijbels, I. (1996). *Local Polynomial Modeling and Its Applications*, Chapman and Hall, London, UK.

Fan, J., Gijbels, I., and King, M. (1997). Local likelihood and local partial likelihood in hazard regression, *Ann. Statist.*, **25**, 1661–1690.

Fan, J., Heckman, N. E., and Wand, M. P. (1995). Local polynomial kernel regression for generalized linear models and quasi-likelihood functions, *J. Amer. Statist. Assoc.*, **90**, 141–150.

Fan, J., and Yao, Q. (1998). Efficient estimation of conditional variance functions in stochastic regression, preprint (to appear in *Biometrika*).

Fan, J., Yao, Q., and Tong, H. (1996). Estimation of conditional densities and sensitivity measures in nonlinear dynamical systems, *Biometrika*, **83**, 189–206.

Fleming, T. R., and Harrington, D. P. (1991). *Counting Processes and Survival Analysis*, Wiley, New York.

Gasser, T., and Müller, H.-G. (1979). Kernel estimation of regression functions, in *Smoothing Techniques for Curve Estimation*, pp. 23–68, Springer–Verlag, New York.

Gasser, T., Müller, H.-G., and Mammitzsch, V. (1985). Kernels for nonparametric curve estimation, *J. Roy. Statist. Soc. B*, **47**, 238–252.

Gentleman, R., and Crowley, J. (1991). Local full likelihood estimation for the proportional hazards model, *Biometrics*, **47**, 1283–1296.

Gijbels, I., Hall, P., and Kneip, A. (1995). *On the Estimation of Jump Points in Smooth Curves*, Discussion Paper #9515, Institute of Statistics, Catholic University of Louvain, Louvain-la-Neuve, Belgium.

Granovsky, B. L. and Müller, H.-G. (1991). Optimizing kernel methods: A unifying variational principle, *Inter. Statist. Rev.*, **59**, 373–388.

Green, P. J., and Yandell, B. (1985). Semiparametric generalized linear models, in *Generalized Linear Models. Proceedings of the GLIM 85 Conference*, pp. 44–55, Springer–Verlag, Berlin, Germany.

Gu, C. (1990). Adaptive spline smoothing in non-Gaussian regression models, *J. Amer. Statist. Assoc.*, **85**, 801–807.

Hall, P., and Jones, M. C. (1990). Adaptive M-estimation in nonparametric regression, *Ann. Statist.*, **18**, 1712–1728.

Härdle, W. (1990)., *Applied Nonparametric Regression*, Cambridge University Press, Boston, MA.

Härdle, W., and Gasser, T. (1984). Robust non-parametric function fitting, *J. Roy. Statist. Soc. B*, **46**, 42–51.

Härdle, W., and Schimek, M. G. (eds.) (1996). *Statistical Theory and Computational Aspects of Smoothing*, Physica-Verlag, Heidelberg, Germany.

Hastie, T. J., and Tibshirani, R. (1990). *Generalized Additive Models*, Chapman and Hall, London, UK.

Henderson, R. (1916). Note on graduation by adjusted average, *Actuarial Soc. Amer.*, **17**, 43–48.

Huang, L.-S. (1995). *On Nonparametric Estimation and Goodness-of-Fit*, Ph.D. thesis, The University of North Carolina, Chapel Hill, NC.

Koenker, R., Portnoy, S., and Ng, P. (1992). Nonparametric estimation of conditional quantile function, in Dodge, Y. (ed.), *Proceedings of the conference on L_1: Statistical Analysis and Related Methods*, pp. 217–229, Elsevier, Amsterdam, The Netherlands.

Macaulay, F. R. (1931). *The Smoothing of Time Series*, National Bureau of Economic Research, New York.

Marron, J. S., and Nolan, D. (1988). Canonical kernels for density estimation, *Statist. Prob. Lett.*, **7**, 195–199.

McCullagh, P., and Nelder, J. A. (1989). *Generalized Linear Models* (2d ed.), Chapman and Hall, London, UK.

Müller, H.-G. (1992). Change-points in nonparametric regression analysis, *Ann. Statist.*, **20**, 737–761.

Parzen, E. (1962). On estimation of a probability density function and mode, *Ann. Math. Statist.*, **33**, 1065–1076.

Ruppert, D. (1995). Empirical-bias bandwidths for local polynomial nonparametric regression and density estimation, *J. Amer. Statist. Assoc.*, **92**, 1049–1062.

Ruppert, D., Sheather, S. J., and Wand, M. P. (1995). An effective bandwidth selector for local least squares regression, *J. Amer. Statist. Assoc.*, **90**, 1257–1270.

Ruppert, D., and Wand, M. P. (1994). Multivariate weighted least squares regression, *Ann. Statist.*, **22**, 1346–1370.

Ruppert, D., Wand, M. P., Holst, U., and Hössjer, O. (1997). Local polynomial variance function estimation, *Technometrics*, **39**, 262–273.

Seifert, B., and Gasser, T. (1996). Finite-sample variance of local polynomials: Analysis and solutions, *J. Amer. Statist. Assoc.*, **91**, 267–275.

Severini, T. A., and Staniswalis, J. (1994). Quasi-likelihood estimation in semiparametric models, *J. Amer. Statist. Assoc.*, **89**, 501–511.

Silverman, B. W. (1986). *Density Estimation for Statistics and Data Analysis*, Chapman and Hall, London, UK.

Simonoff, J. S. (1996). *Smoothing Methods in Statistics*, Springer–Verlag, New York.

Spencer, J. (1904). On the graduation of the rates of sickness and mortality, *J. Inst. Act.*, **38**, 334–347.

Stone, C. J. (1977). Consistent nonparametric regression, *Ann. Statist.*, **5**, 595–645.

Stone, C. J. (1982). Optimal global rates of convergence for nonparametric regression, *Ann. Statist.*, **10**, 1040–1053.

Stone, C. J. (1994). The use of polynomial splines and their tensor products in multivariate function estimation (with discussion), *Ann. Statist.*, **22**, 118–184.

Tibshirani, R., and Hastie, T. J. (1987). Local likelihood estimation, *J. Amer. Statist. Assoc.*, **82**, 559–567.

Truong, Y. K. (1989). Asymptotic properties of kernel estimators based on local medians, *Ann. Statist.*, **17**, 606–617.

Tsybakov, A. B. (1986). Robust reconstruction of functions by the local-approximation method, *Problems of Information Transmission*, **22**, 133–146.

Wand, M. P., and Jones, M. C. (1995). *Kernel Smoothing*, Chapman and Hall, London.

Wang, Y. (1995). Jump and sharp cusp detection by wavelets, *Biometrika*, **82**, 385–397.

Woolhouse, W. S. B. (1870). Explanation of a new method of adjusting mortality tables, with some observations upon Mr. Makeham's modification of Gompertz's theory, *J. Inst. Act.*, **15**, 389–410.

CHAPTER 10

Additive and Generalized Additive Models

Michael G. Schimek
Karl-Franzens-Universität Graz, Graz, Austria

Berwin A. Turlach
The University of Western Australia, Perth, Australia

10.1 INTRODUCTION

This chapter discusses multivariate regression problems evaluated in a nonparametric fashion whose aim is to study the structural relationship between the response variable Y and the vector of d covariates $\mathbf{X} = (X_1, \ldots, X_d)^T$ via

$$m(\mathbf{x}) = E(Y|\mathbf{X} = \mathbf{x})$$

with $\mathbf{x} = (x_1, \ldots, x_d)^T$ and $m(\mathbf{x}) = m(x_1, \ldots, x_d)$.

In the multiple linear regression model, one assumption is that the conditional mean relationship between the response and each of the predictors is linear, that is, that $m(\mathbf{x})$ is linear and additive in the predictors. To gain more flexibility, one can drop the linearity assumption but retain additivity. The result is an additive model in which each explanatory variable can be related to the dependent variable via an individual functional form. As long as these functions obey certain smoothness assumptions, they can be estimated by scatterplot smoothers in a nonparametric fashion. The case of additive models will be studied in Section 10.2, where we consider penalized-least-squares and marginal-integration methods.

In the parametric class of *generalized linear models* (GLMs; McCullagh and Nelder, 1989), the unknown regression function $m(\mathbf{x})$ is modeled linearly via a known link function G. An important example is the logistic regression model

Smoothing and Regression: Approaches, Computation, and Application
Edited by Michael G. Schimek
ISBN 0-471-17946-9

for binary responses. There, the logit of the regression function is represented linearly. Again, we can gain more flexibility by replacing the linear function with some smooth function in a nonparametric setting. Such models have been introduced by Hastie and Tibshirani and summarized in their 1990 monograph. The nonparametric alternative to GLMs is called *generalized additive models* (GAMs). They allow the conditional mean of the response variable to depend, via a fixed link function, on a sum of univariate functions, each function having one component of the vector of explanatory variables as argument. The generalized additive case will be discussed in length in Section 10.3, where the emphasis is again on the penalized-least-squares and the marginal-integration methods.

Section 10.3.3 describes *vector generalized additive models* (VGAMs), which were recently introduced by Yee and Wild (1996). VGAMs handle multivariate (vector) regression problems.

Two further methods, or algorithms named *alternating conditional expectations* (ACE) and *additivity and variance stabilization* (AVAS), are discussed briefly in Section 10.4, because they are (computationally) closely related to GAMs even though they are based on a different working model.

For all models introduced in this chapter, we provide the numerical background (also pointing out alternative estimation concepts), algorithms, software hints (emphasizing S-Plus), and worked-out examples based on real data.

Let us now give some motivation for the use of additive models. The most flexible models one could think of do not have any assumptions about the form of the d-variate function $m(\mathbf{x})$. The problem is to fit a d-dimensional surface to the observed data $\{\mathbf{x}_i^T, Y_i\}$ with $i = (1, \ldots, n)$, where $\mathbf{x}_i = (x_{i1}, \ldots, x_{id})^T$. What comes first to mind is generalizing the univariate smoothing techniques, described in the first chapters of this book, to this multivariate situation. There is, however, a severe problem: it is necessary to define neighborhoods in the d-dimensional space, but under the curse of dimensionality we understand that neighborhoods with a fixed number of points become less local as the dimensions increase (Bellman, 1961).

In the spline world the generalization is restricted to certain cases such as thin-plate splines. (See Chapter 11 for details.) Other problems, and thus restrictions, arise in the domain of kernel smoothers. (See Chapter 12 for a discussion.) The additive model approach is an elegant way around these problems because high dimensionality is tackled via an additive approximation of simple univariate smooth functions, such as cubic smoothing splines or kernel smoothers.

What also remains unsolved is the problem of whether the degree of smoothing should be the same in each dimension. This problem is related to the question of whether the variation in the surface is comparable with respect to all covariates (See Scott, 1992, for further reading.) As will be demonstrated in this chapter, the above problems become less severe when an additive approach is taken, although related questions will emerge. Smoothing-parameter (that is, bandwidth) choice remains a weak point in this class of models. The reason is a lack both of theory and of adequate algorithms. In Sections 10.5 and 10.6, we shall discuss the problem of selecting smoothing parameter(s) in (generalized) additive models, then conclude with a discussion of model diagnostics, another field still open to research.

10.2 THE ADDITIVE MODEL

Consider the general multiple regression model

$$Y = m(\mathbf{X}) + e, \tag{10.1}$$

where we assume that the error e is independent of the vector of explanatory variables $\mathbf{X}$, $E(e) = 0$, and $\mathrm{Var}(e) = \sigma^2$. By contrast with linear regression, in what follows only the additivity property of m is required. The dependent variable Y is approximated by the additive model

$$m(\mathbf{X}) \simeq g(\mathbf{X}) = g_0 + \sum_{j=1}^{d} g_j(X_j), \tag{10.2}$$

where g_0 is a constant and the g_js are univariate smooth functions. To avoid free constants in the functions g_j, we usually require that $E[g_j(X_j)] = 0$ for $1 \le j \le d$ (centering). These identifiability conditions imply that $E(Y) = g_0$.

In the following we shall restrict our interest to linear scatterplot smoothers with an $n \times n$ smoother matrix S. Buja et al. (1989, p. 453) give the following definition for a linear scatterplot smoother: The function estimate $\hat{\mathbf{y}} = (\hat{y}_1, \ldots, \hat{y}_n)^T$ can be written as $\hat{\mathbf{y}} = S\mathbf{y}$, where $\mathbf{y} = (y_1, \ldots, y_n)^T$, and S does not depend on $\mathbf{y}$. Important examples of linear smoothers are kernel smoothers and smoothing splines. Not all scatterplot smoothers have this property.

Under identifiability, if the additive model is correct (not just an approximation), we have

$$E\left[Y - g_0 - \sum_{j \ne k} g_j(X_j)|X_k\right] = g_k(X_k)$$

for $k = 1, \ldots, d$. This relationship suggests that we adopt an iterative procedure for the estimation of the univariate functions $g_1, \ldots, g_d$ corresponding to the explanatory variables. For a known constant g_0 and given functions g_j, $j \ne k$, the function g_k can be estimated by a univariate regression fit based on the measurements (X_{ik}, Y_i) for $i = 1, \ldots, n$ (n observations). Applying a linear smoother S_j, the centered version of the estimator $\hat{g}_j$ of g_j is

$$\hat{g}_j^* = \hat{g}_j - n^{-1} \sum_{i=1}^{n} \hat{g}_j(X_{ij}).$$

This approach is the motivation for an iterative scheme that is now known as *backfitting*. It was introduced and analyzed by Friedman and Stuetzle (1981) and Breiman and Friedman (1985) in the context of projection pursuit regression — the additive model is a special case of the projection pursuit model — and can be described as follows:

Additive Model Algorithm

1. *Initialize*: $\hat{g}_0 = n^{-1}\sum_{i=1}^{n} Y_i$, $\hat{g}_j = g_j^0$, $j = 1, \ldots, d$.
2. *Find new transformations*: For $j = 1, \ldots, d$:
   $\hat{g}_j = S_j\left[Y - \hat{g}_0 - \sum_{j \neq k} \hat{g}_j(X_j) | X_k\right]$;
   $\hat{g}_0 = \hat{g}_0 + n^{-1}\sum_{i=1}^{n} \hat{g}_j(X_{ij})$;
   and $\hat{g}_j^* = \hat{g}_j - n^{-1}\sum_{i=1}^{n} \hat{g}_j(X_{ij})$.
3. Cycle step 2 until convergence.

The idea behind this algorithm is to carry out a fit, calculate partial residuals from that fit, and refit again. That is why the iteration scheme is called *backfitting*. The starting functions $g_1^0, \ldots, g_d^0$ can be obtained in various ways, for example, from a linear regression fit of Y on the predictors X_k. Technical matters concerning the algorithm, such as numerical features, convergence, and statistical implications, will be discussed later in this chapter.

10.2.1 The Normal Equations and Linear Scatterplot Smoothing

In order to justify smoothing in the additive model, we take the L_2-function-space view of Hastie and Tibshirani (1990). Let $\mathcal{H}_j$, for $j = 1, \ldots, d$, denote the Hilbert spaces of measurable functions $\phi_j(X_j)$ with $E[\phi_j(X_j)] = 0$, $E[\phi_j^2(X_j)] < \infty$, and inner product $\langle \phi_j(X_j), \phi_j^*(X_j) \rangle = E[\phi_j(X_j)\phi_j^*(X_j)]$. Furthermore, $\mathcal{H}$ is assumed to be the space of arbitrary centered square-integrable functions of $X_1, \ldots, X_d$. We consider the $\mathcal{H}_j$ as subspaces of $\mathcal{H}$ in a canonical way. $\mathcal{H}^{\text{add}} \subset \mathcal{H}$ describes a closed linear subspace of the additive functions $\mathcal{H}_1 + \cdots + \mathcal{H}_d$. These are all subspaces of $\mathcal{H}_{YX}$, the space of centered square-integrable functions of Y and $X_1, \ldots, X_d$.

The additive model in a population setting amounts to minimizing

$$E[Y - g(\mathbf{X})]^2 \tag{10.3}$$

over

$$g(\mathbf{X}) = \sum_{j=1}^{d} g_j(X_j) \in \mathcal{H}^{\text{add}}.$$

Without the additivity assumption, we would simply obtain $E(Y|\mathbf{X})$. Our goal is to find the closest additive approximation to this function. By the definition of $\mathcal{H}^{\text{add}}$, the above minimum exists and is unique. The same is not true for the individual functions $g_j(X_j)$.

Let P_j denote the conditional expectation operator $E(\cdot|X_j)$ — note that P_j is the orthogonal projection onto H_j. The minimizer $g(\mathbf{X})$ of (10.3) can be represented by residuals $Y - g(\mathbf{X})$ that are orthogonal to the space of fits: $Y - g(\mathbf{X}) \perp \mathcal{H}^{\text{add}}$. $\mathcal{H}^{\text{add}}$ is generated by $\mathcal{H}_j$, so we have (equivalently) $Y - g(\mathbf{X}) \perp \mathcal{H}_j$ for all j, or $P_j[Y - g(\mathbf{X})] = 0$ for all j. If we consider a single component, this can be written as

$$g_k(X_k) = P_k\left[Y - \sum_{\substack{j=1\\ j\neq k}}^{d} g_j(X_j)\right] = E\left[Y - \sum_{\substack{j=1\\ j\neq k}}^{d} g_j(X_j)|X_k\right].$$

The following system of estimating equations is necessary and sufficient for $g = (g_1, \ldots, g_d)$ to minimize (10.3):

$$\begin{pmatrix} I & P_1 & P_1 & \cdots & P_1 \\ P_2 & I & P_2 & \cdots & P_2 \\ \vdots & \vdots & \vdots & \ddots & \vdots \\ P_d & P_d & P_d & \cdots & I \end{pmatrix} \begin{pmatrix} g_1(X_1) \\ g_2(X_2) \\ \vdots \\ g_d(X_d) \end{pmatrix} = \begin{pmatrix} P_1 Y \\ P_2 Y \\ \vdots \\ P_d Y \end{pmatrix}. \tag{10.4}$$

This system can also be expressed as

$$Pg = QY,$$

where P and Q are a matrix and a vector, respectively, of operators. A numerical solution for the additive approximation cannot be obtained from (10.4), because the matrix on the left consists of conditional-expectation operators (not real numbers), but the conditional expectations in our population setting are connected to the sample-data setting via smoothing.

Let us now take a more formal look at linear scatterplot smoothers in additive models. We assume that to compute the fit at design points x_i, in which case we can write a linear smoother as a linear map $S\colon \mathbb{R}^n \mapsto \mathbb{R}^n$ defined by $\hat{\mathbf{y}} = S\mathbf{y}$. Applying a linear-smoothing algorithm, we can produce the corresponding smoother matrix S by smoothing unit basis vectors. Smoothing the ith unit vector results in the ith column of S. The same cannot be done for a nonlinear smoother such as locally weighted scatterplot smoothing (LOWESS), because the estimates depend on y in a nonlinear fashion.

A linear smoother can be written as a smoother matrix S times the response vector $\mathbf{y}$, that is, $\hat{\mathbf{g}} = S\mathbf{y}$. As pointed out earlier, the most prominent linear smoothers are splines and kernel smoothers. Now we can replace the conditional expectation operator P_j by such a smoother with smoother matrix S_j.

Finally, a numerical solution for the additive approximation to the regression curves g_j forming an (nd)-dimensional vector $\mathbf{g}$ can be obtained from the data version of the estimating equations in (10.4), which form an $(nd) \times (nd)$ system:

$$\begin{pmatrix} I & S_1 & S_1 & \cdots & S_1 \\ S_2 & I & S_2 & \cdots & S_2 \\ \vdots & \vdots & \vdots & \ddots & \vdots \\ S_d & S_d & S_d & \cdots & I \end{pmatrix} \begin{pmatrix} \mathbf{g}_1 \\ \mathbf{g}_2 \\ \vdots \\ \mathbf{g}_d \end{pmatrix} = \begin{pmatrix} S_1\mathbf{y} \\ S_2\mathbf{y} \\ \vdots \\ S_d\mathbf{y} \end{pmatrix}. \tag{10.5}$$

In short form we can write

$$A\mathbf{g} = B\mathbf{y},$$

where A and B are block matrices consisting of identity matrices I and linear smoothing operators S_j. This system of equations is known as the normal equations of the additive regression model.

The preceding system can be adapted to handle tied observations; such a modification is important for applied data analysis. Schimek et al. (1993) proposed a weighting scheme (based on permutation and reduction matrix operators) that does not change the structural features of the system matrix. (Later, this approach is discussed in detail for smoothing splines.)

10.2.2 Penalized Least Squares

In this section we shall concentrate on smoothing splines which have the additional property (not shared by, for instance, kernel smoothers) that the smoothing matrix S is symmetric. As a direct consequence of this symmetry, and of the fact that splines can reproduce straight lines, we have

$$HS\mathbf{y} = (S^T H^T)^T \mathbf{y} = (SH)^T \mathbf{y} = H^T \mathbf{y} = H\mathbf{y},$$

where H is the so-called *hat matrix* (often written $\hat{H}$) for linear-least-squares regression. This property will prove useful when studying the formal structure of the backfitting algorithm. Also due to symmetry is the fact that the eigenvectors of the smoother matrix S of a spline resemble those of polynomials of increasing degree (Eubank, 1984). The first two eigenvalues are 1, and correspond to linear functions of the design variable x. The eigendecomposition of S allows us to analyze the smoother by means of the spectrum, analogously to the way we use linear stochastic processes to analyze time series.

For convenience, let us study the popular case of cubic smoothing splines. We have to minimize with respect to g the penalized least-squares criterion

$$\sum_{i=1}^{n} (y_i - g(x_i))^2 + \lambda \int_{-\infty}^{+\infty} [g''(z)]^2 \, dz, \qquad (10.6)$$

where $\lambda (\lambda > 0)$ is a fixed smoothing parameter as introduced by Eubank in Chapter 1 of this book. The solution $\hat{g}$ is a cubic Reinsch spline (Reinsch, 1967) with knots at each distinct x_i.

Following Green and Yandell (1985, p. 46) we introduce $h_i = x_{i+1} - x_i$ for $i = 1, 2, \ldots, n-1$, a tridiagonal $(n-2) \times n$ matrix Δ with $\Delta_{ii} = 1/h_i$, $\Delta_{i,i+1} = -(1/h_i + 1/h_{i+1})$, $\Delta_{i,i+2} = 1/h_{i+1}$, and a symmetric tridiagonal $(n-2) \times (n-2)$ matrix C with $c_{i-1,i} = c_{i,i-1} = h_i/6$, $c_{ii} = (h_i + h_{i+1})/3$. Then the minimization problem in (10.6) can be equivalently expressed by

$$\|\mathbf{y} - \mathbf{g}\|^2 + \lambda \mathbf{g}^T K \mathbf{g} \rightarrow \min, \qquad (10.7)$$

where K denotes a quadratic penalty matrix with

$$K = \Delta^T C^{-1} \Delta.$$

The solution is now

$$\hat{\mathbf{g}} = S\mathbf{y},$$

where the linear smoother matrix S is given by

$$S = (I + \lambda K)^{-1}. \tag{10.8}$$

When smoothing splines are used as scatterplot smoothers, the concept of penalized least squares can be applied to the additive model itself. Further, this concept provides another motivation, apart from that of function spaces, for the application of smoothing in additive models.

Above, we introduced the cubic smoothing spline as the minimizer of the penalized-least-squares criterion (10.6) over all twice-continuously-differentiable functions g. In order to extend this idea to the additive model, we generalize the criterion (10.6) in a straightforward manner. We seek to minimize

$$\sum_{i=1}^{n} \left(y_i - \sum_{j=1}^{d} g(x_{ij}) \right)^2 + \sum_{j=1}^{d} \lambda_j \int_{-\infty}^{+\infty} [g_j''(z)]^2 \, dz \tag{10.9}$$

over all twice-continuously-differentiable functions g_j. Each function in (10.9) is penalized by a separate fixed smoothing parameter λ_j. This in turn determines the smoothness of that function in the solution. There are two extreme cases. One occurs if all the λ_js take the value zero; then the solution to (10.6) is any interpolating set of functions satisfying $y_i = \sum_{j=1}^{d} g(x_{ij})$ for $i = 1, 2, \ldots, n$. The other extreme case occurs if each λ_j goes to infinity; then the penalty term itself goes to infinity unless $g_j''(z) = 0$ for all j (i.e., each g_j is linear as in linear least-squares regression).

The solution of (10.9) is a cubic smoothing spline in each of the predictors. Evaluation at all n observations leads to

$$\left(\mathbf{y} - \sum_{j=1}^{d} \mathbf{g}_j \right)^T \left(\mathbf{y} - \sum_{j=1}^{d} \mathbf{g}_j \right) + \sum_{j=1}^{d} \lambda_j \mathbf{g}_j^T K_j \mathbf{g}_j, \tag{10.10}$$

where the K_js are penalty matrices for each predictor as in the univariate setting of (10.7). Differentiating (10.10) with respect to the function $\mathbf{g}_k$ yields

$$\hat{\mathbf{g}}_k = S_k \left(\mathbf{y} - \sum_{j \neq k} \hat{\mathbf{g}}_j \right), \tag{10.11}$$

where S_k is a smoother matrix given by (compare with (10.8))

$$S_k = (I + \lambda_k K_k)^{-1}. \tag{10.12}$$

(The K_k are the individual penalty matrices.) Writing Equation (10.11) for $k = 1, 2, \ldots, d$ produces the same $(nd) \times (nd)$ system of normal equations as is obtained from the function-space considerations. Again we end up with

$$\begin{pmatrix} I & S_1 & S_1 & \ldots & S_1 \\ S_2 & I & S_2 & \ldots & S_2 \\ \vdots & \vdots & \vdots & \ddots & \vdots \\ S_d & S_d & S_d & \ldots & I \end{pmatrix} \begin{pmatrix} \mathbf{g}_1 \\ \mathbf{g}_2 \\ \vdots \\ \mathbf{g}_d \end{pmatrix} = \begin{pmatrix} S_1\mathbf{y} \\ S_2\mathbf{y} \\ \vdots \\ S_d\mathbf{y} \end{pmatrix}, \tag{10.13}$$

with d different cubic spline smoother matrices S_j. According to Hastie and Tibshirani (1990, p. 111), we can interpret the penalty terms in (10.10) as a down-weighting of each of the components of $\mathbf{g}_j$, the down-weighting determined by the corresponding eigenvalue of that component and by λ_j.

Next we have to cope with the technical problem of tied predictor values, mentioned in Hastie and Tibshirani (1990, p. 74). There is almost no practical situation in multivariate regression where the data within one variable are all different from each other.

The problem of tied observations in additive models is twofold: (i) the predictor variables may require different orderings (this consideration is relevant when, for instance, smoothing splines are adopted); and (ii) in each predictor variable, values may occur with frequency greater than one (tied values). Schimek et al. (1993) put forward a computationally efficient modification of the normal equations in (10.13) that successfully deals with problems (i) and (ii) at the same time.

We apply again the penalized-least-squares criterion for cubic smoothing splines as well as for the additive model. Let us define $x_i = (x_{i1}, \ldots, x_{id})^T$ as the design points (i.e., explanatory observations) for $i = 1, \ldots, n$.

The estimation of model (10.2) requires us to determine the $\mathbf{g}_j$s for each coordinate of the design points x_{ij}. For the distinct design points, we consider the minimization problem (with respect to $\mathbf{g}_j$)

$$\left(\mathbf{y} - \sum_{j=1}^{d} \mathbf{g}_j\right)^T \left(\mathbf{y} - \sum_{j=1}^{d} \mathbf{g}_j\right) + \sum_{j=1}^{d} \lambda_j \mathbf{g}_j^{(l)T} K_j \mathbf{g}_j^{(l)}, \tag{10.14}$$

where the K_j are penalty matrices and the λ_j ($\lambda_j > 0$) are the smoothing parameters of the individual cubic smoothing splines, as before.

The vectors $\mathbf{g}_j^{(l)}$ are of equal or smaller (hence the notation "(l)") dimension than the original $\mathbf{g}$ vectors of estimators (i.e., $n_j \leq n$). They can be obtained as follows. Let O_j be a permutation operator such that $x'_{ij} = O_j x_{ij}$ is ordered ascending, $j = 1, \ldots, d$. Then let R_j be a reduction operator defined by an $n_j \times n$ matrix with elements r_{ts}, where $n_j = \#\{x_{1j}, \ldots, x_{nj}\}$ (# is the number of elements of a set); $r_{ts}:=1$ for $s=1,\dots,n$ and $t=t(s)=\#\{x'_{1j},\dots,x'_{sj}\}$; $r_{ts}:=0$ elsewhere. Let $D_j:=R_jR_j^T$ and $\mathbf{g}_j^{(l)}:=D_j^{-1}R_jO_j\mathbf{g}_j$. The $\mathbf{g}_j^{(l)}$s consist of those remaining values $\mathbf{g}_{ij}$ corresponding to the distinct and ordered design points x'_{ij}, $i=1,\dots,n_j$. Hence we can write

$$\mathbf{g}_j = O_j^{-1}R_j^T\mathbf{g}_j^{(l)}.$$

The least-squares term in (10.14) is equivalently expressed as the square of

$$O_r\mathbf{y} - \left(R_r^T\mathbf{g}_r^{(l)} + \sum_{j\neq r} O_rO_j^{-1}R_j^T\mathbf{g}_j^{(l)}\right), \tag{10.15}$$

for $r=1,\dots,d$. Differentiation of (10.14) using (10.15) with respect to $\mathbf{g}_r^{(l)}$ yields

$$\lambda_rK_r\mathbf{g}_r^{(l)} - R_rO_r\mathbf{y} + R_rR_r^T\mathbf{g}_r^{(l)} + \sum_{j\neq r} R_rO_rO_j^{-1}R_j^T\mathbf{g}_j^{(l)} = 0.$$

$T_{j,l}:=S_jR_jO_jO_l^{-1}R_l^T$, where $S_j=(D_j+\lambda_jK_j)^{-1}$, describes a modified smoother matrix, and $\mathbf{z}_j:=S_jR_jO_j\mathbf{y}$ for $j=1,\dots,d$. Then the normal equations can be expressed by

$$\begin{pmatrix} I_1 & T_{1,2} & \dots & T_{1,d-1} & T_{1,d} \\ T_{2,1} & I_2 & \dots & T_{2,d-1} & T_{2,d} \\ \vdots & \vdots & \ddots & \vdots & \vdots \\ T_{d-1,1} & T_{d-1,2} & \dots & I_{d-1} & T_{d-1,d} \\ T_{d,1} & T_{d,2} & \dots & T_{d,d-1} & I_d \end{pmatrix} \begin{pmatrix} \mathbf{g}_1^{(l)} \\ \mathbf{g}_2^{(l)} \\ \vdots \\ \mathbf{g}_{d-1}^{(l)} \\ \mathbf{g}_d^{(l)} \end{pmatrix} = \begin{pmatrix} \mathbf{z}_1 \\ \mathbf{z}_2 \\ \vdots \\ \mathbf{z}_{d-1} \\ \mathbf{z}_d \end{pmatrix}, \tag{10.16}$$

where I_j is the $n_j \times n_j$ identity matrix. The $\mathbf{g}_j^{(l)}$ are the new estimators of the additive model in (10.2). The system in (10.16) can be solved in the same way as the usual normal equations are. In the next section we learn about the problems and numerical techniques associated with the solution of such systems of the additive model.

10.2.3 Solution of the Normal Equations

10.2.3.1 Statistical Considerations

First we consider consistency of the normal equations in (10.5). Consistency requires that $B\mathbf{y} \in \mathcal{R}(A)$ for arbitrary data $\mathbf{y} \in \mathbb{R}^n$. ($\mathcal{R}$ is the range.) For the d-dimensional smoother case the following results are known for linear scatterplot smoothers with smoother matrices S_j.

Theorem 1. *If each S_j is symmetric with eigenvalues in [0,1], the normal equations are consistent for every* $\mathbf{y}$. (For the proof, see Buja et al., 1989, p. 481.)

For symmetric smoothers with eigenvalues in [0,1), closed formulas for the solutions can be given.

Proposition 1. *If the S_j are symmetric with eigenvalues in [0,1), the solutions of the normal equations can be written as* $\mathbf{g}_j = A_j(I + A)^{-1}\mathbf{y}$, *where* $A_j = (I - S_j)^{-1}S_j$ *and* $A = \sum_j A_j$. (For the proof, see Breiman and Friedman, 1985.)

Less stringent necessary and sufficient conditions on S_j for consistency could be derived by Buja et al. (1989, p. 482f) only for the case of two smoothers, a case which is of limited practical value.

Next we discuss reasons for degeneracy of the normal equations, which results in nonunique solutions. The first reason is collinearity or concurvity. *Collinearity* is well known from linear regression and describes a situation in which predictors are linearly dependent. If nonlinear dependencies are concerned, the term *concurvity* has been established. Both collinearity and concurvity have the same effect on the additive model: degeneracy of the normal equations. In practice, exact singularity of the system matrix is unlikely; however, some degeneration is quite common. There is a second reason for degeneracy of the normal equations: Linear scatterplot smoothers impose certain weighting schemes on the data which can cause near-singularity in some instances, especially in combination with concurvity.

For smoother-based normal equations, exact singularity (concurvity) is defined as the existence of a nonzero solution of the corresponding homogeneous equations

$$A\mathbf{g}^* = \mathbf{0}.$$

If such a $\mathbf{g}^*$ exists, and if $\mathbf{g}$ is a solution to $A\mathbf{g} = B\mathbf{y}$, then $\mathbf{g} + \gamma\mathbf{g}^*$, for arbitrary γ, is also a solution. Hence, if one such $\mathbf{g}^*$ exists, then there are infinitely many solutions.

Theorem 2. *If the smoothers S_j are all symmetric and have eigenvalues in [0,1], then a vector* $\mathbf{g} \neq \mathbf{0}$ *with* $\mathbf{g} \in \mathcal{R}(S_j)$ *represents a concurvity* ($A\mathbf{g}^* = \mathbf{0}$) *iff one of the two following conditions is satisfied:* $B\mathbf{g}^* = \mathbf{0}$ *(i.e.,* $\mathbf{g}$ *minimizes B); or* $\mathbf{g}_j^* \in \mathcal{M}_1(S_j)$ *for* $j = 1, 2, \ldots, d$, *and* $\mathbf{g}_+^* = \mathbf{0}$. $\mathcal{M}_1$ *denotes the eigenspace corresponding to the eigenvalue 1 and* $\mathbf{g}_+^* = \sum_j \mathbf{g}_j^*$. (For the proof, see Buja et al., 1989, p. 485.)

The last condition implies that exact concurvity is exact collinearity if, for instance, all smoothers are of the cubic spline type. Approximate concurvity, however, can be characterized by approximate minimizers of $B\mathbf{g}^*$, leading to approximate nonlinear additive relationships beween the predictors. Another aspect of interest is the following: If the S_j, $j = 1, 2, \ldots, d$, are symmetric with eigenvalues in [0,1), then A is nonsingular. This remark is of practical relevance, as the constant term is usually separated in the additive model and a zero-mean adjustment made for each of the smooth terms. As a matter of fact, doing this redefines our smoother matrices S to S^*, the latter having an eigenvalue of zero for the constant.

Finally, we consider the asymptotic rate of convergence of the nonparametric additive model. Stone (1985) derived the interesting result that the d-dimensional additive model can be estimated with the optimal rate of convergence of one-dimensional smoothing. Under a number of technical conditions and the following definitions—$\mathbf{g}^*$ is the best additive approximation to the true response function $\mathbf{g}$; o is the assumed measure of smoothness of $\mathbf{g}^*$; $\hat{\mathbf{g}}_n$ is the additive (spline) estimator; $\hat{\mathbf{g}}_{n1}, \ldots, \hat{\mathbf{g}}_{nd}$ are the component functions of $\hat{\mathbf{g}}_n$; and $r = o/(2o+1)$—we can obtain the following result.

Theorem 3. $E[\|\hat{\mathbf{g}}_{nj} - \mathbf{g}_j^*\|_j^2 | X_1, \ldots, X_n] = O_p(n^{-2r})$ *for* $1 \le j \le d$.

For the technical aspects and the proof see Stone (1985, p. 693ff). In a corollary, he showed that the rate of convergence does not depend on d the number of dimensions, another surprising result which means that the curse of dimensionality does not affect the asymptotic convergence rate. All these results hold for projection smoothers (e.g., regression splines) and smoothing splines, but not for general linear smoothers. Opsomer and Ruppert (1997) and Opsomer (1999) derived similar results for local linear regression, a nonprojection smoother, under rather strong conditions on the smoothing matrices.

10.2.3.2 Numerical Procedures

The normal equations in (10.5) are a linear system of the form

$$A\mathbf{g} = \mathbf{b},$$

where $\mathbf{b} = B\mathbf{y}$. Let the elements of the matrix A be a_{ij}, and those of the vector $\mathbf{b}$ be b_i. Because of the size of the equation system it is usually solved by iterative numerical techniques.

Here we relate backfitting to two standard iterative procedures (Jacobi and Gauss–Seidel) developed for solving linear equation systems having nonsingular system matrices. After a brief overview we discuss their shortcomings in the context of fitting additive regression models nonparametrically by linear scatterplot smoothers. As was pointed out ealier in this chapter, concurvity is a main source of concern, causing ill-posed system matrices. However, standard iterative procedures are not designed to cope with degeneracy.

Backfitting

The algorithm for the so-called Jacobi procedure is defined by

$$v_i^{(m)} = \left[b_i - \sum_{\substack{j=1 \\ j \neq i}}^{N} a_{ij} v_j^{(m-1)} \right] \Big/ a_{ii},$$

with iterative solutions $v_i^{(m)}$, $i = 1, \ldots, N$, an iteration counter m, and starting values $v_i^{(0)}$ (usually $v_i^{(0)} = 0$). The Gauss–Seidel procedure is characterized by

$$v_i^{(m)} = \left[b_i - \sum_{j=1}^{i-1} a_{ij} v_j^{(m)} - \sum_{j=i+1}^{N} a_{ij} v_j^{(m-1)} \right] \Big/ a_{ii}.$$

How do they differ numerically? Jacobi is a complete-step and Gauss–Seidel a single-step procedure. This difference can be best seen when the information update is studied. It describes how results currently available (estimates $v^{(m-1)}$) are used to obtain the successive results (estimates $v^{(m)}$). The Jacobi algorithm computes the estimates in iteration m on the basis only of estimates from iteration $m-1$. All estimates generated during step m are stored for use in step $m+1$ and remain unused within step m. Information update takes place only when moving from step m to step $m+1$, so the information flow is low. By contrast, the Gauss–Seidel procedure makes use of all the information currently available, no matter which step it comes from (m or $m-1$). There is continual information update and so maximal information flow.

Backfitting was developed in the context of nonparametric multidimensional regression (Friedman and Stuetzle, 1981; Hastie and Tibshirani, 1986). To serve that purpose, it was built upon the idea of determining estimates for the covariates in a successive manner, taking advantage of the specific structural features of the estimation problem. It uses the currently available information from all covariates except the one for which estimates have just been computed. This approach leads to a splitting of the system matrix into d blocks P_j of size $N/d \times N$, where each block corresponds to one of the predictor variables X_j, $j = 1, \ldots, d$. Then an iterative procedure has to be applied to these blocks, resulting in d vectors $\mathbf{v}_j^{(l)}$ (l denoting the last iteration), each of length N/d. The common choice is Gauss–Seidel iterations.

Information update takes place within the iterations, but with a certain lag. Therefore the information flow is higher than that in the Jacobi procedure but lower than that in the Gauss–Seidel procedure. As a consequence, given the block structure imposed by the additive or generalized additive model, backfitting is best viewed as a special case of the Gauss–Seidel algorithm. Moreover, backfitting is most effective in combination with Reinsch splines, because then we can avoid the explicit calculation of the S_k in Equation (10.11). (See Green and Silverman, 1994, p. 19ff for details.) This explains why many software implementers resort to backfitting and the Reinsch algorithm.

Convergence results for the above-mentioned iterative procedures found in the numerical literature were obtained under the assumption of a non-singular system matrix A. Convergence depends on the eigenvalues λ_i of the iteration matrix. (For details, see Hämmerlin and Hoffmann, 1992, p. 372 and p. 376.) Acceleration techniques were developed to improve the speed of convergence. We can write the iterative Jacobi scheme as

$$\mathbf{v}^{(m)} = (I - A)\mathbf{v}^{(m-1)} + \mathbf{b} = \mathbf{v}^{(m-1)} - D^{(m-1)}, \tag{10.17}$$

where $D^{(m-1)} = A\mathbf{v}^{(m-1)} - \mathbf{b}$ is the deficiency of step $m-1$. Equation (10.17) reveals that the procedure may be viewed as an attempt at correcting the previous step's estimate of deficiency. A relaxation parameter ω can be introduced which allows us to control the amount of correction:

$$\mathbf{v}^{(m)} = \mathbf{v}^{(m-1)} - \omega D^{(m-1)}. \tag{10.18}$$

For relaxed Gauss–Seidel iteration, the term *successive overrelaxation* (SOR) is established (Golub and van Loan, 1989, p. 510). After rewriting (10.17) the following relationship holds between the relaxed $\mathbf{v}_{rel}^{(m)}$ and the unrelaxed $\mathbf{v}^{(m)}$:

$$\mathbf{v}_{rel}^{(m)} = (1-\omega)\mathbf{v}^{(m-1)} + \omega\mathbf{v}^{(m)}. \tag{10.19}$$

For $\omega < 1$, we have underrelaxation; for $\omega > 1$, we have overrelaxation; and for $\omega = 1$ we obtain the unrelaxed algorithm. Theoretical results about the admissible range of ω (on one hand) and the optimal choice of ω (on the other) are available in the numerical literature. Under nonsingularity, the optimal choice of ω is as follows: (i) Assuming the Jacobi procedure converges, we have convergence for $0 < \omega \leq 1$, and the optimal parameter is given by $\omega^* = 2/(2 - \lambda_1 - \lambda_n)$, with $1/2 < \omega^* < \infty$. (ii) Assuming the Gauss–Seidel procedure converges, convergence for SOR iteration can be achieved only with $0 < \omega < 2$. There is evidence that this result holds at least for some special matrices, for example, symmetric positive definite matrices (See Gander and Golub, 1989, p. 530.) In such cases, however, the optimal choice ω^* is not known.

For more details we refer to Schimek et al. (1994). We should not close this discussion without mentioning that relaxation concepts have not found their way into commercial software such as S-Plus or XploRe. One obstacle is certainly the need for interaction between the user and the program (activation and selection of a relaxation parameter). Recently, numerical alternatives have been studied.

One such alternative to backfitting and related procedures is a technique based on relaxed iterative projections, proposed by Schimek (1996). Especially in cases where backfitting fails totally to provide appropriate estimation results in an additive or generalized additive model (e.g., when the covariates are substantially correlated), a numerical approach which successfully deals with degeneracy of the normal equations is desirable. In the following section, this new technique is described in brief.

Relaxed Iterative Projections

Let us again consider the linear equations system $A\mathbf{x} = \mathbf{b}$ (with $\mathbf{x} = \mathbf{g}$ and $\mathbf{b} = B\mathbf{y}$ in the additive model). Further, let us have a square $n \times n$ system matrix A (usually large and sparse) and n-dimensional vectors $\mathbf{x}$ and $\mathbf{b}$. In contrast to standard iterative procedures, which are equation-oriented, the iterative projection method is column-oriented.

We assume the matrix $A = (\mathbf{a}_1, \mathbf{a}_2, \ldots, \mathbf{a}_n)$ to consist of column vectors $\mathbf{a}_i$ for $i = 1, 2, \ldots, n$. Let us next have a linear space, span(A), generated by the columns of A, and $\mathbf{b} \in$ span(A). We now define two real sequences. One is (μ_j) with

$$\left(\mathbf{b} - \sum_{i=1}^{j} \mu_i \mathbf{a}_i, \mathbf{a}_j\right) = 0, \quad j = 1, 2 \ldots,$$

where $\mathbf{a}_i$ are the column vectors of A as defined above. The other sequence is (s_{ik}) defined by

$$s_{ik} = \sum_j \mu_j, \quad j = i + nk, \quad k = 1, 2, \ldots \tag{10.20}$$

In the j-th iteration step, μ_j is determined by the orthogonal (perpendicular) projection of the previous "unexplained" residual component $\mathbf{u}_{j-1}$ onto the dimension $\mathbf{a}_j$. As a result, the coefficients μ_j can be calculated by dot (inner) products. Hence

$$\mu_j = \frac{(\mathbf{u}_{j-1}, \mathbf{a}_j)}{(\mathbf{a}_j, \mathbf{a}_j)} \tag{10.21}$$

where

$$\mathbf{u}_{j-1} = \mathbf{b} - \sum_{i=1}^{j-1} \mu_i \mathbf{a}_i.$$

This latter definition makes the geometric interpretation clear. The norm (length) of μ is usually shrinking, and convergence can be expected. Because the s_{ik} from Equation (10.20) tend to the x_i as $k \to \infty$, each element x_i of the solution vector $\mathbf{x} = (x_1, x_2, \ldots, x_n)$ can (for k sufficiently large) be calculated by

$$x_i = \sum_j \mu_j, \quad j = i + nk, \quad k = 1, 2, \ldots.$$

The necessary condition is that the residual components $\mathbf{u}_j$ tend to zero, and that the condition is always true. Indeed, convergence does not depend on special features of the system matrix A, such as positive definiteness or diagonal dominance. Even for singular systems, a solution can be obtained (For the details, see Schimek, 1996, and the references therein.) As a result, we can cope with ill-conditioned systems (which cannot be ruled out in additive models).

The iterative projection method, however, converges more slowly than backfitting does. This drawback can be compensated for through relaxation (Schimek, 1996, p. 456). For that purpose, a relaxation parameter ω is introduced in Equation (10.21), leading to

$$\mu'_j = \frac{(\omega\mathbf{u}_{j-1}, \mathbf{a}_j)}{(\mathbf{a}_j, \mathbf{a}_j)}$$

where

$$\mathbf{u}_{j-1} = \mathbf{b} - \sum_{i=1}^{j-1} \mu_i \mathbf{a}_i.$$

As a matter of fact, one can show that the relaxed version of the iterative projection method maintains all its desirable characteristics. In addition, there is some theory, as well as evidence from simulations, concerning the choice of the relaxation parameter ω.

Recent numerical experiences for nonsingular and near-singular system matrices are quite promising (Schimek, 2000). In all instances, the computational burden can be significantly reduced by relaxation. Degeneracy requires larger ω values. As pointed out earlier, ill-posed linear equation systems should not be solved with classical techniques, because regularity of the system matrix is required throughout. The relaxed iterative projection algorithm has the potential to bridge this latter gap in numerical methodology, but there is a price to be paid: the explicit calculation of the S_k when the method is applied to the system of normal equations in (10.13).

Some Final Remarks on Iterative Solutions

When an additive model is evaluated via an iterative numerical procedure in the sample world of scatterplot smoothers, some questions still remain open. Obviously, small-sample behaviour is different from asymptotic behavior. Even more problematical is that the convergence behavior is not fully understood even for backfitting, despite its being a well-established technique. Statistical considerations in these iterative schemes become really tedious. It is a short while ago that findings were published for certain smoother types.

Until recently, the only findings for backfitting with arbitrary linear scatterplot smoothers have been limited to the case of two dimensions. Buja et al. (1989, pp. 489ff) and Hastie and Tibshirani (1990, pp. 118ff) showed that both the convergence of the backfitting algorithm and the uniqueness of its solution depend on the behaviour of the product of the two smoother matrices S_1 and S_2, which have to be symmetric and shrinking. We call a smoother matrix S *shrinking* if $\|S\mathbf{y}\| \leq \|\mathbf{y}\|$ for all $\mathbf{y}$ (Euclidean or any other matrix norm). This condition will hold if all its singular values are less than or equal to 1. Among smoothers fulfilling this requirement are projection smoothers (expressed in parametric terms, such as regression splines) and smoothing splines. Opsomer (1999) was able to extend the result above to more than two dimensions, by deriving recursive expressions for the backfitting estimator. For $d > 2$, the existence and the uniqueness of this estimator depend additionally on the characteristics of the pairwise products of the smoother matrices: A d-variate additive model with smoother matrices $S_1, \ldots, S_d$ will converge to a unique solution if, for some matrix norm,

$$\max_{\delta \in [2,d]} \left\| \sum_{j=1}^{\delta-1} S_\delta S_j \right\| < 1, \tag{10.22}$$

where δ denotes the δth block consisting of an $n \times n$ identity matrix and zero elements otherwise in a matrix E_δ, so that $\hat{g}_\delta = E_\delta A^{-1} B y$ (provided that the inverse of A for equation system (10.5) exists. (For more details, see Opsomer, 1999.) Thus, the index δ corresponds to the δth block of the solution vector $\hat{g}$. Assume centered smoother matrices (i.e., $S^* \equiv (I - \mathbf{1}\mathbf{1}^T/n)S$). One can prove that the system matrix A is invertible when the spectral radius of $T_\delta = \| \sum_{j=1}^{\delta-1} S_\delta^* S_j^* \|$ is less than one for all $\delta \in [2, d]$. Because of the fact that the ordering of the predictor functions is arbitrary, it is sufficient to show that the condition (10.22) holds for one particular ordering.

Ansley and Kohn (1994) studied the statistical features of d-dimensional backfitting for the special case of smoothing splines. Opsomer and Ruppert (1997) analyzed backfitting for the bivariate additive model, and Opsomer (1999) did the same for the d-variate additive model applying local polynomial regression. Linton, Mammen, and Nielsen (1997) derived the asymptotic properties of a backfitting projection algorithm under weaker conditions for local linear regression and for a special Nadaraya–Watson estimator in an additive model of dimension d. Even more research is desirable into the convergence behaviour of these and other scatterplot smoothers, especially in the case of the iterative-projection algorithm although some results in Opsomer (1999) also apply to the latter.

Recently there has been considerable effort to avoid iterative techniques at all in the evaluation of additive models. Such developments are introduced in the next section.

10.2.4 The Marginal-Integration Method

Here we discuss a direct method, introduced independently by Newey (1994), Tjøstheim and Auestad (1994), and Linton and Nielsen (1995), labelled *marginal integration*. It is based on the fact that, up to a constant, $g_j(x_j)$ is equal to

$$E\{m(X_1, \ldots, X_{j-1}, x_j, X_{j+1}, \ldots, X_d)\},$$

where $m(x) = E(Y|\mathbf{X} = \mathbf{x})$. The estimate of g_j is obtained by marginal integration of an estimate of m. Therefore no iterations are necessary. The explicit definition of the estimation procedure allows a detailed asymptotic analysis, which is easier to carry out than for the backfitting algorithm.

Linton and Nielsen (1995) studied the additive regression model

$$g(x_1, x_2) = g_1(x_1) + g_2(x_2),$$

where $g(x_1, x_2)$ is a bivariate regression function. Let Q be a deterministic (continuous or discrete) weight function with $dQ(x_2) = 1$. Assume a density q of Q with respect to either a Lebesgue or a counting measure. The contrast

$$\alpha_Q(x_1) = \int g(x_1, x_2)\, dQ(x_2) \tag{10.23}$$

is considered, where $\alpha_Q(x_1) = g_1(x_1) + c_1$ with $c_1 = \int g_2(x_2)\, dQ(x_2)$. Thus $\alpha_Q(x_1)$ is identifiable up to a constant and is the univariate component of the additive structure we are interested in.

Let us have observed data $\{x_{1i}, x_{2i}, y_i\}$ for $i = 1, \ldots, n$. Linton and Nielsen (1995) propose, for the additive fit, a local linear smoother with product kernels. (See Chapter 9 in this book). Thus, $\hat{g}(x_1, x_2) = \sum w_j(x_1, x_2) y_j$ (the w_j denoting some weights) is the first element of

$$(D^T K D)^{-1} D^T K y,$$

where $y = (y_1, \ldots, y_n)^T$ and $D = (d_1, \ldots, d_n)^T$ with $d_j = (1, x_{1j} - x_1, x_{2j} - x_2)^T$, while K is a diagonal $n \times n$ matrix with typical diagonal elements $k_{b_1}(x_{1j} - x_1)k_{b_2}(x_{2j} - x_2)$. The b_1 and b_2 are scalar bandwidths, $k_b(\cdot) = b^{-1}k(b^{-1})$ for any b, and k denotes a univariate differentiable density function symmetric about zero. Then $\alpha_Q(x_1)$ is estimated by the sample version of (10.23), namely,

$$\hat{\alpha}_Q(x_1) = \int \hat{g}(x_1, x_2)\, dQ(x_2) = \sum_{j=1}^{n} w_{Qj}(x_1) y_j,$$

where $w_{Qj}(x_1) = \int w_j(x_1, x_2)\, dQ(x_2)$.

We already know from Stone (1985) that both g_1 and g_2 can be estimated with the one-dimensional convergence rate $n^{2/5}$. Under the usual assumptions (such as independent, identically distributed observations; a joint density $f(x_1, x_2)$; marginals $f_{x_1}(x_1)$ and $f_{x_2}(x_2)$ with marginal cumulative distribution functions $F_{x_1}(x_1)$ and $F_{x_2}(x_2)$), Linton and Nielsen (1995, p. 95) proved the following theorem.

Theorem 4. *Assume that g possesses two continuous partial derivatives in each direction, and that f is continuously differentiable. Assume also that $b_1, b_2 \to 0$ and $nb_1b_2^2 \to \infty$. Then, conditional on $\{x_{1i}, x_{2i}\}_{i=1}^n$,*

$$(nb_1)^{1/2}[\hat{\alpha}_Q(x_1) - E[\hat{\alpha}_Q(x_1)]] \to N(0, s^2(x_1)),$$

in distribution, where $s^2(x_1) = v(k)\sigma^2 \int f^{-1}(x_1, x_2) q^2(x_2)\, dx_2$, with $v(k) = \int k(t)^2\, dt$, while

$$E[\hat{\alpha}_Q(x_1)] - \alpha_Q(x_1) = \{b_1^2\beta_1(x_1) + b_2^2\beta_2(x_1)\}(1 + o(1)),$$

where

$$\beta_1(x_1) = \frac{\mu(k)}{2} \int \frac{\partial^2 g}{\partial x_1^2}(x_1, x_2)\, dQ(x_2), \quad \beta_2(x_1) = \frac{\mu(k)}{2} \int \frac{\partial^2 g}{\partial x_2^2}(x_1, x_2)\, dQ(x_2),$$

with $\mu(k) = \int t^2 k(t)\, dt$. Hence, when $b_1, b_2 = O(n^{-1/5})$, then $\hat{\alpha}_Q(x_1)$ of the marginal-integration method converges at $n^{2/5}$, the optimal rate.

Further, Q has to be chosen. Linton and Nielsen (1995) recommend an empirical distribution function as an approximation of Q, one which is optimal with respect to integrated mean-squared error, although in general there is a dependence on both f and g. If g is exactly additive and $b_2 = o(n^{-1/5})$, the bias does not depend on Q.

A main disadvantage compared to the iterative backfitting procedure is the fact that the marginal-integration method as outlined above does not hold true for dimensions (explanatory variables) $d > 2$. Theorem 4 cannot be extended to cope with higher dimensions. Moreover, the local linear pilot estimator is a non-optimal choice. There have been several (as yet most unpublished) attempts to introduce a marginal integration estimator for $d > 2$. I would like to mention a published paper of Severance-Lossin and Sperlich (1999), who also propose a methodology to estimate the derivatives for additive separable models, which are important for certain applications in economics and biology. They also suggest the application of a local polynomial pilot estimator restricted to the direction of interest, with the effect that more information remains in the constant. This approach leads to a faster algorithm with little loss in optimality.

In the marginal-integration method the scatterplot smoothers are restricted to local polynomial and kernel fits. In the next few paragraphs we sketch the marginal-integration algorithm for local polynomials with kernel weights.

The algorithm is represented in the GAUSS language. The input parameters are as follows: `x`, the $n \times d$ matrix of the explanatory variables; `y`, the response variable; `xg`, the design points that is, the grid $(ng \times d)$ on which we want to estimate the additive components; and the bandwidths h $\in \mathbb{R}^d$ for the dimensions (directions of interest and not of interest). The result is written in `fh`.

Marginal Integration Algorithm

```
fh = zeros(ng,d);
p = 1;     @ degree of local polynomial @
pick = 1~zeros(1,p);     @ pick function estimate @
j = 1;
     do until (j>d);
     i = 1;
     do until (i>ng);     @ for dimension of interest @
          hv = g;
          hv[j,.] = h;
          xest = x;
          xest[.,j] = xg[i,j].*ones(n,1);
          weight = zeros(1,n);
          Z = ones(rows(x),1);
          k = p;
          do while (k);     @ create polynomial design @
               Z = ones(n,1)~(Z*~(x[.,j]-xest[i,j]));
               k = k-1;
          endo;
          l = 1;
          do until (l>n);     @ kernel weights @
               dx = kernel((x-xest[l,.])./(hv'));
               zw = Z'.*dx';
               weight = pick*inv(zw*Z)*zw+weight;
               l = l+1;
          endo;
          fh[i,j] = (weight./n)*y;     @ result @
          i = i+1;
     endo;
     j = j+1;
 endo;
```

Fan et al. (1996) extend the marginal-integration method to the estimation of semiparametric additive partial-linear models (For an overview of semiparametric regression models, see Schimek, 1997.) A problem in practice, certainly, is the necessary $O(n^3)$ algorithm. Sperlich, Tjøstheim, and Yang (1999) propose a concept which allows not only for interaction in additive models but also for the estimation and testing of such interactions.

Finally, let us make some comments about the marginal-integration method as compared to backfitting. For the case of two explanatory variables with smoother matrices S_1 and S_2, respectively, the backfitting procedure converges to the n-dimensional vectors

$$\hat{\mathbf{g}}_1^{\infty} = [I - (I - S_1 S_2)^{-1}(I - S_1)]\mathbf{y}$$

and

$$\hat{\mathbf{g}}_2^{\infty} = [I - (I - S_2 S_1)^{-1}(I - S_2)]\mathbf{y},$$

provided that $\|S_1 S_2\| < 1$. (See Hastie and Tibshirani, 1990, p. 118f.) These expressions are quite intractable for general linear smoothers (although their linearity can be exploited to construct, for example, pointwise confidence bounds), and as a consequence the bias and variance cannot be derived except in special cases. (See the final remarks on iterative solutions.) This deficiency is certainly a drawback compared to what marginal integration offers with its result of Theorem 4.

It is also a well-known fact that the marginal-integration method can be very inefficient with respect to the regression function m. For a demonstration of such inefficiency, let us study the marginal-integration approach in terms of L_2 function space. As in Section 10.2.1, assume $\mathcal{H}_j$, for $j = 1, \ldots, d$, denoting the Hilbert spaces of measurable functions $\phi_j(X_j)$ with $E[\phi_j(X_j)] = 0$, $E[\phi_j^2(X_j)] < \infty$, and inner product $\langle \phi_j(X_j)\phi_j^*(X_j)\rangle = E[\phi_j(X_j)\phi_j^*(X_j)]$. Moreover, let $\mathcal{H}^{\text{add}} = \sum_j^d \mathcal{H}_j$ and let all $\mathcal{H}_j$ be subspaces of $\mathcal{H}$, the space of measurable functions of $X_1, \ldots, X_d$. Recall that solving the normal equations means finding the particular member of $\mathcal{H}^{\text{add}}$ which is closest to the regression function $m \in \mathcal{H}$, corresponding to Equation (10.1).

The empirical marginal-integration map $\pi: \mathcal{H} \to \mathcal{H}^{\text{add}}$ is of the form

$$\pi(g)(x) = \int g(x_1, x_2) f_{x_2}(x_2)\, dx_2 + \int g(x_1, x_2) f_{x_1}(x_1)\, dx_1 - \int g(x) f(x_1, x_2)\, dx,$$

where

$$g(x) = g_0 + \sum_{j=1}^{2} g_j(X_j). \tag{10.24}$$

The additive functions are fixed points of π. For that reason, the marginal-integration method estimates g consistently in Equation (10.24), and so π is idempotent. It is also linear: that is, for any $a, b \in \mathcal{H}$, $\pi(a + b) = \pi(a) + \pi(b)$.

However, π is not self-adjoined, and so it is not an orthogonal projection (Linton, 1997, p. 470). As a result, using $E[m(\mathbf{X})] \in \mathcal{H}^{\text{add}}$, when solving the normal equations, provides a better mean-squared-error approximation to the regression function m. A further result is that the marginal-integration method is not very efficient in estimating g and its components.

To overcome this drawback of the marginal-integration method, Linton (1997) suggested calculating starting values via marginal integration and then applying a single-step backfitting iteration. He could prove the efficiency of this technique, in the sense of its being equivalent to a procedure based on knowing the other components of the regression function. The interpretation of these estimation results is not clear, especially when the additivity assumption is violated. Another yet unpublished proposal to improve efficiency, also from a computational point of view, is due to Hengartner (1996). He tackles the problem of pilot-estimator choice, and he comes out favoring a so-called internalized estimator.

Apart from the fact that there is no agreement about appropriate algorithms for marginal integration, there are only two studies comparing it with the backfitting procedure. One is Nielsen and Linton (1998); the other is Sperlich, Linton, and Härdle (1999). According to Nielsen and Linton (1998), both marginal integration and backfitting can be viewed with respect to minimizing an integrated mean-squared-error criterion. Marginal integration optimizes the criterion with weighting given by an independent product measure. This result is correct independently of whether additivity holds or not. Backfitting achieves the same goal with weighting based on a joint empirical measure (joint density). The latter makes sense without prior knowledge of the situation the data come from. The definite advantage of the marginal-integration estimator is that it is explicitly defined. However, it causes a loss of efficiency for non-independent designs. In conclusion, Nielsen and Linton (1998, p. 221) write, "Perhaps the more significant disadvantage of integration, which is specific to this nonparametric setting, is that the curse of dimensionality is completely eliminated. Thus we must use bias reduction arguments to achieve the optimal rate in high dimensions and we might expect poor small sample performance relative to the asymptotics."

Sperlich, Linton, and Härdle (1999) have undertaken the most extensive simulation study up to now, in which they tried to trace down performance differences between the iterative backfitting procedure and the direct marginal-integration method for small samples and $d = 2$ (in one instance, for $d = 4$). They applied local polynomial fitting and Nadaraya–Watson kernel smoothing to a number of simple additive functions. The error assumptions were Uniform and Normal, with constant variance, and with correlation $\rho = \{0.0, 0.4, 0.8\}$ between the covariates (case $d = 2$). The bandwidth was chosen by a rule of thumb due to Linton and Nielsen (1995) and by the plug-in method of Severance-Lossin and Sperlich (1999).

The results can be summed up as follows: There are many similarities between the backfitting algorithm and the marginal-integration algorithm with respect to their statistical performance. Both algorithms run into severe problems in designs with increasing correlation (problems similar to the ones noted in the remarks on ill-posed systems in 10.2.3.2), although backfitting does perform slightly better. At least for $d = 2$, asymptotics hold empirically. The smoothing method adopted generally does not matter much. Backfitting works better at boundary points and under data sparseness; the integration method is more capable of estimating the components as opposed to the function itself (marginal influences). This is especially true for the higher-dimension case ($d = 4$). The results obtained are not conclusive in the sense of favouring one method overall.

Implementations of the marginal-integration procedure are rare, and more or less of prototype nature. The only one publicly available is the macro `intest` in XploRe 4.0 (for Windows 98 or NT, Linux, and UNIX) which is similar to the algorithm presented above. XploRe can be downloaded at `http://www.xplore-stat.de/`.

Finally, we should mention that choosing the bandwidths and smoothing parameters (see Section 10.5 in this chapter) remains a troublesome issue in the context of additive models, regardless of the method of estimation, and it is likely to hamper the interpretation of comparative simulation studies (controlled for in the experiment described above).

10.3 GENERALIZED ADDITIVE MODELS

In this section, we extend the additive model to the class of generalized additive models (GAMs). GAMs were introduced in a series of papers by Hastie and Tibshirani (1986, 1987a, 1987b) and Stone (1986). They are described in detail in Hastie and Tibshirani (1990).

Their purpose is to allow for even more flexibility than additive models do. On the other hand, they retain an important feature of GLMs, namely, additivity of the predictors. However, the predictor effects are generally nonlinear because the functions g_j are arbitrary.

A special case occurs if only one predictor function, say $g_1(x_1)$, is evaluated nonparametrically, while the remaining explanatory variables still enter as a linear combination, say $\tilde{X}^T\beta = x_2\beta_2 + \cdots + x_d\beta_d$. Such semi-parametric models were first considered by Green and Yandell (1985).

Our generalization of the additive model becomes

$$E[Y|\mathbf{X} = x] = G\left(g_0 + \sum_{j=1}^{d} g_j(x_j)\right),$$

where $G(\cdot)$ is a fixed link function and the distribution of Y is assumed to belong to the exponential family as in GLMs. The assumptions concerning identifiability of the functions g_j remain the same (see Section 10.2).

The fitting of a GAM consists of two parts: estimating the additive predictor, and linking it to the function $G(\cdot)$ in an iterative manner. The first part requires solving the system of normal equations as already discussed. For the second part the so-called *local-scoring* algorithm is applied.

The local-scoring algorithm is practically identical with the Fisher scoring algorithm used in GLMs, except that the least-squares step is replaced by the solution step of the normal equations. In GLMs, the least-squares step is used to update the estimate $\hat{\beta}$ for the linear predictor $X^T\beta$. Here we apply the backfitting or iterative-projection algorithm to update the estimates for g_0 and the g_js.

In Section 10.2.1, smoothing in the additive model was motivated in L_2 function space. In Section 10.2.4 the Hilbert-space interpretation was used to interpret its estimation through marginal integration. This view also helps to understand GAMs (Hastie and Tibshirani, 1990, pp. 148ff).

Given $\mathbf{X} = (X_1, \ldots, X_d)$, the response Y has conditional density $h(y, \zeta)$, where $\zeta = \zeta(\mathbf{X})$ is the true regression parameter fulfilling $\zeta \in \mathcal{H}$. The corresponding log likelihood for a single observation is denoted by l. For $\zeta(\mathbf{X})$, we try to obtain the best approximation by maximizing the expected log likelihood

$$E[l(\eta(\mathbf{X}), Y)] \tag{10.25}$$

over $\eta(\mathbf{X}) = \sum_j^d g_j(X_j) \in \mathcal{H}^{\text{add}}$. Stone (1986) gave conditions for the existence and uniqueness of the best additive approximation. The maximum of Equation (10.25) is characterized by a score function $\partial l/\partial\eta$ orthogonal to the space of fits, or equivalently by

$$E\left[\frac{\partial l}{\partial \eta}\bigg| X_j\right] = 0$$

for all j. A solution for these nonlinear equations in η and g_j can be found by a linearization about an approximate η_0. The final result is

$$g_j(X_j) = \frac{E\left[W_0(\mathbf{X})\left\{Z_0 - \sum_{k\neq j} g_k(X_k)\right\}\bigg| X_j\right]}{E[W_0(\mathbf{X}|X_j]}, \tag{10.26}$$

where

$$Z_0 = \eta_0 + (\partial l)/(\partial\eta_0)/(-\partial^2 l)/(\partial\eta_0^2)$$

and

$$W_0(\mathbf{X}) = (-\partial^2 l)/(\partial\eta_0^2).$$

The two operators E in Equation (10.26) are weighted conditional expectations that can be evaluated by scatterplot smoothers.

10.3.1 Penalized Least Squares

Let us consider again the penalized-least-squares concept introduced in Section 10.2.2, which results in cubic smoothing splines. The estimation of the additive model (representing regression with a continuous, usually Gaussian response) was based on Equation (10.10). Now, for the generalized case (representing regression with a noncontinuous response variable), we can apply the penalized log-likelihood criterion (Hastie and Tibshirani, 1986; Fahrmeir and Tutz, 1994)

$$PL(g_1, \ldots, g_d) = \sum_{i=1}^{n} l_i(y_i; \eta_i) - \frac{1}{2} \sum_{j=1}^{d} \lambda_j \mathbf{g}_j^T K_j \mathbf{g}_j,$$

where l denotes the log likelihood (as in GLMs), η_i the additive predictor values

$$\eta_i = \sum_{j=1}^{d} g_j(x_{ij}),$$

and $\mathbf{g}_j = (g_j(x_{1j}), \ldots, g_j(x_{nj}))^T$, $\mathbf{g}_j \in \mathcal{H}_j$, $j = 1, \ldots, d$, the vector of smooth spline functions. The penalty matrices K_j for each predictor X_j are defined as in Equation (10.14) for the additive model.

The Fisher scoring algorithm was originally designed to maximize a (penalized) log likelihood criterion (See McCullagh and Nelder, 1989, for details.) Here we study the Fisher scoring algorithm first, and then later we modify it to what is now known as local-scoring procedure in nonparametric regression.

10.3.1.1 Fisher Scoring and Local Scoring

Let η be the n-dimensional vector of additive predictor values η_i. Differentiation of $\partial PL/\partial \mathbf{g}_j$ for $j = 1, \ldots, d$, yields the likelihood equations

$$s_1 = \lambda_1 K_1 \mathbf{g}_1, \ldots, s_d = \lambda_d K_d \mathbf{g}_d,$$

where the derivative $s = (s_1, \ldots, s_n)$ of the log likelihood is given by

$$s_i = \frac{D_i}{\sigma_i^2}(y_i - \mu_i),$$

with $D_i = \partial G/\partial \eta_i$ being the first derivative of the response function and σ_i^2 being the variance function evaluated at $\mu_i = G(\eta_i)$.

Let $W = \text{diag}(w_1, \ldots, w_n)$, and let $w_i = D_i^2/\sigma_i^2$ be the expected information matrix. Then the Fisher scoring iterations are given by

$$\begin{pmatrix} W^{(k)} + \lambda_1 K_1 & W^{(k)} & \cdots & W^{(k)} \\ W^{(k)} & W^{(k)} + \lambda_2 K_2 & \cdots & W^{(k)} \\ \vdots & \vdots & \ddots & \vdots \\ W^{(k)} & W^{(k)} & \cdots & W^{(k)} + \lambda_d K_d \end{pmatrix} \begin{pmatrix} \mathbf{g}_1^{(k+1)} - \mathbf{g}_1^{(k)} \\ \mathbf{g}_2^{(k+1)} - \mathbf{g}_2^{(k)} \\ \vdots \\ \mathbf{g}_d^{(k+1)} - \mathbf{g}_d^{(k)} \end{pmatrix}$$

$$= \begin{pmatrix} \mathbf{s}^{(k)} - \lambda_1 K_1 \mathbf{g}_1^{(k)} \\ \mathbf{s}^{(k)} - \lambda_2 K_2 \mathbf{g}_2^{(k)} \\ \vdots \\ \mathbf{s}^{(k)} - \lambda_p K_d \mathbf{g}_d^{(k)} \end{pmatrix},$$

where $W^{(k)}$ and $\mathbf{s}^{(k)}$ are W and s evaluated at $\eta^{(k)} = \eta(\mathbf{g}_1^{(k)}, \ldots, \mathbf{g}_d^{(k)})$. By defining the working observation vector

$$\tilde{\mathbf{y}}^{(k)} = \eta^{(k)} + (W^{(k)})^{-1}\mathbf{s}^{(k)}$$

and the smoother matrices

$$S_j^{(k)} = (W^{(k)} + \lambda_j K_j)^{-1} W^{(k)},$$

the iterations can be expressed in the form

$$\begin{pmatrix} I & S_1^{(k)} & \cdots & S_1^{(k)} \\ S_2^{(k)} & I & \cdots & S_2^{(k)} \\ \vdots & \vdots & \ddots & \vdots \\ S_d^{(k)} & S_d^{(k)} & \cdots & I \end{pmatrix} \begin{pmatrix} \mathbf{g}_1^{(k+1)} \\ \mathbf{g}_2^{(k+1)} \\ \vdots \\ \mathbf{g}_d^{(k+1)} \end{pmatrix} = \begin{pmatrix} S_1^{(k)}\tilde{\mathbf{y}}^{(k)} \\ S_2^{(k)}\tilde{\mathbf{y}}^{(k)} \\ \vdots \\ S_d^{(k)}\tilde{\mathbf{y}}^{(k)} \end{pmatrix}. \tag{10.27}$$

This is an $(n \times d)$-dimensional linear system comparable to the normal equations in (10.5). For the same reasons as pointed out there, a direct solution is not feasible, as a method of obtaining the next iterate $\mathbf{g}^{(k+1)} = (\mathbf{g}_1^{(k+1)}, \ldots, \mathbf{g}_d^{(k+1)})$ apart from special instances. Instead, (10.27) is rewritten as

$$\begin{pmatrix} \mathbf{g}_1^{(k+1)} \\ \mathbf{g}_2^{(k+1)} \\ \vdots \\ \mathbf{g}_d^{(k+1)} \end{pmatrix} = \begin{pmatrix} S_1\big(\tilde{y}^{(k)} - \sum_{j \neq 1} \mathbf{g}_j^{(k+1)}\big) \\ S_2\big(\tilde{y}^{(k)} - \sum_{j \neq 2} \mathbf{g}_j^{(k+1)}\big) \\ \vdots \\ S_d\big(\tilde{y}^{(k)} - \sum_{j \neq d} \mathbf{g}_j^{(k+1)}\big) \end{pmatrix}$$

and solved iteratively by the backfitting or iterative-projection algorithm in the inner loop of the local-scoring algorithm. As mentioned earlier, solving the linear system in the core of Fisher scoring iteratively produces a new algorithm, known as local scoring. It can be characterized as follows:

Generalized Additive Model Algorithm

1. *Initialization of outer loop:* $\mathbf{g}_0^{(0)} = F\left(n^{-1}\sum_{i=1}^{n} y_i\right)$, $\mathbf{g}_1^{(0)} = \ldots = \mathbf{g}_d^{(0)} = 0$, $F = G^{-1}$
2. *Scoring steps:* For $k = 0, 1, 2, \ldots$:
   Compute the current working observations

$$\tilde{\mathbf{y}}_i^{(k)} = \eta_i^{(k)} + \frac{\mathbf{y}_i - G(\eta^{(k)})}{D_i^{(k)}}$$

and the weights $w_i^{(k)} = (D_i^{(k)}/\sigma_i^{(k)})^2$ with $\eta_i^{(k)} = \mathbf{g}_0^{(k)} + \sum_{j=1}^{d} \mathbf{g}_j^{(k)}(x_{ij})$, $i = 1, \ldots, n$.

3. *Backfitting* or *iterative projection steps:* Solve (10.27) for $\mathbf{g}^{(k+1)}$:
   (a) Initialization of inner loop:

$$\mathbf{g}_0^{(k)} = n^{-1} \sum_{i=1}^{n} \tilde{\mathbf{y}}_i^{(k)}, \mathbf{g}_j^0 := \mathbf{g}_j^{(k)}, \quad j, p, c = 1, \ldots, d;$$

   (b) Compute updates $\mathbf{g}_j^0 \rightarrow \mathbf{g}_j^1$, $j = 1, \ldots, d$ in each backfitting iteration (index p for "previous" and c for "current")

$$\mathbf{g}_j = S_j^1 \left[\tilde{\mathbf{y}} - \mathbf{g}_0 - \sum_{c<j} \mathbf{g}_c^1 - \sum_{p>j} \mathbf{g}_p^0 \right]$$

   and apply weighted cubic spline fits to the "working residuals" $\tilde{\mathbf{y}} - \mathbf{g}_0 - \sum_{c<j} \mathbf{g}_c^1 - \sum_{p>j} \mathbf{g}_p^0$.
   (Set $\mathbf{g}_j^0 := \mathbf{g}_j^1$ for $j = 1, \ldots, d$ after each iteration.)

   (c) Stop backfitting or iterative projections when $\|\mathbf{g}_j^0 - \mathbf{g}_j^1\|$ for $j = 1, \ldots, d$ is very small.
   Set $\mathbf{g}_j^{(k+1)} := \mathbf{g}_j^1$ after the final iterate.

4. *Iterate steps 2 and 3 until some termination criterion is met.*

As a termination criterion, one could, for instance, use

$$\frac{\sum_{j=1}^{d} \|\mathbf{g}_j^{(k+1)} - \mathbf{g}_j^{(k)}\|}{\sum_{j=1}^{d} \|\mathbf{g}_j^{(k)}\|} \leq \varepsilon$$

for some very small value of ε.

Finally, we should point out again that the constant g_0 was introduced for the purpose of identifiability of the additive approximation. Hence, the local-scoring algorithm is applied to centered working observations, that is, $\tilde{\mathbf{y}}$ is centered to have mean zero and thus ensure that, throughout the iterations, the $\mathbf{g}_j s$ have mean zero.

As far as convergence is concerned, the local-scoring algorithm does not require special attention. Local-scoring iterations correspond to Newton–Raphson steps. (Only when the weighting schema is different these algorithms apply.) Convergence results and methods of step optimization, usually not needed for the estimation of GAMs, are considered in Ortega and Rheinboldt (1970). Crucial in the fitting of GAMs is the inner loop providing the solution of the normal equations, not the outer loop carrying out the local-scoring iterations.

For very large data sets (so-called data-mining tasks) Hegland et al. (1999) proposed a parallel solver which is generally applicable to GAMs.

In this section we emphasized cubic smoothing splines because of their numerical practicability and software availability. Alternatively, any other symmetric linear smoother could be applied. Such smoothers can also be represented by matrices $S_1, \ldots, S_d$ of shapes similar to those of smoothing splines. The only difference is that one has to define the penalty matrices $K_1, \ldots, K_d$ in the penalized log-likelihood criterion by

$$K_j = S_j^- - I,$$

where S_j^- is any generalized inverse of S_j (see Golub and van Loan, 1989, for generalized inverses). Weighted cubic spline smoothing in the inner loop of the algorithm is then substituted by a corresponding symmetric linear smoother, for instance, a running line smoother.

Further, we have to remark that the local-scoring algorithm is not limited to the use of backfitting or iterative projections. Any iterative procedure which solves a linear system would serve this purpose.

Next, we illustrate the fitting of a GAM in the statistical environment S-Plus for UNIX, Linux, and Windows 98 or NT platforms. Alternative software is XploRe (Härdle et al., 1999) for Windows, Linux and UNIX computers. Furthermore, there is a function `addreg` in Douglas Nychka's FUNFITS software, available at `http://www.cgd.ucar.edu/stats/Funfits/index.shtml` for Windows and UNIX platforms with an S-Plus environment. (S/S-Plus functions are required to execute FUNFITS.) The most widely used implementation is actually that of S-Plus. It is based on backfitting in combination with local scoring as described above (see Schimek, 2000). Very useful for GAM fitting in S-Plus are Chambers and Hastie (1992, especially Chapter 7), Venables and Ripley (1999, Chapter 10), and the manual (MathSoft, 1999, Chapter 8).

We will analyze the Irish Educational Transitions Data as described in Raftery and Hout (1985) and distributed freely by StatLib over the Internet. Educational transitions were observed from a sample of 500 Irish school children who in 1967 were aged eleven. Here we use 468 observations (missing-value cases excluded). The binary response variable `Certif` (leaving certificate) has the value 0 if "not taken" and 1 if "taken". In a logistic regression analysis, selected covariates are `Sex` (1 is "male", 2 is "female"), the Drumcondra Verbal Reasoning test score `Score`, and a prestige score for father's occupation, namely `Father`. The aim is prediction of the outcome variable `Certif`. We have reason to assume nonlinear relationships between the variables in

$$P[Y = 1 | \mathbf{X} = \mathbf{x}] = G\left(x_1 + g_0 + \sum_{j=2}^{3} g_j(x_j)\right),$$

where G is the logistic function; x_1 denotes `Sex`, a factor with two values and so not part of the nonparametric fit (the function `gam` in S-Plus allows for this additional feature making this model semiparametric in a strict sense); x_2 denotes

Score; and x_3 denotes Father. The latter two metric explanatory variables can be evaluated nonparametrically in the fashion outlined in this section.

Before we fit a model, we convert the variable Sex into a factor by the assignment

```
Sex <- factor(Sex).
```

Otherwise S-Plus would interpret its values as metric and not as levels (here two levels denoted "0" and "1"). The additive logistic model can now be fitted by the calculation

```
irish.fit1 <- gam(Certif~Sex + s(Score, df=5) + s(Father),
family=binomial).
```

In the GAM formula, s denotes the use of *cubic smoothing splines* as the scatter-plot smoother. The degree of smoothing can be specified only by the (equivalent) *degrees of freedom* df. If not specified, as for the last covariate Father, nonlinear degrees of freedom take the value of three (default suggested by experience). As discussed later in this chapter, smoothing parameter choice remains a weak point in GAMs. The value 5 for Score was found by trial and error in a sequence of fits for the same data. As can be seen, the function gam is used in the same way as any model function (e.g., glm) in S-Plus.

The results of the fit can be obtained by use of the function

```
summary(irish.fit1),
```

and we get

```
Call: gam(formula = Certif~Sex+s(Score, 5)+s(Father),
family = binomial)
Deviance Residuals:
      Min         1Q     Median         3Q       Max
-2.266302 -0.8910529 -0.4777382 0.8948091 2.201811
(Dispersion Parameter for Binomial family taken to be 1)
    Null Deviance: 644.2571 on 467 degrees of freedom
Residual Deviance: 504.7334 on 457.0659 degrees of freedom
Number of Local Scoring Iterations: 4
DF for Terms and Chi-squares for Nonparametric Effects
           Df Npar Df Npar Chisq    P(Chi)
(Intercept) 1
        Sex 1
s(Score, 5) 1       4   6.456659 0.1635463
  s(Father) 1       3   7.594247 0.0543460
```

The S-Plus function gam has converted the five degrees of freedom for the cubic spline smoothing of Score to four nonlinear degrees of freedom (see the above listing of results). Unlike GLMs because of the nonparametric nature of GAMs no "classical" parameter estimates are obtained, thus no exact degrees of freedom.

(Actually three and four are not very much apart in terms of smoothness.) The p-values of the approximate-χ^2-test statistic indicate some importance of the two nonparametrically fitted covariates, although only `Father` is near the usual significance boundary. It should be mentioned that the function `s` activates a cubic smoothing-spline built into a fast backfitting respectively local scoring algorithm, having a relatively small number of knots (controlled by `df`). This algorithm avoids the problem of handling tied observations, but in our own experience, it does not always allow representation of the data in a satisfactory manner.

The results are shown in Figure 10.1. There we see the contributions to the predictor function of the factor `Sex` and the metric variables `Score` and `Father`.

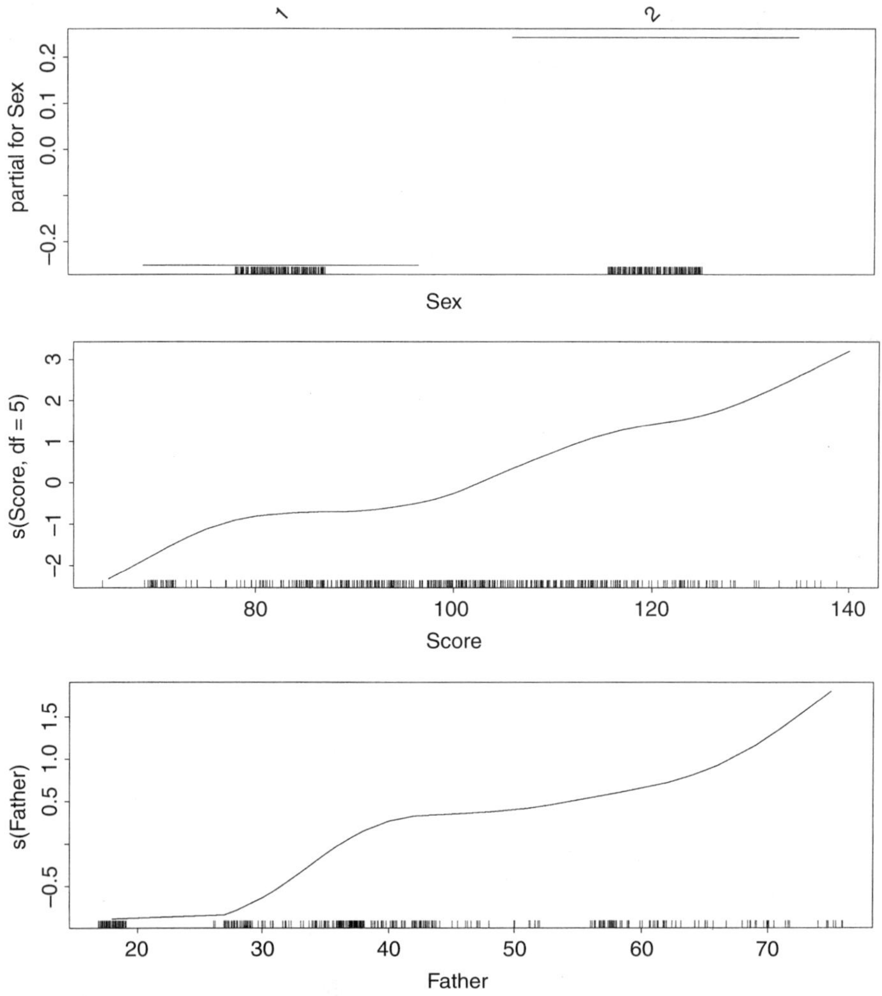

FIGURE 10.1 GAM fitted to Educational Transitions Data.

The functions of the latter two variables are clearly nonlinear. It seems that `Score` has a plateau at around 90 and `Father` has one at around 45. This is indicative of certain requirements (in terms of verbal test score and father's social status) corresponding to specific levels of achievement.

The contribution of `Sex` is questionable. (The values on the y-axis are with ± 0.25 small compared to the range of values of the metric variables.) Hence, we consider instead the model

```
irish.fit2 <- gam(Certif ~ s(Score, df = 5) + s(Father),
family = binomial)
```

The results obtained by `summary(irish.fit2)` are now

```
Call: gam(formula = Certif ~ s(Score, df = 5) + s(Father),
family = binomial)
Deviance Residuals:
     Min          1Q      Median         3Q       Max
-2.289604 -0.9013231 -0.5317411 0.887507 2.313568
(Dispersion Parameter for Binomial family taken to be 1)
    Null Deviance: 644.2571 on 467 degrees of freedom
Residual Deviance: 509.5588 on 458.063 degrees of freedom
Number of Local Scoring Iterations: 4
DF for Terms and Chi-squares for Nonparametric Effects
                Df Npar Df Npar Chisq   P(Chi)
  (Intercept)    1
s(Score,df = 5)  1       4   5.700269 0.2180685
    s(Father)    1       3   7.818090 0.0491733
```

We realize that eliminating the factor `Sex` hardly affects the fit. The difference in deviance between `irish.fit1` and `irish.fit2` is 4.8253 with degrees of freedom approximately one. The approximate χ^2-statistic (see Bowman and Azzalini, 1997, p. 163ff) is not significant. The covariate `Father` now contributes significantly, and (at 5 per cent level) the covariate `Score` has lost predictive power. The fitted functions do not change much (graphical results omitted). Following the concept of model parsimonity, we could reasonably eliminate all covariates but `Father`. However, when `Score` is eliminated too the predictive power of the latter model is considerably reduced. (The difference in deviance takes on a significant value.)

We continue by analyzing a subsample of the Irish Educational Transitions Data. There is a covariate describing the educational level attained. It takes the integer values 1 to 11. Here we are interested only in the vocational schools, where 2 denotes "junior cycle incomplete" versus 4, "terminal leaver." We construct a binary dependent variable `Educat` and fit the model

```
irish.fit3 <- gam(Educat=="4" ~ Sex + s(Scorejv) + s(Father),
family = binomial)
```

where `Scorejv` denotes the Drumcondra Verbal Reasoning test score of junior cycle vocational-school pupils.

The factor level 4 represents "success" (terminal leaver) versus "no success." It turns out that the (previously explanatory) variables `Sex` and `Father` are almost irrelevant. Finally we fit

```
irish.fit4 <- gam(Educat=="4" ~ s(Scorejv), family = binomial).
```

The model yields for its single covariate a χ^2-test statistic of 6.244 and a p-value of 0.094. The null deviance is 151.2648 on 110 degrees of freedom and the residual deviance is 143.5494 on 106.1001 degrees of freedom.

`Score` is the only predictor variable used in the fit to the whole sample (`irish.fit2`) and in the fit to the subsample of junior-cycle vocational-school pupils (`irish.fit4`). In Figure 10.2, the contribution of this variable to the predictor function in the latter fit is displayed. Comparative boxplot of the observations of `Score` in the overall sample and in the subsample are shown in Figure 10.3. The functional form of `Score` (compared with that in Figure 10.1) and its predictive power have changed. In the subsample, `Score` is with no doubt in nonlinear relationship with achievement. The boxplot makes it clear: Median and interquartile distance are smaller for the vocational-school pupils. In terms of the logistic analysis carried out by Raftery and Hout (1985), our results indicate that gender has no effect, social class has an overall effect, and verbal ability has an educational-level-specific effect on success in the Irish educational system.

In concluding this section, we would like to mention that the additive fitting concept can also be applied to the estimation of autoregressive time series (an application beyond the scope of this chapter). Härdle and Chen (1995, p. 379ff) discuss the so-called *nonlinear additive autoregressive models* (NAAR).

After these illustrations of spline smoothing in backfitting with local scoring, our emphasis in the next section will be on kernel smoothing. There we will discuss recent developments in the marginal-integration method which make it possible to apply it to GAMs.

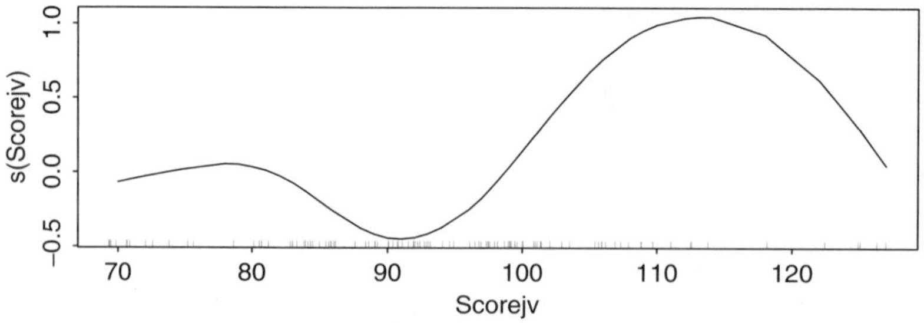

FIGURE 10.2 GAM fitted to subsample of Educational Transitions Data.

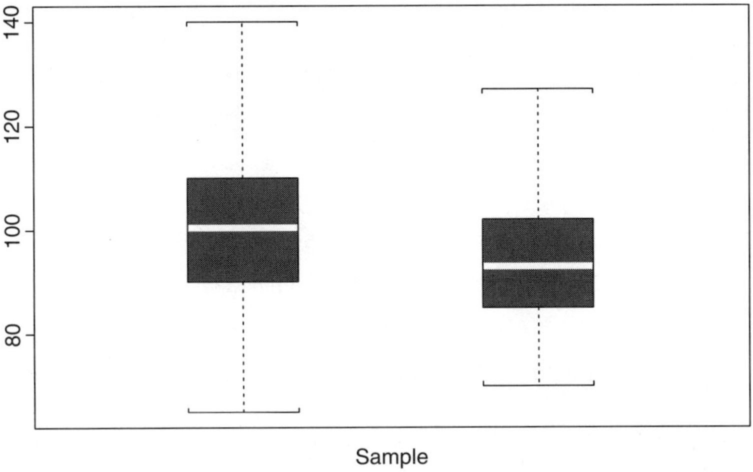

FIGURE 10.3 Comparative boxplot of the Score observations in sample (left) and subsample (right) of the Educational Transitions Data.

10.3.2 The Marginal-Integration Method for Generalized Additive Models

In Section 10.2.4, we described the marginal-integration method for the evaluation of additive models, due to Linton and Nielsen (1995). This direct method (no iterations) is based on the fact that up to a constant, $g_j(x_j)$ is equal to

$$E[\eta(X_1, \ldots, X_{j-1}, x_j, X_{j+1}, \ldots, X_d)],$$

where $\eta(x) = E[Y|\mathbf{X} = \mathbf{x}]$. The estimate of g_j is obtained by marginal integration of an estimate of η. Linton and Härdle (1996) extended this idea to the estimation of

$$G\{\eta(x)\} = g_0 + \sum_{j=1}^{d} g_j(x_j), \tag{10.28}$$

where G is a known link function. They differentiate between a full and a partial specification of the model. Here we are interested in the full specification, which represents generalized additive models with errors following an distribution in the exponential family. This means that the variance is functionally related to the mean. In a partial specification of the model, the variance function is unrestricted. When G is the identity function, we have exactly the situation examined in Section 10.2.4.

Let us assume a partition $X = (X_1, X_2)$, where X_1 is the one-dimensional direction of interest and X_2 a $(d-1)$-dimensional nuisance direction ($d \geq 2$ predictors). Further let $x = (x_1, x_2)$. For any predictor function η and link function G, a functional

$$\gamma_1(x_1) = \int G\{\eta(x_1, x_2)\} p_2(x_2)\, dx_2 \tag{10.29}$$

is defined, where $p_2(x_2)$ is the joint density of x_2. Because of the additive structure of (10.28), γ_1 is g_1 up to a constant g_0. Linton and Härdle (1996) proposed replacing both η and p_2 in (10.29) by estimates. For that purpose they use the Nadaraya–Watson kernel estimator

$$\hat{\eta}(x_1, x_2) := \frac{n^{-1}\sum_{i=1}^{n} K_h(x_1 - X_{1i})L_l(x_2 - X_{2i})Y_i}{n^{-1}\sum_{i=1}^{n} K_h(x_1 - X_{1i})L_l(x_2 - X_{2i})},$$

where K and L are compactly supported Lipschitz-continuous kernels integrating to one (bandwidth h and l respectively). We have $K_h(\cdot) = h^{-1}K(h^{-1}\cdot)$ and $L_l(\cdot) = l^{-(d-1)}L(l^{-1}\cdot)$, and we take K to be a second-order kernel and L to be product of univariate kernels of order q, (i.e., $\int L(u)u^k\, du = 0$ for $k = 1, \ldots, q-1$). For high dimensions d, it is necessary to reduce the bias in the nuisance directions to achieve the optimal one-dimensional rate of convergence for the direction of interest.

The functional $\gamma_1(x_1)$ can be estimated by the sample version of (10.29), that is,

$$\tilde{\gamma}_1(x_1) := \sum_{i=1}^{n} G\{\eta(x_1, X_{2i})\}.$$

Apart from the case where G is the identity function, $\tilde{\gamma}_1(x_1)$ is a nonlinear function of y_i. For the estimation of the regression surface, the above procedure is applied to each direction by redefining in turn the jth covariate to be X_1 and the remainder to be X_2. Estimates are obtained for each γ_j at all the design points. Finally, $\eta(x)$ is reestimated by

$$\tilde{\eta}(x) := F\left\{\sum_{j=1}^{d} \tilde{g}_j(x_j) + \tilde{g}_0\right\},$$

where $F = G^{-1}$,

$$\tilde{g}_0 = d^{-1}n^{-1}\sum_{j=1}^{d}\sum_{i=1}^{n} \tilde{\gamma}_j(X_{ji}),$$

and $\tilde{g}_1(x_1) = \tilde{\gamma}_1(x_1) - \tilde{g}_0$.

In practice, one often has a problem identifying relevant covariates with respect to a response variable. Härdle and Korostelev (1996) address the problem of searching for significant variables.

By contrast with the estimation concept for GAMs worked out in the previous section, the marginal-integration method allows for asymptotic considerations more easily. Linton and Härdle (1996, p. 534) provide the following theorem.

Theorem 5. *Let the order q of L satisfy $q > d - 1$. Let $h = \beta n^{-1/5}$ (β some constant). Assume that $n^{2/5}l^q \to \infty$, that $n^{2/5}l^{d-1} \to \infty$, that F is twice continuously differentiable, and that the additivity assumption holds. Then*

$$n^{2/5}\{\tilde{\eta}(x) - \eta(x)\} \to N\{b(x), v(x)\}$$

in distribution, where

$$b(x) = F'\left\{\sum_{j=1}^{d} g_j(x_j) + g_0\right\}\sum_{j=1}^{d} b_j(x_j)$$

and

$$v(x) = F'\left\{\sum_{j=1}^{d} g_j(x_j) + g_0\right\}\sum_{j=1}^{d} v_j(x_j).$$

The main consequence is that the rate of convergence of $\tilde{\eta}$ is not influenced by the curse of dimensionality. The obtained rate of $n^{2/5}$ is still that derived by Stone (1986) for one-dimensional regression functions. This is a remarkable result.

Despite asymptotic optimality, one has to be cautious about using the algorithm, especially for dimensions $d > 2$. In practice, covariates are more or less correlated, data sparse, and sample sizes small. Extensive simulations for different sample sizes and for different combinations of regression and link functions are necessary before any conclusions can be drawn. Another problem is the lack of software other than the macro `gintest` in XploRe 4, available at `http://www.xplore-stat.de/` based on a prototype algorithm.

In conclusion, the marginal-integration method for generalized additive models is certainly interesting from a theoretical point of view, but it is too new for us to comment on its value for real data analysis.

10.3.3 Vector Generalized Additive Models

Yee and Wild (1996) extended the class of GAMs to handle multivariate (vector) regression problems. Vector generalized additive models (VGAMs) enhance the idea of vector GLMs (not explicitly dealt with in McCullagh and Nelder, 1989).

Suppose that for each measurement unit under study, a q-dimensional response vector $\mathbf{y} = (y_1, \ldots, y_q)^T$ and a d-dimensional covariate vector $\mathbf{x} = (x_1, \ldots, x_d)^T$ are observed. A VGAM is any model for which the conditional distribution of $\mathbf{y}$ given $\mathbf{x}$ is of the form

$$f(\mathbf{y}|\mathbf{x}) = f^*(\mathbf{y}, \eta_1(\mathbf{x}), \ldots, \eta_M(\mathbf{x}))$$

for some function f^*, where (for $j = 1, \ldots, M$)

$$\eta_j(\mathbf{x}) = g_{(j)0} + g_{(j)1}(x_1) + \cdots + g_{(j)d}(x_d).$$

For the evaluation of a VGAM, simultaneous smoothing is required. This condition can be achieved by an interesting generalization (due to Fessler, 1991) of cubic smoothing splines. His vector-spline technique has some nice features, especially for numerical analysis. As a matter of fact, it can be seen as a generalization of univariate Reinsch splines, which computationally are highly efficient. It should be pointed out, however, that the underlying system matrix of such a VGAM is very large. Another feature of vector splines is that they allow for correlated errors ϵ_i, as long as their covariance matrix Σ_i is known and the errors are independent between samples.

Let us suppose a vector response $\mathbf{y}_i$ of dimension M at each value of a scalar x_i, assumed to be a realization from the vector-measurement model in $\mathbb{R}^M$

$$\mathbf{y}_i = \mathbf{g}(x_i) + \epsilon_i$$

for $i = 1, \ldots, n$ with error assumptions $E(\epsilon_i) = 0$ and $E(\epsilon_i \epsilon_i^T) = \delta_{ij} \Sigma_i$, where the Σ_i are known symmetric and positive definite error covariances. The smooth vector function $\mathbf{g}(x) = (g_1(x), \ldots, g_M(x))^T$ can be estimated by minimizing a penalized-least-squares criterion

$$\sum_{i=1}^{n} (\mathbf{y}_i - \mathbf{g}(x_i))^T \Sigma_i^{-1} (\mathbf{y}_i - \mathbf{g}(x_i)) + \sum_{j=1}^{M} \lambda_j \int_a^b [g_j''(t)]^2 \, dt. \tag{10.30}$$

The roughness term penalizes for lack of smoothness in the component functions. Each component demands independent smoothing, not necessarily of the same degree (measured on arbitrary scales). Hence M fixed smoothing parameters $\lambda_j > 0$ are required. We refer to Fessler (1991, p. 855) for the choice of the λ_j in practice, which is nontrivial.

In the special case of a covariance matrix $\mathbf{\Sigma} = I$, the minimization problem of Equation (10.30) is reduced to the unweighted version of a univariate (scalar) spline evaluation.

The solution $\hat{\mathbf{g}}$ can be obtained in a way similar to that for univariate cubic smoothing splines (see Section 10.2.2). For ordered design values $x_1 < x_2 < \cdots < x_n$ and (block) vectors $\mathbf{y} = (\mathbf{y}_1^T, \ldots, \mathbf{y}_n^T)^T$,

$$\mathbf{g} = (f_1(x_1), \ldots, f_M(x_1), \ldots, f_1(x_n), \ldots, f_M(x_n))^T,$$

and $\mathbf{\Sigma} = \mathrm{diag}(\Sigma_1, \ldots, \Sigma_n)$, the penalized-least-squares criterion in (10.30) is equivalent to

$$(\mathbf{y} - \mathbf{g})^T \mathbf{\Sigma}^{-1} (\mathbf{y} - \mathbf{g}) + \mathbf{g}^T K \mathbf{g}$$

for a matrix K of the same structure (not depending on $\mathbf{y}$) as in Section 10.2.2 Actually, there is one difference: here the smoothing parameters λ are part of K. The quadratic in $\mathbf{g}$ is minimized when

$$\hat{\mathbf{g}} = S(\lambda)\mathbf{y},$$

with the smoother matrix being

$$S(\lambda) = (I + \mathbf{\Sigma}K)^{-1}.$$

As a result, the vector spline is a linear smoother. In combination with VGAMs, all the findings for additive models, due to Buja et al. (1989), hold.

The estimation procedure for VGAMs can be developed in the way that was demonstrated in the section about Fisher and local scoring for GAMs. The starting point is again a penalized likelihood criterion. Let us consider solely the kth covariate. Then the minimization problem is

$$\sum_{i=1}^{n}(\mathbf{z}_i - \mathbf{g}_k(x_{ik}))^T W_i(\mathbf{z}_i - \mathbf{g}_k(x_{ik})) + \sum_{j=1}^{M} \lambda_{(j)k} \int_a^b [g''_{(j)k}(t)]^2\, dt, \qquad (10.31)$$

where the M-dimensional vector $\mathbf{z}_i$ represents the adjusted dependent variable. Subject to smoothing (Yee and Wild, 1996, pp. 483ff, give a detailed study for the identity link); $\mathbf{g}_k(x_{ik}) = (g_{(1)k}(x_{ik}), \ldots, g_{(M)k}(x_{ik}))^T$; and each $M \times M$ matrix W_i has elements

$$(w_i)_{jk} = -(\partial^2 l_i)/(\partial\eta_j \partial\eta_k),$$

where l_i is the likelihood of the ith measurement unit. The solution of (10.31) is a vector spline with $\Sigma_i^{-1} = W_i$. For the regression problem with d predictors, we have to apply the backfitting or the iterative-projection algorithm to $\mathbf{z}_i$ with vector spline smoothing. The results obtained are for the vector additive model

$$E(\mathbf{y}_i) = \mathbf{g}_0 + \sum_{j=1}^{d} \mathbf{g}_j(x_{ij}),$$

which is fitted to the vector response $\mathbf{y}$. The backfitting algorithm adapted for vector smoothing is given in Yee and Wild (1996, p. 485). The underlying large linear (block matrix) equation system would in fact be better solved by the relaxed iterative projection algorithm (Schimek, 1996). VGAMs are an application where ill-posed estimation problems, depending primarily on the features of the covariance structure, are likely. The slow convergence of backfitting iterations in such applications is indicative of these problems.

"Generalized" in VGAMs is not used in the sense of directly evaluating arbitrary link functions G (within the exponential-family distribution framework). As a result, VGAM, do not need to develop an alternative to Fisher scoring. (See the previous section on GAMs for comparison.) A generalization, apart from the important concept of vector variables, can be seen in the fact that Yee and Wild (1996) introduced constraints with respect to the covariates. They allow not only for different sets of predictors to be used for each predictor function but also for constraints on how they act. (For details, see Yee and Wild, 1996, p. 486ff.) The example following discusses the proportional-odds model as a typical application. Its parametric version was first introduced by McCullagh (1980). An underlying response variable that is continuous is assumed, but only categorized observations can be observed.

Several types of VGAM models have been implemented in the program VGAM, an S-Plus/ANSI-C implementation for UNIX platforms written by Thomas Yee. The software is available at `http://www.stat.auckland.ac.nz/yee`. The following example illustrates not only the proportional-odds model but also some of the features of the software. The function `vgam()` is very similar to `gam()` of S-Plus.

The data, consisting of a sample of 7882 New Zealand workers, are described in Yee and Wild (1996, p. 489). The ordinal response variable "alcohol consumption" has three levels: $Y = 1$ (less than once a month, on average); $Y = 2$ (once a month to once a week, on average); and $Y = 3$ (more than once a week, on average). There is only one explanatory variable, "age of worker."

In the following, `y` is a 49×3 matrix of counts and `Age` is a vector of length 49 with elements corresponding to the ages 17 to 65. The resulting nonparametric proportional-odds model

$$\operatorname{logit} P(Y \geq j|\text{Age}) = \beta_{(j-1)0} + f(\text{Age}), \quad j = 2, 3,$$

can be fitted by

```
fit1 <- vgam(y~s(Age), cumulative (reverse = T, parallel = T))
```

The function `s()` denotes a vector spline and defaults to four degrees of freedom for each component function. This accommodates a reasonable amount of nonlinearity. Then `plot(fit, se=T)`, where `se` is standard error, will result in Figure 10.4. A proportional-odds model constrains the component functions to be parallel—this is the so-called *proportional-odds assumption*. Thus the two component functions in Figure 10.4 are identical (because they are centered). Specifying `se=T` in `plot()` provides approximate ± 2 standard-error curves about the fitted functions.

In contrast to the above model, a nonparametric non proportional-odds model

$$\operatorname{logit} P(Y \geq j|\text{Age}) = \beta_{(j-1)0} + f_{(j-1)1}(\text{Age}), \quad j = 2, 3,$$

can be fitted by

```
fit2 <- vgam(y~s(Age), cumulative (reverse = T, parallel = F)).
```

The evaluation of such a model allows us to check the proportional-odds assumption. Plotting `fit2` (results shown in Figure 10.5) shows clearly that the proportional-odds assumption is very much in doubt.

In contrast to the two nonparametric models above, one can easily fit a linear model. For example

```
fit3 <- vglm (y~Age, cumulative (reverse = T, parallel = F))
```

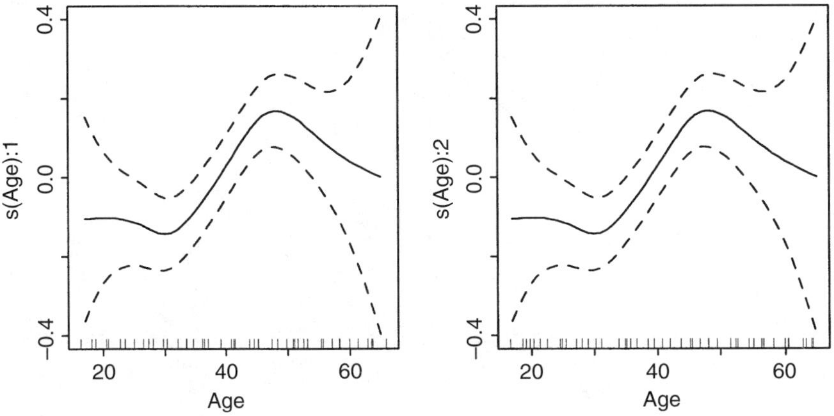

FIGURE 10.4 Nonparametric proportional-odds model fitted to a New Zealand Workforce Data.

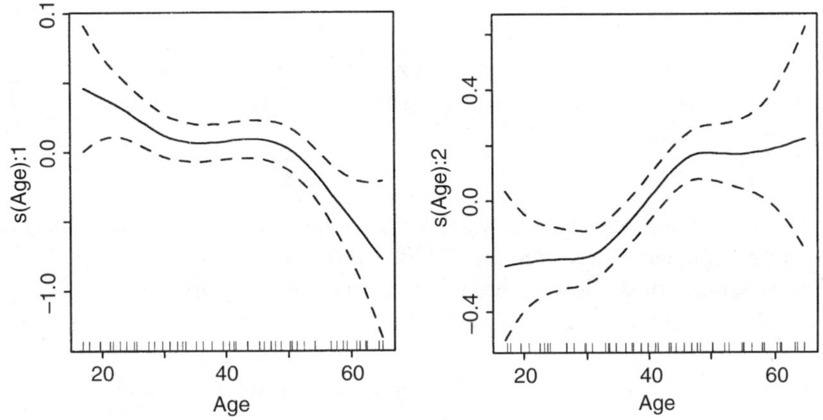

FIGURE 10.5 Nonparametric nonproportional-odds model fitted to a New Zealand Workforce Data.

fits the ordinary (i.e., parametric) proportional-odds model

$$\operatorname{logit} P(Y \geq j|\text{Age}) = \beta_{(j-1)0} + \beta_{(1)1} \times \text{Age}, \quad j = 2, 3.$$

This model is plotted in Figure 10.6.

Most of the features of the object-oriented S functions described in Chambers and Hastie (1992) are available for VGAM objects. For example,

```
> fit2
Call:
vgam(formula = y~s(Age), family = cumulative (reverse = T,
parallel = F))
Degrees of Freedom: 98 total; 88.00039 Residual
```

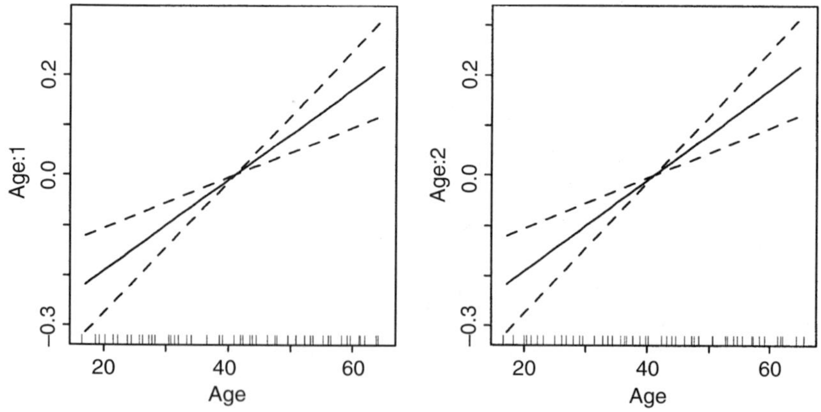

FIGURE 10.6 Proportional odds model fitted to a New Zealand Workforce Data.

```
Residual Deviance: 97.3698
> coef(fit3, matrix=T)
                              (Intercept)          Age
logit(PLY>=2])                  1.7577425  0.009060889
logit(PLY>=3])                 -0.4016867  0.009060889
> deviance(fit3)
[1] 187.5361
```

The latter function, `deviance.vglm()`, enables approximate likelihood-ratio tests to be conducted. (See Yee and Wild, 1996, for details.)

The program VGAM allows users to specify constraint matrices for the purpose of modeling features such as parallelism assumptions, exchangeability, and no-effects of certain variables on certain component functions. In the above, `parallel=T` was a convenient method of setting up the constraint matrices:

```
> fit$constraints
$ (Intercept)
     [,1] [,2]
[1,]    1    0
[2,]    0    1
$ Age
     [,1]
[1,]    1
[2,]    1
```

That is, there are no constraints on the intercepts (the constraint matrix is the identity matrix), but the two component functions of `Age` are constrained to be equal. At this point we mention that Hastie and Tibshirani (1987b) proposed an algorithm for the proportional-odds model; however, they smoothed a (constrained) proportional-odds model, whereas Yee and Wild (1996) applied constrained smoothing to the (unconstrained) non proportional-odds model.

Other important models that can be fitted with the VGAM software are the vector additive model (including seemingly unrelated regression), bivariate logistic models, bivariate probit models, and multinomial logit models.

10.4 ALTERNATING CONDITIONAL EXPECTATIONS, ADDITIVITY, AND VARIANCE STABILIZATION

There are two other methods that we wish to discuss in this chapter, the *alternating conditional-expectation* algorithm (ACE; Breiman and Friedman, 1985) and the *additivity-and-variance-stabilization* algorithm (AVAS; Tibshirani, 1988). Although these methods are based on a different working model than are (generalized) additive models, the estimation process is very similar. Specifically, the working model for these two algorithms is

$$\theta(Y) = \alpha + \sum_{j=1}^{d} \psi_j(X_j) + e, \tag{10.32}$$

where e has mean zero and is independent of the X_js, and $\theta(\cdot)$ and $\psi_j(\cdot)$, $j = 1, \ldots, d$, are unknown (smooth) functions with $E[\psi_j(X_j)] = 0$ for identifiability reasons.

Note that now we assume that the conditional expectation of a transformation of the Y-variable is given by an additive model of the X-variables. By contrast, in a GAM (per Section 10.3), we assume that a transformation of the conditional expectation of Y is given by an additive approximation of the X-variables. Hence, models of the form (10.32) are also know as *transformation models* or *transform-both-sides models* (TBSs) (See, among others, Carroll and Ruppert, 1988; Nychka and Ruppert, 1995.)

10.4.1 Alternating Conditional Expectations

We assumed in (10.32) that $E[\psi_j(X_j)] = 0$ for $j = 1, \ldots, d$. From this assumption, it follows that $E[\theta(Y)] = \alpha$; hence, without loss of generality, we can incorporate the constant α into the function $\theta(\cdot)$ and assume also that $E[\theta(Y)] = 0$. Our working model is now

$$\theta(Y) = \sum_{j=1}^{d} \psi_j(X_j) + e.$$

The ACE algorithm estimates $\theta(\cdot)$ and $\psi_j(\cdot)$, $j = 1, \ldots, d$ by minimizing

$$E\left[\theta(Y) - \sum_{j=1}^{d} \psi_j(X_j)\right]^2.$$

Obviously, this criterion is minimized trivially by choosing $\theta(\cdot) \equiv \psi_j(\cdot) \equiv 0$. Hence, a further normalization is needed, for example, $\mathrm{Var}[\theta(Y)] = 1$. Note that if (10.32) holds, we have, similarly to the additive model in (10.2),

$$E\left[\theta(Y) - \sum_{j \neq k}^{d} \psi_j(X_j) | X_k\right] = \psi_k(X_k)$$

and

$$E\left[\sum_{j=1}^{d} \psi_j(X_j) | Y\right] = \theta(Y).$$

This suggests estimating $\theta(\cdot)$ and $\psi_j(\cdot)$, $j = 1, \ldots, d$, by calculating these conditional expectations instead. Using a univariate scatterplot smoother to approximate these conditional expectations thus leads to the basic ACE algorithm.

ACE Algorithm

1. *Initialize*: Set $\hat{\theta}(\cdot) = (\cdot - \overline{Y})/\hat{\sigma}_Y$ where $\overline{Y} = n^{-1} \sum Y_i$ and $\hat{\sigma}_Y = n^{-1} \sum (Y_i - \overline{Y})^2$.
2. *Find new transformations of Xs*: Fit an additive model to $\hat{\theta}(\cdot)$ to obtain new estimates $\hat{\psi}_j(\cdot)$, $j = 1, \ldots, d$. (See Section 10.2.3.2.)
3. *Find new transformation of Y*: Obtain a new estimate $\hat{\theta}(\cdot)$ by smoothing $\sum_j \hat{\psi}(X_j)$ against Y, and then standardize it in such a way that $\sum_i \hat{\theta}(Y_i) = 0$ and $\sum_i \hat{\theta}(Y_i)^2 = 1$.
4. *Alternate between step 2 and 3 until convergence is reached.*

It can be shown (Breiman and Friedman, 1985) that the solutions of the ACE algorithm are closely related to the solutions for minimizing the correlation between $\theta(Y)$ and $\sum_j \psi_j(X_j)$ under the condition that $\mathrm{Var}[\theta(Y)] = \mathrm{Var}[\sum_j \psi_j(X_j)]$. This relationship implies that ACE is more suitable as a correlation tool than as a regression tool. Indeed, if viewed as a regression tool, ACE has several disturbing features. (See the discussion of Breiman's and Friedman's article, or see Hastie and Tibshirani, 1990, p. 184ff.) One of these features is that even if (10.32) holds, the ACE algorithm may not reconstruct the functions $\theta(\cdot)$ and $\psi(\cdot)$, because the optimal transformations (that minimize the above criterion) depend on the joint distribution of the Xs and e. In the next section we shall describe a modification of ACE that tries to overcome these anomalies.

On the other hand, the original FORTRAN77 implementation of the ACE algorithm of Breiman and Friedman (available from the archive `lib.stat.cmu.edu`) allows the user to specify that some of the transformations should have certain features, such as being monotone or linear. (The same is true for the `ace` function in S-Plus.) Indeed, by restricting $\theta(\cdot)$ to be linear, the ACE algorithm can be used as an effective exploratory tool. (See Raftery and Richardon, 1996, and the references therein.)

10.4.2 Additivity and Variance Stabilization

Tibshirani (1988) proposed a modification of the ACE algorithm to make it more suitable as a regression tool. Essentially, he proposed that instead of calculating conditional expectations to identify the transformations $\theta(\cdot)$ and $\psi_j(\cdot)$, $j = 1, \ldots, d$, a variance-stabilizing transformation should be used to estimate $\theta(\cdot)$. Specifically, we try find a transformation such that

$$E[\theta(Y)|X_1, \ldots, X_d] = \sum_{j=1}^{d} \psi_j(X_j)$$

and

$$\mathrm{Var}\left[\theta(Y)\middle|\sum_{j=1}^{d} \psi_j(X_j)\right] = c,$$

where c is an arbitrary constant. Here, we assume additionally that $\theta(\cdot)$ is strictly monotone — in fact, we may, without loss of generality, assume that $\theta(\cdot)$ is strictly increasing.

Now, if a random variable Z has mean μ and variance $\mathrm{Var}(\mu)$ then the variance-stabilizing transformation for Z is given by

$$h(t) = \int_0^t \frac{1}{\sqrt{\mathrm{Var}(u)}}\, du.$$

(See Serfling, 1980, p. 120f; Sen and Singer, 1993, p. 139f.) That is, $h(Z)$ has approximately constant variance. The proposed algorithm is now as follows:

AVAS Algorithm

1. *Initialize*: Set $\hat{\theta}(\cdot) = (\cdot - \overline{Y})/\hat{\sigma}_Y$ where $\overline{Y} = n^{-1}\sum Y_i$ and $\hat{\sigma}_Y = n^{-1}\sum(Y_i - \overline{Y})^2$.
2. *Find new transformations of Xs*: Fit an additive model to $\hat{\theta}(\cdot)$ to obtain new functions $\hat{\psi}_j(\cdot)$, $j = 1, \ldots, d$. (See Section 10.2.3.2.)
3. *Find the new transformation of Y*: Set $m(\mathbf{X}) = \sum_{j=1}^{d} \psi_j(X_j)$, and compute the variance function $V(u) = \mathrm{Var}[\theta(Y)|m(\mathbf{X}) = u]$. Then calculate the variance-stabilizing transformation

$$h(t) = \int_0^t \frac{1}{\sqrt{V(u)}}\, du,$$

and define the new $\hat{\theta}(\cdot)$ as the transformation $h(\hat{\theta}(\cdot))$ of the old estimate for $\theta(\cdot)$. Finally, renormalize $\hat{\theta}(\cdot)$ such that $E[\theta(Y)] = 0$ and $\text{Var}[\theta(Y)] = 1$.

4. *Alternate steps 2 and 3 until convergence is reached.*

When implementing this algorithm, the user must calculate the variance-stabilizing transformation by means of numerical quadrature and $V(u)$ by means of an appropriate smoothing operation. FORTRAN77 and RATFOR code implementing this algorithm was submitted by Robert Tibshirani to the archive at `lib.stat.cmu.edu` and is available from there. S-Plus, too, has an implementation of this algorithm called `avas`.

Banks et al. (1995) compare several methods for (high-dimensional) nonparametric regression, including ACE and AVAS. Their results seem to indicate that both methods perform similarly and that ACE is often slightly better in the mean integrated-squared-error-sense. This conclusion is somewhat surprising, because AVAS was designed to be a regression tool, whereas ACE is more a correlation tool. Although some theoretical results regarding AVAS are available, several questions remain unsolved, for instance, its global convergence has not yet been established.

To illustrate these two algorithms in S-Plus, we use the Tree Data set from the `sm` library (`http://www.stats.gla.ac.uk/~adrian/sm`) of Bowman and Azzalini (1997). The data record for a sample of black cherry trees, the diameter of the tree trunk (inches), the height of the tree (feet), and the volume of usable timber obtained from each tree (cubic feet). The three corresponding variables are `Diameter`, `Height` and `Volume`. As a first approximation, let us assume a cylindrical shape for the trees. We might expect that the volume of usable timber is proportional to the product of the height of the tree and its squared diameter:

$$\log(\texttt{Volume}) \sim \alpha_0 + \alpha_1 \log(\texttt{Diameter}) + \alpha_2 \log(\texttt{Height}). \qquad (10.33)$$

Rather than fitting this parametric model, let us first use ACE and AVAS to explore these data. Assume that we have read the data into a matrix, say `dat`, such that the first column of the matrix holds the measurements on the diameter of the tree trunk, the second column the height of the tree, and the third the volume of usable timber.

As motivated by (10.33), let us first take the logarithm of each variable. To be able to plot the transformation given by ACE and AVAS later, we also have to know how to sort the data with respect to each regressor variable. All this can be done using the following commands (In `x` the variable `x1` and `x2` are combined.):

```
> y <- log(dat[,3])
> x1 <- log(dat[,1])
> x2 <- log(dat[,2])
> x <- cbind(x1,x2)
> o1 <- order(x1)
> o2 <- order(x2)
```

We may now use ACE and AVAS to fit (10.32) to the data. The degree of smoothness is automatically chosen in S-Plus. To display the result, we use a display with two rows and three columns of panels. In the first row, we shall display the ACE-transformations of log(`Diameter`) and log(`Height`) as solid lines in the first and second panel, respectively. The third panel holds the transformation of log(`Volume`). The second row will hold the corresponding pictures using the transformations given by AVAS. This display is achieved by the following commands, and the result (except for changes to the axes; `type` and `pch` denote line type and plotting symbol) is shown in Figure 10.7.

```
> par(mfrow=c(2,3))
> ace.fit <- ace(x,y)
> plot(x1[o1], ace.fit$tx[o1,1], type="l")
> rug(x1)
> plot(x2[o2], ace.fit$tx[o2,2], type="l")
> rug(x2)
> plot(y, ace.fit$ty, pch=1)
> avas.fit <- avas(x,y)
> plot(x1[o1], avas.fit$tx[o1,1], type="l")
> rug(x1)
> plot(x2[o2], avas.fit$tx[o2,2], type="l")
> rug(x2)
> plot(y, avas.fit$ty, pch=1)
```

Figure 10.7 shows that the ACE algorithm confirms (10.33). AVAS, on the other hand, suggests that log(`Diameter`) and log(`Height`) should have a piecewise linear influence on a transformation of log(`Volume`). Note that the transformation suggested for log(`Volume`) by AVAS does not seem to be linear. The implementation of AVAS allows us to specify that the transformation of a variable should be linear or monotone. For example, to ensure that the transformation of log(`Volume`) is linear, we may use the command

```
> avas.fit2 <- avas(x,y,linear=0).
```

The result is not shown here. The resulting transformation for log(`Diameter`) is much the same as the transformation found by ACE (i.e., nearly linear), and the transformation for log(`Height`) does not change much. Hence, both algorithms suggest that (10.33) holds if we replace α_2 by

$$\alpha_2 + \alpha_3 I(\log(\texttt{Height}) > 4.38),$$

where $I(\cdot)$ is an indicator function. The first AVAS fit suggests that the slope of the transformation for log(`Diameter`) also changes at log(`Diameter`)$\approx$2.45, but this shift is less clear and is not supported by either the ACE fit or the second AVAS fit. However, we may start with such an (exhaustive) parametric model, and then use standard techniques for comparing nested parametric models, to find the model that fits the data best.

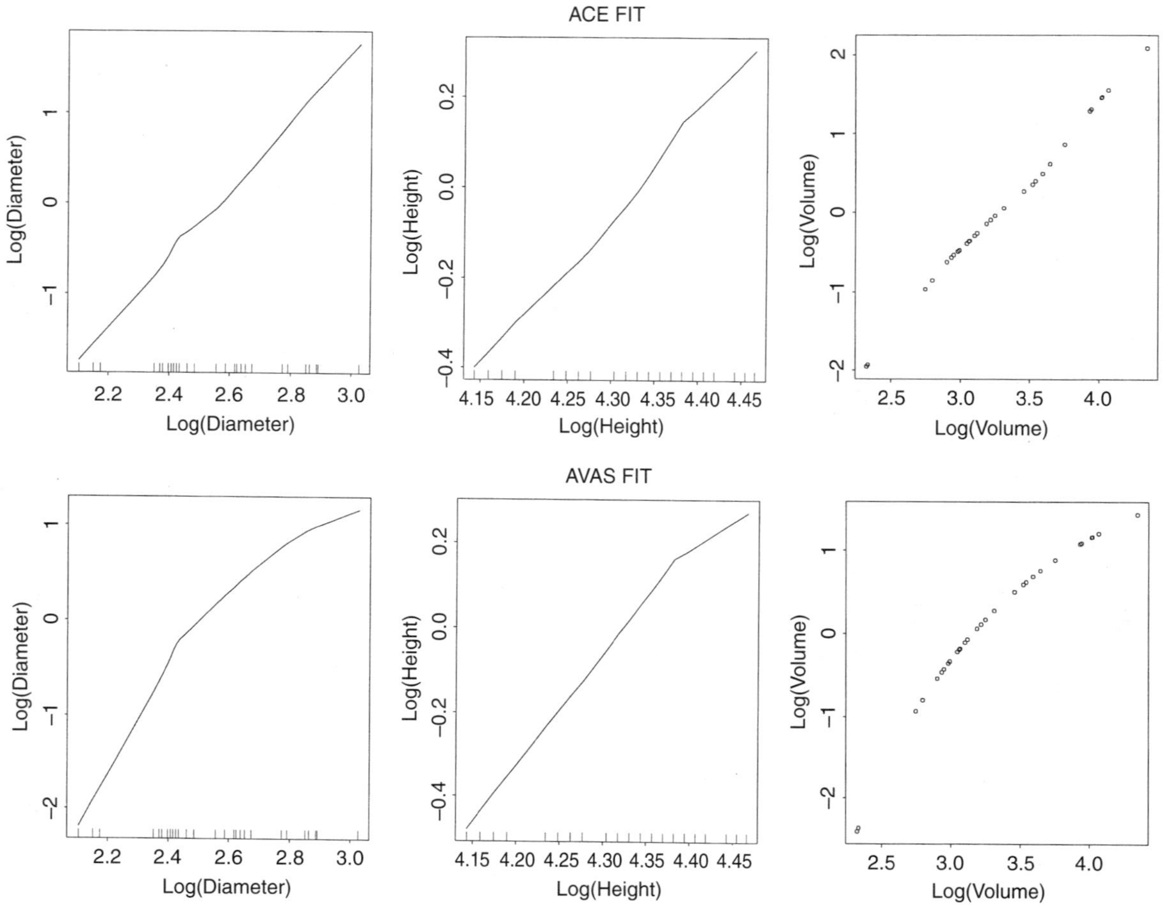

FIGURE 10.7 ACE and AVAS fitted to the Tree Data.

10.5 SMOOTHING PARAMETER AND BANDWIDTH DETERMINATION

All the estimation methods for additive models and GAMs discussed so far work only when smoothing parameter or bandwidth values are chosen for each dimension beforehand. Although univariate in nature, the choice of the degree of smoothing remains a multivariate problem. As a result of that, the univariate techniques (such as generalized cross-validation, introduced in Chapter 2 and in Chapter 4) cannot be applied for this purpose. A data-driven choice of the degree of smoothing (i.e., for $\hat{\lambda} = (\hat{\lambda}_1, \ldots, \hat{\lambda}_d))$ by generalized cross-validation is possible in principle. For additive models the criterion is

$$GCV_{\text{add}}(\lambda) = \frac{1}{n} \sum_{i=1}^{n} \left\{ \frac{(y_i - \hat{\mu}_i)/\hat{\sigma}_i}{1 - \text{tr}(R_\lambda)/n} \right\}^2,$$

where R_λ is the hat matrix that generates the additive predictor $\hat{\mathbf{g}} = R_\lambda \mathbf{y}$ in the last iteration step (at convergence). R_λ can be interpreted as a weighted additive-fit operator. However, optimization of the criterion would require efficient computation of $\text{tr}(R_\lambda)$ in each step of a multidimensional search algorithm. Further, it would be very difficult to find a global minimum in this multidimensional setting. For the generalized case, the problem gets even more complicated, because the link function also plays a role in the search for the optimal degree of smoothing. Taking into account popular link functions, Burman (1990) made a proposal which works for B-splines with equispaced knots. His proposal addresses the question of choosing the correct number of knots when using cross-validation as the criterion. For a pure additive spline model, there is an $O(n^3)$ algorithm available (Gu and Wahba, 1988). Using the marginal-integration method instead of iterative procedures does not circumvent this messy and computationally expensive situation. The standard choice is simple plug-in techniques for bandwidth selection (e.g., as in Severance-Lossin and Sperlich, 1999). There is also a proposal for a plug-in technique for the backfitting procedure when local linear fitting is applied (Opsomer and Ruppert, 1998). All these proposals are limited, because there is theoretical justification neither for the case in which the predictors are correlated nor for the case of dimensions $d > 2$.

We know from univariate function fitting that the number of degrees of freedom is simply $df = \text{tr}(S)$, where S is a linear smoother matrix. As a matter of fact, df amounts to be the sum of the eigenvalues of S and indicates the extent of smoothing. Moreover, $\text{tr}(S)$ is easy to compute because only the diagonal elements of S are required. For certain linear smoothers, there are $O(n)$ algorithms (e.g., Hutchinson and de Hoog, 1985, for smoothing splines).

Just as in univariate function fitting, the overall number of degrees of freedom is $df = \text{tr}(R_\lambda)$ in the additive model (Hastie and Tibshirani, 1990, p. 128f). The problem is that $\text{tr}(R_\lambda)$ is not exactly the sum of the individual R_{λ_j} for $j = 1, \ldots, d$. Buja et al. (1989, p. 496f.) provided some evidence that, under the assumptions of not-too-small smoothing parameters λ_j and not-too-heavily correlated predictors x_j, adding up the individual degrees of freedom gives an upper bound for the overall degree of freedom of R_λ. The degrees of freedom obtained in this way are sometimes called the equivalent degrees of freedom (Hastie, 1992, p. 251). Their prespecification is frequently used to determine the amount of smoothing in an additive model or GAM. Apart from the fact that a prior guess can be wrong, it is often implicitly assumed that the necessary amount of smoothing in each dimension is the same, but this is not true in many instances.

A different approach is feasible. It is based on an improved (correcting for the tendency to undersmooth) Akaike's information criterion and is due to Hurvich et al. (1998). Their proposal is made for univariate smoothing problems, but the idea could be extended to multivariate situations (e.g., for the additive model). Its advantage is that, unlike plug-in techniques, it imposes less limitation on the type of linear smoothers.

In summary, we cannot help saying that smoothing-parameter or bandwidth choice remains a weak point in every kind of additive model discussed so far. It seems that additional research is required for the data-driven selection of the necessary amount of smoothing.

10.6 MODEL DIAGNOSTICS

We recall, at convergence of an additive fit, we can express $\hat{\mathbf{g}}_j$ as $R_{\lambda_j}\mathbf{y}$ for some matrix R_{λ_j} of dimension $n \times n$, because the $\hat{\mathbf{g}}_j$ results from a linear mapping applied to $\mathbf{y}$. Suppose that errors are independent and identically distributed; then the covariance of the estimator is

$$\text{Covar}(\hat{\mathbf{g}}_j) = R_{\lambda_j} R_{\lambda_j}^T \sigma^2,$$

where $\sigma^2 = \text{Var}(Y_i)$. If the system matrix A of the normal equations $A\mathbf{g} = B\mathbf{y}$ has singular values close to zero (i.e., highly correlated covariates), then $\text{Covar}(\hat{\mathbf{g}}_j)$ will be large.

The direct computation of R_{λ_j} is very expensive. Hastie and Tibshirani (1990, pp. 127ff) propose an $O(kn^2)$ backfitting procedure, where $k = dMc$ (with d the number of covariates, M the number of backfitting iterations, and c a linear-smoother-specific constant).

Hastie and Tibshirani (1990, p. 128 and pp. 156ff) also discuss the construction of confidence bands for the estimator. Today it is believed that resampling techniques are the better choice. (See Chapter 14 in this book.)

As in GLMs, the analysis of deviance plays an important role for inference in GAMs. The deviance (or likelihood statistic) for a fitted model represented by $\hat{\eta}$ is defined by

$$D(y; \hat{\eta}) = 2\{l(\eta_{\max}; y) - l(\hat{\eta}; y)\},$$

where $\eta_{\max}$ is the parameter value that maximizes the log likelihood $l(\eta; y)$ over all η (the saturated model). The deviance replaces the residual sum of squares $RSS = \mathbf{y}^T(I - R_\lambda)^T(I - R_\lambda)\mathbf{y}$, which is used in the simple additive model.

For GLMs, there is an asymptotic distribution theory. Consider two linear models, η_1 and η_2, with η_1 nested within η_2. If η_1 is correct, and if some regularity conditions are fulfilled, then $D(\eta_2; \eta_1) = D(y; \eta_1) - D(y; \eta_2)$ has an asymptotic χ^2-distribution with degrees of freedom equal to the difference in the dimensions of the models. The result is usually summarized in an analysis-of-deviance table. For GAMs the deviance remains a sensible means of model assessment. The problem is that it is not even asymptotically χ^2-distributed, although empirical evidence due to Hastie and Tibshirani (1990, pp. 155ff) supports the use of the χ^2-distribution.

Finally, we consider the degrees of freedom for error, df_{err}. In the additive model df_{err} is derived in terms of the expected value of the residual sum of squares RSS and is defined by

$$df_{\text{err}} = n - \text{tr}(2R_\lambda - R_\lambda R_\lambda^T).$$

The quantity in the GAM that is analogous to RSS in the additive model is the deviance D. Starting from an asymptotic approximation to the deviance,

$$D(y; \eta) \approx (\tilde{\mathbf{y}} - \hat{\eta})^T W (\tilde{\mathbf{y}} - \hat{\eta}),$$

we end up with the degrees of freedom for error,

$$df_{\text{err}} = n - \text{tr}(2R_\lambda - R_\lambda^T W R_\lambda W^{-1}),$$

for the GAM obtained from local scoring as described in Section 10.3.1.1. For the use of df_{err}, see Hastie and Tibshirani (1990, pp. 157ff) and Hastie (1992, pp. 302ff).

Similar arguments apply to VGAMs, although with additional complications (see Yee and Wild, 1994, for details).

In conclusion, one must say that since the publication of Hastie's and Tibshirani's 1990 monograph, there have not been substantial new developments in the field of model critique for additive models or GAMs. (However, for testing in nonparametric models, see also Bowman and Azzalini, 1997.) The same must be said about model selection. Current research emphasizes the application of resampling techniques.

10.7 NEW DEVELOPMENTS

There are two most recent developments in the field of GAMs we would like to mention finally. One addesses the estimation of additive models via Markov chain Monte Carlo techniques. This is truly a change of paradigm away from solving the normal equation system (or equivalent concepts). Wong and Kohn (1996) have presented a promising Bayesian estimation technique that can at the same time handle the tedious problem of smoothing parameter selection for the special case of regression splines. This Bayesian approach also facilitates the estimation of diagnostic quantities.

The other new development is based on penalized likelihood. Marx and Eilers (1998) have put forward an estimation concept based on penalized B-splines with equally spaced knots. Unlike the discussion in this chapter, GAMs are evaluated in the GLM framework. The advantage is to get rid of the backfitting or relaxed iterative projections step because smoothing is carried out simultaneously. The performance of this approach in comparison with the classical techniques described in this chapter is studied in Schimek (2000).

Acknowledgements

We thank Thomas Yee for technical help with the VGAM example and Stefan Sperlich for providing the marginal-integration algorithm.

REFERENCES

Ansley, C. F., and Kohn, R. (1994). Convergence of the backfitting algorithm for additive models, *J. Australian Math. Soc., A*, **57**, 316–329.

Banks, D., Maxion, R., and Olszewski, R. (1995). Comparing methods for nonparametric regression, *ASA Proceedings of the Statistical Computing Section*, 136–141.

Bellman, R. E. (1961). *Adaptive Control Processes*. Princeton University Press, Princeton, NJ.

Bowman, A. W., and Azzalini, A. (1997). *Applied Smoothing Techniques for Data Analysis: The Kernel Approach with S-Plus Illustrations*. Oxford University Press, Oxford, UK.

Breiman, L., and Friedman, J. H. (1985). Estimating optimal transformations for multiple regression and correlation, *J. Amer. Statist. Assoc.*, **80**, 580–619.

Buja, A. (1990). Remarks on functional canonical variates, alternating least squares methods, and ACE, *Ann. Statist.*, **18**, 1032–1069.

Buja, A. (1996). What criterion for a power algorithm? in Rieder, H. (ed.), *Robust Statistics, Data Analysis, and Computer Intensive Methods*, pp. 49–61, Springer-Verlag, New York.

Buja, A., Hastie, T. J., and Tibshirani, R. J. (1989), Linear smoothers and additive models, *Ann. Statist.*, **17**, 453–510.

Burman, B. (1990). Estimation of generalized additive models, *J. Multivariate Analysis*, **32**, 230–255.

Carroll, R. J., and Ruppert, D. (1988). *Transformation and Weighting in Regression*, Chapman and Hall, London, UK.

Chambers, J. M., and Hastie, T. J., eds. (1992). *Statistical Models in S*, Wadsworth & Brooks, Pacific Grove, CA.

Eubank, R. L. (1984). The hat matrix for smoothing splines, *Statist. and Prob. Lett.*, **2**, 9–14.

Fahrmeir, L., and Tutz, G. (1994). *Multivariate Statistical Modeling Based on Generalized Linear Models*, Springer-Verlag, New York.

Fan, J., Härdle, W., and Mammen, E. (1996). Direct estimation of additive and linear components for high dimensional data, *Technical Report 96.1, Department of Statistics*, Chinese University of Hong Kong, PRC.

Fessler, J. A. (1991). Nonparametric fixed-interval smoothing with vector splines, *IEEE Trans. Signal Proces.*, **39**, 852–859.

Friedman, J. H., and Stuetzle, W. (1981). Projection pursuit regression, *J. Amer. Statist. Assoc.*, **76**, 817–823.

Gander, W., and Golub, G. H. (1989). Discussion of "Buja, A., Hastie, T., and Tibshirani, R.: Linear smoothers and additive models," *Ann. Statist.*, **17**, 529–532.

Golub, G. H., and van Loan, C. F. (1989). *Matrix Computations*, John Hopkins University Press, Baltimore, MD.

Green, P., and Silverman, B. W. (1994). *Nonparametric Regression and Generalized Linear Models: A Roughness Penalty Approach*, Chapman and Hall, London.

Green, P., and Yandell, B. (1985). Semi-parametric generalized linear models, in Gilchrist, R., Francis, B. J., and Whittaker, J. (eds.), *Generalized Linear Models*, pp. 44–55, Springer-Verlag, Berlin, Germany.

Gu, C., and Wahba, G. (1988). *Minimizing GCV/GML Scores with Multiple Smoothing Parameters via the Newton Method*, Technical Report 847, University of Wisconsin, Madison, WI.

Hämmerlin, G., and Hoffmann, K.-H. (1995). *Numerical Mathematics*, Springer-Verlag, New York.

Härdle, W., and Chen, R. (1995). Nonparametric time series analysis, a selective review with examples, *Bulletin of the International Statistical Institute, Proceedings of the 50th Session, Book 1*, 375–394.

Härdle, W., Klinke, S., and Müller, M. (1999). *XploRe—Academic Edition. The Interactive Statistical Computing Environment* (CD-ROM with Handbook). Springer-Verlag, Berlin, Germany.

Härdle, W., and Korostelev, A. (1996). Search for significant variables in nonparametric additive regression, *Biometrika*, **83**, 541–549.

Hastie, T. J. (1992). Generalized additive models, in Chambers, J. M., and Hastie, T. J. (eds.), *Statistical Models in S*, pp. 249–307, Wadsworth & Brooks, Pacific Grove, CA.

Hastie, T. J., and Tibshirani, R. J. (1986). Generalized additive models. *Statist. Scien.*, **1**, 297–318.

Hastie, T. J., and Tibshirani, R. J. (1987a). Generalized additive models: Some applications, *J. Amer. Statist. Assoc.*, **82**, 371–386.

Hastie, T. J., and Tibshirani, R. J. (1987b). Non-parametric logistic and proportional odds regression, *Appl. Statist.*, **36**, 260–276.

Hastie, T. J., and Tibshirani, R. J. (1990). *Generalized Additive Models*, Chapman and Hall, London, UK.

Hegland, M., McIntosh, I., and Turlach, B. A. (1999). A parallel solver for generalised additive models, *CSDA*, **31**, 377–396.

Hengartner, N. (1996). *Rate Optimal Estimation of Additive Regression via the Integration Method in Presence of Many Covariates*, Statistics Preprint 96, Oct. 1, Yale University, New Haven, CT.

Hurvich, C. M., Simonoff, J. S., and Tsai, C.-L. (1998). Smoothing parameter selection in nonparametric regression using an improved Akaike information criterion, *J. Roy. Statist. Soc. B*, **60**, 271–293.

Hutchinson, M. F., and de Hoog, F. R. (1985). Smoothing noisy data with spline functions, *Numer. Math.*, **47**, 99–106.

Linton, O. B. (1997). Efficient estimation of additive nonparametric regression models, *Biometrika*, **84**, 469–473.

Linton, O. B., and Härdle, W. (1996). Estimation of additive regression models with known links, *Biometrika*, **83**, 529–540.

Linton, O. B., Mammen, E., and Nielsen, J. P. (1997). *The Existence and Asymptotic Properties of a Backfitting Projection Algorithm under Weak Conditions*, Cowles Foundation Discussion Paper No. 1160, Yale University, New Haven, CT.

Linton, O. B., and Nielsen, J. P. (1995). A kernel method of estimating structured nonparametric regression based on marginal integration, *Biometrika*, **82**, 93–100.

Marx, B. D., and Eilers, P. H. C. (1998). Direct generalized additive modeling with penalized likelihood, *Comput. Statist. Data Analysis*, **28**, 193–209.

MathSoft (1999). *S-PLUS 2000: Guide to Statistics*. Vol. I, MathSoft, Inc., Seattle, WA.

McCullagh, P. (1980). Regression models for ordinal data (with discussion), *J. Roy. Statist. Soc. B*, **42**, 109–142.

McCullagh, P., and Nelder, J. A. (1989). *Generalized Linear Models*, 2nd ed., Chapman and Hall, London, UK.

Neubauer, G. P., and Schimek, M. G. (1996). A note on cross-validation for smoothing splines, in Härdle, W., and Schimek, M. (eds.), *Statistical Theory and Computational Aspects of Smoothing*, pp. 165–177, Physica-Verlag, Heidelberg, Germany.

Newey, W. K. (1994). Kernel estimation of partial means, *Econometric Theory*, **10**, 233–253.

Nielsen, J. P., and Linton, O. B. (1998). An optimization interpretation of integration and back-fitting estimators for separable nonparametric models, *J. Roy. Statist. Soc. B*, **60**, 217–222.

Nychka, D., and Ruppert, D. (1995). Nonparametric transformation for both sides of a regression model, *J. Roy. Statist. Soc. B*, **57**, 519–532.

Opsomer, J. D. (2000). Asymptotic properties of backfitting estimators, *J. Multivariate Analysis*, **73**, 166–179.

Opsomer, J. D., and Ruppert, D. (1997). Fitting a bivariate additive model by local polynomial regression, *Ann. Statist.*, **25**, 186–211.

Opsomer, J. D., and Ruppert, D. (1998). A fully automatic bandwidth selector method for fitting additive models, *J. Amer. Statist. Assoc.*, **93**, 605–619.

Ortega, J. M., and Rheinboldt, W. C. (1970). *Iterative Solution of Nonlinear Equations in Several Variables*, Academic Press, New York.

Raftery, A. E., and Hout, M. (1985). Does Irish education approach the meritocratic ideal? A logistic analysis, *The Economic and Social Review*, **16**, 115–140.

Raftery, A. E., and Richardon, S. (1996). Model selection for generalized linear models via GLIB: Application to nutrition and breast cancer, in Berry, D. A., and Stangl, D. K. (eds.), *Bayesian Biostatistics*, pp. 321–354, Marcel Dekker, New York.

Reinsch, C. (1967). Smoothing by spline functions, *Numerische Mathematik*, **10**, 177–183.

Schimek, M. G. (1996). An iterative projection algorithm and some simulation results, in Prat, A. (ed.), *COMPSTAT '96: Proceedings in Computational Statistics*, pp. 453–458, Physica-Verlag, Heidelberg, Germany.

Schimek, M. G. (1997). Non- and semiparametric alternatives to generalized linear models, *Computational Statistics*, **12**, 173–191.

Schimek, M. G. (2000). GAM spline algorithms: A direct comparison, in Bethlehem, S. and van der Heijolen, P. (eds.) *COMPSTAT 2000: Proceedings in Computational Statistics*. Physica-Verlag, Heidelberg, Germany, to appear.

Schimek, M. G., Neubauer, G., and Stettner, H. (1994). Backfitting and related procedures for non-parametric smoothing regression: A comparative view, in Grossmann, W., and Dutter, R. (eds.), *COMPSTAT '94: Proceedings in Computational Statistics*, pp. 63–68, Physica-Verlag, Heidelberg, Germany.

Schimek, M. G., Stettner, H., Haberl, J., and Neubauer, G. P. (1993). Approaches for fitting additive models with cubic smoothing splines and associated problems, *Bulletin of the International Statistical Institute, Contributed Papers, 49th Session, Book 2*, 381–382.

Scott, D. W. (1992). *Multivariate Density Estimation: Theory, Practice and Visualization*, Wiley, New York.

Sen, P. K., and Singer, J. M. (1993), *Large Sample Methods in Statistics: An Introduction with Applications*, Chapman and Hall, New York.

Serfling, R. J. (1980). *Approximation Theorems of Mathematical Statistics*, Wiley, New York.

Severance-Lossin, E., and Sperlich, S. (1999). Estimation of derivatives for additive separable models, *Statistics,* **33**, 241–265.

Sperlich, S., Linton, O. B., and Härdle, W. (1999). A simulation comparison between integration and backfitting methods of estimating separable nonparametric regression models, *Test*, **8**, 419–458.

Sperlich, S., Tjøstheim, D., and Yang, L. (1999). *Nonparametric Estimation and Testing of Interaction in Additive Models*, Working Paper 99–85 (34), Universidad Carlos III de Madrid, Spain.

Stone, C. J. (1985). Additive regression and other nonparametric models, *Ann. Statist.*, **13**, 689–705.

Stone, C. J. (1986). The dimension reduction principle for generalized additive models, *Ann. Statist.*, **14**, 590–606.

Tibshirani, R. (1988). Estimating transformations for regression via additivity and variance stabilization, *J. Amer. Statist. Assoc.*, **83**, 394–405.

Tjøstheim, D., and Auestad, B. (1994). Nonparametric identification of nonlinear time series: Projections, *J. Amer. Statist. Assoc.*, **89**, 1398–1409.

Venables, W. N., and Ripley, B. D. (1999). *Modern Applied Statistics with S-Plus*, 3rd ed., Springer-Verlag, New York.

Wong, C., and Kohn, R. (1996). A Bayesian approach to additive semiparametric regression, *J. Econometrics*, **74**, 209–235.

Yee, T. W. and Wild, C. J. (1994) *Vector Splines and the Vector Additive Model.* Research Report, Department of Statistics, University of Auckland, New Zealand.

Yee, T. W., and Wild, C. J. (1996). Vector generalized additive models, *J. Roy. Statist. Soc. B*, **58**, 481–493.

CHAPTER 11

Multivariate Spline Regression

Chong Gu
Purdue University, West Lafayette, Indiana, USA

11.1 INTRODUCTION

This chapter introduces the smoothing spline approach to multivariate nonparametric regression, with the main focus being on tensor product splines. Instead of the traditional development based on the theory of reproducing kernel Hilbert spaces, which is unfamiliar to most statisticians, our treatment here is based on the more familiar empirical Bayes model. This makes the methodology more accessible at the cost of a slight loss in mathematical rigor.

The basic theoretical framework of the methodology is actually independent of the dimension. In Section 11.2, we discuss the empirical Bayes model using univariate examples. On product domains, structures such as analysis of variance (ANOVA)-like function decompositions can greatly enhance the interpretability of the analysis. A generalization of the ANOVA decomposition to generic product domains is introduced in Section 11.3. Section 11.4 illustrates how such ANOVA decompositions can be built into statistical models using tensor product splines.

A unified computational strategy has been developed for smoothing splines based on the general theoretical framework, and portable code implementing the algorithms is available to the public free of charge. In Section 11.5, the basic computational problem is presented, with a brief discussion of options for smoothing parameter selection. Section 11.6 discusses the calculation of posterior variances under the Bayes model, which are used in the construction of Bayesian confidence intervals. Algorithmic details are omitted in this exposition,

Smoothing and Regression: Approaches, Computation, and Application
Edited by Michael G. Schimek
ISBN 0-471-17946-9

but open-source software implementing the algorithms is described and illustrated in Section 11.7.

The rest of the chapter touches on several related topics. Section 11.8 introduces the cosine diagnostics for ANOVA terms that are negligible or are "collinear" with other terms. Sections 11.9 and 11.10 discuss partial splines and thin-plate splines, respectively. Section 11.11 provides an entry to the field of non-Gaussian regression.

Many of the numerical examples in this chapter include color figures to be produced by the reader using the simple code given in the text. For those without immediate access to the computer, some black-and-white versions of the figures are included.

11.2 SMOOTHING SPLINES AS BAYES ESTIMATES

To illustrate the basic theoretical framework, we shall go over a few univariate examples of smoothing splines and discuss the related empirical Bayes models. Certain quantities appearing in the discussion also serve as building blocks in the construction of tensor product splines in Section 11.4.

Consider the popular cubic smoothing spline. Observing $Y_i = f(x_i) + \epsilon_i$, $i = 1, \ldots, n$, where $x_i \in [0, 1]$ and $\epsilon_i \sim N(0, \sigma^2)$, one estimates $f(\cdot)$ using the minimizer of

$$\frac{1}{n}\sum_{i=1}^{n}(Y_i - f(x_i))^2 + \lambda \int_0^1 (f'')^2. \tag{11.1}$$

The first term of (11.1) penalizes lack of fit, the second term penalizes the roughness of the estimate, and the smoothing parameter λ controls the trade-off between the two conflicting goals. When $\lambda \to \infty$, one has the simple linear regression model $f(x) = \beta_0 + \beta_1(x - 0.5)$, which resides in the null space of the roughness functional $\int_0^1 (f'')^2$, $\{f : \int_0^1 (f'')^2 = 0\}$.

The minimizer of (11.1) is called a cubic spline because it is a piecewise cubic polynomial. It is three times differentiable with the third derivative jumping at the distinctive sampling points known as the knots. Univariate smoothing splines are also known as natural splines in the numerical analysis literature. (See for example, de Boor, 1978 and Schumaker, 1981.)

The functional (11.1) is a specialization of the generic penalized least-squares functional

$$\frac{1}{n}\sum_{i=1}^{n}(Y_i - f(x_i))^2 + \frac{\lambda}{2}J(f), \tag{11.2}$$

which can be used to estimate a regression function $f(\cdot)$ on a generic domain $\mathcal{X}$, possibly multidimensional, where $J(f)$ is a quadratic roughness functional. Using the mathematical theory of reproducing kernel Hilbert spaces, generic theoretical and computational treatments have been developed for the minimizer of (11.2). Specializations of the generic theory to specific multivariate problems result in various multivariate smoothing splines, which is the focus of this chapter. A thorough treatment of the subject can be found in Wahba (1990).

We avoid the theory of reproducing kernel Hilbert spaces in the exposition by explaining the basic construction in a more familiar empirical Bayes setting. The structure can best be explained on a discrete domain, although one rarely calls it smoothing in that context.

Consider the domain $\mathcal{X} = \{1, \ldots, K\}$. A function on such a domain is simply a K-vector, and the regression problem is simply a one-way ANOVA. Setting $J(f) = \sum_{x=1}^{K}(f(x) - \overline{f})^2$, where $\overline{f} = (1/K)\sum_{x=1}^{K} f(x)$, the generic functional (11.2) specializes to

$$\frac{1}{n}\sum_{i=1}^{n}(Y_i - f(x_i))^2 + \lambda\sum_{x=1}^{K}(f(x) - \overline{f})^2, \tag{11.3}$$

which defines a shrinkage estimate shrinking towards a constant. Note that $\sum_{x=1}^{K}(f(x) - \overline{f})^2 = \boldsymbol{f}^T(I - \mathbf{1}\mathbf{1}^T/K)\boldsymbol{f}$, where $\boldsymbol{f}^T = (f(1), \ldots, f(K))$. Write $\boldsymbol{f} = \mu\mathbf{1} + \boldsymbol{\alpha}$, where $\mathbf{1}^T\boldsymbol{\alpha} = 0$. The minimizer of (11.3) is easily seen to be a Bayes estimate with a diffuse prior for μ and a proper prior for $\boldsymbol{\alpha}$, $\boldsymbol{\alpha} \sim N(\mathbf{0}, \sigma^2(n\lambda)^{-1}(I - \mathbf{1}\mathbf{1}^T/K))$. We will refer to a component with a diffuse prior as a fixed effect and that with a proper prior as a random effect. In general, for any nonnegative definite matrix J, it can be shown that the minimizer of

$$\frac{1}{n}\sum_{i=1}^{n}(Y_i - f(x_i))^2 + \lambda\boldsymbol{f}^T J\boldsymbol{f} \tag{11.4}$$

is a Bayes estimate with a fixed effect in the null space of J and a random effect with a Gaussian prior $N(\mathbf{0}, \sigma^2(n\lambda)^{-1}J^+)$, where J^+ is the Moore–Penrose inverse of J. See, for example, Rao (1973), (Sec, 1b.5) for the Moore–Penrose inverse of the matrix.

We now come back to the cubic spline. By a classical result of Kimeldorf and Wahba (1971), the minimizer of (11.1) is also a Bayes estimate under a Gaussian prior for $f(x)$. Specifically, the prior has a fixed effect $f_0(x) = \beta_0 + \beta_1(x - 0.5)$ and a random effect $f_1(x)$, which is a Gaussian stochastic process on [0, 1] with mean zero and covariance function $E[f_1(x_1)f_1(x_2)] = \sigma^2(n\lambda)^{-1}R(x_1, x_2)$, where

$$R(x_1, x_2) = k_2(x_1)k_2(x_2) - k_4(|x_1 - x_2|), \tag{11.5}$$

and

$$\begin{aligned} k_2(x) &= (1/2)(k_1^2(x) - 1/12), \\ k_4(x) &= (1/24)(k_1^4(x) - k_1^2(x)/2 + 7/240), \end{aligned} \tag{11.6}$$

for $k_1(x) = x - 0.5$. It can be shown that $\int_0^1 f_1(x) = \int_0^1 (x - 0.5)f_1(x) = 0$, which are side conditions keeping the random effect "orthogonal" to the fixed effect, similar to the side condition $\mathbf{1}^T\boldsymbol{\alpha} = 0$ on the discrete domain. The nonnegative definite function R in (11.5) and the quadratic functional $J(f) = \int_0^1 (f'')^2$ are in a sense "inverses" of each other. Note also that the fixed effect $f_0(x)$ is in the null space of $J(f) = \int_0^1 (f'')^2$, $\{f : J(f) = 0\}$.

Just as the inverse J^+ of a matrix J can rarely be "seen" through the entries of J, the "inverse" $R(x_1, x_2)$ of $J(f) = \int_0^1 (f'')^2$ as given in (11.5) is not to be perceived intuitively. Mathematically, one has

$$\int_0^1 f'' R_x'' = f(x) \tag{11.7}$$

for f in the space $\{f\colon \int_0^1 f = \int_0^1 f' = 0, \int_0^1 (f'')^2 < \infty\}$, where $R_x(\cdot) = R(x, \cdot)$, which defines the notion of "inverse" in this case. It can be verified that R_x resides in such a space so $R(x_1, x_2) = \int_0^1 R_{x_1}'' R_{x_2}''$. Technical details are to be found in Craven and Wahba (1979). To see why (11.7) defines a notion of "inverse," look no further than functions on the discrete domain $\mathcal{X} = \{1, \ldots, K\}$ with $J(f) = \boldsymbol{f}^T J \boldsymbol{f}$: for $\boldsymbol{f}$ in the column space of J,

$$\boldsymbol{r}_x^T J \boldsymbol{f} = \mathbf{e}_x^T J^+ J \boldsymbol{f} = f(x),$$

where $\mathbf{e}_x$ is the xth unit vector and $\boldsymbol{r}_x = J^+ \mathbf{e}_x$ is the xth column of J^+.

In general, the minimizer of (11.2) is a Bayes estimate with a fixed effect in the null space of $J(f)$ and a Gaussian-process random effect with mean zero and covariance function proportional to the "inverse" of $J(f)$, say, $R_J(x_1, x_2)$. The hyperparameter $b = \sigma^2 (n\lambda)^{-1}$ in the covariance function is to be effectively selected through the choice of λ based on the data, so this is an empirical Bayes setting. While an explicit $J(f)$ provides the most direct intuition for the notion of smoothness, an explicit R_J and the specification of the fixed effect are all that are needed for computation (see Section 11.5).

11.3 ANOVA DECOMPOSITION ON PRODUCT DOMAINS

An important aspect of statistical modeling, which distinguishes it from mere function approximation, is the interpretability of the results. To enhance the interpretability of multidimensional data analysis, we describe below a generalization of the ANOVA decomposition to generic product domains.

First let us fix some notations. Throughout this chapter, we will be using the lowercase letter x, often with a subscript, to denote points on the domain of interest, $\mathcal{X}$. For $x, x_i \in \mathcal{X} = \prod_{l=1}^L \mathcal{X}_l$ a product domain, we write $x = (x_{\langle 1 \rangle}, \ldots, x_{\langle L \rangle})$, $x_i = (x_{i\langle 1 \rangle}, \ldots, x_{i\langle L \rangle})$, where $x_{\langle l \rangle}, x_{i\langle l \rangle} \in \mathcal{X}_l$. In a similar manner, we will also use the bracketed subscript to earmark other entities defined on the marginal domain $\mathcal{X}_l$.

Consider function $f(x)$ on a product domain $\mathcal{X} = \prod_{l=1}^L \mathcal{X}_l$. Let $A_{\langle l \rangle}$ be averaging operators acting on arguments $x_{\langle l \rangle}$ that satisfy $A_{\langle l \rangle}^2 = A_{\langle l \rangle}$. An ANOVA decomposition of the function can be defined as

$$f = \left\{ \prod_{l=1}^L (I - A_{\langle l \rangle} + A_{\langle l \rangle}) \right\} f$$

$$= \sum_{\mathcal{S}\subseteq\{1,\dots,L\}} \left\{ \prod_{l\in\mathcal{S}} (I - A_{\langle l\rangle}) \prod_{l\in\mathcal{S}^c} A_{\langle l\rangle} \right\} f$$

$$= \sum_{\mathcal{S}\subseteq\{1,\dots,L\}} f_{\mathcal{S}}, \tag{11.8}$$

where $\mathcal{S}$ is the index set of active arguments in a component; $f_\emptyset = [\prod_{l=1}^{L} A_{\langle l\rangle}]f$ is a constant, $f_l = f_{\{l\}} = \{(I - A_{\langle l\rangle})\prod_{k\neq l} A_{\langle k\rangle}\}f$ are the $x_{\langle l\rangle}$ main effects, $f_{l,k} = f_{\{l,k\}} = \{(I - A_{\langle l\rangle})(I - A_{\langle k\rangle})\prod_{m\neq l,k} A_{\langle m\rangle}\}f$ are the $x_{\langle l\rangle}$–$x_{\langle k\rangle}$ interactions, and so on. The identifiability of such a decomposition is assured by the side conditions $A_{\langle l\rangle} f_{\mathcal{S}} = 0,\ \forall \mathcal{S} \ni l$.

For $\mathcal{X}_l = \{1, \dots, K\}$ a discrete domain, the choice of

$$A_{\langle l\rangle} f = \left(\frac{1}{K}\right) \sum_{x_{\langle l\rangle}=1}^{K} f(x)$$

defines the standard ANOVA decomposition separating the mean $\overline{f}$ from the contrast $f - \overline{f}$. For $\mathcal{X}_l = [0, 1]$ a real interval, the choice of $A_{\langle l\rangle} f = \int_0^1 f(x)\,dx_{\langle l\rangle}$ similarly decomposes a function into a mean $\int_0^1 f$ and a contrast $f - \int_0^1 f$.

The marginal domains in this construction are generic, and in particular, they can be product domains themselves. The decomposition may also be defined recursively via a series of hierarchical constructions with $L = 2$.

As an example, consider $L = 2$, $\mathcal{X}_l = [0, 1]$, and $A_{\langle l\rangle} f = \int_0^1 f(x)\,dx_{\langle l\rangle}$, $l = 1, 2$. Straightforward calculation yields

$$f_\emptyset = A_{\langle 1\rangle} A_{\langle 2\rangle} f = \int_0^1 \int_0^1 f(x)\,dx_{\langle 2\rangle}\,dx_{\langle 1\rangle},$$

$$f_1 = (I - A_{\langle 1\rangle}) A_{\langle 2\rangle} f = \int_0^1 f(x)\,dx_{\langle 2\rangle} - f_\emptyset,$$

$$f_2 = A_{\langle 1\rangle}(I - A_{\langle 2\rangle}) f = \int_0^1 f(x)\,dx_{\langle 1\rangle} - f_\emptyset,$$

$$f_{1,2} = (I - A_{\langle 1\rangle})(I - A_{\langle 2\rangle}) f = f(x) - \int_0^1 f(x)\,dx_{\langle 1\rangle} - \int_0^1 f(x)\,dx_{\langle 2\rangle} + f_\emptyset.$$

11.4 TENSOR PRODUCT SPLINES

We now describe specializations of (11.2) on product domains with an ANOVA decomposition built in. The terms in the ANOVA decomposition can be individually attached or detached, resulting in a rich variety of models that include the popular main-effect-only additive models as a special case.

First consider the product of two discrete domains, $\mathcal{X} = \{1, \ldots, K_1\} \times \{1, \ldots, K_2\}$. A function f on such a domain can be written as a K_1K_2-vector, say,

$$\boldsymbol{f} = (f(1,1), \ldots, f(1, K_2), \ldots, f(K_1, 1), \ldots, f(K_1, K_2))^T.$$

The construction of the model will be derived from (11.3). Note that the fixed effect in (11.3) can be viewed as the limiting case of a random effect independent of $\boldsymbol{\alpha}$. Specifically, one starts with $\mu\mathbf{1} \sim N(\mathbf{0}, \eta\mathbf{1}\mathbf{1}^T)$ and lets $\eta \to \infty$. The column spaces of $\Sigma_0 = \mathbf{1}\mathbf{1}^T/K$ and $\Sigma_1 = (I - \mathbf{1}\mathbf{1}^T/K)$ span the spaces of the mean and the contrast, respectively, in a one-way ANOVA decomposition.

Using the two covariance matrices on each of the axes, of different sizes when $K_1 \neq K_2$, the four Kronecker products define four covariance matrices for $\boldsymbol{f}$ on the product domain. It is easy to see that an ANOVA decomposition is readily built in, where the column space of $\Sigma_{0\langle 1\rangle} \otimes \Sigma_{0\langle 2\rangle}$ spans the overall mean, that of $\Sigma_{1\langle 1\rangle} \otimes \Sigma_{0\langle 2\rangle}$ spans the $x_{\langle 1\rangle}$ main effect, that of $\Sigma_{0\langle 1\rangle} \otimes \Sigma_{1\langle 2\rangle}$ spans the $x_{\langle 2\rangle}$ main effect, and that of $\Sigma_{1\langle 1\rangle} \otimes \Sigma_{1\langle 2\rangle}$ spans the interaction. The averaging operators are simply those defining the one-way ANOVA decompositions on the marginal domains, namely, $A_{\langle 1\rangle} f = (1/K_1)\sum_{x_{\langle 1\rangle}=1}^{K_1} f(x)$ and $A_{\langle 2\rangle} f = (1/K_2)\sum_{x_{\langle 2\rangle}=1}^{K_2} f(x)$.

Note that a $K \times K$ matrix is simply a bivariate function on the domain $\{1, \ldots, K\}$, so one may write Σ as $\Sigma(x_1, x_2)$. The $K_1K_2 \times K_1K_2$ Kronecker product $\Sigma_{\langle 1\rangle} \otimes \Sigma_{\langle 2\rangle}$ is a bivariate function on the product domain, and it can be verified that $(\Sigma_{\langle 1\rangle} \otimes \Sigma_{\langle 2\rangle})(x_1, x_2) = \Sigma_{\langle 1\rangle}(x_{1\langle 1\rangle}, x_{2\langle 1\rangle})\Sigma_{\langle 2\rangle}(x_{1\langle 2\rangle}, x_{2\langle 2\rangle})$.

For each of the four independent terms, one may choose to include or exclude it from the model. When included, a term may further be classified as a fixed effect or a random effect. Suppose one wants to include all the terms and wants to treat all but the constant as random effects. The joint covariance matrix for the three random effects is proportional to $R_J = \theta_{1,0}\Sigma_{1\langle 1\rangle} \otimes \Sigma_{0\langle 2\rangle} + \theta_{0,1}\Sigma_{0\langle 1\rangle} \otimes \Sigma_{1\langle 2\rangle} + \theta_{1,1}\Sigma_{1\langle 1\rangle} \otimes \Sigma_{1\langle 2\rangle}$, where the θs are to be selected using data-driven procedures. The corresponding $J(f)$ is not needed for computation, but for the record, it is easy to verify that $J(f) = \boldsymbol{f}^T(\theta_{1,0}^{-1}\Sigma_{1\langle 1\rangle} \otimes \Sigma_{0\langle 2\rangle} + \theta_{0,1}^{-1}\Sigma_{0\langle 1\rangle} \otimes \Sigma_{1\langle 2\rangle} + \theta_{1,1}^{-1}\Sigma_{1\langle 1\rangle} \otimes \Sigma_{1\langle 2\rangle})\boldsymbol{f}$.

To fit a main-effect-only additive model, one may simply set $\theta_{1,1} = 0$. The third term in $J(f)$ would then disappear from the formula, but the minimization of (11.2) is restricted to functions satisfying $(I - A_{\langle 1\rangle})(I - A_{\langle 2\rangle})f = 0$. On the other end of the spectrum, we note that a fixed effect actually corresponds to $\theta = \infty$, so it is not penalized by $J(f)$.

Now let us look at a construction on $\mathcal{X} = [0, 1]^2$ using materials from (11.1) as building blocks. The fixed effect in (11.1) again can be viewed as the limit of some zero mean Gaussian process random effect independent of f_1. Actually, one can split the two terms of the fixed effect, with the constant β_0 resulting from a process with the covariance function $\eta_0 R_0$ as $\eta_0 \to \infty$, and $\beta_1(x - 0.5)$ from a process with the covariance function $\eta_1 R_1$ as $\eta_1 \to \infty$, where $R_0(x_1, x_2) = 1$ and $R_1(x_1, x_2) = (x_1 - 0.5)(x_2 - 0.5)$. We shall denote the R function in (11.5) as R_2

below. Note that $\beta_1(x - 0.5) + f_1(x)$ forms the contrast in a one-way ANOVA decomposition with $Af = \int_0^1 f$.

With three covariance functions on each of the axes, one gets nine covariance functions on the product domain by simple multiplication. The ANOVA structures on the marginal domains automatically induce one on the product domain. Specifically, $R_{0\langle 1\rangle}(x_{1\langle 1\rangle}, x_{2\langle 1\rangle})R_{0\langle 2\rangle}(x_{1\langle 2\rangle}, x_{2\langle 2\rangle})$ yields the constant, $R_{1\langle 1\rangle}R_{0\langle 2\rangle}$ and $R_{2\langle 1\rangle}R_{0\langle 2\rangle}$ yield the $x_{\langle 1\rangle}$ main effect, $R_{0\langle 1\rangle}R_{1\langle 2\rangle}$ and $R_{0\langle 1\rangle}R_{2\langle 2\rangle}$ yield the $x_{\langle 2\rangle}$ main effect, and the other four yield the interaction. One then can attach tunable θ's to the nine product covariance functions and add them together to get the joint covariance function R_J. A $\theta = 0$ eliminates the corresponding term from the model, a $\theta = \infty$ sets the term as a fixed effect, and these terms do not show up in the formula for R_J. Note that only the four products with both factors being one of R_0 or R_1 can be set as fixed effects; the other five products are of infinite dimension. The "inverses" of the product covariance functions can be found in Gu and Wahba (1991), which we will omit here, because they do not contribute to the computation of the estimates.

A similar construction is as straightforward on the mixed product domain $\{1, \ldots, K\} \times [0, 1]$. From the marginal covariance functions Σ_0, Σ_1 on $\mathcal{X}_1 = \{1, \ldots, K\}$ and R_0, R_1, and R_2 on $\mathcal{X}_2 = [0, 1]$, one gets six product covariance functions on the product domain by simple multiplication. An ANOVA decomposition is in place with $\Sigma_{0\langle 1\rangle}R_{0\langle 2\rangle}$ yielding the constant, $\Sigma_{1\langle 1\rangle}R_{0\langle 2\rangle}$ yielding the $x_{\langle 1\rangle}$ main effect, $\Sigma_{0\langle 1\rangle}R_{1\langle 2\rangle}$ and $\Sigma_{0\langle 1\rangle}R_{2\langle 2\rangle}$ yielding the $x_{\langle 2\rangle}$ main effect, and $\Sigma_{1\langle 1\rangle}R_{1\langle 2\rangle}$ and $\Sigma_{1\langle 1\rangle}R_{2\langle 2\rangle}$ yielding the interaction. One then attaches θs in the range $[0, \infty]$ to these product covariance functions to preserve (fixed effect with $\theta = \infty$), shrink (random effect with $\theta \in (0, \infty)$), or eliminate (degenerate with $\theta = 0$) the terms. Again, the infinite dimensional terms corresponding to $\Sigma_{0\langle 1\rangle}R_{2\langle 2\rangle}$ and $\Sigma_{1\langle 1\rangle}R_{2\langle 2\rangle}$ cannot be set as fixed effects.

The general construction should be clear at this point. From covariance functions $R_{\langle 1\rangle}$ and $R_{\langle 2\rangle}$ on generic domains $\mathcal{X}_1$ and $\mathcal{X}_2$, one may obtain a covariance function $R_{\langle 1\rangle}R_{\langle 2\rangle}$ on the product domain $\mathcal{X}_1 \times \mathcal{X}_2$ by simple multiplication. When two or more covariance functions define certain function decompositions on the marginal domains, the products of these covariance functions define an induced decomposition on the product domain. The different product covariance functions yield independent modules, which one may choose to use as fixed effects, to use as random effects, or to eliminate in a statistical model. Generalizations to the product of more than two marginal domains is straightforward.

In the above discussion, we have implicitly used an elementary mathematical result, that the product of nonnegative definite functions $R_{\langle 1\rangle}$ on $\mathcal{X}_1$ and $R_{\langle 2\rangle}$ on $\mathcal{X}_2$ is nonnegative definite on $\mathcal{X}_1 \times \mathcal{X}_2$.

At the end of the constructions, one would have a finite number of fixed effects, say, of a total dimension M, and a number of random effects, say, f_β, $\beta = 1, \ldots, p$, with covariance functions $R_\beta(x_1, x_2)$; f_β reside in the linear space spanned by $\{R_\beta(x, \cdot), x \in \mathcal{X}\}$. An ANOVA decomposition on a product domain is defined by a certain regrouping of the terms.

11.5 COMPUTATION

We now describe a general approach to the computation of smoothing splines based on the generic theoretical framework. Related software is described in Section 11.7.

It can be shown that the minimizer of (11.2), while generally over an infinite dimensional function space, has a finite dimensional expression

$$f(x) = \sum_{\nu=1}^{M} d_\nu \phi_\nu(x) + \sum_{i=1}^{n} c_i R_J(x_i, x) = \boldsymbol{\phi}^T(x)\boldsymbol{d} + \boldsymbol{\xi}^T(x)\boldsymbol{c}, \tag{11.9}$$

where $\{\phi_\nu\}_{\nu=1}^M$ span the null space of $J(f)$ and $R_J = \sum_{\beta=1}^{p} \theta_\beta R_\beta$ is the "inverse" of $J(f)$; see Kimeldorf and Wahba (1971). In the empirical Bayes model, $\{\phi_\nu\}_{\nu=1}^M$ are the fixed effects and R_β define the random effects. The work left to be done is the selection of the smoothing parameters λ and θ_β and the calculation of the coefficients d_ν and c_i.

Plugging the expression (11.9) into (11.2), the problem reduces to the minimization of

$$(\boldsymbol{Y} - S\boldsymbol{d} - Q\boldsymbol{c})^T(\boldsymbol{Y} - S\boldsymbol{d} - Q\boldsymbol{c}) + n\lambda\boldsymbol{c}^T Q\boldsymbol{c} \tag{11.10}$$

with respect to $\boldsymbol{d}$ and $\boldsymbol{c}$, where S is $n \times M$ with the (i, ν)th entry $\phi_\nu(x_i)$ and Q is $n \times n$ with the (i, j)th entry $R_J(x_i, x_j)$. The fact that

$$J\left(\sum_{i=1}^{n} c_i R_J(x_i, \cdot)\right) = \boldsymbol{c}^T Q\boldsymbol{c}$$

is a direct consequence of the mathematical definition of R_J as the "inverse" of $J(f)$. It is clear at this point that an explicit expression of $J(f)$ is not needed for computation.

It can be shown that the minimizer of (11.2) is unique provided that S is of full column rank (Gu and Qiu, 1993, Theorem 4.1), whereas (11.10) could have multiple numerical solutions for $\boldsymbol{c}$. This may occur, for example, when one has replicated x_i in the data. All one needs, however, is one solution of (11.10), which can be obtained by solving the well-behaving surrogate linear system

$$\begin{aligned}(Q + n\lambda I)\boldsymbol{c} + S\boldsymbol{d} &= \boldsymbol{Y},\\ S^T\boldsymbol{c} &= 0.\end{aligned} \tag{11.11}$$

Let

$$S = FR^* = (F_1, F_2)\begin{pmatrix} R \\ 0 \end{pmatrix}$$

be the QR-decomposition of S with F orthogonal and R upper-triangular. From $S^T\boldsymbol{c} = 0$, one has $F_1^T\boldsymbol{c} = 0$, so $\boldsymbol{c} = F_2F_2^T\mathbf{c}$. Premultiplying by F_2^T and F_1^T, the first equation of (11.11) gives

$$\begin{aligned} \boldsymbol{c} &= F_2(F_2^T Q F_2 + n\lambda I)^{-1} F_2^T \boldsymbol{Y}, \\ \boldsymbol{d} &= R^{-1}(F_1^T \boldsymbol{Y} - F_1^T Q \boldsymbol{c}). \end{aligned} \tag{11.12}$$

Denote the fitted values by $\hat{\boldsymbol{Y}} = (\hat{f}(x_1), \ldots, \hat{f}(x_n))^T$ and the residuals by $\boldsymbol{e} = \boldsymbol{Y} - \hat{\boldsymbol{Y}}$. Some algebra yields

$$\hat{\boldsymbol{Y}} = Q\boldsymbol{c} + S\boldsymbol{d} = (I - n\lambda F_2(F_2^T Q F_2 + n\lambda I)^{-1} F_2^T)\boldsymbol{Y}. \tag{11.13}$$

The symmetric matrix

$$A(\lambda, \theta) = (I - n\lambda F_2(F_2^T Q F_2 + n\lambda I)^{-1} F_2^T) \tag{11.14}$$

is called the smoothing matrix associated with (11.2), where its dependence on the smoothing parameters λ and θ_β hidden in Q is explicitly noted. It is easy to see from (11.11) that $\boldsymbol{e} = (I - A(\lambda, \theta))\boldsymbol{Y} = n\lambda\boldsymbol{c}$.

The smoothing parameters are yet to be selected. When σ^2 is known, Craven and Wahba (1979) proposed to use the minimizer of

$$U(\lambda, \theta) = n^{-1}\boldsymbol{Y}^T(I - A(\lambda, \theta))^2\boldsymbol{Y} + 2\sigma^2 n^{-1} \operatorname{tr} A(\lambda, \theta), \tag{11.15}$$

and when σ^2 is unknown, they proposed to use the minimizer of

$$V(\lambda, \theta) = \frac{n^{-1}\boldsymbol{Y}^T(I - A(\lambda, \theta))^2\boldsymbol{Y}}{[n^{-1}\operatorname{tr}(I - A(\lambda, \theta))]^2}. \tag{11.16}$$

Consider the mean square error loss at the design points,

$$L(\lambda, \theta) = n^{-1}\sum_{i=1}^{n}(\hat{f}(x_i) - f(x_i))^2. \tag{11.17}$$

Under mild conditions, it can be shown that the minimizers (λ_u, θ_u) of U and (λ_v, θ_v) of V asymptotically minimize $L(\lambda, \theta)$ as $n \to \infty$, in the sense that

$$\begin{aligned} L(\lambda_*, \theta_*)/L(\lambda_u, \theta_u) &= 1 + o_p(1), \\ L(\lambda_*, \theta_*)/L(\lambda_v, \theta_v) &= 1 + o_p(1), \end{aligned}$$

where (λ_*, θ_*) minimizes $L(\lambda, \theta)$ (see Li, 1986). The score U is an unbiased estimate of the relative loss $L(\lambda, \theta) + n^{-1}\epsilon^T\epsilon$, where ϵ is the vector of unobservable noise that does not depend on the smoothing parameters. V is the popular generalized cross-validation score. Note that U and V are derived and justified in a classical setting, where the function $f(x)$ is considered fixed. When V is used to select the smoothing parameters, a good variance estimate is given by

$$\hat{\sigma}_v^2 = \frac{\boldsymbol{Y}^T(I - A(\lambda_v, \theta_v))^2\boldsymbol{Y}}{\text{tr}(I - A(\lambda_v, \theta_v))}. \tag{11.18}$$

(See Wahba, 1983.)

Under the empirical Bayes model, the smoothing parameters λ, θ_β and the sampling variance σ^2 can be estimated by the type II maximum (restricted) likelihood estimate. It can be shown that the smoothing parameters are to be estimated by the minimizer (λ_m, θ_m) of

$$M(\lambda, \theta) = \frac{n^{-1}\boldsymbol{Y}^T(I - A(\lambda, \theta))\boldsymbol{Y}}{|I - A(\lambda, \theta)|_+^{1/(n-M)}}, \tag{11.19}$$

where $|\cdot|_+$ denotes the product of positive eigenvalues of the matrix, and the sampling variance σ^2 is to be estimated by

$$\hat{\sigma}_m^2 = \frac{\boldsymbol{Y}^T(I - A(\lambda_m, \theta_m))\boldsymbol{Y}}{n - M}. \tag{11.20}$$

Details can be found in Wahba (1985); see also Gu and Wahba (1991).

The minimizations of U, V, or M share similar numerical structures. Efficient algorithms were developed in Gu et al. (1989) and Gu and Wahba (1991) and were implemented in a collection of public domain FORTRAN compatible subroutines collectively known as RKPACK. A user-friendly front end to RKPACK is available in the R package GSS, to be discussed in Section 11.7.

The algorithms implemented in RKPACK are $O(n^3)$. For univariate smoothing with the cubic spline (11.1) or the like, $O(n)$ algorithms are available through alternative formulations of the numerical problem. Interested readers may find details about these specialized algorithms in Ansley and Kohn (1985), Hutchinson and de Hoog (1985), O'Sullivan (1985), and Wecker and Ansley (1983).

11.6 BAYESIAN CONFIDENCE INTERVALS

In this section, we will discuss the Bayesian confidence intervals proposed by Wahba (1983), and describe the computation of posterior variances needed in the construction of the intervals.

As mentioned in Section 11.2, the minimizer of (11.2) was shown by Kimeldorf and Wahba (1971) to be the posterior mean $\hat{f}(x) = E(f(x)|\boldsymbol{Y})$ under the Bayes model. Following similar algebra, the posterior variance can be shown to be

$$\text{Var}[f(x)|\boldsymbol{Y}] = b[R_J(x, x) + \boldsymbol{\phi}^T(x)(S^TM^{-1}S)^{-1}\boldsymbol{\phi}(x) - 2\boldsymbol{\phi}^T(x)\boldsymbol{d}_\xi - \boldsymbol{\xi}^T(x)\boldsymbol{c}_\xi], \tag{11.21}$$

where $b = \sigma^2(n\lambda)^{-1}$, $\boldsymbol{\phi}$, and $\boldsymbol{\xi}$ are as in (11.9), $M = Q + n\lambda I$, and

$$\begin{aligned} \boldsymbol{c}_\xi &= F_2(F_2^T Q F_2 + n\lambda I)^{-1} F_2^T \boldsymbol{\xi}(x), \\ \boldsymbol{d}_\xi &= R^{-1}(F_1^T \boldsymbol{\xi}(x) - F_1^T Q \boldsymbol{c}_\xi), \end{aligned} \tag{11.22}$$

where F_1, F_2, and R appeared in (11.12). Comparing (11.22) with (11.12), it can be seen that the computation of $\boldsymbol{c}_\xi$ and $\boldsymbol{d}_\xi$ follows the same procedure as that of $\boldsymbol{c}$ and $\boldsymbol{d}$, except that $\boldsymbol{Y}$ is replaced by $\boldsymbol{\xi}(x)$. Details can be found in Wahba (1983) and Gu and Wahba (1993a).

A Bayesian confidence interval for $f(x)$ can thus be constructed as $\hat{f}(x) \pm z_{\alpha/2}\sqrt{\mathrm{Var}[f(x)|\boldsymbol{Y}]}$. At a sampling point x_i, $\mathrm{Var}[f(x_i)|\boldsymbol{Y}]$ can be shown to be the (i, i)th entry $\sigma^2 a_{i,i}$ of the matrix $\sigma^2 A(\lambda, \theta)$, so the intervals are

$$\hat{f}(x_i) \pm z_{\alpha/2}\sqrt{\sigma^2 a_{i,i}}. \tag{11.23}$$

In practice, one has to select the smoothing parameters λ and θ_β, which determine $\hat{f}(x_i)$ and $a_{i,i}$, and to estimate σ^2.

Wahba (1983) suggested using (λ_v, θ_v) and $\hat{\sigma}_v^2$ in the construction. For such a construction, define

$$\mathrm{ACP}(\alpha) = \#\{i\colon |\hat{f}(x_i) - f(x_i)| < z_{\alpha/2}\hat{\sigma}_v\sqrt{a_{i,i}}\}/n,$$

the *across-the-function coverage percentage* (ACP) of the intervals at the sampling points. Simulation results in Wahba (1983) suggest that, for n large,

$$E[\mathrm{ACP}(\alpha)] \approx 1 - \alpha, \tag{11.24}$$

where the expectation is with respect to ϵ_i in $Y_i = f(x_i) + \epsilon_i$. Heuristic arguments in support of (11.24) can be found in Wahba (1983). A more rigorous treatment for smoothing splines on [0, 1] is given by Nychka (1988). It remains unclear, however, whether (11.24) is a precise statement of the across-the-function coverage property or whether a general treatment is possible. Note also that the derivation of the intervals is under the Bayes model but the property (11.24) is under the classical setting with a fixed smooth f.

To put (11.24) in perspective, consider some parametric model $f(x) = g(x, \beta)$ with $g(x, \beta)$ known up to the parameters β. The standard confidence interval for $f(x)$, $g(x, \hat{\beta}) \pm z_{\alpha/2}\hat{\sigma}_{g(x,\hat{\beta})}$, has the point-wise coverage property

$$P(|g(x, \hat{\beta}) - f(x)| < z_{\alpha/2}\hat{\sigma}_{g(x,\hat{\beta})}) \approx 1 - \alpha. \tag{11.25}$$

The property (11.24) is weaker than (11.25), but (11.25) does imply (11.24). Hence, the intervals satisfying (11.24) can be compared with the standard confidence intervals in parametric models on the basis of across-the-function coverage percentage.

Under the Bayes model of $f(x) = \sum_{\nu=1}^{M} \psi_\nu(x) + \sum_{\beta=1}^{p} f_\beta(x)$, where $\psi_\nu \in \{\phi_\nu\}$ are fixed effects and f_β are random effects with covariance proportional to R_β, one can also calculate the posterior moments of individual terms. Specifically, it can be shown that

$$E[\psi_\nu(x)|\boldsymbol{Y}] = d_\nu \phi_\nu(x),$$

$$E[f_\beta(x)|\boldsymbol{Y}] = \sum_{i=1}^{n} c_i \theta_\beta R_\beta(x_i, x), \tag{11.26}$$

where d_ν and c_i are calculated from (11.12), and

$$b^{-1}\operatorname{Cov}[\psi_\nu(x), \psi_\mu(x)|\boldsymbol{Y}] = \phi_\nu(x)\phi_\mu(x)\mathbf{e}_\nu^T(S^T M^{-1} S)^{-1}\boldsymbol{e}_\mu,$$

$$b^{-1}\operatorname{Cov}[\psi_\nu(x), f_\beta(x)|\boldsymbol{Y}] = -\phi_\nu \boldsymbol{e}_\nu^T \boldsymbol{d}_\beta,$$

$$b^{-1}\operatorname{Cov}[f_\beta(x), f_\gamma(x)|\boldsymbol{Y}] = \theta_\beta R_\beta(x, x) I_{[\beta=\gamma]} - \boldsymbol{c}_\beta^T \boldsymbol{\xi}_\gamma, \tag{11.27}$$

where $\boldsymbol{e}_\nu$ is the νth unit vector, $I_{[\beta=\gamma]}$ is the indicator function,

$$\boldsymbol{\xi}_\beta = \theta_\beta(R_\beta(x_1, x), \ldots, R_\beta(x_n, x))^T,$$

and

$$\boldsymbol{c}_\beta = F_2(F_2^T Q F_2 + n\lambda I)^{-1} F_2^T \boldsymbol{\xi}_\beta,$$

$$\boldsymbol{d}_\beta = R^{-1}(F_1^T \boldsymbol{\xi}_\beta - F_1^T Q \boldsymbol{c}_\beta). \tag{11.28}$$

It is easy to check that (11.21) can be derived from (11.27). The computation of $\boldsymbol{c}_\beta$ and $\boldsymbol{d}_\beta$ again parallels that of $\boldsymbol{c}$ and $\boldsymbol{d}$. Technical details are to be found in Gu and Wahba (1993a). Bayesian confidence intervals for individual ANOVA components can be constructed using (11.26) and (11.27).

For the component-wise version of ACP(α), (11.24) appears less plausible in the simulations of Gu and Wahba (1993a), but the widths of the component-wise intervals are certainly of the proper order of magnitude. A theoretical treatment for the component-wise intervals has yet to be found.

The Bayesian confidence intervals were also suggested by Wecker and Ansley (1983), but with σ^2 and the smoothing parameter λ estimated via the maximum likelihood. It is unclear whether (11.24) holds for such intervals.

11.7 SOFTWARE

As mentioned earlier, efficient algorithms for the computation of smoothing splines as outlined in Sections 11.5 and 11.6 have been developed by Gu et al. (1989) and Gu and Wahba (1991). The algorithms have been implemented in a collection of RATFOR (Kernighan, 1975) routines known as RKPACK, first released in 1989 (Gu, 1989). Routines from public domain linear algebra libraries

BLAS and LINPACK have been used extensively in RKPACK routines as building blocks. The user interface of RKPACK is through two drivers `dsidr`, and `dmudr`, that solve (11.10) with automatically selected single or multiple smoothing parameters, and two utility routines `dsms`, `dcrdr` that compute $(S^T M^{-1} S)^{-1}$ in (11.27) and $\boldsymbol{c}_\beta$, $\boldsymbol{d}_\beta$ in (11.28). RKPACK has been deposited to Netlib and StatLib. For the latest version of RKPACK, check `http://www.stat.purdue.edu/~chong/software.html`.

To further facilitate the application of the techniques developed, a suite of R functions have been developed to serve as a front end to RKPACK. R (Ihaka and Gentleman 1996), an open source clone of S distributed under the GNU General Public License, practices active garbage collecting, which makes it feasible to load and execute the memory hungry RKPACK routines. Add-on modules in R are known as packages as in S. The facilities to be described below comprise part of the GSS package, where GSS abbreviates General Smoothing Splines. The latest public release of the GSS package can be found at any one of the CRAN (Comprehensive R Archive Network) sites along with other R resources. The master CRAN site is at `http://cran.r-project.org`.

We will assume some working knowledge of the modeling facilities in R, which for the purpose is a synonym of S; for a good reference on the subject, see Venables and Ripley (1999). The primary user interface of GSS for penalized least-squares regression is through the fitting function `ssanova` and the associated methods `predict`, `summary`, `residuals`, and `fitted`. The usage of these functions is very similar to that of the `lm` suite for linear models, as can be seen in the following examples.

Example 11.1 (Cubic spline) Assume that the GSS package is installed. At the R prompt, the following command loads the GSS package into R.

```
library(gss)
```

The following sequence generates some synthetic data and fits a cubic spline to the data, with the smoothing parameter selected by $V(\lambda)$.

```
x <- runif (100)
y <- 5 + 3 * sin(2*pi*x) + rnorm(x)
cubic.fit <- ssanova (y~x, type="cubic", method="v")
```

The arguments `type="cubic"` and `method="v"` are the defaults so they can be omitted. The results assigned to `cubic.fit` is a list of objects of class `"ssanova"`. To evaluate the fit on a grid for plotting purposes, one may try the following.

```
range <- max(x) - min(x)
grid <- seq(min(x)-.05*range, max(x)+.05*range, len=51)
est <- predict (cubic.fit, data.frame (x=grid), se.fit=TRUE)
```

The flag `se.fit=TRUE` requests the calculation of the posterior standard deviation corresponding to the evaluated posterior mean. Now `est$fit` and `est$se.fit`

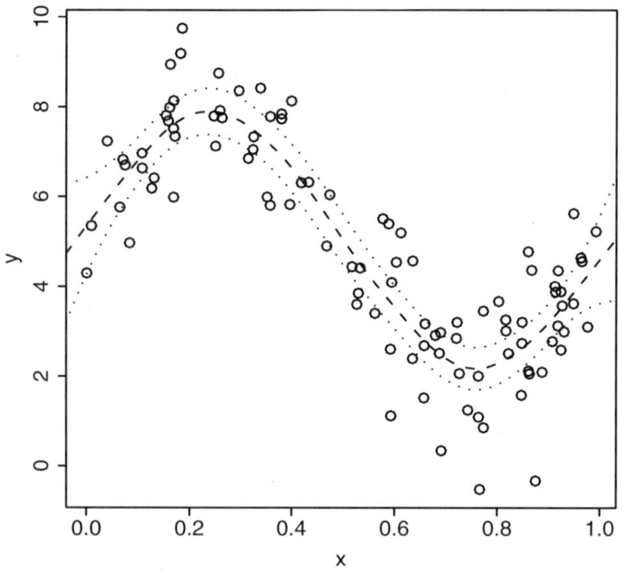

FIGURE 11.1 Cubic spline (Example 11.1).

contain the posterior mean and the posterior standard deviation on the grid, respectively, and the plot containing the data, the fit, and the 95% Bayesian confidence intervals can easily be produced by the following commands.

```
plot(x,y); lines(grid,est$fit,col=2)
lines(grid,est$fit+1.96*est$se.fit, col=3)
lines(grid,est$fit–1.96*est$se.fit, col=3)
```

In case you are not working in front of a color monitor, you may want to replace the `col` (color) options above by some `lty` (linetype) options. The result is displayed in Figure 11.1. □

Example 11.2 (Tensor product cubic spline) The following sequence generates some synthetic data and fits a tensor product cubic spline to the data, with the smoothing parameters selected by $M(\lambda, \theta)$.

```
x1 <- rnorm(100); x2 <- rnorm(100)
y <- 5 + 3 * sin(2*pi*(x1–x2)) + rnorm(x1)
tpcubic.fit <- ssanova (y~x1*x2, method="m",maxiter=50)
```

Note that `type="cubic"` is omitted in the call to `ssanova`. The option `maxiter=50` tells `ssanova` not to give up too soon when the iteration does not converge within 30 steps (the default). The model has four terms labeled `"1"`, `"x1"`, `"x2"`, and `"x1 : x2"`, representing $f_\emptyset$, f_1, f_2, and $f_{1,2}$. To evaluate the fit on a grid, one may try the following, which require sufficient memory be allocated for the R session; check `help (Memory)` to see how to increase memory allocation.

```
range1 <- max(x1) - min(x1)
grid1 <- seq(min(x1)-.05*range1,max(x1)+.05*range1, len=31)
range2 <- max(x2) - min(x2)
grid2 <- seq(min(x2)-.05*range2,max(x2)+.05*range2, len=31)
new <- data.frame(x1=rep(grid1,31), x2=rep(grid2,rep(31,31)))
est <- predict (tpcubic.fit, newdata=new, se.fit=TRUE)
post.mean <- matrix(est$fit,31,31)
post.stdev <- matrix(est$se.fit,31,31)
```

Now let us plot the contours of the posterior mean and the posterior standard deviation, with the data superimposed.

```
filled.contour(grid1,grid2,post.mean, main="post mean")
points(x1,x2)
filled.contour(grid1,grid2,post.stdev, main="post st dev")
points(x1,x2)
```

Note how quickly the posterior standard deviation takes off beyond the data range. By default, `predict` evaluates the overall function, but a partial sum of selected model terms can also be obtained via the specification of an optional argument `include`. For example, the following command returns the interaction on the grid.

```
est.int <- predict(tpcubic.fit, new, se=TRUE, include="x1:x2")
```

One can now plot the contours of the interaction and compare with those of the overall function (see Figure 11.2). □

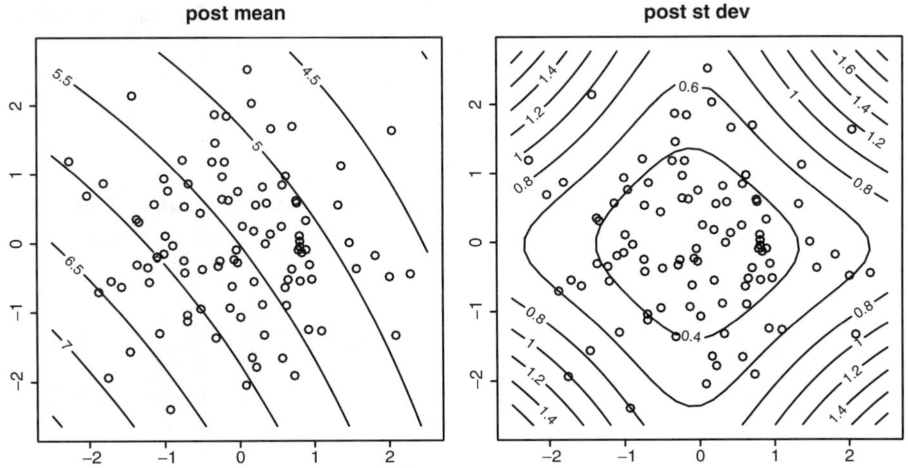

FIGURE 11.2 Tensor product cubic spline (Example 11.2).

The data ranges in the above examples are rather arbitrary whereas the internal calculations are done through model constructions involving (11.5), which is defined on [0, 1]. The program simply maps the marginal domains into [0, 1] and uses the formulas given in Sections 11.2 and 11.4 to calculate the covariance functions for the random effects. The marginal domains are specified via the extension of the data ranges by a certain percentage on both ends; the default is 5% as hinted by the grid span used in the examples. By construction, the averaging operators defining the ANOVA decomposition are integrations over the marginal domains as in the example of Section 11.3. Further details concerning `ssanova` and the affiliated methods can be found in the on-line documentation distributed with the code.

The generic algorithms implemented in RKPACK provide a universal numerical engine for all models sharing the structure of (11.9). For any specific model, all one needs are the null space basis ϕ_ν that generates the S matrix and the reproducing kernels R_β that generate the Q matrix. All that `ssanova` does is to generate the ϕ_ν and R_β functions based on the model formula for certain "types" of marginal construction, to calculate the matrices by evaluating the functions at the data points, and to pass the matrices along with other inputs to the RKPACK drivers. Besides the cubic spline marginal, other "types" supported in `ssanova` include the linear splines with $J(f) = \int (f')^2$ and the thin-plate splines to be discussed in Section 11.10. Clearly, there is no way to round up all useful models in utilities like `ssanova`, and if one desires to fit a model that is not "canned" in `ssanova` or the like, one will have to do things "manually" through the secondary GSS user interface, which simply consists of the R wrappers of the RKPACK user interface; details are of less general interest and hence omitted.

11.8 COSINE DIAGNOSTICS

The ANOVA decomposition is well defined by the side conditions, but due to possible structures in the design, the empirical estimation of individual terms may suffer identifiability problems similar to collinearity in linear models (i.e., concurvity). Also, some estimated terms may contain nothing but noise. Simple geometric diagnostics were proposed in Gu (1992a) to deal with the problems.

Consider $f = \sum_{\beta=0}^{p} f_\beta$, where $f_0 \propto 1$ is the constant and f_β, $\beta > 0$ are other terms in an ANOVA decomposition. Evaluating a fit at the design points, one obtains a retrospective linear model

$$\boldsymbol{Y} = \boldsymbol{f}_0 + \boldsymbol{f}_1 + \cdots + \boldsymbol{f}_p + \boldsymbol{e}, \tag{11.29}$$

where $\boldsymbol{f}_\beta = (f_\beta(x_1), \ldots, f_\beta(x_n))^T$. Removing the constant effect by projecting (11.29) onto $\{\mathbf{1}\}^\perp$, one gets

$$\boldsymbol{Y}^* = \boldsymbol{f}_1^* + \cdots + \boldsymbol{f}_p^* + \boldsymbol{e}^*. \tag{11.30}$$

The diagnostics are based on (11.30). The collinearity indices κ_β of $(\boldsymbol{f}_1^*, \ldots, \boldsymbol{f}_p^*)$ (Stewart, 1987), which are the diagonals of the inverse of the correlation matrix $(\cos(\boldsymbol{f}_\beta, \boldsymbol{f}_\gamma^*))$ of the $\boldsymbol{f}_\beta^*$, measure the identifiability of f_β in the fit. The $\boldsymbol{f}_\beta^*$ are supposed to predict the response $\boldsymbol{Y}^*$, so a near-orthogonal angle between an $\boldsymbol{f}_\beta^*$ and $\boldsymbol{Y}^*$ indicates a noise term. Signal terms should be reasonably orthogonal to the residuals, hence a large cosine between an $\boldsymbol{f}_\beta^*$ and an $\boldsymbol{e}^*$ makes a term suspect. Among informative measures for the signal to noise ratio in the data are $\cos(\boldsymbol{Y}^*, \boldsymbol{e}^*)$ and $R^2 = \|\boldsymbol{Y}^* - \boldsymbol{e}^*\|^2/\|\boldsymbol{Y}^*\|^2$. Finally, a very small Euclidean norm of an $\boldsymbol{f}_\beta^*$ as compared to that of $\boldsymbol{Y}^*$ also indicates a negligible term. These geometric diagnostics are collectively referred to as the cosine diagnostics, since they are largely based on the cosines among the vectors appearing in (11.30).

As illustrations of the use and effectiveness of the cosine diagnostics, we now analyze a few simple synthetic examples on $[0, 1]^3$ using the `ssanova` facilities in the R package GSS.

Example 11.3 (Independent design) First generate some synthetic data as follows:

```
x1 <- runif(100)
x2 <- runif(100)
x3 <- runif(100)
y <- 10*sin(pi*x2)+exp(3*x3)+5*cos(2*pi*(x1-x2))+3*rnorm(x1)
```

A tensor product linear spline model is then fit to the data.

```
fit <- ssanova (y~x1*x2*x3-x1:x2:x3, type="linear", met="v")
```

Note that a tensor product cubic spline is infeasible here as it has way too many random effects (12, to be exact). In comparison, the null space of the linear spline penalty $J(f) = \int (f')^2$ has only one dimension so the above model only involves six random effects. The diagnostics for the fit can be requested from the method `summary`.

```
sum.fit <- summary(fit, diagnostics=TRUE)
```

A look at the κ_β terms confirms that there is no identifiability problem with this fit.

```
round(sum.fit$kappa,2)
#    x1     x2     x3    x1:x2  x1:x3  x2:x3
#  1.07   1.03   1.12    1.08   1.04   1.12
```

Checking the order diagnostics, one finds that the interactions "`x1:x3`" and "`x2:x3`" have effectively been eliminated by the fitting procedure. The main effect "`x1`" is also largely noise, as it is virtually orthogonal to the response it is supposed to predict.

```
round(sum.fit$cosines,2)
#            x1    x2    x3  x1:x2  x1:x3  x2:x3  yhat     y     e
# cos.y -0.02  0.39  0.64   0.57   0.14   0.31  0.99  1.00  0.49
# cos.e  0.19  0.16  0.10   0.35   0.36   0.23  0.38  0.49  0.00
# norm   8.99 29.28 54.11  40.78   0.00   0.00 69.74 73.98 10.41
```

Eliminating "x1:x3" and "x2:x3" but keeping "x1" due to the presence of "x1:x2", a new model is fitted to the data.

```
fit.new <- ssanova (y~x1*x2+x3, type="linear", met="v")
```

A quick check shows that there is virtually no change in the diagnostics associated with the remaining terms. The elimination of "x1:x3" and "x2:x3" knocks out six smoothing parameters from the cubic spline model, so we are tempted to give it a try.

```
fit.new.c <- ssanova (y~x1*x2+x3, type="cubic", met="v")
```

The diagnostics for the cubic spline fit agree well with those for the linear spline fit.

```
round(summary(fit.new.c, TRUE)$cos,2)
#            x1    x2    x3  x1:x2  yhat     y     e
# cos.y -0.05  0.38  0.63   0.50  0.96  1.00  0.38
# cos.e  0.09  0.05  0.02   0.10  0.11  0.38  1.00
# norm   9.44 30.20 55.87  40.13 68.98 73.98 20.17
```

Note that the residuals generally are larger in the cubic spline fit because the fit doesn't bend as easily as the linear spline fit.

Note also that these sequences will show slightly different numbers from one run to the next, since no attempt was made to coordinate the seeds in random number generation. □

Example 11.4 (Simple aliasing design) Instead of an independent design, we now put $x_{i\langle 1\rangle}$ and $x_{i\langle 2\rangle}$ on a curve to create some identifiability problem.

```
x2 <- runif(100)
x3 <- runif(100)
x1 <- sqrt(x2)
y <- 10*sin(pi*x2)+exp(3*x3)+5*cos(2*pi*(x1-x2))+3*rnorm(x1)
```

Fit a tensor product linear spline model to the data and obtain the diagnostics. The κ_β say that "x1" and "x2" are highly concurve, and the norms indicate that the "x1" main effect and the interactions are effectively eliminated by the fitting algorithm. Note that the same fit would look smoother in the "x2" axis since the data spread out more.

```
fit <- ssanova (y~x1*x2*x3-x1:x2:x3, type="linear", met="v")
sum.fit <- summary(fit, diagnostics=TRUE)
```

```
round(sum.fit$kappa,2)
#     x1     x2    x3  x1:x2    x1:x3   x2:x3
# 16.80 21.79 1.41    2.64     2.80    2.41
round(sum.fit$cosines,2)
#           x1     x2     x3 x1:x2 x1:x3 x2:x3  yhat      y     e
# cos.y 0.47   0.53   0.77   0.49   0.45   0.31   0.95   1.00   0.51
# cos.e 0.23   0.24   0.09   0.25   0.26   0.45   0.21   0.51   1.00
# norm  0.00  28.24  44.73   0.00   0.00   0.00  51.90  59.11  19.46
```

Formally removing the terms and refitting, no change is to be found in the diagnostics but in the κ_β terms. We shall skip ahead to the cubic spline fit and get its diagnostics.

```
fit.new.c <- ssanova (y~x2+x3, type="cubic", met="v")
round(summary(fit.new.c, TRUE)$cos,2)
#           x2     x3  yhat      y     e
# cos.y  0.39   0.76   0.87   1.00   0.52
# cos.e  0.05   0.01   0.04   0.52   1.00
# norm  23.91  45.39  50.34  59.11  29.16
```

Once again, the cubic spline fit has larger residuals. □

Example 11.5 (Complex aliasing design) We now change the aliasing pattern to $x_{i\langle 1\rangle} = (x_{i\langle 2\rangle}^2 + x_{i\langle 3\rangle}^2)/2$, and obtain the tensor product linear spline fit and its diagnostics.

```
x2 <- runif(100)
x3 <- runif(100)
x1 <- (x2^2+x3^2)/2
y <- 10*sin(pi*x2)+exp(3*x3)+5*cos(2*pi*(x1–x2))+3*rnorm(x1)
fit <- ssanova (y~x1*x2*x3–x1:x2:x3, type="linear", met="v")
sum.fit <- summary(fit, diagnostics=TRUE)
```

The κ_β indicate that the terms are concurve, but no clear pairing is seen. The fitting procedure effectively eliminated all terms involving "x1".

```
round(sum.fit$kappa,2)
#    x1    x2    x3  x1:x2   x1:x3   x2:x3
# 4.46 3.31 3.88    3.07    2.52    2.64
round(sum.fit$cosines,2)
#           x1     x2     x3 x1:x2 x1:x3  x2:x3  yhat      y     e
# cos.y 0.41   0.46   0.81   0.69   0.53    0.56   0.97   1.00   0.49
# cos.e 0.10   0.15   0.14   0.32   0.29    0.28   0.25   0.49   1.00
# norm  0.00  35.07  63.81   0.00   0.00   20.00  81.51  90.42  23.93
```

Formally removing the terms and refitting a cubic spline model, the diagnostics point to no problems.

```
fit.new <- ssanova (y~x2*x3, type="cubic", met="v")
sum.fit.new <- summary(fit.new, diagnostics=TRUE)
round(sum.fit.new$kappa,2)
#    x2   x3 x2:x3
# 1.01 1.09  1.10
round(sum.fit.new$cosines,2)
#          x2    x3 x2:x3  yhat     y     e
# cos.y  0.44  0.81  0.49  0.95  1.00  0.36
# cos.e  0.06  0.02  0.07  0.06  0.36  1.00
# norm  37.33 67.60 19.21 84.60 90.42 27.59
```

The residuals are again larger in the cubic spline fit. □

11.9 PARTIAL SPLINES

In certain applications, one needs to add a few parametric terms to the otherwise nonparametric estimation. Partial splines provide a convenient tool for such modeling.

Let $\{\zeta_\mu(x)\}_{\mu=1}^q$ be a set of known functions on the domain $\mathcal{X}$. The minimizer of

$$\frac{1}{n}\sum_{i=1}^{n}\left(Y_i - \sum_{\mu=1}^{q}\beta_\mu\zeta_\mu(x_i) - f(x_i)\right)^2 + \lambda J(f), \tag{11.31}$$

with respect to β_μ and $f \in \mathcal{H} = \{f : J(f) < \infty\}$, is called a partial spline. Under the empirical Bayes model, partial splines are easily seen to be trivial extensions of the standard smoothing splines: One simply adds $\{\zeta_\mu(x)\}_{\mu=1}^q$ to the list of fixed effects. The solution expression (11.9) and the computation of the fit remain intact with the augmented fixed effect list.

Example 11.6 (Cubic spline with jump) To estimate a function $\tilde{\eta}$ that has a possible jump at a known location $x = 0.7$ but otherwise believed to be smooth on [0, 1], one may minimize

$$\frac{1}{n}\sum_{i=1}^{n}(Y_i - \beta I_{[x_i>0.7]} - f(x_i))^2 + \lambda\int_0^1 (f'')^2$$

with respect to β and $f \in \{\int_0^1 (f'')^2 < \infty\}$. The solution has an expression

$$\tilde{f}(x) = \sum_{\nu=1}^{3} d_\nu\phi_\nu(x) + \sum_{i=1}^{n} c_i R_J(x_i, x),$$

where $\phi_1(x) = 1$, $\phi_2(x) = (x - 0.5)$, $\phi_3(x) = I_{[x>0.7]}$, and

$$R_J(x_1, x_2) = k_2(x_1)k_2(x_2) - k_4(|x_1 - x_2|).$$

The following R code generates some synthetic data and fits a cubic spline with a jump.

```
x <- (0:100)/100
y <- 5+3*sin(2*pi*x)+2*(x>.7)+rnorm(x)
partial.fit <- ssanova (y~x, partial=as.numeric(x>.7), ext=0)
```

The argument `ext=0` sets the domain to be the data range [0, 1], that is, 0 per cent extension beyond the range. One can evaluate and plot the fit as follows:

```
new <- data.frame(x=x, partial=as.numeric(x>.7))
est <- predict(partial.fit, new, se=TRUE)
plot(x,y); lines(x, est$fit, col=2)
lines(x,est$fit+1.96*est$se, col=3)
lines(x,est$fit-1.96*est$se, col=3)
```

Obviously, one should avoid using the variable name "`partial`" for anything appearing in the model formula. The plot is displayed in Figure 11.3. □

Wahba (1990, Chapter 6) and Green and Silverman (1994, Chapter 4) have more discussions on partial splines.

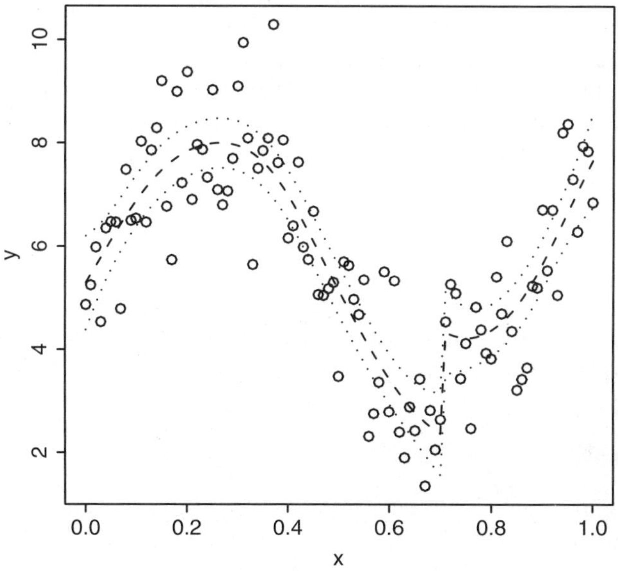

FIGURE 11.3 Cubic spline with jump (Example 11.6).

11.10 THIN-PLATE SPLINES

Before tensor product splines came along, thin-plate splines were the prime tool for multivariate smoothing via the penalized least-squares approach. A thin-plate spline is the minimizer of

$$\frac{1}{n}\sum_{i=1}^{n}(Y_i - f(x_i))^2 + \lambda J_m^d(f) \tag{11.32}$$

on the d-dimensional Euclidean domain $\mathcal{X} = (-\infty, \infty)^d$, where

$$J_m^d(f) = \sum_{\alpha_1+\cdots+\alpha_d=m} \frac{m!}{\alpha_1!\ldots\alpha_d!} \int\cdots\int \left(\frac{\partial^m f}{\partial x_{\langle 1\rangle}^{\alpha_1}\ldots\partial x_{\langle d\rangle}^{\alpha_d}}\right)^2.$$

The null space of J_m^d is composed of polynomials of up to $(m-1)$ total order, which is of dimension

$$M = \binom{d+m-1}{d}.$$

For $J_m^d(f)$ to have an "inverse," it is necessary that $2m - d > 0$; see, for example, Wahba (1990). Note that for $d = 1$ and $m = 2$, a thin-plate spline is simply a cubic spline.

When it does exist, the "inverse" of $J_m^d(f)$ has a rather complicated expression, which requires the knowledge of reproducing kernel Hilbert spaces, hence we omit it here. Interested readers may find related discussion in, for example, Wahba (1990) and Gu and Wahba (1993b).

As a means of estimating multivariate functions, thin-plate splines and tensor product splines target different kinds of applications. Note that $J_m^d(f)$ is invariant under a rotation of the coordinates, so thin-plate splines are natural choices when one needs to estimate spatial effects. When the coordinates are individually interpretable, however, tensor product splines are the ones to use. When spatial effects are to be estimated jointly with other effects, thin-plate splines can be used as marginals in the construction of tensor product splines; see Gu and Wahba (1993b).

Example 11.7 (Tensor product thin-plate spline) To illustrate the use of thin-plate splines, consider the lake acidity data `lake.acid` included in the GSS package. The data, which concern 112 lakes in the Blue Mountains, are extracted from the Eastern Lake Survey of 1984 implemented by the US Enviromental Protection Agency. After loading the GSS package, the data frame `lake.acid` can be loaded through the command.

```
data(lake.acid)
```

The three components of the data frame are `ph`, the pH value of surface water, `cal`, the calcium concentration, and `geog`, the geographic location. The geographic locations are Euclidean coordinates with respect to a local origin, in the unit of the Earth radius, calculated from the longitude and the latitude.

With ph as the response, we have two independent variables cal and geog, where geog is two dimensional. A model of the form

$$Y = f_0 + f_c(c) + f_g(g) + f_{c,g}(c, g) + \epsilon$$

using thin-plate marginals can be fitted by

```
fit <- ssanova (ph~log(cal)*geog, type="tp", data=lake.acid,
                order=2)
```

where cal is used in logarithm. The option order=2 that specifies $m = 2$ is the default for type="tp" and has no effect for other types. The fit is invariant with respect to arbitrary rotations of geog. To evaluate the fitted $f_g(g)$ and the associated posterior standard deviation on a grid, one may use

```
grid <- seq(-0.04, 0.04, len=21)
new <- model.frame(~geog, list(geog=cbind(rep(grid,21),
                     rep(grid, rep(21,21)))))
est <- predict(fit, new, se=TRUE, inc="geog")
```

where include="geog" specifies the components to be included in the evaluation. The fit and the standard error can now be plotted easily by

```
par(pty="s")
contour(grid, grid, matrix(est$fit,21,21), col=1)
contour(grid, grid, matrix(est$se,21,21), add=TRUE, col=2)
points(lake.acid$geog)
```

where the locations are superimposed (see Figure 11.4). □

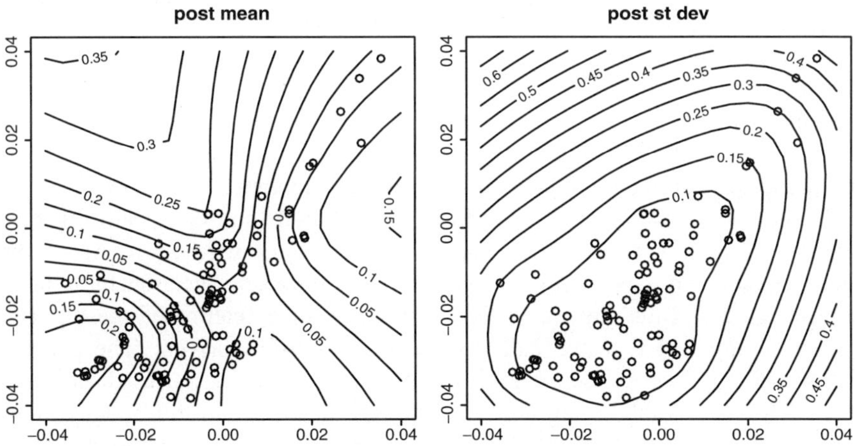

FIGURE 11.4 Tensor product thin-plate spline: geography effect (Example 11.7).

11.11 NON-GAUSSIAN REGRESSION

The penalized least-squares method is actually the specialization to Gaussian regression of the more general penalized likelihood method. Replacing the least-squares functional in (11.2) by other log likelihoods, one may use the technique to estimate functions in other settings.

Consider independent observations $(Y_i|x_i)$, $i = 1, \ldots, n$, from an exponential family distribution with density $\exp\{[yf(x) - b(f(x))]/a(\phi) + c(y, \phi)\}$. The parameter $a(\phi)$, possibly known or otherwise considered as a nuisance parameter, is assumed common to all the observations, finite and positive. Of interest is the estimation of the regression function $f(x)$ on domain $\mathcal{X}$. The penalized likelihood method estimates $f(x)$ by minimizing

$$-\frac{1}{n}\sum_{i=1}^{n}\{Y_i f(x_i) - b(f(x_i))\} + \frac{\lambda}{2}J(f) \tag{11.33}$$

for $f \in \{f : J(f) < \infty\}$, where the first term in (11.33) is a constant away from the minus log likelihood and $J(f)$ is the roughness functional; see, for example, O'Sullivan et al. (1986) and Gu (1990). It is easy to check that (11.33) reduces to (11.2) when the density is Gaussian.

Consider binary responses with $P(Y = 1) = 1 - P(Y = 0) = p$ as an example. The Bernoulli density can be written as $\exp\{yf - \log(1 + e^f)\}$, where $f = \log[p/(1 - p)]$ is the logit. Observing (x_i, Y_i), a nonparametric logistic regression may be performed by minimizing

$$-\frac{1}{n}\sum_{i=1}^{n}\{Y_i f(x_i) - \log(1 + e^{f(x_i)})\} + \frac{\lambda}{2}J(f).$$

The minimizer of (11.33) is the posterior mode under the empirical Bayes model for f as described in Sections 11.2 and 11.4. The expression (11.9) still holds for the minimizer of (11.33), but after plugging it into (11.33), the resulting counterpart of (11.10) is in general nonquadratic. The calculation of the minimizer needs iteration even with fixed smoothing parameters. The empirical selection of the smoothing parameters further complicates the matter.

The development of tools for penalized likelihood (non-Gaussian) regression is nearly as mature as that for penalized least-squares regression. An efficient algorithm with effective empirical smoothing parameter selection built in was proposed and justified in Gu (1990, 1992b). A nice feature of the algorithm is its execution via repeated calls to existing RKPACK drivers. The extension of Bayesian confidence intervals to non-Gaussian regression was discussed by Gu (1992c) and Wahba et al. (1995). Interested readers may find related details directly from the literature. Penalized likelihood regression models can be fitted using the `gssanova` function in GSS, as illustrated in the following example.

Example 11.8 (Logistic regression) The following R commands generate some synthetic data on a grid and fit a cubic spline logistic regression model with $J(f) = \int (f'')^2$.

```
test <- function(x)
        {.3*(1e6*(x^11*(1-x)^6)+1e4*(x^3*(1-x)^10))-2}
x <- (0:100)/100
p <- 1-1/(1+exp(test(x)))
y <- rbinom(x,3,p)
logit.fit <- gssanova(cbind(y,3-y)~x, family="binomial")
```

Note that `type="cubic"` is omitted in the call as it is the default. Equivalently, one may use a one-column response with weights to do the fitting.

```
logit.fit1 <- gssanova(y/3~x, family="binomial", weights=rep
                       (3,101))
```

To evaluate the fit on the grid, use

```
est <- predict(logit.fit, data.frame(x=x), se=TRUE)
```

To plot the fit with the data and the test function superimposed, use

```
plot(x,y/3, ylab="p")
lines(x,p, col=1)
```

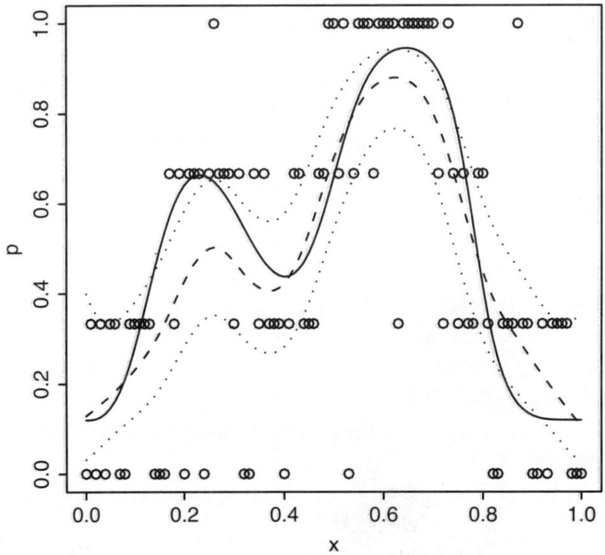

FIGURE 11.5 Logistic regression (Example 11.8).

```
lines(x,1-1/(1+exp(est$fit)), col=2)
lines(x,1-1/(1+exp(est$fit+1.96*est$se)), col=3)
lines(x,1-1/(1+exp(est$fit-1.96*est$se)), col=3)
```

where the vertical scale is in probability. The result is seen in Figure 11.5. □

Finally let me mention that the reader is strongly encouraged to produce the figures (in color) himself using the R code given in the examples.

REFERENCES

Ansley, C. F., and Kohn, R. (1985). Estimation, filtering and smoothing in state space models with incompletely specified initial conditions, *Ann. Statist.*, **11**, 1286–1316.

Craven, P., and Wahba, G. (1979). Smoothing noisy data with spline functions: Estimating the correct degree of smoothing by the method of generalized cross-validation. *Numer. Math.*, **31**, 377–403.

De Boor, C. (1978). *A Practical Guide to Splines*. Springer-Verlag, New York.

Green, P., and Silverman, B. W. (1994). *Nonparametric Regression and Generalized Linear Models*. Chapman and Hall, London, UK.

Gu, C. (1989). RKPACK and its applications: Fitting smoothing spline models. *Proc. Statist. Comput. Sec. Am. Statist. Assoc.*, 42–51.

Gu, C. (1990). Adaptive spline smoothing in non-Gaussian regression models. *J. Am. Statist. Assoc.*, **85**, 801–807.

Gu, C. (1992a). Diagnostics for nonparametric regression models with additive terms, *J. Amer. Statist. Assoc.*, **87**, 1051–1058.

Gu, C. (1992b). Cross validating non-Gaussian data. *J. Comput. Graph. Statist.*, **1**, 169–179.

Gu, C. (1992c). Penalized likelihood regression: A Bayesian analysis. *Statist. Sin.*, **2**, 255–264.

Gu, C., Bates, D. M., Chen, Z., and Wahba, G. (1989). The computation of GCV functions through Householder tridiagonalization with application to the fitting of interaction spline models, *SIAM J. Matrix Anal. Appl.*, **10**, 457–480.

Gu, C., and Qiu, C. (1993). Smoothing spline density estimation: Theory, *Ann. Statist.*, **21**, 217–234.

Gu, C., and Wahba, G. (1991). Minimizing GCV/GML scores with multiple smoothing parameters via the Newton method. *SIAM J. Sci. Statist. Comput.*, **12**, 383–398.

Gu, C., and Wahba, G. (1993a). Smoothing spline ANOVA with component-wise Bayesian confidence intervals, *J. Comput. Graph. Statist.*, **2**, 97–117.

Gu, C., and Wahba, G. (1993b). Semiparametric ANOVA with tensor product thin plate splines, *J. Roy. Statist. Soc. B*, **55**, 353–368.

Hutchinson, M. F., and de Hoog, F. R. (1985). Smoothing noisy data with spline functions, *Numer. Math.*, **47**, 99–106.

Ihaka, R., and Gentleman, R. (1996). R: A language for data analysis and graphics. *J. Comput. Graph. Statist.*, **5**, 299–314.

Kernighan, B. W. (1975). RATFOR—A preprocessor for a rational FORTRAN. In *UNIX Programmer's Manual*, Bell Laboratories.

Kimeldorf, G., and Wahba, G. (1971). Some results on Tchebycheffian spline functions. *J. Math. Anal. Applic.*, **33**, 82–85.

Li, K.-C. (1986). Asymptotic optimality of C_L and generalized cross-validation in the ridge regression with application to spline smoothing, *Ann. Statist.*, **14**, 1101–1112.

Nychka, D. (1988). Bayesian confidence intervals for smoothing splines. *J. Amer. Statist. Assoc.*, **83**, 1134–1143.

O'Sullivan, F. (1985). Comment on "Some aspects of the spline smoothing approach to nonparametric regression curve fitting" by B. W. Silverman. *J. Roy. Statist. Soc. B*, **47**, 39–40.

O'Sullivan, F., Yandell, B., and Raynor, W. (1986). Automatic smoothing of regression functions in generalized linear models. *J. Amer. Statist. Assoc.*, **81**, 96–103.

Rao, C. R. (1973). *Linear Statistical Inference and Its Applications* (2nd ed.). Wiley, New York.

Schumaker, L. (1981). *Spline Functions: Basic Theory*. Wiley, New York.

Stewart, G. W. (1987). Collinearity and least square regression, *Statist. Sci.*, **2**, 68–100.

Venables, W. N., and Ripley, B. D. (1999). *Modern Applied Statistics with S-PLUS* (3rd ed.). Springer-Verlag, New York.

Wahba, G. (1983). Bayesian "confidence intervals" for the cross-validated smoothing spline, *J. Roy. Statist. Soc. B*, **45**, 133–150.

Wahba, G. (1985). A comparison of GCV and GML for choosing the smoothing parameter in the generalized spline smoothing problem. *Ann. Statist.*, **13**, 1378–1402.

Wahba, G. (1990). *Spline Models for Observational Data*. SIAM, Philadelphia, PA.

Wahba, G., Wang, Y., Gu, C., Klein, R., and Klein, B. (1995). Smoothing spline ANOVA for exponential families, with application to the Wisconsin epidemiological study of diabetic retinopathy. *Ann. Statist.*, **23**, 1865–1895.

Wecker, W., and Ansley, C. F. (1983). The signal extraction approach to signal extraction and spline smoothing, *J. Amer. Statist. Assoc.*, **78**, 81–89.

CHAPTER 12

Multivariate and Semiparametric Kernel Regression

Wolfgang Härdle and Marlene Müller
Humboldt-Universität zu Berlin, Berlin, Germany

12.1 INTRODUCTION

Nonparametric smoothing methods serve three essential needs in statistical data analysis. First, they provide a flexible analysis tool, often based on interactive graphical data representation (e.g., Scott 1992). Second, they help in constructing a model from observations — for example, by comparison with concurrent models (e.g., Müller, 1988). Third, they provide pilot estimators in adaptation problems (e.g., Newey and Stoker, 1993.) In this chapter, we present the multivariate kernel smoother, examine the asymptotic properties of both density and regression estimators, and review applications of nonparametric smoothing techniques to semiparametric statistics.

Multivariate nonparametric density estimation is a pilot tool that is often used for examining the structure of data. Regression smoothing helps in investigating the association between covariates and responses. Here, we concentrate on kernel smoothing using local polynomial fitting which includes the Nadaraya–Watson estimator. Some theory on asymptotic behavior and bandwidth selection also is provided. In the applications of the kernel technique, we focus on the semiparametric paradigm. We describe the single-index model (SIM) and the generalized partial linear model (GPLM) in more detail.

12.2 MULTIDIMENSIONAL SMOOTHING WITH KERNELS

In this section, we review kernel-smoothing methods for estimating density and regression functions. Many ideas that we present — in particular, for asymptotics,

Smoothing and Regression: Approaches, Computation, and Application
Edited by Michael G. Schimek
ISBN 0-471-17946-9

bandwidth choice and graphical representation — are similar for both purposes. We can, however, introduce only a small part of the available material. In particular, as regards regression, we restrict the presentation to the random-design case. For a more detailed presentation of the subject, the reader should consult the monographs by Härdle (1990, 1991), Scott (1992), Fan and Gijbels (1995), and Wand and Jones (1995), Kernel regression for univariate data is discussed in detail in Chapter 3 of this volume.

12.2.1 Multivariate Kernel Density Estimation

The goal of multivariate nonparametric density estimation is to approximate the *probability density function* (pdf) $f(t) = f(t_1, \ldots, t_q)$ of the random variables $T = (T_1, \ldots, T_q)^T$. The multivariate kernel density estimator in the q-dimensional case is defined as

$$\hat{f}_h(t) = \frac{1}{n}\sum_{i=1}^{n} \frac{1}{h_1 \ldots h_q}\mathcal{K}\left(\frac{T_{i1} - t_1}{h_1}, \ldots, \frac{T_{iq} - t_q}{h_q}\right), \qquad (12.1)$$

where $\mathcal{K}: \mathbb{R}^q \to \mathbb{R}$ denotes a multivariate kernel function. Note that (12.1) assumes that the bandwidth is a vector of bandwidths $h = (h_1, \ldots, h_q)^T$.

What form will the multidimensional kernel function $\mathcal{K}(u) = \mathcal{K}(u_1, \ldots, u_q)$ take? The easiest way to answer this question is to use a *multiplicative* kernel

$$\mathcal{K}(u) = K(u_1) \cdot \ldots \cdot K(u_q),$$

where K denotes univariate kernel function. For univariate kernels with support $[-1, 1]$ (such as the Epanechnikov kernel $K(u) = 0.75(1 - u^2)I(|u| \leq 1)$), observations in a cube around t are used to estimate the density at the point t. An alternative is to use a genuine multivariate kernel function $\mathcal{K}(u)$, for example, the multivariate Epanechnikov

$$\mathcal{K}(u) \propto (1 - u^T u)I(u^T u \leq 1).$$

This type of multivariate kernel function can be obtained from a univariate one by defining

$$\mathcal{K}(u) \propto K(\|u\|), \qquad (12.2)$$

where $\|u\| = \sqrt{u^T u}$ denotes the Euclidean norm of the vector u. (Note that we use $\propto$ to indicate that $K(\|u\|)$ has to be multiplied by the appropriate constant.) Kernels of the form (12.2) use observations from a ball around t to estimate the pdf at t. This type of kernel is usually called *spherical* or *radially symmetric*, since $\mathcal{K}(u)$ has the same value for all u on a sphere around zero.

Figure 12.1 shows the contour lines from a bivariate product and a bivariate radially symmetric kernel on the left- and right-hand sides respectively. Note that the kernel weights correspond to equal bandwidths in each direction, that is,

$h = (h_1, h_2)^T = (1, 1)^T$. When we employ different bandwidths, the observations around t in the density estimate $\hat{f}_h(x)$ are used with different weights in both dimensions.

Another approach is to use a nonsingular, symmetric *bandwidth matrix* $\mathbf{H}$. The general form for the multivariate density estimator is then

$$\hat{f}_{\mathbf{H}}(t) = \frac{1}{n}\sum_{i=1}^{n} \frac{1}{\det(\mathbf{H})}\mathcal{K}\{\mathbf{H}^{-1}(T_i - t)\} = \frac{1}{n}\sum_{i=1}^{n} \mathcal{K}_{\mathbf{H}}(T_i - t). \tag{12.3}$$

(See Silverman, 1986, and Scott, 1992.) Here we introduce the short notation

$$\mathcal{K}_{\mathbf{H}}(\bullet) = \frac{1}{\det(\mathbf{H})}\mathcal{K}(\mathbf{H}^{-1}\bullet),$$

which is analogous to $K_h = K(\bullet/h)/h$ in the one-dimensional case. A bandwidth matrix includes all simpler cases as special cases. An equal bandwidth h in all dimensions corresponds to $\mathbf{H} = h\mathbf{I}_q$ where $\mathbf{I}_q$ denotes the $q \times q$ identity matrix. Different bandwidths, as in (12.1), are equivalent to $\mathbf{H} = \mathrm{diag}(h_1, \ldots, h_q)$, the diagonal matrix with elements $h_1, \ldots, h_q$.

What effect does the inclusion of off-diagonal elements have? We will see that a good rule of thumb is to use a bandwidth matrix proportional to $\hat{\boldsymbol{\Sigma}}^{-1/2}$, where $\hat{\boldsymbol{\Sigma}}$ is the covariance matrix of the data. Hence, using such a bandwidth corresponds to a transformation of the data to an identity covariance matrix. As a consequence, we can use bandwidth matrices to correct for correlation between the components of T. We have plotted the contour curves of product and radially

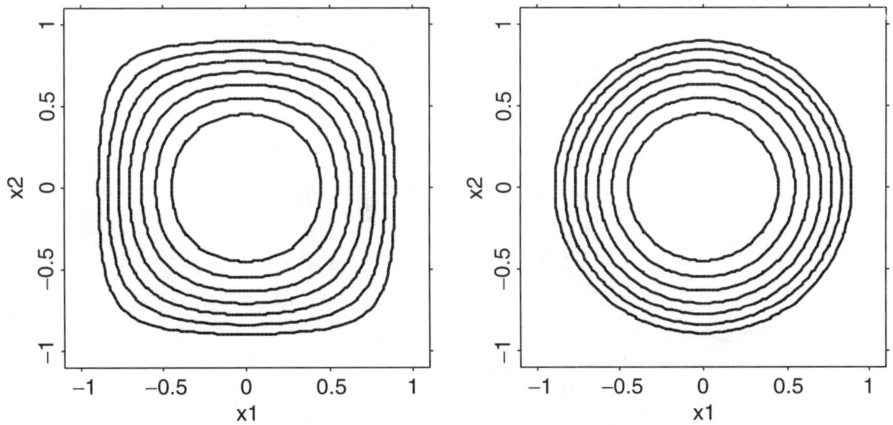

FIGURE 12.1 Contours from bivariate product (left) and bivariate radially symmetric (right) Epanechnikov kernel.

symmetric Epanechnikov weights with bandwidth matrix

$$\mathbf{H} = \begin{pmatrix} 1 & 0.5 \\ 0.5 & 1 \end{pmatrix}^{1/2}$$

(i.e., $\mathcal{K}_{\mathbf{H}}(u) = \mathcal{K}(\mathbf{H}^{-1}u)/\det(\mathbf{H})$) in Figure 12.2.

We next consider statistical properties such as the bias and variance, as well as the issue of bandwidth selection and applications for this estimator. We formulate all results for estimators with bandwidth matrices and multivariate kernel function $\mathcal{K}$.

12.2.1.1 Bias, Variance, and Asymptotics

A consequence of the standard assumption

$$\int \mathcal{K}(u)\,du = 1 \tag{12.4}$$

on the nonnegative kernel $\mathcal{K}$ is that the estimate $\hat{f}_{\mathbf{H}}$ is a density function (i.e., $\int \hat{f}_{\mathbf{H}}(t)\,dt = 1$). The estimate

$$\hat{f}_{\mathbf{H}}(t) = \frac{1}{n}\sum_{i=1}^{n} \mathcal{K}_{\mathbf{H}}(T_i - t) = f(t) + o_p(1) \tag{12.5}$$

is consistent at any point t of continuity of f; that is, if $n \to \infty$, $\det(\mathbf{H}) \to 0$ and $n\det(\mathbf{H}) \to \infty$. (See, e.g., Ruppert and Wand, 1994.) The derivation of the *mean squared error* (MSE) and the *mean integrated squared error* (MISE) is analogous to the one-dimensional case. We will sketch the asymptotic expansions and concentrate on the *asymptotic mean integrated squared error* (AMISE).

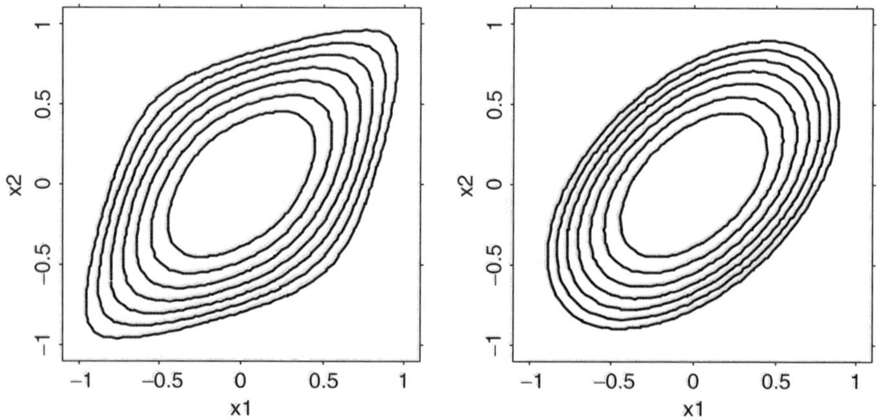

FIGURE 12.2 Contours from bivariate product (left) and bivariate radially symmetric (right) Epanechnikov kernel. Bandwidth matrix $\mathbf{H}$.

As usual, AMISE has a bias part AIB and a variance part AIV. The bias of $\hat{f}_{\mathbf{H}}(t)$ is $E\hat{f}_{\mathbf{H}}(t) - f(t)$, and the *integrated squared bias* (ISB) is

$$\mathrm{ISB}(\mathbf{H}) = \int \{E\hat{f}_{\mathbf{H}}(t) - f(t)\}^2 \, dt.$$

The *asymptotic integrated squared bias* AIB(**H**) is the first-order term of ISB(**H**); that is,

$$\frac{\mathrm{ISB}(\mathbf{H}) - \mathrm{AIB}(\mathbf{H})}{\mathrm{AIB}(\mathbf{H})} = o(1).$$

Note that we assume $\det(\mathbf{H}) \to 0$, $n \to \infty$ and $n \det(\mathbf{H}) \to \infty$. Now we define the *integrated variance* (IV)

$$\mathrm{IV}(\mathbf{H}) = \int E\{\hat{f}_{\mathbf{H}}(t) - E\hat{f}_{\mathbf{H}}(t)\}^2 \, dt$$

and the *asymptotic integrated variance* (AIV) in a manner analogous to AIB. Then the asymptotic mean integrated squared error can be calculated as

$$\mathrm{AMISE}(\mathbf{H}) = \mathrm{AIB}(\mathbf{H}) + \mathrm{AIV}(\mathbf{H}).$$

A detailed derivation of the components of AMISE can be found in Scott (1992) or Wand and Jones (1995) and the references therein. As in the univariate case, we use a second-order Taylor expansion. Here and in what follows, we denote the gradient by ∇ and the Hessian matrix of second-order partial derivatives of a function (here, f) by $\mathcal{H}$. Then the Taylor expansion of $f(\cdot)$ around t is

$$f(t+u) = f(t) + u^T \nabla_f(t) + \tfrac{1}{2} u^T \mathcal{H}_f(t) t + o(u^T u).$$

(See Wand and Jones, 1995, p. 94.) This equation leads to the expression

$$\begin{aligned} E\hat{f}_{\mathbf{H}}(t) &= \int \mathcal{K}_{\mathbf{H}}(u - t) f(u) \, du = \int \mathcal{K}(s) f(t + \mathbf{H}s) \, ds \\ &\approx \int \mathcal{K}(s) \left\{ f(t) + s^T \mathbf{H}^T \nabla_f(t) + \tfrac{1}{2} s^T \mathbf{H}^T \mathcal{H}_f(t) \mathbf{H} s \right\} ds. \qquad (12.6) \end{aligned}$$

If we assume, in addition to (12.4), that

$$\int u \mathcal{K}(u) \, du = 0_q$$

$$\int u u^T \mathcal{K}(u) \, du = \mu_2(\mathcal{K}) \mathbf{I}_q,$$

then (12.6) yields $E\hat{f}_{\mathbf{H}}(t) - f(t) \approx \frac{1}{2}\mu_2(\mathcal{K}) \operatorname{tr}\{\mathbf{H}^T \mathcal{H}_f(t) \mathbf{H}\}$. Hence,

$$\mathrm{AIB}(\mathbf{H}) = \tfrac{1}{4}\mu_2^2(\mathcal{K}) \int [\operatorname{tr}\{\mathbf{H}^T \mathcal{H}_f(t) \mathbf{H}\}]^2 \, dt.$$

As in univariate density estimation, the leading term of the variance part is the second moment of the estimate; that is,

$$\begin{aligned}
\operatorname{Var}\{\hat{f}_{\mathbf{H}}(t)\} &= \frac{1}{n}\int \{\mathcal{K}_{\mathbf{H}}(u-t)\}^2\,du - \frac{1}{n}\{E\hat{f}_{\mathbf{H}}(t)\}^2 \\
&\approx \int \frac{1}{n\det(\mathbf{H})}\mathcal{K}^2(s)f(t+\mathbf{H}s)\,ds \\
&\approx \int \frac{1}{n\det(\mathbf{H})}\mathcal{K}^2(s)\{f(t)+s^T\mathbf{H}^T\nabla_f(t)\}\,ds \\
&\approx \frac{1}{n\det(\mathbf{H})}\|\mathcal{K}\|_2^2 f(t),
\end{aligned}$$

where $\|\mathcal{K}\|_2$ denotes the q-dimensional L_2-norm of $\mathcal{K}$. Hence,

$$\operatorname{AIV}(\mathbf{H}) = \frac{1}{n\det(\mathbf{H})}\|\mathcal{K}\|_2^2,$$

and, in sum, we get the following AMISE formula for the multivariate kernel density estimator:

$$\operatorname{AMISE}(\mathbf{H}) = \frac{1}{4}\mu_2^2(\mathcal{K})\int [\operatorname{tr}\{\mathbf{H}^T\mathcal{H}_f(t)\mathbf{H}\}]^2\,dt + \frac{1}{n\det(\mathbf{H})}\|\mathcal{K}\|_2^2. \tag{12.7}$$

We now turn to the problem of how to choose the AMISE optimal bandwidth. Again, this is the bandwidth that balances the bias–variance trade-off in AMISE. Let h be a scalar such that $\mathbf{H} = h\mathbf{H}_0$ and $\det(\mathbf{H}_0) = 1$. Then AMISE can be written as

$$\operatorname{AMISE}(\mathbf{H}) = \frac{1}{4}h^4\mu_2^2(\mathcal{K})\int [\operatorname{tr}\{\mathbf{H}_0^T\mathcal{H}_f(t)\mathbf{H}_0\}]^2\,dt + \frac{1}{nh^q}\|\mathcal{K}\|_2^2.$$

If we only allow changes in h, the optimal orders for the smoothing parameter h and AMISE are

$$h_0 = O(n^{-1/(4+q)}) \text{ and } \operatorname{AMISE}(h_0\mathbf{H}_0) = O(n^{-4/(4+q)}).$$

Hence, this density estimator has a rather slow rate of convergence, especially if q is large. If we consider $\mathbf{H} = h\mathbf{I}_q$ (the same bandwidth in all q dimensions) and we fix the sample size n, then the AMISE optimal bandwidth has to be considerably larger than in the one-dimensional case in order to make sure that the estimate has reasonably small variability. Some ideas of comparable sample sizes to reach the same quality of density estimates over different dimensions can be found in Silverman (1986, p. 94) and Scott and Wand (1991). Moreover, the computational effort of this technique increases with the number of dimensions, q.

Therefore, multidimensional density estimation is usually not practically applied if $q \geq 5$.

12.2.1.2 Bandwidth Selection and Graphical Representation

The problem of an automatic, data-driven choice of the bandwidth $\mathbf{H}$ is of great importance in the multivariate case. In one or two dimensions, we may choose an "appropriate" bandwidth interactively by looking at the sequence of density estimates for different bandwidths. But how can this be done in three, four, or more dimensions? The problem of graphical representation arises, which we address next.

Theoretically, the bandwidth selection problem can be handled as in the one-dimensional case. Typically, one searches for a global bandwidth $\mathbf{H}$ or a local bandwidth $\mathbf{H}(t)$. Two approaches are frequently used in both cases:

- plug-in bandwidths — in particular, "rule-of-thumb" bandwidths.
- resampling methods — in particular, cross-validation and bootstrap techniques.

We introduce generalizations of Silverman's rule-of-thumb bandwidths and least-squares cross-validation to stress the analogy with one-dimensional bandwidth selectors.

The selection of rule-of-thumb bandwidths proceeds according to a formula that arises from a reference distribution. Obviously, the pdf of a multivariate normal distribution $N_q(\mu, \mathbf{\Sigma})$ is a good candidate for a reference distribution in the multivariate case. Suppose that the kernel $\mathcal{K}$ is Gaussian, that is, it is the pdf of $N_q(0_q, \mathbf{I}_q)$. Note that $\mu_2(\mathcal{K}) = 1$ and $\|\mathcal{K}\|_2^2 = 2^{-q}\pi^{-q/2}$ in this case. Hence, from (12.7) and the fact that

$$\int [\mathrm{tr}\{\mathbf{H}^T \mathcal{H}_f(t)\mathbf{H}\}]^2\,dt$$
$$= \frac{1}{2^{q+2}\pi^{q/2}\det(\mathbf{\Sigma})^{1/2}}[2\,\mathrm{tr}(\mathbf{H}^T\mathbf{\Sigma}^{-1}\mathbf{H})^2 + \{\mathrm{tr}(\mathbf{H}^T\mathbf{\Sigma}^{-1}\mathbf{H})\}^2],$$

we can easily derive rule-of-thumb formulas for different assumptions on $\mathbf{H}$ and $\mathbf{\Sigma}$.

In the simplest case, in which we consider $\mathbf{H}$ and $\mathbf{\Sigma}$ to be diagonal matrices (i.e., $\mathbf{H} = \mathrm{diag}(h_1, \ldots, h_q)$ and $\mathbf{\Sigma} = \mathrm{diag}(\sigma_1^2, \ldots, \sigma_q^2)$), we obtain

$$\tilde{h}_j = \left(\frac{4}{q+2}\right)^{1/(q+4)} n^{-1/(q+4)}\sigma_j. \qquad (12.8)$$

Note that this formula coincides with Silverman's (1986, p. 45) rule of thumb in the case where $q = 1$. Replacing the σ_j's by estimates and noting that the first factor is always between 0.924 and 1.059, we arrive at Scott's (1992, p. 152) rule,

$$\hat{h}_j = n^{-1/(q+4)}\hat{\sigma}_j.$$

It is difficult to derive the rule of thumb for general $\mathbf{H}$ and $\mathbf{\Sigma}$. However, (12.8) shows that it might be a good idea to choose the bandwidth matrix $\mathbf{H}$ proportional to $\mathbf{\Sigma}^{1/2}$. In this case, we get, as a generalization of Scott's rule,

$$\hat{\mathbf{H}} = n^{-1/(q+4)}\hat{\Sigma}^{1/2}. \tag{12.9}$$

This rule is equivalent to applying a Mahalanobis transformation on the data (to transform the estimated covariance matrix to identity), then computing the kernel estimate with equal bandwidths $h = n^{-1/(q+4)}$, and, finally, retransforming the estimated pdf back to the original scale.

But before we go on with applications, let us consider what we can do if we want to use a kernel different from the Gaussian. The idea of canonical kernels (Marron and Nolan, 1988) can be easily extended to the multivariate case. Consider a kernel $\mathcal{K}$ and all equivalent kernel functions $\mathcal{K}_\delta = \delta^{-q}\mathcal{K}(\cdot/\delta)$ with $\delta \geq 0$. Although δ is a scalar, it operates on the q-variate argument of $\mathcal{K}$. Now, we have $\|\mathcal{K}_\delta\|_2^2 = \delta^{-q}\|\mathcal{K}\|_2^2$ and $\mu_2(\mathcal{K}_\delta) = \delta^2\mu_2(\mathcal{K})$. As in the one-dimensional case, we choose δ such that the bias–variance trade-off in AMISE $(\mathbf{H}, \mathcal{K}_\delta)$ is independent of $\mathcal{K}_\delta$. This yields

$$\mu_2^2(\mathcal{K}_{\delta_0}) = \|\mathcal{K}_{\delta_0}\|_2^2 \Longleftrightarrow \delta_0 = \left\{\frac{\|\mathcal{K}\|_2^2}{\mu_2^2(\mathcal{K})}\right\}^{1/(q+4)},$$

where δ_0 again is called *canonical bandwidth* of the kernel $\mathcal{K}$. Now let $\mathcal{K}^A$ be a kernel function with canonical bandwidth δ_0^A, and $\mathcal{K}^B$ a kernel function with canonical bandwidth δ_0^B. Suppose we have used $\mathbf{H}_A$ with kernel $\mathcal{K}^A$ and we want to recompute the kernel density estimate with kernel $\mathcal{K}^B$. Then the approximation

$$\text{AMISE}(\mathbf{H}_A, \mathcal{K}^A) \approx \text{AMISE}(\mathbf{H}_B, \mathcal{K}^B)$$

holds if

$$\mathbf{H}_B = \frac{\delta_0^B}{\delta_0^A}\mathbf{H}_A,$$

which allows us to adjust bandwidths for different kernels, as in the one-dimensional case.

Let us consider an example. Suppose we want to use the product quartic kernel $\mathcal{K}^Q$ instead of the q-dimensional Gaussian $\mathcal{K}^G$, since the former is faster in direct computation because of its compact support on $[-1,1]$. Which rule of thumb is equivalent to (12.9) in this case? Here we have $\delta_0^G = \{1/(2\sqrt{\pi})\}^{q/(q+4)}$ and $\delta_0^Q = (49 \cdot 5^q/7^q)^{1/(q+4)}$, which together give the canonical bandwidths in Table 12.1 for dimensions $q = 1, \ldots, 5$.

The fourth column of the table gives the factor by which the rule-of-thumb bandwidth matrix in (12.9) needs to be multiplied to obtain the rule-of-thumb bandwidth for the multiplicative quartic kernel. Of course, all rule-of-thumb bandwidths for other kernel functions can be calculated in a similar way.

TABLE 12.1 Bandwidth-adjusting Factors for a Gaussian kernel and a Multiplicative Quartic kernel for Different Dimensions q

q	δ_0^G	δ_0^Q	δ_0^Q/δ_0^Q
1	0.7764	2.0362	2.6226
2	0.6558	1.7100	2.6073
3	0.5814	1.5095	2.5964
4	0.5311	1.3747	2.5883
5	0.4951	1.2783	2.5820

For a product kernel $\mathcal{K}$ constructed from an univariate kernel K, $\mu_2(\mathcal{K}) = \mu_2(K)$ and $\|\mathcal{K}\|_2 = \|K\|_2^q$. A table of values of $\mu_2(K)$ for $\|K\|_2^2$ can be found in Härdle (1991, p. 220), for example.

Principally, all plug-in methods for one-dimensional kernel density estimation can be extended to the multivariate case. (See Wand and Jones, 1994, for details.)

As we mentioned before, the cross-validation method is fairly independent of the special structure of the parameter or function estimate. With respect to the bandwidth choice problem, cross-validation techniques allow us to treat a wider class of density functions f than does the rule-of-thumb approach. (Remember that the rule-of-thumb bandwidth is optimal for the reference pdf; hence, it may fail for multimodal densities, for instance.)

Recall that, in contrast to the rule-of-thumb approach, least-squares cross-validation for density estimation aims to estimate the *integrated squared error* (ISE) optimal bandwidth. Here, we approximate the ISE as

$$\begin{aligned} \mathrm{ISE}(\mathbf{H}) &\approx \int \{\hat{f}_{\mathbf{H}}(t) - f(t)\}^2\, dt \\ &= \int \hat{f}_{\mathbf{H}}^2(t)\, dt - 2 \int \hat{f}_{\mathbf{H}}(t) f(t)\, dt + \int f^2(t)\, dt. \end{aligned} \tag{12.10}$$

Apparently, this is the same formula as in the one-dimensional case, and since the last term of (12.10) does not involve $\mathbf{H}$, that term can be ignored. The first term can be easily calculated from the data. Only the second term of (12.10) is unknown and has to be estimated. However, observe that $\int \hat{f}_{\mathbf{H}}(t) f(t)\, dt = E\hat{f}_{\mathbf{H}}(T)$, where the only new aspect now is that T is q dimensional. As in the one-dimensional case, we estimate this term by a leave-one-out estimator

$$\widehat{E\hat{f}_{\mathbf{H}}}(T) = \frac{1}{n} \sum_{i=1}^{n} \hat{f}_{\mathbf{H},-i}(T_i),$$

where

$$\hat{f}_{\mathbf{H},-i}(t) = \frac{1}{n-1} \sum_{i \neq j, j=1}^{n} \mathcal{K}_{\mathbf{H}}(T_j - t).$$

This yields the multivariate cross-validation criterion as a straightforward generalization of CV in the one-dimensional case:

$$\begin{aligned} \mathrm{CV}(\mathbf{H}) = & \frac{1}{n^2 \det(\mathbf{H})} \sum_{i=1}^{n} \sum_{j=1}^{n} \mathcal{K} * \mathcal{K}\{\mathbf{H}^{-1}(T_j - t_i)\} \\ & - \frac{2}{n(n-1)} \sum_{i=1}^{n} \sum_{\substack{j=1 \\ j \neq i}}^{n} \mathcal{K}_{\mathbf{H}}(T_j - T_i). \end{aligned}$$

Here, $\mathcal{K} * \mathcal{K}$ denotes the convolution of $\mathcal{K}$ with itself. The difficulty comes in by the fact that the bandwidth is now a $q \times q$ matrix $\mathbf{H}$. In the most general case, this means that we have to minimize over $q(q+1)/2$ parameters. Still, if we assume $\mathbf{H}$ to be a diagonal matrix, the problem remains a q-dimensional one. This holds as well for other cross-validation approaches. Multivariate resampling methods for bandwidth selection are discussed in more detail in Sain et al. (1994).

Next, we examine the problem of how to graphically display a multivariate density estimate. Assume that $q = 2$. Here we are still able to show the density estimate in a three-dimensional plot. This is particularly useful if the estimated function can be rotated on the computer screen interactively. For a two-dimensional presentation, a contour plot often gives more insight into the structure of the data.

In the following discussion, we use the Credit Scoring Data from Fahrmeir and Hamerle (1984) and Fahrmeir and Tutz (1994) for purposes of illustration. This data set consists of $n = 1000$ clients, of whom 700 paid a credit back without problems and 300 did not. Among a number of categorical variables (running account, previous credits, purpose, personal attributes, etc.), three continuous variables are available: `duration`, `amount` of credit, and `age`.

The upper panels of Figures 12.3 and 12.4 display a two-dimensional density estimate

$$\hat{f}_h(t) = \hat{f}_h(t_1, t_2)$$

for log(`duration`) vs. log(`amount`) and for log(`amount`) vs. log(`age`), respectively. The subscript h indicates that we used a diagonal bandwidth matrix $\mathbf{H} = \mathrm{diag}(h_1, h_2)$.

The lower panels of the two figures give contour plots of the density estimates. Clearly, both distributions are rather symmetric, due to the logarithmic transformation. In the `duration` direction, a typical bimodal structure can be discerned that reproduces slightly in the `amount` direction. Obviously, both variables are related with a positive correlation.

In this case, the bandwidth was chosen according to the general rule-of-thumb (12.9), which tends to oversmooth multimodal structures of the data. In fact, the

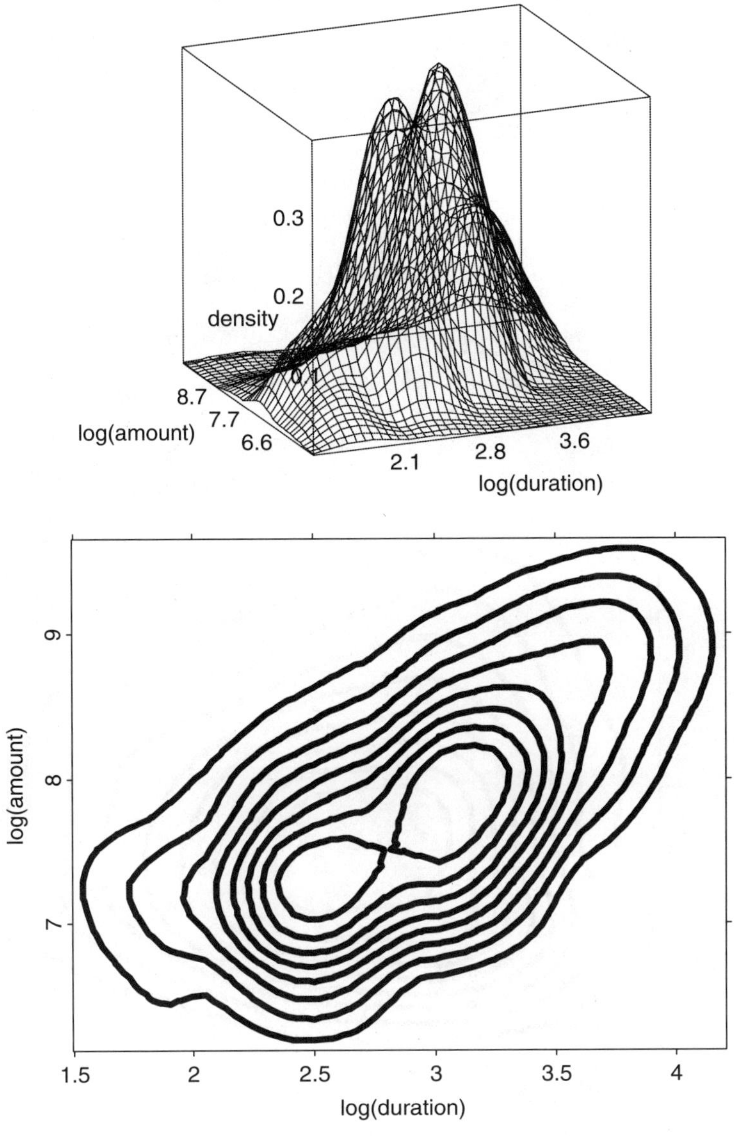

FIGURE 12.3 Two-dimensional density estimate (upper panel) and density contours (lower panel) for duration and amount of Credit Scoring Data. Rule-of-thumb bandwidths $h_1 = 0.48$ and $h_2 = 0.64$.

`durations` of credits are multiples of 6 months in most instances. The two clear modes that we observe are those for `durations` of 12 and 24 months. In all applications discussed in this chapter, we use the quartic (biweight) product kernel. Recall that the univariate quartic kernel is $K(u) = 0.9375(1 - u^2)^2 \cdot I(|u| \leq 1)$.

For three-dimensional density estimates, it is always possible to hold one variable fixed and to plot the density as a function of the other variables.

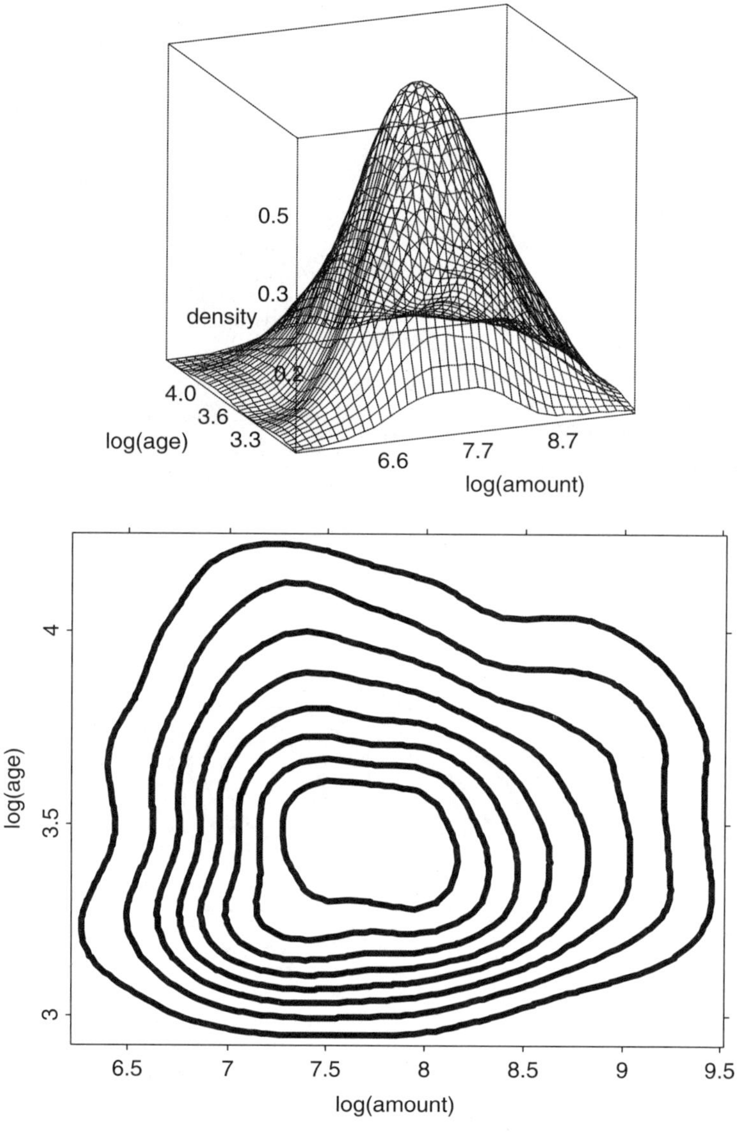

FIGURE 12.4 Two-dimensional density estimate (upper panel) and density contours (lower panel) for amount and age of Credit Scoring Data. Rule-of-thumb bandwidths $h_1 = 0.64$ and $h_2 = 0.25$.

Alternatively, we can again plot contours of the density estimate, which are now three-dimensional surfaces. Figure 12.5 shows this for the credit scoring variables. Colors and the possibility of rotating the contours on the computer screen ease the exploration of the data structures. Of course, we are restricted to two-dimensional plots here. However, one can clearly recognize the ellipsoidal structure of the contour, which indicates a relatively symmetric distribution.

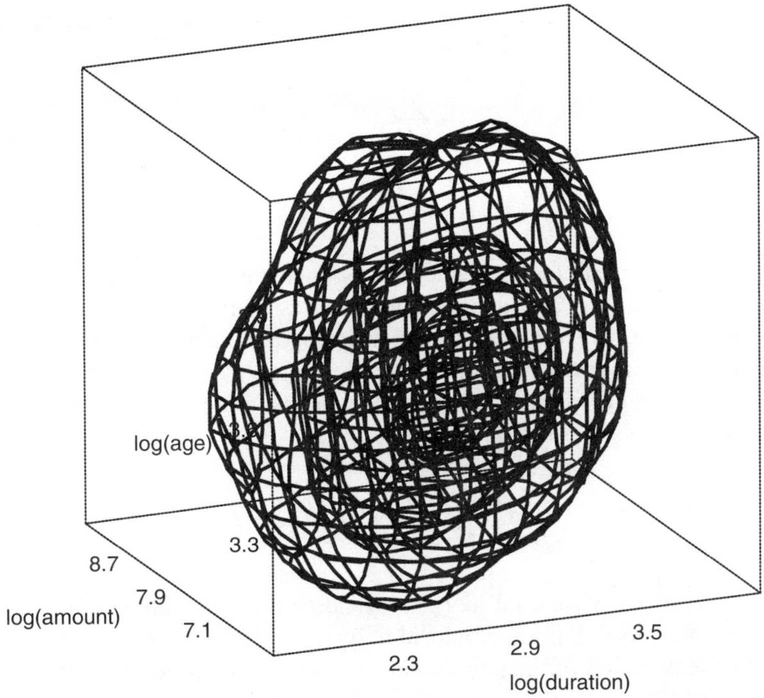

FIGURE 12.5 Three-dimensional density contours for duration, amount, and age of Credit Scoring Data. Rule-of-thumb bandwidths $h_1 = 0.56$, $h_2 = 0.75$, and $h_3 = 0.29$.

12.2.2 Multivariate Kernel Regression

Multivariate nonparametric regression aims to estimate the functional relation between a response variable Y and a multivariate explanatory variable T, that is, the conditional expectation

$$E(Y|T) = E(Y|T_1, \ldots, T_q) = m(T),$$

where, as before, $T = (T_1, \ldots, T_d)^T$. The relation

$$E(Y|T) = \int y f(y|t)\, dy = \frac{\int y f(y, t)\, dy}{f(t)}$$

leads by replacing the multivariate densities $f(y, t)$ by the kernel density estimate

$$\hat{f}_{h,\mathbf{H}}(y, t) = \frac{1}{n} \sum_{i=1}^{n} K_h(Y_i - y) \mathcal{K}_{\mathbf{H}}(t_i - t)$$

and $f(t) = f_T(t)$ by (12.3) to the multivariate generalization of the Nadaraya–Watson estimator,

$$\hat{m}_{\mathbf{H}}(t) = \frac{\sum_{i=1}^{n} \mathcal{K}_{\mathbf{H}}(T_i - t) Y_i}{\sum_{i=1}^{n} \mathcal{K}_{\mathbf{H}}(T_i - t)}.$$

Hence, the multivariate kernel regression estimator is just a weighted sum of the observed responses Y_i. The denominator ensures that the weights sum to unity. Depending on the choice of the kernel, $\hat{m}_{\mathbf{H}}(t)$ is a weighted average of those Y_i such that T_i lies in a ball or cube around t.

Note that the multivariate Nadaraya–Watson estimator is a local constant estimator; that is, it is the solution of

$$\min_{\beta_0} \sum_{i=1}^{n} \{Y_i - \beta_0\}^2 \mathcal{K}_{\mathbf{H}}(T_i - t).$$

Replacing β_0 by a polynomial in $T_i - t$ yields a local polynomial kernel regression estimator. This definition of a local polynomial kernel regression is a straightforward generalization of the univariate case. (For details, see Ruppert and Wand, 1994.) We illustrate this fact with the example of a local linear regression estimate. The minimization problem is

$$\min_{\beta_0, \beta_1} \sum_{i=1}^{n} \{Y_i - \beta_0 - (T_i - t)^T \beta_1\}^2 \mathcal{K}_{\mathbf{H}}(T_i - t).$$

The solution can be written as

$$\hat{\beta} = (\hat{\beta}_0, \hat{\beta}_1^T)^T = (\mathbf{T}^T \mathbf{W} \mathbf{T})^{-1} \mathbf{T}^T \mathbf{W} \mathbf{Y} \tag{12.11}$$

where

$$\mathbf{T} = \begin{pmatrix} 1 & (T_1 - t)^T \\ \vdots & \vdots \\ 1 & (T_n - t)^T \end{pmatrix}, \quad \mathbf{Y} = \begin{pmatrix} Y_1 \\ \vdots \\ Y_n \end{pmatrix},$$

and $\mathbf{W} = \operatorname{diag}(\mathcal{K}_{\mathbf{H}}(T_1 - t), \ldots, \mathcal{K}_{\mathbf{H}}(T_n - t))$. In (12.11), $\hat{\beta}_0$ estimates the regression function itself, whereas $\hat{\beta}_1$ estimates the partial derivatives with respect to the components T. In the discussion that follows, we denote the multivariate local linear estimator as

$$\hat{m}_{1,\mathbf{H}}(t) = \hat{\beta}_0(t).$$

12.2.2.1 Bias, Variance, and Asymptotics

The asymptotic conditional variances of the Nadaraya–Watson estimator $\hat{m}_{\mathbf{H}}$ and the local linear estimator $\hat{m}_{1,\mathbf{H}}$ are identical and equal to

$$\mathrm{Var}\{\hat{m}_{\mathbf{H}}(t)|T_1, \ldots, T_n\} = \frac{1}{n\,\det(\mathbf{H})}\|\mathcal{K}\|_2^2\frac{\sigma^2(t)}{f(t)}\{1 + o_p(1)\},$$

where $\sigma^2(t)$ denotes the variance function in $\mathrm{Var}(Y|t)$. (A derivation of this formula is found in detail in Ruppert and Wand, 1994.)

We sketch the derivation of the asymptotic conditional bias, since we find remarkable differences between both estimators. Let M be the second-order Taylor expansion of $(m(T_1), \ldots, m(T_n))^T$; that is,

$$M \approx m(t)\mathbf{1}_n + L(t) + \tfrac{1}{2}Q(t) = \mathbf{T}\begin{pmatrix} m(t) \\ \nabla_m(t) \end{pmatrix} + \tfrac{1}{2}Q(t), \qquad (12.12)$$

where

$$L(t) = \begin{pmatrix} (T_1 - t)^T\nabla_m(t) \\ \vdots \\ (T_n - t)^T\nabla_m(t) \end{pmatrix}, \quad Q(t) = \begin{pmatrix} (T_1 - t)^T\mathcal{H}_m(t)(T_1 - t) \\ \vdots \\ (T_n - t)^T\mathcal{H}_m(t)(T_n - t) \end{pmatrix}.$$

In addition to (12.5), we can show that

$$\frac{1}{n}\sum_{i=1}^{n}\mathcal{K}_{\mathbf{H}}(T_i - t)(T_i - t) = \mu_2(\mathcal{K})\mathbf{H}\mathbf{H}^T\nabla_f(t) + o_p(\mathbf{H}\mathbf{H}^T\mathbf{1}_d)$$

$$\frac{1}{n}\sum_{i=1}^{n}\mathcal{K}_{\mathbf{H}}(T_i - t)(T_i - t)(T_i - t)^T = \mu_2(\mathcal{K})f(t)\mathbf{H}\mathbf{H}^T\nabla_f(t) + o_p(\mathbf{H}\mathbf{H}^T).$$

(See Ruppert and Wand, 1994.) Therefore, the denominator of the conditional asymptotic expectation of the Nadaraya–Watson estimator $\hat{m}_{\mathbf{H}}$ is approximately $f(t)$. Using $E(\mathbf{Y}|T_1, \ldots, T_n) = M$ and the Taylor expansion for M, we have

$$\begin{aligned} &E\{\hat{m}_{\mathbf{H}}|T_1, \ldots, T_n\} \\ &\approx \{f(t) + o_p(1)\}^{-1}\left\{\frac{1}{n}\sum_{i=1}^{n}\mathcal{K}_{\mathbf{H}}(T_i - t)m(t) + \sum_{i=1}^{n}\mathcal{K}_{\mathbf{H}}(T_i - t)(T_i - t)^T\nabla_m(t)\right. \\ &\quad \left. + \sum_{i=1}^{n}\mathcal{K}_{\mathbf{H}}(T_i - t)(T_i - t)^T\mathcal{H}_m(t)(T_i - t)\right\} \\ &\approx \{f(t)\}^{-1}[f(t)m(t) + \mu_2(\mathcal{K})\nabla_m\mathbf{H}\mathbf{H}^T\nabla_f + \tfrac{1}{2}\mu_2(\mathcal{K})f(t)\,\mathrm{tr}\{\mathbf{H}^T\mathcal{H}_m(t)\mathbf{H}\}], \end{aligned}$$

where "$\approx$" now indicates asymptotic equality. The results for the Nadaraya–Watson estimator are summarized in the following theorem.

Theorem 12.1. *The conditional asymptotic bias and variance of the multivariate Nadaraya–Watson kernel regression estimator are, respectively,*

$$E\{\hat{m}_{\mathbf{H}}|T_1,\ldots,T_n\} - m(t) \approx \mu_2(\mathcal{K})\frac{\nabla_m(t)^T\mathbf{H}\mathbf{H}^T\nabla_f(t)}{f(t)} + \tfrac{1}{2}\mu_2(\mathcal{K})\ \mathrm{tr}\{\mathbf{H}^T\mathcal{H}_m(t)\mathbf{H}\}$$

$$\mathrm{Var}\{\hat{m}_{\mathbf{H}}|T_1,\ldots,T_n\} \approx \frac{1}{n\det(\mathbf{H})}\|\mathcal{K}\|_2^2\frac{\sigma^2(t)}{f(t)}$$

in the interior of the support of f_T.

Recall the notation $e_1 = (1, 0, \ldots, 0)^T$ for the first unit vector in $\mathbb{R}^d$. With this notation, we can write the local linear estimator as

$$\hat{m}_{1,\mathbf{H}}(t) = e_1^T(\mathbf{T}^T\mathbf{W}\mathbf{T})^{-1}\mathbf{T}^T\mathbf{W}\mathbf{Y}.$$

Now, using (12.11) and (12.12), we have

$$\begin{aligned} &E\{\hat{m}_{1,\mathbf{H}}|T_1,\ldots,T_n\} - m(t) \\ &= e_1^T(\mathbf{T}^T\mathbf{W}\mathbf{T})^{-1}\mathbf{T}^T\mathbf{W}\mathbf{T}\left\{\begin{pmatrix} m(t) \\ \nabla_m(t)\end{pmatrix} + \tfrac{1}{2}Q(t)\right\} - m(t) \\ &= \tfrac{1}{2}e_1^T(\mathbf{T}^T\mathbf{W}\mathbf{T})^{-1}\mathbf{T}^T\mathbf{W}Q(t), \end{aligned}$$

since $e_1^T[m(t), \nabla_m(t)^T] = m(t)$. Hence, the numerator of the asymptotic conditional bias depends only on the quadratic term. This is one of the key points in asymptotics for local polynomial estimators. If we would use local polynomials of order d and expand M up to order $d+1$, then only the term of order $d+1$ would appear in the numerator of the asymptotic conditional bias. Of course, the structure of the denominator would then be more complicated.

Theorem 12.2. *The conditional asymptotic bias and variance of the multivariate local linear regression estimator are, respectively,*

$$E\{\hat{m}_{1,\mathbf{H}}|T_1,\ldots,T_n\} - m(t) \approx \tfrac{1}{2}\mu_2(\mathcal{K})\ \mathrm{tr}\{\mathbf{H}^T\mathcal{H}_m(t)\mathbf{H}\}$$

$$\mathrm{Var}\{\hat{m}_{1,\mathbf{H}}|T_1,\ldots,T_n\} \approx \frac{1}{n\det(\mathbf{H})}\|\mathcal{K}\|_2^2\frac{\sigma^2(t)}{f(t)}$$

in the interior of the support of f_T.

For the proof of Theorem 12.2, we refer again to Ruppert and Wand (1994), who also show that the local linear estimate has same order-conditional bias on the boundary of the support of f_T. Fan et al. (1997) point out that the multivariate local linear fit with the Epanechnikov kernel is a best linear estimator and has a minimax efficiency of at least 89.4% among all estimators.

12.2.2.2 Bandwidth Selection and Practical Aspects

The methods used to choose a smoothing parameter in nonparametric regression are essentially the same as those used in density estimation. Again, plug-in and resampling ideas are employed for finding a global bandwidth $\mathbf{H}$ or a local bandwidth $\mathbf{H}(t)$.

In our presentation, we concentrate on the classical cross-validation bandwidth selector. To motivate the discussion, we introduce the *residual sum of squares* (RSS) as a (naive) way to assess goodness of fit. We have

$$\mathrm{RSS}(\mathbf{H}) = n^{-1} \sum_{i=1}^{n} \{Y_i - \hat{m}_{\mathbf{H}}(X_i)\}^2, \tag{12.13}$$

which is also called the resubstitution estimate for the *averaged squared error* (ASE). We concentrate on the Nadaraya–Watson estimator at the moment.

One problem with the RSS is that Y_i is used in $\hat{m}_{\mathbf{H}}(X_i)$ to predict itself. As a consequence, ASE($\mathbf{H}$) can be made arbitrarily small by letting $\mathbf{H} \to 0$ (in which case $\hat{m}_{\mathbf{H}}$ is an interpolation of the Y_i's). This leads to the *cross-validation* (CV) function

$$\mathrm{CV}(\mathbf{H}) = n^{-1} \sum_{i=1}^{n} \{Y_i - \hat{m}_{\mathbf{H},-i}(X_i)\}^2.$$

The function CV replaces $\hat{m}_{\mathbf{H}}(X_i)$ in (12.13) with the *leave-one-out* estimator

$$\hat{m}_{\mathbf{H},-i}(X_i) = \frac{\sum_{j \neq i} \mathcal{K}_{\mathbf{H}}(X_i - X_j) Y_j}{\sum_{j \neq i} \mathcal{K}_{\mathbf{H}}(X_i - X_j)}$$

and is equivalent to multiplying each term in RSS($\mathbf{H}$) by a *penalizing function* that corrects for the downward bias of the resubstitution estimate. For the Nadaraya–Watson estimator,

$$\begin{aligned} \mathrm{CV}(\mathbf{H}) &= \frac{1}{n} \sum_{i=1}^{n} \{Y_i - \hat{m}_{\mathbf{H},-i}(X_i)\}^2 \\ &= \frac{1}{n} \sum_{i=1}^{n} \{Y_i - \hat{m}_{\mathbf{H}}(X_i)\}^2 \left\{ \frac{Y_i - \hat{m}_{\mathbf{H},-i}(X_i)}{Y_i - \hat{m}_{\mathbf{H}}(X_i)} \right\}^2 \end{aligned} \tag{12.14}$$

and

$$\frac{Y_i - \hat{m}_{\mathbf{H}}(X_i)}{Y_i - \hat{m}_{\mathbf{H},-i}(X_i)} = \frac{\sum_j \mathcal{K}_{\mathbf{H}}(X_i - X_j)Y_j - Y_i \sum_j \mathcal{K}_{\mathbf{H}}(X_i - X_j)}{\sum_{j \neq i} \mathcal{K}_{\mathbf{H}}(X_i - X_j)Y_j - Y_i \sum_{j \neq i} \mathcal{K}_{\mathbf{H}}(X_i - X_j)}$$

$$\cdot \frac{\sum_{j \neq i} \mathcal{K}_{\mathbf{H}}(X_i - X_j)}{\sum_j \mathcal{K}_{\mathbf{H}}(X_i - X_j)} = 1 - \frac{\mathcal{K}_{\mathbf{H}}(0)}{\sum_j \mathcal{K}_{\mathbf{H}}(X_i - X_j)}. \qquad (12.15)$$

Note that (12.15) is a function of the ith diagonal element of the smoother matrix. In this case, cross-validation is equivalent to *generalized cross-validation* (Craven and Wahba, 1979). Härdle et al. (1988) showed asymptotic optimality of the selected bandwidth, although the rate of convergence is rather slow. An improved bandwidth selection is discussed in Härdle et al. (1992).

Observe that (12.4) and (12.15) also imply that the computation of $\text{CV}(\mathbf{H})$ does not require more computational effort than the computation of $m_{\mathbf{H}}(X_1), \ldots, m_{\mathbf{H}}(X_n)$. However, the optimization over a matrix $\mathbf{H}$ might be cumbersome; hence, diagonal bandwidth matrices (or even $\mathbf{H} = h\mathbf{I}_q$ with appropriate standardization of the data) are still preferred in practice.

Before we consider CV bandwidth selection in the local linear case, we want to comment on the practical computation of the estimator. Principally, since multivariate kernel regression estimators can be expressed as local polynomial estimators, their computation can be done by any statistical package that is able to run weighted least-squares regression. However, since we estimate a function, this type of regression has to be performed at all observation points or on a grid of points in $\mathbb{R}^q$. Therefore, explicit formulas are useful.

We now give a formula for the multivariate local linear estimator. For a fixed point t, consider the sums

$$\mathcal{S}_0 = \mathcal{S}_0(t) = \sum_{i=1}^{n} \mathcal{K}_{\mathbf{H}}(T_i - t),$$

$$\mathcal{S}_1 = \mathcal{S}_1(t) = \sum_{i=1}^{n} \mathcal{K}_{\mathbf{H}}(T_i - t)(T_i - t),$$

$$\mathcal{S}_2 = \mathcal{S}_2(t) = \sum_{i=1}^{n} \mathcal{K}_{\mathbf{H}}(T_i - t)(T_i - t)(T_i - t)^T,$$

$$\mathcal{T}_0 = \mathcal{T}_0(t) = \sum_{i=1}^{n} \mathcal{K}_{\mathbf{H}}(T_i - t)Y_i,$$

$$\mathcal{T}_1 = \mathcal{T}_1(t) = \sum_{i=1}^{n} \mathcal{K}_{\mathbf{H}}(T_i - t)(T_i - t)Y_i.$$

Note that $\mathcal{S}_1$ and $\mathcal{T}_1$ are q-variate vectors and that $\mathcal{S}_2$ is a $q \times q$ matrix. Then, for the local linear estimate, we can write

$$\hat{\beta} = \begin{pmatrix} \mathcal{S}_0 & \mathcal{S}_1^T \\ \mathcal{S}_1 & \mathcal{S}_2 \end{pmatrix}^{-1} \begin{pmatrix} \mathcal{T}_0 \\ \mathcal{T}_1 \end{pmatrix}.$$

For the regression function, we need only the first component, $e_1^T \hat{\beta}$. Applying blockwise matrix inversion, we obtain

$$e_1^T \begin{pmatrix} \mathcal{S}_0 & \mathcal{S}_1^T \\ \mathcal{S}_1 & \mathcal{S}_2 \end{pmatrix}^{-1} = (\mathcal{S}_0 - \mathcal{S}_1^T \mathcal{S}_2^{-1} \mathcal{S}_1)^{-1} \, (1 - \mathcal{S}_1^T \mathcal{S}_2^{-1}),$$

and hence,

$$\hat{m}_{1,\mathbf{H}}(t) = \frac{\mathcal{T}_0 - \mathcal{S}_1^T \mathcal{S}_2^{-1} \mathcal{T}_1}{\mathcal{S}_0 - \mathcal{S}_1^T \mathcal{S}_2^{-1} \mathcal{S}_1}. \tag{12.16}$$

The CV criterion here is a weighted RSS, as in (12.14). If we denote the leave-one-out estimator by $\hat{m}_{1,\mathbf{H},-i}(t)$ and define its components accordingly, we observe that

$$\mathcal{S}_{0,-i} = \mathcal{S}_0 - \mathcal{K}_{\mathbf{H}}(0), \quad \mathcal{S}_{1,-i} = \mathcal{S}_1, \quad \mathcal{S}_{2,-i} = \mathcal{S}_2,$$
$$\mathcal{T}_{0,-i} = \mathcal{T}_0 - Y_i \mathcal{K}_{\mathbf{H}}(0), \quad \mathcal{T}_{1,-i} = \mathcal{T}_1.$$

This means that

$$\hat{m}_{1,\mathbf{H},-i}(t) = \frac{\mathcal{T}_0 - Y_i \mathcal{K}_{\mathbf{H}}(0) - \mathcal{S}_1^T \mathcal{S}_2^{-1} \mathcal{T}_1}{\mathcal{S}_0 - \mathcal{K}_{\mathbf{H}}(0) - \mathcal{S}_1^T \mathcal{S}_2^{-1} \mathcal{S}_1},$$

which, in analogy to (12.15), yields

$$\frac{Y_i - \hat{m}_{\mathbf{H}}(X_i)}{Y_i - \hat{m}_{\mathbf{H},-i}(X_i)} = 1 - \frac{\mathcal{K}_{\mathbf{H}}(0)}{\mathcal{S}_0 - \mathcal{S}_1^T \mathcal{S}_2^{-1} \mathcal{S}_1}. \tag{12.17}$$

As in the Nadaraya–Watson case, (12.17) is a function of the ith diagonal element of the smoother matrix. A summary of bandwidth selection methods other than CV can be found in Fan and Gijbels (1995) and in Chapter 9 of this book, in which, also rule-of-thumb approaches are covered.

Recall that (12.16) estimates the regression function only at one point t. To estimate the regression plane, we have to apply the equation to a two-dimensional grid of points. The WARPing technique (also called binning) described in Härdle and Scott (1992) and applied to local polynomial kernel regression by Fan and Marron (1994) and Fan and Müller (1995), can be used to speed up calculations. (See also Wand, 1994, for an analysis of fast computation methods for multivariate kernel estimation.)

Figure 12.6 shows the bivariate Nadaraya–Watson and local linear estimates for simulated data. The underlying curve is, in fact, an additive combination of

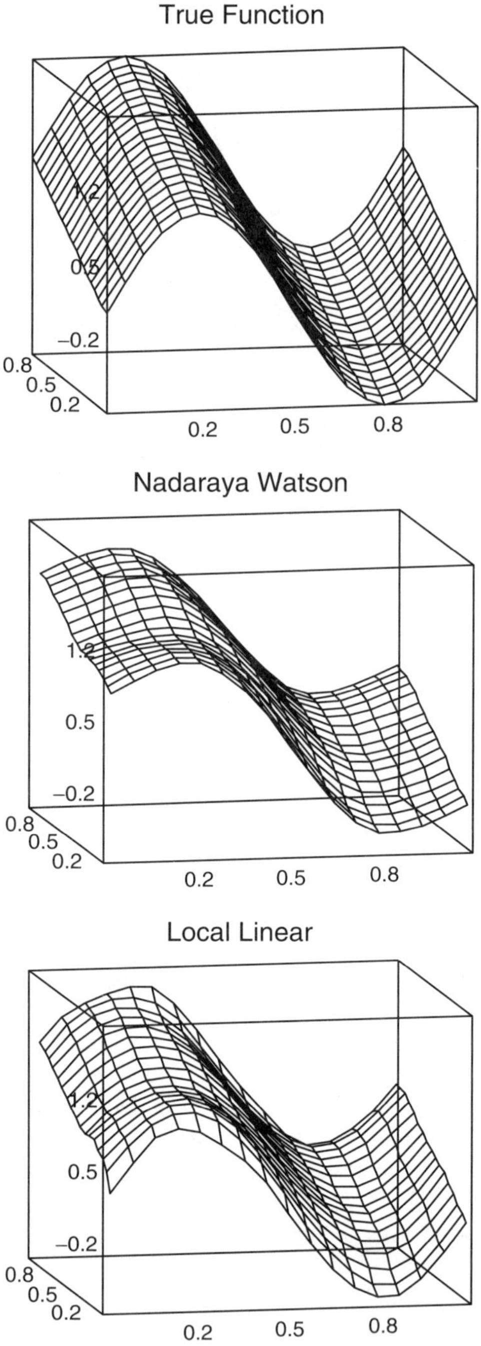

FIGURE 12.6 Two-dimensional Nadaraya–Watson estimates and local linear estimates from simulated data.

a sine function in the first argument and a linear function in the second. Note that we have chosen the same bandwidth in both estimates. The multivariate Nadaraya–Watson regression estimator can be improved in the boundary regions by a boundary correction method. (See Staniswalis et al., 1994, and Staniswalis and Messer, 1996, for more details.)

Of course, nonparametric kernel regression estimation is not limited to bivariate distributions. A practical issue is the graphical display for higher dimensional multivariate functions. This was already considered when we discussed the graphical representation of multivariate density estimates. The corresponding remarks apply here again. The general problem in multivariate nonparametric estimation is the curse of dimensionality. Recall that the nonparametric regression estimators are based on the idea of local (weighted) averaging. In higher dimensions, the observations are usually sparsely distributed for reasonable sample sizes, and consequently, estimators based on local averaging perform unsatisfactorily in this situation.

Technically, one can explain the effect by looking at the AMISE again. Consider a multivariate regression estimator with the same bandwidth h for all components — for example, a Nadaraya–Watson or local linear estimator with bandwidth matrix $\mathbf{H} = h\mathbf{I}_q$. That the AMISE also depends on q is evident from

$$\text{AMISE}(n, h) = \frac{1}{nh^q} C_1 + h^4 C_2,$$

where C_1 and C_2 are constants that depend neither on n nor on h. If we derive the optimal bandwidth, we find that $h_{\text{opt}} \approx n^{-1/(4+q)}$, and hence, the rate of convergence for AMISE is $n^{-4/(4+q)}$. One can clearly see that the speed of convergence decreases dramatically for higher dimensions q.

12.3 SEMIPARAMETRIC GENERALIZED REGRESSION MODELS

As the name suggests, semiparametric models combine two elements, one to be estimated nonparametrically and the other requiring the estimation of a set of finite-dimensional parameters. In this section, we concentrate on *single-index models* (SIM) and *generalized partial linear models* (GPLM).

Often, a canonical partitioning of the explanatory variables exists. In particular, if there are binary or discrete explanatory variables, we keep them separate from the other design variables. In the following we denote a vector of continuous explanatory variables by $T = (T_1, \ldots, T_q)^T$ and refer to $X = (X_1, \ldots, X_p)^T$ as the discrete part of the variables.

Semiparametric generalized linear models are widely used in modeling binary choice — that is, in situations in which the response variable has two alternatives. Recall the example on credit scoring introduced previously. In the analysis of discrete response variables, one typically models the expected value of the response as a nonlinear monotonic function of a linear combination of the explanatory variables. Examples are probit or logit models, wherein the nonlinear (link) function is the cumulative distribution function of a normal and a logistic distribution respectively. (See McCullagh and Nelder, 1989). Then the so-called *generalized linear model* (GLM) has the form

$$E(Y|X,T) = G(X^T\beta + T^T\gamma), \qquad (12.18)$$

with a known monotonic function G and unknown parameters β and γ. The model (12.18) combines computational feasibility (especially for discrete covariates) with good interpretability of the "index" $X^T\beta + T^T\gamma$ and therefore has found wide application in all fields of applied statistics (see, e.g., Fahrmeir and Tutz, 1994, and Maddala, 1983.) However, for some applications, it may be argued that the model is too restrictive (Horowitz, 1993). Indeed, it may not even be clear whether the relationship between the influential variables and the response is monotonic.

Several approaches have been proposed to generalize parametric regression models in order to allow nonmonotonic relationships between explanatory variables and the dependent variable Y. We focus on two classes of semiparametric models that have received a lot of attention:

- A generalization of the known (parametric) link function G to an unknown (nonparametric) link function $g(\cdot)$ yields the SIM given by

  $$E(Y|X,T) = g(X^T\beta + T^T\gamma),$$

  also called a *one-term projection pursuit model* in statistics. Obviously, because of the nonparametric character of the link function, conventional parametric estimation procedures can no longer be applied in this case. Instead, nonparametric estimators are necessary. In this chapter, we give an overview of how this model can be estimated using kernel methods.
- A generalization of the linear form $X^T\beta + T^T\gamma$ to a partial linear form $X^T\beta + m(T)$ yields the GPLM

  $$E(Y|X,T) = G\{X^T\beta + m(T)\},$$

  where G denotes a known link function, as in the GLM model. Here, the $m(\cdot)$ is a multivariate nonparametric function of the variable T.

  In high dimensions of T, the estimate of the nonparametric function $m(\cdot)$ faces the same problems as those of the fully nonparametric multidimensional regression function: the curse of dimensionality and the practical problem of interpretability. Hence, it might be reasonable to think about a lower dimensional nonparametric model for the nonparametric part. A possible alternative is the GPLM with an additive structure in the nonparametric component, that is, the *generalized additive model* (GAM):

  $$E(Y|X,T) = G\{X^T\beta + m_1(T_1) + \cdots + m_d(T_d)\}.$$

  Here, the $m_j(\cdot)$ will be univariate nonparametric functions of the variables T_j.

12.3.1 Generalizing the Link Function: Single-Index Models

SIM derive their name from the term "index", which denotes a summary of different variables into one number. Meanwhile, a number of methods have been proposed do deal with these models. A straightforward semiparametric GLM extension is provided by Weisberg and Welsh (1994), who estimated the unknown link function and its derivative (for the Fisher scoring algorithm) with a kernel smoother. Ichimura (1993) used a similar idea within a least-squares criterion. Klein and Spady (1993) proved an asymptotic efficiency result for a pseudolikelihood binary-choice estimator.

All three methods require the optimization of a pseudolikelihood of possibly complicated structure. We present a direct approach that avoids numerical iterations. The estimation of the SIM

$$E(Y|X,T) = g(X^T\beta + T^T\gamma)$$

is carried out in two steps. First the coefficients vectors β and γ are estimated, and then, using the resulting index values $X_i^T\hat{\beta} + T_i^T\hat{\gamma}$, one can estimate g by the usual univariate nonparametric regression.

12.3.1.1 Average-Derivative Estimation

Consider for a moment only the continuous part of the variables, $T = (T_1, \ldots, T_q)^T$. We denote the regression function to be estimated by $m(\cdot)$; that is, $E(Y|T) = m(T)$. The vector of average derivatives is given by

$$\delta = E\{\nabla_m(T)\} = E\{g'(T^T\beta)\}\beta, \tag{12.19}$$

where $\nabla_m(t)$ is the vector of partial derivatives of $m(\cdot)$ and g' is the derivative of $g(\cdot)$.

Inspection of (12.19) shows that δ equals β up to scale. Hence, any estimate of δ determines β up to scale. The estimation of δ can be carried out by means of several *average-derivative estimation* (ADE) methods. We concentrate on estimators based on the density function of T; however, a variety of other methods exist. (For an overview, see Stoker, 1991.)

The key idea of ADE based on the density $f(\cdot)$ of T lies in "transferring" the derivative of the regression function m on to the derivative of the density f. Observe that

$$\delta = E\{\nabla_m(T)\} = \int \nabla_m(t) f(t)\, dt.$$

If $f(t)m(t) \to 0$ is assumed for $\|t\| \to \infty$, then partial integration yields $E\{\nabla_m(T)\} = -\int m(t)\nabla_f(t)\, dt$. Hence, by introducing the *score vector*

$$\ell(t) = \nabla_{\log f}(t) = \frac{\nabla_f(t)}{f(t)},$$

one arrives at the relationship

$$\delta = -\int \frac{\nabla_f(t)}{f(t)} f(t)m(t)\,dt = \int \ell(t)m(t)f(t)\,dt = E\{\ell(T)m(T)\}.$$

Employing $E\{\ell(T)m(T)\} = E\{\ell(T)Y\}$ immediately allows one to estimate δ by the sample analog

$$\hat{\delta} = n^{-1}\sum_{i=1}^{n} \hat{\ell}_{\mathbf{H}}(T_i)Y_i,$$

where $\ell(t)$ is approximated by $\hat{\ell}_{\mathbf{H}}(t) = -\{\hat{f}_{\mathbf{H}}(t)\}^{-1}(-\partial_1 \hat{f}_{\mathbf{H}}(t), \ldots, -\partial_q \hat{f}_{\mathbf{H}}(t))$. Here, $\hat{f}_{\mathbf{H}}(t)$ is the multivariate kernel density estimator, and the ∂_js are the partial derivatives with respect to the jth dimension (i.e., the $\partial_j \hat{f}_{\mathbf{H}}(t)$s are used for estimating the partial derivatives of the density), so that we have

$$\partial_j \hat{f}_{\mathbf{H}}(t) = \frac{1}{n\,\det(\mathbf{H})}\sum_{j=1}^{n} \partial_j K_{\mathbf{H}}(t - T_j).$$

Due to the sparseness of data in high dimensions, the use of $\hat{f}_{\mathbf{H}}$ can also be problematic, since $\hat{\ell}_{\mathbf{H}}$ might behave badly in regions of low density. Hence, Härdle and Stoker (1989) proposed to use the ADE estimator

$$\hat{\delta} = n^{-1}\sum_{i=1}^{n} \hat{\ell}_{\mathbf{H}}(T_i)Y_i I\{\hat{f}_{\mathbf{H}}(T_i) > b_n\},$$

where $I\{\hat{f}_{\mathbf{H}}(T_i) > b_n\}$ is an indicator that excludes density values which are too small. The trimming bounds b_n are chosen such that $b_n \to 0$ for $n \to \infty$.

Härdle and Stoker (1989) have shown that

$$\sqrt{n}(\hat{\delta} - \delta) \xrightarrow[n\to\infty]{\mathcal{L}} N(\mathbf{0}, \mathbf{\Sigma}_\delta),$$

where $\mathbf{\Sigma}_\delta$ is the covariance matrix of $\ell(T)Y + \{\nabla_m(T) - \ell(T)m(T)\}$. Note that $\hat{\delta}$ achieves $\sqrt{n}$-convergence, a rate that is typical of parametric estimators.

The need for "trimming" the ADE is one of the problems associated with a random denominator, which also complicates the derivation of the distributional properties. These difficulties are overcome by *density-weighted average-derivative estimation* (WADE), proposed by Powell, Stock, and Stoker (1989). Observe that the density-weighted average derivative shares the property that the (unweighted) average derivative has of being proportional to the coefficient vector β in index models:

$$\delta = E\{\nabla_m(T)w(T)\} = E\{g'(T^T\beta)w(T)\}\beta.$$

A "natural" weight function is given by the density f itself. Calculations similar to those for the unweighted ADE with $w(t) = f(t)$ yield

$$\begin{aligned}\delta &= \int \nabla_m(t) f^2(t)\,dt = -2\int m(t)\nabla_f(t) f(t)\,dt \\ &= -2E\{Y\nabla_f(T)\}.\end{aligned}$$

Thus, one may estimate β up to scale by

$$\hat{\delta} = -\frac{2}{n}\sum_{i=1}^{n} Y_i(\partial_1 \hat{f}_{\mathbf{H}}(t), \ldots, \partial_q \hat{f}_{\mathbf{H}}(t))^T. \tag{12.20}$$

The WADE estimator defined in (12.20) shares the desirable distributional features of ADE estimator ($\sqrt{n}$-consistency, asymptotic normality), while not requiring any trimming in practice.

Finally, an estimate for $g(\cdot)$ can be found by applying a univariate estimation method to $\hat{\delta}^T T_i$ and Y_i. For the Nadaraya–Watson estimator, Härdle and Stoker (1989) showed that when $h \approx n^{-1/5}$, the pointwise convergence of the regression function proceeds at the usual rate of $\sqrt{nh}$.

12.3.1.2 Including Discrete Explanatory Variables

By definition, derivatives can be calculated only if the variable under study is continuous. Thus, the method of weighted or unweighted ADE fails when discrete variables $X = (X_1, \ldots, X_d)^T$ need to be included in the model. Before giving a more general solution, let us explain how the coefficient of one dichotomous variable is entered into the model. Recall the SIM

$$E(Y|T, X) = g(X^T\beta + T^T\gamma),$$

with T the continuous and X the discrete part of the covariates. In the simplest case, we suppose that X is binary, that is, either $X = 1$ or $X = 0$.

Then this model can be split into two submodels:

$$\begin{aligned}E(Y|T, X) &= g(T^T\gamma) &&\text{if } X = 0;\\ E(Y|T, X) &= g(T^T\gamma + \beta) &&\text{if } X = 1.\end{aligned}$$

These are, in fact, two models to be estimated, one for $X = 0$ and one for $X = 1$. Note that γ alone could be estimated from the first equation only.

Theoretically, the same T_i can be associated either with $X_i = 0$, yielding an index value of $\gamma^T T_i$, or with $X_i = 1$, leading to an index value of $\gamma^T T_i + \beta$. Thus, the difference between the two indices is exactly β. In practice, finding these horizontal differences will be rather difficult. A very simple estimator is proposed in Korostelev and Müller (1995), using the observation that the integral difference between the two link functions also equals β. Essentially, the coefficient of the binary explanatory variable can be estimated by

$$\hat{\beta} = \hat{J}^{(1)} - \hat{J}^{(0)},$$

where

$$\hat{J}^{(0)} = \sum_{i=0}^{n_0} \gamma^T (T_{i+1}^{(0)} - T_i^{(0)}) Y_i^{(0)} \text{ and } \hat{J}^{(1)} = \sum_{i=0}^{n_1} \gamma^T (T_{i+1}^{(1)} - T_i^{(1)}) Y_i^{(1)},$$

in which the superscripts "0" and "1" denote the observations coming from the subsamples according to $X_i = 0$ and $X_i = 1$. In the simplest case of a binary Y variable, the estimator is $\sqrt{n}$-consistent, and its efficiency can be improved by a one-step estimator (Korostelev and Müller, 1995).

Horowitz and Härdle (1996) extend this approach to multivariate multicategorical X and arbitrary Y. Again, the approach is based on a split of the whole sample into subsamples according to the categories of X. Consider the thresholded link function

$$\tilde{g} = c_0 \operatorname{I}(g < c_0) + g \operatorname{I}(c_0 \le g \le c_1) + c_1 \operatorname{I}(g > c_1).$$

Let $x^{(k)}$ be a possible realization of X. Then the integrated link function conditional on $x^{(k)}$ is

$$J^{(k)} = \int_{v_0}^{v_1} \tilde{g}(v + \beta^T x^{(k)})\, dv,$$

where $[v_0, v_1]$ denotes an appropriately chosen interval. Now compare the integrated link functions for all X-categories $x^{(k)}$ $(k = 1, \ldots, M)$ with that for the first X-category $x^{(0)}$. It holds that

$$J^{(k)} - J^{(0)} = (c_1 - c_0) \left\{ x^{(k)} - x^{(0)} \right\} \beta.$$

Hence, with

$$\Delta J = \begin{pmatrix} J^{(1)} - J^{(0)} \\ \cdots \\ J^{(M)} - J^{(0)} \end{pmatrix} \text{ and } \Delta x = \begin{pmatrix} x^{(1)} - x^{(0)} \\ \cdots \\ x^{(M)} - x^{(0)} \end{pmatrix},$$

one gets $\Delta J = (c_1 - c_0) \Delta x \beta$, which finally yields

$$\beta = (c_1 - c_0)^{-1} (\Delta x^T \Delta x)^{-1} \Delta x^T \Delta J. \tag{12.21}$$

This estimation of β is based on replacing $J^{(k)}$ in (12.21) by

$$\hat{J}^{(k)} = \int_{v_0}^{v_1} \hat{\tilde{g}}(v + \beta^T x^{(k)})\, dv,$$

where $\hat{\tilde{g}}$ is a nonparametric estimate of the thresholded link function $\tilde{g}$. This estimator is obtained by a univariate regression of the estimated "continuous" indices $\hat{\gamma}^T T_i^{(k)}$ on $Y_i^{(k)}$. Using a $\sqrt{n}$-consistent estimate $\hat{\gamma}$ and a Nadaraya–Watson estimator $\hat{\tilde{g}}$ for $\tilde{g}$, Horowitz and Härdle (1996) show that the estimated coefficient $\hat{\beta}$ is itself $\sqrt{n}$-consistent and has an asymptotic normal distribution.

12.3.2 Generalizing the Index: Generalized Partial Linear Models

An alternative way to incorporate a nonmonotonic dependence of the response on the continuous variables is given by the GPLM

$$E(Y|X, T) = G\{X^T \beta + m(T)\}, \tag{12.22}$$

where $\beta = (\beta_1, \ldots, \beta_p)^T$ is a finite-dimensional parameter and $m(\cdot)$ is a smooth function. Models of this type allow a nonparametric inclusion of a part of the explanatory variables. In practice, this might be only those continuous variables that have most influence on the dependent variable Y.

Estimators for β and $m(\cdot)$ have been proposed by Hastie and Tibshirani (1990), Severini and Wong (1992), Severini and Staniswalis (1994), and Hunsberger (1994). Carroll et al. (1997) proposed an extension to the *generalized partial linear single-index model* (GPLSIM), which uses a SIM instead of the fully nonparametric function $m(\cdot)$.

12.3.2.1 Semiparametric Maximum Likelihood

The estimation of model (12.22) can be motivated by the fact that an estimate $\hat{\beta}$ can be found for known m and an estimate $\hat{m}$ can be found for known β. An overview of different algorithms for the GPLM can be found in Müller (1998). We concentrate here on the profile likelihood algorithm proposed by Severini and Wong (1992) and Severini and Staniswalis (1994). An extended presentation of the background of this approach is found in Staniswalis and Thall (1998).

Define

$$\mu = E(Y|X, T) = G\{\beta^T X + m(T)\}$$
$$\sigma^2 V(\mu) = \text{Var}(Y|X, T)$$

and denote the individual log likelihood or quasi-likelihood function by $l(\mu, y)$ (if the distribution of Y does not belong to an exponential family). The "parametric" likelihood function

$$L(\beta) = \sum_{i=1}^{n} l\left[G\{X_i^T \beta + m_\beta(T_i), Y_i\}\right]$$

is used to obtain $\hat{\beta}$. A "smoothed" or "local" likelihood function

$$L^S(\eta) = \sum_{i=1}^{n} \mathcal{K}_{\mathbf{H}}(t - T_i) l \left\{G(X_i^T \beta + \eta, Y_i)\right\} \tag{12.23}$$

is optimized to estimate the smooth function $m_\beta(t) = \eta$ at point t. Note that the use of this smoothed likelihood function leads to the equivalent of the Nadaraya–Watson estimator $\hat{m}_{\mathbf{H}}$ in ordinary regression. To obtain a local polynomial estimator of the nonparametric part $m(\bullet)$, we need to incorporate polynomial terms into the smoothed likelihood function. In the local linear case, we would use

$$L^S(\eta_0, \eta_1) = \sum_{i=1}^{n} \mathcal{K}_{\mathbf{H}}(t - T_i) l \left[G\{X_i^T \beta + \eta_0 + (T_i - t)^T \eta_1, Y_i\}\right]$$

and get $m_\beta(t) = \eta_0$ at point t. Analogous to local linear regression, η_1 points to the gradient of $m(\bullet)$ in t.

The computational algorithm consists in searching maxima of both likelihoods simultaneously. We stay in the framework of a Nadaraya–Watson type estimation of m. Severini and Staniswalis (1994) show that the resulting estimator $\hat{\beta}$ is $\sqrt{n}$-consistent and asymptotically normal and that estimators $\hat{m} = \hat{m}_{\hat{\beta}}$ are consistent in supremum norm. Note that m is estimated as a function of the parametric component β, which yields an asymptotically efficient estimate $\hat{\beta}$ (Severini and Wong, 1992). The possible scale parameter σ can be estimated by

$$\hat{\sigma}^2 = \frac{1}{n} \sum_{i=1}^{n} (Y_i - \hat{\mu}_i)^2 / V(\hat{\mu}_i),$$

where $\hat{\mu}_i = G\{\hat{\beta}^T X_i + \hat{m}(T_i)\}$.

The algorithm we present here corresponds to that proposed in Severini and Staniswalis (1994) for some special cases of link function and distributions of Y. In order to avoid boundary effects, one can use a weight function in the convergence criterion or trimming in the estimation of β.

Define $\hat{\eta}_j(\beta) = \hat{m}_\beta(t_j)$ and $l_i(u) = l\{G(u), Y_i\}$. For example, in a binary response model, we have $l_i(u) = Y_i \log G(u) + (1 - Y_i) \log\{1 - G(u)\}$. In the following, l_i' and l_i'' denote the derivatives of $l_i(u)$ with respect to u. The maximization of the smoothed quasi-likelihood function (12.23) requires that we solve

$$0 = \sum_{i=1}^{n} l_i'\{X_i^T \beta + \hat{\eta}_j(\beta)\} \mathcal{K}_{\mathbf{H}}(T_i - T_j) \tag{12.24}$$

with respect to $\hat{\eta}_j(\beta)$. In some models (in particular, for identity and exponential link functions G), (12.24) can be solved explicitly for $\hat{\eta}_j(\beta)$. Differentiating leads to an estimate for $\hat{\eta}'_j$ as a function of β:

$$\hat{\eta}'_j(\beta) = \frac{-\sum_{i=1}^{n} l''_i\{X_i^T\beta + \hat{\eta}_j(\beta)\}\mathcal{K}_{\mathbf{H}}(T_i - T_j)X_i}{\sum_{i=1}^{n} l''_i\{X_i^T\beta + \hat{\eta}_j(\beta)\}\mathcal{K}_{\mathbf{H}}(T_i - T_j)}. \tag{12.25}$$

For β, we have to solve

$$0 = \sum_{i=1}^{n} l'_i\{X_i^T\beta + \hat{\eta}_i(\beta)\}\{X_i + \hat{\eta}'_i(\beta)\}. \tag{12.26}$$

Equations (12.24)–(12.26) imply the following iterative Newton–Raphson type of algorithm to find $\hat{\beta}$ and $\hat{m}(t_j) = \hat{\eta}_j(\hat{\beta})$ for $j = 1, \ldots, n$:

- *Initialization*
  Different strategies to obtain starting values are possible:
  (i) Start with $\hat{\beta}^{(0)}$ and $\hat{\eta}_j^{(0)}$ from the parametric (GLM) fit. Higher order polynomial terms in T may be included to allow for a nonlinear function $\hat{\eta}_j^{(0)}$.
  (ii) Alternatively, it is possible to use $\hat{\beta}^{(0)} = 0$, and as in GLM, $\hat{\eta}_j^{(0)} = G^{-1}\{(Y_j + \overline{Y})/2\}$. (But $\hat{\eta}_j^{(0)} = G^{-1}\{(Y_j + 0.5)/(m+1)\}$ for binomial responses.)
  (iii) Severini and Staniswalis (1994) propose to start with $\hat{\beta}^{(0)} = 0$ and $\hat{\eta}_j^{(0)} = G^{-1}(Y_j)$ (with an adjustment for binomial responses).
- *Updating step for $\hat{\eta}_j(\beta) = \hat{m}_\beta(T_j)$*
  The function $\hat{\eta}_j(\beta)$ is updated by

$$\hat{\eta}_j^{(k+1)} = \hat{\eta}_j^{(k)} - \frac{\sum_{i=1}^{n} l'_i(X_i^T\hat{\beta}^{(k)} + \hat{\eta}_j^{(k)})\mathcal{K}_{\mathbf{H}}(T_i - T_j)}{\sum_{i=1}^{n} l''_i(X_i^T\hat{\beta}^{(k)} + \hat{\eta}_j^{(k)})K_{\mathbf{H}}(T_i - T_j)}.$$

- *Updating step for β*
  The parameter β is updated by

$$\hat{\beta}^{(k+1)} = \hat{\beta}^{(k)} - B^{-1}\sum_{i=1}^{n} l'_i(X_i^T\hat{\beta}^{(k)} + \hat{\eta}_i^{(k+1)})\tilde{X}_i^{(k)}$$

with a Hessian type of matrix

$$B = \sum_{i=1}^{n} l_i''(X_i^T \hat{\beta}^{(k)} + \hat{\eta}_i^{(k+1)}) \tilde{X}_i^{(k)} \tilde{X}_i^{(k)T}$$

and

$$\tilde{X}_j^{(k)} = X_j - \frac{\sum_{i=1}^{n} l_i''(X_i^T \hat{\beta}^{(k)} + \hat{\eta}_j^{(k+1)}) \mathcal{K}_{\mathbf{H}}(T_i - T_j) X_i}{\sum_{i=1}^{n} l_i''(X_i^T \hat{\beta}^{(k)} + \hat{\eta}_j^{(k+1)}) \mathcal{K}_{\mathbf{H}}(T_i - T_j)}.$$

As an alternative, the functions $l_i''(u)$ can be replaced by their expectations (with respect to Y) in order to obtain a Fisher-scoring type of procedure.

12.4 PRACTICAL APPLICATION AND SOFTWARE HINTS

Let us illustrate semiparametric estimation with the previous example of the Credit Scoring Data (Fahrmeir and Tutz, 1994; Fahrmeir and Hamerle, 1984). Recall that the data set consists of $n = 1000$ clients, among which 700 paid a credit back without problems and 300 did not. We define the binary variable Y with value 1 for those who paid back and 0 for those who did not. The data set contains observations from three continuous variables (`duration` of credit, `amount` of credit, and `age` of client) and 17 discrete variables. Of interest is how the explanatory variables can be used to predict creditworthiness.

A parametric logit model leads to the parameter estimates listed in Table 12.2. We omit the parameter estimates for the discrete explanatory variables. The linear influence of duration is highly significant. By contrast, `amount` and `age` have no

TABLE 12.2 Parametric Logit Coefficients and GPLM Coefficients (t-values in parentheses)

	Coeff.	t-value	Coeff.	t-value	Coeff.	t-value
const.	−17.605	(−1.91)	−34.909	(−2.51)	—	—
duration	−0.036	(−3.85)	−0.033	(−3.48)	−0.037	(−4.23)
log(**amount**)	1.654	(1.41)	4.847	(2.57)	—	—
log(**amount**) square	—	—	−0.229	(−2.26)	—	—
log(**age**)	4.119	(1.59)	6.949	(1.11)	—	—
log(**age**) square	—	—	−0.501	(−0.60)	—	—
log(**age**) * log(**amount**)	−0.484	(−1.47)	−0.384	(−1.17)	—	—
...	...	...	...	...	...	...
	Linear		Quadratic		Partial	Linear

Bandwidths are 40% of range for GPLM.

significant coefficients if we include them linearly. We will see that the insignificant coefficients are a sign of a more complex, structured influence.

Next, we fit a GPLM according to the algorithm just presented. Here, the influence of amount and age is fitted nonparametrically. Figure 12.7 shows the bivariate estimate $\hat{m}$ in the upper panel. The bandwidths were chosen to be 40%

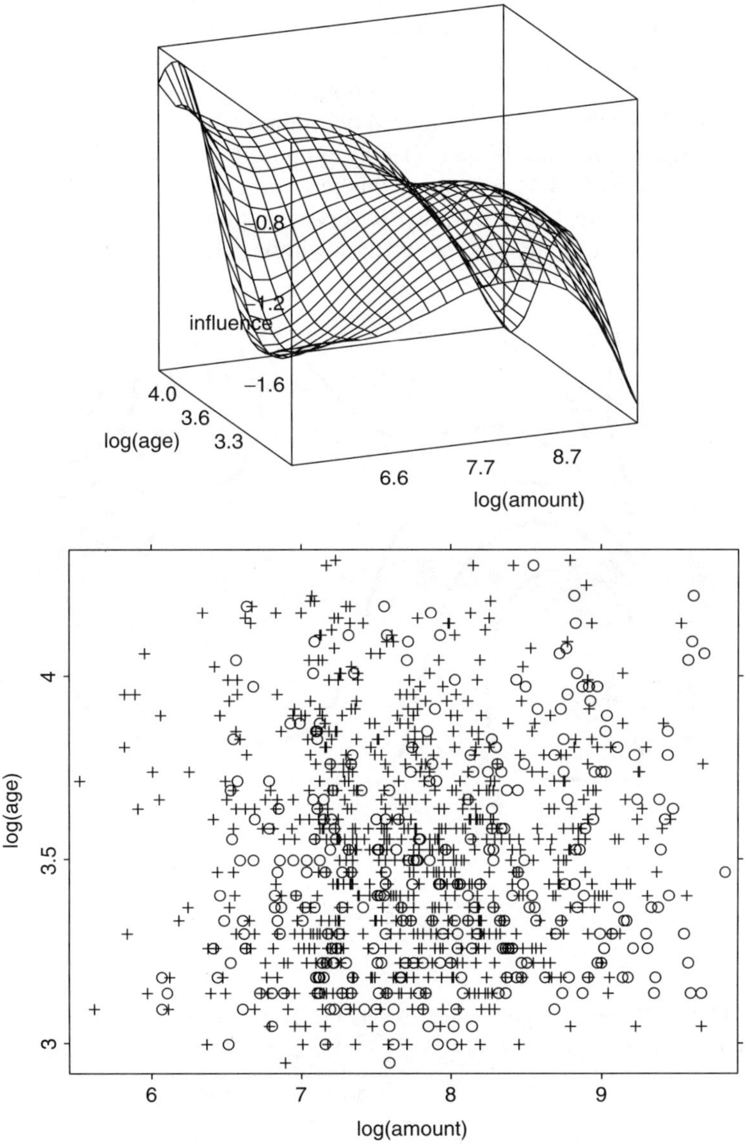

FIGURE 12.7 Credit Scoring Data revisited: Two-dimensional nonparametric function of amount and age in GPLM (upper panel). (Bandwidths are 40% of range.) Scatter plot of amount vs. age (lower panel).

of the range in both dimensions, which gives $h_1 = 1.72$ and $h_2 = 0.55$. A scatter plot of `amount` vs. `age` is given in the lower panel of the figure. The good clients ($Y = 1$) are marked by "+", the bad clients ($Y = 0$) by "o".

We have carried out the analysis for different bandwidths; the nonparametric estimates $\hat{m}$ are obviously nonlinear. However, it is difficult to judge whether a nonparametric estimate gives a significant improvement. The high values of $\hat{m}$ are caused by only a few observations (as can be seen in the scatter plot). Figure 12.8 shows a contour plot of $\hat{m}$. In general, one cannot exclude the possibility that the visual difference between the nonparametric and the linear fit is caused by boundary and bias problems of $\hat{m}$. Additionally, some of the other covariables have a quite dominant influence on creditworthiness.

Härdle, Mammen, and Müller (1998) proposed a bootstrap procedure for testing GLM versus GPLM. We apply this test and compute critical values using their procedure. Table 12.3 shows the observed significance levels for rejection. As before, we report the results for the bandwidths, expressed as a percentage of

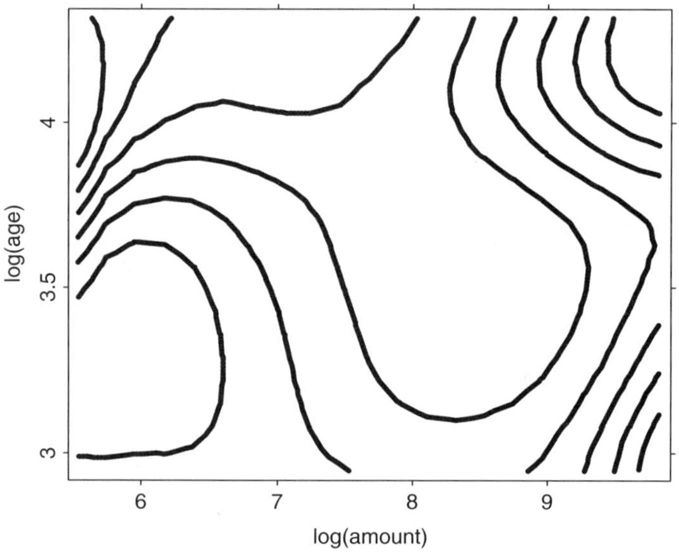

FIGURE 12.8 Credit Scoring Data revisited: Contours for nonparametric function of `amount` and `age` in GPLM. Bandwidths are 40% of range.

TABLE 12.3 Observed Significance Levels for Test of GLM Against GPLM (Bootstrap sample size $n^* = 100$)

h	20%	30%	40%	50%	60%
Linear	<0.01	<0.01	<0.01	0.01	0.29
Linear and interaction	<0.01	<0.01	<0.01	0.07	0.40
Quadratic	<0.01	<0.01	<0.01	0.35	0.55

the range in both dimensions. The decision of the test depends obviously on the bandwidth.

We see from the table that linearity is clearly rejected for bandwidths up to 40%. The significance levels for rejection increase when we include interaction and quadratic terms. But altogether, we conclude that the correct model should be of more complicated structure than the quadratic.

For higher dimensions in T, the possible nonlinearities in (12.22) cannot be graphically displayed and face the above mentioned problems associated with interpretability. Hence, an additive structured partial linear index may be considered, as is done by Hastie and Tibshirani (1990) on the basis of the backfitting algorithm. A variant based on the marginal integration method introduced by Linton and Nielsen (1995) is currently under development. (See Härdle, Huet, Mammen, and Sperlich, 1998.)

Routines for kernel density and kernel regression estimation are included in virtually any modern software package. Semiparametric procedures are typically add-ons in programming environments that allow a user side integration of kernel estimation procedures. XploRe (Härdle, Klinke and Müller, 1999) as used above and S-Plus are such examples.

Acknowledgements

We wish to thank several participants of classes in non- and semiparametric modeling and two referees for valuable proposals that we believe improved the text of this chapter.

REFERENCES

Carroll, R. J., Fan, J., Gijbels, I., and Wand, M. P. (1997). Generalized partially linear single-index models, *J. Amer. Stat. Assoc.*, **92**(438), 477–489.

Craven, P., and Wahba, G. (1979). Smoothing noisy data with spline functions, *Numer. Math.*, **31**, 377–403.

Fahrmeir, L., and Hamerle, A. (1984). *Multivariate Statistische Verfahren*, De Gruyter, Berlin, Germany.

Fahrmeir, L., and Tutz, G. (1994). *Multivariate Statistical Modeling Based on Generalized Linear Models*, Springer-Verlag, New York.

Fan, J., Gasser, Th., Gijbels, I., Brockmann, M., and Engel, J. (1997). Local polynomial regression: optimal kernels and asymptotic minimax efficiency, *Ann. Inst. Statist. Math.*, **49**(1), 79–99.

Fan, J., and Gijbels, I. (1995). *Local Polynomial Modeling and Its Application: Theory and Methodologies*, Chapman and Hall, New York.

Fan, J., and Marron, J. S. (1994). Fast implementations of nonparametric curve estimators, *J. Comput. Graph. Statist.*, **3**(1), 35–56.

Fan, J., and Müller, M. (1995). Density and regression smoothing, in Härdle, W., Klinke, S., and Turlach, B. A. (eds.), *XploRe: An Interactive Statistical Computing Environment*, pp. 77–99, Springer-Verlag, New York.

Härdle, W. (1990). *Applied Nonparametric Regression*, Cambridge University Press, Cambridge, UK.

Härdle, W. (1991). *Smoothing Techniques, with Implementations in S*, Springer-Verlag, New York.

Härdle, W., Hall, P., and Marron, J. S. (1988). How far are automatically chosen regression estimators from their optimum? *J. Amer. Stat. Assoc.*, **83**, 86–97.

Härdle, W., Hall, P., and Marron, J. S. (1992). Regression smoothing estimators that are not far from their optimum, *J. Amer. Stat. Assoc.*, **87**, 227–233.

Härdle, W., Huet, S., Mammen, E., and Sperlich, S. (1998). *Semiparametric Additive Indices for Binary Response*, SFB 373 Discussion Paper 95, Sonderforschungsbereich 373, Humboldt-Universität zu Berlin, Germany.

Härdle, W., Klinke, S., and Müller, M. (eds.) (1999). *XploRe—Academic Edition* (with CD), Springer-Verlag, Berlin, Germany.

Härdle, W., Mammen, E., and Müller, M. (1998). Testing parametric versus semiparametric modeling in generalized linear models, *J. Amer. Stat. Assoc.*, **93**(444), 1461–1474.

Härdle, W., and Scott, D. (1992). Smoothing in by weighted averaging using rounded points, *Computational Statistics*, **7**, 97–128.

Härdle, W., and Stoker, T. (1989). Investigating smooth multiple regression by the method of average derivatives, *J. Amer. Stat. Assoc.*, **84**, 986–995.

Hastie, T. J., and Tibshirani, R. J. (1990). *Generalized Additive Models*, Chapman and Hall, London, UK.

Horowitz, J. L. (1993). Semiparametric and nonparametric estimation of quantal response models, in Madala, G. S., Rao, C. R. and Vinod, H. D. (eds.), *Handbook of Statistics*, pp. 45–72, Elsevier Science Publishers, Netherlands.

Horowitz, J. L., and Härdle, W. (1996). Direct semiparametric estimation of single index models with discrete covariates, *J. Amer. Stat. Assoc.*, **71**, 1632–1640.

Hunsberger, S. (1994). Semiparametric regression in likelihood–based models, *J. Amer. Stat. Assoc.*, **89**, 1354–1365.

Ichimura, H. (1993). Semiparametric least squares (SLS) and weighted SLS estimation of single-index models, *J. Econometrics*, **58**, 71–120.

Klein, R., and Spady, R. (1993). An efficient semiparametric estimator for binary response models, *Econometrica*, **61**, 387–421.

Korostelev, A., and Müller, M. (1995). *Single Index Models with Mixed Discrete–Continuous Explanatory Variables*, Discussion Paper 26, Sonderforschungsbereich 373, Humboldt-Universität zu Berlin, Germany.

Linton, O., and Nielsen, J. P. (1995). A kernel method of estimating structured nonparametric regression based on marginal integration, *Biometrika*, **82**, 93–100.

Maddala, G. S. (1983) *Limited-dependent and qualitative variables in econometrics*, Cambridge University Press, Cambridge, UK.

Marron, J. S., and Nolan, D. (1988). Canonical kernels for density estimation, *Statistics Probability Letters*, **7**, 195–199.

McCullagh, P., and Nelder, J. A. (1989). *Generalized Linear Models*, (2d ed.), Chapman and Hall, London, UK.

Müller, H.-G. (ed.) (1988). *Nonparametric Regression Analysis of Longitudinal Data*, Springer-Verlag, Berlin, Germany.

Müller, M. (1998). Computer-assisted generalized partial linear models, in Scott, D. W. (ed.), *Computing Science and Statistics*, Proceedings of the 29th Symposium of the Interface, pp. 221–230.

Newey, W., and Stoker, T. (1993). Efficiency of weighted average derivative estimators and index models, *Econometrica*, **5**, 1199–1223.

Powell, J. L., Stock, J. H., and Stoker, T. M. (1989). Semiparametric estimation of index coefficients, *Econometrica*, **57**(6), 1403–1430.

Ruppert, D., and Wand, M. P. (1994). Multivariate locally weighted least squares regression, *Annals of Statistics*, **22**(3), 1346–1370.

Sain, S. R., Baggerly, K. A., and Scott, D. W. (1994). Cross-validation of multivariate densities, *J. Amer. Stat. Assoc.*, **89**(427), 807–817.

Scott, D. W. (1992). *Multivariate Density Estimation: Theory, Practice, and Visualization*, Wiley, New York.

Scott, D., and Wand, M. (1991). Feasibility of multivariate density estimates, *Biometrika*, **78**, 197–205.

Severini, T. A., and Staniswalis, J. G. (1994). Quasi-likelihood estimation in semiparametric models, *J. Amer. Stat. Assoc.*, **89**, 501–511.

Severini, T. A., and Wong, W. H. (1992). Generalized profile likelihood and conditionally parametric models, *Annals of Statistics*, **20**, 1768–1802.

Silverman, B. W. (1986). *Density Estimation for Statistics and Data Analysis*, Chapman and Hall, London, UK.

Staniswalis, J. G., and Messer, K. (1996). Addendum to "Kernel estimators for multiple regression," *J. Nonparametric Statistics*, **7**, 67–68.

Staniswalis, J. G., Messer, K., and Finston, D. R. (1994). Kernel estimators for multiple regression, *J. Nonparametric Statistics*, **3**, 103–121.

Staniswalis, J. G., and Thall, P. F. (1998). An explanation of generalized profile likelihood, in Scott, D. W. (ed.), *Computing Science and Statistics*, Proceedings of the 29th Symposium of the Interface, pp. 233–239.

Stoker, T. M. (1991). Equivalence of direct, indirect and slope estimators of average derivatives, in Barnett, W. A., Powell, J., and Tauchen, G. (eds.), *Nonparametric and Semiparametric Methods in Econometrics and Statistics*, Proceedings of the Fifth International Symposium in Economic Theory and Econometrics, pp. 99–118, Cambridge University Press, Cambridge, UK.

Wand, M. P. (1994). Fast computation of multivariate kernel estimators, *J. Comput. Graph. Statist.*, **3**(4), 433–445.

Wand, M. P., and Jones, M. C. (1994). Multivariate plug-in bandwidth selection, *Computational Statistics*, **9**, pp. 97–117.

Wand, M. P., and Jones, M. C. (1995). *Kernel Smoothing*, Vol. 60 of *Monographs on Statistics and Applied Probability*, Chapman and Hall, London, UK.

Weisberg, S., and Welsh, A. H. (1994). Adapting for the missing link, *Annals of Statistics*, **22**, 1674–1700.

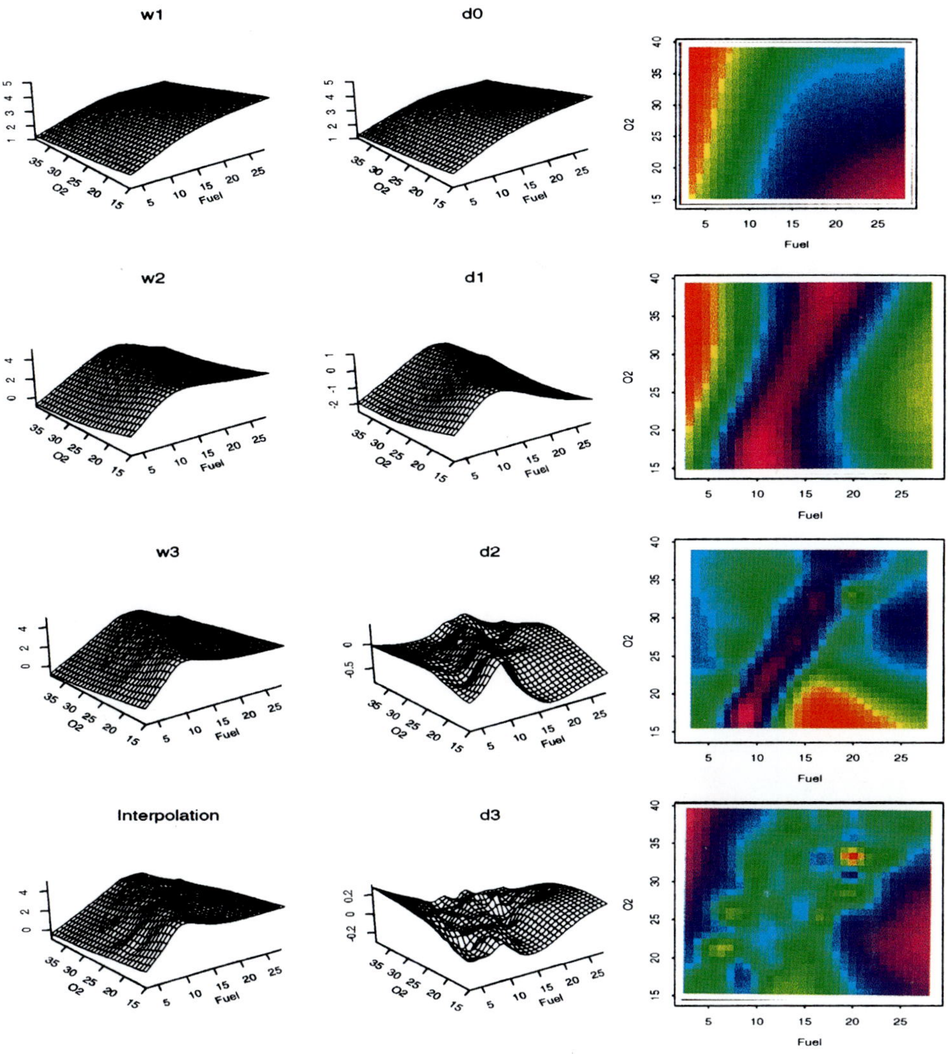

Plate 13.3. Different spline estimates ($m = 3$) and the decomposition of the Flame Data into several detail surfaces. In the first column are the surfaces derived from the three weight functions plotted in Figure 13.2(b), with effective model degrees of freedom 8.0, 43.5, and 60.0. The last plot in this column is an interpolated surface. In the second column are the detail surfaces for the multiresolution decomposition described in Section 13.5.3. The first detail surface is the same as the first smoothed surface. The subsequent surfaces are differences between the smoothed surfaces at two different levels of smoothing. The last column are image plots of the detail surfaces in the second column.

(a)

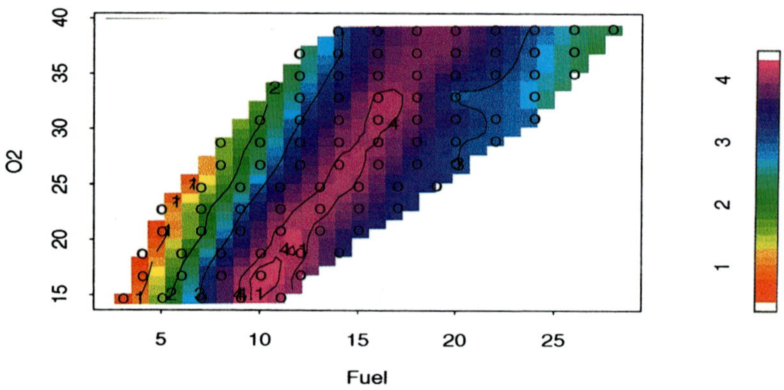

(b)

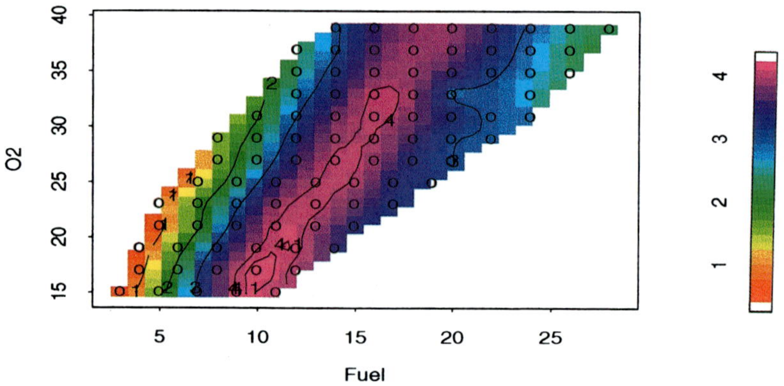

Plate 13.5. (a) Thin-plate spline fitted to the Flame Data with $m = 3$. (b) Kriging estimates for the Flame Data using an exponential-covariance function with range parameter $\theta = 5$. Observation locations are indicated on the contour plots.

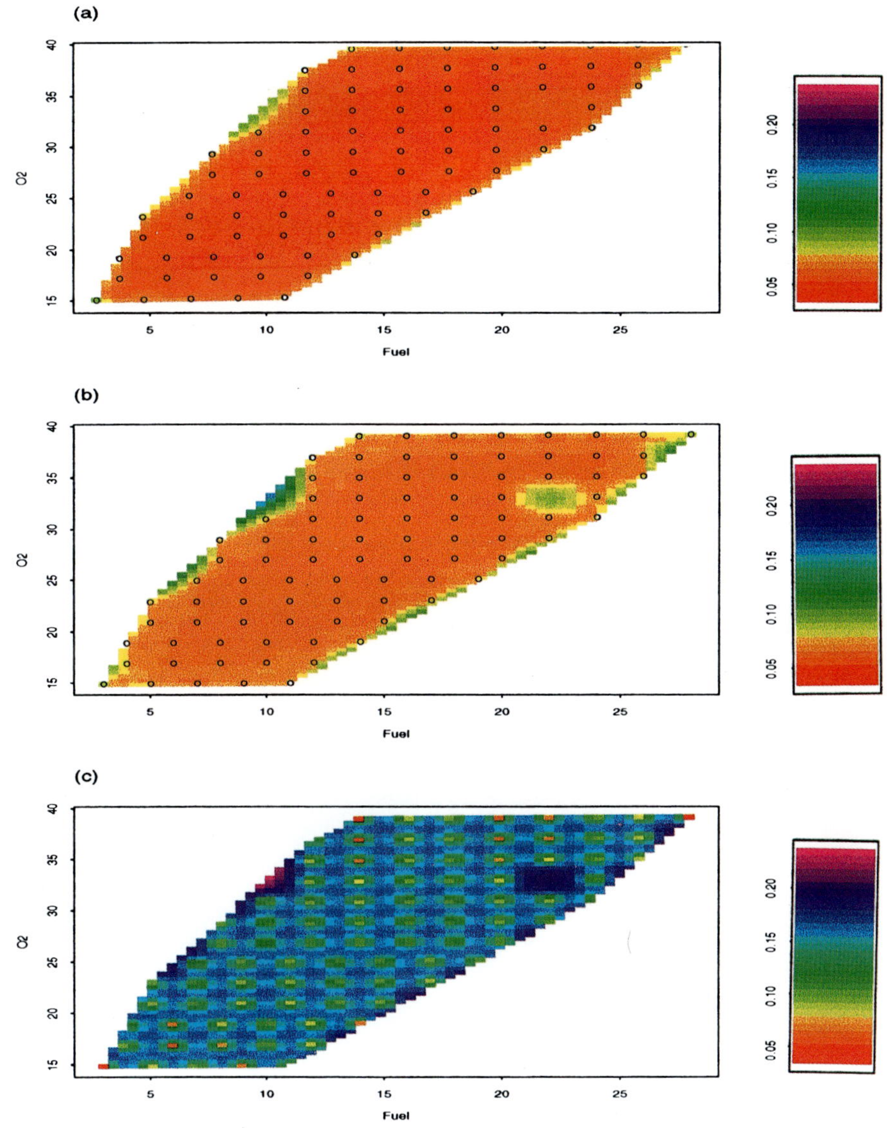

Plate 13.6. Standard errors for thin-plate spline and spatial process estimates for the Flame Data (see Figure 13.5). (a) The standard deviation of the spline estimate that uses a fixed function and no bias. (b) The standard errors associated with the spline estimate that uses the covariance function implied by the radial basis functions. (c) The standard errors associated with the spatial-process estimate that uses an exponential covariance function.

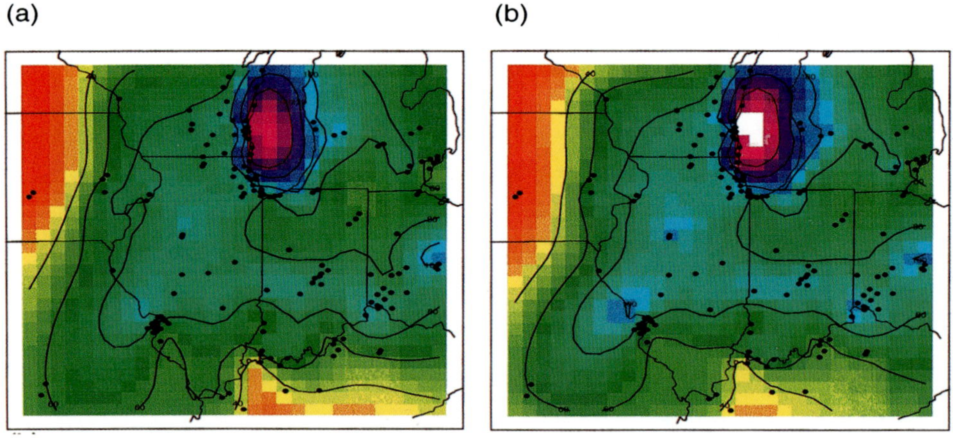

Plate 13.8. Spatial predictions for 8-hour-average ozone in parts per billion (PPB) for June 19, 1987. Results for thin-plate spline (a) and Kriging estimates with an isotropic, exponential correlation function but nonstationary marginal variances (b).

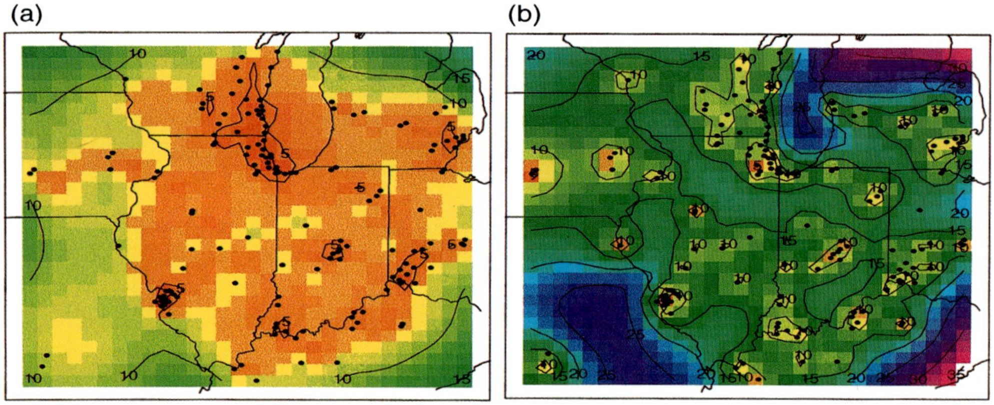

Plate 13.9. Standard errors of prediction (in PPB) for the spline estimate (a) and the spatial process estimate (b) from Figure 13.8.

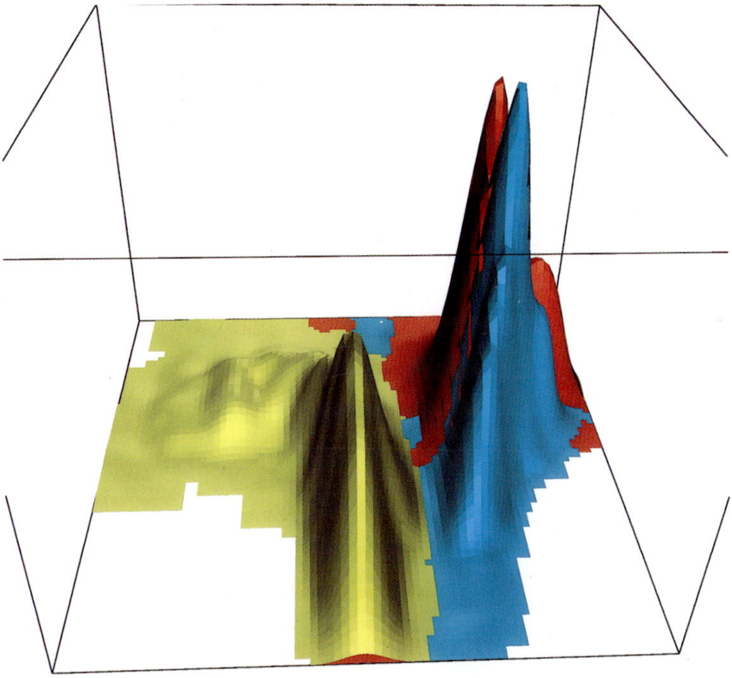

Plate 15.1. Perspective plots for the X-Y variables of three estimated densities for each of the three crops in the Remote-Sensing Data (yellow = sunflower; red = wheat; blue = barley). At each point (x,y) the greatest of the three density values is displayed.

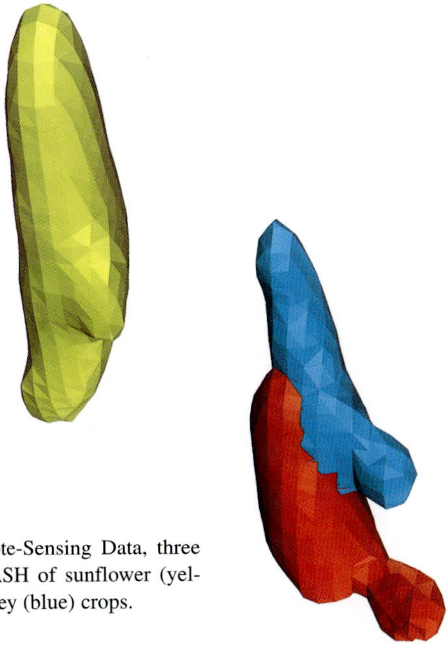

Plate 15.2. For the Remote-Sensing Data, three β-contours (β = 0.5) of ASH of sunflower (yellow), wheat (red), and barley (blue) crops.

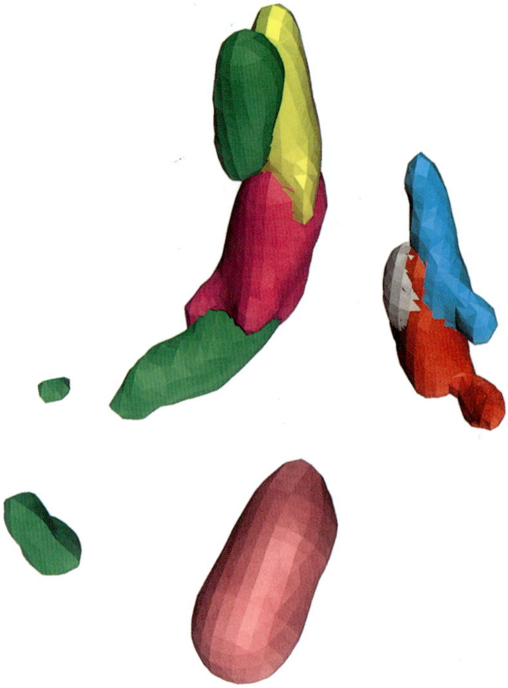

Plate 15.3. For the Remote-Sensing Data, β-contours ($\beta = 0.5$) of ASH for the six major crops. In addition to sunflower (yellow), wheat (red), and barley (blue) crops, we have soybeans (green, $n = 731$), sugar beets (magenta, $n = 506$), spring oats (white, $n = 459$), and peanuts (pink, $n = 306$).

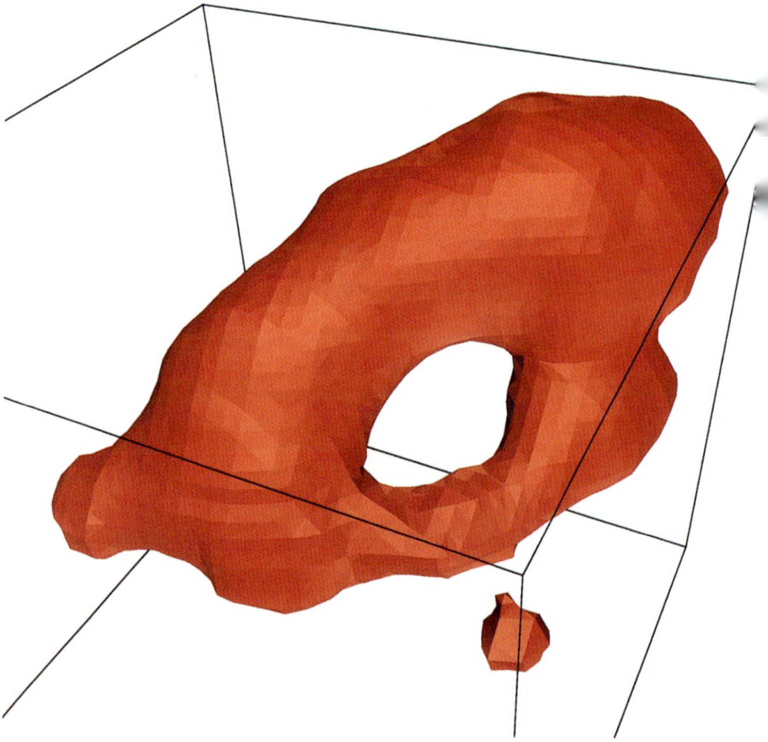

Plate 15.4. A trivariate contour of the lagged log-lynx time series from 1821 to 1934 at the $\alpha = 0.25$ level.

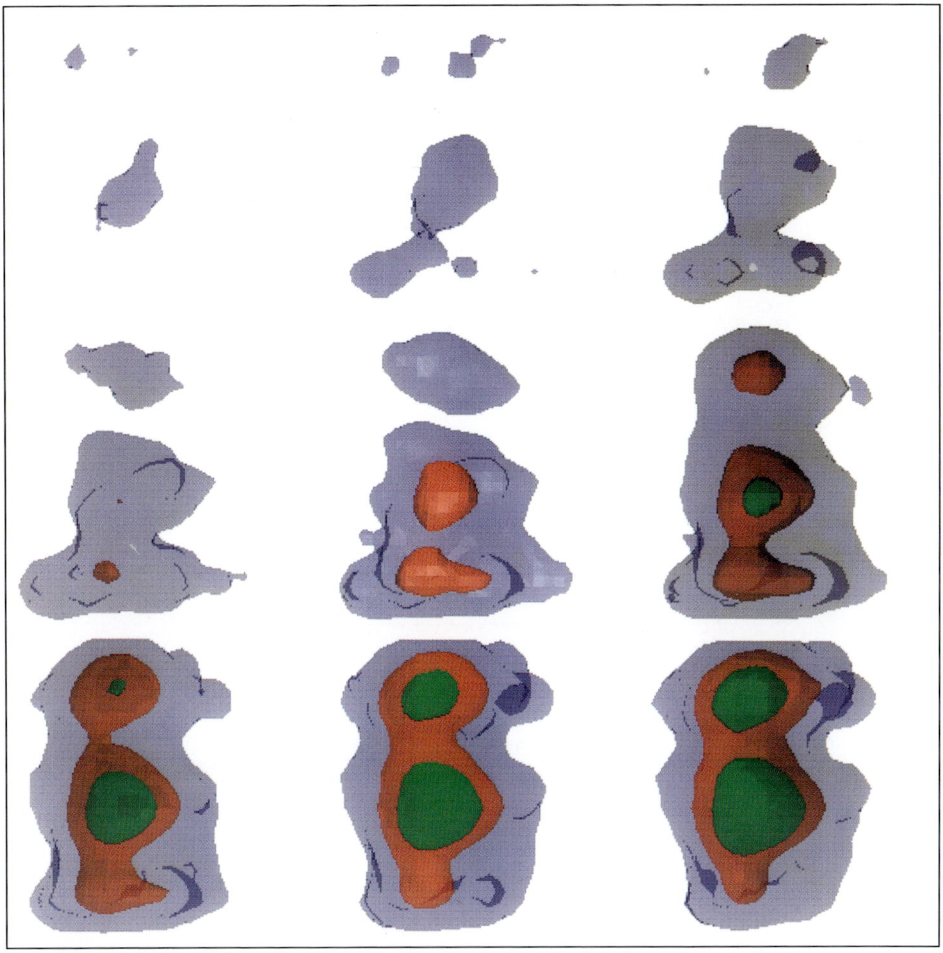

Plate 15.5. Nine frames from a movie of the quadravariate density estimate, $f(x, y, z, t)$, of the Mount St. Helens Data (time moves across and then down). Three α-contours are displayed at each time slice, with $\alpha = 0.02, 0.10, 0.25$.

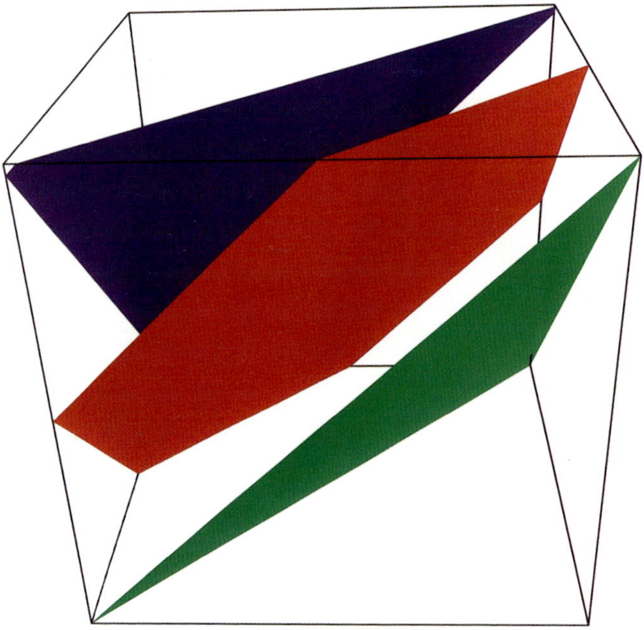

Plate 15.6. Three triumvariate contours of a linear parametric model of experimental design data.

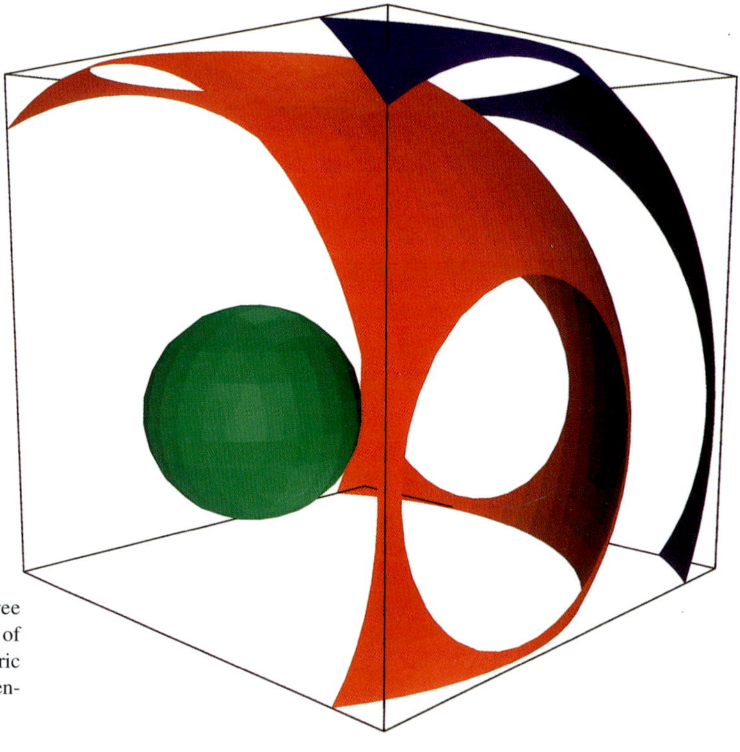

Plate 15.7. Three trivariate contours of a quadratic parametric model of experimental design data.

CHAPTER 13

Spatial-Process Estimates as Smoothers

Douglas W. Nychka
National Center for Atmospheric Research, Boulder, Colorado, USA

13.1 INTRODUCTION

Spatial statistics refers to the class of models and methods for data collected over a region. This region might be a mineral field for a geologist, a quadrat in a forest for an ecologist, or a geographic region for an environmental scientist. A typical problem is to predict values of a measurement at places where it is not observed, or, if the measurements are observed with error, to estimate a smooth spatial process (i.e., a surface) from the data. A family of techniques, known loosely as *Kriging*, was originally developed in geostatistics and are now being widely used for spatial prediction. In this chapter, we will also use the term *spatial-process estimates* to describe these techniques. The goals of Kriging sound very much like those of nonparametric regression, and the understanding of spatial estimates can be enriched through their interpretation as smoothing estimates. In the other direction, the random-process model that is at the heart of spatial methods is also valuable in setting uncertainty estimates for function estimates, especially in low-noise situations. Lying between two standard methods, kernel smoothing and Kriging, are smoothing splines. Because splines have a foot in both of these camps, they aid in understanding the structure common to these approaches. This chapter will present spatial-process estimates and smoothing splines in a common form as least-squares ridge-regression estimates. This approach leads to an interpretation for both of these estimates as a particular kind of orthogonal-series estimate. The orthogonal-series representation is a basis for efficient formulas

Smoothing and Regression: Approaches, Computation, and Application
Edited by Michael G. Schimek
ISBN 0-471-17946-9

for cross-validation and simulation, and also suggests approximate estimators for large data sets.

There are good surveys of spatial estimates (e.g., Cressie, 1991), and in fact the connection between spatial-process models and splines has enjoyed much attention, most recently from Kent and Mardia (1994), and some lively debate (Cressie, 1990, and Wahba, 1990b). So why is another chapter on this subject needed? Most treatments of Kriging do not emphasize the fact that the form of the predictions is a linear combination of basis functions prescribed by the covariance function. The coefficients for this representation are linear functions of the data found by a specific ridge regression. This form makes it easy to see how the Kriging estimator behaves as a smoother, and it illustrates how important parameters of the covariance function can be found by generalized cross-validation or restricted maximum likelihood.

Part of the motivation for this research comes from an interest in the spatial distribution of ambient ozone pollution in the United States. Another less obvious connection comes from the analysis of experimental data for developing new materials and manufacturing processes (O'Connell and Wolfinger, 1997). Here the "spatial" region consists of the variables that can be changed to modify the process. Driven by the applied nature of these problems, we have developed software, FUNFITS[‡] (Nychka et al., 1996), that simplifies the fitting and interpretation of these function estimates. This set of object-oriented software benefits from the numerical algorithms based on the ridge-regression/orthogonal-series representation. The reader is referred to the examples and help files in this software for more supporting information on using these methods and details of computation. One contribution of this chapter is the interweaving of the theoretical development of these estimates with the more pragmatic algorithms for their computation.

13.1.1 Basic Model

The observational model considered in this work has the form

$$Y_i = f(\boldsymbol{x}_i) + \epsilon_i \quad \text{for } 1 \le i \le n, \tag{13.1}$$

where Y_i is the observed response at the ith combination of design variables (or ith location), $\boldsymbol{x}_i \in \mathbb{R}^d$, and f is the function of interest. The random components $\{\epsilon_i\}$ usually associated with the measurement errors are assumed be uncorrelated, zero-mean random variables with variances $\{\sigma^2/W_i\}$.

[‡] FUNFITS is a module of S functions for function fitting and statistical designs that supplement those in S-Plus. A manual and the software for UNIX and Windows can be located from the software subheading on Geophysical Statistics Project home page, `http://www.cgd.ucar.edu/stats`.

One parsimonious strategy for representing f is as the sum of a low-order polynomial and a smooth function:

$$f(\boldsymbol{x}) = P(\boldsymbol{x}) + h(\boldsymbol{x}). \tag{13.2}$$

Typically, in nonparametric regression and smoothing, the assumptions on h are based on higher-order derivatives; in the case of thin-plate splines (Wahba, 1990a; Green and Silverman, 1994), the degree of P and the dimension of $\boldsymbol{x}$ imply a specific roughness penalty on h based on integrated squared (partial) derivatives. In spatial-statistics models, the smoothness assumption is replaced by the assumption that h is a random field. Under the assumption that f is a realization of a spatial process, P can be identified as the "spatial trend" or "drift", the fixed part of the model, and h is modeled as a mean-zero Gaussian process (Cressie, 1991). Let

$$\operatorname{Cov}(h(\boldsymbol{x}), h(\boldsymbol{x}')) = \rho k(\boldsymbol{x}, \boldsymbol{x}') \tag{13.3}$$

be a covariance function, where k is a known function and ρ is an unknown multiplier. Conditions on the covariance function then control the smoothness properties of h.

To simplify the treatment in this chapter, it will be assumed that the covariance function is continuous as $\boldsymbol{x}' \to \boldsymbol{x}$. Thus, if the spatial process has a nugget effect, it is due to the additive errors in (13.1) and not to a discontinuity when $\boldsymbol{x} = \boldsymbol{x}'$ in the covariance function.

13.1.2 The Unifying Theme

Splines and spatial-process estimates are complicated transformations of the data and a careful development of their properties will require some patience. However, it is important not to let the details obscure the basic form of these methods as orthogonal-series estimators. We will construct the set of basis functions $\{g_\nu\}$ for $1 \le \nu \le n$ with the property that if $u_\nu = \sum_{k=1}^{n} g_\nu(\boldsymbol{x}_k) W_k Y_k$, then function $\sum_{\nu=1,n} u_\nu g_\nu(\boldsymbol{x})$ will interpolate the data, Y_k. This construction is possible because of an orthogonality property for the basis. Moreover, the orthogonality will ensure that these interpolation coefficients, u_k, are uncorrelated with each other, and thereby simplify the simulation of these estimators.

In addition to the basis, there is a sequence of increasing nonnegative weights (eigenvalues) $\{D_\nu\}$, so that

$$\hat{f}(\boldsymbol{x}) = \sum_{\nu=1}^{n} \frac{1}{1+\lambda D_\nu} u_\nu g_\nu(\boldsymbol{x}), \tag{13.4}$$

where λ is the usual spline-smoothing parameter. (In terms of the spatial-process estimate, this is the *signal to noise ratio* $\lambda = \sigma^2/\rho$). Thus, we see that both estimates for fixed λ are just tapered versions of the interpolating series. The role of the weighting function is to shrink the magnitude of coefficients where D_ν is large relative to λ.

Where does this magic orthogonal basis come from? For spatial-process models, the usual Kriging surface is a linear combination of low-order polynomial functions, and n functions, $\psi_j(\boldsymbol{x}) = k(\boldsymbol{x}, \boldsymbol{x}_j)$, each obtained by evaluating one of the n arguments of the covariance function at the observed locations. A standard matrix decomposition transforms this basis into the orthogonal one. The fact that the Kriging estimators start with an expansion using the covariance function has important consequences for their behavior away from the observed locations. For splines, one starts with polynomials and radial basis functions and then proceeds in the same way.

In this overview, the case of smoothing has been emphasized, as is appropriate when the function is observed with error ($\sigma > 0$). When measurement error is not present, the situation is the same as setting $\sigma = 0$ (and thus $\lambda = 0$) in the tapering formula in (13.4). The result is a linear combination of the basis functions that interpolates the observed data and can be extrapolated to provide estimates of the surface at unobserved points. Throughout this chapter, the interpolation results can be inferred from the smoothing estimators by considering the limiting case as $\lambda \to 0$.

13.1.3 Chapter Outline

The next section presents the thin-plate spline estimate. Splines are of interest because they have both a spatial-process interpretation and a justification as penalized least-squares estimates independent of a stochastic model for the unknown function. For this reason, it is appropriate that they are discussed first. Section 13.3 develops spatial-process estimates for a random field as the best linear unbiased estimate. The derivation is different from standard treatments, using a QR matrix decomposition to enforce unbiasedness. This approach is included because it parallels the methods of computation.

Section 13.4 gives the common ridge-regression form for Kriging and splines, in order to derive equation (13.4). From this perspective, it is clear how to approximate a Kriging estimate for a large data set from a reduced set of basis functions. Section 13.5 uses a simple response-surface experiment to illustrate how the orthogonal-series representation acts as a smoother and includes a multiresolution decomposition of the data. Section 13.6 presents a spatial-process example based on predicting surface ozone from a limited set of monitoring locations in the Great Lakes region of the US. The last section suggests some areas that need more research effort.

Finally, it should be acknowledged that there is much missing from this treatment. Deeper connections between these two methods, based on Bayesian formalism, have been largely omitted. Also, from a practical viewpoint, the craft of identifying and estimating covariance functions has been alluded to in Sections 13.3.4 and 13.6, but the technology based on variograms has been largely left out.

13.2 THIN-PLATE SPLINES

A spline is defined implicitly as the solution to a variational (minimization) problem. The form of the minimization criterion determines the kind of spline. A thin-plate spline estimator of f is the minimizer of the penalized sum of squares

$$S_\lambda(f) = \frac{1}{n}\sum_{i=1}^{n} W_i(y_i - f(\boldsymbol{x}_i))^2 + \lambda J_m(f) \tag{13.5}$$

for $\lambda > 0$ and for

$$J_m(f) = \int_{\mathbb{R}^d} \sum \frac{m!}{\alpha_1!\ldots\alpha_d!}\left(\frac{\partial^m f}{\partial x_1^{\alpha_1}\ldots\partial x_d^{\alpha_d}}\right)^2 d\boldsymbol{x}.$$

The sum in the integrand is taken over all nonnegative integer vectors, α, such that $\sum \alpha_1 + \cdots + \alpha_d = m$, where $2m > d$. For one dimension and $m = 2$, $J_2(f) = \int (f^{('')}(x))^2\, dx$, giving the standard cubic smoothing-spline roughness penalty. The thin-plate spline is a generalization of the usual cubic smoothing spline (as introduced in Chapter 1) with a "roughness" penalty function, $J_m(f)$, that is rotationally invariant and may involve partial derivatives higher than second-order. An important feature of a spline estimate is the part that is unaffected by the roughness penalty. The set of functions where the roughness penalty is zero is termed the null space, and, for J_m, consists of all polynomials with degree less than or equal to $m - 1$. There is a rich theory that blends the construction of different roughness penalties with statistical models, and the reader is referred to Chapter 11 in this volume for more details on different kinds of spline models.

Minimization of S_λ results in a function that tracks the data, yet is constrained to be smooth. A Bayesian viewpoint is to interpret $-S_\lambda$, to within a constant, as the log of the joint density function for f and Y. Let p be the conditional density of Y given f, and π be the prior density for f; then

$$-S_\lambda = \ln(p(Y|f)\pi(f)) = \ln(p(Y|f)) + \ln(\pi(f)).$$

If one adopts equation (13.5), then the roughness penalty is associated with a log prior distribution for f. Furthermore, this prior can be interpreted as a distribution, such that f is a realization from a smooth (almost m-differentiable) Gaussian process. From Bayes' Theorem, we know that, for fixed Y, the posterior is proportional to the joint density. Thus, maximizing the joint density (i.e., minimizing S_λ) results in a mode of the posterior distribution. The reader is referred to Wahba (1990a) and to Omre and Halvorsen (1989) for additional discussion of this correspondence. Although this brief discussion sketches out a connection between splines and spatial-process estimates, the following sections will pursue a more direct connection. It will be shown that the spline estimate is, in fact, a minimum-variance unbiased estimator under a specific stochastic model. In this way, the interpretation does not depend explicitly on a Bayesian formulation, and it circumvents the measure-theoretic problems of putting prior distributions on function spaces. The next section will set the stage for this (frequentist) interpretation.

13.2.1 Form of the Spline Estimate

The most important step in deriving the spline estimate is identifying the solution as a finite linear combination of basis functions. In this way, the abstract minimization over a function space collapses into minimizing over the coefficients associated with the basis functions.

Thin-plate splines are expressed in terms of radial basis functions. Let $\psi_i(\boldsymbol{x}) = E_{md}(\boldsymbol{x} - \boldsymbol{x}_i)$, where E_{md} are the radial basis functions

$$E_{md}(\boldsymbol{r}) = \begin{cases} a_{md}\|\boldsymbol{r}\|^{(2m-d)}\log(\|\boldsymbol{r}\|), & d \text{ even} \\ a_{md}\|\boldsymbol{r}\|^{(2m-d)}, & d \text{ odd} \end{cases}$$

and where a_{md} is a constant. Also, let $\{\phi_j\}$ be a set of t polynomial functions that span the space of all d-dimensional polynomials with degree less than m. Now consider a function of the form

$$f(\boldsymbol{x}) = \sum_{j=1}^{t} \phi_j(\boldsymbol{x})\beta_j + \sum_{i=1}^{n} \psi_i(\boldsymbol{x})\delta_i,$$

which is parameterized by $\boldsymbol{\beta}$ and $\boldsymbol{\delta}$.

An elegant development of the spline solution can be obtained from Meinguet (1979) and Duchon (1977), with an overview provided by Wahba (1990a). See also Kent and Mardia (1994) for a discussion of generalized functions. For the purposes of this chapter, we will abstract three key results that yield the computational form for the spline. Define T, a $n \times t$ matrix with $T_{k,j} = \phi_j(\boldsymbol{x}_k)$. Let

$$\text{Var} = \{f : J_m(f) < \infty\} \text{ and } \mathcal{M} = \left\{ f : f = \sum_{j=1}^{t} \phi_j\beta_j + \sum_{i=1}^{n} \psi_i\delta_i \text{ and } T^T\boldsymbol{\delta} = 0 \right\}.$$

Then

(1) $\mathcal{M} \subset \text{Var}$
(2) $\min_{f\in\text{Var}} S_\lambda(f) = \min_{f\in\mathcal{M}} S_\lambda(f)$
(3) If T has full rank, then a unique solution exists.

A key point is that $\mathcal{M}$ is spanned by a finite number of basis functions. Moreover, from these three basic properties it follows that the solution to the spline-minimization problem can be reduced to a (quadratic) optimization problem over a finite-dimensional space. With this result, it is useful to rephrase the minimization problem in matrix/vector form, substituting the finite-dimensional form into

$S_\lambda(f)$. Although the spline estimator has its roots in the abstract minimization over a function space, in order to derive computational algorithms we will be using this matrix form for most of the remainder of this chapter.

An important detail in the theory of thin-plate splines is a remarkable simplification for the roughness penalty when the solution is a linear combination of radial basis functions. Let $K_{i,k} = \psi_i(\boldsymbol{x}_k)$; then if $T^T\boldsymbol{\delta} = 0$, it follows that

$$J_m\left(\sum_{j=1}^{t}\phi_j\beta_j + \sum_{i=1}^{n}\psi_i\delta_i\right) = \boldsymbol{\delta}^T K\boldsymbol{\delta}.$$

This expression is very useful toward obtaining a simple form for S_λ.

Assembling all these pieces, we reach

$$S_\lambda\left(\sum_{j=1}^{t}\phi_j\beta_j + \sum_{i=1}^{n}\psi_i\delta_i\right) = (Y - T\boldsymbol{\beta} - K\boldsymbol{\delta})^T W(Y - T\boldsymbol{\beta} - K\boldsymbol{\delta}) + \lambda\boldsymbol{\delta}^T K\boldsymbol{\delta}. \tag{13.6}$$

Now, taking partials with respect to $\boldsymbol{\beta}$ and $\boldsymbol{\delta}$, the solution satisfies the system of equations

$$-2T^T W(Y - T\boldsymbol{\beta} - K\boldsymbol{\delta}) = 0,$$
$$-2KW(Y - T\boldsymbol{\beta} - K\boldsymbol{\delta}) + 2\lambda K\boldsymbol{\delta} = 0.$$

If $\{\boldsymbol{x}_i\}$ are unique locations, then K will have full rank. Multiply the second equation by K^{-1} and substitute this new equation into the first. This operation yields the system of equations

$$T^T\boldsymbol{\delta} = 0, \tag{13.7}$$
$$T\boldsymbol{\beta} + K\boldsymbol{\delta} + \lambda W^{-1}\boldsymbol{\delta} = \boldsymbol{Y}. \tag{13.8}$$

It is easy to verify that the solution is given by

$$\hat{\boldsymbol{\beta}} = (T^T(K + \lambda W^{-1})^{-1}T)^{-1}T^T(K + \lambda W^{-1})^{-1}\boldsymbol{Y},$$
$$\hat{\boldsymbol{\delta}} = (K + \lambda W^{-1})^{-1}(Y - T\hat{\boldsymbol{\beta}}). \tag{13.9}$$

As a link to later discussion, it should be noted that, formally, $\hat{\beta}$ is a generalized least-squares regression estimate, if we identify $(K + \lambda W^{-1})$ as a covariance matrix for Y. This linear system also gives a simple form for the residuals from the splines fit. The predicted values are just $T\hat{\boldsymbol{\beta}} + K\hat{\boldsymbol{\delta}}$, so, based on (13.8), the residual vector must be equal to $\lambda W^{-1}\hat{\boldsymbol{\delta}}$.

13.2.2 The QR Parameterization

The previous derivation produces a solution that is readily interpreted, but it is not a form that is suitable for computation. Also, in terms of making a comparison with Kriging estimates, a different parameterization is required. This computational approach is due to Bates et al. (1987), and the reader is referred to this source and to Gu and Wahba (1991) for further details and software.

To enforce the constraint on δ implied by (13.7), let $T = F_1 R$ denote the QR decomposition of T. $F = [F_1 \mid F_2]$ is an orthogonal matrix. F_1 has columns that span the column space of T, F_2 has columns that span the space orthogonal to T, and R is an upper-triangular matrix. Reparameterizing by $\delta = F_2 \omega_2$ for $\omega_2 \in \mathbb{R}^{n-t}$ now enforces the necessary number of linear constraints on δ and makes the parameters identifiable. Now, setting $\boldsymbol{\beta} \equiv \boldsymbol{\omega}_1$ and $\boldsymbol{\omega}^T = (\boldsymbol{\omega}_1^T, \boldsymbol{\omega}_2^T)$, the system of equations reduces to the form

$$T\boldsymbol{\omega}_1 + KF_2\boldsymbol{\omega}_2 + \lambda W^{-1}F_2\boldsymbol{\omega}_2 = \boldsymbol{Y}.$$

Multiplying through by F_2^T gives

$$\begin{aligned}\hat{\boldsymbol{\omega}}_2 &= (F_2^T(K + \lambda W^{-1})F_2)^{-1}F_2^T\boldsymbol{Y}\\ \hat{\boldsymbol{\omega}}_1 &= R^{-1}F_1^T(\boldsymbol{Y} - (K + \lambda W^{-1})F_2\hat{\boldsymbol{\omega}}_2).\end{aligned} \tag{13.10}$$

We can relate this solution back to the original parameterization by $\hat{\delta} = F_2\hat{\omega}_2$.

13.2.3 The Hat Matrix, Kernels, and Cross-Validation

Examining the equations leading to $\hat{\omega}$, we see that all matrices involved are independent of $\boldsymbol{Y}$. Thus, for fixed λ and fixed locations, splines are linear functions of the data, and there is an $n \times n$ hat matrix, $A(\lambda)$, such that

$$\begin{pmatrix} \hat{f}(\boldsymbol{x}_1) \\ \vdots \\ \hat{f}(\boldsymbol{x}_n) \end{pmatrix} = A(\lambda) \begin{pmatrix} \boldsymbol{Y}_1 \\ \vdots \\ \boldsymbol{Y}_n \end{pmatrix}.$$

By analogy with linear regression, the trace of A will be used as a measure of the effective number of parameters being subsumed by the spline. (More will be said about this topic in Section 13.5.) The ith row of the A matrix gives the weights that are applied to the data to give the function estimate at $\boldsymbol{x}_i$. When one plots these weights, they look like higher-order kernels, although an explicit formula relating λ to a bandwidth is not obvious. In one dimension, the kernel approximation to a smoothing spline has been studied (Nychka, 1995; and Messer, 1991), and for uniformly distributed locations the bandwidth is proportional to $\lambda^{1/2m}$. On the basis of this work, it is not difficult to conjecture, in the multivariate case, that the approximating kernel is a Greens function for a particular $2m$th-order partial-differential operator. An explicit form for the kernel has been derived by Zhang and Stein (1993), but a mathematically rigorous extension to the multivariate case remains to be accomplished.

The hat matrix also yields a simple formula for leave-one-out prediction. Let $\hat{Y}_k$ denote the value predicted for the kth data point by the spline estimate, and let $\hat{Y}_k^*$ be an alternative prediction for this point, based on omitting the kth data point and refitting the spline to the remaining $n - 1$ data values. The difference $Y_k - \hat{Y}_k^*$ will be referred to as the *cross-validated* (CV) residual. Using the properties of linearity and the fact that the spline solves a minimization problem, we have the following compact formula for the CV residuals based on leave-one-out prediction:

$$Y_k - \hat{Y}_k^* = \frac{(Y_k - \hat{Y}_k)}{1 - A_{kk}(\lambda)}.$$

Note that the right side depends on only the full-data spline and the diagonal elements of the A matrix. Thus, the sum of squared CV residuals is a weighted sum using the full-data residuals, and it can be computed without refitting the spline to n different subsets. A further simplification is to substitute the average of the diagonal elements for each of the individual terms in the denominator. These remarks serve as motivation for the *generalized cross-validation* (GCV) function

$$V(\lambda) = \frac{(1/n)\sum_{k=1}^{n} W_k(Y_k - \hat{Y}_k)^2}{(1 - \mathrm{tr}\,A(\lambda)/n)^2} = \frac{(1/n)\|W^{1/2}(I - A(\lambda))\boldsymbol{Y}\|^2}{(1 - \mathrm{tr}\,A(\lambda)/n)^2}.$$

The principle is that good choices of λ will give small CV residuals. So, one might choose λ by minimizing $V(\lambda)$. There is some debate about how much GCV really has to do with cross-validation. However, it can be justified independently on theoretical grounds, and it typically is nearly unbiased as an estimator of the average prediction error of the spline estimate. One pitfall is the variability associated with the estimated smoothing parameter from minimizing the GCV criterion, and a small proportion of estimates may severely undersmooth the data. In practice, such unsuccessful cases are often marked by an unreasonably small estimate of σ.

13.3 SPATIAL-PROCESS ESTIMATES

The derivation of a spline estimate in the previous section assumed that the unknown function is fixed. In contrast, a spatial-process estimate is based on a model for f as a random function. Accordingly, we now assume that f has the form from (13.2) and that h is a mean-zero Gaussian process. We will see that, in this case, the covariance function for h plays the same role for spatial-process estimates as the radial basis functions do for splines.

13.3.1 The Universal Kriging Estimator

Given observed data $\{Y_k\}$, an estimate for $f(x)$ can be found as a linear combination of the observations that minimizes the variance and is unbiased. The standard derivation usually enforces the unbiasedness condition (a linear constraint) using Lagrange multipliers. A more direct approach is to use the QR decomposition to restrict the linear combination to a subspace where unbiasedness holds. In this way we convert the constrained-minimization problem to an unconstrained one that is simple to solve.

For clarity, let $\boldsymbol{x}_0$ be the new point for prediction; then the goal is an optimal estimate for $f(\boldsymbol{x}_0) = \boldsymbol{T}_0^T\boldsymbol{\beta} + h(\boldsymbol{x}_0)$, where $T_{0,j} = \phi_j(\boldsymbol{x}_0)$.

Any linear estimate has the form $\hat{Y}_0 = \boldsymbol{l}^T\boldsymbol{Y}$, with $\boldsymbol{l} \in \mathbb{R}^n$, so our goal is to identify $\boldsymbol{l}$. Now let F and R be the QR decomposition of the T matrix, $(T = F_1R)$, described in the previous section. Equivalently, we can express $\boldsymbol{l}$ as $\boldsymbol{l} = (F\boldsymbol{a})$ for some $\boldsymbol{a} \in \mathbb{R}^n$. Thus, the estimate is $\hat{Y}_0 = (F\boldsymbol{a})^T\boldsymbol{Y}$. Furthermore, $F\boldsymbol{a}$ can be partitioned as $F_1\boldsymbol{a}_1 + F_2\boldsymbol{a}_2$, where we have subdivided the vector $\boldsymbol{a}^T = (\boldsymbol{a}_1^T, \boldsymbol{a}_2^T)$. Although this form may seem awkward at first sight, using this decomposition makes it very easy to enforce the unbiasedness. Determining $\boldsymbol{a}_1$ is easy, and most of our work will focus on finding $\boldsymbol{a}_2$. From the orthogonality of F we have

$$E[(F\boldsymbol{a})^T\boldsymbol{Y}] = (F\boldsymbol{a})^T\boldsymbol{T}\boldsymbol{\beta} = \boldsymbol{a}_1^T R\boldsymbol{\beta}.$$

Note that $E(\hat{Y}_0) = \boldsymbol{T}_0^T\boldsymbol{\beta}$, and so, in order for the estimate to be unbiased for all $\boldsymbol{\beta}$, it must follow that $\boldsymbol{a}_1 = (R^T)^{-1}\boldsymbol{T}_0$.

It remains to determine the second component of $\boldsymbol{a}$. Here we minimize the mean squared error of the estimate. To fix ideas, we start with the elementary expression for the mean squared error,

$$E[(f(x_0) - \hat{Y}_0)^2] = \mathrm{Var}(f(x_0) - \hat{Y}_0). \tag{13.11}$$

Because the estimate is unbiased,

$$\begin{aligned} \mathrm{Var}(f(x_0) - \hat{Y}_0) &= \mathrm{Var}(f(x_0) - \boldsymbol{l}^T\boldsymbol{Y}) & (13.12)\\ &= \mathrm{Var}(f(x_0)) - 2\rho\boldsymbol{l}^T\boldsymbol{k}_0 + \boldsymbol{l}^T(\rho K + \sigma^2 W^{-1})\boldsymbol{l}, & (13.13) \end{aligned}$$

where $\rho K_{ij} = \mathrm{Cov}(h(\boldsymbol{x}_i), h(\boldsymbol{x}_j)) = \rho k(\boldsymbol{x}_i, \boldsymbol{x}_j)$ and $\rho \boldsymbol{k}_{0,j} = \mathrm{Cov}(h(\boldsymbol{x}_0), h(\boldsymbol{x}_j)) = \rho k(\boldsymbol{x}_0, \boldsymbol{x}_j)$. The covariance matrix of the observations is $\rho K + \sigma^2 W^{-1}$.

We now switch to the notation in which $\boldsymbol{l}$ has been partitioned; then we have

$$\begin{aligned} E[(f(x_0) - \hat{Y}_0)^2] &= E[(Z_1 + Z_2)^2] & (13.14)\\ &= E(Z_1^2) + 2E(Z_1Z_2) + E(Z_2^2) & (13.15)\\ &= \mathrm{Var}(Z_1) + 2\,\mathrm{Cov}(Z_1, Z_2) + \mathrm{Var}(Z_2), & (13.16) \end{aligned}$$

where $Z_1 = \boldsymbol{T}_0^T\boldsymbol{\beta} - \boldsymbol{a}_1^T F_1^T\boldsymbol{Y}$ and $Z_2 = h(x_0) - \boldsymbol{a}_2^T F_2^T\boldsymbol{Y}$, using the fact that $E(Z_1) = E(Z_2) = 0$.

The remaining part of the derivation is easy. We will just identify (13.16) as a quadratic function of $\boldsymbol{a}_2$, and set the partial derivatives equal to zero. The solution to this linear system defines the minimizer. Note that, in this process, it is not necessary to work out the term for $\mathrm{Var}(Z_1)$ because it does not depend on $\boldsymbol{a}_2$; thus,

$$\mathrm{Cov}(Z_1, Z_2) = -\boldsymbol{a}_1^T F_1^T \rho \boldsymbol{k}_0 + \boldsymbol{a}_1^T F_1^T (\rho K + \sigma^2 W^{-1}) F_2 \boldsymbol{a}_2$$

and

$$\mathrm{Var}(Z_2) = \mathrm{Var}(h(x_0)) - 2\boldsymbol{a}_2^T F_2^T \rho \boldsymbol{k}_0 + \boldsymbol{a}_2^T F_2^T (\rho K + \sigma^2 W^{-1}) F_2 \boldsymbol{a}_2.$$

Taking (vector) partial derivatives yields

$$\frac{\partial}{\partial \boldsymbol{a}_2} E[(Z_1 + Z_2)^2] = 2F_2^T(\rho K + \sigma^2 W^{-1}) F_1 \boldsymbol{a}_1 - 2F_2^T \rho \boldsymbol{k}_0 + 2F_2^T(\rho K + \sigma^2 W^{-1}) F_2 \boldsymbol{a}_2.$$

Now, divide this equation by ρ, and set $\lambda = \sigma^2/\rho$. Solving for the parameters when the partials are zero yields the solution

$$\boldsymbol{a}_2 = (F_2^T(K + \lambda W^{-1})F_2)^{-1}(F_2^T \boldsymbol{k}_0 - F_2^T(K + \lambda W^{-1})F_1 \boldsymbol{a}_1).$$

Recall that the solution for the first component has already been found to be $\boldsymbol{a}_1 = (R^T)^{-1} T_0$, and so $\boldsymbol{l}$ has been determined.

This solution is complicated; still, much can be gleaned from its form. At this point, the benefit of carrying along the σ^2 and ρ in the derivation is apparent. The estimator depends only on their ratio, λ, not on the separate values. Considering the equations for the two components of $\boldsymbol{a}$, it follows that

$$\hat{f}(\boldsymbol{x}_0) = \boldsymbol{l}^T \boldsymbol{Y} = T_0^T \hat{\boldsymbol{\beta}} + \boldsymbol{k}_0^T \hat{\boldsymbol{\delta}} = \sum_{j=1,t} \phi_j(\boldsymbol{x}_0)\hat{\boldsymbol{\beta}}_j + \sum_{i=1,n} \psi_i(\boldsymbol{x}_0)\hat{\delta}_i,$$

where the basis functions $\{\psi_i\}$ are defined in terms of the covariance function: $\psi_i(\boldsymbol{x}) = \mathrm{Cov}(h(\boldsymbol{x}), h(\boldsymbol{x}_i))$. Here both $\hat{\boldsymbol{\beta}}$ and $\hat{\boldsymbol{\delta}}$ are linear, albeit complicated, combinations of $\boldsymbol{Y}$ that do *not* depend on $\boldsymbol{x}_0$. Thus, the Kriging estimate can be interpreted as a sum of basis functions in which the coefficients are linear functions of the observed $\boldsymbol{Y}$. Taking this interpretation a step further, we can recognize a "hat" matrix, $A(\lambda)$, like the one for splines, such that $\hat{\boldsymbol{Y}} = A(\lambda)\boldsymbol{Y}$.

13.3.2 Kriging Estimates are a Type of Spline

Having noticed the similarity in notation, the astute reader may have already anticipated another key result for this estimate. Symbolically, the form of the spatial-process estimator is identical to that of the spline estimator presented in the previous section. This fact is established by comparing the estimates for $\boldsymbol{\beta}$ and $\boldsymbol{\delta}$ implied in the equations with those in Section 13.2, specifically (13.10). Therefore, the Kriging estimator can be characterized as the solution to the minimization problem at (13.6), provided that the correct identifications are made for the basis functions and the K matrix. Because of this equivalence, the cross-validation formulas presented in Section 13.3 are valid for the Kriging estimate also. This characterization, in the form of a minimization problem, suggests a route to computing spatial-process estimates for large data sets. The minimization of (13.6) can be solved using iterative numerical techniques, such as conjugate-gradient methods.

13.3.3 Thin-Plate Splines as Spatial-Process Estimators

If Kriging estimates also solve a variational problem, going in the other direction, perhaps a thin-plate spline is just a spatial-process estimate? The matrix K, derived from radial basis functions, is not itself a covariance matrix. In fact, its diagonal elements are zero! However, by following the discussion in Wahba (1990a), one can construct a covariance function of the form

$$k(\boldsymbol{x}, \boldsymbol{x}') = E_m(\|\boldsymbol{x}, \boldsymbol{x}'\|) + Q(\boldsymbol{x}, \boldsymbol{x}').$$

The modification Q is symmetric and involves combinations of the radial basis functions with polynomials up to degree $m-1$. It also has the following important property: if $\boldsymbol{u}$ is any vector orthogonal to the columns of T, then

$$\sum_{j=1}^{n} Q(\boldsymbol{x}, \boldsymbol{x}_j)\boldsymbol{u}_j$$

is a polynomial of degree $m-1$ or less in $\boldsymbol{x}$.

Based on this covariance, there is of course a corresponding Kriging estimate. Furthermore, one can show that this estimate reduces to the usual thin-plate spline (Wahba, 1990a). Although the argument will not be reproduced here, the basic ingredient is that K and $\boldsymbol{k}_0$ enter all the expressions for $\boldsymbol{a}_2$ postmultiplied by F_2^T. The columns of F_2 are, of course, orthogonal to T, and so the terms differ from the radial basis function kernel only by a polynomial term of degree $m-1$ or less. These polynomial extra terms are absorbed into the null-space term.

After going to the trouble of constructing a valid covariance function from the radial basis function kernel ($E_m(\|\boldsymbol{x} - \boldsymbol{x}'\|)$, we see that the modification has no impact on the estimator. One can use the radial basis functions alone, pretending that they constitute a valid covariance, and one will recover the same estimate from the Kriging formulas. In fact, the radial basis function kernel does act like a covariance kernel for certain linear combinations of h. This stochastic model has a rigorous foundation, stemming from the work of Matheron (1973) on intrinsic random functions, and the reader is referred to this work and to Wahba (1990a) for the development of this theory. However, in summary, a spline can be identified as a spatial-process estimate with a fairly peculiar covariance function.

13.3.4 Profiling the Likelihood Function for Parameters of the Covariance Function

The QR decomposition also finds use in simplifying the likelihood function in order to estimate parameters of the covariance function. Assuming a Gaussian distribution for h and ϵ in equations (13.1) and (13.2), $F_2^T Y$ will be multivariate normal, with mean zero and with covariance matrix $F_2^T(\rho K + \sigma^2 W^{-1})F_2$. Although a realistic spatial covariance may depend on several parameters, we will emphasize some reductions for the parameters that are linear in the covariance. Recall that we have reparameterized the covariance so that $\sigma^2 = \lambda\rho$. The log likelihood for λ and ρ, up to a constant term, is

$$l(\lambda, \rho) = -(1/2)\log(|M|\rho^{n-t}) - \frac{Y_2^T(M)^{-1}Y_2}{2\rho} + C,$$

where $M = F_2^T(K + \lambda W^{-1})F_2$ and $Y_2 = F_2^T Y$. C is a generic symbol for an expression that depends on neither ρ nor λ. Maximizing over ρ for fixed λ gives

$$\hat{\rho} = Y_2^T(M)^{-1}Y_2/(n-t),$$

and, substituting this back into the likelihood, we have, up to a constant term,

$$l(\lambda, \hat{\rho}) = -(1/2)\log\left((Y_2^T(M)^{-1}Y_2)^{n-t}|M|\right) + C.$$

A useful detail from Section 13.2.1 is that the residual vector, $(I - A(\lambda))\boldsymbol{Y}$, is equal to $\lambda W^{-1}\hat{\delta}$. Now, going back to equation (13.10), and using the fact that $\hat{\boldsymbol{\delta}} = F_2\hat{\omega}_2$, we are led to the relationship that

$$\lambda W^{-1}F_2 M^{-1}F_2^T = I - A(\lambda).$$

Furthermore, one can show that $|M^{-1}|$ is equal to the product of the nonzero eigenvalues of $I - A(\lambda)$. Denote this determinant by $|I - A(\lambda)|_+$; then we have found the profile likelihood with respect to λ,

$$l(\lambda, \hat{\rho}) = -(1/2)\log\left(\frac{(\mathbf{Y}^T W(I - A(\lambda))\mathbf{Y})^{n-t}}{|I - A(\lambda)|_+}\right) + C.$$

Implicit in these expressions is the possibility that the covariance function, k, may depend on parameters other than the leading multiplier ρ. For example, a useful covariance family, analogous to those in the design and analysis of computer experiments (Sacks et al., 1989, and Welch et al., 1992), is

$$k(\boldsymbol{x}, \boldsymbol{x}') = \rho e^{-(\|\boldsymbol{x}-\boldsymbol{x}'\|/\theta)^p}, \tag{13.17}$$

where θ controls the range of correlations and $1 \leq p \leq 2$ determines the smoothness of the random process. Parameters such as the range and smoothness in the example above enter in an inherently nonlinear manner. There is a rich literature on estimating the parameters of the covariance function, and the treatment in Cressie (1991) is a fine introduction. However, given the emphasis in this chapter, we will see that it is possible to evaluate the likelihood or GCV function for different values of λ rapidly by holding other parameters in the covariance fixed. This separation suggests that the nonlinear parameters, such as p and θ in (13.17), could be estimated separately, say by nonlinear least-squares fitting of a variogram. This would be followed by estimating ρ and λ using maximum likelihood, with the formulas derived above. In practice, spatial estimates are often much more sensitive to the choice of λ than to the choice of range or smoothness parameter or both.

13.3.5 Measures of Uncertainty in the Estimates

One advantage of the spatial-process model is that it provides measures of uncertainty for the function estimates. The equations for the prediction variance at (13.13) hold for any unbiased prediction of $f(\boldsymbol{x}_0)$. As a result, the prediction mean squared error can be found by substituting the specific $\boldsymbol{l}$ into this equation. Furthermore, the prediction variance associated with the minimum-variance estimator yields some simplifications. In practice, however, there is an additional uncertainty in the estimate due to estimating ρ, σ, and other covariance parameters. For a more comprehensive treatment including such parameters, the reader is referred to Handcock and Stein (1993), Qian (1997), and Pilz et al. (1997).

The uncertainty measures based on a random field can be very different from the computed variance of a thin-plate spline. The variance associated with the estimate of a fixed function is $\sigma^2 \boldsymbol{l}^t W^{-1} \boldsymbol{l}$; that is only one part of the expression for mean squared error at (13.13). (The remaining part will also be nonnegative, because it is the mean squared error for a estimate when $\sigma = 0$.)

The variance term by itself can be substantially smaller than the actual uncertainty, especially when σ is close to zero. This situation arises not from a defect in the spline estimate but from an inadequate accounting of the estimates error. The problem is that variance alone takes into account only the variability in fitting the spline to the sample. The other component of error arises from approximating the true function in terms of a spline. As a complement to the spatial-process model, we will illustrate this problem in the context of f as a fixed function. In this case, the spline will be a biased estimate of f. It is helpful to study the bias by breaking it into two components. Let f_S be the thin-plate spline that interpolates f at the data points. The bias, is

$$f(\boldsymbol{x}) - E(\hat{f}(\boldsymbol{x})) = (f(\boldsymbol{x}) - f_S(\boldsymbol{x})) + (f_S(\boldsymbol{x}) - E(\hat{f}(\boldsymbol{x}))).$$

The first term will be zero whenever $\boldsymbol{x}$ is a data point. This first term is the interpolation error. It will be zero for *all* $\boldsymbol{x}$ only if the unknown function is *exactly* a spline — a highly unlikely prospect! In practice, it will always represent the *bias caused by interpolation*. The second term will be zero if the spline interpolates the data ($\lambda = 0$), nonzero if some smoothing is done ($\lambda > 0$). Thus, this second term can be thought of as the *bias caused by smoothing*. Both of these biases can cause error in the spline beyond that measured by the variance. In one extreme case, interpolation, the variance of the spline estimate is zero — yet there may exist substantial interpolation error when the estimate is evaluated at points where observations have not been made. One approach to determining the bias is to use bounds based on the approximation properties of splines. Another approach is to fall back on the equivalent spatial model associated with thin-plate splines and to derive uncertainties using the generalized covariance. This latter method is in the spirit of Wahba's "Bayesian"confidence intervals for splines, and it is a practical method even for fixed-function inference (Nychka, 1988).

13.4 RIDGE-REGRESSION ESTIMATES AND SHRINKAGE

The thin-plate spline and the spatial-process estimators can both be expressed in the form of ridge regressions. This general form is at the heart of the numerical procedures for both methods, and it leads to the construction of the orthogonal series and to efficient formulas for determining the smoothing parameter λ.

Assume the model

$$Y = X\omega + \boldsymbol{e},$$

and let H be a nonnegative definite penalty matrix. The general form of the penalized (and weighted) least-squares problem is as minimization of

$$(Y - X\omega)^T W(Y - X\omega) + \lambda\omega^T H\omega \qquad (13.18)$$

over $\omega \in \mathbb{R}^n$. The solution is

$$\omega = (X^T WX + \lambda H)^{-1} X^T WY.$$

13.4.1 A Natural Basis

In this general framework of ridge regression, the specific applications to thin-plate splines and to spatial-process estimates involve identifying the relevant parameterization, in order to find the matrix X and the penalty matrix H. In both cases, the constraint on the parameters, $T^T\boldsymbol{\delta} = 0$, provides a route toward reducing the $t + n$ basis functions to just n. Denote this basis as $\{\eta_i\}$, where the first t functions are equal to the polynomial terms, $\{\phi_i\}$, and the remaining $n - t$ functions are linear combinations of the $\{\psi_j\}$ (based on the columns of F_2). Explicitly, we have for these last members of the basis that $\eta_{i+t} = \sum_{k=1}^{n} F_{2k,i}\psi_i$. We will refer to this as the *natural* basis, because it automatically builds in the constraint necessary for a solution. A more practical feature is that we have reduced an over-determined linear system with linear constraints to a system with as many equations as parameters. To the general reader, there may seem to be nothing really "natural" about this basis; it takes its name from the properties in the one-dimensional case. In the case of a one-dimensional mth-order spline, this constraint forces each of the higher derivatives $j = m, m + 1, \ldots, 2m - 1$ of the spline to be zero at the boundaries. These values can be interpreted, in a physical system, as free or natural boundary conditions. In the case of cubic smoothing splines ($d = 1$, $m = 2$), the estimate will be a linear function at the endpoints of data; as a result, it extrapolates beyond the range of the observed data as a linear function.

Given the natural basis, the representation of the estimate is $f(\boldsymbol{x}) = \sum_{i=1}^{n} \omega_i \eta_i(\boldsymbol{x})$; thus, the relevant regression matrix is $X_{ij} = \eta_j(\boldsymbol{x}_i)$, and the associated penalty matrix is

$$H = \begin{pmatrix} 0 & 0 \\ 0 & F_2^T K F_2 \end{pmatrix}.$$

13.4.2 The Demmler–Reinsch Basis

The final step is to apply a linear transformation to the natural basis, in order to produce an orthogonal basis. This transformation is constructed by finding a matrix, G, that will diagonalize both $X^T W X$ and H. Let B denote the inverse square root of $X^T W X$, and let UDU^T be the singular-value decomposition of BHB^T. Now set $G = BU$, and it is straightforward to verify that $G^T(X^T W X)G = I$ and $G^T(H)G = D$, where D is diagonal. Finally, we define the new, orthogonal basis as

$$g_\nu(\boldsymbol{x}) = \sum_{i=1}^{n} G_{i\nu} \eta_i(\boldsymbol{x}).$$

This basis is known in the context of splines as the *Demmler–Reinsch* (DR) *basis*. Assume that $f(\boldsymbol{x}) = \sum_{\nu=1}^{n} \alpha_\nu g_\nu(\boldsymbol{x})$. We have the following properties for this basis:

(1) $\{g_\nu\}$ spans the same subspace of functions as the natural basis, $\{\eta_\nu\}$.

(2)
$$\sum_{i=1}^{n} g_\nu(\boldsymbol{x}_i) W_i g_\mu(\boldsymbol{x}_i) = \begin{cases} 0 & \mu \neq \nu \\ 1 & \mu = \nu \end{cases}$$

and so

$$\sum_{i=1}^{n} f(\boldsymbol{x}_i)^2 W_i = \sum_{\nu=1}^{n} \alpha_\nu^2.$$

(3) Setting $\boldsymbol{\omega} = G\boldsymbol{\alpha}$ makes

$$f(\boldsymbol{x}) = \sum_{\nu=1}^{n} \omega_\nu \eta_\nu(\boldsymbol{x})$$

and

$$\omega^T H \omega = \boldsymbol{\alpha}^T D \boldsymbol{\alpha} = \sum_{\nu=1}^{n} D_\nu \alpha_\nu^2.$$

These properties can easily be proven by the construction of the transformation matrix G. An important by-product of the second property is a simple representation for the interpolating function. Given the sequences $(\boldsymbol{x}_i, Y_i)$ for $1 \leq i \leq n$, let $u_\nu = \sum_{i=1}^{n} g_\nu(\boldsymbol{x}_i) W_i Y_i$; then the function $\sum_{\nu=1}^{n} u_\nu g_\nu(\boldsymbol{x})$ will interpolate these data.

13.4.3 Simplifications Due to the Demmler–Reinsch Basis

Using the parameterization with respect to the Demmler–Reinsch basis, assume that $f(\boldsymbol{x}) = \sum_{\nu=1}^{n} \alpha_\nu g_\nu(\boldsymbol{x})$, and let $\boldsymbol{u}$ be the coefficients that interpolate the data pairs $(\boldsymbol{x}_i, \boldsymbol{Y}_i)$. Then

$$S_\lambda(f) = \sum_{\nu=1}^{n} (u_\nu - \alpha_\nu)^2 + \lambda \sum_{\nu=1}^{n} D_\nu \alpha_\nu^2.$$

Note that the parameters are now decoupled, and each term can be minimized independently to give the solution

$$\hat{\alpha}_\nu = \frac{u_\nu}{1 + \lambda D_\nu}$$

leading to the function estimate presented in the introduction.

The DR basis also yields simple expressions for the residual sums of squares and for the trace of the hat matrix. The ith residual is

$$Y_i - \hat{f}(\boldsymbol{x}_i) = \sum_{\nu=1}^{n} \left(u_\nu - \frac{u_\nu}{1 + \lambda D_\nu} \right) g_\nu(\boldsymbol{x}_i) = \sum_{\nu=1}^{n} \frac{\lambda D_\nu u_\nu g_\nu(\boldsymbol{x}_i)}{1 + \lambda D_\nu}.$$

Using the orthogonality properties, it follows that

$$\sum_{i=1}^{n} W_i (Y_i - \hat{f}(\boldsymbol{x}_i))^2 = \sum_{\nu=1}^{n} \left(\frac{\lambda D_\nu u_\nu}{1 + \lambda D_\nu} \right)^2.$$

The hat matrix has elements

$$A(\lambda)_{ij} = \sum_{\nu=1}^{n} \frac{g_\nu(\boldsymbol{x}_i) g_\nu(\boldsymbol{x}_j)}{1 + \lambda D_\nu} W_j.$$

Using the orthogonality property again, it follows that the trace of this matrix is

$$\mathrm{tr}(A(\lambda)) = \sum_{\nu=1}^{n} \frac{1}{1+\lambda D_\nu}.$$

In each case, once the DR basis has been computed, the residual sums of squares and the trace can be evaluated rapidly in $O(n)$ operations. Provided that the independent variables are fixed, this feature makes it efficient to estimate the function, to evaluate the cross-validation function, or to evaluate the profiled likelihood for different realizations of the dependent variable.

One important application of the DR form is in the context of resampling. For example, to test for the presence of the nonparametric smooth component or to quantify the uncertainty in estimating λ (see Nychka, 1991), it is necessary to use a bootstrap for approximating the relevant sampling distribution. Under the assumption of normal errors, the interpolation coefficients $\boldsymbol{u}$ can be generated directly. This feature eliminates the need for re-evaluating the basis functions and makes all the computations possible with respect to the DR coefficients. Let Γ denote the matrix with elements $\Gamma_{\nu,i} = g_\nu(\boldsymbol{x}_i)W_i$, so that $\boldsymbol{u} = \Gamma\boldsymbol{Y}$. Under the Gaussian assumptions, $\boldsymbol{Y}$ will be multivariate normal, and so (therefore) will $\boldsymbol{u}$. If f is assumed to be a fixed function, then $E(\boldsymbol{u}) = \boldsymbol{\alpha}$ and $\mathrm{Cov}(\boldsymbol{u}) = \Gamma(\sigma^2 W^{-1})\Gamma^T = \sigma^2 I$. Thus, $\boldsymbol{u} \sim MN(\boldsymbol{\alpha}, \sigma^2 I)$. In the case of the spatial-statistics model with a random function, it is easy to verify that $\{u_\nu\}$ are independent zero-mean normal random variables with variances $\{\rho D_\nu + \sigma^2\}$.

13.4.4 Reducing the Number of Basis Functions

This ridge-regression formalism makes it easy to see how to reduce the size of the basis while still retaining the qualitative, statistical features of these smoothers. There is often strong incentive for doing this. The storage requirements for the estimate involve matrices that are of size $(n-t) \times (n-t)$. Moreover, the operation count for computing the estimate for a fixed λ is $O(n^2)$, (because one must solve an $O(n) \times O(n)$ linear system) and to find the orthogonal basis it increases to $O(n^3)$ (because one must perform an eigenvalue/eigenvector decomposition on a symmetric matrix of size $O(n) \times O(n)$).

Thus, the computation burden increases rapidly as the sample size is increased, and it is difficult to fit these estimates to data sets having more than several hundred unique locations. Working with the complete basis yields estimates with the highest possible spatial resolution; however, for situations in which the function does not have finely detailed structure and in which measurement error is present, one does not need the full resolution provided by an expansion using all of the basis functions. If the weighting function in equation (13.4) is nearly zero for large values of the $\{D_\nu\}$, it is not clear why these terms even need to be computed. Thus, it seems reasonable in some cases to work with a truncated basis.

A natural strategy is to simply restrict the size of the basis, without taking into account the sample size. Start with the full basis $\{\phi_i\}$ and $\{\psi_j\}$, and consider a subset of size N that will give a good approximation to the full expansion.

Because the $\{\psi_j\}$ are generated from the observed locations, one principle is to select a subset derived from locations that are widely separated. One automatic way to accomplish this is selecting the subset of $\{\boldsymbol{x}_k\}$ in a space-filling way. The FUNFITS manual gives an example of this strategy.

Even if only a subset of the natural basis functions is used, the ridge-regression development is largely unchanged, and one will still obtain an orthogonal representation with the same properties. Two differences, however, should be noted: The index for the basis, ν, is restricted to $[1, N]$ with $N < n$; thus, X will be $n \times N$ and H will be $N \times N$. Also, the residual sums of squares should include another term, $\hat{f}_0$, which is associated with the estimate when $\lambda = 0$. (This term is merely the result from the (weighted) regression of $\boldsymbol{Y}$ on X.)

13.5 A RESPONSE-SURFACE EXAMPLE

To help interpret the orthogonal-series estimates, the DR basis and the weighting functions will be computed for a specific example, using thin-plate splines. It is informative to compare the resulting spline to a Kriging estimate and to investigate the effect of changing the null space. The main goal is to give some intuition into the constructions presented in the previous sections.

13.5.1 Demmler–Reinsch Basis Functions for the Flame Data Set

The data set `flame` is a two-dimensional response surface and is included as part of FUNFITS. Its measurements come from an experiment in analytical chemistry exploring the relationship of the intensity of light from the ionization of Lithium ions in aqueous solution to the values of variables controlling the ionizing flame. The independent variables are the flow rates of the fuel and of the oxygen supplied to the burner used to ionize the sample. The data consist of 89 different settings of these two flow rates.

Figure 13.1 illustrates several of the 89 DR basis functions associated with a thin-plate spline with a third-derivative roughness penalty ($m = 3$). The basis functions are in order of increasing D_ν, and thus the first six have zero eigenvalues and span the subspace of second-degree polynomial functions in $\mathbb{R}^2$. Basis functions beyond the null-space terms exhibit increasing numbers of modes, and we informally identify basis functions with larger D_ν as having higher spatial frequencies. This connection leads to associating these higher terms with the more detailed structure in the surface. Figure 13.2 illustrates how the coefficients are filtered to produce a smooth surface. The top plot indicates the weight sequence $1/(1 + \lambda D_\nu)$ for three values of λ, and the middle plot indicates the effect of tapering the interpolating coefficients $\boldsymbol{u}$ by using these weights. Note that the first six weights are always one, because no smoothing is done on the quadratic null space. The resulting smoothed surfaces (and one that interpolates the data) are displayed in the first column of plots in Figure 13.3. The basic feature of this response surface is a broad ridge that is oriented obliquely to the independent variables. There is also some finer structure along the ridge line.

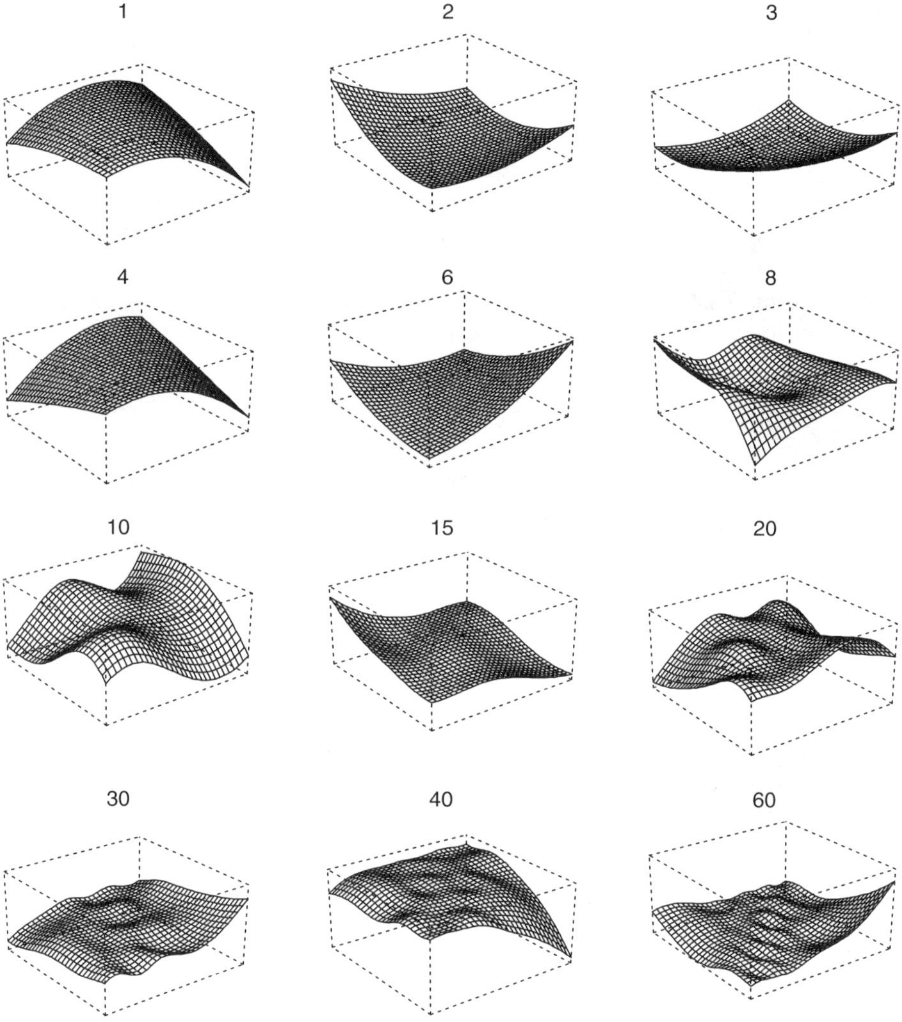

FIGURE 13.1 Some examples of Demmler–Reinsch basis functions for the Flame Data. These are based on a thin-plate spline estimate with $m = 3$, a quadratic-polynomial null space. The number above each plot is the index of the associated eigenvalue when ordered from largest to smallest.

As mentioned in Section 13.2, the amount of smoothing can be estimated from the data using GCV, and Figure 13.4 gives some diagnostic plots from fitting splines with different-sized null spaces ($m = 2, 3, 4$) and roughness penalties. It appears that when $m = 4$, a cubic null model, the spline estimate has a slightly lower GCV value than when $m = 3$ and has fewer degrees of freedom associated with the fitted surface. A smaller number of degrees of freedom might well be preferable, because it gives a more parsimonious representation of the response surface.

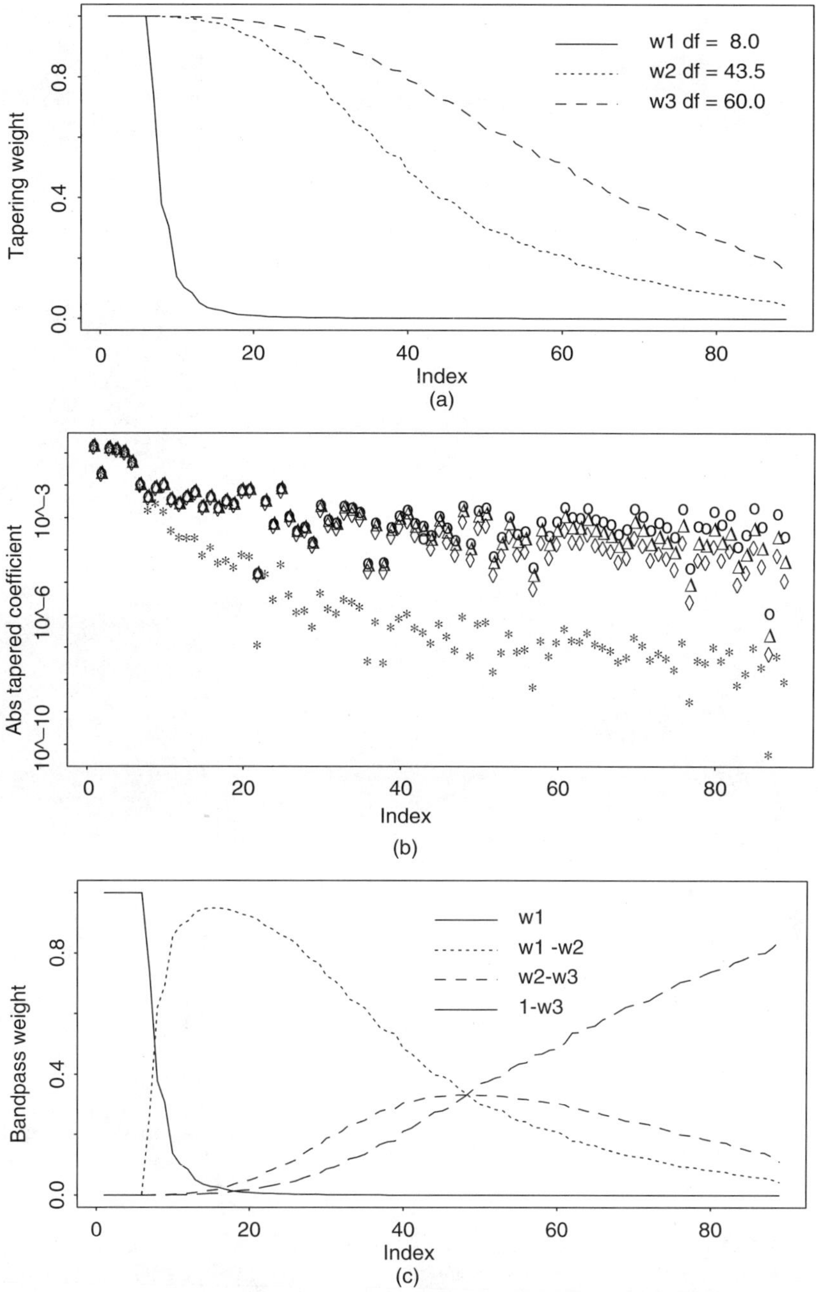

FIGURE 13.2 Tapering functions for the thin-plate spline fitted to the Flame Data. Spline taper functions w_1, w_2, w_3 (for $m = 3$) use the Demmler–Reinsch basis, and follow the form $1/(1 + \lambda D_\nu)$, for three different choices of λ. (a) The effective number of parameters associated with each smoothing tr$A(\lambda)$. (b) The absolute values of the coefficients $\boldsymbol{u}$ and their tapered values, using the weights from (a). (c) Weights used to produce the multiresolution decomposition in Figure 13.3.

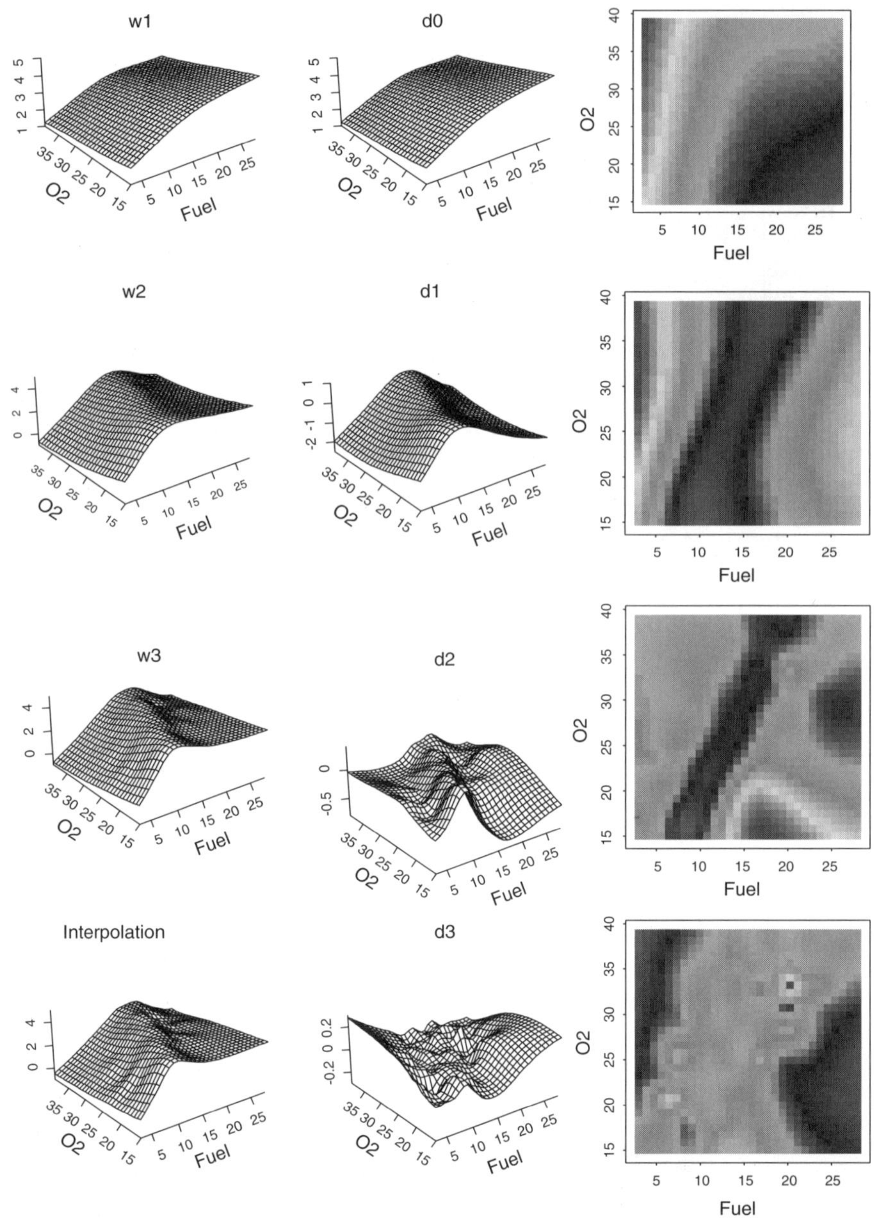

FIGURE 13.3 Different spline estimates ($m = 3$) and the decomposition of the Flame Data into several detail surfaces. In the first column are the surfaces derived from the three weight functions plotted in Figure 13.2(b), with effective model degrees of freedom 8.0, 43.5, and 60.0. The last plot in this column is an interpolated surface. In the second column are the detail surfaces for the multiresolution decomposition described in Section 13.5.3. The first detail surface is the same as the first smoothed surface. The subsequent surfaces are differences between the smoothed surfaces at two different levels of smoothing. The last column are image plots of the detail surfaces in the second column. See insert for color representation of figure.

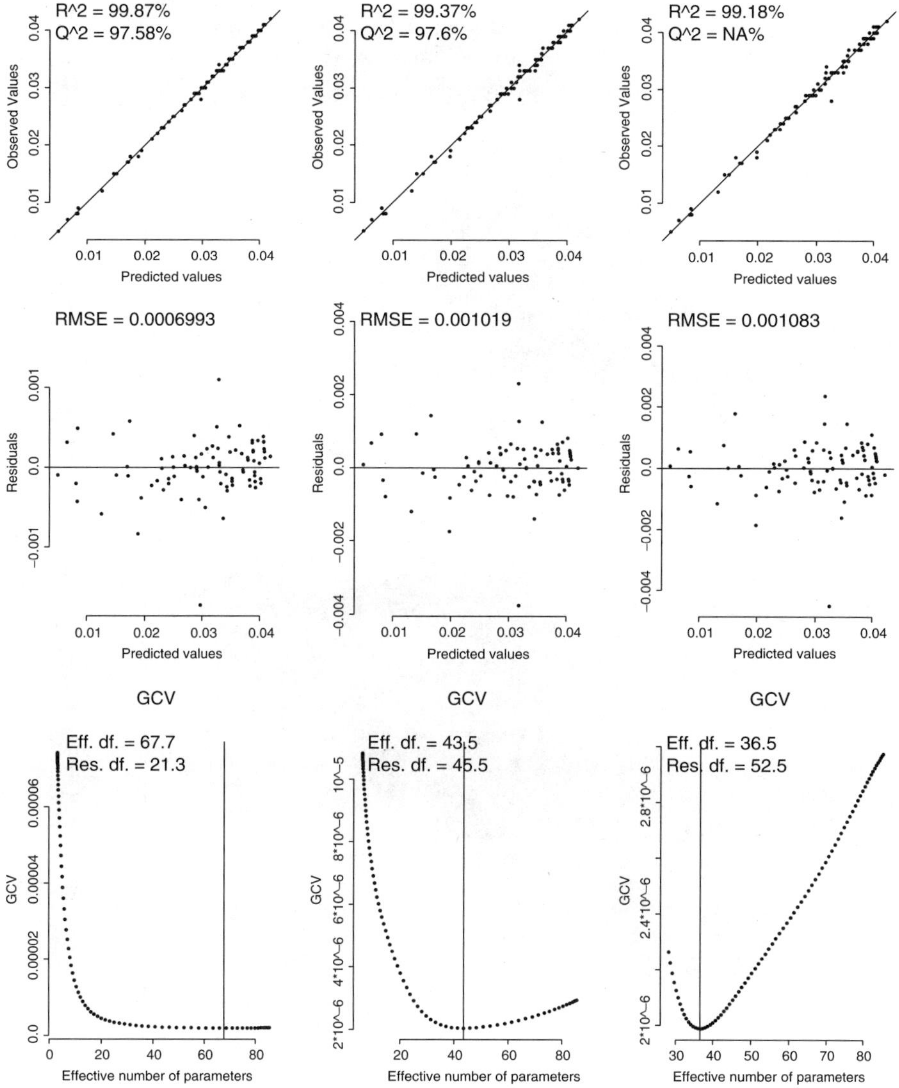

FIGURE 13.4 Diagnostic plots for thin-plate splines fitted to the Flame Data using $m = 2$ (first column), $m = 3$ (second column), and $m = 4$ (third column). In the first row are scatterplots of predicted and fitted values; in the second row are residuals against predicted values; in the third row are the GCV criteria for each model, plotted against the model degrees of freedom.

13.5.2 Comparison with a Spatial-Process Estimate

An exponential-covariance model (Equation 13.17), with $p = 1$) was also estimated for these data, using cross-validation for λ, but fixing the range parameter, θ, at 5 (Figure 13.5(b)). This computation was done largely for comparison. In fact, using cross-validation to choose θ suggests a much larger

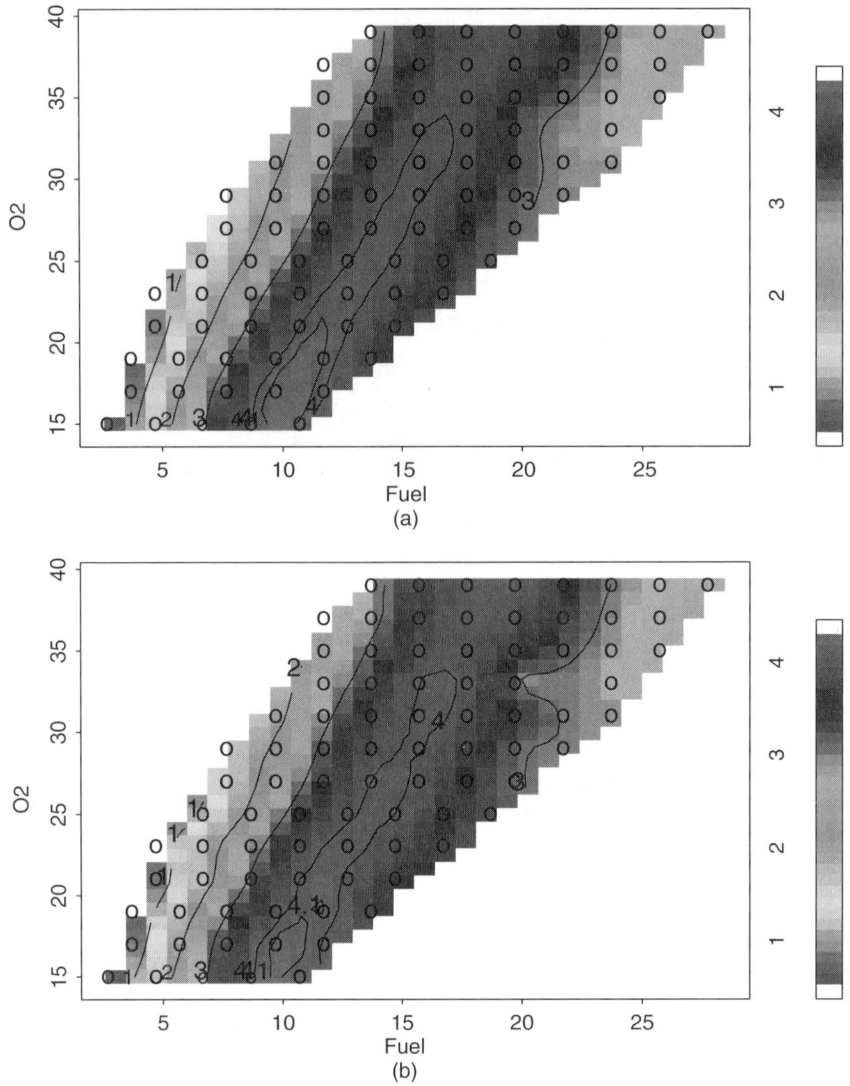

FIGURE 13.5 (a) Thin-plate spline fitted to the Flame Data with $m = 3$. (b) Kriging estimates for the Flame Data using an exponential covariance function with range parameter $\theta = 5$. Observation locations are indicated on the contour plots. See insert for color representation of figure.

value, one that will make the estimator very similar to a spline. Even with this smaller value for the range parameter, there is little difference between the spatial-process estimate and the thin-plate spline. However, if we consider measures of uncertainty, the prediction variance of the spline is substantially different. This difference is due to the fact that under the Kriging model one obtains a much smaller estimate for σ. The two image plots in Figure 13.6 illustrate these

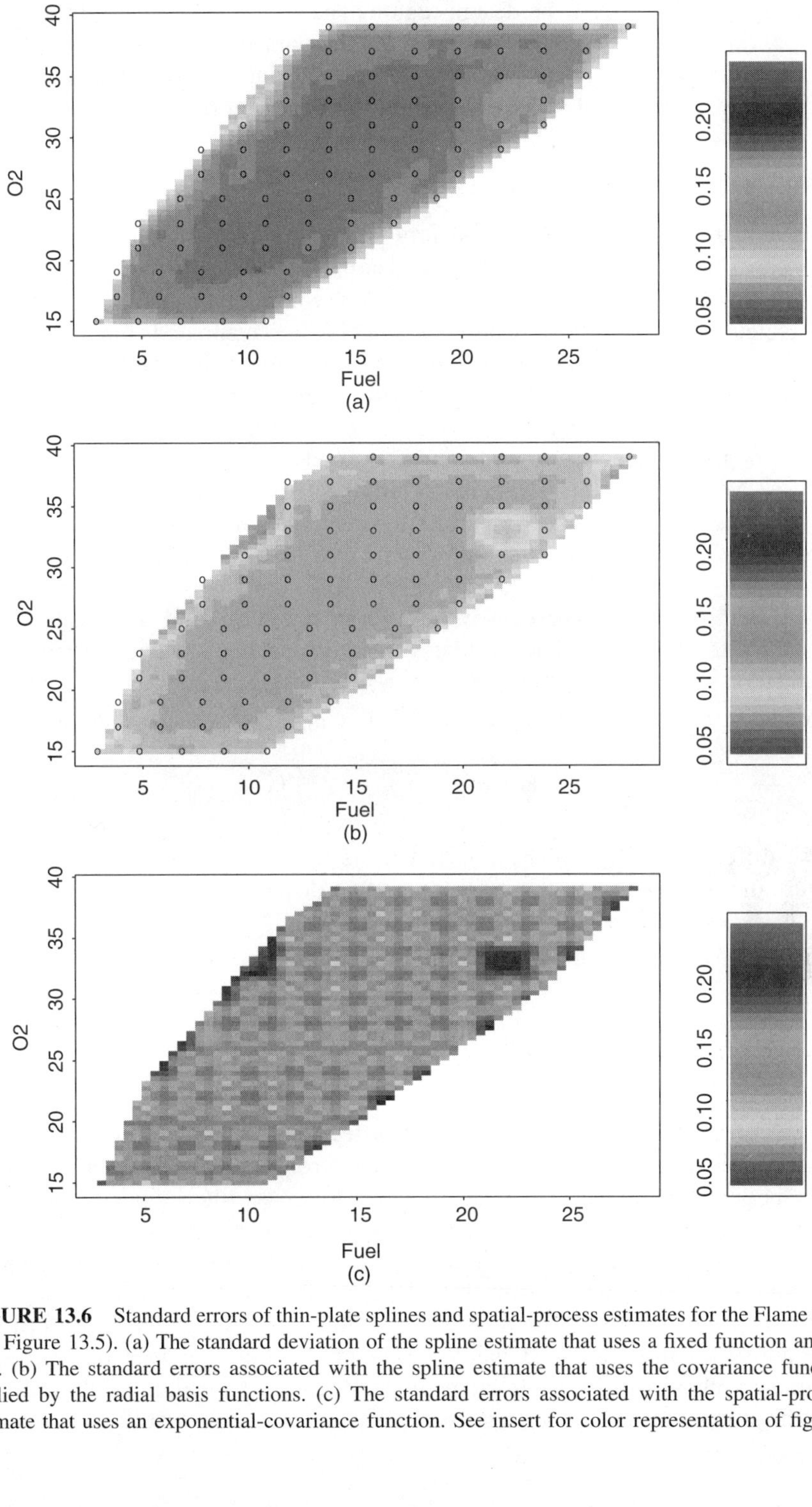

FIGURE 13.6 Standard errors of thin-plate splines and spatial-process estimates for the Flame Data (see Figure 13.5). (a) The standard deviation of the spline estimate that uses a fixed function and no bias. (b) The standard errors associated with the spline estimate that uses the covariance function implied by the radial basis functions. (c) The standard errors associated with the spatial-process estimate that uses an exponential-covariance function. See insert for color representation of figure.

prediction variances over the design region. The spline variances form a smooth surface, reflecting an estimate that smooths the data rather than interpolating it. The spatial-process estimate is much closer to an interpolatory surface, and so prediction variances will be nearly zero at the observed points.

13.5.3 A Multiresolution Analysis

There has been much interest recently in using wavelet (and other related) basis functions for nonparametric regression and smoothing. (See e.g., Nason and Silverman, 1994, and Chapter 7 in this volume.) One contribution from a wavelet analysis is the decomposition of the estimated function into a sequence of curves that represent the structure at different levels of resolution. This is easy in the case of wavelets (for equally spaced data) because the basis functions themselves fall naturally into a hierarchy of resolution levels. The orthogonal-series representation also lends itself to a multi resolution analysis. While it is not as precise or compact as the wavelet representation, it does extend to irregularly spaced observations, and it serves as a useful summary of how the estimated surface changes as a function of the smoothing parameter.

One simple way to use wavelets for smoothing and regression is to consider differences of spline estimates for a decreasing sequence of smoothing parameters. Let $\{\lambda_j\}$, $0 \le j \le K$, be a decreasing sequence of smoothing parameters, where $\lambda_0 = \infty$ and $\lambda_K = 0$. Let $\hat{d}_0(\boldsymbol{x}) \equiv \hat{f}_{\lambda_0}(\boldsymbol{x})$, and, for $j \ge 1$, define

$$\hat{d}_j(\boldsymbol{x}) = \hat{f}_{\lambda_j}(\boldsymbol{x}) - \hat{f}_{\lambda_{(j-1)}}(\boldsymbol{x}).$$

These differences, $\hat{d}_j$, will be referred to as the detail surfaces, because they capture the detail between the estimates at two different levels of smoothing. By the telescoping property of the differences, $Y_i = \sum_{j=0}^{K} \hat{d}_j(\boldsymbol{x}_i)$.

In terms of the orthogonal-series representation,

$$\hat{d}_j(\boldsymbol{x}) = \sum_{\nu=1}^{n} \frac{D_\nu(\lambda_j - \lambda_{j-1})}{(1 + D_\nu\lambda_j)(1 + D_\nu\lambda_{j-1})} u_\nu g_\nu(\boldsymbol{x}) = \sum_{\nu=1}^{n} w_{\nu,j} u_\nu g_\nu(\boldsymbol{x}),$$

so we see that this difference is in effect applying a different type of weighting to the interpolation coefficients. From a signal-processing/image-analysis perspective, the spline estimate is a low-pass filter applied to the data. Thus, the difference of two spline smoothings should behave as a bandpass filter. A filter that passes components in a range of frequencies but omits frequencies that are either much lower or much higher. A bandpass filter can be constructed by the appropriate combination of a low- and a high-pass filter. The third plot in Figure 13.2 illustrates the bandpass weights for a sequence of smoothing parameters chosen to have effective model-degrees-of-freedom of 8, 43.5, and 60. The smoothing-parameter sequence has been chosen so that there is minimal overlap among the weights for the detail surfaces. Because the interpolation coefficients will be independent under a Gaussian model for the errors, the different resolution levels will be only weakly correlated. The second and third columns in Figure 13.3 illustrate the sequence of detail surfaces for the Flame Data. (Note that the image plots are not on a common color scale.) Most of the structure is well explained by the low-resolution surface. The isolated local maxima along the ridge, however, are features that emerge only at the highest resolution levels, but these cause only small adjustments to the overall estimate.

13.6 PREDICTING AMBIENT OZONE

The FUNFITS data set `ozone1` contains the daily 8-hour-average `ozone` measurements during the summer of 1987 for approximately 160 locations in the midwest (Figure 13.7). One use for these data is to compare these observations to the output from a large numerical model for predicting surface ozone (regional oxidant model — ROM). In order to carry out this validation of the ROM, it is important to estimate the ozone concentrations from the observational data at the regular grid points matching with the model output. For the purposes of this

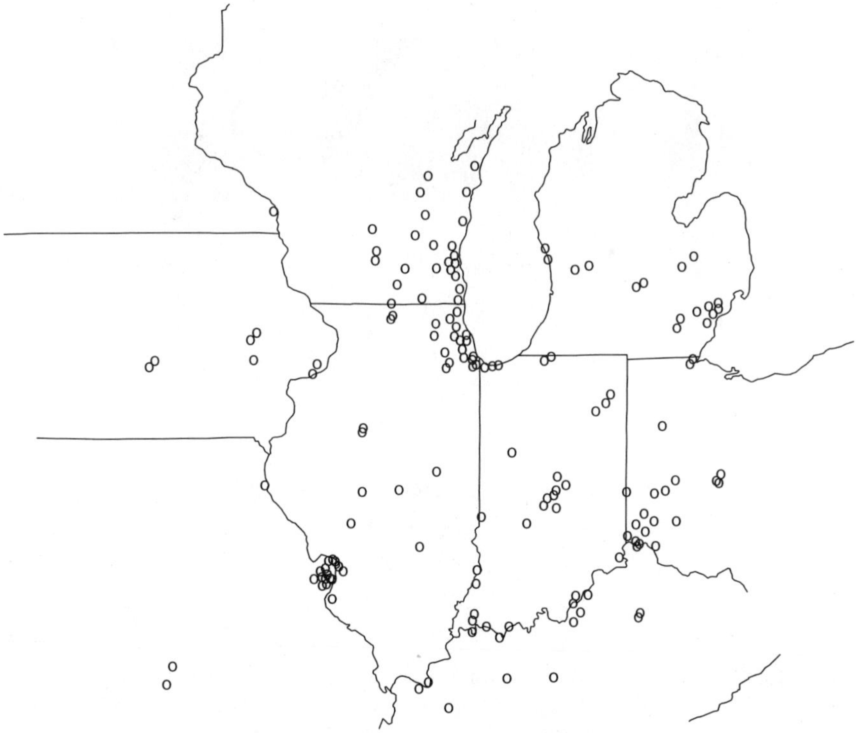

FIGURE 13.7 Locations of the national, state and local (NAMS/SLAMS) ozone-monitoring stations that were operating during 1987 in the Midwestern United States.

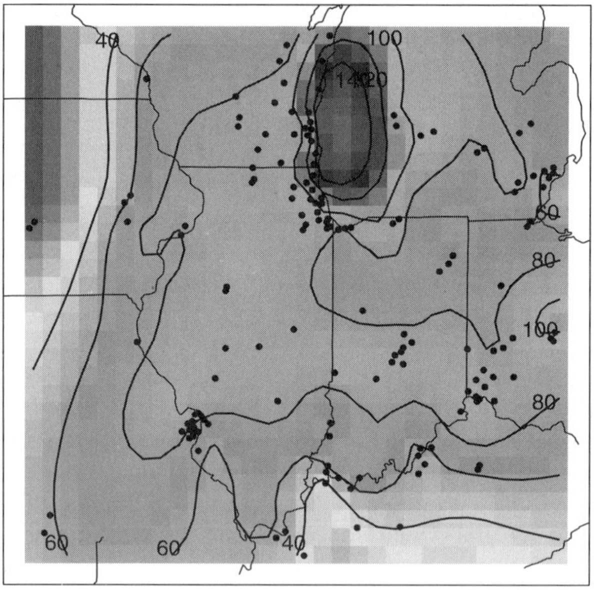

(a)

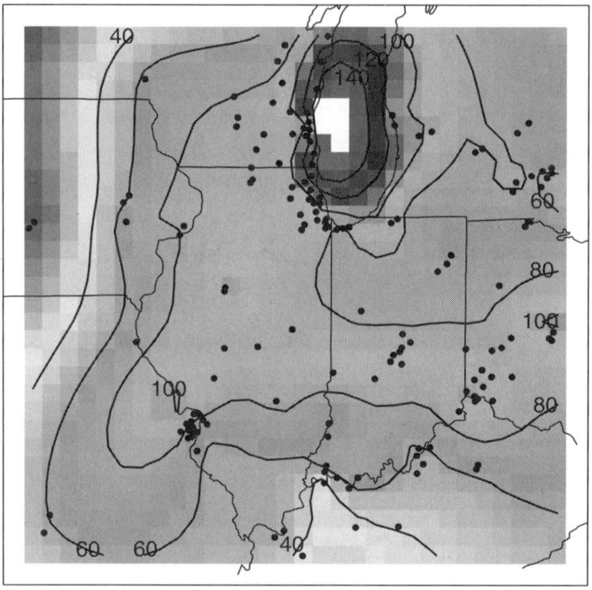

(b)

FIGURE 13.8 Spatial predictions for 8-hour-average ozone in parts per billion (ppb), for June 19, 1987. Results for thin-plate spline (a) and Kriging estimates with an isotropic, exponential correlation function but nonstationary marginal variances (b). See insert for color representation of figure.

chapter, we will focus on the 16th day in this data set (June 19, 1987), and we will compare estimates based on a thin-plate spline and on a more comprehensive spatial model.

The ozone field is not a stationary spatial process. This conclusion is based not only on the physical theories for the formation and transport of ozone but also on the observed data. The covariance function for the ozone field can be estimated empirically under the assumption that each day is an independent realization of the random field. For example, in this data set we have 89 separate measurements for each station. Actually, some days are missing for particular stations, and so 89 is an upper bound on the number of observations.

One simple departure from stationarity is to consider a covariance that has different marginal variances, but isotropic correlations. Specifically, consider

$$\mathrm{Cov}(h(\boldsymbol{x}), h(\boldsymbol{x}')) = \rho\sqrt{w(\boldsymbol{x})w(\boldsymbol{x}')}(e^{||\mathbf{x}-\mathbf{x}'||/\theta})$$

and a nugget (or measurement-error) variance with $\mathrm{Var}(e_k) = \sigma^2 w(\boldsymbol{x}_k)$. As a first step, we can estimate the marginal-standard-deviation surface, $w(\boldsymbol{x})$, using a thin-plate spline fitted to the sample standard deviations for each location over the 89 days of the summer. In the covariance model, $w(\boldsymbol{x})$ is taken to be this spline estimate, and ρ and θ are found by GCV. This covariance model has been set up so that if the marginal variance on this day agrees with the average variance found for the summer, $\sigma^2 + \rho = 1$. Day 16 is apparently different from the summer average, and the estimated variability is larger than expected. The estimated ozone surfaces, one generated by using a thin-plate spline and the other by using a spatial-process estimate, are depicted in Figure 13.8. Note that even though these estimates are based on very different covariance structures, the surfaces are very similar. The standard errors for these estimates are different, as we see in Figure 13.9. The spline has much lower standard errors and this is due to the difference between the generalized covariance (associated with the spline) and the exponential covariance. In particular, the range parameter (θ) can limit the extent of the spatial correlations in the exponential model, implying a rougher surface. In this case, the spatial estimate will have a larger standard error. This example suggests that for inference, one should be careful about modeling the covariance of the surface and should evaluate the sensitivity to different models. If one is forced to choose, the more general spatial model probably gives more believable measures of uncertainty.

13.7 FUTURE DIRECTIONS

All of the smoothing methods discussed in this chapter use a single global smoothing parameter, and so every part of the data is smoothed an equal amount. This constraint may miss local features in data and may cause inferences about the estimate to be unreliable. Wavelet and multiresolution ideas provide one method of local smoothing, and the example in this chapter is one possible

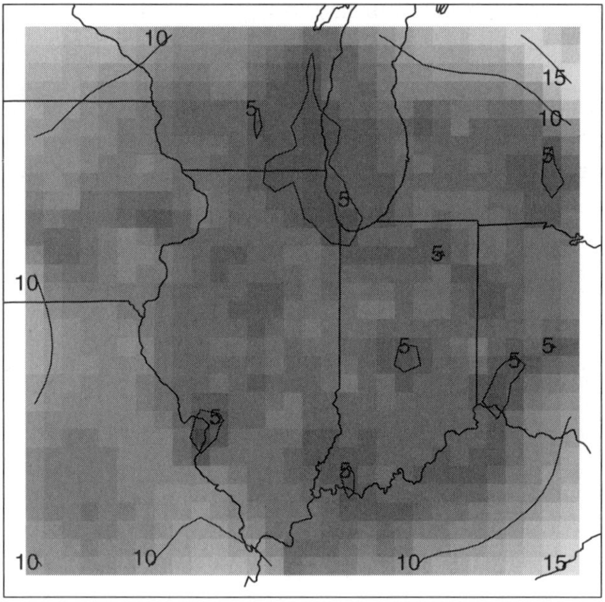

(a)

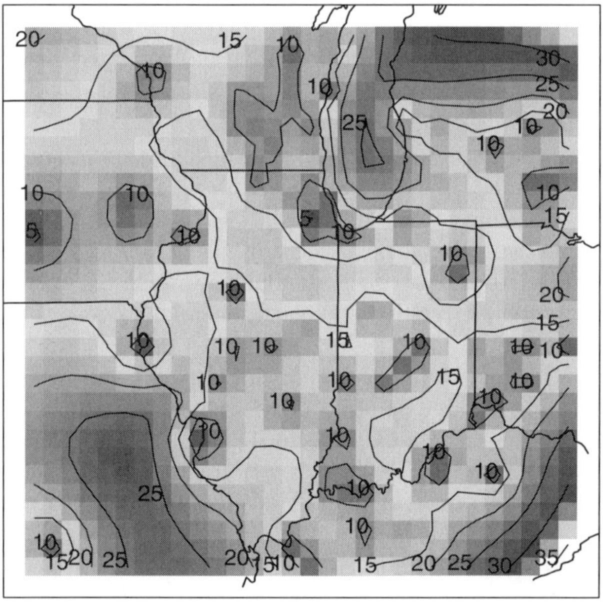

(b)

FIGURE 13.9 Standard errors of prediction (in ppb) for the spline estimate (a) and for the spatial-process estimate (b) from Figure 13.8. See insert for color representation of figure.

implementation for splines. An alternative is to select the basis functions that define the estimate. The idea is to concentrate basis functions in areas where the function has fine structure and to eliminate basis functions where the function does not have features (see, e.g., Luo and Wahba, 1997). In the spatial context, local smoothing is equivalent to a nonstationary covariance function, and the modeling and use of nonstationary spatial covariances is an area of very active research.

Another important area is in nonparametric design. In the context of design for computer experiments, there has been some work on locating points that are efficient for estimating a class of functions and not just a parametric family. There are also opportunities for design in response-surface problems and in the more traditional area of spatial sampling.

In a regression problem, one would like measures of the contribution of different variables to the predictive power of the model. Also, one would like to quantify the importance of the smooth component in the model above the polynomial part. Being able to make inferences about the value of a variable is important in response-surface problems, and identifying key process variables can sharpen the predictions and also make the models easier to interpret. Testing for the significance of a variable usually results in test statistics where the reference distribution must be determined by simulation, and so the computational efficiency of using the DR representation is important. The work by Gu and Wahba (1991) has relevance here, because different smoothing parameters for various parts of a model or for different variables can indicate the amount of structure associated with each component.

Acknowledgments

The author would like to thank Sarah Hardy and Wendy Meiring for a careful reading of this manuscript and for valuable suggestions toward its improvement. This work was sponsored by National Science Foundation grants DMS-9705054 and DMS-93122686.

REFERENCES

Bates, D. M., Lindstrom, M. J., Wahba, G., and Yandell, B. S. (1987). GCVPACK—Routines for generalized cross-validation, *Commun. Stat. Sim. Comput.*, **16**, 263–297.

Cressie, N. A. C. (1990). Reply to Wahba's letter, *Amer. Statistician*, **44**, 256–258.

Cressie, N. A. C. (1991). *Statistics for Spatial Data*, Wiley, New York.

Duchon, J. (1977). Splines minimizing rotation-invariant semi-norms in Sobelev spaces, in Duchon, J. (ed.) *Constructive Theory of Several Variables*, pp. 85–100, Springer–Verlag, Berlin, Germany.

Green, P. J., and Silverman, B. W. (1994). *Nonparametric Regression and Generalized Linear Models*, Chapman and Hall, London, UK.

Gu, C., and Wahba, G. (1991). Minimizing GCV/GML scores with multiple smoothing parameters via the Newton method, *SIAM J. Sci. Stat. Comput.*, **12**, 383–398.

Handcock, M., and Stein, M. (1993). A Bayesian analysis of Kriging, *Technometrics*, **35**, 403–410.

Kent, J. T., and Mardia, K. V. (1994). The link between Kriging and thin plate splines, in F. P. Kelly, ed., *Probability, Statistics and Optimization*, pp. 324–339, Wiley, New York.

Luo, Z., and Wahba, G. (1997). Hybrid adaptive splines, *J. Amer. Statist. Assoc.*, **92**, 107–116.

Matheron, G. (1973). The intrinsic random functions and their applications, *Ad. Appl. Prob.*, **5**, 439–468.

Meinguet, J. (1979). Multivariate interpolation at arbitrary points made simple, *J. Appl. Math. Phys.* (ZAMP), **30**, 292–304.

Messer, K. (1991). A comparison of a spline estimate to its equivalent kernel estimate, *Ann. Statist.*, **19**, 817–829.

Nason, G. P., and Silverman, B. W. (1994). The discrete wavelet transform, *J. Comput. Graph. Statist.*, **3**, 163–191.

Nychka, D. (1988). Bayesian "confidence" intervals for smoothing splines, *J. Amer. Statist. Assoc.*, **83**, 1134–1143.

Nychka, D. (1991). Choosing a range for the amount of smoothing in nonparametric regression, *J. Amer. Statist. Assoc.*, **86**, 653–664.

Nychka, D. (1995). Smoothing splines as locally weighted averages, *Ann. Statist.*, **23**, 1175–1197.

Nychka, D., Bailey, B., Ellner, S., Haaland, P., and O'Connell, M. (1996). *FUNFITS Data Analysis and Statistical Tools for Estimating Functions*. North Carolina Institute of Statistics Mimeoseries, No. 2289, North Carolina State University, Raleigh, NC.

O'Connell, M., and Wolfinger, R. (1997). Spatial regression, response surfaces and process optimization, *J. Computat. Graph. Statist.*, **6**, 224–241.

Omre, H., and Halvorsen, K. (1989). The Bayesian bridge between simple and universal Kriging, *Mathematical Geology*, **21**, 767–786.

Pilz, J., Schimek, M. G., and Spöck, G. (1997). Taking account of uncertainty in spatial covariance estimation, in X. Baafi and N. Schofield (eds.), *Geostatistics Wollongong '96*, pp. 355–358, Kluwer, Dordrecht, The Netherlands.

Qian, S. S. (1997). Estimation of the area affected by phosphorus in an Everglades wetland: A comparison of universal kriging and Bayesian kriging, *Environmental and Ecological Statistics*, **4**, 1–29.

Sacks, J., Welch, W. J., Mitchell, T. J., and Wynn, H. P. (1989). Design and analysis of computer experiments, *Statistical Science*, **4**, 409–435.

Wahba, G. (1990a). *Spline Models for Observational Data*, SIAM, Philadelphia, PA.

Wahba, G. (1990b). Letter to the editor, *American Statistician*, **44**, 256.

Welch, W. J., Buck, R. J., Sacks, J., Wynn, H. P., Mitchell, T. J., and Morris, M. D. (1992). Screening, predicting and computer experiments, *Technometrics*, **34**, 15–25.

Zhang, B., and Stein, M. (1993). Kernel approximations for universal Kriging predictors, *J. Multivariate Analysis*, **44**, 286–313.

CHAPTER 14

Resampling Methods for Nonparametric Regression

Enno Mammen

Ruprecht-Karls-Universität Heidelberg, Heidelberg, Germany

14.1 INTRODUCTION

In this chapter we will discuss resampling methods for nonparametric regression. *Bootstrap* (resampling) approaches have been proposed for model choice, for data-adaptive choice of the smoothing parameter (e.g., bandwidth choice), for testing, for the construction of confidence intervals and bands, and for many other topics. We do not aim at giving a full overview of all these applications here. The aim of this chapter is to give a first impression of the power of resampling methods in the field covered by this book. In particular, we will discuss only nonparametric regression. Applications of bootstrap in nonparametric density estimation will not be described. For these problems, ideas similar to those for the regression setup apply. Furthermore, we will not discuss bootstrap for model choice. That application of bootstrap needs more research.

Before we begin the discussion of bootstrap in nonparametric regression models, we will introduce the basic idea of the bootstrap in Section 14.2, for a general setup. We will discuss why and when bootstrap should be used instead of other classical methods. Bootstrap for nonparametric regression will be described in Section 14.3. The basic idea of bootstrap is to mimic the stochastic structure of the assumed model. In particular, for nonparametric regression, an appropriate choice of the bootstrap method depends on the model (random or deterministic design, homo- or heteroschedastic error).

Smoothing and Regression: Approaches, Computation, and Application
Edited by Michael G. Schimek
ISBN 0-471-17946-9

Some applications of bootstrap will be presented in Sections 14.4–14.7. These include bootstrap confidence bands, bootstrap tests, and bandwidth choice by bootstrap. Some concluding remarks are given in Section 14.8.

14.2 THE IDEA OF BOOTSTRAP

Let us denote the distribution of a sample Z by Q. The bootstrap is a general approach to estimating functionals $T(Q)$. The basic idea of bootstrap is very simple. It consists of two steps: the unknown distribution Q is estimated by an estimate $\tilde{Q}$, and then this estimate is plugged into $T(\cdot)$, that is, the functional $T(Q)$ is estimated by $T(\tilde{Q})$.

We will illustrate this idea by a simple (standard) example. Consider this setup: An i.i.d. real-valued sample $Z = (Z_1, \ldots, Z_n)$ is observed, and one is interested in the distribution of $\overline{Z} - E\,(Z_1)$, where $\overline{Z} = 1/n \sum_{i=1}^{n} Z_i$ denotes the sample mean. Knowledge of this distribution may be used to construct confidence intervals for the mean $\mu = E(Z_1)$. We will discuss this application below.

To begin the description of the bootstrap procedure, we now define the functional T. For a product measure R^n, the functional $T(R^n)$ is given as the distribution of $\sqrt{n}(\overline{Z} - \mu(R))$, where $Z_1, \ldots, Z_n$ is a sample of n i.i.d. observations with distribution R and where $\mu(R) = E_R(Z_1)$ denotes the expectation of Z_1 under the measure R. Here, the factor $\sqrt{n}$ has been introduced for simplification of the notation in later asymptotic considerations.

As a first step, the distribution $Q = P^n$ has to be estimated, for example, by estimating P by the empirical distribution defined by

$$\hat{P}_n(A) = \frac{1}{n} \sum_{i=1}^{n} \mathrm{I}(Z_i \in A).$$

If P is estimated by the empirical distribution $\hat{P}_n$, the bootstrap estimate $T(\hat{P}_n^n)$ is the conditional distribution of $\sqrt{n}(\overline{Z}^* - \overline{Z})$ (given the sample $Z_1, \ldots, Z_n$). Here $\overline{Z}^* = 1/n \sum_{i=1}^{n} Z_i^*$ denotes the sample mean of a conditionally i.i.d. sample $Z_1^*, \ldots, Z_n^*$ having conditional distribution $\hat{P}_n$.

One says that "the bootstrap works" if the bootstrap estimate (of $T(Q)$) is consistent. In general, one expects that bootstrap works under conditions of the following type:

(1) $\tilde{Q}$ is an accurate estimate of Q.

(2) The functional T is continuous in an appropriate sense.

Typically, a mathematical consistency proof for bootstrap is based on some other approach. A popular proof consists of the following two steps, for a sequence of underlying measures Q_n and a sequence of estimates $\tilde{Q}_n$. It is first shown that $T(Q_n)$ has an asymptotic limit τ^∞; then it is proved that $T(\tilde{Q}_n)$ converges in probability to the same limit τ^∞. In our example, one lets the number n of observations tend to infinity, and one sets $Q_n = P^n$. Under weak conditions on the distribution of the sample $Z_1, \ldots, Z_n$ (e.g., that second moments exist), one can show that the conditional distribution $T(\hat{P}_n^n)$ of $\sqrt{n}(\overline{Z}^* - \overline{Z})$ converges in probability to a normal law $N(0, \sigma^2) = \tau^\infty$, and that the distribution $T(P^n)$ of $\sqrt{n}(\overline{Z} - \mu(P))$ converges in distribution to the same limit law. Here σ^2 is the variance of Z_i. The coincidence of the two limiting distributions means that bootstrap works: It gives a consistent estimate of $T(Q_n)$.

In our example, estimates of the functional $T(P^n)$ can be used to construct confidence intervals for the mean $\mu = E_P(Z_1)$. Denote the α quantile of $T(P^n)$ and $T(\hat{P}_n^n)$ by k_α and $\hat{k}_\alpha$, respectively. Then $[\overline{Z} - k_{1-\alpha/2}/\sqrt{n}, \overline{Z} - k_{\alpha/2}/\sqrt{n}]$ defines a confidence interval for μ with exact coverage probability $1 - \alpha$. A bootstrap confidence interval for μ is given by $[\overline{Z} - \hat{k}_{1-\alpha/2}/\sqrt{n}, \overline{Z} - \hat{k}_{\alpha/2}/\sqrt{n}]$. If $T(\hat{P}_n^n)$ converges in probability to the same normal law $N(0, \sigma^2) = \tau^\infty$ as $T(P^n)$, the bootstrap confidence interval has asymptotic coverage probability $1 - \alpha$. More details of this example can be found in every book about the bootstrap; see below for a list of references.

If the functional T is very smooth, bootstrap may work with rough estimates of Q. Let us consider again our simple example, where T is now defined for product measures $Q = R_1 \times \cdots \times R_n$ for which the means of the components are all the same (i.e., $E_{R_1}(Z_1) = \cdots = E_{R_n}(Z_n)$). Here $T(Q)$ is given as the distribution of $\sqrt{n}(\overline{Z} - \mu(R_1))$, and $Z_1, \ldots, Z_n$ is a sample of n independent observations. The observation Z_i has the distribution R_i, and $\mu(R_1)$ denotes the expectation of Z_1 under the measure R_1. Note that $\mu(R_1) = \cdots = \mu(R_n)$. Here, bootstrap requires estimation of n distributions $R_1, \ldots, R_n$. Denote the distribution of $Z_i - \mu(R_1)$ by R_i^c. For this setup, the following resampling plan has been proposed. In this approach the "estimation" of R_i^c makes use only of the single residual $\hat{\varepsilon}_i = Z_i - \overline{Z}$. The estimate $\hat{R}_i^c$ is defined such that

$$E_{\hat{R}_i^c} U_i = 0, \tag{14.1}$$

$$E_{\hat{R}_i^c} U_i^2 = \hat{\varepsilon}_i^2, \tag{14.2}$$

$$E_{\hat{R}_i^c} U_i^3 = \hat{\varepsilon}_i^3. \tag{14.3}$$

For instance, the estimate $\hat{R}_i^c$ can be chosen as a two-point distribution. Then the estimate is uniquely defined by (14.1)–(14.3). In this case, $\hat{R}_i^c$ is the distribution of a binary variable that takes the value $-(\sqrt{5} - 1)\hat{\varepsilon}_i/2$ with probability $(\sqrt{5} + 1)/(2\sqrt{5})$ and the value $(\sqrt{5} + 1)\hat{\varepsilon}_i/2$ with probability $1 - (\sqrt{5} + 1)/(2\sqrt{5})$. In another construction, $\hat{R}_i^c$ is the distribution of $\varepsilon_i V_i$, where $V_1, \ldots, V_n$ is an i.i.d. sample with $E(V_i) = 0, E(V_i^2) = 1$, and $E(V_i^3) = 1$. A possible choice of V_i is given by $V_i = (1 - 2 \cdot 20^{-2/3})^{1/2} W_i + 20^{-1/3}(W_i^2 - 1)$, where the W_i are $N(0,1)$-distributed random variables. For other constructions of $\hat{R}_i^c$ that fulfill (14.1)–(14.3), see Liu (1988) and Mammen (1992b). An estimate $\hat{R}_i$ of R_i is given by the conditional distribution of $U_i + \overline{Z}$ (where U_i has conditional distribution $\hat{R}_i^c$). In this case, the resampling estimate of $T(Q)$ is given by $T(\hat{R}_1^c \times \cdots \times \hat{R}_n^c)$. (Note that in this simple example $T(\hat{R}_1^c \times \cdots \times \hat{R}_n^c) = T(\hat{R}_1 \times \cdots \times \hat{R}_n)$.)

Resampling schemes in which (14.1)–(14.3) hold have been introduced in Wu (1986) for variance estimation in linear models with heteroschedastic errors. It has been remarked by Beran (1986) that they can be used for the estimation of distributions. This application has been called *wild bootstrap* or *external bootstrap*. Condition (14.3) was introduced by Liu (1988) and by Härdle and Mammen (1993), in order to improve the rate of convergence of the wild-bootstrap estimate. Consistency of wild bootstrap in some models has been proven by Liu (1988). In Mammen (1993), it was shown that wild bootstrap achieves better rates of convergence in linear models with random design and increasing dimension. Other comparisons of wild bootstrap against bootstrap in linear models can be found in Liu and Singh (1992). The idea of applying wild bootstrap in nonparametric regression dates back to Härdle and Mammen (1993). They give an example from nonparametric curve estimation for which ordinary bootstrap does not work, yet wild bootstrap is consistent. We will come back to this example in Section 14.5. In our example of non-i.i.d. observations, wild bootstrap does not outperform ordinary bootstrap based on i.i.d. resampling. In this case, wild bootstrap works (i.e., it gives a consistent estimate of $T(Q)$) if and only if the ordinary bootstrap works. This equivalence was shown in Mammen (1992a). So, in this example, it is not true that wild bootstrap works for a larger class of non-i.i.d. models than does the ordinary bootstrap; on the other hand, the ordinary bootstrap works for the broad class of non-i.i.d. models for which the wild bootstrap is consistent. This comparison holds even though the ordinary boostrap is based on i.i.d. resampling. At first sight, this result seems to be counterintuitive, because here wild bootstrap seems to mimic more general models such as the ordinary bootstrap. We will come back to similar findings: We will argue that, in some applications in nonparametric regression, those resampling plans are preferable that mimic a stochastic model that is simpler than the asssumed underlying model and that, in particular, does not coincide with the true model.

Typically, the structure of the bootstrap estimate $T(\tilde{Q})$ is very complex and cannot be calculated explicitly. In such cases, $T(\tilde{Q})$ is often calculated approximately by computer simulations. In our example, an approximate calculation of $T(\hat{P}_n^n)$ can be done by the generation of bootstrap samples $Z_1^*, \ldots, Z_n^*$. In this case, the Z_i^*'s are generated simply by drawing with replacement from the original sample $Z_1, \ldots, Z_n$. A repeated generation of bootstrap samples gives a set of realizations of $\overline{Z}^* - \overline{Z}$. An approximation of $T(\hat{P}_n^n)$ is given by the empirical distribution of these values. For the number of bootstrap replicates, the value 1000 or higher has been proposed. We will use a similar algorithm in our example in Section 14.7. However, it should be remarked that the (approximate) calculation of $T(\tilde{Q})$ is a purely algorithmic point. It is not intrinsic to the statistical idea of the bootstrap. It will not be discussed in this chapter.

14.2.1 Rate of Convergence

For many models, higher-order calculations have been made for the accuracy of the bootstrap estimate. By this approach, the bootstrap estimate can be compared with other methods (e.g., distribution approximations based on asymptotic considerations). In our example, it is well known that bootstrap achieves only the same rate of convergence $1/\sqrt{n}$ as the normal approximation $N(0, \hat{\sigma}^2)$, where $\hat{\sigma}^2$ is the empirical variance $1/n \sum_{i=1}^{n} (Z_i - \overline{Z})^2$.

The situation is quite different for the distribution $S(P^n)$ of the studentized statistic $\sqrt{n}(\overline{Z} - \mu(P))/\hat{\sigma}$. Under weak regularity conditions, the bootstrap estimate $S(\hat{P}_n^n)$ achieves rate of convergence $1/n$, whereas the normal approximation $N(0, 1)$ achieves only $1/\sqrt{n}$. At a heuristic level, it can be easily understood why bootstrap performs better in the latter case. The functional S is "asymptotically constant" (asymptotically equivalent to $N(0, 1)$). The functional T is only "asymptotically continuous" ($T(R^n) = N(0, \sigma^2(R))$, where $\sigma^2(R)$ denotes the variance of the Z_i under R.). For the higher-order performance of the bootstrap, mathematically precise arguments can be constructed by using Edgeworth expansions. In the literature, this construction has been done for several different examples, and it has been shown that bootstrap outperforms normal approximations for studentized statistics—in particular, examples from nonparametric curve estimation. A detailed description of Edgeworth expansions and their applications to the study of bootstrap can be found in Hall (1992b). In the same book, higher-order performance of bootstrap in nonparametric curve estimation is discussed.

14.2.2 Strength of Conditions

For many applications of confidence sets, approximate accuracy of the coverage probability suffices. In these cases (and in other examples where moderate accuracy is sufficient), one is more interested in whether bootstrap works at all. The question of whether bootstrap achieves higher-order accuracy is of less importance in such cases. Mathematically, then, it makes sense to study whether bootstrap works under conditions weaker than those necessary for other methods. Such research was first carried out by Bickel and Freedman (1983). They showed that, for least-squares estimates in linear models with increasing dimension, bootstrap works under weaker conditions than are required for normal approximations. Their results have been generalized to other estimates in Mammen (1989, 1992a, 1992b).

14.2.3 Lack of Alternatives

In some statistical problems, no procedure is available other than bootstrap methods. We will present an example in Section 14.7. There, bootstrap is used to approximate the critical values of a test statistic for multimodality. For this test statistic, no asymptotic theory is known.

14.2.4 Computational Advantage

In other statistical problems, the asymptotic theory is rather complicated. For instance, the calculation of the asymptotic distribution may require computer simulations (e.g., if the limiting distribution is the supremum of a Gaussian process) or the asymptotic formula may be rather complicated (e.g., higher-order Edgeworth expansions). An example for the former case will be given in Section 14.4 (bootstrap confidence bands for nonparametric regression functions). In such cases, one may prefer bootstrap methods, because they require fewer mathematical considerations or because their algorithmic calculation is easier to implement.

14.2.5 Availability of References

Since its introduction (Efron,1979), bootstrap has become a standard tool of modern statistics. It is discussed in many introductory textbooks on statistics. Some books on bootstrap have already appeared (e.g., Efron, 1982; Beran and Ducharme, 1991; Hall, 1992b; Mammen, 1992b; Efron and Tibshirani, 1993; Shao and Tu, 1995; Davison and Hinkley, 1997; see also Le Page and Billard, 1992; Giné, 1997).

14.3 BOOTSTRAP IN NONPARAMETRIC REGRESSION

We will now discuss bootstrap for regression data sets $(X_1, Y_1), \ldots, (X_n, Y_n)$. Here, the X_i are covariables with values in $\mathbb{R}^d$, and the response variable Y_i is assumed to take values in $\mathbb{R}$. A short introduction to bootstrap in nonparametric curve estimation is given in Härdle and Mammen (1991).

For non- and semiparametric modeling, different approaches have been discussed in this book. They include the case of a random covariable X and the case of a nonrandom design. Let us assume that the relation between X and Y can be written as

$$Y = f(X) + \varepsilon, \tag{14.4}$$

where ε is an error variable having $E(\varepsilon|X) = 0$ (case of random design) or $E(\varepsilon) = 0$ (case of nonrandom design).

The function f is assumed to be (partially) unknown. We are thinking here of models for f such as the following:

(i) $f(x) = m(x)$ with an unknown function $m : \mathbb{R}^d \to \mathbb{R}$ (*nonparametric regression*).

(ii) $f(x) = m_1(x_1) + \cdots + m_d(x_d)$ with unknown functions $m_j : \mathbb{R} \to \mathbb{R}$, $j = 1, \ldots, d$ (*additive regression*).

(iii) $f(x) = m_1(x_1) + \cdots + m_p(x_p) + \theta_{p+1}x_{p+1} + \cdots + \theta_d x_d$ with unknown functions $m_j : \mathbb{R} \to \mathbb{R}$, $j = 1, \ldots, p$ and unknown parameters $\theta_{p+1}, \ldots, \theta_d \in \mathbb{R}$ (*partial linear models*).

(iv) $f(x) = G(\rho(x))$, where $G : \mathbb{R} \to \mathbb{R}$ is a known function (*link function*), and $\rho(x)$ is modeled as $f(x)$ in (i), (ii), or (iii) (*nonparametric regression models with link function*).

Up to now, we have not specified the stochastic structure Q of our model (14.4). In this chapter we will consider three different structures:

(a) The pairs (X_i, Y_i) are i.i.d. with $E(Y_i|X_i = x) = f(x)$ (*random design*);
(b) The covariables X_i are deterministic, and the error variables ε_i are i.i.d. with mean 0 (*deterministic design with i.i.d. errors*); and
(c) The covariables X_i are deterministic, and the error variables ε_i are independent with mean 0 (*deterministic design with independent errors*).

The description of model (c) is very rough. Typically, the error distribution is specified more precisely. In particular, it may be assumed that the error variance depends smoothly on the (deterministic) design variable X_i—that is, $\text{Var}(\varepsilon_i) = \sigma^2(X_i)$ has a smooth function $\sigma^2 : \mathbb{R} \to \mathbb{R}$. Alternatively, the error distribution may depend on the unknown function $f(X_i)$. This latter case will occur if the distributions of the response variables Y_i belong to an exponential family that can be parametrized by the expectation $f(X_i)$ of Y_i (e.g., generalized linear models, generalized additive models). An important example are cases in which the response variable Y_i is $\{0, 1\}$-valued (binary-response models). In such cases, $\sigma^2(X_i) = f(X_i)[1 - f(X_i)]$.

Clearly, model (b) is a special case of model (c). Furthermore, after conditioning the covariables $X_1, \ldots, X_n$, model (a) reduces to model (c). Nevertheless, the stochastic structure Q of the models is quite distinct; therefore, quite different estimates $\tilde{Q}$ and quite different bootstrap procedures are motivated by the different model assumptions.

Under model assumption (a), bootstrap samples $(X_1^*, Y_1^*), \ldots, (X_n^*, Y_n^*)$ are generated by i.i.d. sampling from a distribution $\tilde{R}_n$, where $\tilde{R}_n$ is an estimate of the distribution R of (X_1, Y_1). We call this bootstrap procedure *pairwise resampling*. The estimate $\tilde{R}_n$ may be chosen as the empirical distribution $\hat{R}_n$ of $(X_1, Y_1), \ldots, (X_n, Y_n)$. In such a case, the bootstrap sample $(X_1^*, Y_1^*), \ldots, (X_n^*, Y_n^*)$ is generated by drawing with replacement from the original sample. Early references for this approach are McDonald (1982) and Dikta (1990).

This resampling has two disadvantages. For high-dimensional design (d large), the resampling requires the estimation of a distribution on a high-dimensional space, and so it may lead to instabilities. Furthermore, the conditional stochastic nature of the model is not mimicked. In the bootstrap world, we have

$$E^*(Y_i^*|X_i^*) = Y_i^* \tag{14.5}$$

(if the design variables are pairwise different). Here E^* (Var^*) denotes the conditional expectation (variance), given the sample $(X_1, Y_1), \ldots, (X_n, Y_n)$.

Equation (14.5) implies $\mathrm{Var}^*(Y_i^*|X_i^*) = 0$. This outcome causes pairwise resampling to be inconsistent for some applications, as is shown below. The behaviour (14.5) can be avoided by another choice of $\tilde{R}_n$. For instance, suppose that the measure $\tilde{R}_n$ has a density, and that this density converges to the density of (X_1, Y_1). In such a case, the conditional structure of the model is well approximated by resampling from $\tilde{R}_n$. Consistent density estimates can be constructed by smoothing—for instance, a kernel estimate $\tilde{R}_n$ will succeed. In that case, $\tilde{R}_n$ is defined as the *convolution* of the empirical distribution $\hat{R}_n$ with a product kernel

$$\tilde{R}_n(A) = \int_A \frac{1}{n} \sum_{i=1}^{n} K_h(X_i - x) L_g(Y_i - y)\, dx\, dy, \tag{14.6}$$

or

$$\tilde{R}_n(A) = \int_A \frac{1}{n} \sum_{i=1}^{n} K_h(X_i - x) \mathrm{I}[(x, Y_i) \in A]\, dx \tag{14.7}$$

In these cases, $K_h(u) = h^{-d} K(u/h)$ and $L_g(v) = g^{-1} L(v/g)$ are kernels on IR^d or IR, with bandwidths h and g, respectively. Note that (14.7) can be interpreted as the limiting case of (14.6) when $g = 0$. For $g \geq 0$ and $h > 0$, under the assumption that $\int v L(v)\, dv = 0$, we get the following result:

$$E^*(Y_i^*|X_i^* = x) = \frac{\sum_{i=1}^{n} K_h(X_i - x) Y_i}{\sum_{i=1}^{n} K_h(X_i - x)}.$$

This is the *Nadaraya–Watson estimate*. Bootstrap samples with a kernel-smooth $\tilde{R}_n$ can be easily generated. In a first step, a sample based on drawing with replacement is generated. In a second step, i.i.d. variables with density K_h or L_g, respectively, are added to the data. Bootstrap with smooth resampling distributions has also been called *smooth bootstrap*. A discussion of smooth bootstrap in nonparametric regression can be found in Cao-Abad and González-Manteiga (1993).

In the following sections we will discuss two other approaches by means of bootstrap (residual resampling and wild bootstrap).

For bootstrapping model (b), one uses an estimate $\tilde{f}_n$ of the regression function f and an estimate $\tilde{P}_n$ of the distribution P of the error variables ε_i ($i = 1, \ldots, n$). A bootstrap sample $(X_1, Y_1^*), \ldots, (X_n, Y_n^*)$ is generated by first generating (conditionally) i.i.d. errors $\varepsilon_1^*, \ldots, \varepsilon_n^*$ with (conditional) distribution $\tilde{P}_n$ and then setting

$$Y_i^* = \tilde{f}(X_i) + \varepsilon_i^*. \tag{14.8}$$

We call this bootstrap procedure *residual resampling*. In this bootstrap procedure, the covariables X_i are not resampled (i.e., $X_i^* = X_i$), because they are assumed to be deterministic. An estimate $\tilde{P}_n$ of the error distribution is given by the empirical distribution $\hat{P}_n$ of (centered) residuals

$$\overline{\varepsilon}_i = Y_i - \overline{f}(X_i) - \frac{1}{n}\sum_{j=1}^{n}\left[Y_j - \overline{f}(X_j)\right],$$

where $\overline{f}$ is an estimate of f. We remark that three (possibly different) estimates of f are used in this setup. The estimate $\overline{f}$ is used for calculation of residuals, $\tilde{f}$ gives the response variables (see (14.8)), and $\hat{f}$ is the estimate that is analysed by the bootstrap. The choice of $\overline{f}$ will be discussed for bootstrap tests. The estimates $\tilde{f}$ and $\hat{f}$ are used for different purposes: $\hat{f}(x)$ should give an accurate estimate of $f(x)$, whereas $\tilde{f}$ should result in an accurate bootstrap estimate. For this aim, it seems more important that $\tilde{f}$ accurately reflect the curvature of f, as we note below. It is quite natural that, typically, different choices for $\tilde{f}$ and $\hat{f}$ should be used.

A bootstrap of model (c) requires an estimate $\tilde{f}$ of the regression function f and $n(!)$ estimates $\tilde{P}_{n,i}$ of the distributions P_i of the error variables ε_i $(i = 1, \ldots, n)$. A bootstrap sample $(X_1, Y_1^*), \ldots, (X_n, Y_n^*)$ is generated by generating (conditionally) independent errors $\varepsilon_1^*, \ldots, \varepsilon_n^*$ (with (conditional) distributions $\tilde{P}_{n,1}, \ldots, \tilde{P}_{n,n}$ (respectively)) and by once again setting

$$Y_i^* = \tilde{f}(X_i) + \varepsilon_i^*. \tag{14.9}$$

The choice of the estimates $\tilde{P}_{n,i}$ depends on the specifications on the distributions P_i. For instance, the distributions P_i may belong to an exponential family whose parameter depends only on $f(X_i)$, (i.e., $P_i = P_{f(X_i)}$). In such a case, it is appropriate to choose $\tilde{P}_{n,i} = P_{\overline{f}(X_i)}$, where $\overline{f}$ is an estimate of f. As remarked, the estimate $\overline{f}$ may be different from the estimates $\tilde{f}$ and $\hat{f}$. For the case in which not many assumptions are made on the error distributions, it has been proposed in Härdle and Mammen (1993) that one uses wild bootstrap: $\tilde{P}_{n,i}$ are chosen as distributions with (14.1)–(14.3), where $\hat{\varepsilon}_i$ are residuals $Y_i - \overline{f}(X_i)$ (with an estimate $\overline{f}$ of f). For possible choices of $\tilde{P}_{n,i}$, see the discussion in Section 14.2 and the references cited there. Under smoothness assumptions on the error variances (and moments), this approach may be modified by choosing $\tilde{P}_{n,i}$ such that the errors in the wild bootstrap have variance $\hat{\sigma}^2(X_i)$ (and third moment $\hat{\rho}(X_i)$), where $\hat{\sigma}^2$ and $\hat{\rho}$ are smoothing estimates. (See, e.g., Bunke, 1997; and Sommerfeld, 1997).

There is a major difference between using bootstrap in nonparametric curve estimation and using it in other settings. In nonparametric curve estimation, "optimal" choices of estimates $\widehat{m}(x)$ of the underlying curve $m(x)$ are based on balancing the bias and the variance of $\widehat{m}(x)$. So, under typical regularity assumptions for reasonable choices of curve estimates, biases (that do not vanish asymptotically) cannot be avoided. Bootstrap gives an estimate of the distribution of $n^\gamma[\widehat{m}(x) - m(x)]$, where n^γ is the rate of convergence of $\widehat{m}$. Assume that we have proven that bootstrap estimates this distribution consistently; then, in particular, it gives a consistent estimate $\widehat{\text{bias}}_{\text{BOOT}}$ of the bias of $n^\gamma[\widehat{m}(x) - m(x)]$, and so the estimate $\widetilde{m}(x) = \widehat{m}(x) - n^{-\gamma}\widehat{\text{bias}}_{\text{BOOT}}$ has no asymptotic bias. On the other hand, it has the same asymptotic variance as $\widehat{m}(x)$. Therefore $\widetilde{m}(x)$ outperforms $\widehat{m}(x)$. This contradicts the alleged optimality of $\widehat{m}(x)$. This contradiction is typically caused by a mathematical inaccuracy: for the construction of the estimate $\widehat{m}(x)$, weaker model assumptions are used than for the proof of consistency of bootstrap. This contradiction holds not only for the bootstrap but for every procedure that claims to give a consistent estimate of the asymptotic distribution of $\widehat{m}(x)$. At first sight, this contradiction seems to be quite unacceptable. Often, however, this approach seems to be the only possible way to describe asymptotic performance of bootstrap. We will come back to this point in the next section.

14.4 BOOTSTRAP CONFIDENCE INTERVALS AND BANDS

Bootstrap can be used to construct confidence intervals or bands for nonparametric curves m. In this application, m is a nonparametric component of a non- or semiparametric model, as described in Section 14.3. For an estimate $\hat{m}(x)$, the distribution of $\hat{m}(x) - m(x)$ (for fixed x (construction of confidence intervals) or for x in an interval I (construction of confidence bands)) is estimated by the bootstrap distribution $\hat{m}^*(x) - \tilde{m}(x)$. Here, $\hat{m}^*$ is the bootstrap analogue of $\hat{m}$: it is calculated as $\hat{m}$ with the original data $(X_1, Y_1), \ldots, (X_n, Y_n)$ having been replaced by the bootstrap sample $(X_1^*, Y_1^*), \ldots, (X_n^*, Y_n^*)$. The curve estimate $\tilde{m}$ is the nonparametric component of $\tilde{f}$, where $\tilde{f}$ was used in the generation of the bootstrap sample (see for example, (14.8)). We will see below that it makes sense to choose $\tilde{m}$ different from $\hat{m}$.

We discuss now bootstrap confidence intervals for a very simple setup, namely, nonparametric regression models with equidistant design; that is,

$$Y_i = m\left(\frac{i}{n}\right) + \varepsilon_i, \quad i = 1, \ldots, n, \tag{14.10}$$

where $\varepsilon_1, \ldots, \varepsilon_n$ are i.i.d. errors with mean zero. We consider the case where $m(x)$ is estimated by a kernel estimator:

$$\hat{m}(x) = \hat{m}_h(x) = \frac{1}{n}\sum_{i=1}^{n} K_h\left(\frac{i}{n} - x\right) Y_i. \tag{14.11}$$

Here K_h denotes a scaled kernel $K_h(u) = h^{-1}K(h^{-1}u)$ having bandwidth h. It is easy to see, and well known, that under some regularity conditions (m is twice continuously differentiable; K is a symmetric probability density with $\sigma_K^2 = \int u^2 K(u)\,du > 0$; the bandwidth h is of order $n^{-1/5}$, that is, $n^{1/5}h \to c_h > 0$; and

$\sigma^2 = E(\varepsilon_i^2) < \infty$), the distribution of $n^{2/5}[\hat{m}(x) - m(x)]$ converges to a normal distribution with mean $B(x) = \frac{1}{2}c_h^2\sigma_K^2 m''(x)$ and variance $V(x) = c_h^{-1}\rho_K\sigma^2$, where $\sigma_K^2 = \int u^2 K(u)\,du$ and $\rho_K = \int K^2(u)\,du$. We say that bootstrap works if the conditional distribution of $n^{2/5}[\hat{m}^*(x) - \tilde{m}(x)]$ converges (in probability) to the same normal limit $N(B(x), V(x))$.

We will now discuss residual resampling (see Section 14.3). This resampling plan requires two pilot estimates, $\tilde{m}$ and $\overline{m}$. With these estimates, we have $X_i^* = X_i = i/n$ and $Y_i^* = \tilde{m}(i/n) + \varepsilon_i^*$ for $i = 1, \ldots, n$, where $\varepsilon_1^*, \ldots, \varepsilon_n^*$ are (conditionally) i.i.d. variables having distribution $\hat{P}_n$ (the empirical distribution of (centered) residuals $\overline{\varepsilon}_i = Y_i - \overline{m}(X_i) - 1/n \sum_{j=1}^n [Y_j - \overline{m}(X_j)]$). Bootstrap works under the following conditions on the pilot estimates $\tilde{m}$ and $\overline{m}$,

$$\frac{1}{n}\sum_{i=1}^{n}\left[\overline{m}\left(\frac{i}{n}\right) - m\left(\frac{i}{n}\right)\right]^2 \to 0 \text{ in probability.} \tag{14.12}$$

$$n^{-3/5}\sum_{i=1}^{n} K_h\left(\frac{i}{n} - x\right)\left[\tilde{m}\left(\frac{i}{n}\right) - \tilde{m}(x)\right] \to B(x) \text{ in probability.} \tag{14.13}$$

Condition (14.12) is rather weak (e.g., it follows from uniform consistency of $\overline{m}$), and so, typically, the choice $\overline{m} = \hat{m}$ will suffice. There is, however, a problem in choosing $\tilde{m}$ such that (14.13) is fulfilled. Condition (14.13) holds if

$$\tilde{m}''(u) \to m''(u) \quad \text{in probability, for } u \text{ in an } h \text{ neighborhood of } x. \tag{14.14}$$

In particular, for the choice of $\tilde{m} = \hat{m}_h$, with bandwidth h of order $n^{-1/5}$, condition (14.14) is not fulfilled (and neither is condition (14.13)). Several proposals have been made for overcoming this problem:

(i) *Bias correction.* Choose $\tilde{m} = \hat{m}_g$, with bandwidth g of order $n^{-1/5}$ (e.g., $g = h$), and correct the bootstrap estimate by an estimate $\hat{B}(x)$ of the bias $B(x)$. (See, e.g., Härdle and Bowman, 1988.)

(ii) *Oversmoothing in pilot estimation.* Choose $\tilde{m} = \hat{m}_g$, with bandwidth g of order larger than $n^{-1/5}$ (i.e., $n^{1/5}g \to \infty$). Then, under additional smoothness assumptions on m, condition (14.14) holds. This approach has been used in most publications.

(iii) *Choice of nonoptimal h.* Bootstrap the distribution of $\hat{m}_h(x) - m(x)$, with bandwidth h that is of order smaller than $n^{-1/5}$ (i.e., $n^{1/5}h \to 0$). Then $\hat{m}_h(x) - m(x)$ is asymptotically unbiased. Higher-order considerations for bootstrap confidence intervals based on this approach can be found for nonparametric density estimation in Hall (1992a,b) and for nonparametric regression in Hall (1992b).

(iv) *Subsampling.* Use a smaller sample size n' and bandwidth $h^* = (n/n')^{1/5}h$ in the resampling. The distribution of $n^{2/5}[\hat{m}_h(x) - m(x)]$ is estimated by the conditional distribution $(n')^{2/5}[\hat{m}^*_{h^*}(x) - \tilde{m}(x)]$. A general discussion of subsampling can be found in Bickel et al. (1997). Subsampling has been used in Hall (1990) to get bootstrap estimates of the mean-squared error of kernel regression (and density) estimates (see also Section 14.5.) For the use of subsampling to construct bootstrap confidence intervals, see also Politis and Romano (1994).

(v) *Replacing of* $m(x)$ *by* $E[\hat{m}(x)]$. Bootstrap the distribution of $n^{2/5}[\hat{m}_h(x) - E\hat{m}_h(x)]$. Bootstrap estimates of this distribution can be used for confidence intervals for $E[\hat{m}_h(x)]$ (however, not for $m(x)$). A detailed higher-order discussion of this approach can be found in the book Hall (1992b).

Proofs of consistency of the methods (i)–(iv) require additional smoothness conditions on the regression function m (continuity of m'', or existence of higher-order derivatives of m). Under these additional smoothness assumptions, there exist estimates of m that outperform $\hat{m}_h$ (with h of order $n^{-1/5}$) asymptotically (e.g., kernel estimates with higher-order kernels and with bandwidth h of order larger than $n^{-1/5}$). As discussed at the end of the last section, at first sight, this seems to be unacceptable; however, it is intrinsic to every procedure that claims to estimate the distribution of $\hat{m}$. Consistency of such procedures holds only under assumptions that allow more efficient estimation of $m(x)$.

As discussed in the last section for the arithmetic mean (and for other linear statistics), bootstrap achieves higher rates of convergence for studentized statistics. Here, then, it is preferable to bootstrap the statistic $[\hat{m}_h(x) - m(x)]/\hat{S}(x)$ rather than $\hat{m}_h(x) - m(x)$, where $\hat{S}^2(x)$ is an estimate of the variance of $\hat{m}_h(x)$. (A detailed discussion of this point can be found in Hall, 1992b.)

The bootstrap estimates can be used to construct confidence intervals for $m(x)$. Denote the α-quantiles of the bootstrap estimate of the distributions of $\hat{m}_h(x) - m(x)$ and of $[\hat{m}_h(x) - m(x)]/\hat{S}(x)$ by $\hat{k}_\alpha$ and $\hat{k}^S_\alpha$, respectively. $(1-\alpha)$-bootstrap confidence intervals are then defined by

$$[\hat{m}_h(x) - \hat{k}_{1-\alpha/2}, \hat{m}_h(x) - \hat{k}_{\alpha/2}]$$

and

$$\left[\hat{m}_h(x) - \hat{k}^S_{1-\alpha/2}\hat{S}(x), \hat{m}_h(x) - \hat{k}^S_{\alpha/2}\hat{S}(x)\right],$$

respectively.

Similar considerations have been made for wild bootstrap. Consistency of wild bootstrap estimates has been shown in Härdle and Marron (1990). Wild bootstrap can be used to construct (pointwise) confidence intervals, too. Higher-order performance of this method has been considered in Cao-Abad (1991) and in Härdle et al. (1995).

For inference on nonparametric curves, pointwise confidence intervals are often not sufficient. Bootstrap can be used to get confidence bands. $(1-\alpha)$-confidence bands can be constructed by bootstrapping the statistics $\sup_x |\hat{m}_h(x) - m(x)|$ or $\sup_x |\hat{m}_h(x) - m(x)|/\hat{S}(x)$ and by calculating the $(1-\alpha)$-quantile of the bootstrap distribution. This gives bands

$$\{m : |m(x) - \hat{m}_h(x)| \leq \hat{k}_{1-\alpha}^{\text{unif}} \text{ for all } x\},$$

and

$$\{m : |m(x) - \hat{m}_h(x)|/\hat{S}(x) \leq \hat{k}_{1-\alpha}^{S,\text{unif}} \text{ for all } x\},$$

respectively, where $\hat{k}_{1-\alpha}^{\text{unif}}$ and $\hat{k}_{1-\alpha}^{S,\text{unif}}$ are the corresponding estimated quantiles. Consistency of this method can be seen by strong approximations: It suffices to show that there exists a probability space on which are defined two random variables $\Delta_{1,n}$ and $\Delta_{2,n}$, with the following properties:

(i) The (conditional) distribution of $\Delta_{1,n}$ coincides with the (unconditional) distribution of $\sup_x |\hat{m}_h(x) - m(x)|$.
(ii) The (conditional) distribution of $\Delta_{2,n}$ coincides with the conditional (!) distribution of $\sup_x |\hat{m}_h^*(x) - \tilde{m}(x)|$.
(iii) $n^{2/5}[\Delta_{1,n} - \Delta_{2,n}]$ converges to 0 in probability.

In our simple setting, claims (i)–(iii) can be proven by using strong approximations for the partial-sum process of the error variables. Properties (i)–(iii) give a formal mathematical description of why bootstrap works: The stochastic structure of the model is correctly mimicked by the bootstrap. This description is formulated by putting the unconditional stochastic structure of the data and the conditional stochastic structure of the bootstrap resampling onto the same probability space. Mathematically, this approach has the advantage that it does not require knowledge about the asymptotic limit distribution of the statistic $\sup_x |\hat{m}_h(x) - m(x)|$. Although, in our simple setting, an alternative proof could be carried out (by showing that $\sup_x |\hat{m}_h(x) - m(x)|$ and its bootstrap estimate have the same limit distribution), the approach just given is preferable in some settings which are more complicated.

For wild bootstrap, a consistency proof for confidence bands is given in Neumann and Polzehl (1998). Higher-order considerations for bootstrap confidence bands can be found in Hall (1993).

14.5 BOOTSTRAP-BANDWIDTH CHOICE

In this section, we consider applications of bootstrap for choosing smoothing parameters in nonparametric curve estimation. We will discuss this point for bandwidth choice of kernel-regression estimates. Performance of a curve estimate $\hat{m}$ can be measured by the mean integrated squared error

$$\text{MISE}_{n,\hat{m}} = E \int [\hat{m}(x) - m(x)]^2 \, dx. \tag{14.15}$$

Smoothing parameters (such as the bandwidth) can be chosen by minimizing an estimate of the *mean integrated squared error* $\mathrm{MISE}_{n,\hat{m}}$. Bootstrap is one way to estimate $\mathrm{MISE}_{n,\hat{m}}$.

In particular, this approach can be used to choose the bandwidth h of a kernel-regression estimate $\hat{m}_h$. As in the last section, one needs (for a consistent implementation of this method) higher-order smoothness assumptions on the regression function m. One can then show that the method works with "bias correction" (Härdle and Bowman, 1988), with "oversmoothing in the pilot estimation" (Faraway, 1990), or with "subsampling" (Hall, 1990).

The $\mathrm{MISE}_{n,\hat{m}}$ can be decomposed into a bias component and a variance component:

$$\mathrm{MISE}_{n,\hat{m}} = \mathrm{VAR}_n + \mathrm{BIAS}_n^2. \tag{14.16}$$

Here, $\mathrm{VAR}_n = E(\int [\hat{m}(x) - E\hat{m}(x)]^2\, dx)$ and $\mathrm{BIAS}_n^2 = \int [E[\hat{m}(x)] - m(x)]^2\, dx$. For a kernel estimate $\hat{m} = \hat{m}_h$, with bandwidth h, we write $\mathrm{MISE}_n(h) = \mathrm{MISE}_{n,\hat{m}}$, $\mathrm{VAR}_n(h) = \mathrm{VAR}_n$, and $\mathrm{BIAS}_n^2(h) = \mathrm{BIAS}_n^2$. Under regularity conditions, it is well known that (and it can be easily shown that) the variance and bias terms stabilize asymptotically. More precisely,

$$\frac{\mathrm{VAR}_n(h)}{\frac{1}{n}\mathrm{VAR}_\infty(h)} \to 1,$$

$$\frac{\mathrm{BIAS}_n(h)}{\mathrm{BIAS}_\infty(h)} \to 1,$$

for asymptotic expressions $\mathrm{VAR}_\infty(h)$ and $\mathrm{BIAS}_\infty(h)$, depending on h, on the regression function m, and on the design and the error distribution. These convergences hold uniformly over a wide range of sequences of h.

In recent years, it has been proposed that h be chosen by minimizing the formula

$$\frac{1}{n}\widehat{\mathrm{VAR}}_\infty(h) + \widehat{\mathrm{BIAS}}_\infty(h),$$

where $\widehat{\mathrm{VAR}}_\infty$ and $\widehat{\mathrm{BIAS}}_\infty$ are plug-in estimates of the asymptotic terms VAR_∞ and BIAS_∞. Bootstrap bandwidths are based on plug-in estimates of $\mathrm{MISE}_n(h)$ (i.e., of $\mathrm{VAR}_n(h)$ and $\mathrm{BIAS}_n^2(h)$). For an overview and comparison of different bandwidth selectors, we refer to Wand and Jones (1995). (See also Shao and Tu (1995), and Chapter 4 of this book.) A detailed analysis of bootstrap bandwidths and of some related bandwidths is available for density estimation under additional smoothness assumptions on the underlying density. For the model that m has r derivatives (with $r > 2$), it has been shown that a modification of the bootstrap bandwidth (*smoothed cross-validation*) converges with optimal rate to the MISE-optimal deterministic bandwidth. (See Hall, Marron, and Park, 1992; Jones et al., 1991; Marron, 1992; and Cao-Abad, 1993.) In these papers bootstrap (or cross-validation) with oversmoothing in the pilot estimation is considered. The optimality of resampling relies here on the fact that the finite-sample bias term $\text{BIAS}_n^2(h)$ is estimated with higher-order accuracy. This higher-order consistency cannot be achieved by plug-in estimates of the asymptotic bias BIAS_∞.

14.6 BOOTSTRAP TESTS IN NONPARAMETRICS

Bootstrap can be applied to construct tests on different hypotheses of a component m. In this section, and in the next we consider test problems of the following types:

(i) Does a curve m belong to a specified model, or is it of nonparametric form?

(ii) Does a curve m have a specific qualitative shape, such as (piecewise) monotonicity or convexity?

Both questions arise naturally in the interpretation of curve estimates $\hat{m}$. For example, it may be of direct interest whether deviations of $\hat{m}$ from linearity or from monotonicity are caused by random fluctuations or by underlying structure (i.e., by a nonlinear or a nonmonotone function m, respectively). Bootstrap offers a nice data-analytic tool in these cases. One can resample from a linear or a monotone regression function (respectively), and then one can compare the bootstrap curves $\hat{m}^*$ with $\hat{m}$. In particular, one can inspect how significantly the bootstrap curves $\hat{m}^*$ differ from linearity or monotonicity, and one can compare the typical difference with that of $\hat{m}$. For a deeper discussion, we will formalize this approach. The deviance of $\hat{m}$ from linearity (monotonicity) can be measured by the distance of $\hat{m}$ from the class of linear (monotone) functions. This distance can be used as a test statistic. We will discuss whether bootstrap can be used to calculate approximate critical values for this test statistic. We will not address questions about power of the test, which depends mainly on the choice of the distance metric. In this section, we discuss goodness-of-fit tests for parametric modeling of m. Bootstrap tests on the qualitative shape of m will be discussed in the next section.

We consider now a simple regression model with random design. Suppose that $(X_1, Y_1), \ldots, (X_n, Y_n)$ is a set of i.i.d. real-valued tuples. The regression function is defined as $m(x) = E[Y_i|X_i = x]$. The function m can be estimated by the Nadaraya–Watson estimate,

$$\hat{m}_h(x) = \frac{\sum_{i=1}^{n} K_h(X_i - x)Y_i}{\sum_{i=1}^{n} K_h(X_i - x)}.$$

Suppose that we want to discuss a parametric model $\{m_\theta : \theta \in \Theta\}$. Then the Nadaraya–Watson estimate $\hat{m}_h$ can be compared with a parametric estimate $m_{\hat{\theta}}$. For this purpose, we consider the L_2-distance between $\hat{m}_h$ and $m_{\hat{\theta}}$. From a statistical point of view, other distances could also be chosen—these would lead to other power functions. Here, the L_2-distance is chosen mainly for mathematical convenience. With weight function π, we define our test statistic T as

$$T = \int [\hat{m}_h(x) - \mathcal{K}_{h,n} m_{\hat{\theta}}(x)]^2 \pi(x)\, dx, \tag{14.17}$$

where $\mathcal{K}_{h,n}$ denotes the smoothing operator

$$\mathcal{K}_{h,n} g(x) = \frac{\sum_{i=1}^{n} K_h(X_i - x) g(X_i)}{\sum_{i=1}^{n} K_h(X_i - x)}.$$

The smoothing operator has been introduced in the definition of T because $E[\hat{m}_h(x)|X_1, \ldots, X_n] = \mathcal{K}_{h,n} m(x)$. One can show that the test, with test statistic $\int [\hat{m}_h(x) - m_{\hat{\theta}}(x)]^2 \pi(x)\, dx$, shows some erratic power behavior (see Härdle and Mammen, 1993).

The idea of basing tests on nonparametric curve estimates goes back to Bickel and Rosenblatt (1973), where, for a simple hypothesis, the L_2-distance between a kernel-density estimate and its expectation (under the null hypothesis) has been proposed. Application of bootstrap for such tests has been considered in a series of papers (see, e.g., Härdle and Marron, 1990; Härdle and Mammen, 1993; Stute and González-Manteiga, 1996; and Härdle et al., 1998).

We come now to the question of how bootstrap can be applied to get approximate critical values of T. We discuss this for the case of a simple hypothesis, namely, that the parameter set Θ contains only one element. Let us denote the corresponding regression function by m_0. Then we get

$$\begin{aligned} T &= \int [\hat{m}_h(x) - \mathcal{K}_{h,n} m_0(x)]^2 \pi(x)\, dx \\ &= \int \frac{\frac{1}{n} \sum_{i,j=1}^{n} K_h(X_i - x) K_h(X_j - x) \varepsilon_i \varepsilon_j}{\hat{f}_h(x)^2} \pi(x)\, dx, \end{aligned}$$

where $\hat{f}_h(x) = 1/n \sum_{i=1}^{n} K_h(X_i - x)$ is the kernel-density estimate of the density of X_i and where $\varepsilon_i = Y_i - m_0(X_i)$ for $i = 1, \ldots, n$. Under the assumptions that the design variables have a smooth density f and that $\sigma^2(x) = \mathrm{Var}[Y_i|X_i = x] = E[\{Y_i - m(x)\}^2|X_i = x]$ and $\pi(x)$ are smooth functions, one can show that (under the hypothesis $m = m_0$) T is asymptotically equivalent to $T' = T_1' + T_2'$ (much as in Härdle and Mammen, 1993), where

$$T_1' = K_h^{(2)}(0) \int \frac{\sigma^2(x)\pi(x)}{f(x)}\,dx,$$

$$T_2' = \sum_{i,j=1,i\neq j}^{n} W_n(X_i, X_j)\varepsilon_i\varepsilon_j,$$

in which $K_h^{(2)}(w) = \int K_h(w-u)K_h(u)\,du$ is a convoluted kernel and $W_n(u,v) = K_h^{(2)}(v-u)\pi(u)f^{-2}(u)$. Note that T_1' is deterministic. Furthermore, T_2' is a pure U-statistic, that is, $E[W_n(X_i,X_j)\varepsilon_i\varepsilon_j|X_i,\varepsilon_i] = 0$ for $i \neq j$. A first inspection, then, of whether a resampling method works can be based on a check of these same properties in the bootstrap world and, in particular, of whether

$$T_2^* = \sum_{i,j=1,i\neq j}^{n} W_n(X_i^*, X_j^*)\varepsilon_i^*\varepsilon_j^* \tag{14.18}$$

is a pure U-statistic.

We consider now four resampling plans:

(a) *Pairwise resampling*. Generate (conditionally) i.i.d. tuples $(X_{1,A}^*, Y_{1,A}^*), \ldots, (X_{n,A}^*, Y_{n,A}^*)$, with distribution $\hat{R}_n$, where $\hat{R}_n$ is the empirical distribution of $(X_1, Y_1), \ldots, (X_n, Y_n)$. Put $\varepsilon_{i,A}^* = Y_{i,A}^* - m_0(X_{i,A}^*)$.

(b) *Pairwise resampling with adjusted residuals*. Put $X_{i,B}^* = X_{i,A}^*$ and $Y_{i,B}^* = m_0(X_{i,B}^*) + \varepsilon_{i,B}^*$, with $\varepsilon_{i,B}^* = Y_{i,A}^* - \hat{m}_h(X_{i,A}^*)$.

(c) *Pairwise smoothed bootstrap with adjusted residuals*. Generate (conditionally) i.i.d. tuples $(X_{1,C}^*, Y_{1,C}')$, $\ldots$, $(X_{n,C}^*, Y_{n,C}')$, with distribution $\tilde{R}_n$ where $\tilde{R}_n$ is defined in (14.6) or (14.7). Put $\varepsilon_{i,C}^* = Y_{i,C}' - \hat{m}_h(X_{i,C}^*)$ and $Y_{i,C}^* = m_0(X_{i,C}^*) + \varepsilon_{i,C}^*$.

(d) *Wild bootstrap*. Put $X_{i,D}^* = X_i$ and $Y_{i,D}^* = m_0(X_i) + \varepsilon_{i,D}^*$, where $\varepsilon_{1,D}^*, \ldots, \varepsilon_{n,D}^*$ is a conditionally independent sample. The variable $\varepsilon_{i,D}^*$ has distribution $\hat{R}_i$, with properties (14.1)–(14.3), where residuals $\hat{\varepsilon}_i = Y_i - \hat{m}_h(X_i)$ are used.

For further discussion of resampling plans (a), (c), and (d), see Sections 14.2 and 14.3.

Resampling plans (a) and (b) do not work here. They give no consistent estimate of the distribution of T under the hypothesis. As a result, they cannot be used to get approximate critical values of T. The reason for this failure is that for these resampling plans, $E[\varepsilon_i^*|X_1^*, \ldots, X_n^*] = 0$ is not true, in general. In particular, condition (14.18) does not hold. For a formal proof of the inconsistency of bootstrap (a) and (b), we refer to Härdle and Mammen (1993). Their proof is based on the fact that the (conditional) asymptotic variance of $T^* = \int [\hat{m}_h^*(x) - \mathcal{K}_{h,n}m_0(x)]^2\pi(x)\,dx$ does not coincide with the asymptotic variance of

T. (Here, $\hat{m}_h^*(x) = \left[\sum_{i=1}^n K_h(X_i^* - x)Y_i^*\right] / \left[\sum_{i=1}^n K_h(X_i^* - x)\right]$ is the Nadaraya-Watson estimate based on the bootstrap sample).

Resampling plans (c) and (d) do work. Both bootstrap methods are consistent estimates of the distribution of T under the hypothesis. As a result, they give asymptotically correct critical values for T. The reason for consistency is that both bootstrap methods correctly reflect the conditional distribution of $(Y_1, \ldots, Y_n)$ given $(X_1, \ldots, X_n)$. In particular, they fulfil $E[\varepsilon_i^* | X_1^*, \ldots, X_n^*] = 0$ and condition (14.18). For a mathematical proof of consistency, one shows that T_n has a limiting normal distribution and that T_n^* has the same normal limit (in probability). (See Härdle and Mammen, 1993; and Cao-Abad and González-Manteiga, 1993).

This example shows

(i) that a simple i.i.d. resampling may not always work for i.i.d. data (see the breakdown of resampling plans (a) and (b)), and

(ii) that for consistent resampling, the true model need not necessarily be mimicked. In our example, no i.i.d. variables are generated in resampling method (d). This method, then, does not correctly reflect the i.i.d. structure of the model. Nevertheless, this method is consistent. Furthermore, we conjecture that for high-dimensional design variables X_i, method (d) outperforms method (c) even though method (c) correctly mimics the stochastic structure of the underlying model. This conjecture is supported by considerations about high-dimensional linear models (as in Mammen, 1993) and by the experience that it is often better to use an incorrect simple model than to use a correct but too complex model.

As previously mentioned, T has a limiting normal distribution. The accuracy of this normal approximation can be compared with the accuracy of the bootstrap (method (c) or (d)). Simulations suggest that bootstrap is a reliable method here, whereas the normal approximation is rather inaccurate. (See the references mentioned above). This result can be understood by noting that T behaves roughly like a sum of $O(h^{-1})$ independent variables. For instance, for kernels with compact support $[-1, 1]$, say, kernel estimates $\hat{m}_h(x)$ and $\hat{m}_h(y)$ are asymptotically independent if the supports of $K_h(x - \cdot)$ and $K_h(y - \cdot)$ do not overlap. As a result, $\hat{m}_h(x)$ and $\hat{m}_h(y)$ are asymptotically independent if $|x - y| > 2h$. This consideration motivates that the normal approximation has the poor rate of convergence $O(h^{1/2})$.

14.7 BOOTSTRAP INFERENCE ON THE SHAPE OF A CURVE

Interesting shape characteristics may be visible in plots of estimates $\hat{m}$ of nonparametric components m. The estimate may be monotone, piecewise monotone, convex, and so on. Often, the shape of a curve has a direct applied interpretation (see also Chapter 5). The question of interest, then, is whether the shape of a curve estimate reflects the shape of the underlying curve or is caused by random fluctuations. In this section, we focus on the case where one is interested in inference about the number of monotone pieces of a curve. This statistical problem has been addressed in Good and Gaskins (1980), Hartigan and Hartigan (1985), Müller and Sawitzki (1991). A first bootstrap approach has been taken in Silverman (1981, 1983, 1986) for nonparametric density estimation. There, the following bootstrap test for the hypothesis of k monotone pieces has been proposed:

(i) In a first step, a critical bandwidth h_{crit} is determined, for which the kernel-density estimate is just between having k or $k+1$ monotone pieces.
(ii) Bootstrap samples are generated from the density $\hat{f}_{h_{\text{crit}}}$ (where $\hat{f}_h$ is a kernel-density estimate with bandwidth h). These give an estimate of the distribution of h_{crit} under the hypothesis.
(iii) The hypothesis of k monotone pieces is rejected if h_{crit} is larger than the $(1-\alpha)$-quantile of the bootstrap distribution.

Silverman's test was adapted to nonparametric regression models in Bowman et al. (1998). Mammen et al. (1992) and Fisher et al. (1994) present some asymptotic theory and simulations for this test. Their results suggest that Silverman's test achieves its theoretical level as long as the underlying density f has no flat parts and no heavy tails. In particular, in the case of flat parts, the test rejects the hypothesis drastically too often. This over-rejection may be judged a drawback of this procedure.

In what follows, we present another method. We discuss this approach for kernel smoothing in nonparametric regression. For a distance d between functions, we calculate, for fixed bandwidth h, the distance T_k between a kernel estimate $\hat{m}_h$ and the class of functions with k monotone pieces:

$$T_k = \inf\{d^2(\hat{m}_h, m) : m \text{ has } k \text{ monotone pieces}\}.$$

In the following discussion, we use the distance $d^2(m_1, m_2) = \int (m_1 - m_2)^2\, d\pi$, where π is a weighting measure.

This idea is related to Silverman's test, which measures how far mass has to be moved to get a kernel estimate with k monotone pieces. Our test measures how much mass has to be moved to get a function with k monotone pieces. As for Silverman's test, for our test it is difficult to develop an asymptotic theory for T_k. Furthermore, its asymptotic distribution depends strongly on the underlying regression function m. Even the rate of convergence of T_k toward the hypothesis depends on m. As a result, it may be very difficult to get critical values by asymptotic considerations. Again as for Silverman's test, it is natural to try to proceed with bootstrap. In a first step, bootstrap samples are generated under the hypothesis. (This generation requires specification of a regression function

$\tilde{m}$ (with k monotone pieces) that is used in the generation of the resamples.) Bootstrap gives an estimate of the distribution of T_k. Its quantiles can then be used as critical values for T_k.

It is difficult to judge this approach asymptotically. For this reason, we carried out simulations of this approach for two regression functions: a function $m^{(1)}$ with $m^{(1)''} \equiv const \neq 0$, namely,

$$m^{(1)}(x) = x(1-x),$$

and a function $m^{(2)}$ with a flat piece,

$$m^{(2)}(x) = \begin{cases} -\frac{25}{4}x^2 + \frac{5}{2}x & \text{for } x \leq 0.2 \\ \frac{1}{4} & \text{for } 0.2 \leq x \leq 0.8\,. \\ -\frac{25}{4}(1-x)^2 + \frac{5}{2}(1-x) & \text{for } x \geq 0.8 \end{cases}$$

As in Section 14.4, we used a regression model with equidistant deterministic design points $Y_i = m(i/n) + \varepsilon_i$, where the ε_i are i.i.d. normal errors having standard deviation 0.2. The test statistic, T_k, was calculated with π uniform measure on [0, 1] and with kernel estimate $\hat{m}_h$, as defined in (14.11). Observations were generated for $i = 0, \pm 1/n, \pm 2/n, \ldots$ with $n = 100$. (Observations for design points outside of [0, 1] were generated so as to avoid boundary problems. Boundary effects should be discussed separately.) For the bootstrap, the regression function $\tilde{m}$ was chosen to be $\tilde{m} = \arg\min_m \{\int (\hat{m}_g - m)^2\, d\pi :$ m has k monotone pieces$\}$. In this case, $\tilde{m}$ is the projection of a kernel estimate, $\hat{m}_g$, onto the class of functions with k monotone pieces (with respect to the norm of the space $L_2(\pi)$). The bandwidth g was chosen to be $g = h$ and $g = 1.5\,h$. In the simulations, 1000 samples were generated. The bootstrap uses 999 replications. The bootstrap test rejects at the 5 percent level if, in 950 or more of the bootstrap replications, the bootstrap test statistic is smaller than the observed test statistic. (The choice of the number 999 is motivated by the following fact: if one observes i.i.d. variables $V, U_1, \ldots, U_{999}$ with $P(V = U_i) = 0$, then the probability that $\#\{i : U_i \leq V\} \geq 950$ is 0.05. (Here # denotes the number of elements of a set.) The results are summarized in Tables 14.1 and 14.2.

For the choice $g = h$, the bootstrap test maintains its level. It is conservative for the case of function $m^{(1)}$. It should be expected, however, that, for this complex hypothesis, there exists no similar test (i.e., no test that rejects on the whole hypothesis with (approximately) constant probability). Oversmoothing ($g > h$ in the pilot estimation) helps for $m^{(1)}$, but at the cost that then the test rejects too often for $m^{(2)}$. We propose to use the test without oversmoothing ($g = h$).

We have included this example for the following two reasons: It gives a feeling for problems in the qualitative interpretation of nonparametric curve estimates. Furthermore, it shows that, even in cases where no asymptotic theory is available, bootstrap may be a powerful tool.

TABLE 14.1 Results from a Small Simulation Study[a]

Hypothesis		$k = 2$	$k = 3$
$h = 0.1$	$g = 0.1$	0.015	0.019
	$g = 0.15$	0.031	0.039
$h = 0.2$	$g = 0.2$	0.008	0.002
	$g = 0.3$	0.024	0.012
$h = 0.3$	$g = 0.3$	0.036	0.003
	$g = 0.45$	0.044	0.024

[a]Level for testing unimodality and for testing bimodality. Assumed level is 0.05. The underlying function is $m^{(1)}$

TABLE 14.2 Results from a Small Simulation Study[a]

Hypothesis		$k = 2$	$k = 3$
$h = 0.1$	$g = 0.1$	0.070	0.072
	$g = 0.15$	0.095	0.072
$h = 0.2$	$g = 0.2$	0.058	0.052
	$g = 0.3$	0.106	0.062

[a]Level for testing unimodality and for testing bimodality. Assumed level is 0.05. The underlying function is $m^{(2)}$

14.8 EXTENSIONS

In this section we briefly mention two applications that we have not discussed so far.

Bootstrap tests as discussed in Section 14.6 could be used as a first step for model choice procedures. How to do this needs further research.

Another extension concerns dependent observations. Bootstrap for dependent data is an active field of research. Different resampling schemes have been proposed in (parametric) time series analysis:

(1) For ARMA processes, it has been proposed to resample i.i.d. innovations (See, e.g., Kreiss and Franke, 1992).

(2) Künsch (1989) proposed i.i.d. resampling of blocks of consecutive observations.

(3) It has been proposed to use the weak dependence of periodogram ordinates and to generate bootstrap periodograms by i.i.d. resampling of (studentized) ordinates. (See, e.g., Franke and Härdle, 1992; Janas, 1993; and Dahlhaus and Janas, 1996.)

(4) Markov processes can be resampled by a local Markovian bootstrap procedure based on kernel estimates of transition probabilities (see Papavoditis and Politis, 1999a).

These approaches can be used for nonparametric regression models with dependent data. Other related models are nonparametric autoregression models. A simple example for nonparametric autoregression is the model

$$X_{i+1} = m(X_i) + \varepsilon_i, \tag{14.19}$$

where $\varepsilon_1, \ldots, \varepsilon_n$ are i.i.d. mean-zero variables. For a review on nonparametric time series models, see Härdle, Lütkepohl, and Chen (1997). Bootstrap for nonparametric smoothers in nonparametric autoregression models like (14.19) can be done by i.i.d. resampling of the residuals. This method has been considered in Franke and Wendel (1992), Kreutzberger (1993), and Franke et al. (1997, 1998).

The autoregression model (14.19) can be compared with the regression model

$$Y_i = m(X_i) + \varepsilon_i. \tag{14.20}$$

It can be shown that models (14.19) and (14.20) are asymptotically equivalent; see Neumann and Kreiss (1998). This explains why bootstrap methods that generate regression data in the resampling, work for autoregression; see Neumann and Kreiss (1998), Franke et al. (1997), Papavoditis and Politis (1999c), and Ango Nze et al. (1999). For application of local Markovian bootstrap to nonparametric autoregression see Papavoditis and Politis (1999b).

REFERENCES

Ango Nze, P., Bühlmann, P., and Doukhan, P. (1999). *Nonparametric Regression Estimation under Weak Dependence Beyond Mixing and Association*, preprint.

Beran, R. (1986). Discussion to Wu, C. F. J.: Jackknife, bootstrap and other resampling methods in regression analysis, *Ann. Statist.*, **14**, 1295–1298.

Beran, R., and Ducharme, G. R. (1991). *Asymptotic Theory for Bootstrap Methods in Statistics*, Université de Montréal, Montréal, Canada.

Bickel, P. J., and Freedman, D. A. (1983). Bootstrapping regression models with many parameters, in Bickel, P. J., Doksum, K. A., and Hodges, J. C., Jr. (eds.), *A Festschrift for Erich L. Lehmann*, pp. 28–48, Wadsworth, Belmont, CA.

Bickel, P. J., Götze, F., and van Zwet, W. R. (1997). Resampling fewer than n observations: Gains, losses and remedies for losses, *Statistica Sinica*, **7**, 1–32.

Bickel, P. J., and Rosenblatt, M. (1973). On some global measures of the deviations of density function estimates, *Ann. Statist.*, **1**, 1071–1095. [Correction: **3** (1975), 1370.]

Bowman, A. W., Jones, M. C., and Gijbels, I. (1998). Testing monotonicity of regression, *J. Comput. Graph. Statist.*, **7**, 489–500.

Bunke, O. (1997). *Bootstrapping in Heteroscedastic Regression Situations*, Discussion Paper 970004, Sonderforschungsbereich 373, Humboldt Universität, Berlin, Germany.

Cao-Abad, R. (1991). Rate of convergence for the wild bootstrap in nonparametric regression, *Ann. Statist.*, **19**, 2226–2231.

Cao-Abad, R. (1993). Bootstrapping the mean integrated squared error, *J. Multivariate Anal.*, **45**, 137–160.

Cao-Abad, R., and González-Manteiga, W. (1993). Bootstrap methods in regression smoothing, *J. Nonparametric Statistics*, **2**, 379–388.

Dahlhaus, R,. and Janas, D. (1996). A frequency domain bootstrap for ratio statistics in time series analysis, *Ann. Statist.*, **24**, 1934–1963.

Davison, A. C., and Hinkley, D. V. (1997). *Bootstrap Methods and Their Applications*, Cambridge University Press, Cambridge, UK.

Dikta, G.(1990). Approximation of nearest neighbour regression function estimators, *J. Multivariate Anal.*, **32**, 213–229.

Efron, B. (1979). Bootstrap methods: Another look at the jackknife, *Ann. Statist.*, **7**, 1–26.

Efron, B. (1982). *The Jackknife, the Bootstrap, and Other Resampling Plans*. SIAM, Philadelphia, PA.

Efron, B. and Tibshirani, R. J. (1986). *An Introduction to the Bootstrap*, Chapman and Hall, London, UK.

Faraway, J. J. (1990). Bootstrap selection of bandwidth and confidence bands for nonparametric regression, *J. Statist. Comput. Simul.*, **37**, 37–44.

Fisher, N. I., Mammen, E., and Marron, J. S. (1994). Testing for multimodality. *Computat. Statist. Data Analysis*, **18**, 499–512.

Franke, J., and Härdle, W. (1992). On bootstrapping kernel spectral estimates, *Ann. Statist.*, **20**, 121–145.

Franke, J., Kreiss, J. P., and Mammen, E. (1997). *Bootstrap of Kernel Smoothing in Nonlinear Time Series*, Discussion Paper 970020, Sonderforschungsbereich 373, Humboldt Universität, Berlin, Germany.

Franke, J., Kreiss, J. P., Mammen, E., and Neumann, M. H. (1999). *Properties of the Nonparametric Autoregressive Bootstrap*. Discussion Paper 980054, Sonderforschungsbereich 373, Humboldt Universität, Berlin, Germany.

Franke, J., and Wendel, M. (1992). A bootstrap approach for nonlinear autoregressions — some preliminary results, in Jäckel, K.-H., Rothe, G., and Sendler, W., (eds.), *Bootstrapping and Related Techniques*, Springer-Verlag, Berlin, Germany.

Giné, E. (1997). Lectures on some aspects of the bootstrap, in Bernard, P. (ed.), École d'Été de Calcul de Probabilités de Saint-Flour XXVI–1996, pp. 37–151, Springer-Verlag, Berlin, Germany.

Good, I. J., and Gaskins, R. A. (1980). Density estimation and bump-hunting by the penalized likelihood method exemplified by scattering and meteorite data, *J. Amer. Statist. Assoc.*, **75**, 42–73.

Hall, P. (1988). Theoretical comparison of bootstrap confidence intervals, *Ann. Statist.*, **16**, 927–953.

Hall, P. (1990). Using the bootstrap to estimate mean squared error and select smoothing parameter in nonparametric problems, *J. Multivar. Anal.*, **32**, 177–203.

Hall, P. (1992a). On bootstrap confidence intervals in nonparametric regression, *Ann. Statist.*, **20**, 695–711.

Hall, P. (1992b).*The Booststrap and Edgeworth Expansion*, Springer-Verlag, New York.

Hall, P. (1993). On Edgeworth expansion and bootstrap confidence bands in nonparametric curve estimation, *J. Roy. Statist. Soc. B*, **55**, 291–304.

Hall, P., Marron, J. S., and Park, B. U. (1992). Smoothed cross-validation, *Probab. Theory Relat. Fields*, **92**, 1–20.

Härdle, W. (1990). *Applied Nonparametric Regression*, Cambridge University Press, Cambridge, UK.

Härdle, W., and Bowman, A. (1988). Bootstrapping in nonparametric regression: Local adaptive smoothing and confidence bands, *J. Amer. Statist. Assoc.*, **83**, 102–110.

Härdle, W., Huet, S., and Jolivet, E. (1995). Better bootstrap confidence intervals for regression curve estimation, *Statistics*, **26**, 287–306.

Härdle, W., Lütkepohl, H., and Chen, R. (1997). A review of nonparametric time series analysis, *Intern. Statist. Review*, **65**, 49–72.

Härdle, W., and Mammen, E. (1991). Bootstrap methods in nonparametric regression, in Roussas, G. (ed.), *Nonparametric Functional Estimation and Related Topics*, pp. 111–124, Kluwer Academic Publishers, Dordrecht, The Netherlands.

Härdle, W., and Mammen, E. (1993). Testing parametric versus nonparametric regression, *Ann. Statist.*, **21**, 1926–1947.

Härdle, W., Mammen, E., and Müller, M. (1998). Testing parametric versus semiparametric modeling in generalized linear models, *J. Amer. Statist. Assoc.*, **93**, 1461–1474.

Härdle, W., and Marron, J. S. (1990). Semiparametric comparison of regression curves, *Ann. Statist.*, **18**, 63–89.

Härdle, W., and Marron, J. S. (1991). Bootstrap simultaneous error bars for nonparametric regression, *Ann. Statist.*, **19**, 1465–1481.

Hartigan, J. A., and Hartigan, P. M. (1985). The DIP test of unimodality, *Ann. Statist.*, **13**, 70–84.

Janas, D. (1993). *Bootstrap Methods for Time Series*, Shaker, Aachen, Germany.

Jones, M. C., Marron, J. S., and Park, B. (1991). A simple root *n* bandwidth selector, *Ann. Statist.*, **19**, 1919–1932.

Kreiss, J.-P., and Franke, J. (1992). Bootstrapping stationary autoregressive-moving-average models, *J. Time Series Analysis*, **13**, 297–317.

Kreutzberger, E. (1993). Bootstrap für nichtlineare AR(1)-Prozesse, Master Thesis, Universität Kaiserslautern, Kaiserslautern, Germany.

Künsch, H. R. (1989). The jackknife and the bootstrap for general stationary observations, *Ann. Statist.*, **17**, 1217–1241.

LePage, R., and Billard, L. (eds.) (1992). *Exploring the Limits of the Bootstrap*, Wiley, New York.

Liu, R. (1988). Bootstrap procedures under some non i.i.d. models, *Ann. Statist.*, **16**, 1696–1708.

Liu, R., and Singh K. (1992). Efficiency and robustness in resampling, *Ann. Statist.*, **20**, 370–384.

Mammen, E. (1989). Asymptotics with increasing dimension for robust regression with applications to the bootstrap, *Ann. Statist.*, **17**, 382–400.

Mammen, E. (1992a). Bootstrap, wild bootstrap, and asymptotic normality, *Prob. Theory Rel. Fields*, **93**, 439–455.

Mammen, E. (1992b). *When Does the Bootstrap Work? Asymptotic Results and Simulations*, Springer-Verlag, Berlin, Germany.

Mammen, E. (1993). Bootstrap and wild bootstrap for high-dimensional linear models, *Ann. Statist.*, **21**, 255–285.

Mammen, E., Marron, J. S., and Fisher, N. I. (1992). Some asymptotics for multimodality tests based on kernel density estimates, *Prob. Theory Rel. Fields*, **91**, 115–132.

Marron, J. S. (1992). Bootstrap bandwidth selection, in LePage, R., and Billard, L. (eds.), *Exploring the Limits of the Bootstrap*, pp. 249–262, Wiley, New York.

McDonald, J. A. (1982). *Projection Pursuit Regression with the ORION I Workstation: A 20 Minute, 16 mm Color Sound Film*, Stanford University, Stanford, CA.

Müller, D. W., and Sawitzki, G. (1991). Excess mass estimates and tests for multimodality, *J. Amer. Statist. Assoc.*, **86**, 738–746.

Neumann, M. H., and Kreiss, J. P. (1998). Regression-type inference in nonparametric autoregression, *Ann. Statist.*, **26**, 1570–1613.

Neumann, M. H., and Polzehl, J. (1998). *Simultaneous Bootstrap Confidence Bands in Nonparametric Regression*, preprint.

Papavoditis, E., and Politis, D. N. (1999a). The local bootstrap for Markov processes, *J. Statist. Pl. Int.*, to appear.

Papavoditis, E., and Politis, D. N. (1999b). A Markovian local resampling scheme for nonparametric estimators in time series analysis, *Econometric Theory*, to appear.

Papavoditis, E., and Politis, D. N. (1999c). The local bootstrap for kernel estimators under general dependence conditions, *Ann. Inst. Statist. Math.*, to appear.

Politis, D. N., and Romano, J. P. (1994). Large sample confidence regions based on subsamples under minimal assumptions, *Ann. Statist.*, **22**, 2031–2050.

Shao, J., and Tu, T. (1995). *The Jackknife and Bootstrap*, Springer-Verlag, New York.

Silverman, B. W. (1981). Using kernel estimates to investigate multimodality, *J. Roy. Statist. Soc. B*, **43**, 97–99.

Silverman, B. W. (1983). Some properties of a test for multimodality based on kernel density estimates, in Kingman, J. F. C., and Reuter, G. E. H. (eds.), *Probability, Statistics and Analysis*, pp. 248–259, Cambridge University Press, Cambridge, UK.

Silverman, B. W. (1986). *Density Estimation for Statistics and Data Analysis*, Chapman and Hall, London, UK.

Sommerfeld, V. (1997). *Construction of Automatic Confidence Intervals in Nonparametric Heteroscedastic Regression by a Moment Oriented Bootstrap*, Discussion Paper 970022, Sonderforschungsbereich 373, Humboldt Universität, Berlin, Germany.

Stute, W., and González-Manteiga, W. (1996). NN goodness-of-fit tests for linear models, *J. Statist. Plann. Infer.*, **53**, 75–92.

Wand, M. P., and Jones, M. C. (1995). *Kernel Smoothing*, Chapmann and Hall, London, UK.

Wu, C. F. J. (1986). Jackknife, bootstrap, and other resampling methods in regression analysis (with discussion), *Ann. Statist.*, **14**, 1261–1295.

CHAPTER 15

Multidimensional Smoothing and Visualization

David W. Scott
Rice University, Houston, Texas, USA

15.1 INTRODUCTION

Visualization of data and visualization of functionals of data are often treated separately. Exploratory visualization of data is often an end unto itself. Sometimes, exploratory data may not meet the criteria for stochastic modeling; in such cases, no move towards estimation is contemplated, and further functional visualization is inappropriate. Even when exploratory visualization is undertaken as a data preparation step for parametric estimation, multivariate functional visualization is rare. Statistical data, however, points to statistical functions, and the opportunities for functional visualization are growing rapidly. This chapter describes some techniques and provides examples of functional visualization. For simplicity, the discussion will be limited to continuous functions. Whenever possible, data and functional visualization will be presented in parallel. Data visualization without functional visualization, or vice versa, represents an incomplete picture and a missed opportunity.

Parametric modeling in classical statistics relies on tabular display of model coefficients, of error bands, and of p-values, rather than on graphical display. A major component of model-building is the process of reducing the complexity of the statistical model to its most parsimonious form. The very simplicity of parametric models aids in understanding the data, even in high-dimensional settings. Graphical display plays a secondary role, beginning with simple one- and two-dimensional data presentation, proceeding through histograms and scatter plots, and ending with a diagnostic role using residual plots. Parametric models are

Smoothing and Regression: Approaches, Computation, and Application
Edited by Michael G. Schimek
ISBN 0-471-17946-9

preferred for their low variance, but they are susceptible to high bias due to model misspecification.

Nonparametric approaches to density and regression estimation do not have such a parsimonious representation; as a result, graphical display is the most efficient means of understanding a nonparametric estimate. Nonparametric estimates are neither unbiased nor minimum-variance, but they are not as susceptible to high model bias, at least until the curse of dimensionality takes effect. As the number of variables increases, the number of visualization options quickly decreases, and the statistical efficiency of nonparametric methods also decreases rapidly. However, the more common availability of massive data sets and of sophisticated graphical engines represents a clear opportunity for modern computational statistics.

A glance at most statistical journals and textbooks reveals a surprising lack of sophistication in common data and model presentation. Almost all figures are flat. Almost all are either line graphs (including contours) or point graphs. Color is rarely employed; when it is, there is often insufficient impact or importance to support or justify its use. The tendency is to present results of the form $f(x)$ and to include graphical presentation, but to switch to tabular form with additional variables. When a bivariate function, $f(x, y)$, is displayed, contour and perspective presentations are about equally frequent, as these capabilities are now widely available. In many sciences, display of trivariate functions, $f(x, y, z)$, has become common, in as much as the data represent real spatial objects. The more general nature of statistical data and the common use of parametric models have made the use of volumetric visualization relatively rare among statisticians.

Several authors have written generally about statistical graphics. Cleveland (1985) describes the details that make a scatter diagram faithfully convey the structure in the data. Tufte (1983) and Wainer (1997) have written more generally about visualization and presentation. Both focus on charts and maps as much as on function visualization. These books do not include volumetric representation. Many collections do; one is that of Wolff and Yaeger (1993). However, these tend to be computer-graphics efforts at representing "real" things as graphical objects that can be manipulated. Statistical functions, such as a regression plane or a density surface, are not real-world objects; instead, they are mathematical abstractions. However, the same tools apply, and statisticians may benefit from a more hands-on, exploratory view of nonparametric functionals. Some of the newer technologies in virtual reality, among them the data cave (Cook et al., 1997) and the virtual workbench, offer unexplored opportunities for both data exploration and function exploration.

15.2 DATA-POINT VISUALIZATION

Data-point visualization is a well-explored endeavor, whether in one, two, three, or more dimensions. The most common figure by far is the ordinary (x_i, y_i) bivariate scatter diagram. For univariate data, a scatter diagram (commonly

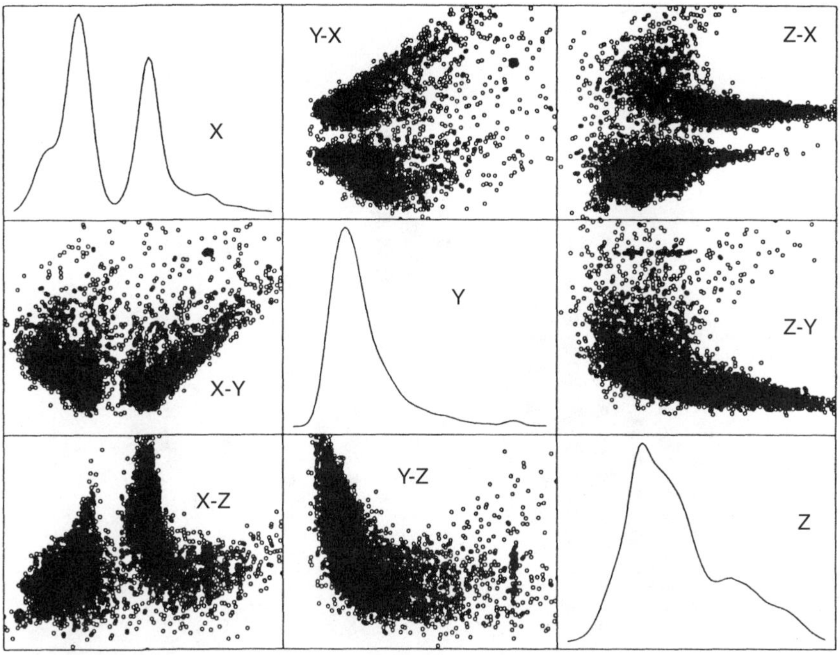

FIGURE 15.1 Scatter-plot matrix of Remote-Sensing Data ($n = 8088$). Smoothed histograms (ASH, see Section 15.4) of the variables are shown in the diagonal frames.

referred to as a dot plot) is seldom presented; rather, a histogram is the more common choice. A hybrid display is Tukey's (1977) stem-and-leaf plot, which displays individual data in such a manner that the frequency of the data in bins is apparent. For higher-dimensional data, these ideas can be assembled for all pairs of variables into a scatter-plot matrix.

In Figure 15.1, we display an enhanced 3×3 scatter-plot matrix of $n = 8397$ trivariate points from a remote-sensing satellite data set, which will be described below. (Actually, only 8088 points are displayed, as 3.7% of the data were truncated because outside the viewing panel.) In the diagonal frames, smoothed histograms of the three variables are displayed. The off-diagonal frames contain the pairwise scatter plots. Each pair of variables appears twice. A first impression is that there are two dominant clusters, with possibly two other smaller bumps at least suggested in the *X* histogram. In fact, these data were extracted from Landsat images (of size 117×196 pixels) of land in agricultural production. Gray-scale and false-color images from these data may be found in Scott (1986, 1992), and the data are available at `ftp.stat.rice.edu`. Of the 22932 pixels, we selected a subset for the crops sunflower, spring wheat, and spring barley, for use in Figure 15.1. We will examine these data further below. The scatter diagrams in Figure 15.1 illustrate a common problem: overplotting when the sample size exceeds one thousand. The outliers are overemphasized in such a scatter plot.

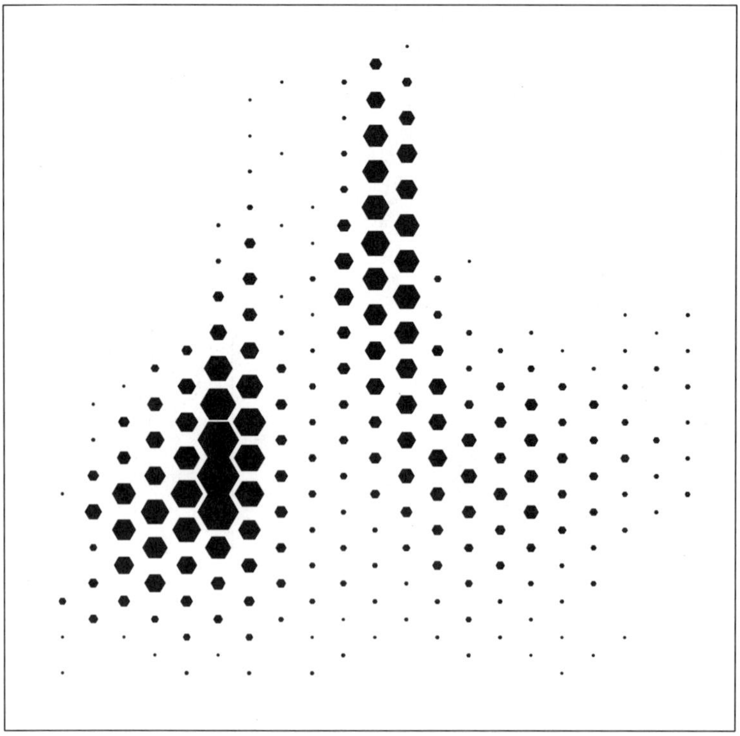

FIGURE 15.2 Hexagonal histogram glyph of the X–Z scatter diagram of the Remote-Sensing Data. The size of each hexagonal glyph is approximately proportional to the number of data points in the hexagonal bin.

Carr et al. (1992) have suggested a hybrid figure called a glyph histogram, which provides better balance between the two views. Figure 15.2 displays a hexagonal histogram glyph of the X-Z scatter diagram from Figure 15.1. This representation largely confirms that there are no hidden clusters in the high-density region of data.

For trivariate data in particular, rotating scatter plots have been popular since the mid-1980's with MacSpin (Donoho et al., 1988), and they are available in most packages today. Animation control is quite sophisticated these days. Alternatively, stereo representation provides a static 3-D view of trivariate data (Carr and Nicholson, 1988). Early and crude red-green stereo techniques and their cardboard glasses have given way to advanced LCD glasses. With virtual reality and data caves, true stereo representation can be merged with animation. High-resolution stereo representation seems to be more effective than rotation for displaying larger data sets. Given the number of individuals with various vision deficiencies, such as color blindness or depth-perception limitations, the newer LCD technologies are almost foolproof. Personal experience in a data cave (Loftin, 1997) suggests that total immersion can be disorienting and tiring; however, the capabilities are absolutely unique and powerful.

For higher-dimensional data, an effective exploration tool is the grand tour (Asimov, 1985). Rather than a selection of projections presented in some unmotivated order, a smooth set of projections is viewed, and in such a manner that eventually all two-dimensional projections are produced. Huber (1985) has provided estimates of how long that might take as the dimension increases, and the increase is rapid. The marriage of the grand tour and an "optimal" projection selection via the grand tour (Friedman and Tukey, 1974) has proven highly successful. Not only can the grand tour provide a set of different starting views for the nonlinear optimization inherent with projection pursuit; it can also provide search directions. A particularly effective implementation of the grand tour and projection pursuit may be found in the larger XGobi program (Swayne et al., 1991). The idea of displaying frequency curves rather than data points has been investigated by Scott (1995), who described the resulting representation as the density grand tour. An implementation in C using `geomview` as the display engine has been described by Salch and Scott (1997). `Geomview` is distributed by The Geometry Center, at `www.geom.umn.edu`. The density grand tour is available at `ftp.stat.rice.edu`.

With exceptions, such as some work by Nason (1995), most projection-pursuit indices are two dimensional. There has been some discussion of the advantage, if any, of viewing three full coordinates in a grand tour and of distinguishing the extra dimension from the movement through the data space. Now, however, with the capability of displaying trivariate density estimates (described below) in the density grand tour, the advantage is clear. The same advantage should be apparent in totally immersive environments, such as the data cave.

Other glyph representations are available for high-dimensional data. Perhaps the best known is the Chernoff face (Chernoff, 1973), in which variables are associated with facial features and determine the size or other attribute of those features. Because each data point is represented individually, sample size is a limiting factor. This representation attempts to display all variables simultaneously; although the assignment of variables to particular features is largely subjective.

A relatively simple device that also displays all variables, but in a more symmetric manner, is the parallel-coordinate plot (Inselberg, 1985; Wegman, 1990). Axes are drawn in parallel, equally spaced. The coordinate values of a data point are marked and connected by line segments. The proposers note a number of interesting dualities between points and lines in this representation. Parallel-coordinate plots cannot handle as many points as a scatter diagram, but Wegman has partially addressed this limitation by plotting the intensity (or density) of ink on the diagram. This will reveal the patterns in these diagrams for much larger data sets.

15.3 FUNCTIONAL VISUALIZATION IN ONE AND TWO VARIABLES

Suppose $f(\cdot)$ represents a generic statistical functional, whether estimated parametrically or nonparametrically, where the dependence on the data $\{x_1, x_2, \ldots, x_n\}$ has been suppressed. High-quality graphical display of $f(x)$ and $f(x, y)$ in postscript is routinely available. In the bivariate setting, as opposed to the univariate setting, the user has a number of choices to make — contour versus perspective representation, number of contour levels versus surface viewing angle, use of contour labels versus choice of wireframe or shaded surface, and so on.

In Figure 15.3, we continue our examination of the Remote-Sensing Data. Extending the idea of a scatter-plot matrix, a contour-plot matrix is displayed. Furthermore, because the actual classification ("ground truth" in remote-sensing jargon), is known for these data, each frame can include three density estimates, one for each crop. The original image contained 3694 pure pixels of sunflowers, 3811 of spring wheat, and 892 of spring barley. When we compare Figures 15.1 and 15.3, the mixing of the three univariate densities into one becomes apparent

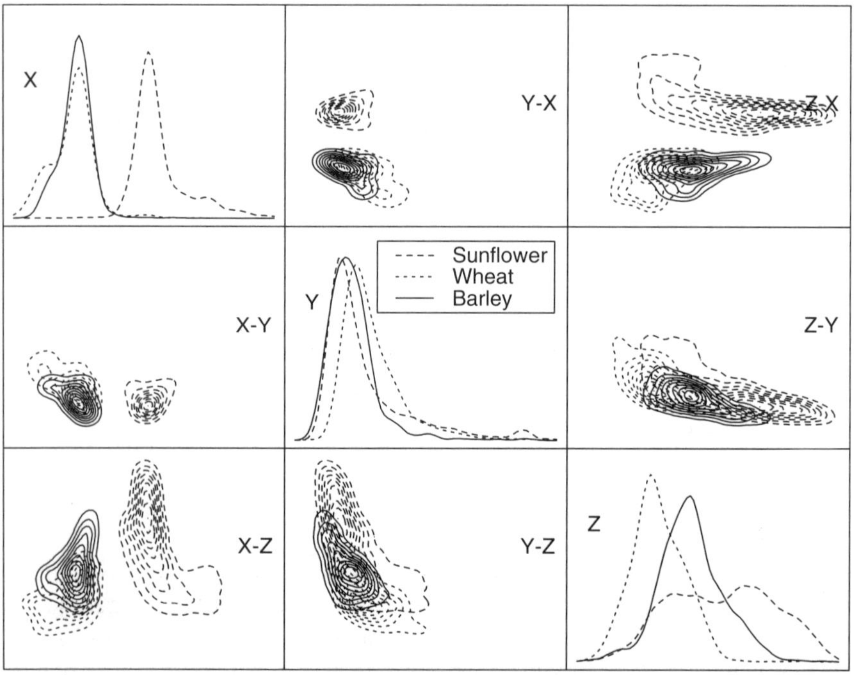

FIGURE 15.3 Contour-plot matrix of Remote-Sensing Data as in Figure 15.1. Contours of bivariate smoothed histograms (ASH., see Section 15.4) for each of three crops (sunflower, spring wheat, and spring barley) are shown for each pair of variables. Smoothed histograms (ASH., see Section 15.4) of the variables for each crop are shown in the diagonal frames.

(diagonal frames). The two clusters observed before — a well-separated sunflower cluster and one spring grain cluster — really represent three classes. Spring wheat and spring barley are not sufficiently separated to manifest two modes.

When we examine the off-diagonal frames, the earlier remark that scatterplots of large data sets tend to overemphasize low-density regions is reinforced. To be more specific, a definition of an individual contour of a multivariable function $f(\mathbf{x})$ for $\mathbf{x} \in \mathbb{R}^d$ is required:

$$C_\alpha = \left\{\mathbf{x} : f(\mathbf{x}) = \alpha \max_{\mathbf{x}} f(\mathbf{x})\right\}, \quad \text{where } 0 \le \alpha \le 1.$$

We refer to C_α as an α-contour. Thus, C_1 gives the location of the (largest) mode, while C_0 would delineate the support of the function f. In Figure 15.3, nine equally spaced contours are displayed for each density estimate; that is, C_{α_k} is drawn in different line types for each of the $\alpha_k = k/10$, for $k = 1, \ldots, 10$.

The α-contour concept may be applied to any (preferably continuous) function, not just to density functions. A regression example is considered below. Contours of a density function are usually closed, a circumstance which aids and facilitates interpretation. In such cases, the equivalent concept of a β-contour, $\mathcal{C}_\beta$, may be considered, where again $0 \le \beta \le 1$. The fraction of the probability density contained in $\mathcal{C}_\beta$ defines the constant β, and $\mathcal{C}_\beta$ is the (unique) contour among C_α satisfying

$$\int_{\mathcal{C}_\beta} f(\mathbf{x})\, d\mathbf{x} = \beta.$$

Here, $\mathcal{C}_0$ represents the largest mode, and $\mathcal{C}_1$ represents the support of f. Note that α and β are negatively correlated. It happens that, in the bivariate setting, most practical experience with contours comes from reading topographical maps, which are of the C_α category. In practice, β is usually estimated for a given α-contour by computing the density at each data point and counting the number of density values greater than or equal to α. If a particular value of β is required, then a simple line search over α may be used.

The choice between a contour plot and a perspective representation of $f(x, y)$ is somewhat subjective. Personal experience suggests that the mathematical notion of equivalence between the two representations does not translate into practice. Indeed, trying to imagine a surface perspective view by studying a contour plot is rarely successful. A perspective plot provides a more complete view of a function. A contour plot provides only a portion, perhaps half, of the extra dimension. On the other hand, part of a perspective plot is likely to be hidden behind a peak, or the height of the function may vary so widely that small features cannot be simultaneously depicted, as they can in a contour plot. Nevertheless, the essential difference in the use of continuity between the contour and the surface representations favors the latter. Although individual contours are continuous, the jump from contour to contour is discontinuous. Furthermore, surfaces that undulate require special care in reading them in contour space, because few contour representations indicate whether nested levels are increasing or decreasing in value.

In Color Plate 15.1, the bivariate X-Z variable density estimates for the three crops are shown; they should be compared to the lower-left frame in Figure 15.3, rotated about 180°. The software program `geomview` was used to generate this display. The density estimates are based on smoothed histograms and have finite support. The greatest of the three densities is displayed at each point. If a nonparametric Bayesian classifier with equal priors was of interest, the highest color displayed would give nonparametric decision regions for classification. `Geomview` supports real-time visualization, including rotation, shading, and transparency, so that just a few manipulations will suffice to eliminate any hidden regions of the surface. The perspective plot more clearly indicates the relative height of the three densities (compared to the contour picture), and provides much greater detail in the tails and in the regions of low density.

15.4 AVERAGED SHIFTED HISTOGRAMS

15.4.1 Univariate Data

Plotting routines require values of the function on a mesh—$\{f_i\}$, $i = 1, \ldots, m_1$ for a single-variable function, and $\{f_{ij}\}$, $i = 1, \ldots, m_1$ and $j = 1, \ldots, m_2$ for a two-variable function. Similarly, $\{f_{ijk}\}$ and $\{f_{ijkl}\}$ are required for three- and four-variable function visualization. Depending on the sophistication of the software, these mesh values may be interpolated by linear functions or even by specialized spline approximations.

When f is a density function, the most natural estimator that provides a grid of values is the classical histogram with equally spaced bins. Generally, the bin widths in a histogram are selected to minimize bias and variance, and not to match the desired graphical resolution of a plotting routine. Optimal bin widths are certain to be much wider than any desired graphical resolution. The *averaged shifted histogram* (ASH) of Scott (1985), however, is well suited to both tasks. The ASH retains much of the computational advantage of the histogram for massive data sets, while computing precisely the array of estimates required for visualization.

The construction of the ASH is easily described in one variable and generalized to d variables. Consider not one but m equally spaced histograms, $\hat{p}_1$, $\hat{p}_2, \ldots,$ $\hat{p}_m$, of a data set. Each histogram employs the same bin width, h, but the meshes are shifted by a fixed amount, h/m, from the preceding mesh. Define δ to be this shift, that is, $\delta = h/m$. The ordinary ASH, $\hat{f}(x)$, is formed by simply averaging the m different shifted histogram levels covering the point x:

$$\hat{f}(x) = \frac{1}{m} \sum_{k=1}^{m} \hat{p}_k(x).$$

Because each histogram, $\hat{p}_k(\cdot)$, is piecewise constant, so is the ASH estimator, $\hat{f}(\cdot)$. The "apparent bin width" of the ASH is, however, not h, but rather h/m, or δ. Thus h may be selected appropriately to minimize statistical error, while m

(and hence δ) are selected to control the mesh size. In one dimension, 100–400 mesh points might be desirable; in three dimensions, however, 100^3 mesh points is beyond the comfortable range for most computers, but 50^3 might be acceptable.

An explicit formula for the ASH can be derived. Focusing on the narrow bin intervals of width δ, let ν_k denote the bin counts. Without loss of generality, consider the value of the ASH over the bin $[0, h)$, which we label bin 0. Then the m histograms covering this interval are given by

$$\begin{aligned}
\hat{p}_m &= (\nu_0 + \nu_1 + \cdots + \nu_{m-1})/nh; \\
\hat{p}_{m-1} &= (\nu_{-1} + \nu_0 + \cdots + \nu_{m-2})/nh; \\
&\ \vdots \\
\hat{p}_1 &= (\nu_{1-m} + \nu_{2-m} + \cdots + \nu_0)/nh.
\end{aligned}$$

Therefore, the ASH is a weighted average of these bin counts, namely,

$$\hat{f}(x) = \frac{\nu_{1-m} + 2\nu_{2-m} + \cdots + m\nu_0 + \cdots + 2\nu_{m-2} + \nu_{m-1}}{mnh},$$

for $0 \le x < \delta$. In general, for x in bin B_ℓ,

$$\hat{f}(x) = \frac{1}{nh} \sum_{k=1-m}^{m-1} \left(1 - \frac{|k|}{m}\right) \nu_{l+k}.$$

The weight function in the ordinary ASH is sampled from the isosceles triangle, which is called a kernel function, K. Here, $K(t) = 1 - |t|$ for $-1 \le t \le 1$. The ASH may be generalized to allow for unequal weighting of the shifted histograms. Let this new set of weights be denoted by $\{w_m(k)\}$; then

$$\hat{f}(x) = \frac{1}{nh} \sum_{k=1-m}^{m-1} w_m(k)\, \nu_{l+k},$$

for $x \in B_l$. Given a kernel, K, defined on the interval $[-1, 1]$, the weights may be obtained by the formula

$$w_m(k) = \frac{mK(k/m)}{\sum_{i=1-m}^{m-1} K(i/m)},$$

for $k = 1 - m, \ldots, m - 1$. Other popular choices for the kernel are the Epanechnikov function, $3/4(1 - t^2)$, and the biweight function, $15/16(1 - t^2)^2$. The normal function, $\exp(-\pi x^2)$, does not have finite support, and so it cannot be used without further approximation.

As the number of shifted histograms grows, the ASH converges to a well-known kernel estimator,

$$\hat{f}(x) = \frac{1}{nh} \sum_{i=1}^{n} K\left(\frac{x - x_i}{h}\right).$$

The ASH may, therefore, be thought of as a statistical approximation to a finite-support kernel estimator. With this understanding, the choice of the smoothing parameter h need not be an integer multiple of δ in the ASH, because the normalization in the definition of the weight function works generally.

Alternatively, a numerical approximation to the kernel estimator may be considered. Let t_k denote the center of bin B_k, with bin counts ν_k as above; then consider rounding all data values to bin centers. The kernel estimate becomes

$$\hat{f}(x) = \frac{1}{nh} \sum_{k=-\infty}^{\infty} \nu_k K\left(\frac{x - t_k}{h}\right).$$

The desired grid of density values, $\{f_j\}$, is obtained by

$$f_j = \hat{f}(t_j) = \frac{1}{nh} \sum_{k=-\infty}^{\infty} \nu_k K\left(\frac{t_j - t_k}{h}\right).$$

Silverman (1982) has shown how the fast Fourier transform may be used to compute these grid values efficiently when the kernel is the normal density. To avoid aliasing problems, a large number of empty bin counts must be appended onto the end of the bin-count vector. In several variables, therefore, if padding doubles the number of bins required in each dimension, considerable extra storage and computation may be required.

15.4.2 Multivariate Data

The extension of the ASH to bivariate and multivariate data is straightforward. In the bivariate setting, a collection of shifted histograms is produced. Suppose the histogram bin is a rectangle of size $h_1 \times h_2$. Then construct $m_1 m_2$ histograms, by performing m_1 equal shifts along the x-axis and m_2 equal shifts along the y-axis, and average; in other words,

$$\hat{f}(x, y) = \frac{1}{m_1 m_2} \sum_{k_1=1}^{m_1} \sum_{k_2=1}^{m_2} \hat{p}_{k_1 k_2}(x, y),$$

for $(x, y) \in B_{k_1 k_2}$. The resulting estimate with general weights can be shown to be a simple product of two univariate weight functions (as defined earlier), applied to the bivariate bin counts:

$$\hat{f}(x, y) = \frac{1}{nh_1 h_2} \sum_{k_1=1-m_1}^{m_1-1} \sum_{k_2=1-m_2}^{m_2-1} w_{m_1}(k_1)\, w_{m_2}(k_2)\, \nu_{l_1+k_1, l_2+k_2},$$

for $(x, y) \in B_{l_1, l_2}$. Further generalization to trivariate data is immediate.

Software to compute these estimates in S–Plus is available at `ftp.stat.rice.edu`. Three separate functions with interfaces to Fortran have been written: `ash1`, `ash2`, and `ashn`. The latter has been written to return a trivariate slice of up to ten dimensions of data. As mentioned above, mesh size grows exponentially; as a result, in core, meshes beyond four or five dimensions are not generally feasible.

15.4.3 WARPing: ASHs for Regression

A multivariate ASH can be used to obtain the corresponding regression function, by formally computing

$$\hat{r}(\mathbf{x}) = \frac{\int_y y\hat{f}(\mathbf{x}, y)\,dy}{\int_y \hat{f}(\mathbf{x}, y)\,dy}.$$

The well-known Nadarya–Watson regression estimator,

$$\hat{r}(x) = \frac{\sum_{i=1}^{n} y_i K((x - x_i)/h)}{\sum_{i=1}^{n} K((x - x_i)/h)},$$

falls out from this expression when $\hat{f}$ is a product kernel estimator.

Whittaker and Scott (1994) and Scott and Whittaker (1996) discuss the derivation of the ASH regression estimator and its application to mapping of and to spatial smoothing of complex survey data. The additional information required is the bin response averages. In the case of bivariate data, these would be the $\overline{y}_i$; they would be of the form $\overline{z}_{ij}$ for trivariate data. The two ASH regression estimates for these two cases are

$$\hat{r}_i = \frac{\sum_a \overline{y}_{i+a} w_a \nu_{i+a}}{\sum_a w_a \nu_{i+a}}$$

$$\hat{r}_{ij} = \frac{\sum_a \sum_b \overline{z}_{i+a,j+b} w_a w_b \nu_{i+a,j+b}}{\sum_a \sum_b w_a w_b \nu_{i+a,j+b}}.$$

For the mapping applications above, the ordinary bin counts were replaced by the sampling-frame weights. Taking account of such complicated features of real data turned out to be a trivial modification in the ASH setting, but for other estimators.

The extension to regression and to *average derivative estimation* (ADE) of the binning technology fundamental to the ASH was described by Härdle and Scott (1992). They coined the acronym WARPing, which stands for *weighted average of rounded points*. Improvements to approximation by using linear binning counts rather than raw bin counts have been described by Hall and Wand (1996).

15.5 FUNCTIONAL VISUALIZATION IN THREE VARIABLES AND BEYOND

We return to our discussion of functional visualization. The ASH we have described is the basis of our color figures. An option in the `ashn` software outputs a full object-oriented file describing the contours, which may be piped into the `geomview` program for visualization. On a high-performance workstation, such as the Silicon Graphics High Impact Indigo 2 Model used to produce these figures, real-time rotation, animation, and contour transparency are supported. `Geomview` does provide postscript output for our figures, but a screen-capture program was used to capture the figures using transparency, because postscript does not support that capability.

15.5.1 Trivariate Functionals

Estimating and visualizing trivariate densities, $f(x, y, z)$, has progressed rapidly since I wrote Scott (1992). Widely available software packages, such as, for example, Mathematica and MATLAB, provide such capabilities. The two options, of contouring and of perspective plotting, available for bivariate densities are, however, reduced to the contour representation in this case. I prefer to use the `ashn` software on account of its richness in control and design parameters (described below).

Three-dimensional contours corresponding to level sets of a four-dimensional surface may not be familiar to everyone. For a trivariate normal density, the contours are "egg-shaped" surfaces (hyperellipses). The shape of one of these hyperellipses (determined by the eigenvalues and eigenvectors of the covariance matrix) does not change as α or β varies, only the size changes. Using transparency techniques, several contours can be visualized simultaneously. Trivariate examples may be found in Scott (1992); a quadrivariate example is given below.

The `ashn` software uses the "marching cubes" algorithm (Lorensen and Cline, 1987). The values at the eight vertices of a cube (or hyper-rectangle) are used to construct a number of triangles, if the contour intersects the cube. This computation can be performed without reference to neighboring information; hence, it is fast. For many applications, further smoothing by Gouraud shading may be applied. We prefer to leave the underlying binning and interpolation information as part of the visualization, because the surfaces represent estimates, not noiseless functions. The details of bivariate and trivariate contouring are interesting in their own right and are covered in Scott (1992, Appendix A.1) and other references.

Returning to the Landsat remote-sensing example: Trivariate ASHs for each of the three crops were computed, using all three variables. A number of α-contours were computed to find the one containing half the estimated probability density. Thus, a β-contour at the level $\beta = 0.5$ was computed for each crop density. These three contours, $C_{0.5}$, are displayed in Color Plate 15.2. The β-contours for wheat and barley are nearly disjoint. The sequence of figures of the Landsat data, beginning with one variable at a time and progressing to all three simultaneously, emphasizes the value of multivariate analysis over univariate analysis. Adding variables to an analysis may well improve its results, by creating more separation among the classes than additional noise. This feature space is not perfect for these classes, but clearly the trivariate feature space will outperform any one- or two-dimensional subspace of features.

The $\beta = 0.5$ contour has special appeal. This object is one candidate for a generalization of the median to more than one dimension. The β-contours may be obtained directly in the ashn software, through option 18. When a value of β is entered, a simple line search is performed over α-contours to find the one with the correct probability mass (to four significant digits).

The image from which the Remote-Sensing Data were extracted is more complex. There are at least six significant crops identified in the ground truth; approximately half of the acre-sized pixels are identified as "mixed" (and represent boundary pixels). The median contour, $C_{0.5}$, is displayed for all six ASHs. The impression of the quality of the feature space is changed radically. It is interesting to see that spring oats are located in the middle of the spring wheat and spring barley portion of the feature space. The peanut contour in the bottom middle — well — really looks appropriate. The lower half of the sunflower contour is hidden by the β-contour of sugar beets. Of special interest is the disjointed nature of the β-contour for soybean crops. Given a random sample of a relatively homogeneous crop such as soybean, the nonconnectedness of the contour is rather unexpected. A standard use of such "training data" is to fit a discriminant function, such as Fisher's linear discriminant function. Implicit in that parametric model is that the covariance matrix for all the groups is the same. Color Plate 15.3 gives a clear impression that not only the *size* of the covariance matrices but also the *orientation* of the major axes of the covariance matrices, varies from class to class. Nonparametric analysis and visualization serve a diagnostic function, and are also a basis for a nonparametric classification algorithm.

15.5.2 Time-series Data

The study of univariate time series leads naturally to multivariate analysis as a means of understanding the autocorrelation structure in the data. Univariate data, x_t, become bivariate data, (x_t, x_{t+1}), and then trivariate data, (x_t, x_{t+1}, x_{t+2}). Nonparametric density-estimation algorithms may be applied to correlated data, although care must be taken with regard to selection of the smoothing parameter. (For a discussion, see Section 15.6).

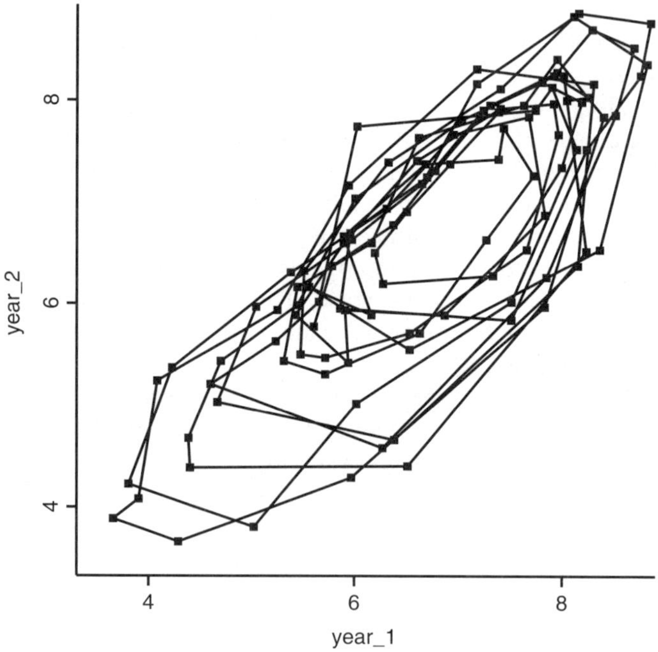

FIGURE 15.4 Lagged values (x_t, x_{t+1}) of the log–lynx time series from 1821 to 1934. Successive points are connected by line segments.

For a short example, consider the annual Canadian Lynx Trapping Data from 1821 to 1934 (Elton and Nicholson, 1942), a data set in S–Plus. A time-series plot (not shown) shows a fairly regular sinusoidal pattern. The lagged data, (x_t, x_{t+1}, x_{t+2}), transformed to a logarithmic scale, were fed into XGobi. A scatterplot of the first two dimensions is shown in Figure 15.4. Without the line segments connecting successive points, a fairly standard normal density seems apparent. However, the flow of data is not random; it proceeds in clockwise fashion, taking 9–12 years to complete a cycle. A number of density estimates were examined informally, and none conformed to the doughnut shape that is apparent from inspection.

Viewing all three dimensions in XGobi (after a rotation) gives the view shown in Figure 15.5. Rotating this point cloud reveals that the data have little variation in the direction orthogonal to the plot. The "hole" in the data also becomes more prominent. Indeed, the ashn view of these data (see Color Plate 15.4) shows the hole quite clearly now. It is instructive to look at the difference in the diameters of this $\alpha = 0.25$ contour at several cross-sections. The wider cross-section at around 10:30 is the result of more (not less) data concentration; that is, the density function gets much higher inside this region. The lowest density cross-section occurs at about 4:30. The orientation of the views in Figure 15.5 and in Color Plate 15.4 are nearly identical, so a comparison is possible. The separate extra little mass at 5:00 can be traced to the presence of four points that

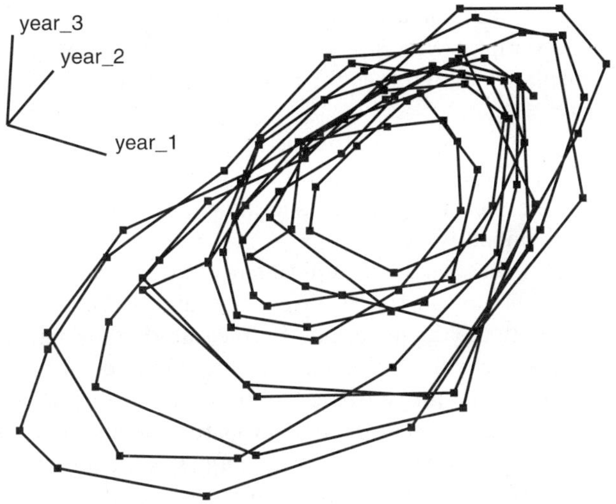

FIGURE 15.5 Lagged values (x_t, x_{t+1}, x_{t+2}) of the log–lynx time series from 1821 to 1934. Successive points are connected by line segments.

are quite close there; their proximity result in a little local peak in the estimate. Notice that the density is not to be taken as zero outside the red surface—rather, the density estimate goes smoothly to zero. There is one strong mode, located in the middle of the mass at around 1:00.

15.5.3 Quadrivariate Functionals and Beyond

Trivariate contouring is a powerful tool, but not radical. Biochemists have used this tool to represent "real" pictures of molecules and bonding forces. Furthermore, trivariate contours live in our natural dimension of space. When we get to four variables and beyond, however, we leave the comfort of our real world and must try to do the best we can at understanding the structure in higher dimensions.

In effect, we earlier proposed going beyond three dimensions, by drawing several trivariate contours at different levels and then making the outer surface transparent, in order to visualize all contours simultaneously. Just as the set of nine bivariate contours took us part way to the full three-dimensional function surface, so does the set of three to five trivariate contours take us partway to the full four-dimensional function surface of $\hat{f}(x, y, z)$. These surfaces must be distinguished, and color provides a simple (discontinuous) method. By using lighting models, we can have gradual shading of a single color over a single contour, while different colors are used to indicate different α or β values.

The only other major cue left is motion (or animation), so now we propose to use that prominently in visualizing the quadrivariate density $\hat{f}(x, y, z, t)$. By varying the fourth variable, t, we obtain a smoothly varying density surface and, hence, smoothly varying density contours as well. The grand tour, which is composed of many discontinuous points, takes advantage of the continuity of the trajectory of each individual point along the way. Earlier slicing algorithms (in MacSpin, for example) were less effective, because points jumped into and then out of the display. Continuity is the most important link for viewing beyond three dimensions.

Therefore, we propose to make a movie of $\hat{f}(x, y, z, t)$, by choosing three to five fixed levels of α for contours. Each frame of the movie will have a slightly different value for the fourth variable, t, and the movie will show contours which are continuous and smoothly deforming.

As a specific example, consider the epicenters of the 500 earthquakes preceding the 1982 eruption of Mount St. Helens (Weaver et al., 1983). These data follow a space-time format, with the depth variable, z, measured in log-units, and with the time variable spanning the five weeks before the eruption. In Color Plate 15.5, a sample of nine out of the thirty frames in the movie is displayed, from left to right and then down. The z-axis or depth is properly oriented. If we first examine the bottom set of frames, three clusters of earthquakes are apparent. Taking into account the logarithmic depth transformation (more details in Scott, 1992), the highest cluster is near the surface, the middle cluster is about 1–2 km deep, and the bottom cluster is approximately 6–8 km deep. Note that the same absolute α levels are maintained across the animation. Thus, it is not until the fourth frame that the red contour level is pierced, and the sixth frame for the green level. The initial three frames show a low but increasing intensity of earthquake activity. The deepest earthquakes appear first, and a close examination of the progression from frame four to the final frame (nine) reveals a surge in activity at the middle and then the top levels, as the deepest activity tails off.

It is important to note that the animation is *not* of the conditional density, $\hat{f}(x, y, z|t)$, but rather of its unnormalized form, $\hat{f}(x, y, z, t = t_k)$. With the normalization implicit in conditioning, the overall level of the function is missing, and the animation is less effective. In particular, the animation of Color Plate 15.5 would completely miss the dramatic conclusion. If the upper-left frame is normalized by conditioning, then it will contain contours at the red and green levels. There is, however, almost no data there, and the estimates would be useless. In fact, if conditioning were applied, the blue contours might reasonably approximate the green level.

With the use of color for the three different contour levels and of animation for the fourth variable, then, we are well along the way to visualizing a five-dimensional function surface. The data cave holds the promise of introducing other senses into the mix, including smell, sound, and touch (changes in air pressure and wind). Each of these latter cues is less precise than the ones already employed, but the goal of higher-dimensional function visualization is not precision but, rather, an overall understanding of the structure and of its position in the full-dimensional space.

At the American Statistical Association's Data Exposition in 1986, I attempted a visualization of a five-variable density, $\hat{f}(x, y, z, u, v)$, in the following manner. A 7×7 grid in the (u, v) space was constructed by looking at the bivariate ASH, $\hat{f}(u, v)$. Then 49 six-dimensional ASH surfaces were estimated using `ashn`, and figures depicting $\hat{f}(x, y, z, u = u_i, v = v_j)$, similar to those in Color Plate 15.5, were placed at the appropriate position in (u, v) space. Actually, 49 photographs were made and then pinned into a 7×7 array. The data had had a chunk removed by David Coleman, and points forming the word EUREKA inserted in its place. The density estimate was able to detect the hole in the data by an unusual nesting of contours (see Scott, 1992). Of course, it could not spell or detect the word inserted in the hole.

For six-variable data, this dual visualization-frame and reference-frame model will still work (especially in a data cave). Appropriate paths in the three-dimensional reference space are not obvious. Without such deliberate paths, one could resort to random paths, much as the grand tour does, and we could then return to our earlier discussion of the density grand tour. Extreme detail cannot, however, be the goal of visualization beyond four to six dimensions.

Projection algorithms, including projection pursuit (see Chapter 16 in this book), may be able to take us to three to five dimensions; they should be explored carefully. In fact, the remote-sensing Landsat example represents such a careful projection, in that case from 24 dimensions to 3. The 24 original dimensions were built up by combining six Landsat IV satellite images taken during the summer of 1977. Each image was four-dimensional, because the early Landsat passive sensors recorded measurements in four spectral bands. An agronomist (Badhwar et al., 1982) spent two years constructing a three-parameter growth model that was applied to the 24-dimensional data pixel by pixel. Statistical techniques, such as principal components and projection pursuit, can often accomplish the same goal much faster, but most scientists resist the use of purely "statistical" techniques in favor of models that incorporate the known science of the experiment. The original data are probably lost, but it would be interesting to see how the results of modern statistical algorithms would compare to the work of one knowledgeable scientist.

15.6 VISUALIZATION OF REGRESSION FUNCTIONS

We will conclude this chapter by briefly indicating how visualization techniques, which so far have been applied primarily to density functions, can be applied equally well to regression functions. For simplicity, we will present this application in the setting of an experimental design in three variables. We will consider a linear and a quadratic response function fitted to data collected over a cube according to some design.

If we restrict our visualization to the cube itself, and do not extrapolate outside the support of the data, we have some interesting results. With a linear model, the best response will, of course, occur in one corner of the cube. The level sets of a linear function are just planes. In Color Plate 15.6, we show three contours of an example of a linear response function. It perhaps serves to remind us of the shapes possible when intersecting a plane with a solid cube (the hexagon always seems the most surprising).

A more interesting model to consider is a fully quadratic fit. The example shown in Color Plate 15.7 turned out to have a definite rather than indefinite covariance, with an optimal setting at the middle of the green contour. In fact, all of the contours are spherical in this parametric model, but (again) the three contours shown were truncated to the cube in which data were measured.

Similar pictures may be constructed using nonparametric regression algorithms, including local polynomial algorithms. Such displays do not offer that certain precision in location that is an important component of regression predictions. Such precision can be obtained by plotting low-dimensional slices of the regression function, in much the same way as was proposed above for higher-dimensional density visualization. At least one program, `coplot` (in the `trellis` package found in S–Plus), presents a conditioning view of a response surface. `trellis` is based on the ideas of Cleveland (1993).

15.7 SUMMARY

We have presented a number of examples and algorithms for visualizing data (and functionals of data) in one to five dimensions. Most nonparametric modeling currently takes place in one or two variables, and it is informative even at those low levels. The capability to attempt to extend that experience easily to the next several dimensions is at hand, because of advances in both software and hardware. Presentation is an important part of science, and we can learn much about effective presentation in several variables by collaborating with our colleagues in the experimental sciences, computer science, and psychology.

All software for generating these examples is available from the author at `scottdw@rice.edu`.

Acknowledgments

This research was supported in part by the National Science Foundation under grant DMS-9306658 and by the Department of Defense under contract MDA 904-95-C-2203.

REFERENCES

Asimov, D. (1985). The grand tour: A tool for viewing multidimensional data, *SIAM J. Sci. Statist. Comp.*, **6**, 128–143.

Badhwar, G. D., Carnes, J. G., and Austin, W. W. (1982). Use of Landsat-derived temporal profiles for corn–soybean feature extraction and classification, *Remote Sensing of Environment*, **12**, 57–79.

Carr, D. W., and Nicholson, W. L. (1986). Explor4: A program for exploring four-dimensional data, in Wegman, E. J., and Depriest, D. J. (eds.), *Statistical Image Processing and Graphics*, pp. 215–250, Marcel Dekker, New York.

Carr, D. B., Olsen, A. R., and White, D. (1992). Hexagon mosaic maps for the display of univariate and bivariate geographical data, *Cartograph. Geograph. Information Systems*, **19**, 228–231.

Chernoff, H. (1973). The use of faces to represent points in k-dimensional space graphically, *J. Amer. Statist. Assoc.*, **68**, 361–368.

Cleveland, W. S. (1985). *The Elements of Graphing Data*, Wadsworth, Monterey, CA.

Cleveland, W. S. (1993). *Visualizing Data*, Hobart Press, Summit, NJ.

Cook, D., Cruz-Neira, C., Kohlmeyer, B. D., Lechner, U., Lewin, N., Elson, L., Olsen, A., Pierson, S., and Symanzik, J. (1997). Exploring environmental data in a highly immersive virtual reality environment, *Inter. J. Environ. Monitoring and Assessment*, **51**, 441–450.

Donoho, A. W., Donoho, D. L., and Gasko, M. (1988). MacSpin: Dynamic graphics on a desktop computer, in Cleveland, W. S., and McGill, M. E. (eds.), *Dynamic Graphics for Statistics*, pp. 331–351, Wadsworth & Brooks/Cole, Pacific Grove, CA.

Elton, C., and Nicholson, M. (1942). The ten-year cycle in numbers of lynx in Canada, *J. Animal Ecology*, **11**, 215–244.

Friedman, J. H., and Tukey, J. W. (1974). A projection pursuit algorithm for exploratory data analysis, *IEEE Transactions in Computers*, **C-23**, 881–890.

Hall, P., and Wand, M. P. (1996). On the accuracy of binned kernel density estimators, *J. Multivariate Analysis*, **56**, 165–184.

Härdle, W., and Scott, D. W. (1992). smoothing in low and high dimensions by weighted averaging using rounded points, *Computational Statistics*, **7**, 97–128.

Huber, P. J. (1985). Projection pursuit (with discussion), *Ann. Statist.*, **13**, 435–525.

Inselberg, A. (1985). The plane with parallel coordinates, *The Visual Computer*, **1**, 69–91.

Loftin, R. B. (1997). *NASA/UH Virtual Environment Technology Laboratory*, brochure, Department of Computer Science, University of Houston, Houston, TX.

Lorensen, W. E., and Cline, H. E. (1987). Marching cubes: A high resolution 3D surface construction algorithm, *Computer Graphics*, **21**, 163–169.

Nason, G. (1995). Three-dimensional projection pursuit, *Applied Statistics*, **44**, 411–430.

Salch, J. D., and Scott, D. W. (1997). Data exploration with the density grand tour. *ASA Statistical Graphics and Computing Newsletter*, **8**, 7–11.

Scott, D. W. (1985). Averaged shifted histograms: Effective nonparametric density estimators in several dimensions, *Ann. Statist.*, **13**, 1024–1040.

Scott, D. W. (1986). Data analysis in three and four dimensions with nonparametric density estimation, in Wegman, E. J., and DePriest, D. (eds.), *Statistical Image Processing and Graphics*, pp. 291–305, Marcel Dekker, New York.

Scott, D. W. (1992). *Multivariate Density Estimation: Theory, Practice, and Visualization*, John Wiley, New York.

Scott, D. W. (1995). Incorporating density estimation into other exploratory tools, *Proceedings of the Section on Statistical Graphics*, ASA, Alexandria, VA, 28–35.

Scott, D. W., and Whittaker, G. (1996). Multivariate Applications of the ASH in Regression, *Communications in Statistics*, **25**, 2521–2530.

Silverman, B. W. (1982). Algorithm AS176: Kernel density estimation using the fast Fourier transform, *Appl. Statist.* **31**, 93–99.

Swayne, D., Cook, D., and Buja, A. (1991). XGobi: Interactive dynamic graphics in the X Window system with a link to S, *Proceedings of the Section on Statistical Graphics*, ASA, Alexandria, VA, 1–8.

Tufte, E. R. (1983). *The Visual Display of Quantitative Information*, Graphics Press, Cheshire, CT.

Tukey, J. W. (1977). *Exploratory Data Analysis*, Addison-Wesley, Reading, MA.

Wainer, H. (1997). *Visual Revelations*, Springer-Verlag, New York.

Weaver, C. S., Zollweg, J. E., and Malone, S. D. (1983). Deep earthquakes beneath Mount St. Helens: Evidence for magmatic gas transport? *Science*, **221**, 1391–1394.

Wegman, E. J. (1990). Hyperdimensional data analysis using parallel coordinates, *J. Amer. Statist. Assoc.*, **85**, 664–675.

Whittaker, G., and Scott, D. W. (1994). Spatial estimation and presentation of regression surfaces in several variables via the averaged shifted histogram, *Computing Science and Statistics*, **26**, 8–17.

Wolff, R. S., and Yaeger, L. (1993). *Visualization of Natural Phenomena*, Springer-Verlag, New York.

CHAPTER 16

Projection Pursuit Regression

Sigbert Klinke
Humboldt-Universität zu Berlin, Berlin, Germany

Janet Grassmann
SONEM, Würzburg, Germany

16.1 INTRODUCTION

In statistics, many methods are available to model a response function

$$E(Y|X=x) = m(x_1, \ldots, x_p) + \epsilon,$$

where $X \in \mathbb{R}^p$ and $Y \in \mathbb{R}$ are a pair of random variables, m a p-dimensional regression function, and ϵ some error. These methods range from parametric approaches such as linear regression to some of the non- and semiparametric approaches described in previous chapters. All of them can be characterized by specific advantages and disadvantages. In a parametric model, the structure is completely fixed. In terms of "interpretation" and "reporting of results," this is certainly a strong point. In case the parametric model does not match the underlying true model, we estimate the "best" approximation in the space of the chosen function class. However, this approximation can be quite far from the true model, and the conclusion from the estimation may be wrong.

Data-driven nonparametric models are indeed highly flexible, but at the cost of prespecification of one or more parameters, known as smoothing parameters or bandwidths. These parameters have to be determined from the data, mostly in computer-intensive routines, before estimation via the model can be carried out. The next problem arises from the multivariate space itself: The so-called *curse of dimensionality* describes the property of data to become sparse the

Smoothing and Regression: Approaches, Computation, and Application
Edited by Michael G. Schimek
ISBN 0-471-17946-9

more the dimension of the space grows. For a fixed number n of observations and a volume $V(p) \approx \mathcal{D}^p$, depending on the diameter $\mathcal{D}$, it follows that the average distance $\mathcal{D}_p$ between two data points is

$$\lim_{p\to\infty} \mathcal{D}_p = \lim_{p\to\infty} \sqrt[p]{\frac{V(p)}{n}} \approx \mathcal{D}.$$

How severe this problem is, can be seen from an estimation point of view in Silverman (1986, p. 94, Table 4.2) and from a visualization point of view in Asimov (1985, Table 1). Last, but not least, the interpretation of the fit of a nonparametric model is far from trivial. The local averaging property of a smoother does not provide a general structure of the data. The inclusion of a parametric part in a nonparametric model, which makes it semiparametric, can bring about more structure in the model and, hence, better interpretability.

In this chapter, we assume that the explanatory variable X is of high dimension. As a result, the application of the previously discussed non- and semiparametric models cannot be that promising. Now we have to attack the dimensionality problem in a more direct manner, compared with how other regression methods do so. For that purpose, we introduce an exploratory methodology, named *projection pursuit regression* (PPR). This approach applies an additive model to projected variables. The high-dimensional X is projected in M adequately chosen directions. PPR is nonparametric because of the nonparametric estimation techniques applied to find these directions. The idea of "projection pursuit" goes back to Kruskal (1969, 1972). The term *projection* implies that we are looking at projected data, and the term *pursuit* means that we must find a "good" projection for the purpose of regression.

PPR has been successfully implemented for exploratory purposes by many authors (Friedman and Tukey, 1974; Jee, 1985; Huber, 1985; Jones and Sibson, 1987; Friedman, 1987; Hall, 1989a; Cook et al., 1993; Posse, 1995; Nason, 1995). The idea has been applied to location and symmetry estimation (Blough, 1989; Maller, 1989), density estimation (Friedman et al., 1984; Chen, 1987; Rejtö and Walter, 1990, 1992), classification (Friedman and Stuetzle, 1981a; Portier et al., 1987; Flick et al., 1990), and discriminant analysis (Posse, 1992; Ahn and Rhee, 1992; Polzehl, 1995). Good overviews of projection pursuit are Huber (1985) and Jones and Sibson (1987).

We begin by introducing the basic algorithm. Then we discuss the problems associated with the quality of the approximation, the number of terms in the regression function, and the interpretation of the function. We then give convergence rates and numerous modifications of standard PPR. Finally, we consider an interesting relationship between PPR and *feed-forward neural networks* (FNNs). (For a general overview of FNNs, see Bishop (1995) and Ripley (1996).) We conclude by describing the implementation of PPR in S-Plus and XploRe illustrated on the well-known Boston Housing Data set.

16.2 THE BASIC PPR ALGORITHM

A pure nonparametric approach can lead to strong oversmoothing, since the sparseness of the space requires a lot of space and many observations for a reliable estimate when local averaging is adopted. To estimate the response function $m(x)$ from the data $\{(x_{1,1}, \ldots, x_{1,M}, y_1), \ldots, (x_{n,1}, \ldots, x_{n,M}, y_n)\}$, Friedman and Stuetzle (1981b) have suggested the following algorithm:

1. *Set* $r_i^{[0]} = y_i$.
2. *For* $j = 1, \ldots$ *maximize*

$$R_{[j]}^2 = 1 - \frac{\sum_{i=1}^{n} \left(r_i^{[j-1]} - \hat{m}_{[j]}(\hat{\alpha}_{[j]}^T x_i) \right)^2}{\sum_{i=1}^{n} \left(r_i^{[j-1]} \right)^2}$$

   by varying over the parameters $\hat{\alpha}_{[j]} \in \mathbb{R}^p (\|\alpha_{[j]}\| = 1)$ *and a univariate regression function* $\hat{m}_{[j]}$.
3. *Define* $r_i^{[j]} = r_i^{[j-1]} - \hat{m}_{[j]}(\hat{\alpha}_{[j]}^T x_i)$ *and repeat step 2 until* $R_{[j]}^2$ *becomes small. A small* $R_{[j]}^2$ *implies that* $\hat{m}_{[j]}(\hat{\alpha}_{[j]}^T x_i)$ *is approximately the zero function and that we can not find another useful direction.*

This algorithm leads to an estimation of the response function by

$$\hat{m}_M(x) = \sum_{j=1}^{M} \hat{m}_{[j]}(\hat{\alpha}_{[j]}^T x). \tag{16.1}$$

The advantages of estimating via this response function are as follows:

(i) We can use univariate regression functions instead of their multivariate analogs, hence circumventing the curse of dimensionality.
(ii) Univariate regressions are simple and fast to calculate.
(iii) In contrast to *generalized additive models* (GAMs), PPR is able to approximate a much richer class of functions.
(iv) In comparison to local averaging methods, such as the k-nearest neighborhood estimator, we are able to ignore variables that contain little or no information about m.

Of course, there are also some disadvantages:

(a) We have to examine an M-dimensional parameter space to estimate $\hat{\alpha}_{[j]}$.
(b) We have to solve the problem of selecting a smoothing parameter if we use nonparametric smoothers for $\hat{m}_{[j]}$.
(c) The interpretation of a single term may not be easy.

For simplicity, let us assume that $E(X_i) = 0$, $\mathrm{Var}(X_i) = 1$, $E(Y) = 0$, and $\mathrm{Var}(Y) = 1$. Friedman and Stuetzle (1981b) constructed a special smoother for estimating the unknown regression function $\hat{m}_M$, similar to the supersmoother (Friedman, 1984b). Moreover, they suggested the use of backfitting to improve the quality of the estimate. In particular, we have

$$E\left(Y - \sum_{\substack{j=1\\ j\neq k}}^{M} m_{[j]}(\alpha_{[j]}^T X) \,\middle|\, \alpha_{[k]}^T X\right) = m_{[k]}(\alpha_{[k]}^T X).$$

Now we cycle through $k = 1, \ldots, M, 1, \ldots, M, \ldots$ and update either $\alpha_{[k]}$ or $m_{[k]}$ or both of them by

$$\hat{m}_{[k]}(\hat{\alpha}_{[k]}^T x_i) = y_i - \sum_{\substack{j=1\\ j\neq k}}^{M} \hat{m}_{[j]}(\hat{\alpha}_{[j]}^T x_i).$$

16.3 QUALITY OF APPROXIMATION

Till now, we have described how to perform PPR. But which kind of function can be picked up by PPR and which not? To answer this question, Donoho and Johnstone (1989) examined the approximation of bivariate polynomials by PPR (in comparison to kernel-based methods), studying the complex function $f(x, y) = \Re(z^8) = x^8 - 28x^6y^2 + 70x^4y^4 - 28x^2y^6 + y^8$ evaluated on $[-1, 1] \times [-1, 1]$ ($\Re$ denotes the real part).

Figure 16.1 shows the general problem of estimating a multivariate function: The corners contain the structure. We require either a lot of data points over the whole area or data points in the "right" place. In practice, both situations are rare. Moreover, Donoho and Johnstone (1989) showed that radial functions (e.g., $f(x, y) = g(\sqrt{x^2 + y^2})$) can be better approximated by PPR than can functions for which the oscillation is averaged out locally.

To overcome the last problem, we may use a local PPR algorithm. The idea is to find a point $\hat{\mu}_{[j]}$ and use a neighborhood around the point to estimate a local regression function $\hat{m}_{[j]}(x) = \hat{m}_{[j]}(\hat{\mu}_{[j]} + \hat{\alpha}_{[j]}^T x)$. For a unique location estimate, we may choose $|\hat{\mu}_{[j]}|$ to be minimal. Since $\hat{\mu}_{[j]} + \hat{\alpha}_{[j]}^T x$ defines a line in $\mathbb{R}^M$, we can compute the distances from the line for each observation. Distance can then be used to define the weight of an observation for the regression.

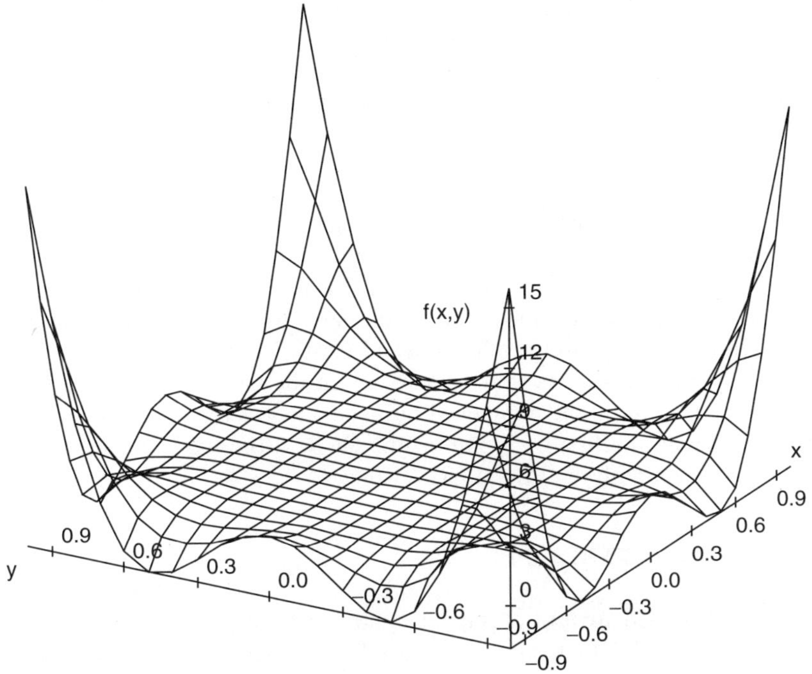

FIGURE 16.1 Real part of the complex polynomial z^8 plotted in $[-1, 1] \times [-1, 1]$.

This approach has some serious drawbacks: The interpretation of the location parameters $\hat{\mu}_{[j]}$ and the projection vectors $\hat{\alpha}_{[j]}$ is difficult. Also, optimization involves a search in a much larger space than $\mathbb{R}^{M-1}$.

Another way to handle the problem might be to use a different method for decomposing the space orthogonal to $\hat{\alpha}_{[j]}$. Let us assume that the model is

$$m(x_1, x_2) = \sum_{i=1}^{M} G_{[j]}(f_{[j,1]}(x_1) + f_{[j,2]}(x_2)),$$

where $G_{[j]}$ is a nonparametric estimate and $f_{[j,\bullet]}$ is a parametric transformation of the x-coordinates (e.g., polynomials of fixed degree). As in *generalized partial linear models* (GPLM), we can choose a metric other than Euclidean.

16.4 NUMBER OF TERMS TO CHOOSE

The previous section made clear that the number of terms we have to choose depends strongly on the underlying regression function. In the case of $\Re(z^8)$, we needed a lot of projections (at least eight) to model the regression function appropriately.

A method similar to choosing the number of components in principal-component analysis can be applied in PPR, too. Computing $\beta_{[j]} = \sqrt{\text{Var}(r_{[j]})}$, we get

$$\text{Var}(r_{[j]}/\beta_{[j]}) = \text{Var}(r_{[j]})/\beta^2_{[j]} = 1..$$

We can construct a plot of $(j, \beta_{[j]})$ and stop if the decrease in the $\beta_{[j]}$ becomes too small.

Other criteria (e.g., standard model selection criteria such as *generalized cross-validation* (GCV)) have been used in Hwang et al. (1994) and Roosen and Hastie (1994). Friedman (1984a) has remarked that adding more terms than necessary and using a backward stepwise model selection helps to avoid getting stuck in local maxima. In the spline setting of Roosen and Hastie (1994), the authors suggested adding terms until two consecutive increases in the GCV are noticed. Then one should prune back until the number of terms is one less than the number of terms in the model with the first minimum of the GCV. From the sequence of models, that one is chosen with the lowest GCV value.

16.5 INTERPRETABLE PPR

Morton (1989) tried to find a general approach to improve the interpretability of a PPR model. The first question to ask is, what are interpretable models? A simple answer is to choose α_j equal to one of the standard unit vectors which allows a quite easy interpretation. But then we are fitting an additive model to the data. Suppose that the number of terms M is fixed. Then we have projection vectors $\hat{\alpha}_{[1]}, \ldots, \hat{\alpha}_{[M]}$. A goal could be to obtain as many zeros as possible in each vector $\hat{\alpha}_{[j]} = (\hat{\alpha}_{[j]1}, \ldots, \hat{\alpha}_{[j]p})$ such that only a few variables contribute to the current fit. The vector

$$\eta_i = \sum_{j=1}^{M} \hat{\alpha}^2_{[j]i}$$

contains the sum of each term's relative contribution to the ith variable ($i = 1, \ldots, p$). Now, let us define an index

$$D(\eta) = \frac{p}{p-1} \sum_{i=1}^{p} \left(\frac{\eta_i}{M} - \frac{1}{p} \right)^2$$

that measures the individual interpretability of each combination $\hat{\alpha}_{[1]}, \ldots, \hat{\alpha}_{[M]}$. A large value of D means that the different variables contribute to different projection vectors.

The second question is, can we use this index to choose M in practice? Morton (1989) gave the following example:

$$\hat{m}_M(x_1, x_2) = \begin{cases} \hat{m}_{[1]} \left(\dfrac{x_1 + x_2}{\sqrt{2}} \right) & M = 1 \\ \hat{m}_{[1]}(x_1) + \hat{m}_{[2]}(x_2) & M = 2 \end{cases}$$

Which model should be chosen? She constructed three different indices based on D, but none of them satisfied her completely. A solution would be to return a sequence of models and let the user with special knowledge about the underlying topic decide on a model. Some practical experience shows that the interpretability drops down when we improve the approximation. One possibility might be to use the index as a penalization during the search for $\hat{\alpha}_{[j]}$.

16.6 CONVERGENCE RATES

Hall (1988) stated that the convergence rate, the bias, and the variance are affected by two issues: the estimation of α_j and the estimation of m_j. He showed that the estimation of α_j can be done in order $n^{-1/2}$, whereas the nonparametric estimation of m_j has an order worse than $n^{-1/2}$. Since the standard algorithm cannot achieve an order $n^{-1/2}$, he proposed a two-stage estimator:

(1) Use an undersmoothed regression estimator to find $\hat{\alpha}_{[j]}$.
(2) Apply the obtained $\hat{\alpha}_{[j]}$ in the regression smoother with a correct amount of smoothing.

Hall (1988) considered the case of projection pursuit density estimation, whereas Hall (1989b) worked on the PPR case. He proved that

$$\sqrt{nh}(\hat{m}_{[j]}(\hat{\alpha}_{[j]}^T x) - m_j(\alpha_j^T x)) = \tau(x)N(0, 1) + \sqrt{nh}h^r(\beta_1 + \beta_2 + \beta_3) + o_p(1),$$

in which $r > 1$ is the number of smooth derivatives of $m_{[j]}$. Thus, the bias is determined by the parameters β_1, β_2, and β_3, where β_1 and β_2 are the contribution to the bias resulting from maximizing the direction and β_3 is the contribution to the bias from the nonparametric estimator (e.g., the Nadaraya–Watson estimator). If we choose the bandwidth (window) proportional to $n^{-1/3}$ to $n^{-1/4}$, then we achieve a parametric rate and get

$$\sqrt{nh}(\hat{m}_{[j]}(\hat{\alpha}_{[j]}^T x) - m_{[j]}(\alpha_{[j]}^T x)) = \tau(x)N(0, 1) + \sqrt{nh}h^2\beta_3 + o_p(1)$$

under the assumption that the smoothing parameter $h \approx n^{-1/5}$. Of course, an undersmoothed estimate causes a wiggly error function with a lot of local minima.

Huber (1985) established weak L_2-convergence for PPR. Jones (1992) proved the norm convergence and established an $O(1/\sqrt{M})$ nonsampling convergence rate. He used a modification of the PPR algorithm that he called relaxed PPR. In this model

$$\hat{m}_M(x) = (1 - \lambda_{[M]})\hat{m}_{M-1}(x) - \lambda_{[M]}\hat{m}_{[M]}(\hat{\alpha}_{[M]}^T x)$$

is chosen such that $0 < \lambda_{[M]} < 1$ and $\hat{\alpha}_{[M+1]}$ minimizes the L_2-norm of

$$e_M = |\hat{m}_M(x) - m(x)|.$$

Under sufficient regularity conditions for m and the assumption that the L_2-norm of $|m_{[M]}| \leq C$ and $\overline{C} = C + |m|$ the error can be estimated by

$$e_M^2 \leq \frac{\overline{C}^2}{M + \overline{C}^2/e_0^2}.$$

e_0 and $\overline{C}$ can be estimated from the data if the d-dimensional Fourier transform of $m(x)$ exists and has a finite $L_1(\mathbb{R}^M)$ norm. Obviously, the dimension of the search space will be increased by one to M. This finding also applies to FFN with one hidden layer or sigmoidal activation functions.

Retjö and Walter (1992) generalized PPR to Hilbert spaces. Assuming that in each step the regression function is chosen to be

$$|m_{[j]}(\alpha_{[j]}^T x)| \geq \rho \sup_{\alpha} |m_{[j]}(\alpha^T x)|,$$

where $0 < \rho < 1$, the authors were able to prove strong convergence for this algorithm.

16.7 MODIFICATIONS

16.7.1 Other Smoothing Methods

Other smoothing methods are available as well. Linear (order $k = 1$) and polynomial (order $k > 1$) regression would lead to an approximation of the unknown regression surface by polynomials of order k given by

$$\hat{m}_M(x) = \sum_{j_1,\ldots,j_p} c_{j_1\ldots j_p} x_1^{j_1} \ldots x_p^{j_p}$$

and used by Hwang et al. (1994). In the study of Roosen and Hastie (1994), three different smoothers (smoothing spline, supersmoother, and polynomial) are compared with known functions m in terms of the *fraction of variance* that is *unexplained* FVU,

$$\mathrm{FVU}_M = \frac{\sum_{i=1}^{n} (\hat{m}_M(x_i) - m(x_i))^2}{\sum_{i=1}^{n} (m(x_i) - \overline{m}(x_i))^2}$$

where $\overline{m}(x_i) = 1/n \sum_{i=1}^{n} m(x_i)$. These researchers employed the same bivariate functions as did Hwang et al. (1994), namely, the following:

- Simple interaction function:

$$m^{(1)}(x, y) = 10.391((x - 0.4)(y - 0.6) + 0.36).$$

- Radial function:

$$m^{(2)}(x, y) = 24.234(r^2(0.75 - r^2)),\ r^2 = (x - 0.5)^2 + (y - 0.5)^2.$$

- Harmonic function:

$$m^{(3)}(x, y) = 42.659\left(\frac{2 + x}{20 + \Re(z^5)}\right),\ z = x + iy - 0.5(1 + i).$$

- Additive function:

$$m^{(4)}(x, y) = 1.3356(1.5(1 - x) + \exp(2x - 1)\sin(3\pi(x - 0.6)^2) + \exp(3(y - 0.5))\sin(4\pi(y - 0.9)^2).$$

- Complicated interaction function:

$$m^{(5)}(x, y) = 1.9(1.35 + \exp(x)\sin(13(x - 0.6)^2)\exp(y)\sin(7y)).$$

With noiseless data, the polynomials perform best for the simple interaction function, for the harmonic function, and for the complicated interaction function; the smoothing splines perform well for the complicated interaction function and slightly better for the radial function. With noisy data, the smoothing splines perform better on all functions except the harmonic.

Roosen and Hastie (1994) used a modified GCV criterion to determine the amount of smoothing in their smoothing spline approach. Let

$$r_{ij} = y_i - \sum_{\substack{k=1 \\ k \neq j}}^{M} \hat{m}_{[k]}(\hat{\alpha}_{[k]}^T x_i)$$

be the partial residual and

$$d_j(\lambda) = Mp + \sum_{j=1}^{M} \text{tr } (S_{[j]}(\lambda)),$$

where λ is the smoothing parameter and $S_{[j]}(\lambda)$ is the smoother matrix. Now define

$$GCV_j(\lambda) = \frac{\sum_{i=1}^{n}(r_{ij} - \hat{m}_{[j]}(\hat{\alpha}_{[j]}^T x_i))^2/n}{(1 - d_j(\lambda)/n)^2}$$

and find the "optimal" directions and smoothing parameters by cycling through $j = 1, \ldots, M, 1, \ldots, M, \ldots$. Hastie and Tibshirani (1990) stated that the GCV tends to undersmooth. Because we fit residuals in each step of the PPR, the autocorrelation may be interpreted as structure and may lead to substantial overfitting. To overcome this problem, they counted the number of local maxima. If the number was above some threshold, then $d_j(\lambda)$ was replaced by $4 + Mp$ to remove the structure that confuses the GCV criterion.

16.7.2 Optimization Criteria

As a modification of the basic PPR algorithm one can apply another optimization criterion. In fact, $R^2_{[j]}$ can be viewed as the well-known R^2 criterion used in linear regression. From linear regression, we also know other criteria — for example, the minimization of the *sum of the absolute deviation* (SAD),

$$\mathrm{SAD}_{[j]} = \sum_{i=1}^{n} |r^{[j-1]} - \hat{m}_{[j]}(\hat{\alpha}^T_{[j]} x_i)|,$$

or the *median absolute deviation* (MAD),

$$\mathrm{MAD}_{[j]} = \text{median}\ (|r^{[j-1]} - \hat{m}_{[j]}(\hat{\alpha}^T_{[j]} x_i)|),$$

for robust estimation. Obviously, all these criteria can be reformulated for weighted observations.

For a large number of variables, as in speech and image recognition (which typically require several thousand variables), Intrator (1990), suggested adding not only a smoothness criterion, but also a projection pursuit index from exploratory projection pursuit to the error function. So, apparently, one can decouple the search for an interesting projection from the smoothing part of the optimization.

16.7.3 PPR for Moderate Nonlinearities

In practice, the data we work with often do not have strong nonlinearities. Thus, Aldrin et al. (1993) suggested the use of the *ordinary least-squares* (OLS) estimator to estimate both the direction and the function nonparametrically. Under the condition that the regressors have an elliptically contoured distribution and the function is approximately monotonic, it follows that

$$\alpha_{[j],OLS} = E(\alpha^T_{[j]} X m_{[j]}(\alpha^T_{[j]} X))\alpha_{[j]}. \tag{16.2}$$

A more complicated method can be used if the expectation in (16.2) is near zero (e.g., if $m_{[j]}$ is a symmetric quadratic function). In that case, we have

$$\begin{aligned}
\alpha_{[j],OLS} &= \lambda \alpha_{[j]} + B, \\
\lambda &= E(\alpha^T_{[j]} X m_{[j]}(\alpha^T_{[j]} X)) \\
B &= \Sigma_x^{-1} E(m_{[j]}(\alpha^T_{[j]} X)(X - \left(\Sigma_x^{-1}\alpha_{[j]}\right) \alpha^T_{[j]} X),
\end{aligned}$$

in which Σ_x is the covariance matrix of the regressors. A sample version is obtained from

$$\hat{\alpha}_{[j],OLS} = \hat{\lambda}\hat{\alpha}_{[j]} + \hat{B} + \hat{\Sigma}_x^{-1}\hat{\Sigma}_{x\epsilon},$$

$$\hat{\lambda} = \frac{1}{n}\sum_{i=1}^{n} \hat{\alpha}_{[j]}^T X \hat{m}_{[j]}(\hat{\alpha}_{[j]}^T X))$$

$$\hat{B} = \hat{\Sigma}_x^{-1}\frac{1}{n}\sum_{i=1}^{n} \hat{m}_{[j]}(\hat{\alpha}_{[j]}^T x_i)(x_i - (\hat{\Sigma}_x^{-1}\hat{\alpha}_{[j]})\hat{\alpha}_{[j]}^T x_i),$$

with $\hat{\Sigma}_x$ the sample covariance matrix and $\hat{\Sigma}_{x\varepsilon}$ the sample cross-covariance matrix of x and the residuals. The iterative algorithm for finding a direction will then be:

1. *Estimate $\hat{\alpha}_{[j,0]}$ via OLS and compute a smooth $\hat{m}_{[j,0]}$.*
2. *Compute*

$$\hat{\alpha}_{[j,k+1]} = \frac{\hat{\alpha}_{[j,k]} - \hat{B}}{\hat{\lambda}}$$

and standardize by $\hat{\alpha}_{[j,k+1]}^T \hat{\Sigma}_x \hat{\alpha}_{[j,k+1]} = 1$.

3. *Smooth again and obtain a new $\hat{m}_{[j,k+1]}$.*
4. *Repeat steps 2 and 3 until*

$$D_k = \left| \frac{\hat{\alpha}_{[j,k]}}{\sqrt{\hat{\alpha}_{[j,k]}\Sigma_x\hat{\alpha}_{[j,k]}}} - \frac{\hat{\alpha}_{[j,k+1]}}{\sqrt{\hat{\alpha}_{[j,k+1]}\Sigma_x\hat{\alpha}_{[j,k+1]}}} \right|$$

becomes arbitrarily small.

Aldrin et al. (1993) applied this technique to the Boston Housing Data (see later).

16.7.4 PPR and Sliced Inverse Regression

Li (1991) proposed a new technique for reducing dimensions and for regression analysis called *sliced inverse regression* SIR (see Chapter 17 in this volume). The technique can be used to determine an interesting direction for PPR. Assume that the dependent variable Y will be cut into $S = n^q (0.5 \le q < 1)$ non-overlapping slices, so that we can define

$$h(y) = \begin{cases} c_k & c + k(d-c)/S \le y < c + (k+1)(d-c)/S \quad (k = 0, \ldots, S-1), \\ 0 & \text{otherwise} \end{cases}$$

where $c < \min(y_i)$ and $d > \max(y_i)$ are constants and $c_k = \text{mean}(y_i)$ in the slice $k(y_i)$. Choose a projection vector $\hat{\alpha}_{[j,0]}$, and minimize the c_ks by

$$\frac{\sum_{i=1}^{n}(h_{[0]}(y_i) - \hat{\alpha}_{[j,0]}^T X_i)^2}{n \operatorname{Var}(h_{[0]}(y_i))} \tag{16.3}$$

with the constraint $\mathrm{Var}(h_{[l]}(y_i)) > v$ (v fixed) to avoid fitting a constant transformation h. The next step is a linear regression to compute a projection vector $\hat{\alpha}_{[j,l]}$. Then we again compute a function $h_{[l]}$, a projection vector $\hat{\alpha}_{[j,l+1]}$, and so on.

Minimizing over c_k is equivalent to solving a set of S quadratic equations in $c_0, \ldots, c_{S-1}$ that can be solved by iterative techniques. Note that this method has the same problems as SIR. For example, it will not provide a good $\hat{\alpha}_{[j]}$ if the data are clustered. In fact, (16.3) is an approximation to $1 - R^2$, where R^2 is the coefficient of determination known from ordinary regression and the minimization is equivalent to finding the best linear predictor and a nonlinear transformation function h. Zhu and Li (1991) showed that this method leads to a consistent estimate for $1 - R^2$.

In the case of two slices ($S = 2$), this method may provide a quick way to find a reasonable starting vector $\hat{\alpha}_{[j]}$, rather than starting with the linear regression projection.

16.7.5 Projection Pursuit-like Regression

Although PPR is designed to overcome the problem of the curse of dimensionality, Donoho and Johnstone (1989) showed that the bias of (16.1) depends on the dimension of X if the regression function is harmonic. Chen (1991) proposed a modification of the PPR model that is not as flexible as PPR, but whose rate of convergence is not affected by the dimensionality of X.

For a given $M < \dim(X)$ and a given set of projection vectors $\alpha_1, \ldots, \alpha_M$, we can solve the (constrained) minimization problem

$$\min \sum_{i=1}^{n} (y_i - \hat{m}(x_i))^2 \mathrm{I}_U(x_i), \tag{16.4}$$

where U is an open set that contains the unit ball C and $\mathrm{I}_U(x)$ is the indicator function. To avoid several fits along the same direction, we choose a constant $0 < A \leq \pi/2$, so that the minimal angle between α_i and span $(\alpha_1, \ldots, \alpha_{i-1}, \ \alpha_{i+1}, \ldots, \alpha_M)$ is larger than A. A choice of $A = \pi/2$ would lead to an orthogonal system of projection vectors α_i.

If we use polynomial splines of degree q to approximate m_j, then the minimization problem has a unique solution, and we can apply model selection criteria such as final prediction error

$$\mathrm{FPE} = \frac{n_C + \mathrm{tr}(P_n I_C)}{n_C - \mathrm{tr}\ (P_n I_C)} \frac{\mathrm{RSS}}{n_C}$$

or

$$\mathrm{GCV} = \left(\frac{n_C}{n_C - \mathrm{tr}(P_n I_C)} \right)^2 \frac{\mathrm{RSS}}{n_C}$$

where

$$\mathrm{RSS} = \sum_{i=1}^{n} (y_i - \hat{m}(x_i))^2 I_C(x_i)$$

is the residual sum of squares, n_C is the number of observations, P_n is the smoother matrix that minimizes (16.4) ($y = P_n x$, $y = (y_1, \ldots, y_n)$, $x = (x_1, \ldots, x_n)$), and I_C is a diagonal matrix with a one in the diagonal if $x_i \in C$ and a zero otherwise.

16.7.6 Adjoint PPR

Duan (1990) proposed an algorithm, called "adjoint" PPR, that is related to SIR. In linear regression, the least-squares solution $L(Y|\alpha^T X)$ solves the equation

$$\text{Cov}(X, Y - L(Y|\alpha^T X)) = 0. \tag{16.5}$$

In comparison, each step in PPR solves the equivalent nonparametric problem

$$\text{Cov}(X, R - E(R|\alpha^T X)) = 0$$

for the residual random variable R instead of Y. From linear algebra (e.g., in Halmos, 1958), we have the notation of adjoint linear operators. For linear least-squares regression, it follows that α solves not only (16.5), but also the equations

$$\text{Cov}(X - L(X|\alpha^T X), Y) = 0$$

and

$$\text{Cov}(X - L(X|\alpha^T X), Y - L(Y|\alpha^T X)) = 0.$$

Duan (1990) formulated the adjoint problem for the nonparametric case where $\text{Cov}(X - E(X|\alpha^T X), R) = 0$. The advantage of estimating $E(X|\alpha^T X)$ rather than $E(R|\alpha^T X)$ is that we can delete or modify outliers in a realization of X. This moves the $E(X|\alpha^T X)$ more to $L(X|\alpha^T X)$ and does not affect the estimation of α, whereas a modification of y affects the estimation of α.

The disadvantage is that the computation will be more complex than in PPR. To ensure a unique solution for α, it is necessary that the true model is a PPR model *and* that $m_{[j]}$ be strictly monotonic. Donoho and Johnstone (1989) showed that any reasonable function can be approximated in this way if M tends to infinity. The close relationship of PPR to neural networks with monotonic activation functions implies that monotonicity is not too strong an assumption. Neural networks with one hidden layer can also approximate any reasonable function when the number of hidden units tends to infinity. (Section 16.8 examines the relationship between neural networks and PPR more closely.) Additionally, Duan (1990) established a fixed-point theorem under the same conditions. The estimation process works as follows:

1. *Choose an initial estimate $\alpha_{[j,0]}$ (e.g., the least-squares estimate).*
2. *Compute the "modified regressor"*

$$\tilde{x}_{[i]} = L(x|\alpha_{[j,i]}^T x) + x - E(x|\alpha_{[j,i]}^T x)$$

by linear and nonparametric regression. (Duan (1990) sliced $\alpha_{[j,i]}^T x$ and took the sample mean over each slice in order to estimate $E(x|\alpha_{[j,i]}^T x)$.)

3. *Compute the new* $\alpha_{[j,i+1]}$ *by solving the linear regression of* $r_{[j]}$ *on* $\tilde{x}_{[i]}$.
4. *Repeat the last steps until some convergence criterion is fulfilled—for example, when the angle between* $\alpha_{[j,i]}$ *and* $\alpha_{[j,i+1]}$ *falls below some threshold value.*

16.7.7 Generalized PPR

Roosen and Hastie (1993) developed a framework for *generalized projection pursuit regression* (GPPR). In a classification framework, PPR is often used as a regression method; however, it cannot take into account the binary structure of the problem. Analogously to generalized linear models (GLMs) in McCullagh and Nelder (1989) and *generalized additive models* (GAMs) in Hastie and Tibshirani (1990), we fit

$$\sum_{i=1} \hat{m}_{[j]}(\hat{\alpha}_{[j]}^T x_i) = \hat{\eta}_i = G(y_i),$$

where G is a known link function and y has a distribution in the exponential family.

If G is the identity function, we get the standard PPR. To compute the solution of the system, we have to combine the PPR algorithm with the *iteratively reweighted least-squares* (IRLS) algorithm.

Roosen and Hastie (1993) compared three possibilities of fitting binary data by PPR: with an identity link function and weights equal to unity; with a logistic link and a binomial random component (called *logistic projection pursuit* (LPP)); and with a logistic link and a Gaussian random component (termed *squared-error logistic response* (SELR)). The five functions in Section 16.7.1 were used together with a logistic function with a scale parameter that describes the "pureness" of the data. To have purer data means that the regions of zero and one are better separable. Roosen and Hastie (1993) noticed numerical problems with SELR in the case of pure data. However, LPP seems to perform slightly better than PPR with pure data. Otherwise, PPR performs somewhat better, but not significantly so. The GPPR framework has been fully developed by Lingjærde and Liestøl (1998).

16.8 PPR AND NEURAL NETWORKS

To compare neural networks with PPR, we examine the single-hidden-layer *feed-forward neural network* (FNN). To estimate the regression function

$$E(Y|X = x) = m(x_1, \ldots, x_p, \Omega), \tag{16.8}$$

with weights $\Omega = (\omega_1, \ldots, \omega_m, \ldots)$ we can build a neural network consisting of p input units, one layer of a certain number of hidden units, and one output unit forming the output layer. There are weighted connections between the layers in the direction from the input to the output. The input units represent the input variables, or predictor variables in statistical notation. The input of the hidden units is then usually the weighted sum of the values of the input variables, whereas the output units generally have the weighted sum of the outputs of the hidden units as input. The output of each unit is determined by a prespecified activation function that is often taken to be the identity function for the input units or a linear, logistic, or threshold function for hidden units and output units. Using the "tanh" function with range $(-1,1)$ for the hidden units, instead of the logistic function with range $(0,1)$, is sometimes said to be more convenient by resulting in faster convergence. But, theoretically, there is no difference, because the two functions differ only by a linear transformation. Taking m logistic hidden units and a linear output unit, we get

$$m(x, \Omega) = \omega_0^{(2)} + \sum_{j=1}^{m} \omega_j^{(2)} \Lambda \left(\omega_{j0}^{(1)} + \sum_{i=1}^{p} \omega_{ji}^{(1)} x_i \right)$$

as the model equation for our regression function, where $\Lambda(z) = [\exp(z)]/[1 + \exp(z)]$ the logistic function and $\Omega = (\omega_{00}^{(1)}, \ldots, \omega_{mp}^{(1)}, \omega_0^{(2)}, \ldots, \omega_m^{(2)})$. The absolute parameters ω_0 and ω_{j0} are called *bias parameters* or *threshold parameters.* The similarity of an FNN to a PPR model is that both project the input vector by the first layer of weights onto a one-dimensional hyperplane followed by a nonlinear transformation by the activation functions of the hidden units and a final linear combination giving the output value. Thus, both reduce the dimension of the feature space by taking the sum of univariate nonlinear functions of linear combinations of the input variables to overcome the well-known curse of dimensionality. One main difference is that the activation functions in a PPR model can be different for each hidden unit and are estimated nonparametrically, whereas they are prespecified parametric functions (e.g., logistic) in FNNs.

Typically, the parameters in a neural network are estimated simultaneously, whereas those in a PPR model are fitted successively for each hidden unit. After the projection direction represented by the weights between the input and the hidden layer of a PPR model is found, the activation function of the current hidden unit is adapted by some smoothing technique. Then the weights from the hidden units to the output unit can be estimated by a least-squares procedure. Finally, with the use of backfitting, all weights in the model can be updated together. Hidden units are added to the model until a certain fitting criterion is reached.

Common error functions for FNNs are the mean squared error

$$E_S = \sum_{i=1}^{n} \|y_i - m(x_i, \Omega)\|^2$$

and the Kullback–Leibler distance

$$E_{\text{KL}} = \sum_{i=1}^{n}\sum_{k=1}^{K}\left\{y_{i,k}\log\left(\frac{y_{i,k}}{m_k(x_i,\Omega)}\right) + (1-y_{i,k})\log\left(\frac{1-y_{i,k}}{1-m_k(x_i,\Omega)}\right)\right\}.$$

Obviously, the minimization of the latter is equivalent to that of the negative log-likelihood function

$$E_{ML} = -\log(L) = -\sum_{i=1}^{n}\sum_{k=1}^{K}\{y_{i,k}\log(m_k(x_i,\Omega)) + (1-y_{i,k})\log(1-m_k(x_i,\Omega))\}.$$

For the minimization, we have to choose an appropriate optimization method that depends on the smoothness of the activation functions.

The best-known algorithm for training networks with hidden layers is the classical *backpropagation* algorithm (generalized delta rule) that was popularized by Rumelhart et al. (1986).

Both PPR models and FNNs can approximate any continuous function to any degree, provided that the model is complex enough. The complexity of a model is given by the effective number of its parameters, which, in the FFNs, grows with an increasing number of units in the hidden layer. Analogously, the complexity of a PPR model becomes larger as the number of its nonlinear terms increases. But the "universal approximator" property does not say anything about the statistical properties of the approximation based on noisy data of limited sample size. White (1990) showed how the complexity of an FFN must grow in relation to the size of the data set in order to be consistent. But in practice, we want to find the optimal complexity, given limited and noisy data. In the statistical literature, this quest often focuses on the trade-off between bias and variance. That is, the mean squared error can be divided into a bias and a variance component of the estimated regression function, and both can contribute to poor performance. Asymptotically, both should approach zero for a consistent estimate and increasing sample size. But with finite samples, balancing the two is a complicated issue (Geman et al., 1992). In nonparametric regression analysis, that is, kernel regression, this is done by adjusting a bandwidth parameter of the fitted function. If the bandwidth is too large, the fitted function will be very smooth and will not be able to detect details of the real regression function. Thus, a bias component is introduced. For a too small bandwidth, the estimator will follow closely the noise in the data, which increases the variance component of the error function, and thus will not be able to generalize well.

For FNNs, a number of heuristic methods is available for finding the optimal complexity. One such method used to handle the bias-variance trade-off is *regularization*—that is, the application of a penalty term in the error function. In the neural network literature, weight decay is one of the most often used regularization approaches. A fraction d of the sum of squared weights of the current model is added to the error function, and one obtains

$$E_d = E_S + \delta \sum_{ij} w_{ij}^2,$$

where w_{ij} are the weights used in the network to penalize large weights during the training process. Before applying this method, the input variables should be normalized. A popular way to choose the weight decay parameter δ is cross-validation. (For details, see Stone, 1974; Efron and Tibshirani, 1993.)

Thus, FNNs provide an alternative to PPR and can, basically be applied to the same type of data. Of course, there are differences regarding optimization and the interpretation of the results.

16.9 OPTIMIZATION METHODS FOR PPR AND NEURAL NETWORKS

The estimation of the parameters of a nonlinear regression function

$$f(x, \omega), \quad \omega = (\omega_1, \ldots, \omega_d)$$

typically requires the minimization of some kind of an error function $h(\omega)$, such as the weighted or unweighted squared error or the negative log-likelihood function. In most instances, these minimization problems cannot be solved analytically, and an iterative method has to be applied. Usually, the error function that is involved represents a "landscape of hills and valleys", and we hope to find at least one of the many local minima. Generally, we decide among three approaches for the iterative minimization of a function $h(\omega)$:

(i) Methods using second derivatives of h in terms of ω (e.g., Newton's method, or modified Newton methods).

(ii) Methods using the first derivatives of h (e.g., simple gradient descent, discrete Newton methods, quasi-Newton methods, or conjugate gradient methods).

(iii) Methods using just function values and no derivatives of h (direct-search methods, e.g., the simplex algorithm).

The choice of the optimization method plays a key role in implementing neural network and PPR methods. For neural networks, this point is stressed by LeCun (1996); for PPR, it is emphasized by Posse (1995).

Apparently, Newtonian procedures perform quite well for neural networks, whereas simpler numerical techniques, like the simplex algorithm, simulated annealing, or even random searches, are preferred for projection pursuit models.

For the numerical details of Newtonian procedures (e.g., a modified or quasi-Newtonian technique, see Seber and Wild (1989). The simplex direct-search technique, also called a polytope algorithm (see Nelder and Mead, 1965) should not be confused with the simplex method known from linear optimization (i.e., linear programming).

16.10 THE IMPLEMENTATION OF PPR IN S-PLUS, R, AND XPLORE

S-Plus offers an intrinsic function for PPR (see also Venables and Ripley, 1999)

```
ppr <-  ppreg (x, y, termno)
```

where x is a high-dimensional variable, y is a one-dimensional response, and termno gives the minimal number of terms that will be used. ppr itself is a list of the components:

TABLE 16.1 S-Plus PPR Components

Component	Description
ypred	Residuals of the fitted model
fl2	Squared residuals divided by the corrected sums of squares
alpha	Matrix of the used projection directions
beta	Matrix of the weights for every observation y_i
z	Matrix that contains $\alpha_i^T X$
zhat	Matrix that contains the smoothed values $\sum_{i=1}^{\text{termno}} G_i(\alpha_i^T X)$
allalpha	Three-dimensional array that contains the fitted α values for each step
allbeta	Three-dimensional array that contains the fitted β values for each step
esq	Fraction of unexplained variance
esqrsp	Fraction of unexplained variance for each observation

The algorithm is based on the work of Friedman (1985) and uses backfitting. Since ppreg is an intrinsic function, we do not have any insight into the workings of the algorithm.

The programming language R (can be obtained via the World Wide Web from http://www.stat.auckland.ac.nz/r/r.html) uses basically the same code as S-Plus. Since R has been developed later than S-Plus (and S) its functionality is wider. As an additional smoother we find smoothing splines with and without automated (cross-validation) smoothing parameter selection. Another extension is the type of backfitting that is used: none; backfit only smoothers; and backfit smoothers with projection directions. The general strategy is to first fit a rather comprehensive model to the data and then drop "less important" terms later. Interesting is the note in the R documentation that "the code is extremely sensitive to the compiler used".

Let us now consider XploRe. PPR is part of the ADDMOD (additive modeling) library of XploRe 3.2 (Härdle et al., 1995). Two macros are available: PPR, to perform the regression, and PPRINTER, to analyze the projections and fits that are obtained. The calls are

```
(xs ys vs fs) = PPR (x cmd)
```

and

```
PPRINTER (x vs fs).
```

The PPR macro allows an interactive fitting of a PPR model with different smoothers. The code for the macro call is as follows:

```
LIBRARY("addmod")            ; Load the addmod library
x = READ("djppr")            ; Read a data set generated from
                             ; Donoho and Johnstone (1989)
(xs ys vs fs) = PPR(x)       ; Do the PPR
```

From the input variable x, it is expected that the last column contains the response. The outputs xs and ys contain the standardized data, vs the projection vectors found by an m-term fit, and fs the values of the fitted function. When the user activates the macro, he or she can choose among different smoothing methods and smoothing parameters. A sequence of pictures, as in Figure 16.2, will be the output. There we can see the current fit and the projected data in the large window. The window underneath the large one displays how the error function will increase by changing the projection vector (the solid line). The dotted line

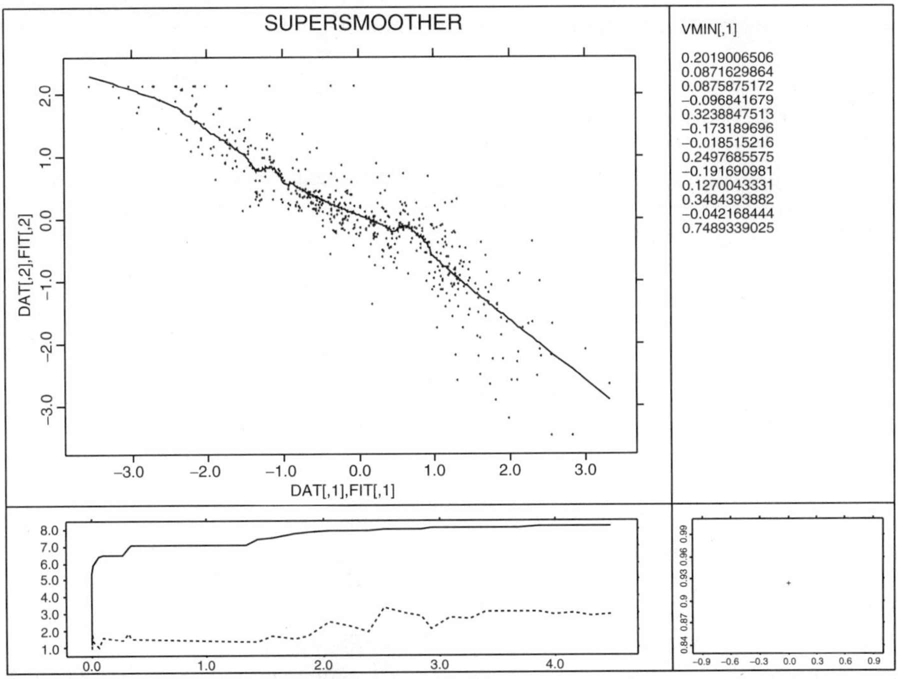

FIGURE 16.2 The first term of an eight term PPR fit of the Boston Housing Data. In the upper large window we see the current fit and the projected data. The window below the large window displays how the error function will increase by changing the projection vector. The dotted line indicates the value of the interpretability index due to Morton (1989). In the right lower window the decreasing error with increasing number of terms can be seen and the upper window shows the current projection.

shows the value of some interpretability index. The right lower window illustrates how the error in each term decreases, and the upper window shows the actual projection values.

The algorithm in `S-Plus` was designed to achieve a quick result. As a consequence, an intrinsic function had been implemented for PPR. `XploRe`, however, allows the user to have a look at the algorithm, since everything is written in macro language. Of course, the disadvantage is, that `XploRe` takes minutes instead of seconds to fit one term for instance in an example as described in the next section.

16.11 AN EXAMPLE

As an example for an application of PPR, we analyze the famous Boston Housing Data. Aldrin et al. (1993) applied their PPR algorithm for moderate nonlinearities to these data. For the purpose of demonstration we applied both, S-Plus and XploRe 3.2. Harrison and Rubinfeld (1978) collected 15 variables (see Table 16.2) for 506 districts in the Boston standard metropolitan statistic area.

After they transformed the variables, they performed a linear regression to estimate the relationship between MEDV (median value of owner occupied homes)

TABLE 16.2 Variables of the Boston Housing Data

Symbol	Definition
MEDV	Median price of owner-occupied houses (in thousands of US dollars)
LMV	Logarithm of the median value of owner-occupied homes
CRIM	Per capita crime rate, by town
ZN	Proportion of a town's residential land zoned for lots greater than 25,000 square feet
IND	Proportion of nonretail business acres per town
CHAS	Charles river dummy variable with value 1 if tract has a boundary on the Charles river
NOX	Nitrogen oxide concentration (parts per hundred million) squared
RM	Average number of rooms squared
AGE	Proportion of owner-occupied units built prior to 1940
DIS	Logarithm of the weighted distances to five employment centers in the Boston region
RAD	Logarithm of index of accessibility to radial highways
TAX	Full-value property-tax rate (per 10,000 US dollars)
PTR	Pupil–teacher ratio
B	$B \equiv (\mathrm{Bk} - 0.63)^2$, where Bk is the proportion of blacks in the population
LSTA	Logarithm of the proportion of the population that is of lower status

and the other variables ($R^2 = 0.806$). The data are known to contain outliers and clustered structures in some variables. The "helix" structure in the data was detected by Li (1991) with sliced inverse regression (see Chapter 17).

The standardized coefficients of the linear regression (first line) and the coefficients of the projection vectors of the PPR model fitted in S-Plus and XploRe can be found in Table 16.3 (see also Klinke, 1997). We have here an 8-term fit in XploRe, a 4-term fit and a 8-term fit in S-Plus. Since XploRe does not use backfitting the 4-term fit is identical to the first four projections of the 8-term fit.

The first projection in Table 16.3 found by XploRe is nearly the same as the one found by linear regression. The values coincide, only the signs are reverted. Due to the stochastic nature of the optimization algorithm used in XploRe, we found other coefficients in Figure 16.2 (right upper part) for the first projection. In S-Plus, we got a similar result for the first projection, but closer to the result obtained by linear regression. We guess that this might be due to the fact that XploRe starts with a random projection and S-Plus starts with the coefficients of the linear regression.

The fact that the variable LSTA (proportion of population that is lower status) is most important is grasped by both routines. Consequently, a picture is produced, $\alpha_{[1]}^T x_i$ against $m_{[1]}(\alpha_{[1]}^T x)$ (upper left part in Figure 16.2), which looks as it was obtained by linear regression. We see a side view of the "helix" structure mentioned above. Since we have restricted ourselves to one-dimensional projections, this is the best view which can be found. Of course, one could also consider bivariate projections.

Although the supersmoother, a local linear regression with varying window width (choosen by cross-validation) is a nonlinear regression method, we can use the following definition of the $R^2 = 1 - \sum_{i=1}^n (\hat{y}_i - y_i)^2 / \sum_{i=1}^n (y_i - \overline{y})^2$ criterion known from linear regression. As shown in the lower right part of Figure 16.2, we obtained a value above 0.9. The second projection explains 60 per cent of the unexplained variance of y, or about 6 per cent on top of the first.

As a consequence, a one term PPR would be sufficient to fit the Boston Housing Data. For that we can use the projection direction given by the linear regression.

But Table 16.3 also reveals the influence of backfitting in S-Plus. The first projection of the 8-term fit in S-Plus is nearly identical with the result from the linear regression. This implies that it is necessary to fit much more terms than usually applied (see also Friedman 1984a; or Roosen and Hastie, 1994).

Since the first projection of the 4- as well as the 8-term fit of S-Plus differs considerably, it raises the question about the role of B, some kind of pureness measure for the population of the school districts. Harrison and Rubinfeld (1981, p. 96) state that: "At a low to moderate level of Bk, an increase in Bk should have a negative influence on housing value if Blacks regarded as undesirable neighbors by Whites. However, market discrimination means that housing values are higher at very high levels of Bk. One expects, therefore, a parabolic relationship between proportion Blacks in a neighborhood and housing values."

TABLE 16.3 Projection Vectors from PPR and the Linear Regression for the Boston Housing Data*

Algorithm	Terms	CRIM	ZN	IND	CHAS	NOX	RM	AGE	DIS	RAD	TAX	PTR	B	LSTA
Lin.Reg.		−0.02	+0.00	+0.00	+0.16	−0.01	+0.01	+0.00	−0.33	+0.17	−0.00	−0.05	+0.64	−0.65
XploRe	1	+0.01	+0.00	+0.00	−0.25	+0.01	−0.01	−0.00	+0.31	−0.14	+0.00	+0.10	−0.31	+0.83
XploRe	2	−0.00	+0.00	+0.02	−0.63	−0.00	−0.00	+0.00	−0.64	+0.20	+0.00	−0.01	−0.35	−0.07
XploRe	3	+0.00	−0.00	+0.00	+0.13	−0.00	−0.03	+0.00	+0.28	+0.10	+0.00	−0.01	−0.66	−0.67
XploRe	4	−0.00	−0.00	−0.00	−0.76	+0.00	−0.00	−0.00	−0.13	+0.16	+0.00	−0.01	−0.59	−0.11
XploRe	5	+0.02	−0.00	+0.00	−0.06	−0.01	−0.08	+0.00	−0.05	−0.28	−0.00	−0.08	+0.02	−0.95
XploRe	6	+0.00	+0.00	+0.02	+0.11	−0.00	+0.00	+0.00	+0.07	+0.02	+0.00	+0.02	+0.99	+0.02
XploRe	7	+0.00	−0.00	+0.01	+0.37	+0.00	−0.01	+0.00	+0.05	+0.02	−0.00	−0.04	−0.92	−0.08
XploRe	8	−0.00	−0.01	+0.01	+0.06	−0.01	+0.00	+0.01	+0.41	+0.20	−0.00	+0.04	−0.89	−0.05
S-Plus	1	−0.02	−0.00	+0.00	+0.11	−0.01	+0.01	−0.00	−0.21	+0.22	−0.00	−0.08	+0.44	−0.82
S-Plus	2	+0.00	+0.00	−0.01	−0.01	−0.02	−0.00	+0.00	−0.92	+0.04	−0.00	+0.01	−0.36	−0.01
S-Plus	3	+0.01	+0.00	−0.00	−0.03	+0.00	+0.02	−0.00	−0.42	−0.33	−0.00	+0.02	+0.51	+0.65
S-Plus	4	−0.02	+0.00	−0.01	−0.24	−0.00	+0.04	−0.00	−0.37	+0.30	+0.00	+0.03	−0.34	+0.76
S-Plus	1	−0.02	−0.00	−0.00	+0.11	−0.01	+0.02	+0.00	−0.27	+0.14	−0.00	−0.08	+0.67	−0.66
S-Plus	2	+0.00	+0.00	−0.02	+0.01	−0.02	−0.00	+0.00	−0.97	+0.07	−0.00	+0.02	−0.25	+0.02
S-Plus	3	+0.00	+0.00	−0.01	+0.09	−0.00	+0.02	−0.01	−0.33	−0.20	−0.00	+0.01	+0.70	+0.59
S-Plus	4	−0.02	+0.00	−0.02	−0.28	−0.01	+0.05	+0.00	−0.41	+0.31	+0.00	+0.03	−0.40	+0.70
S-Plus	5	+0.00	−0.02	+0.11	+0.08	+0.02	+0.04	−0.02	+0.82	+0.24	−0.00	+0.02	−0.43	+0.23
S-Plus	6	−0.01	+0.01	−0.01	+0.68	−0.02	−0.02	+0.00	−0.21	−0.60	+0.00	+0.11	+0.29	−0.21
S-Plus	7	+0.03	−0.00	−0.03	−0.38	−0.00	−0.03	+0.00	−0.35	+0.40	−0.00	−0.00	+0.25	−0.71
S-Plus	8	−0.01	−0.00	+0.15	+0.42	−0.08	−0.01	+0.01	−0.61	+0.13	+0.00	−0.17	−0.12	−0.60

*The coefficients of the linear regression (first line) are rescaled so that the Euclidean norm is one. Further the projections of a 8-term-PPR-fit in XploRe and the projections of a 4-term- and 8-term-fit in S-Plus are displayed.

Estimating MEDV by the supersmoother only by projections of B and LSTA reveals in fact two maxima of R^2. A broad global maximum ($R^2 = 0.68$) with almost only LSTAT contributing to the projection and a smaller local maximum ($R^2 = 0.56$) with LSTAT and B contributing equally to the projection. Obviously, the 8-term fit of XploRe (without backfitting) and the 4-term fit of S-Plus (with backfitting) prefer the global maximum for the first projection whereas the 8-term fit of S-Plus prefers the second maximum and places more explanation of the variance on other projections. Here an expert has to be called in to decide which model is best.

16.12 CONCLUSION

Although PPR has been available for nearly 20 years, we have to consider it still as an exploratory regression tool. Its promise is a solution to the problem of the identifcation of information-poor variables. It also addresses, to some extent, the problem of multicollinearity in the data. But the issues of interpreting projections and choosing the number of terms are still unsolved. The theoretical mathematical treatment is very difficult, since it involves the optimization of a projection vector and the estimation of a nonparametric function.

Consequently, PPR should be applied to check results from other models, like, for example, in Beebe and Kowalski (1988) or van Leeuwen et al. (1994). Moreover, in a comparison of different nonparametric and semiparametric techniques applied to real and simulated data in Sekulic et al. (1993) all methods performed equally well. But they also state critically that, especially for neural networks and PPR, no guidelines are at hand.

As a consequence, software implementers are required to offer much more than the bare algorithms. It is most desirable to have an output of a sequence of models which gives the user the chance to evaluate them based on their own background knowledge. This holds true for models obtained from various variable selection methods as well as different smoothing techniques.

Acknowledgments

We thank Jerome H. Friedman and Jörg Polzehl for helpful discussions and hints. Part of the second authors work was carried out during a stay as a visiting scholar sponsored by the German Academic Exchange Service (DAAD) in the Department of Statistics at Stanford University.

REFERENCES

Ahn, Y., and Rhee, S. (1992). A simulation study on projection pursuit discriminant analysis (in Korean), *Korean J. Appl. Statist.* **5**, 103–111.

Aldrin, M., Bølviken, E., and Schweder, T. (1993). Projection pursuit regression for moderate non-linearities, *Computat. Statist. Data Analysis*, **16**, 379–403.

Blough, D. (1989). Multivariate symmetry via projection pursuit, *Annals of the Institute of Statistical Mathematics* **41**, 461–475.

Chen, J. -H. (1987). A result on the projection pursuit density estimation (in Chinese), *J. Systems Sci. Math. Sci.*, **7**, 183–192.

Chen, H. (1991). Estimation of a projection-pursuit type regression model, *Ann. Statist.*, **19**, 142–157.

Cook, D., Buja, A., and Cabrera, J. (1993). Projection pursuit indexes based on orthonormal function expansions, *J. Comput. Graph. Statist.*, **2**, 225–250.

Donoho, D., and Johnstone, I. (1989). Projection-based approximation and a duality with kernel methods, *Ann. Statist.*, **17**, 58–106.

Duan, N. (1990). The adjoint projection pursuit regression, *J. Amer. Statist. Assoc.*, **85**, 1029–1038.

Efron, B., and Tibshirani, R. (1993). *An Introduction to the Bootstrap*, Chapman and Hall, London, UK.

Flick, T., Jones, L., Priest, R., and Herman, C. (1990). Pattern classification using projection pursuit, *Pattern Recognition*, **23**, 1367–1376.

Friedman, J. (1984a). Smart User's Guide, *Technical Report No. 1*, Department of Statistics, Stanford University, CA.

Friedman, J. (1984b). A Variable Span Smoother, *Technical Report LCS005*, Department of Statistics, Stanford University, CA.

Friedman, J. (1985). Classification and Multiple Regression through Projection Pursuit, *Technical Report LCS012*, Department of Statistics, Stanford University, CA.

Friedman, J. (1987). Exploratory projection pursuit, *J. Amer. Statist. Assoc.*, **82**, 249–266.

Friedman, J., and Stuetzle, W. (1981a). Projection pursuit classification. (unpublished manuscript).

Friedman, J., and Stuetzle, W. (1981b). Projection pursuit regression, *J. Amer. Statist. Assoc.*, **76**, 817–823. (also `http://www.slac.standford.edu/pubs/slacpubs/2000/slac-pub-2466.html`)

Friedman, J., Stuetzle, W., and Schroeder, A. (1984). Projection pursuit density estimation, *J. Amer. Statist. Assoc.*, **79**, 599–608.

Friedman, J., and Tukey, J. (1974). A projection pursuit algorithm for exploratory data analysis, *IEEE Transactions on Computers C*, **23**, 881–889.

Geman, S., Bienenstock, E., and Doursat, R. (1992). Neural networks and the bias/variance dilemma, *Neural Computation*, **4**, 1–58.

Härdle, W., Klinke, S., and Turlach, B. A. (1995). *XploRe—An Interactive Statistical Computing Environment*, Springer-Verlag, Berlin, Germany.

Hall, P. (1988). Estimating the direction in which a data set is most interesting, *Probab. Theory Related Fields*, **80**, 51–77.

Hall, P. (1989a). On polynomial-based projection indices for exploratory projection pursuit, *Ann. Statist.*, **17**, 589–605.

Hall, P. (1989b). On projection pursuit regression, *Ann. Statist.*, **17**, 573–588.

Halmos, P. (1958). *Finite-Dimensional Vector Spaces*, Van Nostrand, Princeton, NJ.

Harrison, D., and Rubinfeld, D. (1978). Hedonic prices and the demand for clean air, *J. Environ. Economics Manage.* **5**, 81–102.

Hastie, T., and Tibshirani, R. (1990). *Generalized Additive Models*, Chapman and Hall, London, UK.

Huber, P. (1985). Projection pursuit, *Ann. Statist.*, **13**, 435–475.

Hwang, J., Lay, S., Maechler, M., Martin, D., and Schimert, J. (1994). Regression modeling in backpropagation and projection pursuit learning networks, *IEEE Transactions on Neural Networks*, **5**, 342–353.

Intrator, N. (1990). Combining exploratory projection pursuit and projection pursuit regression with applications to neural networks, *Neural Computation*, **5**, 443–455.

Jee, J. (1985). Exploratory projection pursuit using nonparametric density estimation, *Proceedings of the Statistical Computing Section, ASA*, pp. 335–339, Alexandria, VA.

Jones, L. (1992). A simple lemma on greedy approximation in Hilbert space and convergence rates for projection pursuit regression and neural network training, *Ann. Statist.*, **20**, 608–613.

Jones, M., and Sibson, R. (1987). What is projection pursuit?, *J. Roy. Statist. Soc., A*, **150**, 1–18.

Klinke, S. (1997). *Data Structures for Computational Statistics*, Physica-Verlag, Heidelberg, Germany.

Kruskal, J. (1969). Toward a practical method which helps uncover the structure of a set of multivariate observations by finding the linear transformation which optimizes a new "index of condensation," in Milton, R. C. and Nelder, J. A. (eds.), *Statistical Computation*, pp. 427–440, Academic Press, New York.

Kruskal, J. (1972). Linear transformation of multivariate data to reveal clustering, in Shepard, R. N., Romney, A. K., and Nerlov, S. B. (eds.), *Multidimensional Scaling: Theory and Applications in the Behavioural Sciences*, Vol. 1, pp. 179–191, Seminar Press, London, UK.

LeCun, Y. (1996). Efficient backprop, `ftp://www.first.gmd.de/persons/Mueller.Klaus-Robert/cookbook.ps.gz`.

Li, K. (1991). Sliced inverse regression for dimension reduction (with discussion), *J. Amer. Statist. Assoc.*, **86**, 316–342.

Lingjærde, O. C., and Liestøl, K. (1998). Generalized projection pursuit regression, *SIAM Journal on Scientific Computing*, **20**, 844–857.

Maller, R. (1989). Some consistency results on projection pursuit estimators of location and scale, *Canad. J. Statist.* **17**, 81–90.

McCullagh, P., and Nelder, J. (1989). *Generalized Linear Models*, Chapman and Hall, London, UK.

Morton, S. (1989). *Interpretable Projection Pursuit*, Doctoral Thesis, Department of Statistics, Stanford University, CA.

Nason, G. (1995). Three-dimensional projection pursuit, *Applied Statistics*, **4**, 411–430.

Nelder, J., and Mead, R. (1965). A simplex method for functional minimization, *The Computer Journal*, **7**, 308–313.

Polzehl, J. (1995). Projection pursuit discriminant analysis, *Computat. Statist. Data Analysis*, **20**, 141–157.

Portier, K., Dippon, D., and Hetrick, V. R. (1987). An implementation of projection pursuit for classification, *Proceedings Statistical Computing Section, ASA*, pp. 340–344, Alexandria, VA.

Posse, C. (1992). Projection pursuit discriminant analysis for two groups, *Communications in Statistics A–Theory Methods*, **21**, 1–19.

Posse, C. (1995). Projection pursuit exploratory data analysis, *Computat. Statist. Data Analysis*, **20**, 669–887.

Rejtö, L., and Walter, G. (1990). Generalized projection pursuit regression and density approximation, in Page, C. and LePage, R. (eds.), *Computing Science Statistics: Proceedings of the Symposium on the Interface*, pp. 501–503, Springer-Verlag, Berlin, Germany.

Rejtö, L., and Walter, G. (1992). Remarks on projection pursuit regression and density estimation, *Stochastic Anal. Appl.*, **10**, 213–222.

Ripley, B. (1996). *Pattern Recognition and Neural Networks*, Cambridge University Press, Cambridge, UK.

Roosen, C., and Hastie, T. (1993). Logistic response projection pursuit, *Document No. bl011214-930806-09tm*, AT&T Bell Laboratories, Murry Hill, NJ.

Roosen, C., and Hastie, T. (1994). Automatic smoothing spline projection pursuit, *J. Computat. Graph. Statist.*, **3**, 235–248.

Rumelhart, D., Hinton, G., and Williams, R. (1986). Learning internal representations by error propagation, in D. Rumelhart, J. McClelland and the PDP Research Group (eds.), *Parallel Distributed Processing: Explorations in the Microstructure of Cognition, Volume 1: Foundations*, pp. 318–362, MIT Press, Cambridge, MA.

Seber, G. A. F., and Wild, C. J. (1989). *Nonlinear Regression*, Wiley, New York.

Stone, M. (1974). Cross-validation choice and assessment of statistical predictions, *J. Roy. Statist. Soc., B*, **36**, 111–147.

White, H. (1990). Connectionist nonparametric regression: Multilayer feed-forward networks can learn arbitrary mappings, *Neural Networks*, **3**, 535–549.

Venables, W. N., and Ripley, B. D. (1999). *Modern Applied Statistics with S-Plus*, Springer-Verlag, New York.

Zhu, L., and Li, G. (1991). A consistent estimator for the supremum of the projection index based on sliced inverse regression, *Communications in Statistics A—Theory Methods*, **20**, 5–16.

CHAPTER 17

Sliced Inverse Regression

Thomas T. Kötter
SAP AG, Walldorf, Germany

17.1 INTRODUCTION

In this chapter, we examine the relationship between a univariate response variable and a multidimensional regressor variable. In contrast to regression analysis, the aim is not the estimation of an unknown regression function, but that of the influence of the regressor space on the response variable. That means that we hope to find (linear) combinations of the regressors that can describe the dependency between the regressors and the response variable. Note that the dependency itself (i.e., the regression function) is not searched for! If only a few linear combinations of the explanatory variable are necessary in comparison to the dimensionality of the regressor variable, the aim of reducing dimensions is reached.

Sliced inverse regression (SIR) also known as *slicing regression* and *sliced regression*, is a technique for such a reduction in dimensionality. The concept was developed jointly by Li and Duan (1987) and was described in detail by Duan and Li (1991) and Li (1991). Meanwhile, many researchers have contributed further results pertaining to the technique. Some of these extensions and modifications are presented in the sections that follow. The rather general model, the moderate assumptions, and the fast algorithms make this procedure interesting for data analysts.

The advantage of having only a few regressor variables, which are linear combinations of the original exogenous variable, is that the application of nonparametric regression techniques to the reduced data becomes possible. If the dimensionality of the original regressor is larger than three or four, problems with such techniques are likely because of the so-called *curse of dimensionality*. That is, nonparametric regression is based on local averaging of the data or

Smoothing and Regression: Approaches, Computation, and Application
Edited by Michael G. Schimek
ISBN 0-471-17946-9

similar techniques, taking into account points in the neighborhood of a given observation. Unfortunately, however, the volume of a d-dimensional space grows exponentially in d. Therefore, the property of being a neighborhood vanishes swiftly in higher dimensions.

Of course, there are other methods such as *projection pursuit regression* (PPR) and *average derivative estimation* (ADE) that handle high-dimensional data non-parametrically. However, they require certain assumptions on the dependency or on the underlying model, in contrast to the quite general assumptions of SIR.

Example. Given an interval with width 1 and uniform distributed observations, in order to get a slice of approximately 10 percent of the observations, we need a subinterval of width 0.1. If we have a cube with volume 1, we would require an inner cube with edge length $\sqrt[3]{0.1} \approx 0.464$ to get about 10 percent of the observations. Actually, this is nearly the half of the edge length of the whole cube.

What are standard approaches to cope with high-dimensional data? Imagine that the unknown regression function could be considered as sum of univariate functions that depend on a certain projection of the regressor variable. The corresponding model is known as PPR model (Friedman and Stuetzle, 1981, and Chapter 16, in this volume) given by

$$Y = \sum_{i=1}^{d} f_i(\beta_i^T X) + \epsilon. \tag{17.1}$$

Clearly, in this case a very specific reduction in dimensionality is assumed: Only one-dimensional projections are considered. Moreover, for the estimation of the directions β_i, the accompanying functions $f_i(\cdot)$ have to be estimated too.

The second approach is ADE (Powell et al., 1989, and Chapters 10 and 16 in this volume) and avoids the problem of estimating an unknown function along a dimension-reduction direction β. In order to find the direction β, the expectation of the derivative of the unknown $f(\cdot)$ is considered. This expectation is

$$\begin{aligned} E\left[f\left(\frac{\partial}{\partial x}\beta^T X\right)\right] &= E[f'(\beta^T X)]\beta \\ &= c\beta, \end{aligned}$$

where $c \in \mathbb{R}$, and is proportional to the unknown vector β. Anyway, only the direction, and not the length, of β can be estimated. Hence, the sample version of the average derivative can be used to find an estimate $\hat{\beta}$.

17.1.1 A More General Model

For SIR, an even more general model than (17.1) is used,

$$Y = m(\beta_1^T X, \ldots, \beta_K^T X, \epsilon), \tag{17.2}$$

where ϵ is independent of X and the covariance of the exogeneous variables X Σ_X exists. Furthermore, $1 \leq K \leq d$ is unknown, $m : \mathbb{R}^{d+1} \to \mathbb{R}$ is unknown, Y is a random variable, and X is a d-dimensional random vector.

Before we start with the SIR approach, let us make some general observations about the model (17.2). First, apparently, it contains (17.1), as well as the common linear model. But in contrast to PPR, in which only the lengths of the β_is are not identifiable, with (17.2) the directions also cannot be identified. Only the span of these vectors can be determined. Now, we can restate (17.2) as

$$y = m_0(\gamma_1^T x, \ldots, \gamma_K^T x, \epsilon),$$

where $\text{span}(\gamma_1, \ldots, \gamma_K) = \text{span}(\beta_1, \ldots, \beta_K)$. Therefore, the projectors of span $(\gamma_1, \ldots, \gamma_K)$ and $\text{span}(\beta_1, \ldots, \beta_K)$ must be the same.

Another consequence is that Y depends on X only through the linear combination $D^T X$, where $D = (\beta_1, \ldots, \beta_K)$ is defined as the matrix of the directions and the conditional distribution functions of $Y|X$ and $Y|D^T X$ are identical, that is,

$$F_{Y|X} = F_{Y|D^T X}.$$

Then, conditional on $D^T X$, the random variables X and Y are independent, of course.

Li (1991) called the vectors $\beta_1, \ldots, \beta_K$ the *effective dimension reduction* (edr) vectors and the space $B := \text{span}(\beta_1, \ldots, \beta_K)$ edr-*space*. Without loss of generality, we can assume that the vectors $\beta_1, \ldots, \beta_K$ are linearly independent and B is K-dimensional, respectively. Otherwise the representation of $m(\cdot)$ would be redundant, and a lower dimensional function $m_l(\cdot) = m(\cdot)$ could be determined.

17.2 THE IDEA

SIR avoids the curse of dimensionality by applying the inverse regression of X given Y. That means, instead of a single d-dimensional regression, d one-dimensional regressions are carried out. The question is what we can learn from the inverse regression curve $E[X|y]$. Actually, we need an additional assumption in order to derive a relation between the inverse regression and the unknown β_i. This assumption is

$$E[b^T X|\beta_1^T x, \ldots, \beta_K^T x] = c_0 + \sum_{i=1}^{K} c_i \beta_i^T x, \tag{17.3}$$

where $c_i \in \mathbb{R}$. Accordingly, given a point in the $\text{span}(\beta_1, \ldots, \beta_K)$, the expectation of any linear combination of the regressor X is linear in the conditional coordinates.

Li (1991) mentioned that elliptically symmetric distributions fulfil this condition, and Cook and Weisberg (1991) pointed out that assumption (17.3) is also sufficient for this kind of distribution. Hence, (17.3) is equivalent to supposing that X follows an elliptically symmetric distribution. Although this seems to be a severe restriction, Hall and Li (1993) reported that the assumption needs to hold only for the true β_is and is often valid in practical situations. Moreover, most low-dimensional projections of high-dimensional data are close to a normal distribution (Diaconis and Freedman, 1984), which belongs to the class of elliptically symmetric distributions. Furthermore, Duan and Li (1991) investigated deviations from (17.3).

But how can we exploit this assumption to find the unknown $\beta_1, \ldots, \beta_K$? The answer is given in the following theorem (Li, 1991).

Theorem 17.2.1. *Let Z_c be the centered random vector $Z_c = X - E[X]$. Then the K-dimensional space* $\text{span}(\Sigma_X\beta_1, \ldots, \Sigma_X\beta_K)$ *contains the inverse regression curve $E[Z_c|y]$.*

Corollary 17.2.1. *Let Z be the standardized random vector $Z = \Sigma_X^{-1/2} Z_c = \Sigma_X^{-1/2}(X - E[X])$. Then the conditional expectation $E[Z|y]$ lies in the K-dimensional space* $\text{span}(\eta_1, \ldots, \eta_K)$, *where the $\eta_i = \Sigma_X^{1/2}\beta_i$ are called standardized* edr-*directions.*

The theorem and the corollary enable us to estimate the unknown edr-directions $\beta_1, \ldots, \beta_K$. The idea is to investigate the inverse regression curve $E[Z_c|y]$ or $E[Z|y]$. An appropriate measure finding this space is the covariance, which tells us in which directions the inverse regression has high variability and in which directions it has low or no variability. More formally, we can write

$$\text{Cov}[E[Z|y]] = E[E[Z|y]E[Z^T|y]].$$

Now, let $b \in \mathbb{R}^d$ be orthogonal to all $\eta_i = \Sigma_X^{1/2}\beta_i$ and therefore orthogonal to the standardized edr-space. Then

$$\text{Cov}[E[Z|y]]b = E[E[Z|y]E[Z^T b|y]] = 0,$$

because $E[Z^T b|y] = 0$. The same argument holds for Z_c, where the standardized edr-directions η_i are replaced by $\Sigma_X\beta_i$. We can take advantage of this property and construct an algorithm for estimating the edr-directions $\beta_1, \ldots, \beta_K$ because $\text{Cov}[E[Z|y]]$ contains all necessary information. By the decomposition of this matrix into eigenvalues and eigenvectors, we can directly see how the variance is spread over the d-dimensional space. Since, being inner products, covariance matrices are usually positive semidefinite, their eigenvalues are positive real numbers, or zero. Eigenvectors γ_i with small associated eigenvalues λ_i characterize uninteresting directions because $\text{Cov}[E[Z|y]]\gamma_i = \lambda_i\gamma_i \approx 0$. Conversely, eigenvectors with large eigenvalues mark important directions wherein $E[Z|y]$ is highly variable.

Hence, we can get estimates for a basis of the edr-space B by the eigenvalue and eigenvector decomposition of an estimate of $\text{Cov}[E[Z|y]]$. Fortunately, a very simple estimate for $E[Z|y]$ is sufficient to derive reasonable results from the estimates. As an estimate of $E[Z|y]$, Duan and Li (1991) proposed the average of the standardized observations in certain intervals around y (like a step function). The sample covariance of these estimates then serves as estimator of the matrix $\text{Cov}[E[Z|y]]$.

In the original algorithm of Duan and Li (1991), $\{(x_i, y_i)\}_{i=1}^n$ is the sample, $\overline{x}$ denotes the mean vector, and $\hat{\Sigma}_X$ is the sample covariance of the exogeneous variables. The range of the endogenous variable, $[\min(y_i), \max(y_i)]$, is separated into S nonoverlapping subintervals H_s called *slices*. The indicator $\text{I}_{H_s}(y_i)$ is unity if $y_i \in H_s$ and zero otherwise. The number of observations in each slice is $n_s = \Sigma_{i=1}^n \text{I}_{H_s}(y_i)$. The algorithm is as follows:

1. *Center and standardize the exogenous variable:*

$$z_i = \hat{\Sigma}_X^{-1/2}(x_i - \overline{x}), \qquad \forall i \in \{1, \ldots, n\}.$$

2. *Estimate* $E[Z|y]$ *by*

$$\overline{z}_s = n_s^{-1} \sum_{i=1}^{n} z_i \text{I}_{H_s}(y_i), \qquad \forall s \in \{1, \ldots, S\}.$$

3. *Estimate* $\text{Cov}[E[Z|y]]$ *by*

$$\hat{V} = n^{-1} \sum_{s=1}^{S} n_s \overline{z}_s \overline{z}_s^T.$$

4. *Compute* $\hat{\Gamma}^T \hat{\Lambda} \hat{\Gamma} = \hat{V}$, *where* $\hat{\Lambda}$ *contains the eigenvalues* $\hat{\lambda}_i$ *and* $\hat{\Gamma}$ *the eigenvectors* $\hat{\eta}_i$. (The latter are estimators for the standardized edr-directions, while the $\hat{\Sigma}_X^{-1/2}\hat{\eta}_i = \hat{\beta}_i$s estimate the original edr-directions.)

Actually, this algorithm is very simple: No complicated numerical tasks are involved, so it can be implemented without problems in any programming language that provides matrix algebra or at least an eigenvalue–eigenvector decomposition. We consider the topic in more detail in Section 17.6. What is remarkable is that no explicit iteration is required in certain programming languages.

17.3 STATISTICAL PROPERTIES

Without going into detail, it can be shown that the estimators of the algorithm are $\sqrt{n}$-consistent. Even if the number of elements in each slice is fixed, a consistent estimate for $\mathrm{Cov}[E[Z|y]]$ can be derived from $\hat{V}$. The consistency can be transfered to the eigenvalues $\hat{\lambda}_i$ and eigenvectors $\hat{\eta}_i$, so that the estimators for the edr-directions $\hat{\beta}_i = \hat{\Sigma}_X^{1/2}\hat{\eta}_i$ are also consistent.

These results follow mainly from the consistency of $\hat{V}$, which is composed of consistent estimates of $E[Z|y]$, the probability $P(y \in H_s)$, and the covariance.

Under the condition that n_s is constant for all slices (the case $n_s = 2$ is treated in detail by Hsing and Carroll, 1992) and the number of slices is increasing at the same rate as the sample size n, it can be shown that the vector of estimated eigenvalues $(\hat{\lambda}_i)_{i=1}^d$ is asymptotically normally distributed (Kötter, 1996). Using the same technique as in principal-component analysis (PCA), we can establish the asymptotic normality of

$$\hat{\Psi}_K = \frac{\sum_{i=1}^{K} \hat{\lambda}_i}{\sum_{i=1}^{d} \hat{\lambda}_i}.$$

However, the computation of the covariance structure is quite complicated and requires a separation of the given sample. This kind of asymptotics is found in different versions of SIR, too, and is presented in the sections that follow.

17.4 THE UNKNOWN DIMENSIONALITY

The foregoing result is strongly related to the question of which value for K has to be chosen—in other words, how many directions carry the information between X and Y. In a manner similar to that of PCA, the plot of the sorted eigenvalues of $\hat{V}$ (the so-called *scree plot*) can be used with the elbow criterion to determine K. Additionally, Li (1991) defined the scale-invariant measure

$$R^2(b) = \max_{\beta \in B} \frac{\left(b^T \Sigma_X \beta\right)^2}{b^T \Sigma_X b \; \beta^T \Sigma_X \beta}. \tag{17.4}$$

Ferre (1996b) generalized this measure and investigated its properties on the basis of elliptically symmetric distributions and allowing K to be an arbitrary dimension parameter. He also investigated the behavior for the Hsing–Carroll case (Ferre, 1996a). Under the assumption that X is normal, Li (1991) found that, asymptotically,

$$n\lambda_{d-u}^* \approx \chi^2_{(d-u)(S-u-1)}, \quad \text{with } \lambda_{d-u}^* = \sum_{j=u+1}^{d} \hat{\lambda}_j, \tag{17.5}$$

where the eigenvalues $\hat{\lambda}_i$ are in descending order ($\hat{\lambda}_1 \geq \cdots \geq \hat{\lambda}_d$) and λ_{d-u}^* is the sum of the $d-u$ smallest eigenvalues. Moreover, Duan and Li (1991) investigated the efficiency of SIR under this assumption. Note that when $n\lambda_{d-u}^*$ is larger than the quantiles of the corresponding χ^2-distribution, we can assume that the edr-space is at least $K = (u+1)$-dimensional.

Schott (1994) extended this result not only to elliptically symmetric distributions, but to a larger class of estimates as well. He also gave asymptotically χ^2-distributed test statistics for estimates based on higher conditional moments than the inverse regression curve (e.g., Cov[$Z|y$], as in SIR II; see Section 17.7.1.) However, he reported that, for moments larger than two, the statistic's complexity increases considerably.

All these tests presuppose that the number of elements in each slice goes to infinity. Another asymptotic behavior is used for the above mentioned result for $\hat{\Psi}_K$. Here, the number of slices increases with n, and asymptotic normality follows for the statistics $\hat{\Psi}_K$ and can be interpreted as the ratio of the variance explained by the first K directions to the total variance. With this approach, the stochastic independence of at least two estimates for Σ_X is required if the estimation of $\hat{V}$ is based on Z. For Z_c, this condition can be relaxed (Kötter, 1996).

17.5 SLICING STRATEGIES

The SIR algorithm makes no assumption about how to separate the range of the response variable. Apparently, there are two possibilities:

- Fix the width of the slices (similar to a bandwidth in kernel regression).
- Fix the number of observations per slice (corresponding to a k-nearest neighbor approach).

In their investigations of SIR, numerous researchers have preferred the second approach to slicing. Hsing and Carroll (1992) defined a special version of SIR with exactly two elements per slice. They considered not only the original nonoverlapping slicing, but also an overlapped version. For both methods, they derived $\sqrt{n}$-consistency and asymptotic normality for the covariance structure. This is similar to the asymptotic result for $\hat{\Psi}_K$, where the elements in each slice must also be constant.

Already Li (1991, Remark 4.3) mentioned that the number of slices is less crucial than the choice of a bandwidth, as in kernel-based methods. Aragon and Saracco (1997) could overcome this problem by introducing *pooled slicing* (PSIR). Because the relevant properties of $\hat{V}$ do not depend on S, these authors proposed the estimator

$$\hat{V}_P = \sum_{r \in R} w_r \hat{V}_r, \quad w_r \geq 0, \quad \sum_{r \in R} w_r = 1,$$

where $\hat{V}_r$ is calculated by r elements per slice and R is a set of integers with recommended values larger than two. The observations that are left are equally distributed over the existing slices, so that the number of elements between the slices can differ only by one. For the weights w_r, Aragon and Saracco suggest weights either equal or proportional to the largest eigenvalue $\hat{\lambda}_1^r$ of the rth estimate. A simulation study showed a good performance of the PSIR estimates in comparison to the usual SIR and both Hsing–Carroll variants. Additionally, Aragon and Saracco (1997) considered kernel-based modifications. Instead of merely averaging the slice observations, one can apply nonparametric methods to get estimates of $E[Z|y]$ or of $\mathrm{Cov}[E[Z|y]]$ directly. Zhu and Ng (1995) studied the use of weighted methods for the inverse regression and also achieved asymptotic normality for the elements of $\hat{V}$.

17.6 IMPLEMENTATION

As already mentioned in Section 17.2, the implementation of SIR is not complicated. However, for different programming languages, efficient solutions are distinct. The reason is that using the features of a matrix or array language results in another approach to implementation than using a common language like C. Therefore, this section comprises two more or less different implementations. The first is based on the matrix language S-Plus (Kötter, 1995) and is quite similar to one given by Kötter and Turlach (1995). The second uses the special features of the new version of the array-based language XploRe (Härdle, Klinke, and Turlach, 1995). An implementation in a common high-level language like C is not harder if one has a numerical library at his or her disposal. But it surely needs more lines of code and more loops.

Before we start with the programming work, let us examine the algorithm thoroughly. The first step is clear: We must calculate the sample mean $\bar{x}$ and the covariance $\hat{\Sigma}_X$; determine $\hat{\Sigma}_X^{-1/2}$; and standardize all values. This computation is dominated by the sample size n.

The second step actually describes a nested loop. We have to run over all S slices and sum up the values belonging to the current slice. This means that we need $S \times n$ operations for step 2. Clearly, it is not necessary to run S times over the whole sample: When we sort it according to y, the X-observations already lie ordered in slices, although the slicing itself is not fixed yet. Now, the observations of a single slice can be collected by running only once through the data, and step 2 requires n operations. Ordering requires only $O(n \log(n))$ operations, so the algorithm is faster if the number of slices grows with n, as in the Hsing–Carroll case or the asymptotic results for $\hat{\Psi}_K$.

Calculating $\hat{V}$ is done in S operations, and eventually, the decomposition into eigenvectors and eigenvalues needs a constant amount of steps if d is fixed (actually, $O(d^3)$). Hence, the crucial improvement is the sorting. However, in the implementation that follows, the fact that the X-observations are actually packed into slices is more important than the possible improvement in running time.

Now we have to define the slicing strategy for our procedure. We focus on the second approach to slicing and use slices with approximately equal numbers of elements. Then we can define the number of slices or the number of slice elements as the slicing parameter. In the code that follows, the latter is used. The procedure's interface in the S-Plus language looks like

```
sir <- function (x, y, h) {
```

where h $= n_s$. Now we check the input arguments and standardize, as well as sort, the data:

```
n    <- nrow (x)
ndim <- ncol (x)
if (n != length (c (y))) {
  stop ("length of y doesn't match the rows of x") }
if (h>n) {
  stop ("Number of slice elements exceeds number of data") }

xb  <- apply (x, 2, mean)
si2 <- solve (chol (var (x)))
xt  <- (x - matrix (xb, nrow (x), ncol (x), byrow=T)) %*% si2
xt  <-  xt [order (y),]
```

In matrix languages, it is useful to exploit the implicit loops of the software. Given the index vector of the observations $(1, \ldots, n)^T$, we can find the endpoints of the slices with the vector $(n_1, n_1 + n_2, \ldots, \Sigma n_s = n)^T$. The endpoints can be easily calculated with the following code:

```
ns     <- floor (n / h)        # the number of slices
condit <- 1:n
slends <- (1:ns)*h
```

If n is not a multiplicative of n_s, the remaining elements have to be distributed over the slices. Let us spread these observations onto the first and last slices according to the following code:

```
if (h*ns != n) {
  hk     <- floor ((n-h*ns)/2)
  slends <- slends + c(1:hk,rep(hk,ns-2*hk-2),hk:(2*hk + 1))
  slends <- slends + hk }
slends[ns] <- n
```

The last line corrects an odd number of remaining elements. Before the loop over the number of slices can begin, three variables have to be introduced:

```
v   <- matrix (0, ndim, ndim) # estimate for Cov(E[z|y])
ind <-  rep (T, n)            # for already sliced elements
j   <-  1                     # loop counter
```

The matrix $\hat{V}$ is stored in v, ind serves as an indicator vector for the values that are not yet selected, and j denotes the current slice. Now we execute the loop, select the slice observations, and update v and ind:

```
while (j<=ns)  {
  sborder <– (condit <=slends[j]) & ind    # index of slice j
  if (any (sborder))  {
   ind    <– ind - sborder
   xslice <– xt [sborder,]
   if (sum (sborder) == 1)  {             # xslice is a vector
     xmean <– xslice
     v     <– v + outer(xmean, xmean, "*")  }
   else  {
     xmean <–  apply(xslice, 2, mean)
     v     <–  v + outer(xmean, xmean, "*") * nrow(xslice) } }
j <–  j+1 }
```

Finally, the eigenvalue–eigenvector decomposition has to be carried out, and the standardized edr-directions have to be rescaled. In addition, the latter are normalized to length unity:

```
eig <– eigen (v/n)
b   <– si2 %*% eig$vectors # estimates for edr-directions
b   <– t( t( b) / sqrt (apply (b*b, 2, sum)))
return (list (edr = b, evalues = eig$values))  }
```

17.6.1 Implementation without a Loop

In XploRe*, it is possible to implement the SIR algorithm without any explicit loop. XploRe offers a very flexible kind of aggregation of array structures that is used for the slice means later on. Among other things this ability supports a rapid implementation of array-based algorithms. Furthermore, in array languages, intrinsic loops are much more efficient than are explicit loops.

The above code is essentially the same in XploRe, except that an assignment is denoted by "=" instead of "<–". The while loop is replaced by the following lines:

```
k   = sum (condit > slends', 2)
nns = slends [1] | slends [2:ns] - slends[1:ns-1]
```

The variable nns consists of the numbers of elements in each slice, and, for each observation, k counts how many times the different slice ends are larger than the current observation. The argument of the function sum is a kind of indicator

* XploRe can be downloaded from http://www.xplore-stat.de.

matrix containing the results of all combinations of the comparison. Therefore, `sum` counts the appropriate numbers of slice elements. All values of `nns` lie between 0 and (`ns` – 1). Further we need:

```
sm = mean (xt, 1, k.*matrix (n, d))
v  = (nns.*sm)' * sm
```

In `sm` the mean for all slices is row-wise computed, as a function of each different value of `k`. The matrix `v` is then the weighted outer product of all slice means. The entire implementation can be found in Kötter (1998).

Finally, note that, although the data are sorted, both implementations require $O(nS)$ operations, which is equal to $O(n^2)$ if S grows proportionally with n. However, for $n < 10000$ and $S < 100$, which includes usual sample sizes, the run time is rather short.

17.7 MODIFICATIONS

Imagine that the response variable is discrete (e.g., a categorical variable). Then the algorithm can be easily applied, except that we cannot slice arbitrarily as before. In some sense, the categories determine natural slices, wherein all regressor variables with the same response are gathered. Subsequently, we have at most as many slices as distinct values for Y. Of course, we can join neighboring slices. Apparently, SIR is able to handle categorical response variables without any modifications. However, the rank of the matrix $\hat{V}$ depends on the number of slices S and therefore also on the number of categories. The rank of $\hat{V}$ is less than or equal to S, because each slice contributes a matrix of rank 1 to $\hat{V}$. Consequently, for a binary-choice regression, we can find only a single edr-direction. In this case, model (17.2) reduces to a *single-index model* (see, e.g., Chapter 12, in this volume) like that for ADE.

Another conclusion is that Y need not be a univariate variable. If the response is a multidimensional variable, the slices can be formed as before. This means that we define S cubes in the response space. If the dimension of Y is larger than two, the number of slices is affected by the curse of dimensionality. On the one hand, the number of observations might be too small to fill the slices with a sufficient number of observations; on the other hand, it might be difficult to define the slices in a way that they contain roughly the same number of elements. For example, given a three-dimensional response and five slices in each dimension, we obtain $5^3 = 125$ slices. Depending on the distribution of $X|Y$, some of these slices could be quite empty. It is much more complicated to define slices that contain essentially the same number of observations. In such a situation, Aragon et al. (1994) proposed *alternating* SIR (see also Kaddeche, 1995), in which the algorithm of SIR is applied iteratively, while the roles of X and Y alternate after each iteration.

17.7.1 SIR II

In fact, SIR can miss edr-directions even if assumption (17.3) is valid. The reason is that it is "blind" for symmetric dependencies. Li (1992) motivates this fact as follows:

Example. Let X follow a bivariate normal distribution, $(X_1, X_2)^T \approx N(0, I_2)$, $\beta = (1, 0)^T$, and $Y = X_1^2$. Then $E[X_2|y] = 0$ because of the independence and $E[X_1|y] = 0$ because of the symmetry.

Therefore, the inverse regression does not contain any information about the edr-direction β. Li (1992) modified SIR so that not the inverse regression $E[Z|y]$, but the conditional covariance $\text{Cov}[Z|y]$, is considered. He showed that, under the additional assumption that X follows a normal distribution, the eigenvectors with the largest eigenvalues of the matrix

$$E\left[(\text{Cov}[Z|y] - E[\text{Cov}[Z|y]])^2\right]$$

are estimates for the standardized edr-directions. The implementation of this modification is quite easy. Instead of computing $E[Z|y]$, one estimates the covariance $\text{Cov}[Z|y]$. The rest of the procedure remains practically the same. Earlier, Cook and Weisberg (1991) had addressed the problem of missed edr-directions in proposing *slice average variance estimates* (SAVE), which are actually a combination of SIR and SIR II. In the more general case of elliptically symmetric distributions, a similar asymptotic χ^2-test for the unknown K is available as before (Schott, 1994).

17.8 AN EXAMPLE

Finally, SIR should be applied to a dimension-reduction problem. Let x be a five-dimensional regressor that is normally distributed with identity matrix as covariance, and let $\beta_1 = (1, 0, 1, 0, 1)^T/2$ and $\beta_2 = (1, 1, 0, 0, 0)^T$ be directions. The error ϵ is independently standard normal, and the response is computed by

$$y_i = \sin(\beta_1^T x_i) \exp(\beta_2^T x_i - 1) + 1/2\epsilon_i.$$

$n = 500$ observations were simulated, and SIR was performed with 20, 30, and 40 observations per slice, which correspond to 25, 16, and 12 slices, respectively. Figure 17.1 shows both indices, $\beta_1^T x$ and $\beta_2^T x$. The left panel illustrates the first index $\beta_1^T x$ vs. y and the right panel the second index $\beta_2^T x$ vs. y. The shape of the exponential function is visible. Although the range of y seems to be quite large (the maximum is approximately 15), the minimum is -2.77, and the 95 percent quantile is 2.84. Hence, the mass of the data lies in a rather small band.

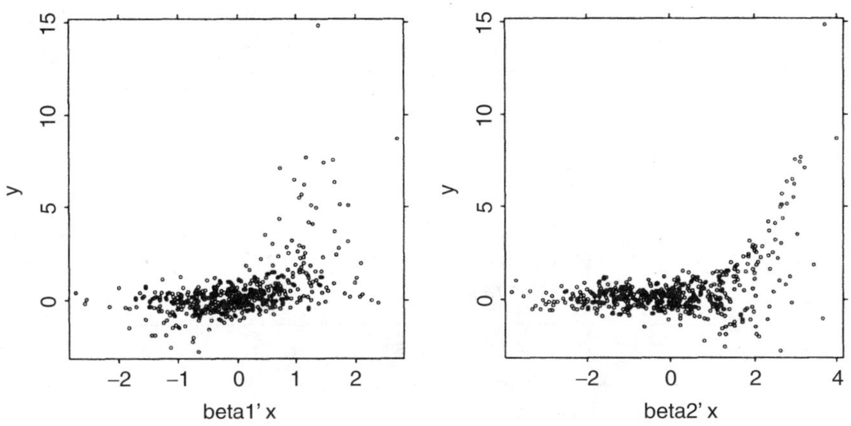

FIGURE 17.1 Plot of the indices $\beta_1^T x$ and $\beta_2^T x$ vs. y.

TABLE 17.1 The Eigenvalues for the Three Different Slicing Parameters

Observations per Slice	$\hat{\lambda}_1$	$\hat{\lambda}_2$	$\hat{\lambda}_3$	$\hat{\lambda}_4$	$\hat{\lambda}_5$
20	0.432	0.251	0.058	0.045	0.017
30	0.425	0.224	0.044	0.030	0.011
40	0.409	0.222	0.032	0.017	0.005

Table 17.1 shows the results for the eigenvalues for the three different slicing parameters.

The estimates according to the first two eigenvalues are displayed in Table 17.2, with the first direction belonging to the higher eigenvalue. The estimates $\hat{\beta}_1$ and $\hat{\beta}_2$ are quite similar, no matter the number of observations per slice — 20, 30, or 40. In order to judge them, we could consider the projection matrices of the true edr-space B and its estimate $\hat{B}$. In Table 17.3, we compare the projector based on the first two edr-directions with 20 elements per slice with the true projector. Apparently, they are very similar. For 30 and 40 elements per slice, the results are also nearly the same. For the three different estimates, Li's measure (17.4) was computed for $\hat{\beta}_1$ and $\hat{\beta}_2$ (see Table 17.4.).

Now let us check how well the estimates are judged. For each different slicing and for each possible dimension, we can compute the χ^2-test statistic (17.5). Here, we are in a simulated setting and know the edr-directions. In a real situation, one should apply this test with care, because the error level of the sequential application of the test is unknown. Table 17.5 displays the values of the statistic for each of the three numbers of slices — 25, 16, and 12.

TABLE 17.2 The First Two Directions Estimated by SIR with 20, 30, and 40 Observations per Slice

Observations per Slice		x_1	x_2	x_3	x_4	x_5
20	$\hat{\beta}_1$	0.765	0.540	0.177	0.018	0.234
	$\hat{\beta}_2$	−0.177	0.653	−0.543	−0.058	−0.493
30	$\hat{\beta}_1$	0.743	0.542	0.213	0.041	0.266
	$\hat{\beta}_2$	−0.174	0.678	−0.487	−0.039	−0.520
40	$\hat{\beta}_1$	0.774	0.528	0.191	0.022	0.226
	$\hat{\beta}_2$	−0.164	0.647	−0.504	−0.010	−0.561

TABLE 17.3 The True and the Estimated Projection Matrices for the edr-Space *B*

0.60	0.40	0.20	0.00	0.20
0.40	0.60	−0.20	0.00	−0.20
0.20	−0.20	0.40	0.00	0.40
0.00	0.00	0.00	0.00	0.00
0.20	−0.20	0.40	0.00	0.40
0.64	0.31	0.24	0.03	0.28
0.31	0.73	−0.26	−0.03	−0.19
0.24	−0.26	0.32	0.04	0.31
0.03	−0.03	0.04	0.00	0.03
0.28	−0.19	0.31	0.03	0.30

TABLE 17.4 Li's Measure (17.4) for $\hat{\beta}_1$ and $\hat{\beta}_2$ for all Applied Slicings

	20	30	40
$\hat{\beta}_1$	0.961	0.960	0.964
$\hat{\beta}_2$	0.955	0.960	0.982

In Table 17.6 the p-values and the degrees of freedom (in parentheses) of the corresponding χ^2-distribution are presented. Regardless of the number of slices, the test statistics suggest a two-dimensional dependency. Comparing these results with the estimates of Ψ_K supports the hypothesis that K is 2, as Table 17.7 suggests.

TABLE 17.5 Values of the Test Statistic (17.5) for $u = 0, \ldots, 4$ and the Number of Slices

	25	16	12
$n\ \lambda_5^*$	8.29	5.47	2.66
$n\ \lambda_4^*$	30.57	20.55	11.12
$n\ \lambda_3^*$	59.67	42.69	27.10
$n\ \lambda_2^*$	185.15	154.66	138.32
$n\ \lambda_1^*$	401.11	367.25	342.57

TABLE 17.6 p-Values and Degrees of Freedom (in parentheses) of the Test Statistic (17.5) for $u = 0, \ldots, 4$ and the Number of Slices

	25		16		12	
$n\ \lambda_5^*$	0.010	(20)	0.094	(11)	0.085	(7)
$n\ \lambda_4^*$	0.095	(42)	0.335	(24)	0.198	(16)
$n\ \lambda_3^*$	0.305	(66)	0.684	(39)	0.542	(27)
$n\ \lambda_2^*$	1	(92)	1	(56)	1	(40)
$n\ \lambda_1^*$	1	(120)	1	(75)	1	(55)

TABLE 17.7 $\hat{\Psi}_K$ for all Dimensions and Slicings

	25	16	12
$\hat{\Psi}_1$	0.538	0.579	0.596
$\hat{\Psi}_2$	0.851	0.884	0.921
$\hat{\Psi}_3$	0.924	0.944	0.968
$\hat{\Psi}_4$	0.980	0.985	0.992
$\hat{\Psi}_5$	1	1	1

The second dimension raises the value of $\hat{\Psi}$ for all numbers of slices to approximately 90 percent. The gain from the first to the second direction is in all cases above 30 percent, while for the third direction it is only about 5 percent.

REFERENCES

Aragon, Y., Li, K. C., and Thomas-Agnan, C. (1994). Modeling income distributions using multivariate sliced inverse regression. (unpublished).

Aragon, Y., and Saracco, J. (1997). Sliced inverse regression: An appraisal of small sample alternatives to slicing, *Computational Statistics*, **12**, 109–130.

Cook, R., and Weisberg, S. (1991). Comment on "Sliced inverse regression for dimension reduction," *J. Amer. Statist. Assoc.*, **86**, 328–332.

Diaconis, P., and Freedman, D. (1984). Asymptotics of graphical projection pursuit, *Ann. Statist.*, **12**, 793–815.

Duan, N., and Li, K. (1991). Slicing regression: A link-free regression method, *Ann. Statist.*, **19**, 505–530.

Ferre, L. (1996a). Choix de Dimension en Régression Inverse par Tranches (SIR), *C.R. Acad. Sci.*, Paris, France.

Ferre, L. (1996b). *Determining the Dimension of Sliced Inverse Regression and Related Methods*, Technical Report, Université Paul Sabatier, Toulouse, France.

Friedman, J., and Stuetzle, W. (1981). Projection pursuit regression, *J. Amer. Statist. Assoc.*, **76**, 817–823.

Hall, P., and Li, K. (1993). On almost linearity of low dimensional projections from high dimensional data, *Ann. Statist.*, **21**, 867–889.

Härdle, W., Klinke, S., and Turlach, B. A. (1995). *XploRe: An Interactive Statistical Computing Environment*, Springer-Verlag, Berlin, Germany.

Hsing, T., and Carroll, R. (1992). An asymptotic theory for sliced inverse regression, *Ann. Statist.*, **20**, 1040–1061.

Kaddeche, N. (1995). Computation of sliced inverse regression in XploRe. (unpublished).

Klinke, S. (1997). *Data Structures in Computational Statistics*, Physica-Verlag, Heidelberg, Germany.

Kötter, T. (1995). S-plus function for sliced inverse regression, Statlib Archive, Carnegie Mellon University, `http://lib.stat.cmu.edu/S/sir`.

Kötter, T. (1996). An asymptotic result for sliced inverse regression, *Computational Statistics*, **11**, 113–136.

Kötter, T. (1998). *Entwicklung statistischer Software*, Physica-Verlag, Heidelberg, Germany.

Kötter, T., and Turlach, B. A. (1995). Additive modeling, in W. Härdle, S. Klinke, and B. A. Turlach (eds.), *XploRe: An Interactive Statistical Computing Environment*, pp. 223–250, Springer-Verlag, Berlin, Germany.

Li, K. (1991). Sliced inverse regression for dimension reduction (with discussion), *J. Amer. Statist. Assoc.*, **86**, 316–342.

Li, K. (1992). Sliced inverse regression-II. (unpublished).

Li, K., and Duan, N. (1987). Distribution-free and link-free estimation for the sample selection model, *J. Econometrics*, **35**, 25–35.

Powell, J. L., Stock, J., and Stoker, T. M. (1989). Semiparametric estimation of index coefficients, *Econometrica*, **57**, 1403–1430.

Schott, J. (1994). Determining the dimensionality in sliced inverse regression, *J. Amer. Statist. Assoc.*, **89**, 141–148.

Zhu, L., and Ng., K. W. (1995). Asymptotics of sliced inverse regression, *Statistica Sinica*, **5**, 727–736.

CHAPTER 18

Dynamic and Semiparametric Models

Ludwig Fahrmeir and Leonhard Knorr-Held
Ludwig-Maximilians-Universität München, Munich, Germany

18.1 INTRODUCTION

This chapter surveys dynamic or state-space models and their relationship to non- and semiparametric models that are based on the roughness-penalty approach. We concentrate on recent advances in dynamic modeling of non-Gaussian time series and longitudinal data, with special emphasis on discrete responses. Furthermore, we describe how ideas from dynamic models can be adopted for Bayesian semiparametric inference in generalized additive and in varying coefficient models. Basic tools for corresponding inference techniques are penalized likelihood estimation, Kalman filtering and smoothing, and Markov chain Monte Carlo (MCMC) simulation. Similarities and relative merits, advantages, and disadvantages of these methods are illustrated by means of several applications.

Section 18.2 gives a short introductory review of results for the classical situation of Gaussian time series observations. We start with Whittaker's (1923) "method of graduation" for estimating trends and show that it is equivalent to the posterior mean estimate from a linear Kalman filter model with known smoothing or variance parameters. We sketch extensions to general Gaussian linear dynamic or state-space models and to continuous-time analogues, such as the Bayesian version of cubic spline smoothing (Wahba, 1978). For more detailed expositions of the equivalence between Bayesian smoothness priors and penalized least squares, we refer the reader to Kohn and Ansley (1988) and previous work cited there and to van der Linde (1995, and Chapter 2 in this

Smoothing and Regression: Approaches, Computation, and Application
Edited by Michael G. Schimek
ISBN 0-471-17946-9

book) for a thorough discussion of splines from a Bayesian point of view. This equivalence also suggests alternative ways of estimating unknown smoothing or variance parameters: Within a semiparametric approach, estimation by optimizing some cross-validated criterion is a common choice. Empirical Bayes models, also treating hyperparameters as fixed or unknown, lead to marginal likelihood estimation. Maximization can be done by EM-type algorithms. Fully Bayesian models put a weakly informative prior on the hyperparameters and make feasible a complete posterior analysis with MCMC techniques.

We then turn briefly to so-called conditionally Gaussian dynamic models that are still linear, but with errors distributed as scale mixtures of normals. Even with this seemingly moderate generalization, penalized least squares and posterior mean estimates are no longer equivalent. Beyond various approximate Kalman filters and smoothers, fully Bayesian approaches based on MCMC are available that make efficient use of the conditionally Gaussian structure.

Fundamentally non-Gaussian time series and longitudinal data, in particular for categorical and count data, are considered in Section 18.3. Dynamic binomial and Poisson models are important members of the family of dynamic generalized linear models. Semiparametric counterparts based on penalized likelihood estimation can be derived as posterior mode estimators, with extended or iterative Kalman-type smoothing algorithms as efficient computational tools (Fahrmeir, 1992; Fahrmeir and Tutz, 1994, chapter 8; Fahrmeir and Wagenpfeil, 1997), but the equivalence between posterior mean and penalized likelihood estimation is lost. Fully Bayesian inference is possible with recently developed MCMC techniques for non-Gaussian dynamic models, an area of intensive current research. In Section 18.3.2, we outline the ideas for Metropolis–Hastings algorithms suggested by Knorr–Held (1999). These algorithms are used for the applications; they are generalized in Section 18.3.3 to nonnormal longitudinal data having additional unobserved population heterogeneity across units.

Ideas from non-Gaussian dynamic modeling, in particular for unequally-spaced or continuous-time parameter models, can be transferred to semiparametric regression models for cross-sectional data (Section 18.4). This transfer leads to Bayesian spline-type smoothing for generalized additive and varying coefficient models using MCMC techniques as a supplement and an alternative to penalized likelihood or, equivalently, to posterior mode estimation (Hastie and Tibshirani, 1990, 1993).

Finally, Section 18.5 summarizes conclusions, and then it indicates extensions to other data situations and statistical models.

18.2 LINEAR DYNAMIC MODELS AND OPTIMAL SMOOTHING FOR TIME SERIES DATA

This section gives a brief survey on the correspondence between linear dynamic or state-space models and semiparametric optimal-smoothing methods based on the roughness-penalty approach. We illustrate this correspondence by some simple and commonly-used examples, and then we review more general and recent work.

18.2.1 Gaussian Models

In the classical smoothing problem treated by Whittaker (1923), time series observations $y = (y_1, \ldots, y_T)$ are assumed to be the sum

$$y_t = \tau_t + \epsilon_t, \quad t = 1, \ldots, T \tag{18.1}$$

of a smooth trend function τ and an irregular noise component ϵ. Whittaker suggested estimating τ by minimizing the *penalized least squares* (PLS) criterion,

$$\text{PLS}(\tau) = \sum_{t=1}^{T} (y_t - \tau_t)^2 + \lambda \sum_{t=3}^{T} (\tau_t - 2\tau_{t-1} + \tau_{t-2})^2, \tag{18.2}$$

with respect to $\tau = (\tau_1, \ldots, \tau_T)$. Minimization of the PLS criterion (18.2) tries to balance two desirable criteria: close fit to the data, as measured by the first term in (18.2); and the smoothness of the estimate of the trend, as measured by the second term in (18.2). The smoothing parameter λ controls the trade-off between these two goals.

The trend model (18.1) and the PLS criterion (18.2) can be generalized in a flexible way. Inclusion of a seasonal component $\gamma = (\gamma_1, \ldots, \gamma_T)$ with period m, leads to the additive trend-seasonal model

$$y_t = \tau_t + \gamma_t + \epsilon_t, \quad t = 1, \ldots, T \tag{18.3}$$

and to the PLS criterion

$$\begin{aligned} \text{PLS}(\tau, \gamma) = &\sum_{t=1}^{T} (y_t - \tau_t - \gamma_t)^2 + \lambda_1 \sum_{t=3}^{T} (\tau_t - 2\tau_{t-1} + \tau_{t-2})^2 \\ &+ \lambda_2 \sum_{t=m}^{T} (\gamma_t + \gamma_{t-1} + \cdots + \gamma_{t-m+1})^2. \end{aligned} \tag{18.4}$$

The trend function τ and the seasonal component γ are estimated by minimizing (18.4). More generally, the influence of covariates can be taken into account by extending the additive predictor $\tau_t + \gamma_t$ to

$$\eta_t = \tau_t + \gamma_t + x_t^T \beta_t + w_t^T \delta, \tag{18.5}$$

with time-varying effects β_t for x_t and constant effects δ for w_t.

This penalized-least-squares approach is reasonable if time series observations are (at least approximately) Gaussian. This approximation is made explicit by assuming that the errors ϵ_t in (18.1) and (18.3) are i.i.d. $N(0, \sigma^2)$ random variables. Then the fit term in (18.4) corresponds to the log-likelihood of the additive Gaussian observation model (18.3), and the PLS approach appears as a semiparametric method for estimating the fixed, unknown sequences $\tau = (\tau_1, \ldots, \tau_T)$ and $\gamma = (\gamma_1, \ldots, \gamma_T)$.

The dynamic model corresponding to (18.3) and (18.4) considers τ and γ as sequences of random variables. It is hierarchical and consists of two stages. The first stage is the Gaussian observation model (18.3) for y, given τ and γ. In the second stage, a transition model corresponding to the roughness-penalty term in (18.4) is given by the difference equations

$$\tau_t - 2\tau_{t-1} + \tau_{t-2} = u_t, \qquad t = 3, \ldots, T \tag{18.6}$$

$$\gamma_t + \gamma_{t-1} + \cdots + \gamma_{t-m+1} = w_t, \qquad t = m, \ldots, T. \tag{18.7}$$

The errors u_t and w_t are distributed i.i.d. $N(0, \sigma_u^2)$ and $N(0, \sigma_w^2)$, respectively. Initial values are specified—for example, by

$$(\tau_1, \tau_2)^T \sim N(0, k_\tau I), \qquad (\gamma_1, \ldots, \gamma_{m-1})^T \sim N(0, k_\gamma I). \tag{18.8}$$

All errors and initial values are assumed to be mutually independent. The difference equation (18.6), also called a random walk of second order, penalizes deviations from the linear trend $\tau_t = 2\tau_{t-1} - \tau_{t-2}$. Similarly, the model (18.7) penalizes deviations from the fixed seasonal pattern $\gamma_t + \gamma_{t-1} + \cdots + \gamma_{t-m+1} = 0$.

For the moment, we will consider the hyperparameters (i.e., the variances σ^2, σ_u^2, σ_w^2) as known or given constants, and we will suppress them in our notation. From a Bayesian point of view, (18.6) and (18.7), together with (18.8), define a multivariate normal prior [$\mathrm{p}(\tau, \gamma) = \mathrm{p}(\tau)\mathrm{p}(\gamma)$, for $(\tau, \gamma) = (\tau_1, \ldots, \tau_T, \gamma_1, \ldots, \gamma_T)$], the so-called smoothness prior. Also, the observation model (18.3) defines a multivariate normal distribution $\mathrm{p}(y|\tau, \gamma)$ for the data y, given τ and γ. Therefore, the posterior

$$\mathrm{p}(\tau, \gamma | y) = \frac{\mathrm{p}(y|\tau, \gamma)\mathrm{p}(\tau)\mathrm{p}(\gamma)}{\mathrm{p}(y)} \propto \mathrm{p}(y|\tau, \gamma)\mathrm{p}(\tau)\mathrm{p}(\gamma) \tag{18.9}$$

is also normal and is characterized by the posterior expectation $E(\tau, \gamma|y)$ and the covariance $\mathrm{Var}(\tau, \gamma|y)$. Because of normality, the posterior expectation and the posterior mode (i.e., the maximizer of (18.9)) coincide. By taking the logarithm, using suitable conditional-independence assumptions, and ignoring constant factors, the posterior mode can be calculated by minimizing

$$\frac{1}{\sigma^2}\sum_{t=1}^{T}(y_t - \tau_t - \gamma_t)^2 + \frac{1}{k_\tau}(\tau_1^2 + \tau_2^2) + \frac{1}{\sigma_u^2}\sum_{t=3}^{T}(\tau_t - 2\tau_{t-1} + \tau_{t-2})^2$$
$$+ \frac{1}{k_\gamma}(\gamma_1^2 + \cdots + \gamma_{m-1}^2) + \frac{1}{\sigma_w^2}\sum_{t=m}^{T}(\gamma_t + \cdots + \gamma_{t-m+1})^2 \tag{18.10}$$

with respect to τ and γ. Setting $\lambda_1 = \sigma^2/\sigma_u^2$ and $\lambda_2 = \sigma^2/\sigma_w^2$, and choosing diffuse priors (18.8) with $k_\tau \to \infty$, $k_\gamma \to \infty$, the criteria (18.4) and (18.10) are identical.

As a result, the semiparametric PLS estimate $\widehat{\tau}$, $\widehat{\gamma}$ is identical to the posterior mode estimate and (because of posterior normality) the posterior mean:

$$(\widehat{\tau}, \widehat{\gamma}) = E(\tau, \gamma | y). \tag{18.11}$$

This equivalence remains valid for more general linear Gaussian observation and transition models; see Kohn and Ansley (1988) for a thorough treatment. By collecting the trend, the seasonal, and other parameters, as in (18.5), into a so-called state vector α_t, for example,

$$\alpha_t^T = (\tau_t, \tau_{t-1}, \gamma_t, \dots, \gamma_{t-m+1}, \beta_t, \delta),$$

most linear dynamic models can be put into the form of Gaussian linear state-space models

$$y_t = z_t^T \alpha_t + \epsilon_t, \quad \epsilon_t \sim N(0, \sigma^2), \tag{18.12}$$

$$\alpha_{t+1} = F_t \alpha_t + v_t, \quad v_t \sim N(0, Q_t), \tag{18.13}$$

by appropriate definition of design vectors z_t and transition matrices F_t, and covariance matrices Q_t. (See, for example, Harvey (1989), West and Harrison (1989), or Fahrmeir and Tutz, 1994, chapter 8.1.) The well-known classical Kalman filter and smoother, or recent variants like the diffuse filter (de Jong, 1991), can be used for efficiently computing posterior expectations $\widehat{\alpha}_t = E(\alpha_t | y)$ and variances $\text{Var}(\alpha_t | y)$. Because of the equivalence with a corresponding PLS criterion, the Kalman filter and smoother can also be regarded as an algorithmic tool for computing semiparametric PLS estimates, without any need for a Bayesian interpretation. (See Fahrmeir and Tutz, 1994, chapter 8.1.) Using Kalman smoothers for semiparametric additive models like (18.3) avoids backfitting, and it provides diagonal bands of smoother matrices as a by-product. However, forcing dynamic models into state-space form can result in high-dimensional state vectors with singular multivariate priors, results which cause unnecessary algorithmic complications.

Up until now, we have tacitly assumed that the time series is spaced equally in time. Extensions to unequally spaced data are possible, either by using modified difference priors or by using continuous-time models. For example, a first-order random walk [$\tau_t - \tau_{t-1} = u_t$, $u_t \sim N(0, \sigma_u^2)$] is generalized to

$$\tau_t - \tau_{t-1} = u_t, \quad u_t \sim N(0, \delta_t \sigma_u^2), \tag{18.14}$$

where δ_t is the time between observation y_{t-1} and y_t. Similarly, a generalized second-order random walk can be defined by

$$\tau_t - \left(1 + \frac{\delta_t}{\delta_{t-1}}\right) \tau_{t-1} + \frac{\delta_t}{\delta_{t-1}} \tau_{t-2} = u_t, \quad u_t \sim N(0, k_t \sigma_u^2), \tag{18.15}$$

where the left-hand side is in the form of second differences for unequally spaced data. The factor k_t on the right hand side is a weight function that depends on δ_t and δ_{t-1}.

A simple and straightforward ad-hoc choice is $k_t = \delta_t$, as was used for first-order random-walk priors. In our experience, this choice is quite reasonable from a practical point of view. More elaborate forms can be derived by different arguments: for example, by using second-order differences for unequally-spaced data, to approximate second derivatives; or by using consistency arguments, as in Knorr–Held (1997). As in Harvey (1989, chapter 9) or in Kohn and Ansley (1987), one could also derive discrete-time versions of the continuous-time models described in the following paragraph.

For continuous-time models, the trend, the seasonal, and other time-varying parameters are considered smooth functions of time. With a slight change in notation, the simple trend model (18.1) becomes

$$y_s = \tau(t_s) + \epsilon_s, \qquad s = 1, \ldots, T,$$

having observation times $t_1 \leq \cdots \leq t_s \leq \cdots \leq t_T$, a smooth trend function $\tau(t)$, and i.i.d. errors $\epsilon_s \sim N(0, \sigma^2)$. A continuous-time version of the PLS criterion (18.2) is also easily constructed: Find a twice-differentiable function $\widehat{\tau}$ that minimizes

$$\sum_{s=1}^{T} (y_s - \tau(s))^2 + \lambda \int (\tau''(t))^2 \, dt \tag{18.16}$$

with respect to τ. The minimizing function is a cubic smoothing spline; see Green and Silverman (1994) and Chapter 1 of this volume.

Wahba (1978) showed that (18.16) has a Bayesian justification, by placing the solution of the stochastic differential equation

$$\frac{d^2\tau(s)}{ds^2} = \lambda^{-1/2} \sigma \frac{dW(s)}{ds}, \qquad s \geq t_1 \tag{18.17}$$

as a smoothness prior over τ. Such a differential equation of order two is the continuous-time version of a second-order random walk (18.6). Here $W(s)$ is a standard Wiener process, with $W(t_1) = 0$, independent of the errors ϵ_s. For the diffuse initial conditions

$$(\tau(t_1), \tau^T(t_1)) \sim N(0, kI), \qquad k \to \infty,$$

the cubic smoothing spline $\widehat{\tau}(s)$ at s coincides with the posterior expectation of $\tau(s)$ given the data—that is,

$$\widehat{\tau}(s) = E(\tau(s)|y).$$

This equivalence can be established for more general types of splines also, ones where second derivatives are replaced by linear differential operators. Ansley and Kohn (1983) give a simple proof of the smoothness properties of conditional means of stochastic differential equations. Kohn and Ansley (1987, 1988) also derive a discrete-time stochastic difference equation from (18.17), and they use state-space techniques for computation of the smoothing spline. Again, pointwise Bayesian confidence bands can be computed as a by-product. For a recent discussion of splines from a Bayesian point of view, we refer to van der Linde (1995).

In practice, smoothing parameters λ or such hyperparameters as σ^2, σ_u^2, and σ_w^2 are usually unknown. Within the semiparametric roughness-penalty approach, a data-driven choice of smoothing parameters is often made by cross-validated optimization of some selection criterion. Even for a small number of smoothing parameters, problems may occur, because the selection criterion can be a rather flat function of $\lambda = (\lambda_1, \lambda_2, \ldots)$. Within an empirical Bayes approach, hyperparameters in dynamic models are treated as unknown constants; in that situation, the method of maximum likelihood is a natural choice, and it was suggested by Wecker and Ansley (1983). A comparison of direct numerical optimization routines with indirect maximization via the EM algorithm is given in Harvey (1989, chapter 4). If the likelihood is rather flat, then ML estimation also performs poorly. Fully Bayesian approaches can avoid these problems by providing additional information about hyperparameters in the form of "hyperpriors". A traditional approach, using discrete priors, leads to multiprocess Kalman filters (Harrison and Stevens, 1976). More recently, Markov chain Monte Carlo (MCMC) techniques have been developed to estimate hyperparameters by simulation from their posteriors (Carlin, et al. 1992; Carter and Kohn, 1994; Frühwirth-Schnatter, 1994; de Jong and Shephard, 1995). An advantage of these simulation methods is that their basic concepts are useful also in conditionally non-Gaussian situations, such as the ones in the following sections.

18.2.2 Conditionally Gaussian Models

Gaussian models are not robust against outliers in the observation errors, change points in the trend function, or other unobserved components. One way to robustify linear dynamic models is to assume that error distributions are scale mixtures of normals. For given values of the mixture variables, the linear dynamic model is then conditionally Gaussian. Mixture variables may be discrete or continuous. A popular choice are χ^2-mixture variables, leading to t-distributions for the errors. A conditionally Gaussian version of the simple trend model (18.1) with a second-order random-walk model for the trend is

$$\begin{aligned} y_t &= \tau_t + \epsilon_t, & \epsilon_t|\omega_{1t} &\sim N(0, \sigma^2/\omega_{1t}), \\ \tau_t - 2\tau_{t-1} + \tau_{t-2} &= u_t, & u_t|\omega_{2t} &\sim N(0, \sigma_u^2/\omega_{2t}). \end{aligned}$$

Assuming $\omega_{1t}\nu_1$ and $\omega_{2t}\nu_2$ to be independently χ^2-distributed with ν_1 and ν_2 degrees of freedom, then ϵ_t and u_t are independently $t(\nu_1)$- and $t(\nu_2)$-distributed. Although Kalman filters and smoothers are still best linear estimators, they perform poorly for small values of ν_1 and ν_2. Various approximate filtering and smoothing algorithms have, therefore, already been given, in early work on robustified state-space modeling (Masreliez, 1975; Masreliez and Martin, 1977; Martin and Raftery, 1987). More recently, fully Bayesian MCMC methods have been developed to tackle this problem. Carlin, Polson, and Stoffer (1992) suggest a Gibbs sampling algorithm by adding the mixture variables ω_{1t} and ω_{2t} to the set of unknown parameters. Their approach applies to rather general nonnormal dynamic models, but it can be inefficient with respect to mixing and convergence properties. Carter and Kohn (1994) and Frühwirth-Schnatter (1994) propose a modified Gibbs sampling algorithm for Gaussian state-space models, one that updates the "state vector" $\tau = (\tau_1, \ldots, \tau_T)$ all at once. This modification makes the algorithm much more efficient. The parameters $\tau_1, \ldots, \tau_T$ are often highly correlated, so updating τ_t, $t = 1, \ldots, T$ one at a time, as is done in Carlin, Polson, and Stoffer (1992), often results in poor mixing (i.e., the corresponding Markov chain is not moving rapidly throughout the support of the posterior distribution). Consequently, Monte Carlo standard errors of sample averages will be large. The idea of updating the whole sequence τ at once was adapted to the more general class of conditionally Gaussian models in Shephard (1994) and in Carter and Kohn (1996a, 1996b).

As an alternative to these fully Bayesian methods, one may also consider posterior mode estimation. Let $\rho_1(\epsilon_t)$ and $\rho_2(u_t)$ denote the negative log-densities of the i.i.d. errors ϵ_t and u_t. Taking logarithms and using (conditional) independence assumptions, a robustified version of the PLS criterion (18.2), namely,

$$\sum_{t=1}^{T} \rho_1(y_t - \tau_t)^2 + \sum_{t=3}^{T} \rho_2(\tau_t - 2\tau_{t-1} + \tau_{t-2})^2, \tag{18.18}$$

can be derived. Computation of the minimizer $\widehat{\tau}$ of (18.18) can be carried out by iterative Kalman-type algorithms; see Künstler (1996). An advantage of posterior mode estimation is that it can be extended to other ρ-functions, for example, Huber functions, or $\rho(u) = |u|$. Also, one may work directly from criterion (18.18), without Bayesian interpretation, to obtain robust semiparametric estimators, and one may transfer this approach to robust continuous-time spline-type estimation. It should be noted, however, that, even for conditionally Gaussian dynamic linear models, posterior mean estimates (obtained from a fully Bayesian approach) and posterior mode or spline-type estimators are no longer equivalent. This property holds only for linear Gaussian models with known hyperparameters, as in Section (18.2.1).

18.3 NON-GAUSSIAN OBSERVATION MODELS

This section deals with fundamentally non-Gaussian time-series and longitudinal data. We progress from simple examples for discrete-valued time series to general non-Gaussian situations.

18.3.1 Non-Gaussian Time Series

Figure 18.1 displays the number, y_t, of occurrences of rainfall exceeding 1 mm in the Tokyo area, for each calendar day during the years 1983–1984. The data, presented in Kitagawa (1987) and reanalyzed later by several authors, constitute an example of a discrete-valued time series. Responses y_t, $t = 1, \ldots, 366$, are assumed binomial; thus,

$$y_t \sim B(n_t, \pi_t) \text{ with } \begin{cases} n_t = 2 \text{ for } t \neq 60, \\ n_t = 1 \text{ for } t = 60 \quad \text{(February 29)} \end{cases},$$

where π_t is the probability of rainfall on calendar day t. To compare it to similar data from other areas or other years, and to see some seasonal pattern, the probabilities $\pi = (\pi_1, \ldots, \pi_T)$, $T = 366$, will be estimated as a smooth curve. We reparameterize π_t by a logit link to τ_t:

$$\tau_t = \log \frac{\pi_t}{1 - \pi_t}, \quad \text{or} \quad \pi_t = \pi(\tau_t) = \frac{\exp(\tau_t)}{1 + \exp(\tau_t)}.$$

A semiparametric discrete-time roughness-penalty approach will work from a *penalized log-likelihood* (PL) criterion

$$\text{PL}(\tau) = \sum_{t=1}^{T} \{y_t \log \pi(\tau_t) + (n_t - y_t) \log(1 - \pi(\tau_t))\} - \lambda \sum_{t=3}^{T} (\tau_t - 2\tau_{t-1} + \tau_{t-2})^2 \tag{18.19}$$

in order to obtain smooth estimates $\hat{\tau} = \arg\max \text{PL}(\tau)$ and $\hat{\pi} = \pi(\hat{\tau})$ of the fixed, unknown sequences τ and π. Comparison with the penalized-least-squares criterion, (18.2), shows that, essentially, the Gaussian log-likelihood of the observation

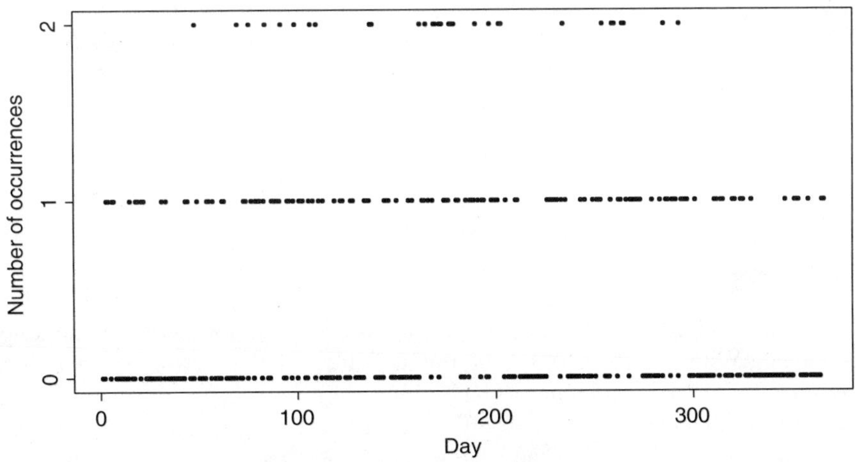

FIGURE 18.1 Tokyo Rainfall Data.

model (18.1) is replaced by the sum of binomial log-likelihood contributions. Instead of second-order differences, one might use a sum $\Sigma(\tau_t - \tau_{t-1})^2$ of squared first-order differences.

In terms of the notation in Section 18.2, the continuous-time version of (18.19) is constructed as follows: Find a twice-differentiable function $\widehat{\tau} = \arg\max \text{PL}(\tau)$, where

$$\text{PL}(\tau) = \sum_{s=1}^{T} \{y_s \log \pi(\tau(t_s)) + (n_s - y_s) \log(1 - \pi(\tau(t_s)))\} - \lambda \int (\tau''(t))^2 \, dt. \tag{18.20}$$

For a given smoothing parameter λ, the solution is again a cubic smoothing spline; see Hastie and Tibshirani (1990) and Green and Silverman (1994). For equally-spaced data, as in the example, the discrete- and continuous-time spline solutions to (18.19) and (18.20) are usually in quite close agreement. Efficient algorithmic solutions of the high-dimensional nonlinear optimization problems (18.19) and (18.20) are usually carried out by iteratively applying smoothers for penalized-least-squares estimation to working observations.

For a Bayesian version of the semiparametric approach (18.19) in the form of a non-Gaussian dynamic model, we take

$$y_t|\tau_t \sim B(n_t, \pi(\tau_t)), \quad \pi(\tau_t) = \frac{\exp(\tau_t)}{1 + \exp(\tau_t)} \tag{18.21}$$

as the observation model. We supplement it, as we did in (18.6), by a random-walk model of first or second order

$$\tau_t = \tau_{t-1} + u_t \text{ or } \tau_t = 2\tau_{t-1} - \tau_{t-2} + u_t, \tag{18.22}$$

which is used as a smoothness prior for τ. The errors u_t are i.i.d. $N(0, \sigma_u^2)$-distributed, and the initial values τ_1 (resp. τ_1, τ_2) are specified as in (18.8). Variances are assumed to be known. In addition, conditional independence is assumed among all $y_t|\tau_t$.

In contrast to what happens in Gaussian models, the posterior

$$\text{p}(\tau|y) = \frac{\text{p}(y|\tau)\text{p}(\tau)}{\text{p}(y)} \propto \text{p}(y|\tau)\text{p}(\tau) \tag{18.23}$$

is now nonnormal. Thus, posterior expectations and posterior modes are no longer equivalent. With a diffuse prior for initial values, the posterior mode $\widehat{\tau}$ is the maximizer of (18.19), and it has its smoothing parameter, λ, equal to $1/2\sigma_u^2$. Algorithmic solutions can be obtained efficiently by the implementation of extended or iterative Kalman filtering and smoothing. (See Fahrmeir, 1992; Singh and Roberts, 1992; Fahrmeir and Tutz, 1994; and Fahrmeir and Wagenpfeil, 1997).

As in the Gaussian case, these techniques may also be viewed as convenient computational tools for computing penalized likelihood estimators, without a Bayesian interpretation. For a fully Bayesian analysis, including computation of posterior moments and quantiles, simulation-based estimation methods, in particular MCMC methods, are generally the most appropriate. Details are given in Section 18.3.2.

A continuous-time dynamic model corresponding to (18.20) is obtained by placing the stochastic differential equation (18.17) as a smoothness prior over τ. Once again, posterior modes are still equivalent to cubic smoothing splines but are different from posterior expectations. Fully Bayesian spline-type smoothing will also be based on MCMC for dynamic models. For this purpose, it is useful to rewrite the continuous-time prior (18.17) as a stochastic difference equation for the state vector

$$\alpha_t = (\tau(t),\ d\tau(t)/dt)$$

of the trend, τ, and its derivative. Following Kohn and Ansley (1987), we note that the sequence $\alpha_s := \alpha(t_s)$ of evaluations at t_s, $s = 1, \ldots, T$ obeys the stochastic difference equation

$$\alpha_{s+1} = F_s \alpha_s + u_s, \qquad s = 1, \ldots, T, \tag{18.24}$$

with transition matrices

$$F_s = \begin{pmatrix} 1 & \delta_{s+1} \\ 0 & 1 \end{pmatrix}, \qquad \delta_{s+1} = t_{s+1} - t_s$$

and independent errors

$$u_s \sim N(0, \sigma^2 U_s), \qquad U_s = \begin{pmatrix} \delta_{s+1}^3/3 & \delta_{s+1}^2/2 \\ \delta_{s+1}^2/2 & \delta_{s+1} \end{pmatrix}.$$

Putting (18.24) together with the observation model

$$y_s | \tau(t_s) \sim B(n_s, \pi(t_s)),$$

we obtain a binomial dynamic, or state-space, model. Higher-order splines can also be written in state-space form; see Kohn and Ansley (1987).

As a second example, we consider a time series of counts y_t of the weekly incidence of *acute hemorrhagic conjunctivitis* (AHC) in Chiba prefecture in Japan during 1987. Kashiwagi and Yanagimoto (1992) analyzed these data, assuming a log-linear Poisson model

$$y_t | \lambda_t \sim Po(\lambda_t), \qquad \lambda_t = \exp(\tau_t)$$

and a first-order random-walk prior for τ. They obtain a posterior mean estimate based on numerical integrations, which are similar to that in Kitagawa (1987).

Of course, other smoothness priors as second-order random walks and as the continuous-time analogue (18.17) can be used as well. A PL approach would start directly from

$$\mathrm{PL}(\tau) = \sum_{t=1}^{T} (y_t \log \lambda_t - \lambda_t) - \lambda \sum_{t=2}^{T} (\tau_t - \tau_{t-1})^2$$

(or with other forms of the roughness-penalty term). Once again, the penalized-likelihood estimators $\widehat{\tau} = \arg\max \mathrm{PL}(\tau)$ are equivalent to posterior mode estimators but different from corresponding posterior means.

Both examples belong to the class of dynamic generalized linear models. The general *observation model* is constructed as follows.

The conditional density of y_t, given the unknown state vector α_t, is of the linear exponential-family type, and has the conditional expectation

$$E(y_t|\alpha_t) = \mu_t = h(\eta_t),$$

which is related to the linear predictor $\eta_t = z_t'\alpha_t$ by a suitable link h. As in the Gaussian case, the components of α_t may consist of trend τ_t, seasonal γ_t, and some (possibly time-varying) effects β_t of the covariates x_t, and z_t is a suitable design vector. For example, an additive predictor,

$$\eta_t = \tau_t + \gamma_t + x_t^T \beta_t,$$

can be written in this form. Although time-constant effects δ can be incorporated formally by setting $\delta_t = \delta_{t-1}$, it is often advantageous to split up the predictor into

$$\eta_t = \tau_t + \gamma_t + x_t^T \beta_t + w_t^T \delta.$$

For the second stage, smoothness priors $\mathrm{p}(\alpha)$ are put on the sequence $\alpha = (\alpha_1, \ldots, \alpha_T)$, in the form of a *transition model.* Linear Gaussian transition models, such as the difference equations (1.6) and (1.7) or such as the state-space form $\alpha_{t+1} = F\alpha_t + u_t$, are often retained as a common choice, but we will also use priors for unequally-spaced observations and for continuous-time priors.

As we did for the examples, we can always write down a corresponding semiparametric model and an associated PL criterion

$$\mathrm{PL}(\alpha) = \sum_{t=1}^{T} l_t(y_t|\alpha_t) - \sum_{j=1}^{p} \lambda_j I(\alpha_j). \tag{18.25}$$

Here, $\alpha_j = (\alpha_{j1}, \ldots, \alpha_{jT})$ is the j-th component of α, $I(\alpha_j)$ is a penalty function, and λ_j is a smoothing parameter. For given smoothing parameters λ_j, estimates $\widehat{\alpha} = \arg\max \mathrm{PL}(\alpha)$ are obtained by iterative smoothing (with backfitting in an inner loop); see Hastie and Tibshirani (1990, 1993), Green and Silverman (1994), or Fahrmeir and Tutz (1994, chapter 5). As in the examples, (18.25) can always be derived from the corresponding dynamic model by using the principle of posterior mode estimation.

Estimation of unknown smoothing parameters λ_j or of corresponding hyperparameters σ_j^2 can be based on the same principles as for Gaussian models. Using the roughnes-penalty approach, smoothing parameters are selected as the minimizer of a generalized cross-validation criterion; see O'Sullivan et al. (1986) or Wahba et al. (1995). Empirical Bayes approaches consider hyperparameters σ_j^2 as unknown but fixed and then use (approximate) maximum-likelihood estimation, for example, an approximate EM-type algorithm, such as is suggested in Fahrmeir (1992). Wagenpfeil (1996) compares some of these approaches. In a fully Bayesian setting, hyperparameters σ_j^2 are considered to be random, and independent inverse-gamma priors

$$\sigma_j^2 \sim \text{IG}(a_j, b_j), \qquad j = 1, \ldots, p \tag{18.26}$$

are a common choice for hyperpriors. By appropriate choices of a_j and b_j, these priors can be made more or less informative.

18.3.2 MCMC in Non-Gaussian Dynamic Models

The design of efficient MCMC algorithms in dynamic models with non-Gaussian observation model is currently an active research area. For easier presentation, we first discuss several MCMC algorithms for simple non-Gaussian dynamic-trend models, such as the dynamic-binomial or Poisson models in the examples in preceding sections. Extensions to the general case are outlined at the end of this subsection. When we supplement the model (18.21) and (18.22) with a hyperprior $\text{p}(\sigma_u^2)$ for σ_u^2—for example, an inverse-gamma prior as in (18.26)—the posterior distribution of the parameters τ and σ_u^2 is then given by

$$\text{p}(\tau, \sigma_u^2|y) \propto \text{p}(y|\tau)\text{p}(\tau|\sigma_u^2)\text{p}(\sigma_u^2). \tag{18.27}$$

MCMC methods construct Markov chains that converge to a given distribution (here, the posterior). Once the chain has reached equilibrium, it provides (dependent) samples from that posterior distribution. Quantities of interest, such as the posterior mean or median, can now be estimated by the appropriate empirical versions.

The well-known Gibbs sampling algorithm (e.g., Gelfand and Smith, 1990) is based on samples from the full conditional distributions of all parameters. Any full conditional distribution is proportional to the posterior (18.27), but often considerable simplification can be done. To implement the Gibbs sampler in dynamic trend models, we have to sample from

$$\text{p}(\tau_t|\tau_{s\neq t}, \sigma_u^2, y) \propto \text{p}(y_t|\tau_t)\text{p}(\tau_t|\tau_{s\neq t}, \sigma_u^2) \tag{18.28}$$

and

$$p(\sigma_u^2|\tau, y) \propto p(\tau|\sigma_u^2)p(\sigma_u^2). \tag{18.29}$$

If inverse gamma priors are assigned to σ_u^2, (18.29) is still inverse-gamma, and samples can be generated easily using standard algorithms.

Suppose we could also easily generate samples from (18.28), $t = 1, \ldots, T$. The Gibbs sampling algorithm iteratively updates $\tau_1, \ldots, \tau_T$ and σ_u^2 by samples from their full conditionals. Markov-chain theory shows that the sequence of random numbers so generated converges to the posterior (18.27) for any starting value of the Markov chain. Such an algorithm is proposed in Fahrmeir et al. (1992), following suggestions of Carlin et al. (1992). However, there are some drawbacks to pure Gibbs sampling in non-Gaussian dynamic models. First, samples from (18.28), which is nonstandard for non-Gaussian observation models, can be obtained only by a carefully designed rejection algorithm, one which itself may already require a considerable amount of computation time. Fortunately, instead of sampling from the full conditional distribution, a member of the more general class of Hastings algorithms (Hastings, 1970) can be used to update τ_t, $t = 1, \ldots, T$. Here so-called *proposals* are generated from an arbitrary distribution, and a specific accept-reject step is added. Such a Hastings step is typically both easier to implement and more efficient in terms of computer time. A thorough discussion of the Hastings algorithm is given in Tierney (1994) and in Besag et al. (1995).

For example, to update (18.28), it is sufficient to generate a proposal τ_t^* from the conditional prior distribution $p(\tau_t|\tau_{s\neq t}, \sigma_u^2)$ and then accept that proposal as the new state of the Markov chain, with probability

$$\delta = \min\left\{1, \frac{p(y_t|\tau_t^*)}{p(y_t|\tau_t)}\right\},$$

where τ_t denotes the current state of the chain. The resulting algorithm requires less computation time than pure Gibbs sampling, because the conditional prior distribution is Gaussian with known moments and so proposals are easy to generate.

The generated Markov chain might, however, show signs of slow convergence, and it does not mix rapidly—that is, the Markov chain is not moving rapidly throughout the support of the posterior distribution. As a result, subsequent samples are highly dependent, and Monte Carlo estimates become imprecise. This inefficiency is a consequence of the underlying single-move strategy (i.e., the parameters τ_t, $t = 1, \ldots, T$, are updated one by one). Various attempts have been made to design algorithms that converge quickly and mix rapidly. A fruitful idea is the use of blocking: Blocks of parameters, say $\tau_{(a,b)} = (\tau_a, \tau_{a+1}, \ldots, \tau_{b-1}, \tau_b)$, can be updated simultaneously, rather than step by step. Such a blocking strategy is a compromise between updating τ all at once (a method infeasible for fundamentally non-Gaussian time series) and updating τ one element at a time. The algorithms of Shephard and Pitt (1997) and of Knorr–Held (1999) are based on blocking. Knorr–Held generalizes the conditional-prior proposal earlier in this section to *block-move* algorithms: Generate a proposal $\tau^*_{(a,b)}$ from the conditional prior distribution $p(\tau_{(a,b)}|\tau_{(1,a-1)}, \tau_{(b+1,T)}, \sigma^2_u)$, and accept that proposal as the new state of the Markov chain, with probability

$$\delta = \min\left\{1, \frac{\prod_{t=a}^{b} p(y_t|\tau^*_t)}{\prod_{t=a}^{b} p(y_t|\tau_t)}\right\}.$$

Generation of the proposal $\tau^*_{(a,b)}$ involves an inversion of a precision matrix and a Cholesky decomposition. Standard algorithms can be used as long as the block size is not too large. For large block size, the block-diagonal structure of the precision matrix can be used to implement a version more efficient numerically.

One of the advantages of MCMC is the possibility of calculating exact posterior distributions of functionals of parameters. For the Tokyo Rainfall Data, the posterior estimates of the probabilities

$$\pi_t = \frac{\exp(\tau_t)}{1 + \exp(\tau_t)} \tag{18.30}$$

are of main interest. Instead of plugging an estimate for $\{\tau_t\}$ into (18.30), we calculate posterior samples from $\{\pi_t\}$, using the original samples from $p(\tau|y)$. The posterior distributions $p(\pi|y)$ can now be explored in detail, without any approximation. In contrast, posterior mode or spline estimation does not have this feature. In these cases, plug-in estimates, especially confidence bands, typically are biased, because of the non-linearity in (18.30). Similar considerations apply to the AHC example, where $\lambda_t = \exp(\tau_t)$ is to be estimated.

Figure 18.2 shows the posterior estimates of the probabilities $\{\pi_t\}$ for the Tokyo Rainfall Data, calculated by a conditional-prior block-MCMC algorithm. A highly dispersed but proper inverse-gamma hyperprior (18.26), with $a = 1$ and $b = 0.00005$, was assigned to σ^2_u. This prior has a mode at 0.000025. The estimated posterior median was 0.0001. The pattern in Figure 18.2, with peaks for wet seasons, nicely reflects the climate in Tokyo. It would be difficult to see this by looking at only the raw data (Figure 18.1). In Fahrmeir and Tutz (1994, section 5.3), the probabilities $\{\pi_t\}$ are fitted by a cubic smoothing spline, with the smoothing parameter estimated by a generalized cross-validation criterion. This criterion had two local minima, one at $\lambda = 32$ and one at $\lambda = 4064$. The smoothing spline for $\lambda = 4064$ is quite close to the posterior median fitted in Figure 18.2, while the smoothing spline for $\lambda = 32$ is much rougher. Such rough posterior median estimates are also obtained if the parameter b for the inverse-gamma prior is set to higher values. For example, with $a = 1$ and $b = 0.005$, the prior mode equals 0.0025. This prior is in favor of larger values for σ^2_u; hence,

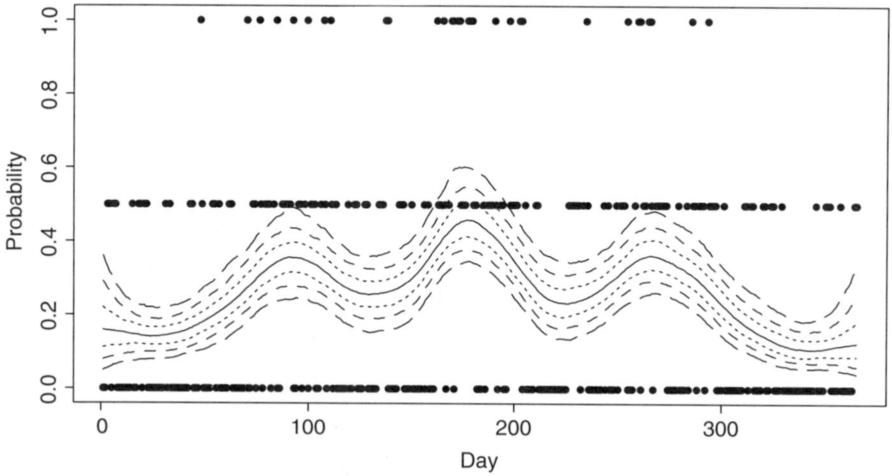

FIGURE 18.2 Tokyo Rainfall Data. Observations and fitted probabilities (posterior median within 50, 80 and 95 percent credible regions). The data is reproduced as relative frequencies with values 0, 0.5 and 1.

posterior median estimates for $\{\pi_t\}$ become rougher. As a third approach, posterior mode estimation, with an approximate EM-type algorithm for estimating σ_u^2 by maximization of the marginal likelihood, also gives estimates that are in good agreement. These results correspond to empirical evidence experienced in other applications: If smoothing and variance parameters are properly adjusted, posterior means and medians are often rather close to posterior modes and penalized-likelihood estimates. Also, estimation of hyperparameters by cross-validation or by marginal likelihood can be helpful for the choice of parameters of the hyperprior in a fully Bayesian model.

Estimates for the AHC data are shown, in Figure 18.3(a) and (b), for both first- and second-order random-walk priors. The posterior distribution of the intensities $\{\lambda_t\}$ shows a peak around week 33 similar to the results of Kashiwagi and Yanagimoto (1992). Compared with those from the model with second-order random-walk priors, estimates in Figure 18.3(a) are somewhat rougher, and the peak around week 33 is lower and flatter. This reflects the fact that first-order random-walk priors are in favor of horizontal, locally straight lines. Figure 18.3(c) shows Bayesian cubic spline-type estimates with the continuous-time prior (18.17). As was to be expected with equally spaced observations, these estimates are in very close agreement with those in Figure 18.3(b). Figure 18.3(d) displays the cubic smoothing spline, which is the posterior mode estimator from the Bayesian point of view. As in the rainfall data example, it is quite close to the posterior median in 18.3(c).

In more general dynamic models, response y_t is related to some unknown parameter vector α_t; see, for example, the state-space representation (18.24) of the spline-type prior (18.17). MCMC simulation in dynamic models can be

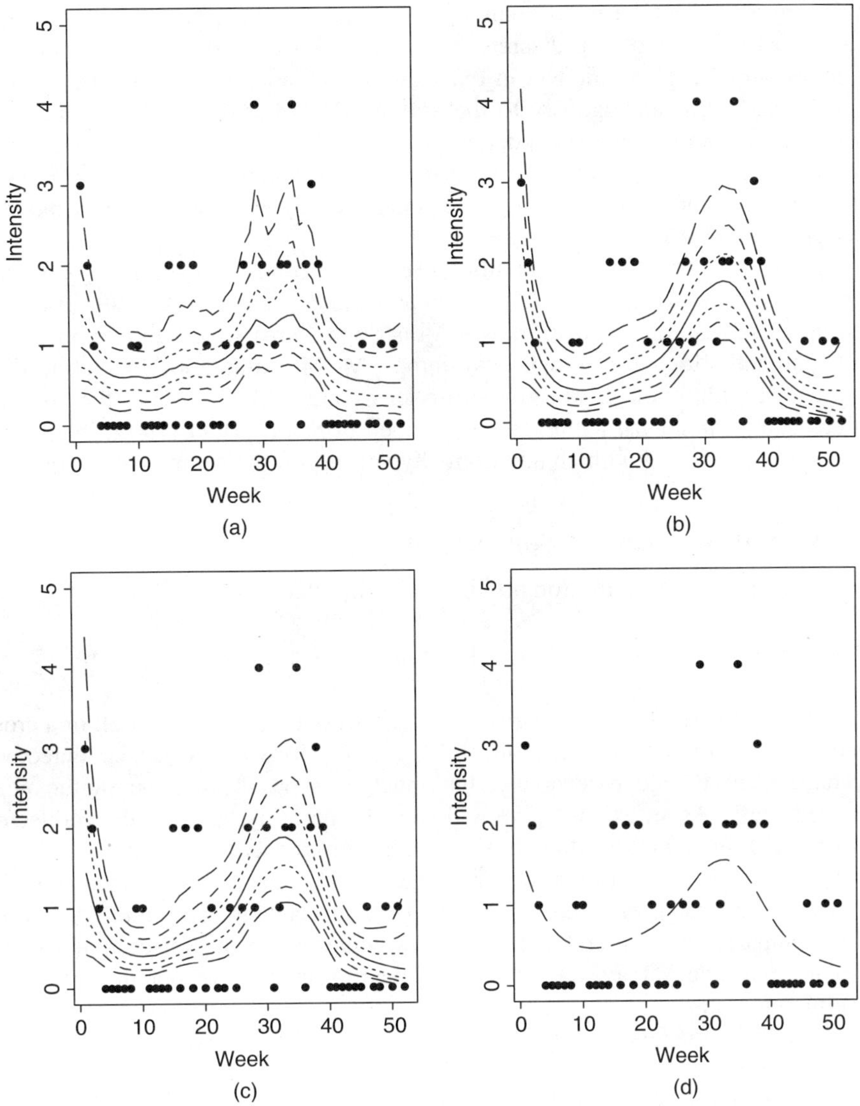

FIGURE 18.3 AHC Data. Observations and fitted intensities (posterior median within 50, 80 and 95 percent credible regions); (a) First-order random walk prior, (b) second-order random walk prior, (c) stochastic differential equation prior, (d) posterior mode estimator.

performed (as it is for the simple dynamic-trend model, where $\alpha_t = \tau_t$ is a scalar) by single- or block-move algorithms. Shephard and Pitt (1997) make specific Fisher scoring-type steps in order to construct a proposal that approximates the full conditional distribution, taking the observation into account. In contrast, conditional-prior proposals are built independent of the observation and are, therefore, easier to construct. Their performance is good for situations in which the posterior is not very different from the conditional prior. Such situations are typically the case for discrete-valued observations, such as the binomial or multinomial logistic models in our examples. Sometimes, components α_j of $\alpha = (\alpha_1, \ldots, \alpha_T)$ (analogously to the notation in (18.25)) are a priori independent, and a componentwise updating strategy with conditional-prior proposals can have advantages. Componentwise updating becomes inevitable in problems involving multiple time scales or, more generally, in generalized additive models. (See Section 18.4.)

Finally, we note that it is possible to combine robust-transition models with non-Gaussian observation models, much as we did Section 18.2.2. For example, one may use random-walk priors with t-distributed errors for trend components; this approach allows for abrupt, large jumps. MCMC simulation in such models is often straightforward, because error terms are still Gaussian (the mixture values being unknown). An example is given in Knorr–Held (1999), with $t(\nu)$-distributed errors and with an additional hyperprior on the degrees of freedom, ν.

18.3.3 Non-Gaussian Longitudinal Data

In this section we consider longitudinal data, in which the observations

$$(y_{ti}, x_{ti}), \quad t = 1, \ldots, T, \quad i = 1, \ldots, n$$

on a response variable y and on a vector x of covariates are made for all of a cross section of n units at each time point $t = 1, \ldots, T$. Models for Gaussian outcomes y_{ti} have been treated extensively, but much less has been done in the non-Gaussian case. As an example, we will consider monthly business data collected by the IFO institute in Munich for a large cross-section of firms. Answers given in a monthly questionnaire are categorical, most of them trichotomous (typically with categories like "increase" (+), "no change" (=), and "decrease" (−)); compare Fahrmeir and Tutz (1994, examples 6.3, 8.5). Selecting a specific response variable y (say, answers on production plans), we obtain categorical longitudinal data.

Observation models for longitudinal data can be defined by appropriate extensions of models for time-series data. A straightforward generalization within the exponential-family framework is as follows: For given covariates x_{ti} and for a possibly time-varying parameter vector α_t, the q-dimensional response y_{ti} comes from a linear exponential density having the conditional mean

$$E(y_{ti}|x_{ti}, \alpha_t) = h(\eta_{ti}) \tag{18.31}$$

and the linear predictor

$$\eta_{ti} = Z_{ti}\alpha_t. \tag{18.32}$$

Here $h: \mathrm{IR}^q \rightarrow \mathrm{IR}^q$ is a q-dimensional link, and the matrix Z_{ti} is a function of the covariates x_{ti} (and possibly of past responses). Individual responses are assumed to be conditionally independent. A dynamic model for longitudinal data is obtained by supplementing the observation model (18.31) and (18.32) with transition models as smoothness priors for α, as was done in Section 18.3.1. Just as for time series, some subvector $\widetilde{\alpha}_t$ of α_t may, indeed, be time-constant. Such partially dynamic models are formally covered by (18.32) with the addition of the restriction $\widetilde{\alpha}_t = \widetilde{\alpha}_{t-1}$, or they can be covered by making this condition explicit by rewriting the predictor in the additive form $\eta_{ti} = Z_{ti}\alpha_t + V_{ti}\beta$.

The posterior-mode or penalized-likelihood approach maximizes

$$\mathrm{PL}(\alpha) = \sum_{t=1}^{T}\sum_{i=1}^{n} l_{ti}(\alpha_t) - \sum_{j=1}^{p} \lambda_j I(\alpha_j) \tag{18.33}$$

with respect to α. Here, $l_{ti}(\alpha_t) = \log f(y_{ti}|x_{ti}, \alpha_t)$ is the likelihood contribution of observation y_{ti}. Computationally efficient solutions can be obtained by, for example, extended or iterative Kalman-type smoothers; see Fahrmeir and Tutz (1994, chapter 8.4) and Wagenpfeil (1996).

Observation models of the form (18.31), (18.32) may be appropriate if heterogeneity among units is sufficiently described by observed covariates. This sufficiency will not always be the case, in particular for larger cross-sections. A natural way to deal with this problem is an additive extension of the linear predictor to

$$\eta_{ti} = Z_{ti}\alpha_t + W_{ti}b_i,$$

where b_i are unit-specific parameters and W_{ti} is an appropriate design matrix. A *dynamic mixed* model is obtained from a usual transition model for α and a *random effects* model for the unit-specific parameter. A common assumption is to assume that the b_i s are i.i.d. Gaussian,

$$b_i \sim N(0, D), \tag{18.34}$$

with covariance matrix D. For posterior-mode or penalized-likelihood estimation of $\alpha = (\alpha_1, \ldots, \alpha_T)$ and $b = (b_1, \ldots, b_n)$, a further penalty term

$$I(b) = \sum_{i=1}^{n} b_i^T D b_i,$$

corresponding to the Gaussian prior (18.34), is added to (18.33). An algorithmic solution for the resulting joint posterior-mode or penalized-likelihood estimates $(\widehat{\alpha}, \widehat{b})$ is worked out in Biller (1997), which uses an EM-type algorithm for estimation of smoothing parameters. However, computation times become large for multicategorical responses. Moreover, serious bias may occur; see Breslow and Clayton (1993) and Breslow and Lin (1995).

MCMC techniques are more attractive for dynamic mixed models on account of their model flexibility. The parameters $b_1, \ldots, b_n$ are added to the set of unknown parameters, and they are updated with some well-designed proposals — for example, with Metropolis random-walk proposals — in every MCMC cycle. Additionally, a hyperprior for D has to be introduced. The usual choice is the inverted Wishart distribution

$$\mathrm{p}(D) \propto |D|^{-\zeta-(m+1)/2} \exp(-\operatorname{tr}(BD^{-1})),$$

with parameters $\zeta > (m-1)/2$ and $|B| > 0$; here, m is the dimension of b_i. A Gibbs step can then be used to update D.

Turning to the IFO business example, we investigate the dependency of current production plans on demand and on orders in hand, in the specific branch "Vorprodukte Steine und Erden". We have complete longitudinal observations of 51 firms, for the period from 1980 to 1994. Our model allows for time-changing effects of covariates and for trend and seasonal variation of threshold parameters (which represent corresponding probabilities of the response categories). Additional unit-specific parameters, b_i, are introduced to allow for firm-specific differences of these probabilities.

The response variable "production plans" is given in three ordered categories: "increase" (+), "no change" (=), and "decrease" (−). Its conditional distribution is assumed to depend on the covariates "orders in hand" and "expected business conditions" and on the production plans of the previous month. All these covariates are trichotomous. We used a dummy-coding approach for comparison with previous analyses; we took the category (−) as the reference category. The corresponding dummies are denoted by A^+, $A^=$ (orders in hand), G^+, $G^=$ (expected business conditions) and P^+, $P^=$ (production plans of the previous month); they define the covariate vector x_{ti}. The inclusion of P as a covariate reduces the panel length (by 1) to $T = 179$ (February 1980 to December 1994).

A cumulative logistic model (e.g., Fahrmeir & Tutz, 1994a, chapter 3) was used, because of the ordinal nature of the response variable: Let $\widetilde{y}_{ti} = 1$ and $\widetilde{y}_{ti} = 2$ denote the response categories "increase" and "no change", respectively. Then

$$P(\widetilde{y}_{ti} \le j) = F(\theta_{tij} + x_{ti}^T \beta_t), \qquad j = 1, 2$$

is assumed, with $x_{ti} = (G^+, G^=, P^+, P^=, A^+, A^=)'$ and $F(x) = 1/(1 + \exp(-x))$.

We decompose both threshold parameters θ_{ti1} and θ_{ti2} into trend parameters τ_t, seasonal parameters γ_t, and unit-specific parameters b_i, one for each threshold:

$$\theta_{tij} = \tau_{tj} + \gamma_{tj} + b_{ij}, \qquad j = 1, 2.$$

Note that the threshold parameters have to follow the restriction $\theta_{ti1} < \theta_{ti2}$ for all combinations of t and i. A seasonal model (18.7) with period $m = 12$ was chosen for the seasonal parameters of both thresholds. First-order random-walk priors are assigned to all covariate effect parameters β_t and to both trend parameters θ_{t1}, θ_{t2}. All time-changing parameters are assumed to be mutually independent, with proper but highly dispersed inverse-gamma hyperpriors ($a = 1, b = 0.005$).

The firm-specific parameters $b_i = (b_{i1}, b_{i2})^T$ are assumed to follow a Gaussian distribution having mean zero and dispersion D. We used the parameter values $\zeta = 1$ and $B = \mathrm{diag}(0.005, 0.005)$ for the inverted Wishart hyperprior specification for D.

This model can be written as a dynamic mixed model with

$$\pi_{ti} = h(\eta_{ti}) = h(Z_{ti}\alpha_t + W_{ti}b_i),$$

where $\alpha_t^T = (\tau_{t1}, \gamma_{t1}, \tau_{t2}, \gamma_{t2}, \beta_t^T)$,

$$W_{ti} = \begin{pmatrix} 1 & 0 \\ 0 & 1 \end{pmatrix},$$

and

$$Z_{ti} = \begin{pmatrix} 1 & 1 & 0 & 0 & x_{ti}^T \\ 0 & 0 & 1 & 1 & x_{ti}^T \end{pmatrix}.$$

The response variable y_{ti} is multinomially distributed; that is,

$$y_{ti} \sim M_2(1, \pi_{ti}),$$

where $y_{ti} = (1, 0)^T$, $(0, 1)^T$, or $(0, 0)^T$ if the first (+), second (=), or third (−) category, respectively, is observed. The link function h is given by

$$h(\eta_{ti}) = \begin{pmatrix} F(\eta_{ti1}) & & \\ F(\eta_{ti2}) & - & F(\eta_{ti1}) \end{pmatrix}.$$

Figure 18.4 displays the temporal pattern of the trend parameters τ_{tj}, $j = 1, 2$ and of both threshold parameters $\theta_{tj} = \tau_{tj} + \gamma_{tj}$, $j = 1, 2$. The first trend parameter is slightly decreasing, while the second remains constant over the whole period. A distinct seasonal pattern can be seen, with higher probabilities of positive response in spring and negative response in fall. Firm-specific deviations from this pattern are, however, substantial. They can be seen in Figure 18.5. Here, posterior median estimates of the first and second firm-specific parameter b_{i1} and b_{i2} are plotted against each other for all 51 firms. Interestingly, these two parameters are often highly negatively correlated. The estimated dispersion matrix of the random effect distribution is

$$\widehat{D} = \begin{pmatrix} 0.78 & -0.28 \\ -0.28 & 0.23 \end{pmatrix};$$

the estimated correlation, based on posterior samples of the corresponding functional of D, is -0.67. Both estimates are posterior median estimates. Some firms

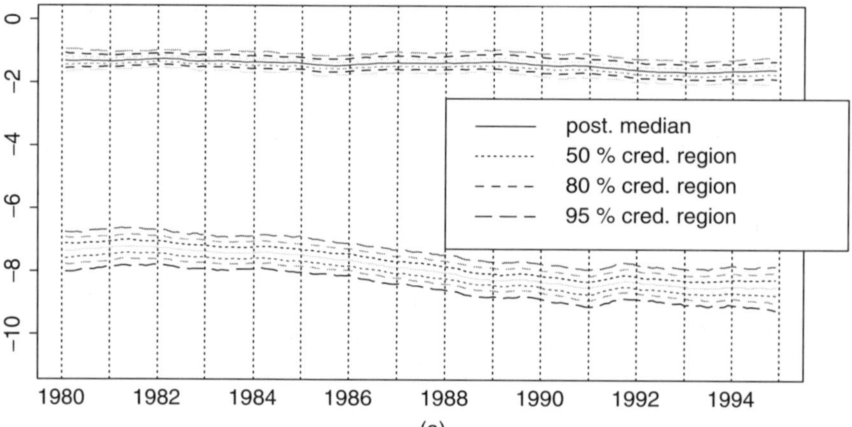

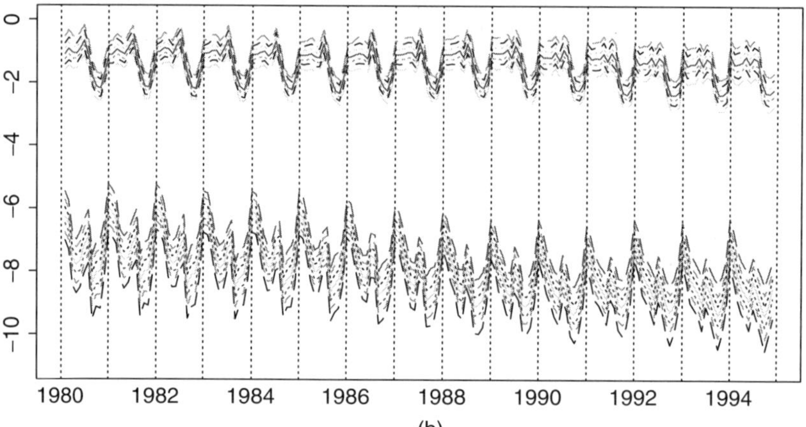

FIGURE 18.4 Estimates of trends (a) and thresholds (b). Dashed vertical lines represent the month January of each year.

seem more "conservative" in their answers; they choose "no change" for the response variable often, relative to the overall frequencies. Such firms have negative values for b_{i1} and positive values for b_{i2}. Other firms seem to avoid the category "no change"; they often answer more extremely, with "decrease" or "increase". For these firms, b_{i1} is positive, and b_{i2} is negative.

The estimated patterns of time-dependent covariate effects (Figure 18.6) show an interesting temporal pattern (in particular, the effect of the dummy $G+$ (Figure 18.7), which stands for expected improved business conditions, relative to $G-$). A distinct low can be seen at the end of 1981, when the German economy was shaken by a recession. In 1982, a new government, under the leadership of Chancellor Helmut Kohl, was established. From that time onward, the effect increases until 1989/1990 (with some additional variation); it can be interpreted as a growing trust in the government.

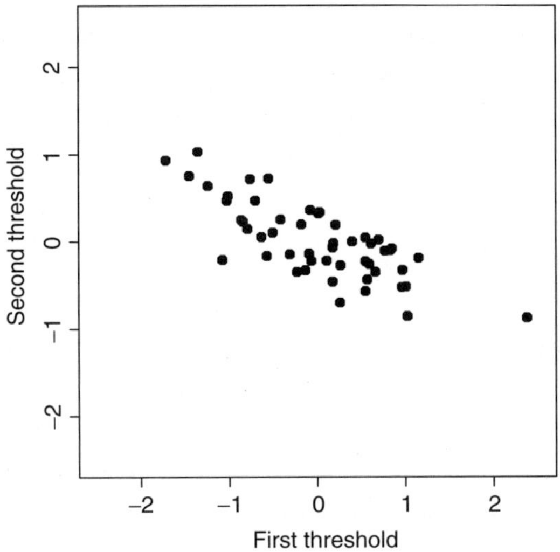

FIGURE 18.5 Plot of the estimates of b_{i1} against b_{i2} for each unit.

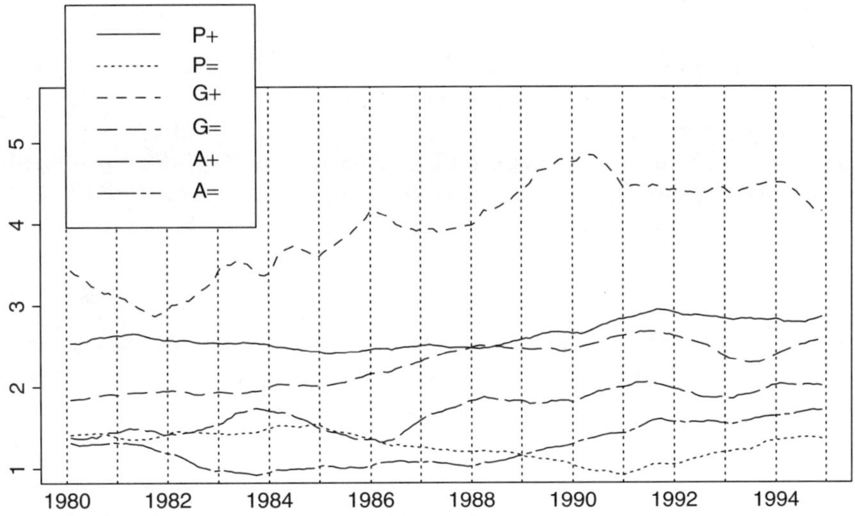

FIGURE 18.6 Estimated time–changing covariate effects (point estimates). Dashed vertical lines represent the month January of each year.

The peak in 1989/1990 coincidences with the German reunification, which was expected to have a catalytic effect on the economy through the sudden opening of the market in the former East Germany. In the years 1986, 1990, and 1994, parliamentary elections were held in the fall. In these years, the effect is always decreasing towards the end of the year; that pattern may be due to uncertainty about the election results.

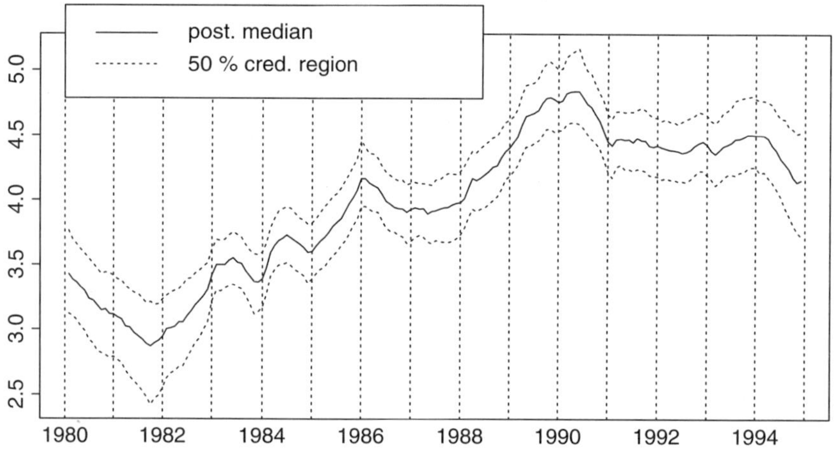

FIGURE 18.7 Estimated time–changing covariate effect of G^+. Dashed vertical lines represent the month January of each year.

18.4 GENERALIZED ADDITIVE AND VARYING COEFFICIENT MODELS

Let us now turn to a cross-sectional regression situation, in which observations $(y_i, x_{i1}, \ldots, x_{ip}), i = 1, \ldots, n$ on a response y and a vector $(x_1, \ldots, x_p)$ of covariates are given. Generalized additive models (GAMs; Hastie and Tibshirani, 1990, and Chapter 10 in this book) assume that, given $x_i = (x_{i1}, \ldots, x_{ip})$, the distribution of y_i belongs to an exponential family having mean $\mu_i = E(y_i|x_i)$ linked to an additive predictor η_i by

$$\mu_i = h(\eta_i), \quad \eta_i = f_1(x_{i1}) + \cdots + f_p(x_{ip}). \tag{18.35}$$

Here $f_1, \ldots, f_p$ are unknown, smooth functions of continuous covariates $x_1, \ldots, x_p$. If some covariates are assumed to have a linear effect on the predictor, then semiparametric or generalized partial linear models like

$$\eta_i = f_1(x_{i1}) + \cdots + \beta_p x_{ip} \tag{18.36}$$

are appropriate modifications of (18.35). In (18.36), x_{ip} could also be a binary or categorical covariate. In the following, we will focus on model (18.35).

Nonparametric estimates of the functions $f_1, \ldots, f_p$ are found by maximizing the PL criterion

$$\text{PL}(f_1, \ldots, f_p) = \sum_{i=1}^{n} l_i(y_i|\eta_i) - \sum_{j=1}^{p} \lambda_j \int (f_j''(u))^2 \, du, \tag{18.37}$$

with individual likelihood contributions l_i from $y_i|x_i$. The maximizing functions are cubic smoothing splines $\widehat{f}_1, \ldots, \widehat{f}_p$. Other types of penalty terms are possible — for example, replacing second derivatives by mth-order derivatives $f_j^{(m)}(u)$, or using discretized penalty terms. Computation is usually carried out by backfitting algorithms; see Hastie and Tibshirani (1990) or Fahrmeir and Tutz (1994, chapter 5). One drawback is that construction of confidence bands relies on the approximate normality of the penalized-likelihood estimators; a rigorous proof of that property is still missing. Also, data-driven choice of smoothing parameters can be problematic. (See Chapter 10 in this volume.)

Bayesian inference in GAMs, as outlined in what follows, uses ideas from dynamic models for time-series data. Basically, time is replaced by metrical covariates; each different covariate x_j, $j = 1, \ldots, p$ corresponds to a different time scale t_j.

For each covariate, let

$$t_{j1} < \cdots < t_{js} < \cdots < t_{jT_j}, \qquad T_j \leq n$$

denote the strictly-ordered, different values of observations x_{ij}, $i = 1, \ldots, n$. Bayesian smoothness priors for the unknown values

$$f(t_{j1}), \ldots, f(t_{js}), \ldots, f(t_{jT_j})$$

can now be defined by adapting random-walk models (18.5) and (18.12) for unequally spaced time-points, or continuous-time priors like (18.17), to the present situation. Setting $\alpha_{js} = f_j(t_{js})$ and $\delta_{js} = t_{js} - t_{j,s-1}$, first- and second-order random-walk priors are given by

$$\alpha_{js} - \alpha_{j,s-1} = u_{js}, \qquad u_{js} \sim N(0, \delta_{js}\sigma_j^2)$$

and

$$\alpha_{js} - \left(1 + \frac{\delta_{js}}{\delta_{j,s-1}}\right)\alpha_{j,s-1} + \frac{\delta_{js}}{\delta_{j,s-1}}\alpha_{j,s-2} = u_{js}, \quad u_{js} \sim N(0, k_{js}\sigma_j^2), \quad (18.38)$$

with mutually independent errors u_{js}. Priors for Bayesian cubic spline-type smoothing, corresponding to the penalized log-likelihood (18.37), are given by the stochastic differential equations

$$\frac{d^2 f_j(s)}{ds^2} = \sigma_j \frac{dW_j(s)}{ds}, \qquad s \geq t_{j1}, \quad (18.39)$$

having mutually independent standard Wiener processes, $W_j(t_{j1}) = 0$, and diffuse initial conditions for

$$\alpha_{js} = (f_j(s), f_j^T(s)).$$

In complete analogy to Section 18.3.1, the priors (18.39) can be written in state-space form (18.24), and some hyperpriors are assigned to σ_j.

The likelihood $p(y|\alpha, \sigma)$, the priors $p(\alpha)$ and $p(\sigma)$, and, as a consequence, the posterior $p(\alpha|y)$ all have the same structure as in Section (18.3.2). Therefore, the single- and block-move schemes outlined there can be used to simulate from the posterior. Details and some generalizations are given in Lang (1996) for random-walk priors and in Biller and Fahrmeir (1997) for stochastic differential-equation priors. Another spline-based approach is given in Wong and Kohn (1996).

As an application, we consider the credit-scoring problem described in Fahrmeir and Tutz (1994, chapter 2.1). In the credit business, banks are interested in estimating the risk that a consumer will not pay back credit advances in accord with contract. The aim of credit-scoring is to model or predict the probability that a client with certain covariates ("risk factors") is a risk. The data set consists of 1000 consumer credit records from a South German bank. The response variable of interest is "creditability"; it is given in dichotomous form ($y = 0$ for creditworthy, $y = 1$ for not creditworthy). In addition, 20 covariates that are assumed to influence creditability were collected. As in Fahrmeir and Tutz, we will use a subset of these data, containing only the following covariates (which are partly metrical and partly categorical).

x_1	`running account`; trichotomous, with the categories "no running account" (=1),"good running account" (=2), and "medium running account" ("less than 200 DM" = 3 = reference category)
x_3	`duration of credit` in months; metrical
x_4	`amount of credit` in DM; metrical
x_5	`payment of previous credits`; dichotomous, with categories "good" and "bad" (=reference category)
x_6	`intended use`; dichotomous, with categories "private" and "professional" (=reference category)
x_8	`marital status`; (reference category "living alone")

A parametric logit model for the probability $P(y = 1|x)$ of being not creditworthy leads to the somewhat surprising conclusion that the covariate `amount of credit` has no significant influence on the risk. Here, we reanalyze the data with a partial-linear logit model

$$\log \frac{P(y = 1|x)}{1 - P(y = 1|x)} = \beta_0 + \beta_1 x_1^1 + \beta_2 x_1^2 + f_3(x_3) + f_4(x_4) + \beta_5 x_5 + \beta_6 x_6 + \beta_8 x_8.$$

Here x_1^1 and x_1^2 are dummies for the categories "good" and "medium" `running account`, respectively. The predictor has semiparametric or partial-linear form: The smooth functions $f_3(x_3)$ and $f_4(x_4)$ of the metrical covariates `duration of credit` and `amount of credit`, respectively, are estimated by the usual cubic splines and by Bayesian spline-type smoothing, using second-order random-walk models (18.38) for unequally-spaced observations. The constant β_0 and the effects $\beta_1, \ldots, \beta_8$ of the remaining categorical covariates are considered fixed for PL estimation; they are estimated jointly with the curves f_3 and f_4. For Bayesian estimation, diffuse priors are chosen for β_0, β_1, β_2, β_5, β_6, and β_8, and additional Metropolis-Hastings-steps with random-walk proposals are included for MCMC simulation.

Figure 18.8 shows the estimates for the curves f_3 and f_4. Again, the posterior mean of the spline-type smoother (dotted line) and of the posterior-mode or PL estimator (solid line) are not far away from each other. The effect of the variable `duration of credit` is not far away from linearity. The effect of `amount of credit` is clearly nonlinear; its curve has "bathtub" shape and demonstrates that not only high credits, but also low credits, increase the risk, relative to that from "medium" credits (between 3000-6000 DM). If the influence is misspecified by assuming a linear function $\beta_4 x_4$ instead of $f_4(x_4)$, the estimated effect $\widehat{\beta}_4$ will clearly be near zero and will be represented by an almost horizontal line $\widehat{\beta}_4 x_4$

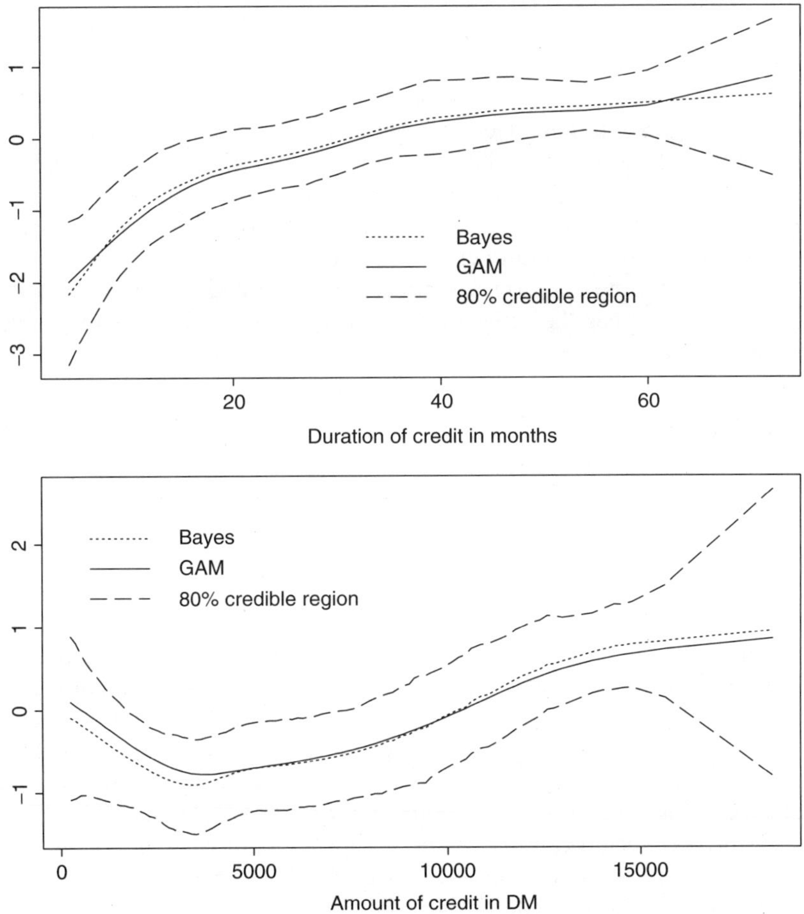

FIGURE 18.8 Estimated functions of the covariate `duration of credit` and `amount of credit`.

TABLE 18.1 Estimates of Constant Parameters in the Credit-Scoring Data

Covariate	Posterior Mean	80 percent CI		ML Estimator
x_1^1	0.662	−0.004	1.335	0.633
x_1^2	−1.468	−2.243	−0.733	−1.324
x_5	−1.085	−2.051	−0.135	−0.998
x_6	−0.442	−1.035	0.209	−0.440
x_8	−0.578	−1.180	0.016	−0.516

near zero, so x_4 will be falsely considered nonsignificant. Table 18.1 gives the posterior means, together with 80 percent *credible intervals* (CI) and maximum-likelihood estimates of the remaining effects. The two estimates are in good agreement.

Finally, we note that the whole approach can be extended to varying coefficient models (Hastie and Tibshirani, 1993) in which the predictor has the form

$$\eta_i = f_1(x_{i1})z_{i1} + \cdots + f_p(x_{ip})z_{ip}, \tag{18.40}$$

with $z_{i1}, \ldots, z_{ip}$ as further "factors". For the special case $z_{i1} = \ldots = z_{ip} = 1$, (18.40) reduces to a generalized additive model (18.35). If $z_1, \ldots, z_p$ are further covariates, possibly including some components of x, then a term $f_j(x_{ij})z_{ij}$ can be interpreted as an interaction term between x and z, or $f_j(x_{ij})$ can be considered as an effect of z, one varying over the "effect-modifier" $f_j(x_j)$. For $x_j \equiv t$ (i.e., if covariate x_j is time t), $f_j(t)$ is a time-varying effect; for $x_1 = \ldots = x_p = t$, the linear predictor has the same form as in dynamic models for time-series or longitudinal data.

18.5 CONCLUSIONS

In this chapter, we have shown that dynamic models with Bayesian smoothness priors and semiparametric models based on the roughness-penalty approach both provide a supplementary method for nonparametric function estimation. Semiparametric Bayesian smoothing has some attractive features: It provides a natural framework for Bayesian analysis beyond posterior-mode or *maximum a posteriori* estimation, and recent advances in MCMC techniques make it possible to estimate posterior means, medians, quantiles, and other functionals of regression functions or of other parameters. No approximation based on a conjecture of asymptotic normality need be made. Bayesian data-driven choice of smoothing parameters is automatically incorporated into the model. Because of its hierarchical model formulation and its modular estimation techniques, the Bayesian approach offers great flexibility in modifying methods and in extending them to other situations—for example, to dynamic mixed models for longitudinal data (Section 18.3.3), to generalized additive and varying coefficient models (Section 18.4) and to data with missing values (an issue not treated here).

To some extent, of course, one has to pay for these advantages: MCMC techniques produce a rich output, but computation time can be quite high. Metropolis–Hastings algorithms provide a wide variety of possibilities for the updating steps, but the rates of convergence and mixing of the Markov chain so constructed also need to be checked empirically. Careful convergence diagnostics deserve close attention in particular applications. Above all, the choice of reasonable priors on the unknown functions remains subjective, and it may not be easily accepted.

Semiparametric models based on the roughness-penalty approach are useful supplementary tools for data analysis: Roughness penalties corresponding to smoothness priors can be interpreted without any underlying Bayesian framework. Thus, if the roughness penalty looks reasonable, then it supports the choice of the smoothness prior. As we have shown, the PL estimator can always be interpreted as a corresponding posterior-mode estimator, from a Bayesian point of view. Computation is done by numerically efficient solutions of a nonlinear maximization problem, in the process employing commonly accepted and well-understood optimization routines. As we demonstrated by examples, the posterior mode is often quite near to posterior means or medians; therefore, it can be quite useful for checking convergence of MCMC simulations.

We focused on non-Gaussian models for time series, longitudinal, and regression data within the setup of generalized linear models having a prespecified link function of known parametric form, such as, for example, the logistic or the exponential function. This restriction could be relaxed by defining a generalized parametric family of link functions (as for example in Stukel, 1988; and Czado, 1992) and by estimating unknown parameters in the link function jointly with unknowns in the predictor. A nonparametric Bayesian approach, avoiding any parametric specification of a link function, has been proposed by Arjas and Gasbarra (1994) and by Arjas and Liu (1996), in the related context of hazard regression. Generally, we believe that, in situations with many covariates, flexible non- or semiparametric modeling and exploration of the predictor are more worthwhile than making a nonparametric choice of the link function while retaining linear parametric predictors. For Gaussian models, Bayesian analysis of regression splines with adaptive knot selection has been recently proposed by Smith and Kohn (1996), Smith et al. (1998), and Denison et al. (1998). For two-dimensional surface estimation, a related Bayesian approach is applied in the following Chapter. For a non-Bayesian spline-based alternative see Chapter 11. It would be interesting to adapt these methods for non-Gaussian regression models.

Extensions to other data structures are made possible by choosing other observation models and smoothness assumptions. In particular, event-history analysis and spatial statistics are a wide and promising field of research; see, e.g., Fahrmeir and Knorr–Held (1997), Arjas and Liu (1996), Arjas and Heikkinen (1997), Besag et al. (1991), and Knorr–Held and Besag (1998). Also, problems of model diagnostics and model choice have to be dealt with convincingly. Here again, Bayesian and non-Bayesian data analyses could complement one another in a productive way.

REFERENCES

Ansley, C. F., and Kohn, R. (1983). On the smoothness properties of the best linear unbiased estimate of a stochastic process observed with noise, *Ann. Statist.*, **11**, 1011–1017.

Arjas, E., and Gasbarra, D. (1994). Nonparametric Bayesian inference from right censored survival data, using the Gibbs sampler, *Statistica Sinica*, **4**, 505–524.

Arjas, E., and Heikkinen, J. (1997). An algorithm for nonparametric Bayesian estimation of a Poisson intensity, *Computational Statistics*, **12**, 385–402.

Arjas, E., and Liu, L. (1996). A non-parametric Bayesian approach to hazard regression: A case study with a large number of missing covariate values, *Statistics in Medicine*, **15**, 1757–1770.

Besag, J. E., Green, P. J., Higdon, D., and Mengersen, K. (1995). Bayesian computation, and stochastic systems (with discussion), *Statistical Science*, **10**, 3–66.

Besag, J. E., York, J., and Mollie, A. (1991). Bayesian image restoration with two applications in spatial statistics (with discussion), *Annals Institute of Statistical Mathematics*, **43**, 1–59.

Biller, C. (1997). *Posterior Mode Estimation in Dynamic Generalized Linear Mixed Models*, Discussion Paper No. 70, Sonderforschungsbereich 386 der Ludwig-Maximilians-Universität München, Munich (also `http://www.stat.uni-muenchen.de/sfb386/publikation.html`).

Biller, C., and Fahrmeir, L. (1997). Bayesian spline-type smoothing in generalized regression models, *Computational Statistics*, **12**, 135–151.

Breslow, N. E., and Clayton, D. G. (1993). Approximate inference in generalized linear mixed models, *J. Amer. Statist. Assoc.*, **88**, 9–25.

Breslow, N. E., and Lin, X. (1995). Bias correction in generalised linear mixed models with a single component of dispersion, *Biometrika*, **82**, 81–91.

Carlin, B. P., Polson, N. G., and Stoffer, D. S. (1992). A Monte Carlo approach to nonnormal and nonlinear state-space-modeling, *J. Amer. Statist. Assoc.*, **87**, 493–500.

Carter, C. K., and Kohn, R. (1994). On Gibbs sampling for state space models, *Biometrika*, **81**, 541–553.

Carter, C. K., and Kohn, R. (1996a). Markov chain Monte Carlo in conditionally Gaussian state space models, *Biometrika*, **83**, 589–601.

Carter, C. K., and Kohn, R. (1996b). Robust Bayesian nonparametric regression, in Härdle, W., and Schimek, M. G. (eds.), *Statistical Theory and Computational Aspects of Smoothing*, pp. 128–148, Physica-Verlag, Heidelberg, Germany.

Czado, C. (1992). On link selection in generalized linear models, in Fahrmeir, L., Francis, B., Gilchrist, R., and Tutz, G. (eds.), *Advances in GLIM and Statistical Modeling*, pp. 60–65, Springer-Verlag, Heidelberg, Germany.

De Jong, P. (1991). The diffuse Kalman filter, *Ann. Statist.*, **19**, 1073–1083.

De Jong, P., and Shephard, N. (1995). The simulation smoother for time series models, *Biometrika*, **82**, 339–350.

Denison, D. G. T., Mallick, B. K., and Smith, A. F. M. (1998). Automatic Bayesian curve fitting, *J. Roy. Statist. Soc. B*, **60**, 333–350.

Fahrmeir, L. (1992). Posterior mode estimation by extended Kalman filtering for multivariate dynamic generalized linear models, *J. Amer. Statist. Assoc.*, **87**, 501–509.

Fahrmeir, L., Hennevogl, W., and Klemme, K. (1992). Smoothing in dynamic generalized linear models by Gibbs sampling, in Fahrmeir, L., Francis, B., Gilchrist, R., and Tutz, G. (eds.), *Advances in GLIM and Statistical Modeling*, pp. 85–90, Springer-Verlag, Heidelberg, Germany.

Fahrmeir, L., and Knorr–Held, L. (1997). Dynamic discrete-time duration models: Estimation via Markov chain Monte Carlo, *Sociological Methodology*, **27**, 417–452.

Fahrmeir, L., and Tutz, G. (1994). *Multivariate Statistical Modeling Based on Generalized Linear Models*, Springer-Verlag, New York.

Fahrmeir, L., and Wagenpfeil, S. (1997). Penalized likelihood estimation and iterative Kalman smoothing for non-Gaussian dynamic regression models, *Computat. Statist. Data Analysis*, **24**, 295–320.

Frühwirth-Schnatter, S. (1994). Data augmentation and dynamic linear models, *J. Time Series Analysis*, **15**, 183–202.

Gelfand, A. E., and Smith, A. F. M. (1990). Sampling-based approaches to calculating marginal densities, *J. Amer. Statist. Assoc.*, **85**, 398–409.

Green, P. J., and Silverman, B. W. (1994), *Nonparametric Regression and Generalized Linear Models*, Chapman and Hall, London, UK.

Harrison, P. J., and Stevens, C. F. (1976). Bayesian forecasting (with discussion), *J. Roy. Statist. Soc. B*, **38**, 205–247.

Harvey, A. C. (1989), *Forecasting, Structural Time Series Models, and the Kalman Filter*, Cambridge University Press, Cambridge, UK.

Hastings, W. K. (1970). Monte Carlo sampling methods using Markov chains and their applications, *Biometrika*, **57**, 97–109.

Hastie, T., and Tibshirani, R. (1990), *Generalized Additive Models*, Chapman and Hall, London, UK.

Hastie, T., and Tibshirani, R. (1993). Varying coefficient models, *J. Roy. Statist. Soc. B*, **55**, 757–796.

Kashiwagi, N., and Yanagimoto, T. (1992). Smoothing serial count data through a state-space model, *Biometrics*, **48**, 1187–1194.

Kitagawa, G. (1987). Non-Gaussian state-space modeling of non-stationary time series (with discussion), *J. Amer. Statist. Assoc.*, **82**, 1032–1063.

Knorr–Held, L. (1997). *Hierarchical Modeling of Discrete Longitudinal Data: Applications of Markov Chain Monte Carlo*, Herbert Utz Verlag Wissenschaft, Munich, Germany.

Knorr–Held, L. (1999). Conditional prior proposals in dynamic models, *Scand. J. Statist.*, **26**, 129–144.

Knorr–Held, L., and Besag, J. (1998). On modeling disease risk data in time and space, *Statistics in Medicine*, **17**, 2045–2060.

Kohn, R., and Ansley, C. (1987). A new algorithm for spline smoothing based on smoothing a stochastic process, *SIAM J. Scientific Statistical Computing*, **8**, 33–48.

Kohn, R., and Ansley, C. (1988). Equivalence between Bayesian smoothing priors and optimal smoothing for function estimation, in Spall, J. C. (ed.), *Bayesian Analysis of Time Series and Dynamic Models*, pp. 393–430, Dekker, New York.

Künstler, R. (1996). *Robuste Zustandsraummodelle*, Doctoral Dissertation, Universität München, Munich, Germany (also `http://www.stat.uni-muenchen.de/sfb386/publikation.html`).

Lang, S. (1996). *Bayesianische Inferenz in Modellen mit variierenden Koeffizienten*, Master Thesis, Universität München, Munich, Germany.

van der Linde, A. (1995). Splines from a Bayesian point of view, *Test*, **4**, 63–81.

Martin, R. D., and Raftery (1987). Robustness, computation, and non-Euclidean models (comment), *J. Amer. Statist. Assoc.*, **82**, 1044–1050.

Masreliez, C. J. (1975). Approximate non-Gaussian filtering with linear state and observation relations, *IEEE Trans. Automatic Control*, **AC-20**, 107–110.

Masreliez, C. J., and Martin, R. D. (1977). Robust Bayesian estimation for the linear model and robustifying the Kalman filter, *IEEE Trans. Automatic Control*, **AC-22**, 361–371.

O'Sullivan, F., Yandell, B., and Raynor, W. (1986). Automatic smoothing of regression functions in generalized linear models, *J. Amer. Statist. Assoc.*, **81**, 96–103.

Shephard, N. (1994). Partial non-Gaussian state space, *Biometrika*, **81**, 115–131.

Shephard, N., and Pitt, M. K. (1997). Likelihood analysis of non-Gaussian measurement time series, *Biometrika*, **84**, 653–667.

Singh, A. C., and Roberts, G. R. (1992). State space modeling cross-classified time series and counts, *International Statistical Review*, **60**, 321–335.

Smith, M., and Kohn, R. (1996). Nonparametric regression using Bayesian variable selection, *J. Econometrics*, **75**, 317–343.

Smith, M., Wong, C., and Kohn, R. (1998). Additive nonparametric regression with autocorrelated errors, *JRSS, B*, **60**, 311–331.

Stukel, T. A. (1988). Generalized logistic models, *J. Amer. Statist. Assoc.*, **83**, 426–431.

Tierney, L. (1994). Markov chains for exploring posterior distributions (with discussion), *Annal. Statist.*, **22**, 1701–1762.

Wagenpfeil, S. (1996), *Dynamische Modelle zur Ereignisanalyse*. Herbert Utz Verlag Wissenschaft, Munich, Germany.

Wahba, G. (1978). Improper priors, spline smoothing and the problem of guarding against model errors in regressions, *J. Roy. Statist. Soc. B*, **40**, 364–372.

Wahba, G., Wang, Y., Gu, C., Klein, R., and Klein, B. (1995). Smoothing spline ANOVA for exponential families, with application to the Wisconsin epidemiological study of diabetic retinopathy, *Annal. Statist.*, **23**, 1865–1896.

Wecker, W., and Ansley, C. F. (1983). The signal extraction approach to nonparametric regression and spline smoothing, *J. Amer. Statist. Assoc.*, **78**, 81–89.

West, M., and Harrison, P. J. (1989), *Bayesian Forecasting and Dynamic Models*, Springer-Verlag, New York.

Whittaker, E. T. (1923). On a new method of graduation, *Proc. Edinborough Math. Assoc.*, **78**, 81–89.

Wong, C., and Kohn, R. (1996). A Bayesian approach to additive semiparametric regression, *J. Econometrics*, **74**, 209–235.

CHAPTER 19

Nonparametric Bayesian Bivariate Surface Estimation

Michael Smith
University of Sydney, Sydney, Australia

Robert Kohn and Paul Yau
Australian Graduate School of Management, Sydney, Australia

19.1 INTRODUCTION

This chapter discusses how to estimate, in a flexible manner, the bivariate surface f in the regression model

$$y_i = f(x_i, z_i) + e_i \quad \text{for} \quad i = 1, \ldots, n. \tag{19.1}$$

Here, y_i is the ith observation of the dependent variable, x_i and z_i are the corresponding observations of two independent variables, the errors e_i are independent and identically distributed $N(0, \sigma^2)$, and f is a general function of the independent variables. A popular way of estimating f is to assume that its form is known *a priori*, except for a small number of parameters. For example, one might assume that f is a linear or quadratic function of x and z. However, if the true regression function is not of the form assumed, then the parametric approach can be grossly biased. In this chapter, we just assume that the regression surface is smooth, with the estimate of the function adapting to the true underlying functional form. This is called a *nonparametric* approach to regression.

There are a number of approaches to estimating a bivariate surface in a nonparametric manner. In what follows, we explicitly model the bivariate surface as a linear combination of a set of known bivariate functions $\mathcal{S} = \{B_i(x, z), i \in I\}$,

Smoothing and Regression: Approaches, Computation, and Application
Edited by Michael G. Schimek
ISBN 0-471-17946-9

where I is an indexing set, so that

$$f(x, z) = \sum_{i \in I} \beta_i B_i(x, z). \tag{19.2}$$

That is, we assume that $f(x, z)$ belongs to the vector space spanned by the elements of $\mathcal{S}$. Usually, we take the elements of $\mathcal{S}$ to be linearly independent, in which case they form a basis of the resulting vector space, and we refer to the $B_i(x, z)$ as *basis elements*. Popular choices for the basis functions B_i are B-forms (de Boor, 1987), tensor products of univariate B-splines (de Boor, 1978), multivariate regression spline bases (Stone, 1994; Silverman and Friedman, 1989), and multivariate radial function bases (Powell, 1987). For a given set $\mathcal{S}$ of basis functions and the approximation of f by (19.2), the regression equation (19.1) becomes a linear model in the coefficients β_i. To make (19.2) a good approximation to a large number of bivariate functions, it is necessary that the number of functions in $\mathcal{S}$ be large. However, if the number of functions in $\mathcal{S}$ is large, then simultaneously estimating *all* the β_i by least squares will result in an estimate of the function that has an unacceptably high variance. In particular, if the number of linearly independent functions in $\mathcal{S}$ is equal to the sample size n (and in the absence of replicated observations), then the estimate of the function will interpolate the data. For example, regressing on all the basis functions with $\mathcal{S}$ a radial function basis or a wavelet basis will result in interpolation. This suggests that it is necessary to select some data-driven subset of the elements of $\mathcal{S}$ in order to obtain a useful estimate of the unknown regression function.

Smith and Kohn (1996) propose a Bayesian approach to univariate nonparametric regression using a univariate regression spline basis containing a large number of elements. They consider two estimators of the regression function. The first estimate is based on the subset of basis functions with the highest posterior probability. The second estimates the posterior mean of the regression function, averaging over the distribution of all possible subsets of the basis functions. Both estimates are obtained by the use of a carefully constructed Markov chain Monte Carlo procedure. Smith and Kohn (1996) show that such an approach works well on both real and simulated data, comparing favorably with a local linear approach using a direct plug-in estimate of the bandwidth. They also extend their analysis to additive and robust nonparametric regression.

As in Smith and Kohn (1997), this chapter extends the procedure to bivariate surface estimation. We apply the Bayesian subset selection procedure first introduced in Smith and Kohn (1996) to a function basis of bivariate regression splines. In doing so, we attempt to demonstrate some important properties of the estimator. First, even though the Bayesian approach selects terms from a bivariate regression spline basis similar to that of MARS (Friedman, 1991), it appears to do so in a substantially more efficient manner. Second, as a data-driven surface estimator, the approach is competitive with some of the other procedures discussed in the literature on surface estimation, including multivariate smoothing splines (Wahba, 1990; Gu et al., 1989) and variable bandwidth based local bivariate regression (see Section 19.3.4). Third, one of the most striking features of the Bayesian approach is how it can be easily extended to allow bivariate surface estimation in situations that are difficult to deal with using traditional approaches. Two such problems are addressed here: bivariate surface estimation in the presence of outlying values of the dependent variable, y, and bivariate surface estimation in the presence of an autocorrelated error structure. Finally, the estimation procedure can be applied to other bivariate basis families, such as B-forms (de Boor, 1987) and radial bases, and not just to those families that have the simple structure of multivariate regression splines.

To establish the preceding points, the chapter is organized as follows: Section 19.2 outlines Bayesian model selection, including the specific case of Bayesian subset selection. A comprehensive discussion of how to implement such a procedure in an efficient manner is also presented. Section 19.3 illustrates how such subset selection can be coupled with a bivariate regression spline basis to provide a bivariate surface estimator. Extensive empirical comparisons with other methodologies are undertaken, along with an analysis of a real bivariate data set. Section 19.4 extends Bayesian model selection, and hence surface estimation, to the case where there are outlying values in the dependent variable; the resulting surface estimate is "robust" to such observations. In the discussion, we indicate how the associated algorithm can be implemented in a fast and stable manner. Section 19.5 discusses how the surface estimator can be extended to the case where the data are collected over time and the errors are modeled as an autoregression of unknown order. Here, the bivariate surface, autoregressive order, and autoregressive parameters are simultaneously estimated. Section 19.6 discusses how the estimation procedure can be generalized to basis families that do not have the simple structure of multivariate regression splines.

19.2 BAYESIAN MODEL AND SUBSET SELECTION

19.2.1 Introduction

This section discusses Bayesian model selection (and hence the special case of Bayesian subset selection) for linear regression models. The procedure forms the basis for the nonparametric estimator of a bivariate surface presented in Section 19.3.

Let Γ be a family of linear regression models, all having the same dependent variable and such that, for $\boldsymbol{\gamma} \in \Gamma$,

$$\boldsymbol{y} = X_\gamma \boldsymbol{\beta}_\gamma + \boldsymbol{e}. \tag{19.3}$$

Here, $\boldsymbol{y} = (y_1, \ldots, y_n)^T$ is the vector of dependent variables, X_γ is an $n \times q_\gamma$ matrix of independent variables that is of full column rank, $\boldsymbol{\beta}_\gamma$ is a q_γ-vector of regression parameters, and the errors $\boldsymbol{e} \sim N(0, \sigma^2 I_n)$. Bayesian model selection for such linear models usually consists of selecting those members of Γ with the highest posterior probability.

Subset selection is a special case of this model selection problem in which there exists an $n \times q$ matrix of independent variables X, such that Γ is the set of all 2^q possible subsets of the columns of X. In this case, we can write the model parameter as $\boldsymbol{\gamma} = (\gamma_1, \ldots, \gamma_q)$ and set $\gamma_i = 1$ if the ith column of X is in the model and $\gamma_i = 0$ if it is not. If $\boldsymbol{\beta} = (\beta_1, \ldots, \beta_q)$ is a coefficient vector corresponding to the full design X, then $\boldsymbol{\beta}_\gamma$ consists of all elements β_i such that $\gamma_i = 1$.

Some approaches to nonparametric regression correspond to such a subset selection problem; for example, the regression spline formulation focused on in this chapter and the use of radial basis functions as in Powell (1987) and Holmes and Mallick (1998) are of this variety. However, for other approaches, it is more useful to think in terms of model selection in which Γ need not have the special structure outlined above. An example of this is the use of B-splines and B-forms as in Stone et al. (1996), where dropping one column of X results in changes to several other columns of the design matrix.

To tackle model selection, including the special case of subset selection, the following hierarchical prior, similar to that proposed by Smith and Kohn (1996), will be used to provide a Bayesian analysis:

(i) The prior for $\beta_\gamma|\sigma^2, \boldsymbol{\gamma}$ is

$$\boldsymbol{\beta}_\gamma|\sigma^2, \boldsymbol{\gamma} \sim N(\boldsymbol{\mu}_\gamma, c\sigma^2(X_\gamma^T X_\gamma)^{-1}), \tag{19.4}$$

where $\boldsymbol{\mu}_\gamma = (X_\gamma^T X_\gamma)^{-1} X_\gamma^T \mathbf{y}$. In (19.4), the variance matrix of $\boldsymbol{\beta}_\gamma|\sigma^2, \boldsymbol{\gamma}$ is the variance matrix of the least-squares estimate of $\boldsymbol{\beta}_\gamma$, scaled by an inflation factor c. Here, $c > 1$ is set large enough that this conditional prior of $\boldsymbol{\beta}_\gamma$ contains less information about $\boldsymbol{\beta}_\gamma$ than the likelihood $p(\mathbf{y}\,|\,\boldsymbol{\beta}_\gamma, \sigma^2, \gamma)$. For example, Smith and Kohn (1997) used $c = 100{,}000$ and found this worked well for selecting variables from a bivariate regression spline basis. All the empirical work in this chapter results from taking c to be 100,000.

(ii) The prior for $\sigma^2|\boldsymbol{\gamma}$ is $p(\sigma^2|\boldsymbol{\gamma}) \propto 1/\sigma^2$, which means that $\log \sigma^2$ has a flat prior (Box and Tiao, 1973).

(iii) Following Smith and Kohn (1996, 1997), we choose $p(\gamma) = 1/\#(\Gamma)$, where $\#(\Gamma)$ is the number of elements in Γ, making each model in Γ equally likely *a priori*. In the case of subset selection, the binary variables γ_i are taken to be *a priori* independent with

$$p(\gamma_i) = 1/2 \quad \text{and} \quad p(\boldsymbol{\gamma}) = 2^{-q}. \tag{19.5}$$

19.2.2 Calculating the Posterior Probability of a Model

Using the preceding framework, the posterior probability of any model $\boldsymbol{\gamma}$ is given by

$$\begin{aligned} p(\boldsymbol{\gamma}|\mathbf{y}) &\propto p(\mathbf{y}\,|\,\boldsymbol{\gamma})p(\boldsymbol{\gamma}) \\ &\propto \int\!\!\int \underbrace{p(\mathbf{y}\,|\,\boldsymbol{\beta}_\gamma, \sigma^2, \boldsymbol{\gamma})}_{\text{likelihood}}\, \underbrace{p(\boldsymbol{\beta}_\gamma|\sigma^2, \boldsymbol{\gamma})p(\sigma^2|\boldsymbol{\gamma})}_{\text{given priors}}\, d\boldsymbol{\beta}_\gamma\, d\sigma^2\, p(\boldsymbol{\gamma}) \end{aligned}$$

$$\propto \int\int (\sigma^2)^{-(n+q_\gamma+2)/2}(2\pi c)^{-q_\gamma/2}|X_\gamma^T X_\gamma|^{1/2} \times \exp\left\{\frac{-1}{2\sigma^2}\,[\mathbf{y}^T\mathbf{y}\right.$$

$$\left. + (\boldsymbol{\beta}_\gamma - \boldsymbol{\mu}_\gamma)^T \Sigma_\gamma^{-1}(\boldsymbol{\beta}_\gamma - \boldsymbol{\mu}_\gamma) - \boldsymbol{\mu}_\gamma^T \Sigma_\gamma^{-1}\boldsymbol{\mu}_\gamma\right\} d\boldsymbol{\beta}_\gamma d\sigma^2 p(\boldsymbol{\gamma}),$$

where

$$\boldsymbol{\mu}_\gamma = (X_\gamma^T X_\gamma)^{-1} X_\gamma^T \mathbf{y}$$

and

$$\Sigma_\gamma = \frac{c}{c+1}(X_\gamma^T X_\gamma)^{-1}.$$

Integrating out a $N(\boldsymbol{\mu}_\gamma, \Sigma_\gamma)$ density in $\boldsymbol{\beta}_\gamma$, we obtain

$$p(\boldsymbol{\gamma}|\mathbf{y}) \propto (1+c)^{-q_\gamma/2}\int (\sigma^2)^{-(n/2+1)} \exp\left\{\frac{-S(\boldsymbol{\gamma})}{2\sigma^2}\right\} d\sigma^2 p(\boldsymbol{\gamma}),$$

where $S(\boldsymbol{\gamma}) = \mathbf{y}^T\mathbf{y} - \mathbf{y}^T X_\gamma (X_\gamma^T X_\gamma)^{-1} X_\gamma^T \mathbf{y} = \mathbf{y}^T\mathbf{y} - \boldsymbol{\mu}_\gamma^T \Sigma_\gamma^{-1} \boldsymbol{\mu}_\gamma$ is the residual sum of squares for model $\boldsymbol{\gamma}$. By integrating out an inverse gamma density in σ^2, the posterior density can be derived as

$$\begin{aligned} p(\boldsymbol{\gamma}|\mathbf{y}) &\propto (1+c)^{-q_\gamma/2} S(\boldsymbol{\gamma})^{-n/2} p(\boldsymbol{\gamma}) \\ &\propto (1+c)^{-q_\gamma/2} S(\boldsymbol{\gamma})^{-n/2}, \end{aligned} \tag{19.6}$$

where we have employed the prior $p(\boldsymbol{\gamma})$ specified earlier in (iii). Notice that this posterior probability can be equated with popular information criteria via specific choices for the inflation factor c. These criteria include the *Schwarz information criterion* (SIC; Schwarz, 1978) and the *Akaike information criterion* (AIC; Akaike, 1973), which are detailed in Table 19.1.

19.2.3 The Gibbs Sampler for Subset Selection

In the special case of subset selection, the model variable can be given the specific binomial form $\boldsymbol{\gamma} = (\gamma_1, \gamma_2, \ldots, \gamma_q)$, where each γ_i is a binary variable. However, when there are a large number of models, it is computationally impractical to

TABLE 19.1 Different Choices for c that Result in Posterior Probabilities Corresponding to Popular Information Criteria

Criterion	Value of c	$\log_e(p(\boldsymbol{\gamma}|\mathbf{y}))$ up to a constant
AIC	$e^2 - 1$	$-\frac{n}{2}\log_e(S(\boldsymbol{\gamma})) - q_\gamma$
SIC	$n - 1$	$-\frac{n}{2}\log_e(S(\boldsymbol{\gamma})) - \frac{1}{2}q_\gamma \log_e(n)$

consider each one separately, and some search strategy is necessary to find the most promising models. This is particularly important when the regression function (19.1) is approximated as a linear combination of basis terms, as indicated in (19.2) and discussed in Section 19.3, because the number, q, of basis terms will be large. For example, the bases discussed in Sections 19.3.1 and 19.3.2 and applied to problems in Section 19.3.5, have $q = 169$.

A number of deterministic search strategies, such as stepwise regression, have been suggested to solve the problem. The Gibbs sampler (Gelfand and Smith, 1990) provides a stochastic alternative to stepwise regression in which a Monte Carlo sample from the posterior distribution $\boldsymbol{\gamma}|\mathbf{y}$ is eventually produced. As a search strategy, it can be used with any model selection criterion, such as adjusted R^2, Mallows C_p, or SIC, although we concentrate on the posterior probability $p(\boldsymbol{\gamma}|\mathbf{y})$ itself.

The following Gibbs sampler for subset selection was proposed by Smith and Kohn (1996):

(i) Choose an initial value $\boldsymbol{\gamma}^{[0]} = (\gamma_1^{[0]}, \ldots, \gamma_q^{[0]})$ of $\boldsymbol{\gamma}$.
(ii) Successively generate from $p(\gamma_i|\mathbf{y}, \boldsymbol{\gamma}_{j\neq i}), \quad i \in \{1, \ldots, q\}$.

It can be shown (Gelfand and Smith, 1990) that the iterates of $\boldsymbol{\gamma}$ produced by the Gibbs sampler converge to samples from the posterior distribution $p(\boldsymbol{\gamma}|\mathbf{y})$. Step (ii) is carried out many times and in two stages. The first stage is a warm-up period, at the end of which it is assumed that the sampler has converged to the joint distribution $p(\boldsymbol{\gamma}|\mathbf{y})$. The second stage is a sampling period, and the $\boldsymbol{\gamma}$s generated during this period are used for inference.

The conditional posterior density of γ_i can be deduced from (19.6) as

$$\begin{aligned} p(\gamma_i|\mathbf{y}, \boldsymbol{\gamma}_{j\neq i}) &\propto p(\mathbf{y}|\boldsymbol{\gamma})p(\gamma_i|\boldsymbol{\gamma}_{j\neq i}) \\ &\propto (1+c)^{-q_\gamma/2} S(\boldsymbol{\gamma})^{-n/2} p(\gamma_i). \end{aligned} \tag{19.7}$$

As we do throughout the chapter, taking the prior for $\boldsymbol{\gamma}$ given at (19.5), the exact conditional density $p(\gamma_i|\mathbf{y}, \boldsymbol{\gamma}_{j\neq i})$ can be obtained by evaluating (19.7) for $\gamma_i = 0$ and $\gamma_i = 1$ and normalizing, so that

$$p(\gamma_i = 1|\mathbf{y}, \boldsymbol{\gamma}_{j\neq i}) = 1/(1+h), \tag{19.8}$$

where

$$h = (c+1)^{1/2} \left(\frac{S(\boldsymbol{\gamma}^1)}{S(\boldsymbol{\gamma}^0)}\right)^{n/2},$$

$$\boldsymbol{\gamma}^1 = (\gamma_1, \ldots, \gamma_{i-1}, \gamma_i = 1, \gamma_{i+1}, \ldots, \gamma_q),$$

and

$$\boldsymbol{\gamma}^0 = (\gamma_1, \ldots, \gamma_{i-1}, \gamma_i = 0, \gamma_{i+1}, \ldots, \gamma_q).$$

Section 19.2.5 details how generation from this distribution can be undertaken in an efficient manner.

19.2.4 Estimation with Subset Selection

This section discusses estimating parameters under the assumption of subset uncertainty. We assume the use of the Gibbs sampler for subset selection outlined earlier and extend it to the more general case of model selection in Section 19.6.

In most approaches to nonparametric regression that use a linear model, such as MARS (Friedman, 1991), a model is first selected and the coefficients are then estimated, usually by least squares. In a similar manner, by examining the posterior probabilities $p(\boldsymbol{\gamma}^{[j]}|\mathbf{y})$ of the J iterates of the converged sampling sequence $\{\boldsymbol{\gamma}^{[1]}, \dots, \boldsymbol{\gamma}^{[J]}\}$, estimates of the mode of the model space Γ are obtainable, as in Smith and Kohn (1996). This allows the selection of an "optimal" model which can be conditioned on to obtain coefficient estimates. However, because the cardinality $\#(\Gamma)$ is often very large, a long sampling period may be required to uncover the model with the very best posterior probability. Therefore, we focus on the posterior mean $E(\boldsymbol{\beta}|\mathbf{y}) = \sum_{\boldsymbol{\gamma}\in\Gamma} E(\boldsymbol{\beta}|\mathbf{y}, \boldsymbol{\gamma})p(\boldsymbol{\gamma}|\mathbf{y})$, which is approximated by using the following mixture estimate proposed in Smith and Kohn (1996):

$$\hat{\boldsymbol{\beta}} = \frac{1}{J}\sum_{j=1}^{J} E(\boldsymbol{\beta}|\mathbf{y}, \boldsymbol{\gamma}^{[j]}). \tag{19.9}$$

Here, the calculations are completed by using the iterates of the converged sampling sequence $\{\boldsymbol{\gamma}^{[1]}, \dots, \boldsymbol{\gamma}^{[J]}\}$. The conditional expectation at (19.9) is evaluated exactly at each iteration because

$$\boldsymbol{\beta}_{\gamma}|\mathbf{y}, \boldsymbol{\gamma} \sim T\left[n, \boldsymbol{\mu}_{\gamma}, \frac{S(\boldsymbol{\gamma})}{n}\Sigma_{\gamma}\right],$$

while the other elements of $\boldsymbol{\beta}$ are exactly zero. Here, $T[d, \mu, M]$ denotes a multivariate T distribution with d degrees of freedom, location μ, and scale matrix M (Box and Tiao, 1973).

Raftery et al. (1997) call (19.9) *model averaging*, and it has the advantage that when the true function is not exactly a linear combination of the basis terms being considered, it appears to outperform the posterior mode. (See Smith and Kohn, 1996, for examples in the univariate case.) It also usually requires far fewer iterations to obtain a good mixture estimate than it takes to search for a single mode through the 2^q possible models in Γ. This difference in the number of iterations is more pronounced in the robust case (see Section 19.4), in which the multinomial parameter space of the outlier process and possible subsets has cardinality 2^{n+q}. In our experience, converged sampling runs of less than 1,000 iterations are more than adequate to get an excellent estimate of the posterior mean. Throughout the chapter, we use 300 iterations to ensure convergence and a further $J = 500$ for estimation procedures. Finally, we note that mixture estimates are available only in a Bayesian framework and not in frequentist approaches.

19.2.5 Efficient Implementation of the Sampling Scheme

The Gibbs sampler requires the generation of γ_i, for $i \in \{1, 2, \ldots, q\}$, from the conditional posterior density at equation (19.8). To induce numerical stability in these calculations, we follow Dongarra et al. (1979) in standardizing the matrix X, by setting $\tilde{X} = XK^{-1}$, where $K = \operatorname{diag}(\|\boldsymbol{x}_1\|, \|\boldsymbol{x}_2\|, \ldots, \|\boldsymbol{x}_q\|)$, $\|\boldsymbol{x}_i\| = (\sum_{j=1}^{n} x_{ij})^{1/2}$, and x_{ij} is the element in the jth row and ith column of X. Note that the calculation of the sum of squares, $S(\boldsymbol{\gamma})$, in the posterior probability given in (19.8) is unaffected by this standardization, since

$$S(\boldsymbol{\gamma}) = \boldsymbol{y}^T\boldsymbol{y} - \boldsymbol{y}^T X_\gamma (X_\gamma^T X_\gamma)^{-1} X_\gamma^T \boldsymbol{y} = \boldsymbol{y}^T\boldsymbol{y} - \boldsymbol{y}^T \tilde{X}_\gamma (\tilde{X}_\gamma^T \tilde{X}_\gamma)^{-1} \tilde{X}_\gamma^T \boldsymbol{y}.$$

The sampling scheme can be made independent of n after an initial $O(nq^2)$ calculation by precalculating $\boldsymbol{y}^T\boldsymbol{y}$, $\tilde{X}^T\tilde{X}$, and $\tilde{X}^T\boldsymbol{y}$. Then, the vector

$$\boldsymbol{v}_\gamma = \tilde{X}_\gamma^T \boldsymbol{y}$$

and matrix

$$M_\gamma = \tilde{X}_\gamma^T \tilde{X}_\gamma$$

are simply subelements of these precalculated quantities.

To generate γ_i efficiently, we make use of a number of definitions. Let $\boldsymbol{\gamma}^1 = (\gamma_1, \ldots, \gamma_{i-1}, \gamma_i = 1, \gamma_{i+1}, \ldots, \gamma_q)$ and $\boldsymbol{\gamma}^0 = (\gamma_1, \ldots, \gamma_{i-1}, \gamma_i = 0, \gamma_{i+1}, \ldots, \gamma_q)$, and define the terms R_1, R_0, S_1, and S_0 according to Table 19.2.

The following algorithm, with the order of each major computation noted on the right-hand side, indicates how $\boldsymbol{\gamma}$ can be generated, one element at a time, in a fast and stable manner:

```
do i in {1,2,...,q}
   if (γi=1) then
      (1a) Obtain R0 from R1 using Cholesky column
           extraction (Dongarra et al., 1979, p. 10.2)      O(qγ²)
```

TABLE 19.2 Definition of Some Terms

Term	Description
R_1	The upper triangular Cholesky factor such that $R_1^T R_1 = M_{\gamma^1}$
R_0	The upper triangular Cholesky factor such that $R_0^T R_0 = M_{\gamma^0}$
S_1	The sum of squares, $S(\boldsymbol{\gamma}^1) = \boldsymbol{y}^T\boldsymbol{y} - \boldsymbol{v}_{\gamma^1}^T M_{\gamma^1}^{-1} \boldsymbol{v}_{\gamma^1}$
S_0	The sum of squares, $S(\boldsymbol{\gamma}^0) = \boldsymbol{y}^T\boldsymbol{y} - \boldsymbol{v}_{\gamma^0}^T M_{\gamma^0}^{-1} \boldsymbol{v}_{\gamma^0}$

```
      (1b) Solve R_0 z = v_γ0 for z                                  O(q_γ^2)
      (1c) Set S_0 = y^T y - z^T z                                   O(q_γ)
   else
      (2a) Obtain R_1 by adding an extra column to the
           Cholesky factor R_0                                       O(q_γ^2)
      (2b) Solve R_1 z = v_γ1 for z                                  O(q_γ^2)
      (2c) Set S_1 = y^T y - z^T z                                   O(q_γ)
   endif
      (3a) Set h = exp{ (n/2)(log(S_1) - log(S_0)) + log(c)/2 }
      (3b) Evaluate p(γ_i = 1|γ_{j≠i}, y) = 1/(1 + h), and generate γ_i
   if (γ_i=1) then
      (4) R_0 = R_1 and S_0 = S_1                                    O(q_γ^2)
   else
      (5) R_1 = R_0 and S_1 = S_0                                    O(q_γ)
   endif
enddo
```

It is recommended that the outer loop be traversed in random order, so that for each iteration of the Gibbs sampler the order in which the γ_i are generated is different.

Overall, the generation of a single iterate of $\boldsymbol{\gamma}$ is an $O(q_\gamma^2 q)$ operation, as there are q elements in the vector $\boldsymbol{\gamma}$. In the problem of nonparametric regression examined in this chapter, q is likely to be fairly large ($q = 169$ in the examples in Section 19.3.5), because flexible bivariate bases often involve a large number of terms. However, q_γ is usually much smaller than q, because most smooth surfaces can be approximated parsimoniously with only a small number of terms. For example, q_γ was typically in the range of $\{2, 3, \ldots, 15\}$ at any iteration of the samplers used to estimate the surfaces in Section 19.3.5.

Given the initialization, together with the fact that this sampling scheme needs to be run for J iterations, the algorithm requires $O(Jq_\gamma^2 q + nq^2)$ floating-point operations. Comparative timings are provided in Section 19.3.5 and illustrate that the speed of the procedure is highly competitive in practice.

19.2.6 Brief Review of Related Literature

The literature on Bayesian subset selection is extensive; George and McCulloch (1993) provide a useful set of references. Of particular interest is the work of Mitchell and Beauchamp (1988), who use point null priors to calculate the posterior probabilities of different regression models. By a point null prior, we mean that if a variable is not in the regression, then the corresponding coefficient is set identically to zero. Such a point null prior is utilized in the approach discussed here, with $\beta_i = 0$ if $\gamma_i = 0$.

George and McCulloch (1993) were the first to suggest using a Gibbs sampling approach for subset selection to compute the posterior mode estimate; but they do not use point null priors, or integrate out $\boldsymbol{\beta}$ from the resulting sampling scheme, which converges only slowly for our applications. Smith and Kohn (1996) proposed the Gibbs sampler presented in this section and show that it takes fewer iterations to converge and requires less computation per iteration than the Gibbs sampler of George and McCulloch (1993). The conditional prior of (19.4) is similar to Zellner's (1986) g-prior, but Zellner does not discuss subset selection. This prior is also similar to the family of priors in O'Hagan (1995). Raftery et al. (1997) discuss model selection and model mixing, but neither use the g-prior structure nor offer a fast sampling algorithm for selection. A comprehensive discussion of current sampling approaches to subset selection is provided by George and McCulloch (1997).

19.3 BIVARIATE SURFACE ESTIMATION

19.3.1 Tensor-Product Regression Splines

The bivariate surface at (19.2) can be modeled using a tensor product of two univariate functional bases, so that

$$f(x, z) \in \text{ span } \left[\{1, b_j^1(x) | j \in I_1\} \otimes \{1, b_j^2(z) | j \in I_2\}\right]. \tag{19.10}$$

Here, I_1 and I_2 are indexing sets, $\{1, b_j^1(x) | j \in I_1\}$ and $\{1, b_j^2(z) | j \in I_2\}$ are univariate function bases corresponding to the two independent variables x and z, and the tensor product $U \otimes V := \{uv | u \in U, v \in V\}$. Writing f as a linear combination of the elements of such a tensor-product basis, we obtain

$$f(x, z) = \beta_1 + \sum_{j \in I_1} \beta_j^1 b_j^1(x) + \sum_{j \in I_2} \beta_j^2 b_j^2(z) + \sum_{i \in I_1} \sum_{j \in I_2} \beta_{i,j}^{1,2} b_i^1(x) b_j^2(z), \tag{19.11}$$

or

$$f(x, z) = \beta_1 + \sum_{j \in I} \beta_j B_j(x, z). \tag{19.12}$$

where I is an indexing set, the β_j's are the regression parameters, and the functions $B_j(x, z)$ are bivariate basis functions. Notice that this bivariate basis can be decomposed into main effects f_1 and f_2 in x and z, along with an interaction part f_{12}, so that

$$f_1(x) = \sum_{j \in I_1} \beta_j^1 b_j^1(x), \; f_2(z) = \sum_{j \in I_2} \beta_j^2 b_j^2(z), \; f_{12}(x, z) = \sum_{i \in I_1} \sum_{j \in I_2} \beta_{i,j}^{1,2} b_i^1(x) b_j^2(z).$$

The univariate function bases we use in this chapter are the $(m-1)$-times differentiable regression spline bases, where

$$\{1, b_j^1(x) | j \in I_1\} = \{1, x, x^2, \ldots, x^m, (x - k_1^1)_+^m, \ldots, (x - k_{K_1}^1)_+^m\}$$
$$\{1, b_j^2(z) | j \in I_2\} = \{1, z, z^2, \ldots, z^m, (z - k_1^2)_+^m, \ldots, (z - k_{K_2}^2)_+^m\},$$

in which $(k_1^1, k_2^1, \ldots, k_{K_1}^1)$ and $(k_1^2, k_2^2, \ldots, k_{K_2}^2)$ are knot sequences for x and z, respectively, the function $(\cdot)_+^m = (\max\{\cdot, 0\})^m$, $I_1 = \{2, 3, \ldots, m + K_1 + 1\}$, and $I_2 = \{2, 3, \ldots, m + K_2 + 1\}$.

When the independent variable is continuous, we use the cubic regression spline basis ($m = 3$) with the knots placed so that they follow the variable's observed density. That is, one knot is placed every ath value of the sorted independent variable. Here, a is chosen so that there are enough knots to be reasonably sure that a knot pair (k_i^1, k_j^2) is at, or near, every position required to capture the fluctuations in the surface. Unless the true underlying function is highly oscillatory, the configuration of 81 knot pairs (a nine-by-nine grid) used throughout this chapter will usually suffice. When the independent variable is highly discrete, we use linear splines ($m = 1$) with knots placed at internal values of x and z, as in Smith et al. (1996).

19.3.2 Design-Adaptive Regression Splines

For the many designs in x and z, most knot pairs will fall inside the convex hull of the data; the few that do not are removed automatically. However, to ensure that knot pairs follow the bivariate density of x and z in cases where there are strong relationships between the independent variables, we alter part of the interaction terms in (19.11) by replacing each term of the form $(x - k_i^1)_+^m (z - k_j^2)_+^m$ with $(x - k_i^1)_+^m (z - k_j^{2,i})_+^m$. The other terms, including the main effects, remain unchanged, while the knot pairs $(k_i^1, k_j^{2,i})$ are now allowed to adapt to the design.

For each i, the knots $k_1^{2,i}, \ldots, k_{K_2}^{2,i}$ are chosen to follow the conditional observed density of z in the locality of $x = k_i^1$, rather than the marginal density of z. Figure 19.1 gives an example of how such adaptive placing guards against initial knot placement outside the observed bivariate density of the design.

Although the resulting bivariate basis is no longer a tensor-product basis as in (19.10) and (19.11), the decomposition into main and interaction effects and the ability to write the model as a linear combination of bivariate basis functions as in (19.12) still apply. To differentiate between the two bivariate bases, we call the first a *tensor-product regression spline basis* and the second a *design-adaptive regression spline basis*.

19.3.3 Estimating the Surface by Bayesian Subset Selection

In the previous two sections, we constructed a bivariate regression spline basis with a large number of terms. Equation (19.12) expresses the function f as a linear combination of basis functions, making the nonparametric regression problem a subset selection problem in which we expect most of the variables to be redundant. The coefficient vector $\boldsymbol{\beta}$ is estimated by the mixture estimate $\hat{\boldsymbol{\beta}}$ of (19.9), and the estimate of the regression surface is therefore

$$\hat{f}(x, z) = \hat{\beta}_1 + \sum_{i \in I} \hat{\beta}_i B_i(x, z).$$

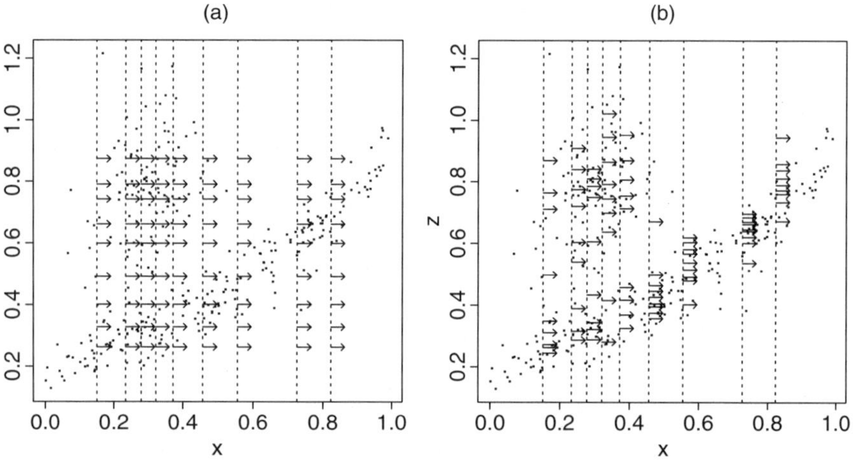

FIGURE 19.1 (a) A highly dependent design with knots placed using the tensor-product regression spline basis. The lines indicate the positioning of the knots in the x-dimension. The base of each arrow is the location of the knot pairs (k_i^1, k_j^2). The direction of the arrows denotes the direction of the nonzero contribution of the term $(x - k_i^1)_+^m(z - k_j^2)_+^m$. (b) A similar plot for the design-adaptive regression spline basis. The locations of the knots $k_j^{2,i}$ follow the observed density of $z|x \in (k_i^1, k_{i+1}^1]$.

In addition, we can extract the columns and estimated coefficients corresponding to the main effects and the pure interaction terms and write them as $\hat{f}_1(x)$, $\hat{f}_2(z)$, and $\hat{f}_{12}(x, z)$.

19.3.4 Discussion of Other Surface Estimators

There are a number of different approaches to nonparametric bivariate regression. A popular method is kernel-weighted local polynomial regression, discussed by Cleveland and Grosse (1991) and Ruppert and Wand (1994). Data-driven estimates of the smoothing parameters crucial to these estimators can, in principle, be obtained using cross-validation, direct plug-in, or alternative methodologies. However, these are often computationally difficult to implement. Nevertheless, one recent procedure is Clive Loader's `locfit` estimator, available at the time of writing on the World Wide Web at `http://cm.bell-labs.com/ms/departments/sia/project/locfit/index.html`.

A second approach to nonparametric bivariate surface estimation is to use multivariate smoothing splines. This approach was developed by Wahba and her coworkers, who use two distinct types of smoothing splines. The first is called a thin-plate spline and involves a single smoothing parameter, while the second is called a tensor-product spline and involves more than one smoothing parameter (Gu et al., 1989). Usually, data-driven estimates of the smoothing parameters are obtained by either generalized cross-validation (GCV) or generalized maximum-likelihood techniques (Wahba, 1990). Both methods require $O(n^3)$ computations, although the tensor-product multivariate smoothing spline has a much larger multiplier than that of the thin-plate multivariate smoothing spline. Both procedures also require $O(n^2)$ storage locations, which, as Section 19.3.5 demonstrates, can make their application to large data sets difficult. However, the computational burden can be reduced via subsampling of the data at some cost in efficiency.

A number of other approaches to nonparametric regression consist effectively of basis selection. While the bases used are often multivariate regression splines, similar to that discussed in Sections 19.3.1 and 19.3.2, the search procedures and the method of estimating the function are radically different from that discussed here. They include the use of deterministic search procedures, such as a stepwise algorithm (Silverman and Friedman, 1989) and regression trees (Friedman, 1991; Stone et al., 1996), to locate a single best model. Once this model has been selected, the coefficients of the linear model are then estimated, usually by least squares. In direct contrast, the search algorithm in this chapter is stochastic (see Section 19.2), and the entire distribution of models (parameterized by γ) is accounted for via the use of mixture estimates that smooth over the distribution of $\gamma \in \Gamma$. (See Section 19.2.4.)

19.3.5 Simulations

Simulations were performed to compare the performance of the proposed nonparametric Bayesian estimator with that of some other surface estimators in the literature. The following five examples were considered for the bivariate regression model in (19.1).

Example 1. The independent variable x was uniformly distributed on the interval [0,1] independently of z, which is normally distributed with a mean of 0.5 and variance of 0.0625. The function is

$$f(x, z) = \frac{(\phi(x, 0.25, 0.1) + \phi(x, 0.6, 0.2))\, z}{4},$$

where $\phi(t, \mu, \sigma)$ is a normal density with mean μ and standard deviation σ evaluated at t.

Example 2. The independent variable x was uniformly distributed on [0,1] independently of z, which is normally distributed with a mean of 0.5 and variance of 0.0625. The function is

$$f(x, z) = \begin{cases} -z/2 & \text{if } x < 0.5 \\ z/2 & \text{if } x > 0.5. \end{cases}$$

Example 3. The independent variables x and z were normally distributed with a mean of 0.5, variance of 0.05, and covariance of -0.025. The function is $f(x, z) = 3$.

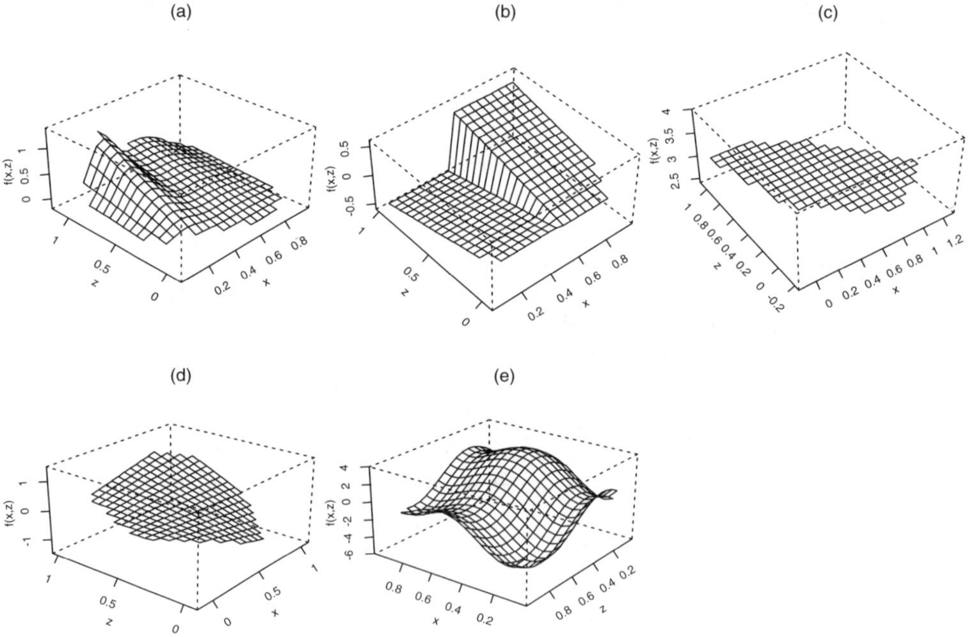

FIGURE 19.2 (a)–(e) Plots of the five functions over a typical convex hull resulting from the designs of the five examples presented in Section 19.3.5. (Panel (a) is for Example 1, panel (b) for Example 2, and so on.)

Example 4. The independent variables x and z are normally distributed with a mean of 0.5, variance of 0.05, and covariance of 0.0125. The function is $f(x, z) = 2x(z - 0.5)$.

Example 5. The independent variables x and z are independent and uniformly distributed on the interval [0,1]. The function is $f(x, z) = 1 - 48x + 218x^2 - 315x^3 + 145x^4 + 2\sin(2\pi z)$.

Perspectival plots of all five functions over a typical convex hull from the designs in each example are given in Figures 19.2(a)–(e).

Three hundred observations of x, z, and y were generated for each of the five examples. The signal-to-noise ratio for Examples 1, 2, 4 and 5 was $\frac{1}{4}$; that is, $\sigma = \frac{1}{4} \times \text{range}(f)$ in (19.1). Because the range of the function in Example 3 is zero, we instead set $\sigma = \frac{1}{4}$. A design-adaptive regression spline basis was used with $K_1 = K_2 = 9$, resulting in $q = (9 + 4)^2 = 169$ potential terms in the design matrix X. Along with the *Bayesian subset selection* based procedure (BSS) outlined in this chapter, the following estimators were used to fit the generated data:

(1) MARS (Friedman, 1991). Version 3.6 was used, with the number of possible terms restricted to 169, as the program default of 15 appeared purely arbitrary. Rarely were there more than 15 terms selected anyway.

(2) LFAF. Clive Loader's `locfit` program was applied with a variable bandwidth, which requires an initial estimate of σ^2. Following Loader's advice, we obtained this estimate from a nearest-neighbor fit with a fixed global bandwidth. The estimated values $\hat{\sigma}^2$ were typically very close to the true value of σ^2.

(3) TPMSS. A bivariate cubic thin-plate spline with a single smoothing parameter was estimated by generalized cross validation (GCV) as in Gu et al. (1989).

(4) TENSMSS. This estimator is a tensor-product cubic smoothing spline which uses five smoothing parameters that were estimated by GCV as in Gu et al. (1989).

(5) ACE. Breiman and Friedman's (1985) additive backfitting routine based on "supersmoother" as implemented in S-Plus with the data transformation turned off, was used.

(6) LSP. A least-squares fit of the parametric linear interaction model $y_i = \beta_0 + \beta_1 x_i + \beta_2 z_i + \beta_3 x_i z_i + e_i$ was applied.

One hundred replications of this simulation were carried out for each example and for each of the estimators. The performance of all the estimators was measured for each of the five test functions by using an approximation to the integrated squared error given by

$$\text{AISE }(\hat{f}) = \frac{1}{N}\sum_{i=1}^{N}\left\{(f_i - \hat{f}_i)^2\right\}.$$

Here, $\{f_i\}_{i=1}^N$ and $\{\hat{f}_i\}_{i=1}^N$ are the true and estimated function values, respectively, over some N points. In this simulation, we used the $n = 300$ design locations.

Box plots of the results for the five examples are given in Figures 19.3(a)–(e). Perspectival plots of the surface estimates corresponding to the upper tenth, median, and lower tenth decile of the AISE for each of the five examples are presented in Figures 19.4 through 19.8. The plots are for the BSS and MARS estimators, together with the better performer of LFAF, TPMSS, and TENSMSS for each example.

The simulation results provide a range of interesting insights into the relative strengths and weaknesses of the various estimators. We stress the following points:

(i) The excellent performance of ACE on the additive function in Example 5 is matched by the more general bivariate estimators BSS, TPMSS, and TENMSS. In the case of BSS, this is attributable to the fact that the underlying signal is largely captured by the main-effect components $f_1(x)$ and $f_2(z)$ of the basis.

(ii) MARS often does not select appropriate terms from the bivariate regression spline basis it considers. This is apparent in Examples 1, 2, 4, and 5,

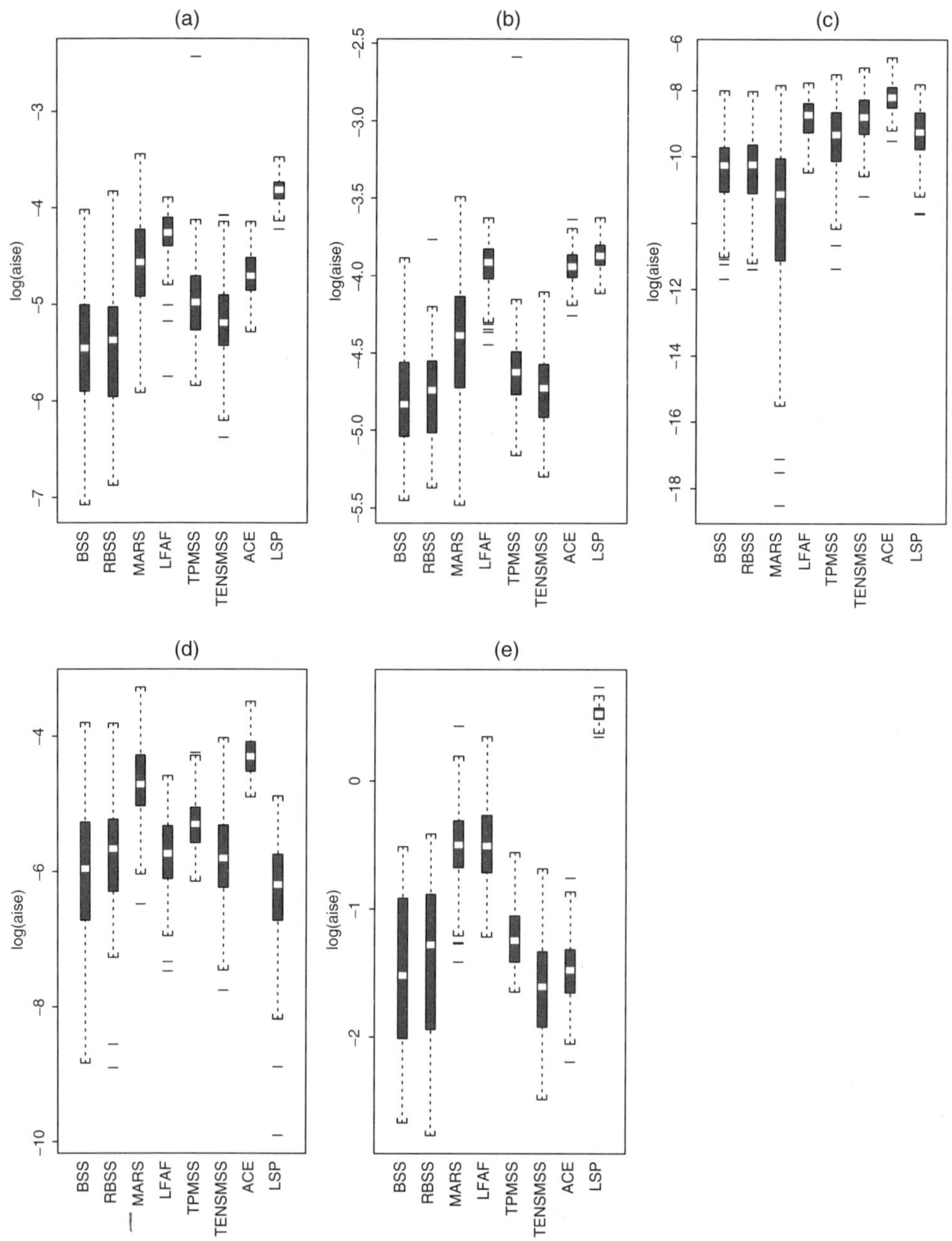

FIGURE 19.3 (a)–(e) Box plots of log(AISE) for the various estimators (see the text for their description) fitted to the 100 replicated data sets corresponding to the five examples. (Panel (a) is for Example 1, panel (b) for Example 2, and so on.) From left to right, the estimators used in each panel are BSS, RBSS, MARS, LFAF, TPMSS, TENSMSS, ACE, and LSP.

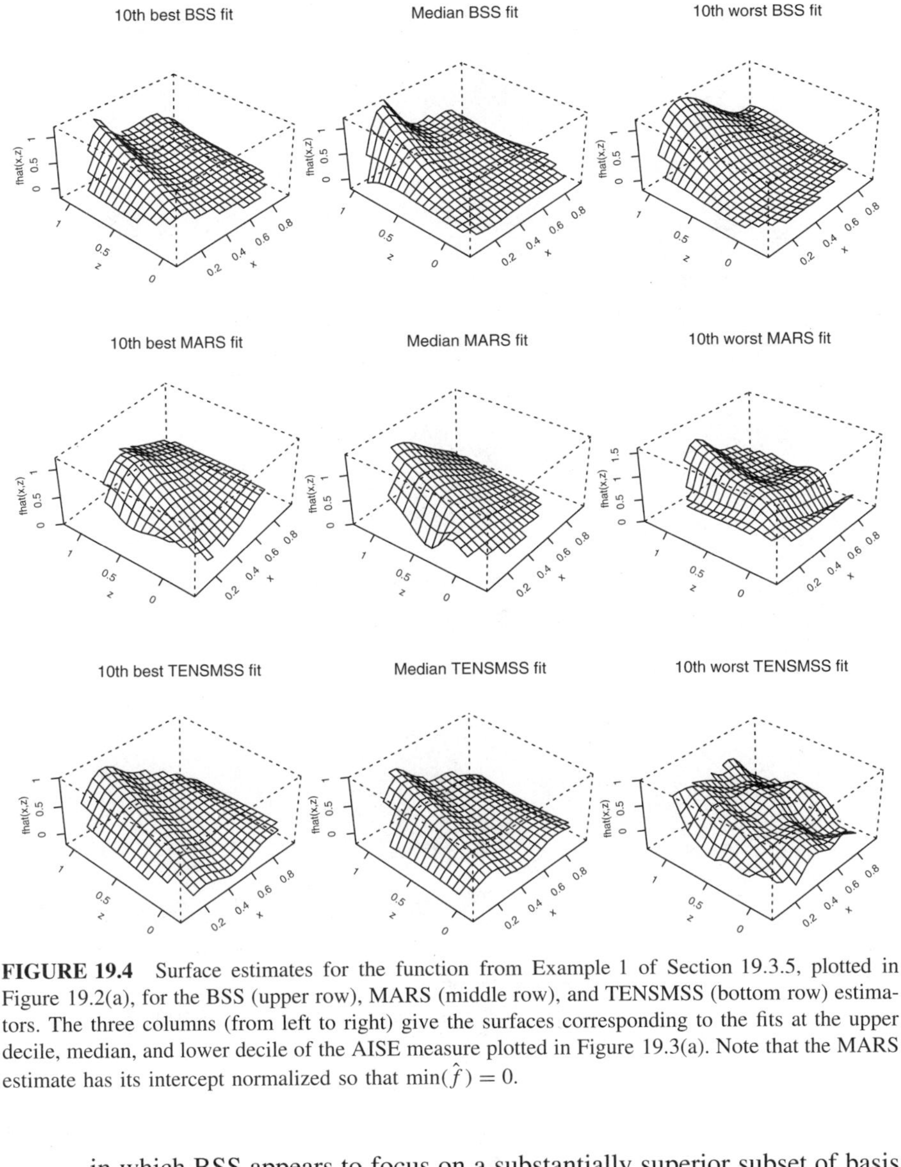

FIGURE 19.4 Surface estimates for the function from Example 1 of Section 19.3.5, plotted in Figure 19.2(a), for the BSS (upper row), MARS (middle row), and TENSMSS (bottom row) estimators. The three columns (from left to right) give the surfaces corresponding to the fits at the upper decile, median, and lower decile of the AISE measure plotted in Figure 19.3(a). Note that the MARS estimate has its intercept normalized so that $\min(\hat{f}) = 0$.

in which BSS appears to focus on a substantially superior subset of basis terms than that of MARS. However, in Example 3, MARS outperforms the other procedures in terms of AISE. Nevertheless, an examination of the actual surface estimates in Figure 19.6 reveals that all procedures appear to identify the underlying function $f(x, z) = 3$ well.

(iii) In direct contrast, the BSS estimator performs well in all five examples. It almost matches the LSP estimator in Example 4, which is very encouraging, as the parametric model fitted under LSP is close to the true model. This near agreement can be attributed to the fact that the models

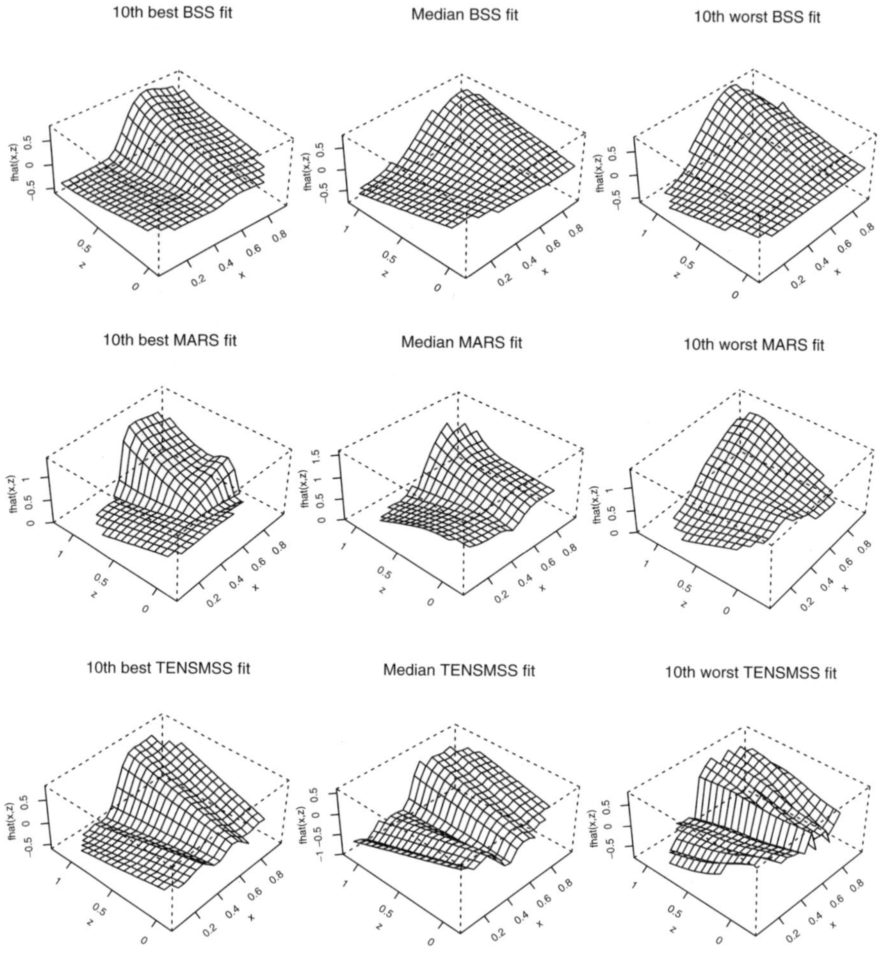

FIGURE 19.5 Surface estimates for the function from Example 2 of Section 19.3.5, plotted in Figure 19.2(b), for the BSS (upper row), MARS (middle row), and TENSMSS (bottom row) estimators. The three columns (from left to right) give the surfaces corresponding to the fits at the upper decile, median, and lower decile of the AISE measure plotted in Figure 19.3(b). Note that the MARS estimate has its intercept normalized so that $\min(\hat{f}) = 0$.

$\boldsymbol{\gamma}^{[j]}$ generated under BSS are often exactly the true model, which is in the regression spline basis.

(iv) Of the two multivariate smoothing splines, TENSMSS is consistently more variable than TPMSS, because more smoothing parameters are estimated. In Example 5, TENSMSS performs much better than TPMSS because the later function has very different curvature in the x- and z-directions and so requires more than one smoothing parameter. In Example 3 (where $f(x, z) = 3$), a single bandwidth is appropriate, and therefore, TPMSS outperforms TENSMSS.

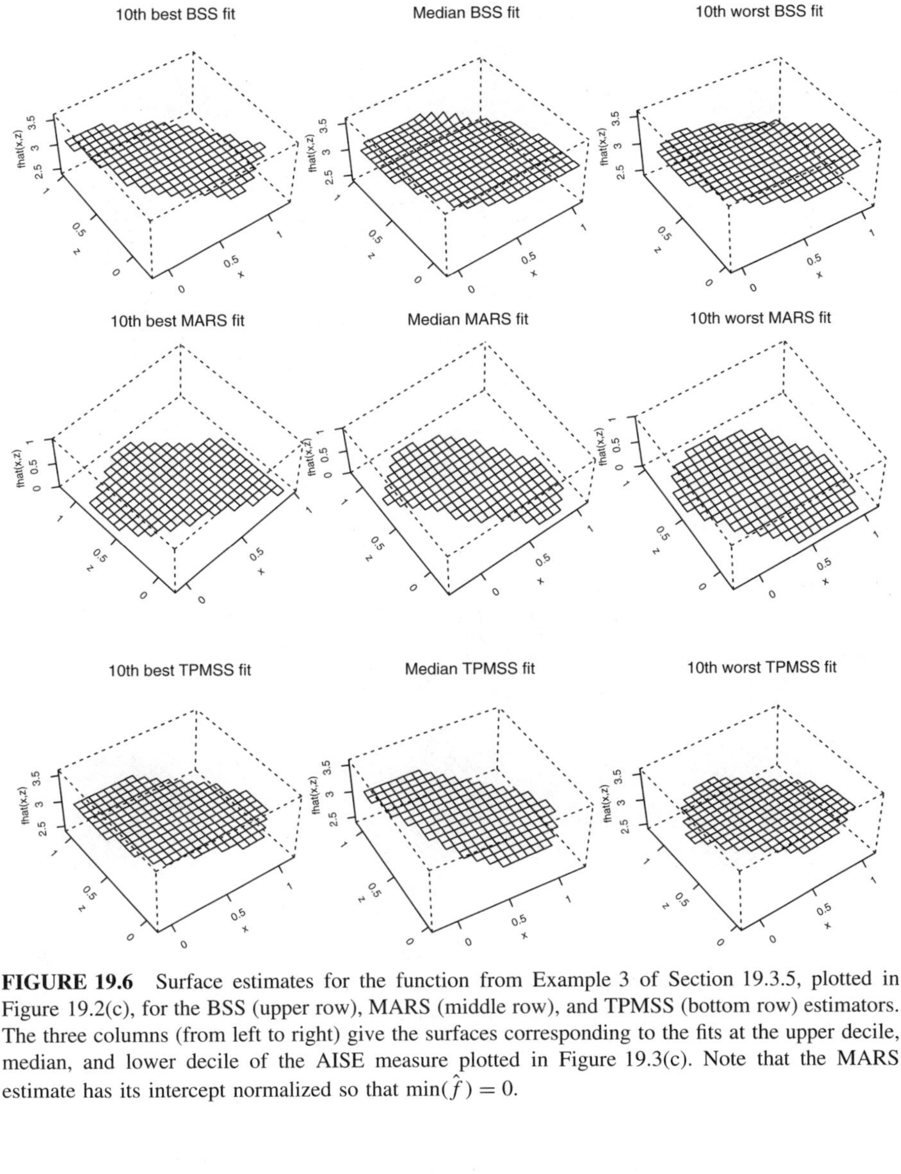

FIGURE 19.6 Surface estimates for the function from Example 3 of Section 19.3.5, plotted in Figure 19.2(c), for the BSS (upper row), MARS (middle row), and TPMSS (bottom row) estimators. The three columns (from left to right) give the surfaces corresponding to the fits at the upper decile, median, and lower decile of the AISE measure plotted in Figure 19.3(c). Note that the MARS estimate has its intercept normalized so that $\min(\hat{f}) = 0$.

(v) It is important to note that the bivariate regression spline basis used here may not provide good approximations to certain functions outside the polynomial space it spans. Nevertheless, the excellent results for the functions in Examples 1, 2, and 5, which are not in that space, illustrates that good results often are obtained.

(vi) The simulations are, of course, necessarily limited. However, we hope that they at least display the competitive nature of the bivariate surface estimator BSS in the standard bivariate regression problem. The real

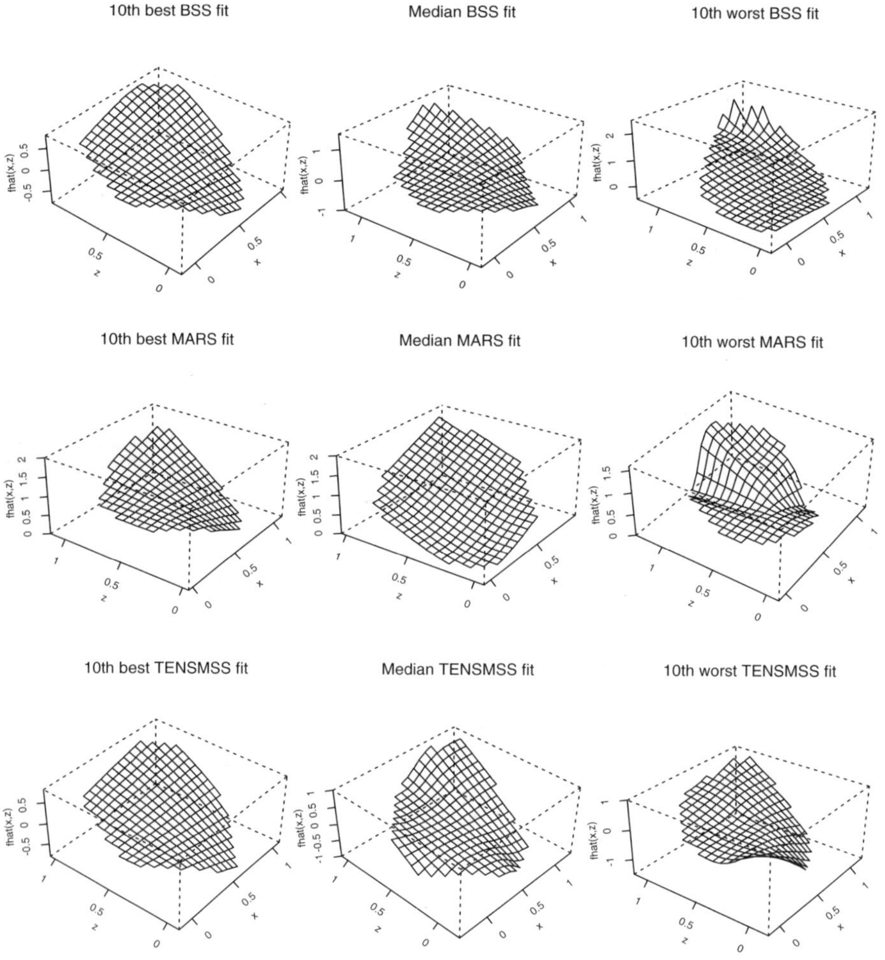

FIGURE 19.7 Surface estimates for the function from Example 4 of Section 19.3.5, plotted in Figure 19.2(d), for the BSS (upper row), MARS (middle row), and TENSMSS (bottom row) estimators. The three columns (from left to right) give the surfaces corresponding to the fits at the upper decile, median, and lower decile of the AISE measure plotted in Figure 19.3(d). Note that the MARS estimate has its intercept normalized so that $\min(\hat{f}) = 0$.

strength of the estimator is the ease with which it can be extended to what have previously been highly problematic settings. This extensibility is discussed at greater length in Sections 19.4 and 19.5.

Observe that the MARS code normalizes its graphical surface estimates so that $\min(\hat{f}) = 0$. However, this is not a hindrance, because in the surface plots we are concerned just with the shape of the surface. Also, the surfaces are plotted only over the convex hull of x and z, as this represents the support of the design.

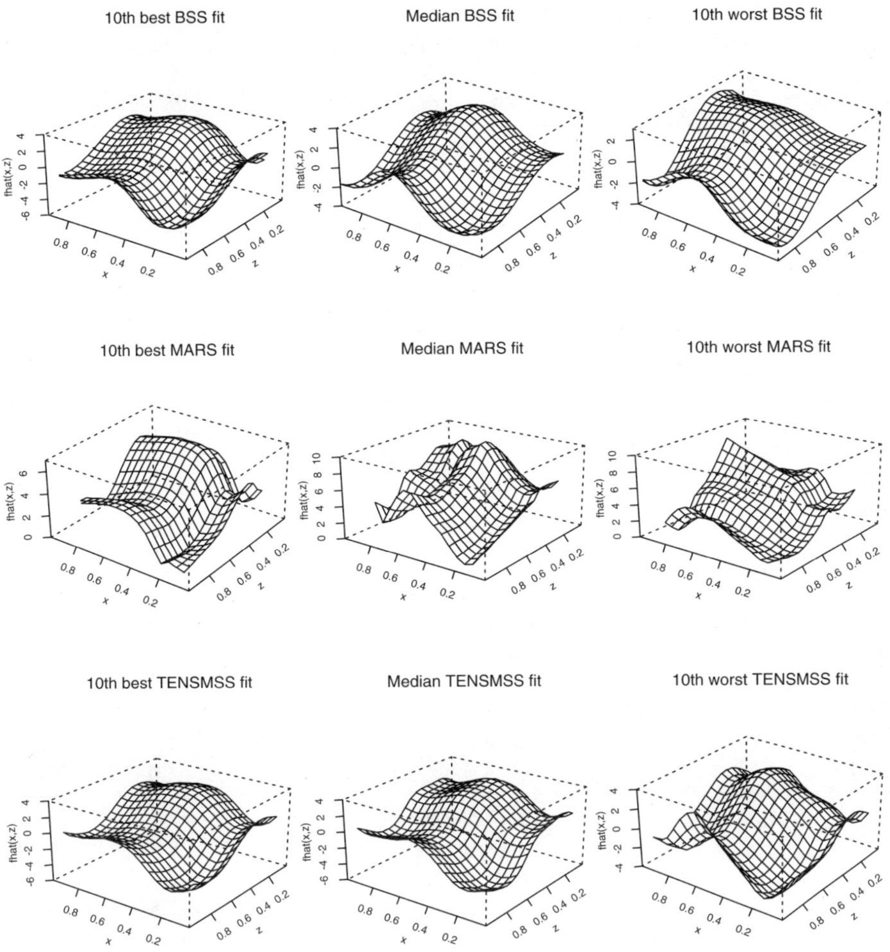

FIGURE 19.8 Surface estimates for the function from Example 5 of Section 19.3.5, plotted in Figure 19.2(e), for the BSS (upper row), MARS (middle row), and TENSMSS (bottom row) estimators. The three columns (from left to right) give the surfaces corresponding to the fits at the upper decile, median, and lower decile of the AISE measure plotted in Figure 19.3(e). Note that the MARS estimate has its intercept normalized so that $\min(\hat{f}) = 0$.

The times taken to estimate the surfaces in each example at a variety of observation levels are reported in Table 19.3; they demonstrate that the sampling-based solution proposed here is computationally competitive with the other estimators. The figures are both machine and implementation dependent and are meant only as an indicator of the viability of the procedures. Nevertheless, the Bayesian estimator BSS is fairly insensitive to the sample size, as there is only a single $O(nq)$ matrix operation (and q is fixed with respect to n when the bases outlined in this section are used). The multivariate smoothing splines, however, are not so insensitive. The local regression procedure LFAF is not

TABLE 19.3 Times (in Seconds) Required to Run the Various Programs on Data Generated from the Five Examples of Section 19.3.5 with Various Sample Sizes[a]

Smoother	(a)	(b)	(c)	(d)	(e)
			Sample size of $n = 300$		
BSS	11.5	17.0	14.0	15.5	28.5
RBSS	164.0	167.0	150.0	151.0	174.5
MARS	12.0	11.0	11.0	10.5	12.0
TPMSS	4.0	4.0	4.0	4.0	4.0
TENSMSS	319.5	300.0	213.0	196.0	439.0
ACE	0.5	0.5	0.5	0.5	1.0
			Sample size of $n = 600$		
BSS	29.0	25.0	20.0	23.5	32.5
RBSS	330.5	329.5	308.5	302.0	349.0
MARS	33.5	32.5	33.5	33.5	32.5
TPMSS	26.0	24.5	25.5	26.0	26.0
TENSMSS	2541.0	2993.5	1146.5	1307.5	3489.5
ACE	2	1.5	1.5	2.0	1.0
			Sample size of $n = 1200$		
BSS	39	36.5	30.5	32.0	45.0
RBSS	648	628.0	599.0	618.5	707.5
MARS	67	68.0	60.0	65.0	61.0
TPMSS	195	195.0	195.0	195.0	195.0
TENSMSS	NEM	NEM	NEM	NEM	NEM
ACE	3.5	4.0	4.0	4.0	4.0
			Sample size of $n = 2400$		
BSS	70.0	68.0	58.5	58.5	75.5
RBSS	1515.5	1609.0	1443.0	1475.5	1674.0
MARS	167.5	152.5	153.5	147.5	155.5
TPMSS	26,723.0	24,535.0	27,598.5	25,788.0	26,207.5
TENSMSS	NEM	NEM	NEM	NEM	NEM
ACE	4.0	4.0	4.5	4.0	4.0

[a]NEM denotes that there was not enough (physical or virtual) memory to run the program because the algorithm required $O(n^2)$ storage locations. The machine used was a low-end workstation (IBM Power PC) with 32 MB of memory and 200 MB of paging space. Note that TPMSS with $n = 2400$ paged heavily, while RBSS is the robust Bayesian estimator introduced in Section 19.4.

included here, as we ran it on another machine, and it would be misleading to compare running times across machines. However, in general, LFAF is a fast approach with computing times slightly lower than those of MARS.

19.3.6 Body Fat Data

To demonstrate the bivariate surface estimation procedure, we apply it to Body Fat Data discussed by Johnson (1996). Two potentially important predictors,

X = age in years and Z = weight in pounds, are considered for a body fat percent measure Y, calculated by using the procedure found in Siri (1956). The data include measurements of X, Z, and Y for $n = 248$ males; four observations that were regarded as inconsistent by Johnson were removed.

The design-adaptive regression spline basis was used with $q = 169$ terms corresponding to a 9-by-9 grid of knot locations. Figure 19.9(a) plots the overall surface estimate $\hat{f}$, while Figures 19.9(b) and (c) give the estimated main effect of age ($\hat{f}_1$) and weight ($\hat{f}_2$), respectively. The plots reveal that, while weight is a significant determinant of body fat percentage, age has an almost negligible main effect. Nevertheless, age does appear to affect the percentage of body fat in the form of the nonlinear interaction effect $\hat{f}_{12}$, given in Figure 19.9(d).

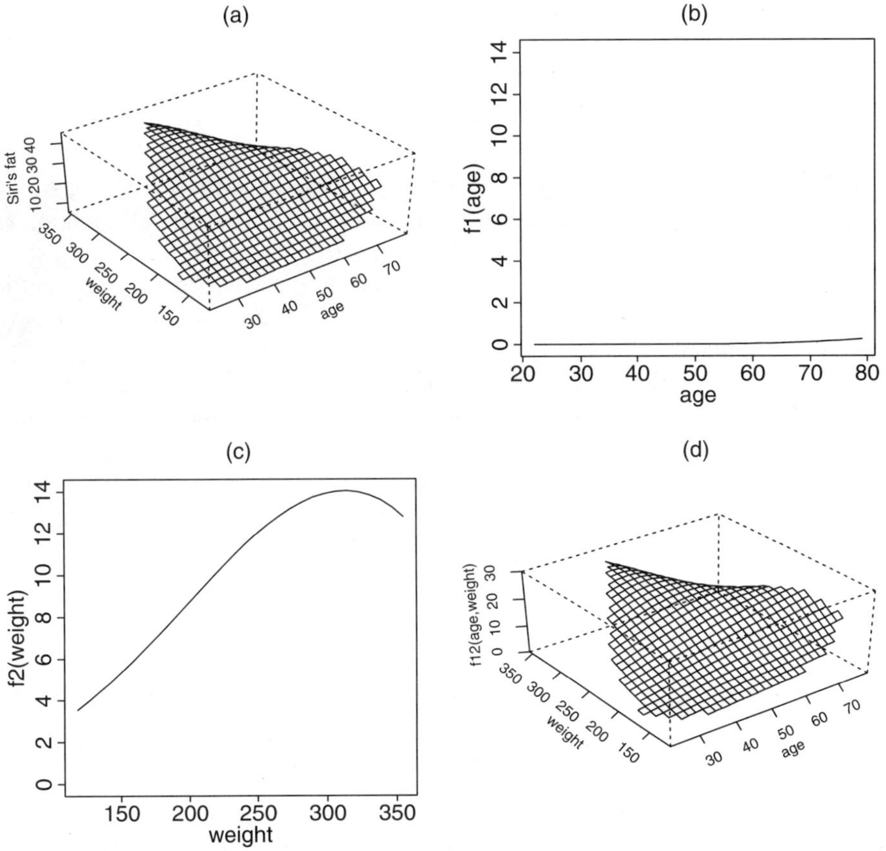

FIGURE 19.9 (a) The overall surface estimate ($\hat{f}$) of the effect of weight and age on Siri's percentage-fat measure. (b)–(c) The main effects of age ($\hat{f}_1$) and weight ($\hat{f}_2$) on Siri's percentage-fat measure, plotted on the same scale to demonstrate the marginal main effect of age compared with that of weight. (d) The pure interaction effect $\hat{f}_{12}$, which is the most significant component of the overall surface estimate $\hat{f}$.

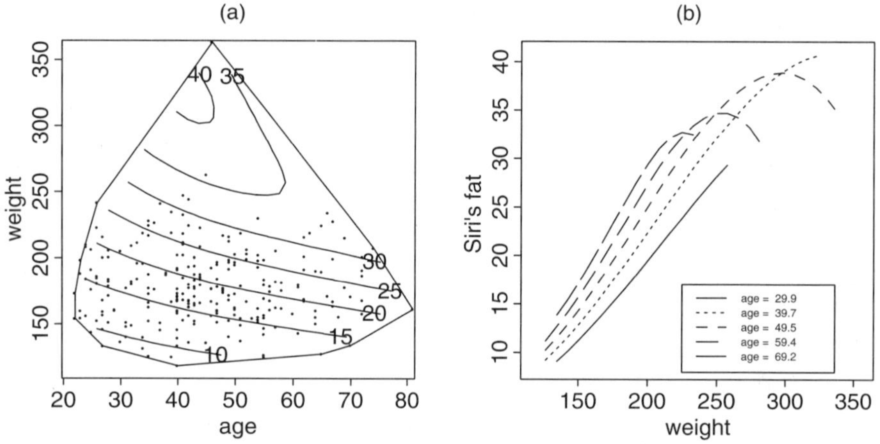

FIGURE 19.10 (a) The scatter plot is of age against weight—that is, the bivariate "design". The convex hull of the data is outlined, while the contour plot is of the estimated regression surface $\hat{f}$. (b) Cross-sections of the regression surface $\hat{f}$ for given values of age.

Figure 19.10(a) is a scatter plot of X against Z, along with the outline of the convex hull of the data and a plot of the contours of $\hat{f}$. A single observation of a very heavy middle-aged man sits a long way from the bulk of the data. In a parametric regression this observation has high leverage, while in a nonparametric regression the effect on the estimated response $\hat{f}$ over the domain of the rest of the data is minimal.

Figure 19.10(b) plots cross sections of the overall bivariate surface estimate $\hat{f}$ for five different values of age. The estimates highlight the changing profile of the relationship between body fat percentage and body weight, conditional on age. However, it is important to note that much of the nonlinear nature of this interaction estimate arises from the aforementioned single anomalous observation in the design.

19.4 ROBUST SURFACE ESTIMATION

19.4.1 Introduction

This section shows how to make Bayesian model selection, and hence bivariate surface estimation, robust to outliers in the observations y. We are unaware of any other robust bivariate nonparametric estimator in the literature that is fully data driven. We follow Smith and Kohn (1996) and Smith et al. (1996) and model the errors in the regression equation (19.1) as an independent and identically distributed mixture of two normals, so that

$$e_i \sim N(0, \omega_i \sigma^2) \quad \text{with } \omega_i = \begin{cases} \kappa > 1 & \text{if the } i\text{th observation is an outlier} \\ 1 & \text{otherwise} \end{cases}$$

We take $\kappa = 100$ to capture the notion that outliers have a much greater variance than that of standard observations.

The use of a mixture of normals to model an outlier process in a linear regression without model selection is a popular approach. See, for example, Box and Tiao (1968), and Verdinelli and Wasserman (1991). Smith et al. (1996) compared this type of approach with that of traditional estimators and demonstrated that it is highly competitive. They generated outliers from error processes that are not distributed as the mixture of normals above and showed that the approach can nevertheless identify and down-weight any outliers.

19.4.2 Robust Bayesian Model Selection

The extension of the work presented in Section 19.4.1 to robust model selection is fairly straightforward and was first discussed in Smith and Kohn (1996) and Smith et al. (1996). If $\boldsymbol{\omega} = (\omega_1, \ldots, \omega_n)^T$ is a vector that parameterizes the outlier process for the n observations, then the linear model (19.3), remains valid, except now the errors $e \sim N(0, \sigma^2\Omega_\omega)$, where $\Omega_\omega = \text{diag}(\boldsymbol{\omega})$. The priors that determine the hierarchical model in Section 19.2 can be amended as follows:

(1) The conditional prior for $\boldsymbol{\beta}_\gamma$ is adjusted to incorporate the error variance from the outlier process, so that

$$\boldsymbol{\beta}_\gamma|\sigma^2, \boldsymbol{\gamma}, \boldsymbol{\omega} \sim N\left((X_\gamma^T\Omega_\omega^{-1}X_\gamma)^{-1}X_\gamma^T\Omega_\omega^{-1}\mathbf{y}, c\sigma^2(X_\gamma^T\Omega_\omega^{-1}X_\gamma)^{-1}\right). \quad (19.13)$$

(2) The prior for σ^2 is unchanged, except that it is also assumed that σ^2 is *a priori* independent of $\boldsymbol{\omega}$, so that $p(\sigma^2|\boldsymbol{\gamma}, \boldsymbol{\omega}) \propto 1/\sigma^2$.

(3) The ω_i are *a priori* independent and identically distributed with $p(\omega_i = \kappa) = \pi_e$. We set $\pi_e = 0.05$ in our numerical work to represent the notion that outliers are rare and find that this works well in practice.

(4) The parameters $\boldsymbol{\gamma}$ and $\boldsymbol{\omega}$ are assumed to be *a priori* independent, and $p(\boldsymbol{\gamma})$ is unchanged from Section 19.2.

19.4.3 The Gibbs Sampler for Robust Subset Selection

In the special case of subset selection, the Gibbs sampler in Section 19.2.3 can be generalized as follows:

(i) Choose initial values $\boldsymbol{\gamma}^{[0]}$ and $\boldsymbol{\omega}^{[0]}$ for $\boldsymbol{\gamma}$ and $\boldsymbol{\omega}$, respectively.

(ii) Successively generate from $p(\gamma_i|\mathbf{y}, \boldsymbol{\gamma}_{j\neq i}, \boldsymbol{\omega})$, $i \in \{1, \ldots, q\}$.

(iii) Successively generate from $p(\omega_i|\mathbf{y}, \boldsymbol{\gamma}, \boldsymbol{\omega}_{j\neq i})$, $i \in \{1, \ldots, n\}$.

Here, steps (ii) and (iii) are iterated repeatedly. Eventually, after an initial warm-up period required for convergence, the iterates that are generated will form observations from the joint distribution of $\boldsymbol{\omega}, \boldsymbol{\gamma}|\mathbf{y}$.

The conditional density $p(\gamma_i|\mathbf{y}, \boldsymbol{\gamma}_{j\neq i}, \boldsymbol{\omega})$ can be calculated in a manner similar to that in Section 19.2, where

$$p(\gamma_i|\boldsymbol{\gamma}_{j\neq i}, \boldsymbol{\omega}, \mathbf{y}) \propto p(\mathbf{y}|\boldsymbol{\gamma}, \boldsymbol{\omega})p(\gamma_i)$$
$$\propto \iint \underbrace{p(\mathbf{y}|\boldsymbol{\beta}_\gamma, \sigma^2, \boldsymbol{\gamma}, \boldsymbol{\omega})}_{\text{likelihood}} \underbrace{p(\boldsymbol{\beta}_\gamma|\sigma^2, \boldsymbol{\gamma}, \boldsymbol{\omega})p(\sigma^2|\boldsymbol{\gamma}, \boldsymbol{\omega})}_{\text{given priors}}\, d\boldsymbol{\beta}_\gamma\, d\sigma^2\, p(\gamma_i).$$

Plugging in the likelihood and prior densities, we can integrate out a normal density in $\boldsymbol{\beta}_\gamma$ followed by an inverse gamma in σ^2, in much the same way as in Section 19.2.2. The resulting conditional posterior is

$$p(\gamma_i|\boldsymbol{\gamma}_{j\neq i}, \boldsymbol{\omega}, \mathbf{y}) \propto (1+c)^{-q_\gamma/2} S(\boldsymbol{\gamma}, \boldsymbol{\omega})^{-n/2} p(\gamma_i), \tag{19.14}$$

where $S(\boldsymbol{\gamma}, \boldsymbol{\omega}) = \mathbf{y}^T\Omega_\omega^{-1}\mathbf{y} - \mathbf{y}^T\Omega_\omega^{-1}X_\gamma\left(X_\gamma^T\Omega_\omega^{-1}X_\gamma\right)^{-1} X_\gamma^T\Omega_\omega^{-1}\mathbf{y}$. Using this, the exact density can be obtained by calculating (19.14) for both $\gamma_i = 1$ and $\gamma_i = 0$ and then normalizing.

The conditional density $p(\omega_i|\mathbf{y}, \boldsymbol{\gamma}, \boldsymbol{\omega}_{j\neq i})$ can be calculated by means of the proportionality

$$p(\omega_i|\mathbf{y}, \boldsymbol{\gamma}, \boldsymbol{\omega}_{j\neq i}) \propto p(\mathbf{y}|\boldsymbol{\gamma}, \boldsymbol{\omega})p(\omega_i)$$
$$\propto \iint \underbrace{p(\mathbf{y}|\boldsymbol{\gamma}, \boldsymbol{\omega}, \boldsymbol{\beta}_\gamma, \sigma^2)}_{\text{likelihood}} \underbrace{p(\boldsymbol{\beta}_\gamma|\sigma^2, \boldsymbol{\gamma}, \boldsymbol{\omega})p(\sigma^2|\boldsymbol{\gamma}, \boldsymbol{\omega})}_{\text{given priors}}\, d\boldsymbol{\beta}_\gamma\, d\sigma^2\, p(\omega_i).$$

By integrating out a normal density in $\boldsymbol{\beta}_\gamma$ and an inverse gamma in σ^2 in exactly the same manner as before, we obtain the conditional posterior

$$p(\omega_i|\boldsymbol{\omega}_{j\neq i}, \boldsymbol{\gamma}, \mathbf{y}) \propto (\omega_i)^{-1/2} S(\boldsymbol{\gamma}, \boldsymbol{\omega})^{-n/2} p(\omega_i). \tag{19.15}$$

The exact conditional density can then be obtained by calculating (19.15) for both $\omega_i = 1$ and $\omega_i = \kappa$ and then normalizing.

After a warm-up period, J iterates $\{(\boldsymbol{\gamma}^{[1]}, \boldsymbol{\omega}^{[1]}), (\boldsymbol{\gamma}^{[2]}, \boldsymbol{\omega}^{[2]}), \ldots, (\boldsymbol{\gamma}^{[J]}, \boldsymbol{\omega}^{[J]})\}$ can be collected to form a Monte Carlo sample from the posterior distribution of $\boldsymbol{\gamma},\boldsymbol{\omega}|\mathbf{y}$. Using these iterates, we can calculate mixture estimates of $E(\boldsymbol{\beta}|\mathbf{y})$ and $p(\omega_i = \kappa|\mathbf{y})$ as

$$\hat{\boldsymbol{\beta}} = \frac{1}{J}\sum_{j=1}^{J} E(\boldsymbol{\beta}|\mathbf{y}, \boldsymbol{\gamma}^{[j]}, \boldsymbol{\omega}^{[j]}), \qquad \hat{p}(\omega_i = \kappa|\mathbf{y}) = \frac{1}{J}\sum_{j=1}^{J} p(\omega_i = \kappa|\mathbf{y}, \boldsymbol{\gamma}^{[j]}, \boldsymbol{\omega}_{k\neq i}^{[j]}).$$

Note that conditional expected value of $\boldsymbol{\beta}$ in the equation on the left can be calculated exactly, because

$$\boldsymbol{\beta}_\gamma|\mathbf{y}, \boldsymbol{\gamma}, \boldsymbol{\omega} \sim T\left[n,\ (X_\gamma^T\Omega_\omega^{-1}X_\gamma)^{-1}X_\gamma^T\Omega_\omega^{-1}\mathbf{y},\ \frac{cS(\boldsymbol{\gamma}, \boldsymbol{\omega})}{(1+c)n}(X_\gamma^T\Omega_\omega^{-1}X_\gamma)^{-1}\right],$$

with the remaining elements of $\boldsymbol{\beta}$ exactly zero. These quantities give robust estimates of the regression parameters (and hence the resulting bivariate surface) and the posterior probability of the ith observation being an outlier.

19.4.4 Efficient Implementation of the Sampling Scheme

19.4.4.1 Generation of γ_i

Given $\boldsymbol{\omega}$ and $\boldsymbol{\gamma}_{j\neq i}$, the generation of γ_i, $i \in \{1, 2, \ldots, q\}$, is the same as is outlined in Section 19.2.5, except that the precalculations are now of $\boldsymbol{y}^T\Omega_\omega^{-1}\boldsymbol{y}$, $\tilde{X}^T\Omega_\omega^{-1}\boldsymbol{y}$, and $\tilde{X}^T\Omega_\omega^{-1}\tilde{X}$, the active subvector is now $\boldsymbol{v}_\gamma = \tilde{X}_\gamma^T\Omega_\omega^{-1}\boldsymbol{y}$, the active submatrix is $M_\gamma = \tilde{X}_\gamma^T\Omega_\omega^{-1}\tilde{X}_\gamma$, and $\boldsymbol{y}^T\boldsymbol{y}$ is replaced with $\boldsymbol{y}^T\Omega_\omega^{-1}\boldsymbol{y}$ wherever it appears in the algorithm.

19.4.4.2 Generation of ω_i

Given $\boldsymbol{\gamma}$ and $\boldsymbol{\omega}_{j\neq i}$, the efficient generation of ω_i, for $i \in \{1, 2, \ldots, n\}$, is as outlined in Smith et al. (1996). Let $\boldsymbol{\omega}^\kappa = (\omega_1, \omega_2, \ldots, \omega_{i-1}, \kappa, \omega_{i+1}, \ldots, \omega_n)$, $\boldsymbol{\omega}^1 = (\omega_1, \omega_2, \ldots, \omega_{i-1}, 1, \omega_{i+1}, \ldots, \omega_n)$, $a_\omega = \boldsymbol{y}^T\Omega_\omega^{-1}\boldsymbol{y}$, $\boldsymbol{v}_\omega = \tilde{X}_\gamma^T\Omega_\omega^{-1}\boldsymbol{y}$, and $M_\omega = \tilde{X}_\gamma^T\Omega_\omega^{-1}\tilde{X}_\gamma$. Also, let $\boldsymbol{x}_i$ be the ith row of $\tilde{X}_\gamma$ and R be the upper triangular Cholesky factor of M_ω. Notice that a change in the value of an individual element ω_i does not require the complete recalculation of a_ω, $\boldsymbol{v}_\omega$, and M_ω. Fast updating can be provided by the three identities

$$a_{\omega^\kappa} = a_{\omega^1} - \left(1 - \frac{1}{\kappa}\right) y_i^2, \qquad \boldsymbol{v}_{\omega^\kappa} = \boldsymbol{v}_{\omega^1} - \left(1 - \frac{1}{\kappa}\right) \boldsymbol{x}_i y_i,$$

$$M_{\omega^\kappa} = M_{\omega^1} - \left(1 - \frac{1}{\kappa}\right) \boldsymbol{x}_i\boldsymbol{x}_i^T.$$

The calculation of the quadratic form $\boldsymbol{v}_\omega^T M_\omega^{-1}\boldsymbol{v}_\omega$ is the backbone of the evaluation of the conditional density for ω_i in (19.15). Fast updating of this quadratic form for a change in a single element ω_i is possible with the use of the upper triangular Cholesky factor R of the previous matrix M_ω. For example, if $R^TR = M_{\omega^1}$, then

$$\begin{aligned}
\boldsymbol{v}_{\omega^\kappa}^T M_{\omega^\kappa}^{-1}\boldsymbol{v}_{\omega^\kappa} &= \boldsymbol{v}_{\omega^\kappa}^T \left[M_{\omega^1} - (1 - 1/\kappa)\boldsymbol{x}_i\boldsymbol{x}_i^T\right]^{-1} \boldsymbol{v}_{\omega^\kappa} \\
&= \boldsymbol{v}_{\omega^\kappa}^T \left[R^T\left(I - (1 - 1/\kappa)(R^T)^{-1}\boldsymbol{x}_i\boldsymbol{x}_i^T R^{-1}\right)R\right]^{-1} \boldsymbol{v}_{\omega^\kappa} \\
&= \boldsymbol{\xi}^T(I + \boldsymbol{\delta}\boldsymbol{\delta}^T)^{-1}\boldsymbol{\xi} \\
&= \boldsymbol{\xi}^T\boldsymbol{\xi} + \frac{(\boldsymbol{\xi}^T\boldsymbol{\delta})^2}{1 - \boldsymbol{\delta}^T\boldsymbol{\delta}},
\end{aligned}$$

where $\boldsymbol{\xi} = (R^T)^{-1}\boldsymbol{v}_{\omega^\kappa}$ and $\boldsymbol{\delta} = (1 - 1/\kappa)^{1/2}(R^T)^{-1}\boldsymbol{x}_i$. However, if $R^TR = M_{\omega^\kappa}$, then similar manipulations lead to the equation

$$\boldsymbol{v}_{\omega^1}^T M_{\omega^1}^{-1}\boldsymbol{v}_{\omega^1} = \boldsymbol{\xi}^T\boldsymbol{\xi} - \frac{(\boldsymbol{\xi}^T\boldsymbol{\delta})^2}{1 + \boldsymbol{\delta}^T\boldsymbol{\delta}},$$

where $\boldsymbol{\xi} = (R^T)^{-1}\boldsymbol{v}_{\omega^1}$.

Using these definitions and fast updating identities, we can employ the following algorithm to generate the ω_i for i in $\{1, 2, \ldots, n\}$, one element at a time (the order of each operation is noted on the right-hand side):

```
do i in {1, 2, ..., n}
   if ω_i=1 then
      (1a) Update a_{ω^κ} = a_{ω^1} − (1 − 1/κ) y_i^2
      (1b) Update v_{ω^κ} = v_{ω^1} − (1 − 1/κ) x_i y_i                         O(q_γ)
      (1c) Solve the lower triangular system R^T ξ = v_{ω^κ} for ξ           O(q_γ^2)
      (1d) Solve the lower triangular system R^T δ = (sqrt(1 − 1/κ)) x_i
           for δ                                                              O(q_γ^2)
      (1e) Calculate S(γ, ω^κ) = a_{ω^κ} − ξ^T ξ − (ξ^T δ)^2 / (1 − δ^T δ)   O(q_γ)
   else
      (2a) Update a_{ω^1} = a_{ω^κ} + (1 − 1/κ) y_i^2
      (2b) Update v_{ω^1} = v_{ω^κ} + (1 − 1/κ) x_i y_i                         O(q_γ)
      (2c) Solve the lower triangular system R^T ξ = v_{ω^1} for ξ
                                                                              O(q_γ^2)
      (2d) Solve the lower triangular system R^T δ = (sqrt(1 − 1/κ)) x_i
           for δ                                                              O(q_γ^2)
      (2e) Calculate S(γ, ω^1) = a_{ω^κ} − ξ^T ξ + (ξ^T δ)^2 / (1 + δ^T δ)   O(q_γ)
   endif
   (3a) h = exp{ n/2 [log(S(γ, ω^κ)) − log(S(γ, ω^1))]
               + log(1 − π_e) − log(π_e) + 1/2 log(κ) }
   (3b) Evaluate p(ω_i = κ|γ, ω_{j≠i}, y) = 1/(1 + h) and generate ω_i^new
   if ω_i^new = 1 and ω_i = κ then
        (4) Update R using Cholesky updating
            (Dongarra et al., 1979, 10.1)                                     O(q_γ^2)
   else if ω_i^new = κ and ω_i = 1 then
        (5) Update R using Cholesky updating
            (Dongarra et al., 1979, 10.1)                                     O(q_γ^2)
   endif
   Set ω_i = ω_i^new
enddo
```

Note that the updating of the Cholesky factor R at steps (4) and (5) need be undertaken only if the value of ω_i that is generated is different from that of the previous iterate. This is usually not the case, so these operations are undertaken infrequently. Overall, the generation of the $\boldsymbol{\omega}$ iterate is an $O(nq_\gamma^2)$ algorithm.

However, to synchronize the generation of $\boldsymbol{\omega}$ with that of $\boldsymbol{\gamma}$ requires the updating of $\mathbf{y}^T\Omega_{\omega}^{-1}\mathbf{y}$, $\tilde{X}^T\Omega_{\omega}^{-1}\mathbf{y}$, and $\tilde{X}^T\Omega_{\omega}^{-1}\tilde{X}$, as $\boldsymbol{\omega}$ has now changed. If there are, say, k altered elements of $\boldsymbol{\omega}$, then the synchronization can be done in $O(kq^2)$ operations using the fast-updating identities previously discussed.

19.4.5 Simulations

Smith et al. (1996) presented evidence which suggests that modeling errors as a mixture of normals when fitting a linear model without subset selection is competitive with several traditional robust estimators arising out of a frequency view of probability. Smith and colleagues also undertook simulations indicating that Bayesian subset selection, combined with the mixture-of-normals approach to robustness, compares favorably with robust quantile smoothing spline techniques (Koneker et al. 1994) in univariate nonparametric regression. Smith and Kohn (1996) also undertake a comparison with local polynomial robust regression (Cleveland, 1979) and again show that the Bayesian approach to robust nonparametric additive regression is competitive.

To extend these empirical results, the simulation experiments of the previous section are repeated, but with 20 of the 300 observations generated being potential outliers. These 20 outlying observations were restricted to fall within the convex hull of the data and with an error standard deviation equal to $\frac{15}{4} \times \text{range}(f)$. The error standard deviation of the "clean" observations remained $\frac{1}{4} \times \text{range}(f)$. Example 3 is a special case where, because the range of the function is zero, the clean observations were given an error standard deviation of 1/4 and the potential outliers an error standard deviation of 15/4. *Robust Bayesian subset selection* (RBSS) was applied to select terms from the same design-adaptive regression spline basis as was used in Section 19.3.2. The data were also fitted by the same non-Bayesian estimators used in Section 19.3.5 to see how sensitive they were to outliers.

Figures 19.11 gives box plots of log(AISE) for all the estimators over 100 replicated data sets. The nonrobust techniques, of course, perform poorly. However, it is interesting to note that the nonparametric procedures appear to perform worse than the parametric model LSP, suggesting that if outliers appear prominent in the data, a parametric model should be considered instead of the nonrobust nonparametric estimators.

Nevertheless, the RBSS estimator appears highly successful in down-weighting the outlying observations, even though the method of generating the errors is not the same as that of generating the prior for the outliers. This feature is further discussed by Smith and Kohn (1996) and Smith et al. (1996), who found that outliers generated from nonnormal distributions were also well captured by using a simple mixture of two normals.

The resulting surface fits appear highly robust. A comparison of the box plots for RBSS in Figure 19.11 and those for BSS in Figure 19.3 illustrates that the robust fit to the outlier-contaminated data is only slightly less efficient than the nonrobust fit to the "clean" data. Moreover, the robust Bayesian estimator RBSS

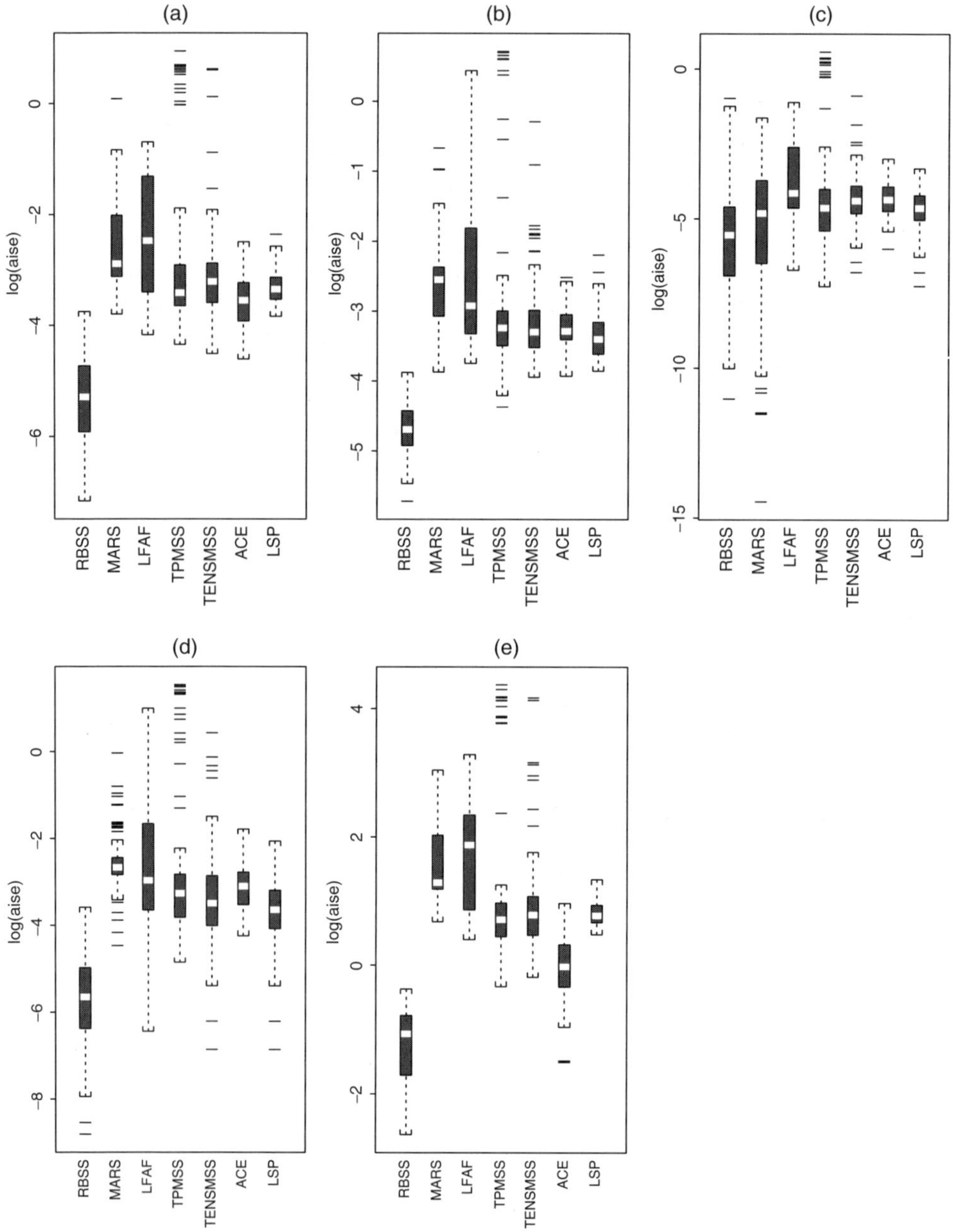

FIGURE 19.11 (a)–(e) Box plots of log (AISE) for the various estimators (see the text for their description) fitted to 100 replicated data sets corresponding to the five examples with outliers present. From left to right, the estimators are RBSS, MARS, LFAF, TPMSS, TENSMSS, ACE, and LSP.

was also fitted to the uncontaminated data from Section 19.3 in which there were no outliers, in order to find out whether the procedure was identifying outliers where they did not exist. The box plots for the RBSS and BSS estimators in Figure 19.3 indicate that no such identification occurs and that there is little loss in efficiency in using a robust estimator when no outliers are present in the data.

19.5 SURFACE ESTIMATION FOR TIME SERIES DATA

19.5.1 Bayesian Model Selection with Autocorrelated Errors

If observations y_t are collected in time order and equally spaced over the interval $t = 1, \ldots, n$, then the errors that arise are potentially autocorrelated. If ignored, autocorrelation can result in very inefficient estimates of the underlying regression surface. This is demonstrated for the univariate case in Chapter 6 of this volume and in Smith et al. (1998) and for the bivariate case in Smith and Kohn (1997). Smith et al. (1998) show how to handle autocorrelated errors in additive nonparametric regression; here, we extend the approach to bivariate surface estimation. As far as we are aware, this chapter is one of only a few papers that deal with estimating a bivariate regression surface when the errors are correlated over time. For example, Wang (1996) extends the multivariate smoothing spline methodology to regression models in which the errors have a parametric correlation matrix that depends on a parsimonious set of parameters. The required smoothing parameters can be estimated simultaneously with these correlation parameters with the use of several popular procedures, including GCV and generalized maximum-likelihood techniques. Our approach to bivariate surface estimation with autocorrelated errors is outlined next; a detailed exposition is given in Smith et al. (1998).

We model the errors u_t as a stationary autoregressive process of maximal order s, so that the surface estimation problem at (19.1) can be rewritten as

$$y_t = f(x_t, z_t) + u_t,$$

where

$$u_t = \sum_{j=1}^{s} \theta_j u_{t-j} + e_t, \tag{19.16}$$

in which $\boldsymbol{\theta} = (\theta_1, \ldots, \theta_s)^T$ are the autoregressive parameters and can be reparameterized in terms of the partial autocorrelations $\boldsymbol{\psi} = (\psi_1, \ldots, \psi_s)^T$ (Monahan, 1984; Smith et al., 1998). The errors $e_t \sim N(0, \sigma^2)$ are independent.

Following the exposition in previous sections, we model the function f as a sum of bivariate basis terms, as in (19.2). Again, we use either the tensor product or design-adaptive regression spline bases discussed in Sections 19.3.1 and 19.3.2, although there is no reason that other bases could not be employed. Again, a large number of potential knot positions are considered, with the regression parameters $\boldsymbol{\beta}$ estimated so that the uncertainty regarding their significance is taken into account. As in previous sections, this uncertainty is parameterized by $\boldsymbol{\gamma}$, and the regression parameters $\boldsymbol{\beta}$ are estimated under the assumption of subset uncertainty.

The major problem in extending Bayesian model selection (and therefore the special case of subset selection) to linear models in which the errors are dependent over time is to develop a conditional prior for $\boldsymbol{\beta}_\gamma$ that is analogous to those found in (19.4) and (19.13). We first note that $\text{var}(\boldsymbol{u}) = \sigma^2\Omega_\psi$, where the matrix Ω_ψ depends only on the partial autocorrelation parameters $\boldsymbol{\psi}$. In a manner similar to that exhibited by (19.13), we take the prior for $\boldsymbol{\beta}_\gamma$, conditional on σ^2, $\boldsymbol{\gamma}$, and $\boldsymbol{\psi}$ as

$$\boldsymbol{\beta}_\gamma|\sigma^2, \boldsymbol{\gamma}, \boldsymbol{\psi} \sim N((X_\gamma^T\Omega_\psi^{-1}X_\gamma)^{-1}X_\gamma^T\Omega_\psi^{-1}\boldsymbol{y}, c\sigma^2(X_\gamma^T\Omega_\psi^{-1}X_\gamma)^{-1}), \tag{19.17}$$

where the conditional prior mean and variance matrix of $\boldsymbol{\beta}_\gamma$ have been adjusted for the $\text{Var}(\boldsymbol{u})$.

In addition to estimating the autocorrelation process (i.e., $\boldsymbol{\psi}$) it is possible to estimate the order of the autocorrelation, l, given the prespecified maximal order $s \geq l$. To undertake this, we follow a procedure first discussed in Barnett et al. (1996) and applied to univariate nonparametric regression with autocorrelated errors in Smith et al. (1998). The procedure involves the introduction of a vector of binary indicator variables $\kappa = (\kappa_1, \ldots, \kappa_s)^T$ such that $\kappa_i = 0$ if ψ_i is identically zero and $\kappa_i = 1$ otherwise. The true order $l \leq s$ of the autoregression can then be determined from the posterior distribution of κ, because, by Monahan (1984), $\theta_{l+1} = \cdots = \theta_s = 0$ if and only if $\kappa_{l+1} = \cdots = \kappa_s = 0$.

The sampling scheme results in mixture and histogram estimates $\hat{f}$, $\hat{\boldsymbol{\psi}}$, and $\hat{\kappa}$ of the posterior means $E(f|\mathbf{y})$, $E(\boldsymbol{\psi}|\mathbf{y})$, and $E(\kappa|\mathbf{y})$, respectively. That is, simultaneous estimates of the bivariate surface f, autocorrelation process $\boldsymbol{\psi}$, and autocorrelation order l are obtained. Smith and Kohn (1997) provide empirical examples of such surface estimation and show that taking account of autocorrelation results in extensive improvement in surface estimates from data for which autocorrelation exists.

19.6 ALTERNATIVE BASES AND MODEL MIXING

The focus of this chapter has been on the multivariate regression spline bases outlined in Sections 19.3.1 and 19.3.2 and on estimating $\boldsymbol{\beta}$ under the assumption of subset uncertainty. The two are a natural combination, as the removal of an element from a regression spline basis corresponds exactly to the removal of a column from the associated design matrix. This correspondence permits the fast updating of the Cholesky factors in steps (1a) and (2a) in the algorithm for generating $\boldsymbol{\gamma}$ presented in Section 19.2.5.

However, other bases can be more complex and do not correspond exactly to subset selection. In this case, not only are the fast Cholesky updates no longer applicable, but the sampling schemes in Sections 19.2.3 and 19.4.3 may not be appropriate, because they rely on the binomial structure of the model parameterization $\boldsymbol{\gamma} \in \Gamma$. Nevertheless, the formulation of the hierarchical model for general model selection found in Sections 19.2.3, 19.4.3 and 19.5 does extend to the selection from a general family of bases. For example, in the plain model-selection case (that is, that without an outlier or autoregressive error process), the extension can be accomplished in the following manner: Let $\{\mathcal{S}_\gamma; \boldsymbol{\gamma} \in \Gamma\}$ be a family of bivariate bases, where $\mathcal{S}_\gamma = \{B_{i,\gamma}(x, z); i \in I_\gamma\}$ is the basis for model $\boldsymbol{\gamma}$. Then, for $\boldsymbol{\gamma} \in \Gamma$, we approximate the surface f as a linear combination of basis terms, so that

$$f(x, z) = \sum_{i \in I_\gamma} \beta_{i,\gamma} B_{i,\gamma}(x, z).$$

This can be written as

$$f(x, z) = \boldsymbol{x}_\gamma(x, z)^T \boldsymbol{\beta}_\gamma,$$

where

$$\boldsymbol{x}_\gamma(x, z) = (B_{1,\gamma}(x, z), B_{2,\gamma}(x, z), \ldots, B_{q_\gamma,\gamma}(x, z))^T,$$
$$\boldsymbol{\beta}_\gamma = (\beta_{1,\gamma}, \beta_{2,\gamma}, \ldots, \beta_{q_\gamma,\gamma})^T,$$

and $q_\gamma = \#(\mathcal{S}_\gamma)$. Thus, the nonparametric regression can be expressed in the linear regression form given in (19.3), and the posterior probability of any model, $p(\boldsymbol{\gamma}|\boldsymbol{y})$, can be calculated as in (19.6). The posterior mean of $f(x, z)$ can be expressed as

$$E[f(x, z)|\boldsymbol{y}] = \sum_{\gamma \in \Gamma} E[f(x, z)|\boldsymbol{y}, \boldsymbol{\gamma}] p(\boldsymbol{\gamma}|\boldsymbol{y}) = \sum_{\gamma \in \Gamma} x_\gamma(x, z)^T E[\boldsymbol{\beta}_\gamma|\boldsymbol{\gamma}, \boldsymbol{y}] p(\boldsymbol{\gamma}|\boldsymbol{y}). \tag{19.18}$$

Now, if an appropriate sampling scheme can be constructed to generate a sample $\{\boldsymbol{\gamma}^{[1]}, \ldots, \boldsymbol{\gamma}^{[J]}\}$ from $\boldsymbol{\gamma}|\boldsymbol{y}$, then the posterior mean in (19.18) can be estimated using the mixture estimate

$$\hat{f}(x, z) = \frac{1}{J} \sum_{j=1}^{J} \boldsymbol{x}_{\gamma^{[j]}}(x, z)^T E[\boldsymbol{\beta}_{\gamma^{[j]}}|\boldsymbol{\gamma}^{[j]}, \boldsymbol{y}].$$

The estimates for the robust and autocorrelated error model selections are similar.

The construction of an appropriate sampling scheme requires knowledge of the form of the parameter $\boldsymbol{\gamma}$, which in turn depends on the basis families $\{\mathcal{S}_\gamma; \boldsymbol{\gamma} \in \Gamma\}$ considered. Such schemes therefore should be constructed on a case-by-case basis. For example, two popular bivariate basis families are tensor-product B-splines (de Boor, 1978) and B-forms (de Boor, 1987). Here, the removal of a knot corresponds to the removal of a term in a manner similar to that used with regression splines, but also requires the recalculation of several adjacent basis terms found in the design matrix X_γ. In this case, the binomial structure of γ remains appropriate (because a decision on the inclusion or exclusion of a finite set of knots is still being made), and therefore, the sampling schemes in Sections 19.2.3 and 19.4.3 can still be used. However, the fast implementation details given in Section 19.2.5 no longer apply, as they relied on the fact that removing terms from X_γ did not alter any adjacent basis terms in the resulting design matrix.

Acknowledgments

We would like to thank Grace Wahba and Chong Gu for extensive help in implementing their smoothing spline routines, Jerome Friedman for sending us his MARS program, and Clive Loader for helping us implement his `locfit` program. We would also like to thank two anonymous referees for helpful comments.

REFERENCES

Akaike, H. (1973). Information theory and an extension of the maximum likelihood principle, in Petrov, B. N. and Csàki, F., (eds.), *Second International Symposium on Information Theory*, pp. 267–281, Akademia Kiad, Budapest, Hungary.

Barnett, G., Kohn, R., and Sheather, S. (1996). Bayesian estimation of an autoregressive model using Markov chain Monte Carlo, *J. Econometrics*, **74**, 237–254.

Box, G., and Tiao, G. (1968). A Bayesian approach to outlier detection and residual analysis, *Biometrika*, **55**, 119–129.

Box, G., and Tiao, G. (1973). *Bayesian Inference in Statistical Analysis*, Addison-Wesley, Reading, MA.

Breiman, L., and Friedman, J. (1985). Estimating optimal transformations for multiple regression and correlation, *J. Amer. Statist. Assoc.*, **80**, 580–598.

Cleveland, W. (1979). Robust locally weighted regression and smoothing scatterplots, *J. Amer. Statist. Assoc.*, **74**, 828–836.

Cleveland, W., and Grosse, E. (1991). Computational methods for local regression, *Statistics and Computing*, **1**, 47–62.

De Boor, C. (1978). *A Practical Guide to Splines*, Springer-Verlag, New York.

De Boor, C. (1987). B-form basics, in Farin, G., (ed.), *Geometric Modeling*, pp. 131–148, SIAM, Philadelphia, PA.

Dongarra, J., Moler, C., Bunch, J., and Stewart, G. (1979). *Linpack User's Guide*, SIAM, Philadelphia, PA.

Friedman, J. (1991). Multivariate adaptive regression splines, *Ann. Statist.*, **19**, 1–141.

Friedman, J. H., and Silverman, B. W. (1989). Flexible parsimonious smoothing and additive modeling, *Technometrics*, **31**, 3–39.

Gelfand, A., and Smith, A. (1990). Sampling based approaches to calculating marginal densities, *J. Amer. Statist. Assoc.*, **85**, 398–409.

George, E., and McCulloch, R. (1993). Variable selection via Gibbs sampling, *J. Amer. Statist. Assoc.*, **88**, 881–889.

George, E., and McCulloch, R. (1997). Approaches for Bayesian variable selection, *Statistica Sinica*, **7**(2), 339–373.

Gu, C., Bates D., Chen, Z., and Wahba, G. (1989). The computation of GCV functions through Householder tridiagonalization with application to the fitting of interaction spline models, *SIAM J. Matrix Analysis and Applications*, **10**, 457–480.

Holmes, C., and Mallick, B. (1999). Bayesian radial basis functions of variable order, *Neural Computation*, **10**, 1217–1233.

Johnson, R. (1996). Fitting percentage of body fat to simple body measurements, *J. Stats. Education* (World Wide Web journal), 4.

Koneker, R., Ng, P., and Portnoy, S. (1994). Quantile smoothing splines, *Biometrika*, **81**, 673–680.

Mitchell, T., and Beauchamp J. (1988). Bayesian variable selection in linear regression, *J. Amer. Statist. Assoc.*, **83**, 1023–1036.

Monahan, J. (1984). Full Bayesian analysis of ARMA time series models, *J. Econometrics*, **21**, 307–331.

O'Hagan, A. (1995). Fractional Bayes factors for model comparison (with discussion), *J. Roy. Statist. Soc., B*, **57**, 99–138.

Powell, M. J. D. (1987). Radial basis functions for multivariate interpolation: A review, in Mason, J., and Cox, M. (eds.), *Algorithms for Approximation*, pp. 143–167, Oxford University Press, Oxford, UK.

Raftery, A. E., Madigan, D., and Hoeting, J. (1997). Bayesian model averaging for linear regression models, *J. Amer. Statist. Assoc.*, **92**, 179–191.

Ruppert, R, and Wand, M. (1994). Multivariate locally weighted least squares regression, *Ann. Statist.*, **22**, 1346–1370.

Schwarz, G. (1978). Estimating the dimension of a model, *Ann. Statist.*, **6**, 461–464.

Silverman, B. W., and Friedman, J. (1989). Flexible parsimonious smoothing and additive modeling, *Technometrics*, **31**, 3–21.

Siri, W. E. (1956). Gross composition of the body, in Lawrence, J. H., and Tobias, C. A. (eds.). *Advances in Biological and Medical Physics*, IV, Academic Press, New York.

Smith, M., and Kohn, R. (1996). Nonparametric regression via Bayesian variable selection, *J. Econometrics*, **75**, 317–344.

Smith, M., and Kohn R. (1997). A Bayesian approach to nonparametric bivariate regression, *J. Amer. Statist. Assoc.*, **92**, 1522–1535.

Smith, M., Sheather, S. J., and Kohn, R. (1996). Finite sample performance of robust Bayesian regression, *Computational Statistics*, **11**, 269–301.

Smith M., Wong, C., and Kohn, R. (1998). Additive nonparametric regression with autocorrelated errors, *J. Roy. Statist. Soc., B*, **60**, 311–331.

Stone, C. J. (1994). The use of polynomial splines and their tensor products in multivariate function estimation (with discussion), *Ann. Statist.*, **22**, 118–184.

Stone, C., Hansen, M., Kooperberg, C., and Truong, Y. (1996). *Polynomial Splines and Their Tensor Products in Extended Linear Modeling*, Technical Report 437, Department of Statistics, University of California, Berkeley, CA.

Verdinelli, I., and Wasserman, L. (1991). Bayesian analysis of outlier problems using the Gibbs sampler, *Statistics and Computing*, **1**, 105–117.

Wahba, G. (1990). *Spline Models for Observational Data*, SIAM, Philadelphia, PA.

Wang, Y. (1996). *Smoothing Spline Models with Correlated Random Errors*, Technical Report 966, Department of Statistics, University of Wisconsin, Madison, WI.

Zellner, A. (1986). On assessing prior distributions and Bayesian regression analysis with g-prior distributions, in Goel, P., and Zellner, A. (eds.). *Bayesian Inference and Decision Techniques: Essays in Honor of Bruno de Finetti*, pp. 233–243, North-Holland, Amsterdam, The Netherlands.

Index

WILEY SERIES IN PROBABILITY AND STATISTICS
ESTABLISHED BY WALTER A. SHEWHART AND SAMUEL S. WILKS

Probability and Statistics Section

*ANDERSON · The Statistical Analysis of Time Series
ARNOLD, BALAKRISHNAN, and NAGARAJA · A First Course in Order Statistics
ARNOLD, BALAKRISHNAN, and NAGARAJA · Records
BACCELLI, COHEN, OLSDER, and QUADRAT · Synchronization and Linearity: An Algebra for Discrete Event Systems
BARNETT · Comparative Statistical Inference, *Third Edition*
BASILEVSKY · Statistical Factor Analysis and Related Methods: Theory and Applications
BERNARDO and SMITH · Bayesian Statistical Concepts and Theory
BILLINGSLEY · Convergence of Probability Measures, *Second Edition*
BOROVKOV · Asymptotic Methods in Queuing Theory
BOROVKOV · Ergodicity and Stability of Stochastic Processes
BRANDT, FRANKEN, and LISEK · Stationary Stochastic Models
CAINES · Linear Stochastic Systems
CAIROLI and DALANG · Sequential Stochastic Optimization
CONSTANTINE · Combinatorial Theory and Statistical Design
COOK · Regression Graphics
COVER and THOMAS · Elements of Information Theory
CSÖRGŐ and HORVÁTH · Weighted Approximations in Probability Statistics
CSÖRGŐ and HORVÁTH · Limit Theorems in Change Point Analysis
*DANIEL · Fitting Equations to Data: Computer Analysis of Multifactor Data, *Second Edition*
DETTE and STUDDEN · The Theory of Canonical Moments with Applications in Statistics, Probability, and Analysis
DEY and MUKERJEE · Fractional Factorial Plans
*DOOB · Stochastic Processes
DRYDEN and MARDIA · Statistical Shape Analysis
DUPUIS and ELLIS · A Weak Convergence Approach to the Theory of Large Deviations
ETHIER and KURTZ · Markov Processes: Characterization and Convergence
FELLER · An Introduction to Probability Theory and Its Applications, Volume I, *Third Edition,* Revised; Volume II, *Second Edition*
FULLER · Introduction to Statistical Time Series, *Second Edition*
FULLER · Measurement Error Models
GHOSH, MUKHOPADHYAY, and SEN · Sequential Estimation
GIFI · Nonlinear Multivariate Analysis
GUTTORP · Statistical Inference for Branching Processes
HALL · Introduction to the Theory of Coverage Processes
HAMPEL · Robust Statistics: The Approach Based on Influence Functions
HANNAN and DEISTLER · The Statistical Theory of Linear Systems
HUBER · Robust Statistics

*Now available in a lower priced paperback edition in the Wiley Classics Library.

Probability and Statistics (Continued)

HUSKOVA, BERAN, and DUPAC · Collected Works of Jaroslav Hajek—with Commentary
IMAN and CONOVER · A Modern Approach to Statistics
JUREK and MASON · Operator-Limit Distributions in Probability Theory
KASS and VOS · Geometrical Foundations of Asymptotic Inference
KAUFMAN and ROUSSEEUW · Finding Groups in Data: An Introduction to Cluster Analysis
KELLY · Probability, Statistics, and Optimization
KENDALL, BARDEN, CARNE, and LE · Shape and Shape Theory
LINDVALL · Lectures on the Coupling Method
McFADDEN · Management of Data in Clinical Trials
MANTON, WOODBURY, and TOLLEY · Statistical Applications Using Fuzzy Sets
MORGENTHALER and TUKEY · Configural Polysampling: A Route to Practical Robustness
MUIRHEAD · Aspects of Multivariate Statistical Theory
OLIVER and SMITH · Influence Diagrams, Belief Nets and Decision Analysis
*PARZEN · Modern Probability Theory and Its Applications
PRESS · Bayesian Statistics: Principles, Models, and Applications
PUKELSHEIM · Optimal Experimental Design
RAO · Asymptotic Theory of Statistical Inference
RAO · Linear Statistical Inference and Its Applications, *Second Edition*
RAO and SHANBHAG · Choquet-Deny Type Functional Equations with Applications to Stochastic Models
ROBERTSON, WRIGHT, and DYKSTRA · Order Restricted Statistical Inference
ROGERS and WILLIAMS · Diffusions, Markov Processes, and Martingales, Volume I: Foundations, *Second Edition;* Volume II: Îto Calculus
RUBINSTEIN and SHAPIRO · Discrete Event Systems: Sensitivity Analysis and Stochastic Optimization by the Score Function Method
RUZSA and SZEKELY · Algebraic Probability Theory
SCHEFFE · The Analysis of Variance
SEBER · Linear Regression Analysis
SEBER · Multivariate Observations
SEBER and WILD · Nonlinear Regression
SERFLING · Approximation Theorems of Mathematical Statistics
SHORACK and WELLNER · Empirical Processes with Applications to Statistics
SMALL and McLEISH · Hilbert Space Methods in Probability and Statistical Inference
STAPLETON · Linear Statistical Models
STAUDTE and SHEATHER · Robust Estimation and Testing
STOYANOV · Counterexamples in Probability
TANAKA · Time Series Analysis: Nonstationary and Noninvertible Distribution Theory
THOMPSON and SEBER · Adaptive Sampling
WELSH · Aspects of Statistical Inference
WHITTAKER · Graphical Models in Applied Multivariate Statistics
YANG · The Construction Theory of Denumerable Markov Processes

Applied Probability and Statistics Section

ABRAHAM and LEDOLTER · Statistical Methods for Forecasting
AGRESTI · Analysis of Ordinal Categorical Data
AGRESTI · Categorical Data Analysis

*Now available in a lower priced paperback edition in the Wiley Classics Library.

Applied Probability and Statistics (Continued)

ANDERSON, AUQUIER, HAUCK, OAKES, VANDAELE, and WEISBERG · Statistical Methods for Comparative Studies
ARMITAGE and DAVID (editors) · Advances in Biometry
*ARTHANARI and DODGE · Mathematical Programming in Statistics
ASMUSSEN · Applied Probability and Queues
*BAILEY · The Elements of Stochastic Processes with Applications to the Natural Sciences
BARNETT and LEWIS · Outliers in Statistical Data, *Third Edition*
BARTHOLOMEW, FORBES, and McLEAN · Statistical Techniques for Manpower Planning, *Second Edition*
BASU and RIGDON · Statistical Methods for the Reliability of Repairable Systems
BATES and WATTS · Nonlinear Regression Analysis and Its Applications
BECHHOFER, SANTNER, and GOLDSMAN · Design and Analysis of Experiments for Statistical Selection, Screening, and Multiple Comparisons
BELSLEY · Conditioning Diagnostics: Collinearity and Weak Data in Regression
BELSLEY, KUH, and WELSCH · Regression Diagnostics: Identifying Influential Data and Sources of Collinearity
BHAT · Elements of Applied Stochastic Processes, *Second Edition*
BHATTACHARYA and WAYMIRE · Stochastic Processes with Applications
BIRKES and DODGE · Alternative Methods of Regression
BLISCHKE AND MURTHY · Reliability: Modeling, Prediction, and Optimization
BLOOMFIELD · Fourier Analysis of Time Series: An Introduction, *Second Edition*
BOLLEN · Structural Equations with Latent Variables
BOULEAU · Numerical Methods for Stochastic Processes
BOX · Bayesian Inference in Statistical Analysis
BOX and DRAPER · Empirical Model-Building and Response Surfaces
*BOX and DRAPER · Evolutionary Operation: A Statistical Method for Process Improvement
BUCKLEW · Large Deviation Techniques in Decision, Simulation, and Estimation
BUNKE and BUNKE · Nonlinear Regression, Functional Relations and Robust Methods: Statistical Methods of Model Building
CHATTERJEE and HADI · Sensitivity Analysis in Linear Regression
CHERNICK · Bootstrap Methods: A Practitioner's Guide
CHILÈS and DELFINER · Geostatistics: Modeling Spatial Uncertainty
CHOW and LIU · Design and Analysis of Clinical Trials: Concepts and Methodologies
CLARKE and DISNEY · Probability and Random Processes: A First Course with Applications, *Second Edition*
*COCHRAN and COX · Experimental Designs, *Second Edition*
CONOVER · Practical Nonparametric Statistics, *Second Edition*
CORNELL · Experiments with Mixtures, Designs, Models, and the Analysis of Mixture Data, *Second Edition*
*COX · Planning of Experiments
CRESSIE · Statistics for Spatial Data, *Revised Edition*
DANIEL · Applications of Statistics to Industrial Experimentation
DANIEL · Biostatistics: A Foundation for Analysis in the Health Sciences, *Sixth Edition*
DAVID · Order Statistics, *Second Edition*
*DEGROOT, FIENBERG, and KADANE · Statistics and the Law
DODGE · Alternative Methods of Regression
DOWDY and WEARDEN · Statistics for Research, *Second Edition*
DUNN and CLARK · Applied Statistics: Analysis of Variance and Regression, *Second Edition*
*ELANDT-JOHNSON and JOHNSON · Survival Models and Data Analysis
*FLEISS · The Design and Analysis of Clinical Experiments

*Now available in a lower priced paperback edition in the Wiley Classics Library.

Applied Probability and Statistics (Continued)

FLEISS · Statistical Methods for Rates and Proportions, *Second Edition*
FLEMING and HARRINGTON · Counting Processes and Survival Analysis
GALLANT · Nonlinear Statistical Models
GLASSERMAN and YAO · Monotone Structure in Discrete-Event Systems
GNANADESIKAN · Methods for Statistical Data Analysis of Multivariate Observations, *Second Edition*
GOLDSTEIN and LEWIS · Assessment: Problems, Development, and Statistical Issues
GREENWOOD and NIKULIN · A Guide to Chi-Squared Testing
*HAHN · Statistical Models in Engineering
HAHN and MEEKER · Statistical Intervals: A Guide for Practitioners
HAND · Construction and Assessment of Classification Rules
HAND · Discrimination and Classification
HEIBERGER · Computation for the Analysis of Designed Experiments
HEDAYAT and SINHA · Design and Inference in Finite Population Sampling
HINKELMAN and KEMPTHORNE: · Design and Analysis of Experiments, Volume 1: Introduction to Experimental Design
HOAGLIN, MOSTELLER, and TUKEY · Exploratory Approach to Analysis of Variance
HOAGLIN, MOSTELLER, and TUKEY · Exploring Data Tables, Trends and Shapes
HOAGLIN, MOSTELLER, and TUKEY · Understanding Robust and Exploratory Data Analysis
HOCHBERG and TAMHANE · Multiple Comparison Procedures
HOCKING · Methods and Applications of Linear Models: Regression and the Analysis of Variables
HOGG and KLUGMAN · Loss Distributions
HOSMER and LEMESHOW · Applied Logistic Regression
HØYLAND and RAUSAND · System Reliability Theory: Models and Statistical Methods
HUBERTY · Applied Discriminant Analysis
HUNT and KENNEDY · Financial Derivatives in Theory and Practice
JACKSON · A User's Guide to Principle Components
JOHN · Statistical Methods in Engineering and Quality Assurance
JOHNSON · Multivariate Statistical Simulation
JOHNSON and KOTZ · Distributions in Statistics
JOHNSON, KOTZ, and BALAKRISHNAN · Continuous Univariate Distributions, Volume 1, *Second Edition*
JOHNSON, KOTZ, and BALAKRISHNAN · Continuous Univariate Distributions, Volume 2, *Second Edition*
JOHNSON, KOTZ, and BALAKRISHNAN · Discrete Multivariate Distributions
JOHNSON, KOTZ, and KEMP · Univariate Discrete Distributions, *Second Edition*
JUREČKOVÁ and SEN · Robust Statistical Procedures: Aymptotics and Interrelations
KADANE · Bayesian Methods and Ethics in a Clinical Trial Design
KADANE AND SCHUM · A Probabilistic Analysis of the Sacco and Vanzetti Evidence
KALBFLEISCH and PRENTICE · The Statistical Analysis of Failure Time Data
KELLY · Reversability and Stochastic Networks
KHURI, MATHEW, and SINHA · Statistical Tests for Mixed Linear Models
KLUGMAN, PANJER, and WILLMOT · Loss Models: From Data to Decisions
KLUGMAN, PANJER, and WILLMOT · Solutions Manual to Accompany Loss Models: From Data to Decisions
KOTZ, BALAKRISHNAN, and JOHNSON · Continuous Multivariate Distributions, Volume 1, *Second Edition*
KOVALENKO, KUZNETZOV, and PEGG · Mathematical Theory of Reliability of Time-Dependent Systems with Practical Applications
LACHIN · Biostatistical Methods: The Assessment of Relative Risks

*Now available in a lower priced paperback edition in the Wiley Classics Library.

Applied Probability and Statistics (Continued)

LAD · Operational Subjective Statistical Methods: A Mathematical, Philosophical, and Historical Introduction

LANGE, RYAN, BILLARD, BRILLINGER, CONQUEST, and GREENHOUSE · Case Studies in Biometry

LAWLESS · Statistical Models and Methods for Lifetime Data

LEE · Statistical Methods for Survival Data Analysis, *Second Edition*

LEPAGE and BILLARD · Exploring the Limits of Bootstrap

LINHART and ZUCCHINI · Model Selection

LITTLE and RUBIN · Statistical Analysis with Missing Data

LLOYD · The Statistical Analysis of Categorical Data

MAGNUS and NEUDECKER · Matrix Differential Calculus with Applications in Statistics and Econometrics, *Revised Edition*

MALLER and ZHOU · Survival Analysis with Long Term Survivors

MANN, SCHAFER, and SINGPURWALLA · Methods for Statistical Analysis of Reliability and Life Data

McLACHLAN and KRISHNAN · The EM Algorithm and Extensions

McLACHLAN · Discriminant Analysis and Statistical Pattern Recognition

McNEIL · Epidemiological Research Methods

MEEKER and ESCOBAR · Statistical Methods for Reliability Data

*MILLER · Survival Analysis, *Second Edition*

MONTGOMERY and PECK · Introduction to Linear Regression Analysis, *Second Edition*

MYERS and MONTGOMERY · Response Surface Methodology: Process and Product in Optimization Using Designed Experiments

NELSON · Accelerated Testing, Statistical Models, Test Plans, and Data Analyses

NELSON · Applied Life Data Analysis

OCHI · Applied Probability and Stochastic Processes in Engineering and Physical Sciences

OKABE, BOOTS, and SUGIHARA · Spatial Tesselations: Concepts and Applications of Voronoi Diagrams

PANKRATZ · Forecasting with Dynamic Regression Models

PANKRATZ · Forecasting with Univariate Box-Jenkins Models: Concepts and Cases

PIANTADOSI · Clinical Trials: A Methodologic Perspective

PORT · Theoretical Probability for Applications

PUTERMAN · Markov Decision Processes: Discrete Stochastic Dynamic Programming

RACHEV · Probability Metrics and the Stability of Stochastic Models

RÉNYI · A Diary on Information Theory

RIPLEY · Spatial Statistics

RIPLEY · Stochastic Simulation

ROLSKI, SCHMIDLI, SCHMIDT, and TEUGELS · Stochastic Processes for Insurance and Finance

ROUSSEEUW and LEROY · Robust Regression and Outlier Detection

RUBIN · Multiple Imputation for Nonresponse in Surveys

RUBINSTEIN · Simulation and the Monte Carlo Method

RUBINSTEIN and MELAMED · Modern Simulation and Modeling

RYAN · Statistical Methods for Quality Improvement, *Second Edition*

SCHIMEK · Smoothing and Regression: Approaches, Computation, and Application

SCHUSS · Theory and Applications of Stochastic Differential Equations

SCOTT · Multivariate Density Estimation: Theory, Practice, and Visualization

*SEARLE · Linear Models

SEARLE · Linear Models for Unbalanced Data

SEARLE, CASELLA, and McCULLOCH · Variance Components

SENNOTT · Stochastic Dynamic Programming and the Control of Queueing Systems

STOYAN, KENDALL, and MECKE · Stochastic Geometry and Its Applications, *Second Edition*

*Now available in a lower priced paperback edition in the Wiley Classics Library.

Applied Probability and Statistics (Continued)

STOYAN and STOYAN · Fractals, Random Shapes and Point Fields: Methods of Geometrical Statistics
THOMPSON · Empirical Model Building
THOMPSON · Sampling
THOMPSON · Simulation: A Modeler's Approach
TIJMS · Stochastic Modeling and Analysis: A Computational Approach
TIJMS · Stochastic Models: An Algorithmic Approach
TITTERINGTON, SMITH, and MAKOV · Statistical Analysis of Finite Mixture Distributions
UPTON and FINGLETON · Spatial Data Analysis by Example, Volume 1: Point Pattern and Quantitative Data
UPTON and FINGLETON · Spatial Data Analysis by Example, Volume II: Categorical and Directional Data
VAN RIJCKEVORSEL and DE LEEUW · Component and Correspondence Analysis
VIDAKOVIC · Statistical Modeling by Wavelets
WEISBERG · Applied Linear Regression, *Second Edition*
WESTFALL and YOUNG · Resampling-Based Multiple Testing: Examples and Methods for p-Value Adjustment
WHITTLE · Systems in Stochastic Equilibrium
WOODING · Planning Pharmaceutical Clinical Trials: Basic Statistical Principles
WOOLSON · Statistical Methods for the Analysis of Biomedical Data
*ZELLNER · An Introduction to Bayesian Inference in Econometrics

Texts and References Section

AGRESTI · An Introduction to Categorical Data Analysis
ANDERSON · An Introduction to Multivariate Statistical Analysis, *Second Edition*
ANDERSON and LOYNES · The Teaching of Practical Statistics
ARMITAGE and COLTON · Encyclopedia of Biostatistics: Volumes 1 to 6 with Index
BARTOSZYNSKI and NIEWIADOMSKA-BUGAJ · Probability and Statistical Inference
BENDAT and PIERSOL · Random Data: Analysis and Measurement Procedures, *Third Edition*
BERRY, CHALONER, and GEWEKE · Bayesian Analysis in Statistics and Econometrics: Essays in Honor of Arnold Zellner
BHATTACHARYA and JOHNSON · Statistical Concepts and Methods
BILLINGSLEY · Probability and Measure, *Second Edition*
BOX · R. A. Fisher, the Life of a Scientist
BOX, HUNTER, and HUNTER · Statistics for Experimenters: An Introduction to Design, Data Analysis, and Model Building
BOX and LUCEÑO · Statistical Control by Monitoring and Feedback Adjustment
BROWN and HOLLANDER · Statistics: A Biomedical Introduction
CHATTERJEE and PRICE · Regression Analysis by Example, *Third Edition*
COOK and WEISBERG · Applied Regression Including Computing and Graphics
COOK and WEISBERG · An Introduction to Regression Graphics
COX · A Handbook of Introductory Statistical Methods
DILLON and GOLDSTEIN · Multivariate Analysis: Methods and Applications
*DODGE and ROMIG · Sampling Inspection Tables, *Second Edition*
DRAPER and SMITH · Applied Regression Analysis, *Third Edition*
DUDEWICZ and MISHRA · Modern Mathematical Statistics
DUNN · Basic Statistics: A Primer for the Biomedical Sciences, *Second Edition*
EVANS, HASTINGS, and PEACOCK · Statistical Distributions, *Third Edition*
FISHER and VAN BELLE · Biostatistics: A Methodology for the Health Sciences

*Now available in a lower priced paperback edition in the Wiley Classics Library.

Texts and References (Continued)

FREEMAN and SMITH · Aspects of Uncertainty: A Tribute to D. V. Lindley
GROSS and HARRIS · Fundamentals of Queueing Theory, *Third Edition*
HALD · A History of Probability and Statistics and their Applications Before 1750
HALD · A History of Mathematical Statistics from 1750 to 1930
HELLER · MACSYMA for Statisticians
HOEL · Introduction to Mathematical Statistics, *Fifth Edition*
HOLLANDER and WOLFE · Nonparametric Statistical Methods, *Second Edition*
HOSMER and LEMESHOW · Applied Logistic Regression, *Second Edition*
HOSMER and LEMESHOW · Applied Survival Analysis: Regression Modeling of Time to Event Data
JOHNSON and BALAKRISHNAN · Advances in the Theory and Practice of Statistics: A Volume in Honor of Samuel Kotz
JOHNSON and KOTZ (editors) · Leading Personalities in Statistical Sciences: From the Seventeenth Century to the Present
JUDGE, GRIFFITHS, HILL, LÜTKEPOHL, and LEE · The Theory and Practice of Econometrics, *Second Edition*
KHURI · Advanced Calculus with Applications in Statistics
KOTZ and JOHNSON (editors) · Encyclopedia of Statistical Sciences: Volumes 1 to 9 wtih Index
KOTZ and JOHNSON (editors) · Encyclopedia of Statistical Sciences: Supplement Volume
KOTZ, REED, and BANKS (editors) · Encyclopedia of Statistical Sciences: Update Volume 1
KOTZ, REED, and BANKS (editors) · Encyclopedia of Statistical Sciences: Update Volume 2
LAMPERTI · Probability: A Survey of the Mathematical Theory, *Second Edition*
LARSON · Introduction to Probability Theory and Statistical Inference, *Third Edition*
LE · Applied Categorical Data Analysis
LE · Applied Survival Analysis
MALLOWS · Design, Data, and Analysis by Some Friends of Cuthbert Daniel
MARDIA · The Art of Statistical Science: A Tribute to G. S. Watson
MASON, GUNST, and HESS · Statistical Design and Analysis of Experiments with Applications to Engineering and Science
MURRAY · X-STAT 2.0 Statistical Experimentation, Design Data Analysis, and Nonlinear Optimization
PURI, VILAPLANA, and WERTZ · New Perspectives in Theoretical and Applied Statistics
RENCHER · Linear Models in Statistics
RENCHER · Methods of Multivariate Analysis
RENCHER · Multivariate Statistical Inference with Applications
ROSS · Introduction to Probability and Statistics for Engineers and Scientists
ROHATGI · An Introduction to Probability Theory and Mathematical Statistics
RYAN · Modern Regression Methods
SCHOTT · Matrix Analysis for Statistics
SEARLE · Matrix Algebra Useful for Statistics
STYAN · The Collected Papers of T. W. Anderson: 1943–1985
TIERNEY · LISP-STAT: An Object-Oriented Environment for Statistical Computing and Dynamic Graphics
WONNACOTT and WONNACOTT · Econometrics, *Second Edition*
WU and HAMADA · Experiments: Planning, Analysis, and Parameter Design Optimization

*Now available in a lower priced paperback edition in the Wiley Classics Library.

WILEY SERIES IN PROBABILITY AND STATISTICS
ESTABLISHED BY WALTER A. SHEWHART AND SAMUEL S. WILKS

Survey Methodology Section

BIEMER, GROVES, LYBERG, MATHIOWETZ, and SUDMAN · Measurement Errors in Surveys
COCHRAN · Sampling Techniques, *Third Edition*
COUPER, BAKER, BETHLEHEM, CLARK, MARTIN, NICHOLLS, and O'REILLY (editors) · Computer Assisted Survey Information Collection
COX, BINDER, CHINNAPPA, CHRISTIANSON, COLLEDGE, and KOTT (editors) · Business Survey Methods
*DEMING · Sample Design in Business Research
DILLMAN · Mail and Telephone Surveys: The Total Design Method, *Second Edition*
GROVES and COUPER · Nonresponse in Household Interview Surveys
GROVES · Survey Errors and Survey Costs
GROVES, BIEMER, LYBERG, MASSEY, NICHOLLS, and WAKSBERG · Telephone Survey Methodology
*HANSEN, HURWITZ, and MADOW · Sample Survey Methods and Theory, Volume 1: Methods and Applications
*HANSEN, HURWITZ, and MADOW · Sample Survey Methods and Theory, Volume II: Theory
KISH · Statistical Design for Research
*KISH · Survey Sampling
KORN and GRAUBARD · Analysis of Health Surveys
LESSLER and KALSBEEK · Nonsampling Error in Surveys
LEVY and LEMESHOW · Sampling of Populations: Methods and Applications, *Third Edition*
LYBERG, BIEMER, COLLINS, de LEEUW, DIPPO, SCHWARZ, TREWIN (editors) · Survey Measurement and Process Quality
SIRKEN, HERRMANN, SCHECHTER, SCHWARZ, TANUR, and TOURANGEAU (editors) · Cognition and Survey Research

*Now available in a lower priced paperback edition in the Wiley Classics Library.